GENOMES 2
SECOND EDITION

GENOMES 2

SECOND EDITION

T.A. BROWN

Department of Biomolecular Sciences, UMIST, Manchester, M60 1QD, UK

WILEY-LISS

A JOHN WILEY & SONS, INC., PUBLICATION

Published by John Wiley & Sons, Inc., by arrangement with BIOS Scientific
Publishers Limited, 9 Newtec Place, Magdalen Road, Oxford OX4 1RE, UK

This edition published in the United States of America, its dependent territories, Central and
South America, Canada, Australia, Brunei, Cambodia, Hong Kong, India, Indonesia, Laos, Macau,
Malaysia, New Zealand, People's Republic of China, Philippines, Singapore, South Korea, Taiwan,
Thailand, the South Pacific and Vietnam only and not for export therefrom.

First published 1999
Second Edition 2002

A CIP catalogue record for this book is available from the British Library.

ISBN 0-471-25046-5

Library of congress Cataloging-in-Publication Data

Genomes / edited by Terence A. Brown. -- 2nd ed.
        p. ; cm.
Rev. ed. of: Genomes / T.A. Brown. 1999.
Includes bibliographical references and index
        ISBN 0-471-25046-5 (alk. paper)
    1. Genomes.
        [DNLM: 1. Genome. QH 477 G33605 2002] 1. Brown., T.A. (Terence A.)
        QH447 .B76 2002
        572.8′6--dc21

                                                            2002003471

USA
John Wiley & Sons Inc.,
605 Third Avenue, New York,
NY 10158–0012, USA

Canada
John Wiley & Sons (Canada) Ltd,
22 Worcester Road, Rexdale,
Ontario M9W 1L1, Canada

Project Manager: Helen Barham PhD
Production Editor: Sarah Carlson
Designed, typeset and illustrated by J&L Composition Ltd, Filey, North Yorkshire, UK
Printed by Ajanta Offset, New Delhi, India

Contents in Brief

PART 1   Genomes, Transcriptomes and Proteomes   1

Chapter 1   The Human Genome   3

Chapter 2   Genome Anatomies   29

Chapter 3   Transcriptomes and Proteomes   69

PART 2   Studying Genomes   93

Chapter 4   Studying DNA   95

Chapter 5   Mapping Genomes   125

Chapter 6   Sequencing Genomes   163

Chapter 7   Understanding a Genome Sequence   187

PART 3   How Genomes Function   219

Chapter 8   Accessing the Genome   221

Chapter 9   Assembly of the Transcription Initiation Complex   239

Chapter 10   Synthesis and Processing of RNA   273

Chapter 11   Synthesis and Processing of the Proteome   313

Chapter 12   Regulation of Genome Activity   347

PART 4   How Genomes Replicate and Evolve   381

Chapter 13   Genome Replication   383

Chapter 14   Mutation, Repair and Recombination   417

Chapter 15   How Genomes Evolve   459

Chapter 16   Molecular Phylogenetics   483

Contents

Abbreviations   xvii

Preface to the Second Edition   xxi

Preface to the First Edition   xxiii

An Introduction to *Genomes*   xxv

PART 1   **Genomes, Transcriptomes and Proteomes**   1

Chapter 1   **The Human Genome**   3

   1.1   **DNA**   5
   1.1.1   Genes are made of DNA   6
     Bacterial genes are made of DNA   6
     Virus genes are made of DNA   8
   1.1.2   The structure of DNA   8
     Nucleotides and polynucleotides   9
     RNA   10
   1.1.3   The double helix   11
     The evidence that led to the double helix   13
     The key features of the double helix   14
     Box 1.1: Base-pairing in RNA   14
     The double helix has structural flexibility   14
     Box 1.2: Units of length for DNA molecules   16

   1.2   **The Human Genome**   16
   1.2.1   The content of the human nuclear genome   18
     Genes and related sequences   19
     The functions of human genes   21
     Box 1.3: How many genes are there in the human genome?   22
     Pseudogenes and other evolutionary relics   22
     Genome-wide repeats and microsatellites   22
     Box 1.4: The organization of the human genome   23
   1.2.2   The human mitochondrial genome   24

   1.3   **Why is the Human Genome Project Important?**   24

   Study Aids   26

Chapter 2   **Genome Anatomies**   29

   2.1   **An Overview of Genome Anatomies**   30
   2.1.1   Genomes of eukaryotes   31
   2.1.2   Genomes of prokaryotes   33

   2.2   **The Anatomy of the Eukaryotic Genome**   35
   2.2.1   Eukaryotic nuclear genomes   35
     Packaging of DNA into chromosomes   35
     Technical Note 2.1: Agarose gel electrophoresis   37

The special features of metaphase chromosomes 37
Box 2.1: Unusual chromosome types 39
Where are the genes in a eukaryotic genome? 41
What genes are present in a eukaryotic genome? 41
Technical Note 2.2: Ultracentrifugation techniques 42
Families of genes 43
Box 2.2: Two examples of unusual gene organization 44
2.2.2   Eukaryotic organelle genomes 46
Physical features of organelle genomes 46
The genetic content of organelle genomes 46
The origins of organelle genomes 46

2.3   **The Anatomy of the Prokaryotic Genome** 49
2.3.1   The physical structure of the prokaryotic genome 50
The traditional view of the bacterial 'chromosome' 50
Complications on the *E. coli* theme 50
Research Briefing 2.1: Supercoiled domains in the *Escherichia coli* nucleoid 52
2.3.2   The genetic organization of the prokaryotic genome 54
Operons are characteristic features of prokaryotic genomes 55
Prokaryotic genomes and the species concept 56
Box 2.3: Mechanisms for gene flow between prokaryotes 57
Speculation on the minimal genome content and the identity of distinctiveness genes 58

2.4   **The Repetitive DNA Content of Genomes** 59
2.4.1   Tandemly repeated DNA 59
Satellite DNA is found at centromeres and elsewhere in eukaryotic chromosomes 59
Minisatellites and microsatellites 60
2.4.2   Interspersed genome-wide repeats 60
Transposition via an RNA intermediate 61
DNA transposons 63

Study Aids 66

Chapter 3   **Transcriptomes and Proteomes** 69

3.1   **Genome Expression in Outline** 70
Box 3.1: Cross-references to Part 3 of *Genomes* 71

3.2   **The RNA Content of the Cell** 72
3.2.1   Coding and non-coding RNA 72
Box 3.2: Non-coding RNA specified by the human genome 74
3.2.2   Synthesis of RNA 74
Processing of precursor RNA 74
3.2.3   The transcriptome 78
Studies of the yeast transcriptome 78
The human transcriptome 80

3.3   **The Protein Content of the Cell** 80
3.3.1   Protein structure 80
The four levels of protein structure 80
Amino acid diversity underlies protein diversity 82
Box 3.3: Non-covalent bonds in proteins 82
3.3.2   The link between the transcriptome and the proteome 83
The genetic code specifies how an mRNA sequence is translated into a polypeptide 84
Research Briefing 3.1: Elucidation of the genetic code 85
The genetic code is not universal 86
Box 3.4: The origin and evolution of the genetic code 87
3.3.3   The link between the proteome and the biochemistry of the cell 88
The amino acid sequence of a protein determines its function 88
The multiplicity of protein function 89

Study Aids 90

PART 2   **Studying Genomes**   93

Chapter 4   **Studying DNA**   95

        4.1   **Enzymes for DNA Manipulation**   97
               Technical Note 4.1: DNA labeling   98
        4.1.1   DNA polymerases   98
               The mode of action of a template-dependent DNA polymerase   98
               The types of DNA polymerases used in research   100
        4.1.2   Nucleases   102
               Restriction endonucleases enable DNA molecules to be cut at defined positions   102
               Examining the results of a restriction digest   103
        4.1.3   DNA ligases   105
        4.1.4   End-modification enzymes   107

        4.2   **DNA Cloning**   108
        4.2.1   Cloning vectors and the way they are used   109
               Vectors based on *E. coli* plasmids   109
               Technical Note 4.2: DNA purification   110
               Cloning vectors based on *E. coli* bacteriophage genomes   112
               Vectors for longer pieces of DNA   113
               Technical Note 4.3: Working with a clone library   117
               Cloning in organisms other than *E. coli*   117

        4.3   **The Polymerase Chain Reaction (PCR)**   119
               Technical Note 4.4: Techniques for studying RNA   120
        4.3.1   Carrying out a PCR   120
        4.3.2   The applications of PCR   121

        Study Aids   123

Chapter 5   **Mapping Genomes**   125

        5.1   **Genetic and Physical Maps**   128

        5.2   **Genetic Mapping**   128
        5.2.1   Genes were the first markers to be used   129
        5.2.2   DNA markers for genetic mapping   129
               Restriction fragment length polymorphisms (RFLPs)   130
               Simple sequence length polymorphisms (SSLPs)   130
               Single nucleotide polymorphisms (SNPs)   130
               Box 5.1: Why do SNPs have only two alleles?   131
               Technical Note 5.1: DNA microarrays and chips   133
        5.2.3   Linkage analysis is the basis of genetic mapping   134
               The principles of inheritance and the discovery of linkage   134
               Partial linkage is explained by the behavior of chromosomes during meiosis   137
               From partial linkage to genetic mapping   139
        5.2.4   Linkage analysis with different types of organism   140
               Linkage analysis when planned breeding experiments are possible   140
               Box 5.2: Multipoint crosses   141
               Gene mapping by human pedigree analysis   142
               Genetic mapping in bacteria   144

        5.3   **Physical Mapping**   145
        5.3.1   Restriction mapping   145
               The basic methodology for restriction mapping   147
               The scale of restriction mapping is limited by the sizes of the restriction fragments   147
               Direct examination of DNA molecules for restriction sites   150
        5.3.2   Fluorescent *in situ* hybridization (FISH)   152
               In situ hybridization with radioactive or fluorescent probes   152
               FISH in action   153
        5.3.3   Sequence tagged site (STS) mapping   153
               Any unique DNA sequence can be used as an STS   154

Fragments of DNA for STS mapping 155
Research Briefing 5.1: The radiation hybrid map of the mouse genome 156
A clone library can also be used as the mapping reagent for STS analysis 159

Study Aids 161

Chapter 6 **Sequencing Genomes** 163

6.1 **The Methodology for DNA Sequencing** 164
6.1.1 Chain termination DNA sequencing 165
Chain termination sequencing in outline 165
Technical Note 6.1: Polyacrylamide gel electrophoresis 165
Box 6.1: DNA polymerases for chain termination sequencing 166
Chain termination sequencing requires a single-stranded DNA template 168
The primer determines the region of the template DNA that will be sequenced 169
Thermal cycle sequencing offers an alternative to the traditional methodology 170
Fluorescent primers are the basis of automated sequence reading 170
Box 6.2: The chemical degradation sequencing method 170
6.1.2 Departures from conventional DNA sequencing 171

6.2 **Assembly of a Contiguous DNA Sequence** 172
6.2.1 Sequence assembly by the shotgun approach 173
The potential of the shotgun approach was proven by the *Haemophilus influenzae* sequence 173
6.2.2 Sequence assembly by the clone contig approach 176
Clone contigs can be built up by chromosome walking, but the method is laborious 176
Newer more rapid methods for clone contig assembly 178
6.2.3 Whole-genome shotgun sequencing 179
Key features of whole-genome shotgun sequencing 180

6.3 **The Human Genome Projects** 181
6.3.1 The mapping phase of the Human Genome Project 181
6.3.2 Sequencing the human genome 182
6.3.3 The future of the human genome projects 182

Study Aids 184

Chapter 7 **Understanding a Genome Sequence** 187

7.1 **Locating the Genes in a Genome Sequence** 188
7.1.1 Gene location by sequence inspection 188
The coding regions of genes are open reading frames 189
Simple ORF scans are less effective with higher eukaryotic DNA 189
Homology searches give an extra dimension to sequence inspection 191
7.1.2 Experimental techniques for gene location 191
Hybridization tests can determine if a fragment contains transcribed sequences 192
cDNA sequencing enables genes to be mapped within DNA fragments 192
Methods are available for precise mapping of the ends of transcripts 193
Exon–intron boundaries can also be located with precision 194

7.2 **Determining the Functions of Individual Genes** 195
7.2.1 Computer analysis of gene function 196
Homology reflects evolutionary relationships 196
Homology analysis can provide information on the function of an entire gene or of segments within it 196
Homology analysis in the yeast genome project 197
7.2.2 Assigning gene function by experimental analysis 198
Functional analysis by gene inactivation 198
Individual genes can be inactivated by homologous recombination 198
Gene inactivation without homologous recombination 199
Box 7.1: The phenotypic effect of gene inactivation is sometimes difficult to discern 200
Gene overexpression can also be used to assess function 200
Research Briefing 7.1: Analysis of chromosome I of *Caenorhabditis elegans* by RNA interference 202
7.2.3 More detailed studies of the activity of a protein coded by an unknown gene 204
Directed mutagenesis can be used to probe gene function in detail 204

Technical Note 7.1: Site-directed mutagenesis 205
Reporter genes and immunocytochemistry can be used to locate where and when genes are expressed 206

7.3   Global Studies of Genome Activity 207
7.3.1   Studying the transcriptome 207
The composition of a transcriptome can be assayed by SAGE 207
Using chip and microarray technology to study a transcriptome 207
7.3.2   Studying the proteome 208
Proteomics – methodology for characterizing the protein content of a cell 208
Identifying proteins that interact with one another 211
Protein interaction maps 211

7.4   Comparative Genomics 213
7.4.1   Comparative genomics as an aid to gene mapping 214
7.4.2   Comparative genomics in the study of human disease genes 215

Study Aids 217

PART 3   How Genomes Function 219

Chapter 8   Accessing the Genome 221

8.1   Inside the Nucleus 222
8.1.1   The internal architecture of the eukaryotic nucleus 222
Box 8.1: Accessing the prokaryotic genome 223
Technical Note 8.1: Fluorescence recovery after photobleaching (FRAP) 224
8.1.2   Chromatin domains 224
Functional domains are defined by insulators 226
Some functional domains contain locus control regions 227

8.2   Chromatin Modifications and Genome Expression 228
8.2.1   Activating the genome 228
Histone modifications determine chromatin structure 228
Nucleosome remodeling influences the expression of individual genes 229
Box 8.2: Chromatin modification by the HMGN proteins 231
8.2.2   Silencing the genome 231
Histone deacetylation is one way of repressing gene expression 231
Research Briefing 8.1: Discovery of the mammalian DNA methyltransferases 232
Genome silencing by DNA methylation 234
Methylation is involved in imprinting and X inactivation 235

Study Aids 237

Chapter 9   Assembly of the Transcription Initiation Complex 239

9.1   The Importance of DNA-binding Proteins 241
9.1.1   Locating the positions of DNA-binding sites in a genome 242
Gel retardation identifies DNA fragments that bind to proteins 242
Protection assays pinpoint binding sites with greater accuracy 242
Modification interference identifies nucleotides central to protein binding 244
9.1.2   Purifying a DNA-binding protein 245
9.1.3   Studying the structures of proteins and DNA–protein complexes 246
X-ray crystallography has broad applications in structure determination 246
NMR gives detailed structural information for small proteins 248
9.1.4   The special features of DNA-binding proteins 248
The helix–turn–helix motif is present in prokaryotic and eukaryotic proteins 249
Box 9.1: RNA-binding motifs 250
Zinc fingers are common in eukaryotic proteins 250
Other DNA-binding motifs 251
Box 9.2: Can sequence specificity be predicted from the structure of a recognition helix? 252
9.1.5   The interaction between DNA and its binding proteins 252
Direct readout of the nucleotide sequence 252

The nucleotide sequence has a number of indirect effects on helix structure — 253
Contacts between DNA and proteins — 253

9.2 **DNA–Protein Interactions During Transcription Initiation** — 254
9.2.1 RNA polymerases — 254
Box 9.3: Mitochondrial and chloroplast RNA polymerases — 255
9.2.2 Recognition sequences for transcription initiation — 255
Bacterial RNA polymerases bind to promoter sequences — 255
Eukaryotic promoters are more complex — 256
9.2.3 Assembly of the transcription initiation complex — 257
Transcription initiation in *E. coli* — 257
Transcription initiation with RNA polymerase II — 258
Box 9.4: Initiation of transcription in the archaea — 258
Transcription initiation with RNA polymerases I and III — 259
Research Briefing 9.1: Similarities between TFIID and the histone core octamer — 260

9.3 **Regulation of Transcription Initiation** — 261
9.3.1 Strategies for controlling transcription initiation in bacteria — 261
Promoter structure determines the basal level of transcription initiation — 261
Regulatory control over bacterial transcription initiation — 262
Box 9.5: *Cis* and *trans* — 263
9.3.2 Control of transcription initiation in eukaryotes — 265
Activators of eukaryotic transcription initiation — 265
Contacts between activators and the pre-initiation complex — 266
Repressors of eukaryotic transcription initiation — 266
Box 9.6: The modular structures of RNA polymerase II promoters — 267
Controlling the activities of activators and repressors — 268

Study Aids — 271

Chapter 10 **Synthesis and Processing of RNA** — 273

10.1 **Synthesis and Processing of mRNA** — 274
10.1.1 Synthesis of bacterial mRNAs — 274
The elongation phase of bacterial transcription — 274
Termination of bacterial transcription — 275
Control over the choice between elongation and termination — 277
Research Briefing 10.1: The structure of the bacterial RNA polymerase — 278
Box 10.1: Antitermination during the infection cycle of bacteriophage λ — 280
10.1.2 Synthesis of eukaryotic mRNAs by RNA polymerase II — 283
Capping of RNA polymerase II transcripts occurs immediately after initiation — 283
Elongation of eukaryotic mRNAs — 284
Termination of mRNA synthesis is combined with polyadenylation — 286
10.1.3 Intron splicing — 287
Conserved sequence motifs indicate the key sites in GU–AG introns — 288
Outline of the splicing pathway for GU–AG introns — 288
snRNAs and their associated proteins are the central components of the splicing apparatus — 289
Alternative splicing is common in many eukaryotes — 291
AU–AC introns are similar to GU–AG introns but require a different splicing apparatus — 294

10.2 **Synthesis and Processing of Non-coding RNAs** — 295
10.2.1 Transcript elongation and termination by RNA polymerases I and III — 295
10.2.2 Cutting events involved in processing of bacterial and eukaryotic pre-rRNA and pre-tRNA — 295
10.2.3 Introns in eukaryotic pre-rRNA and pre-tRNA — 296
Eukaryotic pre-rRNA introns are self-splicing — 296
Eukaryotic tRNA introns are variable but all are spliced by the same mechanism — 298

10.3 **Processing of Pre-RNA by Chemical Modification** — 298
10.3.1 Chemical modification of pre-rRNAs — 299
Box 10.2: Other types of intron — 302

10.3.2 RNA editing   303

10.4   **Degradation of mRNAs**   304
        Box 10.3: More complex forms of RNA editing   305
10.4.1 Bacterial mRNAs are degraded in the 3′→5′ direction   305
10.4.2 Eukaryotes have more diverse mechanisms for RNA degradation   306

10.5   **Transport of RNA Within the Eukaryotic Cell**   308

        Study Aids   310

Chapter 11 Synthesis and Processing of the Proteome   313

11.1   **The Role of tRNA in Protein Synthesis**   314
11.1.1 Aminoacylation: the attachment of amino acids to tRNAs   315
        All tRNAs have a similar structure   315
        Aminoacyl-tRNA synthetases attach amino acids to tRNAs   316
11.1.2 Codon–anticodon interactions: the attachment of tRNAs to mRNA   317

11.2   **The Role of the Ribosome in Protein Synthesis**   319
11.2.1 Ribosome structure   321
        Ultracentrifugation was used to measure the sizes of ribosomes and their components   321
        Probing the fine structure of the ribosome   322
11.2.2 Initiation of translation   323
        Initiation in bacteria requires an internal ribosome binding site   324
        Initiation in eukaryotes is mediated by the cap structure and poly(A) tail   325
        Initiation of eukaryotic translation without scanning   328
        Regulation of translation initiation   328
11.2.3 Elongation of translation   328
        Elongation in bacteria and eukaryotes   328
        Frameshifting and other unusual events during elongation   330
        Research Briefing 11.1: Peptidyl transferase is a ribozyme   332
11.2.4 Termination of translation   333
        Box 11.1: Translation in the archaea   334

11.3   **Post-translational Processing of Proteins**   335
11.3.1 Protein folding   335
        Not all proteins fold spontaneously in the test tube   335
        In cells, folding is aided by molecular chaperones   337
11.3.2 Processing by proteolytic cleavage   339
        Cleavage of the ends of polypeptides   339
        Proteolytic processing of polyproteins   339
11.3.3 Processing by chemical modification   340
11.3.4 Inteins   342

11.4   **Protein Degradation**   343

        Study Aids   346

Chapter 12 Regulation of Genome Activity   347

12.1   **Transient Changes in Genome Activity**   350
12.1.1 Signal transmission by import of the extracellular signaling compound   351
        Lactoferrin is an extracellular signaling protein which acts as a transcription activator   352
        Some imported signaling compounds directly influence the activity of pre-existing protein factors 352
        Some imported signaling compounds influence genome activity indirectly   354
12.1.2 Signal transmission mediated by cell surface receptors   354
        Signal transduction with one step between receptor and genome   356
        Signal transduction with many steps between receptor and genome   358
        Signal transduction via second messengers   359

12.2   **Permanent and Semipermanent Changes in Genome Activity**   360
12.2.1 Genome rearrangements   360

Yeast mating types are determined by gene conversion events 360
Genome rearrangements are responsible for immunoglobulin and T-cell receptor diversities 361
12.2.2 Changes in chromatin structure 362
12.2.3 Genome regulation by feedback loops 363
Research Briefing 12.1: Unraveling a signal transduction pathway 364

12.3 **Regulation of Genome Activity During Development** 365
12.3.1 Sporulation in *Bacillus* 366
Sporulation involves coordinated activities in two distinct cell types 366
Special σ subunits control genome activity during sporulation 366
12.3.2 Vulva development in *Caenorhabditis elegans* 369
C. elegans is a model for multicellular eukaryotic development 369
Determination of cell fate during development of the *C. elegans* vulva 369
Research Briefing 12.2: The link between genome replication and sporulation in *Bacillus* 370
12.3.3 Development in *Drosophila melanogaster* 372
Maternal genes establish protein gradients in the *Drosophila* embryo 372
A cascade of gene expression converts positional information into a segmentation pattern 374
Box 12.1: The genetic basis of flower development 375
Segment identity is determined by the homeotic selector genes 375
Homeotic selector genes are universal features of higher eukaryotic development 376

Study Aids 378

PART 4 **How Genomes Replicate and Evolve** 381

Chapter 13 **Genome Replication** 383

13.1 **The Topological Problem** 384
13.1.1 Experimental proof for the Watson–Crick scheme for DNA replication 385
The Meselson–Stahl experiment 386
13.1.2 DNA topoisomerases provide a solution to the topological problem 388
13.1.3 Variations on the semiconservative theme 389

13.2 **The Replication Process** 391
13.2.1 Initiation of genome replication 391
Initiation at the *E. coli* origin of replication 391
Origins of replication in yeast have been clearly defined 392
Replication origins in higher eukaryotes have been less easy to identify 393
13.2.2 The elongation phase of replication 393
The DNA polymerases of bacteria and eukaryotes 395
Discontinuous strand synthesis and the priming problem 396
Events at the bacterial replication fork 397
The eukaryotic replication fork: variations on the bacterial theme 400
13.2.3 Termination of replication 401
Replication of the *E. coli* genome terminates within a defined region 402
Little is known about termination of replication in eukaryotes 403
Box 13.1: Genome replication in the archaea 403
13.2.4 Maintaining the ends of a linear DNA molecule 404
Telomeric DNA is synthesized by the telomerase enzyme 405
Telomere length is implicated in senescence and cancer 405
Box 13.2: Telomeres in *Drosophila* 408

13.3 **Regulation of Eukaryotic Genome Replication** 409
13.3.1 Coordination of genome replication and cell division 409
Establishment of the pre-replication complex enables genome replication to commence 409
Research Briefing 13.1: Replication of the yeast genome 410
Regulation of pre-RC assembly 412
13.3.2 Control within S phase 412
Early and late replication origins 412
Checkpoints within S phase 413

Study Aids 415

Chapter 14 Mutation, Repair and Recombination   417

Box 14.1: Terminology for describing point mutations   419

14.1   Mutations   420
14.1.1   The causes of mutations   420
Errors in replication are a source of point mutations   420
Replication errors can also lead to insertion and deletion mutations   423
Mutations are also caused by chemical and physical mutagens   424
14.1.2   The effects of mutations   428
The effects of mutations on genomes   428
Technical Note 14.1: Mutation detection   429
The effects of mutations on multicellular organisms   431
The effects of mutations on microorganisms   432
14.1.3   Hypermutation and the possibility of programmed mutations   433

14.2   DNA Repair   434
14.2.1   Direct repair systems fill in nicks and correct some types of nucleotide modification   434
Research Briefing 14.1: Programmed mutations?   435
14.2.2   Excision repair   437
Base excision repairs many types of damaged nucleotide   437
Nucleotide excision repair is used to correct more extensive types of damage   437
14.2.3   Mismatch repair: correcting errors of replication   440
14.2.4   Repair of double-stranded DNA breaks   441
14.2.5   Bypassing DNA damage during genome replication   442
14.2.6   Defects in DNA repair underlie human diseases, including cancers   444

14.3   Recombination   444
14.3.1   Homologous recombination   444
The Holliday model for homologous recombination   445
Proteins involved in homologous recombination in *E. coli*   446
The double-strand break model for recombination in yeast   447
Box 14.2: The RecE and RecF recombination pathways of *Escherichia coli*   447
14.3.2   Site-specific recombination   448
Integration of λ DNA into the *E. coli* genome involves site-specific recombination   448
14.3.3   Transposition   450
Box 14.3: DNA methylation and transposition   450
Replicative and conservative transposition of DNA transposons   451
Transposition of retroelements   454

Study Aids   456

Chapter 15 How Genomes Evolve   459

15.1   Genomes: the First 10 Billion Years   460
15.1.1   The origins of genomes   460
The first biochemical systems were centered on RNA   461
The first DNA genomes   462
How unique is life?   463

15.2   Acquisition of New Genes   465
15.2.1   Acquisition of new genes by gene duplication   465
Whole-genome duplications can result in sudden expansions in gene number   465
Research Briefing 15.1: Segmental duplications in the yeast and human genomes   468
Duplications of individual genes and groups of genes have occurred frequently in the past   470
Box 15.1: Gene duplication and genetic redundancy   471
Genome evolution also involves rearrangement of existing genes   472
15.2.2   Acquisition of new genes from other species   473

15.3   Non-coding DNA and Genome Evolution   476
15.3.1   Transposable elements and genome evolution   476
Box 15.2: The origin of a microsatellite   476
15.3.2   The origins of introns   477

'Introns early' and 'introns late': two competing hypotheses 477

The current evidence disproves neither hypothesis 478

Box 15.3: The role of non-coding DNA 479

15.4 The Human Genome: the Last 5 Million Years 479

Study Aids 482

Chapter 16 Molecular Phylogenetics 483

16.1 The Origins of Molecular Phylogenetics 484

Box 16.1: Phenetics and cladistics 486

16.2 The Reconstruction of DNA-based Phylogenetic Trees 487

16.2.1 The key features of DNA-based phylogenetic trees 487

Gene trees are not the same as species trees 488

Box 16.2: Terminology for molecular phylogenetics 488

16.2.2 Tree reconstruction 489

Sequence alignment is the essential preliminary to tree reconstruction 490

Converting alignment data into a phylogenetic tree 491

Technical Note 16.1: Phylogenetic analysis 491

Assessing the accuracy of a reconstructed tree 493

Molecular clocks enable the time of divergence of ancestral sequences to be estimated 493

16.3 The Applications of Molecular Phylogenetics 494

16.3.1 Examples of the use of phylogenetic trees 494

DNA phylogenetics has clarified the evolutionary relationships between humans and other primates 494

The origins of AIDS 495

16.3.2 Molecular phylogenetics as a tool in the study of human prehistory 496

Intraspecific studies require highly variable genetic loci 496

The origins of modern humans – out of Africa or not? 496

Box 16.3: Genes in populations 497

The patterns of more recent migrations into Europe are also controversial 498

Research Briefing 16.1: Neandertal DNA 499

Prehistoric human migrations into the New World 502

Study Aids 504

Appendix 507

Keeping up to Date 507

Keeping up to Date by Reading the Literature 507

Keeping up to Date using the Internet 507

Glossary 511

Index 551

5-bU	5-bromouracil
A	adenine; alanine
ABF	ARS binding factor
Ac/Ds	activator/dissociation
ADAR	adenosine deaminase acting on RNA
ADP	adenosine 5'-diphosphate
AIDS	acquired immunodeficiency syndrome
ala	alanine
AMP	adenosine 5'-monophosphate
ANT-C	Antennapedia complex
AP	apurinic/apyrimidinic
arg	arginine
ARS	autonomously replicating sequence
asn	asparagine
ASO	allele-specific oligonucleotide
asp	aspartic acid
ATP	adenosine 5'-triphosphate
ATPase	adenosine 5'-triphosphatase
BAC	bacterial artificial chromosome
bis	N, N'-methylenebisacrylamide
bp	base pair
BSE	bovine spongiform encephalopathy
BX-C	Bithorax complex
C	cysteine; cytosine
cAMP	cyclic AMP
CAP	catabolite activator protein
CASP	CTD-associated SR-like protein
cDNA	complementary DNA
CEPH	Centre d'Études du Polymorphisme Humaine
cGMP	cyclic GMP
CHEF	contour-clamped homogeneous electric fields
Col	colicin
CPSF	cleavage and polyadenylation specificity factor
CRM	chromatin remodeling machine
CstF	cleavage stimulation factor
CTAB	cetyltrimethylammonium bromide
CTD	C-terminal domain
CTP	cytidine 5'-triphosphate
cys	cysteine
D	aspartic acid
DAG	diacylglycerol
Dam	DNA adenine methylase
DAPI	4, 6-diamino-2-phenylindole dihydrochloride
DASH	dynamic allele-specific hybridization
dATP	2'-deoxyadenosine 5'-triphosphate

DBS	double-stranded DNA binding site
Dcm	DNA cytosine methylase
dCTP	2'-deoxycytidine 5'-triphosphate
ddATP	2', 3'-dideoxyadenosine 5'-triphosphate
ddCTP	2', 3'-dideoxycytidine 5'-triphosphate
ddGTP	2', 3'-dideoxyguanosine 5'-triphosphate
ddNTP	2', 3'-dideoxynucleoside 5'-triphosphate
ddTTP	2', 3'-dideoxythymidine 5'-triphosphate
dGTP	2'-deoxyguanosine 5'-triphosphate
DNA	deoxyribonucleic acid
DNase	deoxyribonuclease
Dnmt	DNA methyltransferase
dNTP	2'-deoxynucleoside 5'-triphosphate
dsRAD	double-stranded RNA adenosine deaminase
dsRBD	double-stranded RNA binding domain
dTTP	2'-deoxythymidine 5'-triphosphate
E	glutamic acid
EDTA	ethylenediamine tetraacetate
eEF	eukaryotic elongation factor
EEO	electroendosmosis value
EF	elongation factor
eIF	eukaryotic initiation factor
EMS	ethylmethane sulfonate
eRF	eukaryotic release factor
ERV	endogenous retrovirus
ES	embryonic stem
ESE	exonic splicing enhancer
ESS	exonic splicing silencer
EST	expressed sequence tag
F	fertility; phenylalanine
FEN	flap endonuclease
FIGE	field inversion gel electrophoresis
FISH	fluorescent in situ hybridization
G	glycine; guanine
G1	gap phase 1
G2	gap phase 2
GABA	γ-aminobutyric acid
GAP	GTPase activating protein
Gb	gigabase pair
GDP	guanosine 5'-diphosphate
GFP	green fluorescent protein
gln	glutamine
glu	glutamic acid
gly	glycine
GMP	guanosine 5'-monophosphate
GNRP	guanine nucleotide releasing protein
GTF	general transcription factor
GTP	guanosine 5'-triphosphate

H	histidine
HAT	hypoxanthine + aminopterin + thymidine
HBS	heteroduplex binding site
HDAC	histone deacetylase
his	histidine
HIV	human immunodeficiency virus
HLA	human leukocyte antigen
HMG	high mobility group
HNPCC	hereditary non-polyposis colorectal cancer
hnRNA	heterogenous nuclear RNA
HOM-C	homeotic gene complex
HPLC	high-performance liquid chromatography
HPRT	hypoxanthine phosphoribosyl transferase
HTH	helix–turn–helix
I	isoleucine
ICF	immunodeficiency, centromere instability and facial anomalies
IF	initiation factor
IHF	integration host factor
ile	isoleucine
Inr	initiator
Ins(1,4,5)P$_3$	inositol-1,4,5-trisphosphate
IRE-PCR	interspersed repeat element PCR
IRES	internal ribosome entry site
IS	insertion sequence
ITR	inverted terminal repeat
JAK	Janus kinase
K	lysine
kb	kilobase pair
kDa	kilodalton
L	leucine
LCR	locus control region
leu	leucine
LINE	long interspersed nuclear element
LTR	long terminal repeat
lys	lysine
M	methionine; mitosis phase
MALDI-TOF	matrix-assisted laser desorption ionization time-of-flight
MAP	mitogen activated protein
MAR	matrix-associated region
Mb	megabase pair
MeCP	methyl-CpG-binding protein
met	methionine
MGMT	O^6-methylguanine-DNA methyltransferase
mRNA	messenger RNA
Myr	million years
N	2′-deoxynucleoside 5′-triphosphate; asparagine
NAD	nicotinamide adenine dinucleotide
NADH	reduced nicotinamide adenine dinucleotide
NHEJ	non-homologous end joining
NJ	neighbor-joining
NMD	nonsense-mediated RNA decay
NMR	nuclear magnetic resonance
NTP	nucleoside 5′-triphosphate

OFAGE	orthogonal field alternation gel electrophoresis
ORC	origin recognition complex
ORF	open reading frame
OTU	operational taxonomic unit
P	proline
PAC	P1-derived artificial chromosome
PADP	polyadenylate binding protein
PAUP	Phylogenetic Analysis Using Parsimony
PCNA	proliferating cell nuclear antigen
PCR	polymerase chain reaction
phe	phenylalanine
PHYLIP	Phylogeny Inference Package
PIC	pre-initiation complex
PNA	peptide nucleic acid
PNPase	polynucleotide phosphorylase
pro	proline
PtdIns(4,5)P$_2$	phosphatidylinositol-4,5-bisphosphate
PTRF	polymerase I and transcript release factor
Pu	purine
Py	pyrimidine
Q	glutamine
R	arginine; purine
RACE	rapid amplification of cDNA ends
RAM	random access memory
RBS	RNA binding site
RC	replication complex
RF	release factor
RFC	replication factor C
RFLP	restriction fragment length polymorphism
RHB	Rel homology domain
RLF	replication licensing factor
RMP	replication mediator protein
RNA	ribonucleic acid
RNase	ribonuclease
RNP	ribonucleoprotein
RPA	replication protein A
RRF	ribosome recycling factor
rRNA	ribosomal RNA
RT-PCR	reverse transcriptase-PCR
RTVL	retroviral-like element
S	serine; synthesis phase
SAGE	serial analysis of gene expression
SAP	stress activated protein
SAR	scaffold attachment region
SCAF	SR-like CTD-associated factor
scRNA	small cytoplasmic RNA
SCS	specialized chromatin structure
SDS	sodium dodecyl sulfate
SeCys	selenocysteine
ser	serine
SINE	short interspersed nuclear element
SIV	simian immunodeficiency virus
snoRNA	small nucleolar RNA
SNP	single nucleotide polymorphism
snRNA	small nuclear RNA
snRNP	small nuclear ribonucleoprotein
SRF	serum response factor

SSB	single-strand binding protein	TPA	tissue plasminogen activator
SSLP	simple sequence length polymorphism	TRAP	*trp* RNA-binding attenuation protein
STAT	signal transducer and activator of transcription	tRNA	transfer RNA
		trp	tryptophan
STR	simple tandem repeat	tyr	tyrosine
STS	sequence tagged site	U	uracil
T	threonine; thymine	UCE	upstream control element
TAF	TBP-associated factor	UTP	uridine 5'-triphosphate
TBP	TATA-binding protein	UTR	untranslated region
TEMED	N, N, N', N'-tetramethylethylenediamine	UV	ultraviolet
TF	transcription factor	val	valine
thr	threonine	VNTR	variable number of tandem repeats
Ti	tumor inducing	W	adenine or thymine; tryptophan
TIC	TAF and initiator-dependent cofactor	X-gal	5-bromo-4-chloro-3-indolyl-β-D-galactopyranoside
TK	thymidine kinase		
T_m	melting temperature	Y	pyrimidine; tyrosine
tmRNA	transfer-messenger RNA	YAC	yeast artificial chromosome
Tn	transposon	YIp	yeast integrative plasmid
TOL	toluene		

Preface to the Second Edition

Three exciting years have elapsed since publication of the first edition of *Genomes*. Draft sequences have appeared for the fruit fly, *Arabidopsis* and human genomes, and prokaryotic genome sequences are now published at the rate of two or three per month. Experimental techniques for studying the transcriptome and proteome have begun to mature and are providing novel insights into genome expression. And as well as these new directions, the genome expression and replication processes continue to be described in ever-increasing detail. All of these advances have been incorporated into this second edition of *Genomes*. The human genome is now the central feature of Chapter 1, followed immediately by a survey of the physical and genetic organizations of genomes in general, with Part 1 completed by an overview of the transcriptome and proteome. Part 2, on the methods used to study the genome, has been supplemented by the addition of an entirely new chapter on cloning techniques and PCR, which were interspersed in a rather unsatisfactory manner throughout the first edition. The chapters on sequencing and functional analysis have been updated and extended to reflect changes in technology since 1999. Part 3, describing genome expression, has been given a thorough update, as has Part 4 on genome replication and evolution. A number of readers commented on how up-to-date the first edition of *Genomes* was, and I hope that I have been able to retain this quality in the new edition.

Other changes have been designed to make the book more user friendly. The reorganization of material in Part 1 gives a more gentle introduction for students who are encountering molecular biology for the first time, and each chapter now ends with a series of study aids that I hope will be useful both as a guide to revision and in directing supplementary tutorial work. I have also prefaced each chapter with a set of learning outcomes, these being perhaps the most useful of the teaching innovations forced on UK universities by the quality-assessment initiatives of recent years.

I would like to say a general thank you to the many people who have been kind enough to send me comments and suggestions for the second edition of *Genomes*. I hope that you will recognize the changes, large and small, that I have made in response to your feedback. Also I thank Jonathan Ray and Simon Watkins of BIOS for the tremendous support that they provided when I was writing *Genomes*, and Sarah Carlson and Helen Barham for ensuring that the production phase was not a stressful experience. Finally, this second edition of *Genomes* would not have appeared without the support of my wife, Keri. In the Acknowledgements to the First Edition I wrote, 'if you find this book useful then you should thank Keri, not me, because she is the one who ensured that it was written', and I am pleased that one or two people actually took me up on this.

T.A. Brown
Manchester

Preface to the First Edition

Genomes attempts to bring a fresh approach to the teaching of undergraduate molecular biology. It starts with the premise that the syllabus for a university course in molecular biology should reflect the major research issues of the new millennium rather than those topics that were in vogue during the 1970s and 1980s. The book is therefore centered on genomes, not genes, in recognition of the fact that today's molecular biology is driven less by research into the activities of individual genes and more by genome sequencing and functional analysis. Many of today's molecular biology undergraduates will be involved in genome research when they begin their graduate careers and all of them will find their work influenced in one way or another by genome projects. If the objective of undergraduate teaching is to prepare students for their future careers then they must be taught about genomes!

It would of course be foolish to suggest that genes are no longer important. The major challenge that I faced when writing *Genomes* was to combine the essential elements of the traditional molecular biology syllabus with the new material relating to genomes. It is not yet possible to describe adequately the events leading from DNA to protein entirely in terms of 'genome to proteome', hence a substantial part of *Genomes* is devoted to the expression pathways of individual genes. This book differs from many others in that it attempts to describe these expression pathways in the context of the activity and function of the genome as a whole. Similarly, DNA replication, mutation and recombination are dealt with largely in terms of their effects on the genome, and not simply as processes responsible for the replication and alteration of genes.

My belief that molecular biology teaching should be centered on genomes grew as I wrote this book and discovered how much more satisfying and informative the approach is compared with the traditional syllabus. A number of topics that in the past have seemed to me to be of peripheral interest have fallen into place and taken on new relevance. I hope that at least some of the excitement that I felt while writing *Genomes* is conveyed to the reader.

T.A. Brown
Manchester

An Introduction to *Genomes*

I have tried to make the second edition of *Genomes* as user friendly as possible. The book therefore includes a number of devices intended to help the reader and to make the book an effective teaching aid.

Organization of the Book

Genomes is divided into four parts:

- **Part 1 – Genomes, Transcriptomes and Proteomes** introduces the central concepts of modern molecular biology. Chapter 1 begins with DNA and then summarizes the key features of the human genome, with Chapter 2 extending the survey to the genomes of eukaryotes and prokaryotes in general. Chapter 3 then uses the new concepts of the transcriptome and the proteome to introduce the basic steps in genome expression. By the end of Part 1 the reader will have acquired a good working knowledge of the structures and organizations of genomes and will understand, in outline, how the information contained in the genome is released and made available to the cell.

- **Part 2 – Studying Genomes** begins with an orientation chapter that introduces the reader to the methods, centered on cloning and PCR, that were used in the pre-genome era to examine individual genes. The techniques that are more specifically used for studying genomes are then described in the order in which they would be used in a genome project: methods for constructing genetic and physical maps (Chapter 5); DNA sequencing methodology and the strategies used to assemble a contiguous genome sequence (Chapter 6); and methods for identifying genes in a genome sequence and determining the functions of those genes in the cell (Chapter 7). The Human Genome Project forms a continuous thread throughout Part 2, but this is not to the exclusion of all else and I have tried to give adequate coverage to the strategies that have been used, and are being used, to understand the genomes of other organisms.

- **Part 3 – How Genomes Function** covers the material that in the past has been described (inadequately in my opinion) as 'DNA goes to RNA goes to protein'. Chapter 8 addresses the increasingly important issue of how chromatin structure influences genome expression. Chapter 9 then describes the assembly of the transcription initiation complexes of prokaryotes and eukaryotes, and includes a fairly detailed discussion of DNA-binding proteins, these playing the central roles in the initial stages of genome expression. Chapters 10 and 11 give details of the synthesis of RNA and protein, and Chapter 12 surveys the regulation of genome activity. Keeping Chapter 12 to a manageable length was difficult, as many different topics are relevant to genome regulation, but I hope that by using specific examples to illustrate general themes I have managed to achieve a satisfactory balance between conciseness and breadth of coverage.

- **Part 4 – How Genomes Replicate and Evolve** links DNA replication, mutation and recombination with the gradual evolution of genomes over time. In Chapters 13 and 14 the molecular processes responsible for replication, mutation, repair and recombination are described, and in Chapter 15 the ways in which these processes are thought to have shaped the structures and genetic contents of genomes over evolutionary time are considered. Finally, Chapter 16 is devoted to the increasingly informative use of molecular phylogenetics to infer the evolutionary relationships between DNA sequences.

Organization of Chapters

Learning outcomes

Each chapter starts with a set of learning outcomes. These have been phrased very carefully. They are not merely a series of synopses of the factual content of each chapter, but instead indicate the level and type of knowledge that the student should gain from reading the chapter. Therefore, the learning outcomes state what the student should be able to describe, draw, discuss, explain, evaluate, etc., each verb having been selected to convey precisely what it is that the student is expected to be able to do. The intention is that the student is left in no doubt about what they should get out of each chapter, and hence is in no doubt about whether they have dealt satisfactorily with the material.

Figures

A good diagram is certainly worth a thousand words but a bad one can confuse the reader and a superfluous one is merely distracting. I have therefore tried to ensure that every figure is necessary and fulfils a purpose beyond simply breaking up the text and making the book look pretty. I have also tried to make figures reproducible because in my opinion this makes them much more useful as a learning aid for the student. I have never understood the penchant for making textbook diagrams into works of art because if the student cannot redraw a diagram then it is merely an illustration and does not help the student learn the information that it is designed to convey. The figures in *Genomes* are as clear, simple and uncluttered as possible.

Boxes, Technical Notes and Research Briefings

The main text in each chapter is supported and extended by additional information, separated into three distinct categories:

- **Boxes** contain discrete packages of information that I have taken out of the main text, either for emphasis or to avoid disrupting the flow of the text. Some boxes summarize the key points regarding a topic that is described at length in the text, or provide a pointer towards a later topic that has a bearing on the issues being discussed. Other boxes are used to give a more extended coverage of interesting topics, and some describe current speculation regarding areas that have not yet been resolved.
- Each **Technical Note** is a self-contained description of a technique or a group of techniques important in the study of genomes. The Technical Notes are designed to be read in conjunction with the main text, each one being located at the place in the book where an application of that technique is described for the first time.
- **Research Briefings** are designed to illustrate some of the strategies that are used to study genomes. Each Briefing is based on one or a few research papers and explains the background and rationale of a research project, describes how the resulting data were analyzed, and summarizes the conclusions that were drawn. The objective is to illustrate the way in which real research is conducted and to show how research into molecular biology has established the 'facts' about genomes.

Reading lists

The reading lists at the end of each chapter are divided into two sections:

- **References** are lists of articles that are cited in the text. *Genomes* is not itself a research publication and the text is not referenced in the way that would be appropriate for a review or scientific paper. Many points and facts are not referenced at all, and those citations that are given are often review articles rather than the relevant primary research papers. In several cases, for example, I have referred to a *Science* Perspective or *Nature* News and Views article, rather than a research paper, because these general articles are usually more helpful in explaining the context and relevance of a piece of work. My intention throughout *Genomes* has been that the reference lists should be as valuable as possible to students writing extended essays or dissertations on particular topics.
- **Further Reading** contains books and review articles that are not referred to directly in the main text but which are useful sources of additional material. In most cases I have appended a short summary stating the particular value of each item to help the reader decide which ones he or she wishes to seek out. The lists are not all-inclusive and I encourage readers to spend some time searching the shelves of their own libraries for other books and articles. Browsing is an excellent way to discover interests that you never realized you had!

Study aids

Each set of study aids is divided into three sections: key terms, self study questions and problem-based learning.

Key terms

This is a list of the important words and short phrases that the student will have encountered for the first time when reading the chapter. A short definition is required for each one. All of the terms are highlighted in the text and defined in the Glossary, so the student can check the accuracy of their answers after they have completed the exercise. Short definitions of this kind are a useful type of revision aid: if a student can accurately define every key term then they almost certainly have an excellent knowledge of the factual content of the chapter.

Self study questions

These require 100–500 word answers, or occasionally ask for an annotated diagram or a table. The questions cover the entire content of each chapter in a straightforward manner, and they can be marked simply by checking each answer against the relevant part of the text. A student can use the self study questions to work systematically through a chapter, or can select individual ones in order to evaluate their ability to answer questions on specific topics. The self study questions could also be used in closed-book examinations.

Problem-based learning

This is a student-centered activity in which a group of students research a problem and, through their studies, obtain the information more normally delivered by a teacher-centered activity such as a lecture. Most students and teachers who have adopted this educational tool believe that it is a more effective means of learning than the traditional approaches, and is also more fun. The questions vary in nature and in difficulty. Some are reasonably straightforward and merely require a literature survey, the intention with these problems being that the students take their learning a few stages on from where *Genomes* leaves off. Some problems require that the students evaluate a statement or a hypothesis, which could be done by reading around the subject but which, hopefully, will engender a certain amount of thought and critical awareness. A few problems are very difficult, to the extent that there is no solid answer to the question posed. These are designed to stimulate debate and speculation, which stretches the knowledge of each student and forces them to think carefully about their statements. Ideally, problem-based learning is conducted as a group exercise, each group comprising 5–10 students, with an exercise lasting 1–2 weeks and being carried out through a series of meetings between the group and a facilitator, interspersed with meetings that the students conduct on their own. The facilitator helps the students to organize their thoughts, steers them away from unproductive lines of research, and points out any serious omissions in their approach. The output from the exercise is a written report, a poster, an oral presentation, or a combination of these things. Most of the problems given in *Genomes* are suitable for any type of output, and many can also be adapted for use as discussion topics in tutorials. There are no answers at the back of book! To provide answers would defeat the purpose – the intention is that the students discover a solution for themselves.

Appendix – Keeping up to Date

The Appendix gives the reader advice regarding the best way to keep up to date with the latest research discoveries. It is divided into two sections. The first section covers the various journals and other publications that include reviews and news articles on genome research, and the second section contains a list of some of the many Internet sites that contain relevant information.

Glossary

I am very much in favor of glossaries as learning aids and I have provided an extensive one for this second edition of *Genomes*. Every term that is highlighted in bold in the text is defined in the Glossary, along with a number of additional terms that the reader might come across when referring to books or articles in the reading lists. Each term in the Glossary also appears in the index, so the reader can quickly gain access to the relevant pages where the Glossary term is covered in more detail.

1

Genomes, Transcriptomes and Proteomes

Part 1 of *Genomes* introduces the central concepts of modern molecular biology. In these three chapters you will become familiar with the genomes of the different types of organism that live on planet Earth and you will begin to understand the link between the genome, transcriptome and proteome.

Chapter 1 explains how the structure of DNA enables this molecule to perform its role as a store of biological information, and also describes the construction of the human genome, as revealed by the draft DNA sequence that was completed in 2001.

In Chapter 2 we broaden our focus by surveying the genomes of other organisms, including plants, insects, yeast, bacteria and archaea. As you read this chapter, you will realize that our perception of the basic principle of life – how an organism's biological information is stored in its genome – is undergoing rapid change as more and more genome sequences are being completed.

Chapter 3 explains that these conceptual changes encompass not only the genome itself but also the processes by which the information contained in the genome is utilized by the cell. You will learn how the old-fashioned notions of *gene expression* are now being redefined as *genome expression*, and you will discover how the flow of biological information from genome to cell involves synthesis and maintenance of the transcriptome and the proteome.

Chapter 1

The Human Genome

Chapter 2

Genome Anatomies

Chapter 3

Transcriptomes and Proteomes

The Human Genome

Chapter Contents

1.1	*DNA*	5
	1.1.1 Genes are made of DNA	6
	1.1.2 The structure of DNA	8
	1.1.3 The double helix	11
1.2	*The Human Genome*	16
	1.2.1 The content of the human nuclear genome	18
	1.2.2 The human mitochondrial genome	24
1.3	*Why is the Human Genome Project Important?*	24

Learning outcomes

When you have read Chapter 1, you should be able to:

■ Describe the two experiments that led molecular biologists to conclude that genes are made of DNA, and state the limitations of each experiment

■ Give a detailed description of the structure of a polynucleotide and summarize the chemical differences between DNA and RNA

■ Discuss the evidence that Watson and Crick used to deduce the double helix structure of DNA and list the key features of this structure

■ Explain why the DNA double helix has structural flexibility

■ Describe in outline the content of the human nuclear genome

■ Draw the structure of an 'average' human gene

■ Categorize the human gene catalog into different functional classes

■ Distinguish between conventional and processed pseudogenes and other types of evolutionary relic

■ Give specific examples of the repetitive DNA content of the human genome

■ Give an outline description of the structure and organization of the human mitochondrial genome

■ Discuss the importance of the Human Genome Project

LIFE AS WE KNOW IT is specified by the **genomes** of the myriad organisms with which we share the planet. Every organism possesses a genome that contains the **biological information** needed to construct and maintain a living example of that organism. Most genomes, including the human genome and those of all other cellular life forms, are made of **DNA** (deoxyribonucleic acid) but a few viruses have **RNA** (ribonucleic acid) genomes. DNA and RNA are polymeric molecules made up of chains of monomeric subunits called **nucleotides**.

The human genome, which is typical of the genomes of all multicellular animals, consists of two distinct parts (*Figure 1.1*):

■ The **nuclear genome** comprises approximately 3 200 000 000 nucleotides of DNA, divided into 24 linear molecules, the shortest 50 000 000 nucleotides in length and the longest 260 000 000 nucleotides, each contained in a different chromosome. These 24 chromosomes consist of 22 autosomes and the two sex chromosomes, X and Y.

■ The **mitochondrial genome** is a circular DNA molecule of 16 569 nucleotides, multiple copies of which are located in the energy-generating organelles called mitochondria.

Each of the approximately 10^{13} cells in the adult human body has its own copy or copies of the genome, the only exceptions being those few cell types, such as red blood cells, that lack a nucleus in their fully differentiated state. The vast majority of cells are **diploid** and so have two copies of each autosome, plus two sex chromosomes, XX for females or XY for males – 46 chromosomes in all. These are called **somatic cells**, in contrast to **sex cells** or **gametes**, which are **haploid** and have just 23 chromosomes, comprising one of each autosome and one sex chromosome. Both types of cell have about 8000 copies of the mitochondrial genome, 10 or so in each mitochondrion.

This book is about genomes. It explains what genomes are (Part 1), how they are studied (Part 2), how they function (Part 3), and how they replicate and evolve (Part 4). We begin our journey with our own genome, which is quite naturally the one that interests us the most. Later in this chapter we will examine how the human genome is constructed, some of this information dating from the old days when biologists studied genes rather than genomes, but much of it revealed only since the Human Genome Project was completed in the first year of the new millennium. First, however, we must understand the structure of DNA.

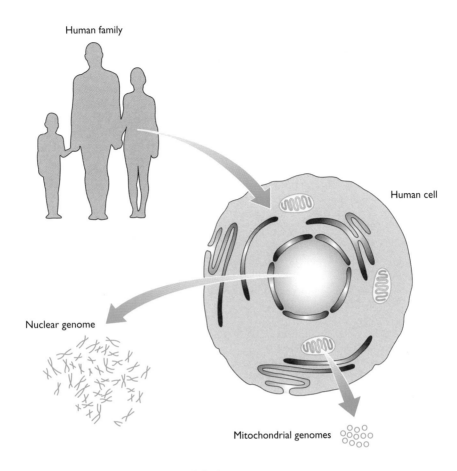

Human family

Human cell

Nuclear genome

Mitochondrial genomes

Figure 1.1   The nuclear and mitochondrial components of the human genome.

For more details on the anatomy of the human genome, see Section 1.2.

1.1 DNA

DNA was discovered in 1869 by Johann Friedrich Miescher, a Swiss biochemist working in Tübingen, Germany. The first extracts that Miescher made from human white blood cells were crude mixtures of DNA and chromosomal proteins, but the following year he moved to Basel, Switzerland (where the research institute named after him is now located) and prepared a pure sample of **nucleic acid** from salmon sperm. Miescher's chemical tests showed that DNA is acidic and rich in phosphorus, and also suggested that the individual molecules are very large, although it was not until the 1930s when biophysical techniques were applied to DNA that the huge lengths of the polymeric chains were fully appreciated.

Three years before Miescher discovered DNA, Gregor Mendel had published the results of his breeding experiments with pea plants, carried out in the monastery gardens at Brno, a central european city some 550 km from Tübingen in what is now the Czech Republic. Mendel's paper in the *Proceedings of the Society of Natural Sciences in Brno* describes his hypothesis that inheritance

is controlled by **unit factors**, the entities that geneticists today call **genes**. It is very unlikely that Miescher and Mendel were aware of each other's work, and if either of them had happened to read about the other's discoveries then they certainly would not have made any connection between DNA and genes. To make such a connection – to infer that genes are made of DNA – would have been quite illogical in the late 19th century or indeed for many decades afterwards. The precise biological function of DNA was not known, and the supposition that it was a store of cellular phosphorus seemed entirely reasonable at the time. The chemical nature of genes was equally unknown, and indeed was an irrelevance for most geneticists, who in the years immediately after 1900, when Mendel's work was rediscovered, were able to make remarkable advances in understanding heredity without worrying about what genes were actually made of.

It was not until the 1930s that scientists began to ask more searching questions about genes. In 1944, Erwin Schrödinger, more famous for the wave equation which still terrifies many biology students taking introductory courses in physical chemistry, published a book entitled *What is Life?*, which encapsulated a variety of issues that were being discussed not only by geneticists but also by physicists such as Niels Bohr and Max Delbrück. These scientists were the first **molecular biologists** and the first to suggest that 'life' could be explained in molecular terms; our current knowledge of how the genome functions stems directly from their pioneering work. The starting point for the new molecular biology was to discover what genes are made of.

1.1.1 Genes are made of DNA

How could the molecular nature of the genetic material be determined? Back in 1903, WS Sutton had realized that the inheritance patterns of genes paralleled the behavior of chromosomes during cell division. This observation led to the proposal that genes are located in chromosomes and by the 1930s it was universally accepted that the **chromosome theory** was correct. Examination of cells by **cytochemistry**, after staining with dyes that bind specifically to just one type of biochemical, had shown that chromosomes are made of DNA and protein, in roughly equal amounts. Some biologists looked on the combination between the two ('nucleoprotein') as the genetic material, but others argued differently. From today's perspective it can be difficult to understand why these arguments favored the notion that genes were made, not of DNA, but of protein. The explanation is that, at the time, many biochemists thought that all DNA molecules were the same, which meant that DNA did not have the immense variability that was one of the postulated features of the genetic material. Billions of different genes must exist and for each one to have its own individual activity, the genetic material must be able to take many different forms. If every DNA molecule were identical then DNA could not satisfy this requirement and so genes must be made of protein. This assumption made perfect sense because proteins were known, correctly, to be highly variable polymeric molecules, each one made up of a different combination of 20 chemically distinct amino-acid monomers (Section 3.3.1).

The errors that had been made in understanding DNA structure lingered on until the late 1930s. Gradually, however, it was accepted that DNA, like protein, has immense variability. Could DNA therefore be the genetic material? The results of two experiments performed during the middle decades of the 20th century forced biologists to take this possibility seriously.

Bacterial genes are made of DNA

The first molecular biologists realized that the most conclusive way to identify the chemical composition of genes would be to purify some and subject them to chemical analysis. But nothing like this had ever been attempted and it was not clear how it could be done. Ironically, the experiment was performed almost unwittingly by a group of scientists who did not look upon themselves as molecular biologists and who were not motivated by a curiosity to know what genes are made of. Instead, their objective was to find a better treatment for one of the most deadly diseases of the early 20th century, pneumonia.

Before the discovery of antibiotics, pneumonia was mainly controlled by treating patients in the early stages of the disease with an antiserum prepared by injecting dead cells of the causative bacterium (now called *Streptococcus pneumoniae*) into an animal. In order to prepare more effective antisera, studies were made of the immunological properties of the bacterium. It was shown that there is a range of different types of *S. pneumoniae*, each characterized by the mixture of sugars contained in the thick capsule that surrounds the cell and elicits the immunological response (*Figure 1.2A*). In 1923, Frederick Griffith, a British medical officer, discovered that as well as the virulent strains, there were some types of *S. pneumoniae* that did not have a capsule and did not cause pneumonia. This discovery was not a huge surprise because other species of pathogenic bacteria were known to have avirulent, unencapsulated forms. However, 5 years later Griffith obtained some results that were totally unexpected (Griffith, 1928). He performed a series of experiments in which he injected mice with various mixtures of bacteria (*Figure 1.2B*). He showed that, as anticipated, mice injected with virulent *S. pneumoniae* bacteria developed pneumonia and died, whereas those injected with an avirulent type remained healthy. What he did not anticipate, however, was what would happen when mice were injected with a mixture made up of live avirulent bacteria along with some virulent cells that had been killed by heat treatment. The only live bacteria were the harmless ones, so surely the mice would remain healthy? Not so: the mice died. Griffith carried out biopsies of these dead mice and discovered that their respiratory tracts contained virulent bacteria, which were always of the same immunological type as the strain that had been killed by heat treatment before injection. Somehow the living harmless bacteria had acquired the ability to make the capsule sugars of the dead bacteria. This process – the conversion of living harmless bacteria into virulent cells – was called **transformation**. Although not recognized as such by Griffith, the **transforming principle** – the component of the dead cells that conferred on the live cells the ability to make the capsular sugars – was genetic material.

Oswald Avery, together with his colleagues Colin MacLeod and Maclyn McCarty, of Columbia University, New York, set out to determine what the transforming principle was made of. The experiments took a long time to carry out and were not completed until 1944 (Avery *et al.*, 1944). But the results were conclusive: the transforming principle was DNA. The transforming principle behaved in exactly the same way as DNA when subjected to various biophysical tests, it was inactivated by enzymes that degraded DNA, and it was not affected by enzymes that attacked protein or any other type of biochemical (*Figure 1.3*).

Avery's experiments were meticulous but, because of several complicating factors, they did not immediately lead to acceptance of DNA as the genetic material. It was

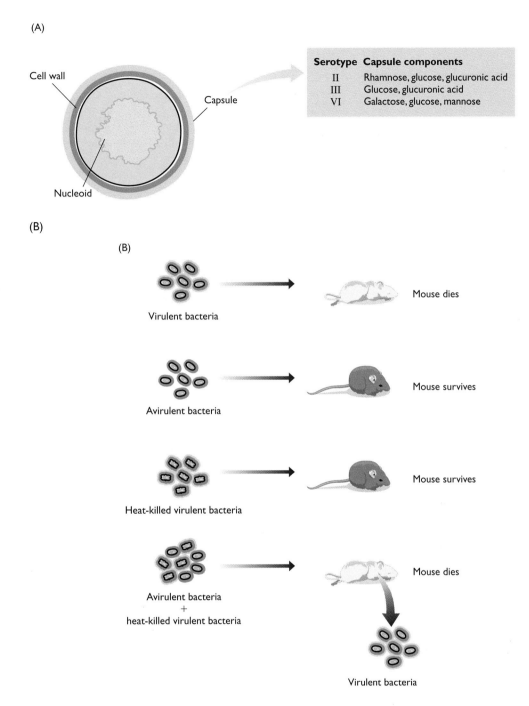

Figure 1.2   Griffith's experiments with virulent and avirulent *Streptococcus pneumoniae* bacteria.

(A) Representation of a *S. pneumoniae* bacterium. A serotype is a bacterial type with distinctive immunological properties, conferred in this case by the combination of sugars present in the capsule. Avirulent types have no capsule. (B) The experiments which showed that a component of heat-killed bacteria can transform living avirulent bacteria into virulent cells. Griffith showed that the avirulent bacteria were always transformed into the same serotype as the dead cells. In other words, the living bacteria acquired the genes specifying synthesis of the capsule of the dead cells.

not clear in the minds of all microbiologists that transformation really was a genetic phenomenon, and few geneticists really understood the system well enough to be able to evaluate Avery's work. There was also some

doubt about the veracity of the experiments. In particular, there were worries about the specificity of the **deoxyribonuclease** enzyme that he used to inactivate the transforming principle. This result, a central part of the

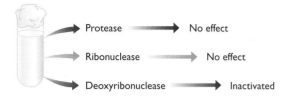

Figure 1.3 The transforming principle is DNA.

Avery and his colleagues showed that the transforming principle is unaffected by treatment with a protease or a ribonuclease, but is inactivated by treatment with a deoxyribonuclease.

evidence for the transforming principle being DNA, would be invalid if, as seemed possible, the enzyme contained trace amounts of a contaminating **protease** and hence was also able to degrade protein. These uncertainties meant that a second experiment was needed to provide more information on the chemical nature of the genetic material.

Virus genes are made of DNA

The second experiment was carried out by two *bona fide* molecular biologists, Alfred Hershey and Martha Chase, at Cold Spring Harbor, New York, in 1952 (Hershey and Chase, 1952). Like the work on the transforming principle, the Hershey–Chase experiment was not done specifically to determine the chemical nature of the gene. Hershey and Chase were two of several biologists who were studying the infection cycle of **bacteriophages** (or '**phages**') – viruses that infect bacteria. Phages are relatively simple structures made of just DNA and protein, with the DNA contained inside the phage, surrounded by a protein **capsid** (*Figure 1.4A*). To replicate, a phage must enter a bacterial cell and subvert the bacterial enzymes into expressing the information contained in the phage genes, so that the bacterium synthesizes new phages. Once replication is complete, the new phages leave the bacterium, possibly causing its death as they do so, and move on to infect new cells (*Figure 1.4B*). The objective of Hershey and Chase's experiment was to determine if the entire phage particle entered the bacterium at the start of the infection cycle, or if part of the phage stayed outside. If only one component of the phage – the DNA or the protein – entered the cell then that component must be the genetic material.

Their experimental strategy was based on the use of radioactive labels, which had recently been introduced into biology. DNA contains phosphorus, which is absent from protein, so DNA can be labeled specifically with the radioactive phosphorus isotope, ^{32}P. Protein, on the other hand, contains sulfur, which is absent from DNA, and so protein can be labeled with ^{35}S. Hershey and Chase were not the first to use radiolabeling to try to determine which part of the phage entered the cell, but previous experiments by James Watson, Ole Maaløe and others had been

inconclusive because of the difficulty in distinguishing between phage material that was actually inside a bacterium and a component that did not enter the cell but remained attached to the outer cell surface. To get round this difficulty, Hershey and Chase made an important modification to the previous experiments. They infected *Escherichia coli* bacteria with radiolabeled T2 phages but, rather than allowing the infection process to go to completion, they left the culture for just a few minutes and then agitated it in a blender. The idea was that the blending would detach the phage material from the surface of the bacteria, enabling this component to be separated from the material inside the cells by centrifuging at a speed that collected the relatively heavy bacteria as a pellet at the bottom of the tube but left the detached material in suspension (*Figure 1.5*). After centrifugation, Hershey and Chase examined the bacterial pellet and found that it contained 70% of the ^{32}P-labeled component of the phages (the DNA) but only 20% of the ^{35}S-labeled material (the phage protein). In a parallel experiment, the bacteria were left for 20 minutes, long enough for the infection cycle to reach completion. With T2 phage, the cycle ends with the bacteria bursting open and releasing new phages into the supernatant. These new phages contained almost half the DNA from the original phages, but less than 1% of the protein.

Hershey and Chase's results suggested that DNA was the major component of the infecting phages that entered the bacterial cell and, similarly, was the major, or perhaps only, component to be passed on to the progeny phages. These observations lent support to the view that DNA is the genetic material, but were they conclusive? Not according to Hershey and Chase (1952) who wrote 'Our experiments show clearly that a physical separation of phage T2 into genetic and non-genetic parts is possible . . . . The chemical identification of the genetic part must wait, however, until some questions . . . have been answered.' Even if the experiment had provided compelling evidence that the genetic material of phages was DNA, it would have been erroneous to extrapolate from these unusual life forms (which some biologists contend are not really 'living') to cellular organisms. Indeed, we know that some phage genomes are made of RNA. The Hershey–Chase experiment is important, not because of what it tells us, but because it alerted biologists to the fact that DNA *might* be the genetic material and was therefore worth studying. It was this that influenced Watson and Crick to study DNA and, as we will see below, it was their discovery of the **double helix** structure, which solved the puzzling question of how genes can replicate, that really convinced the scientific world that genes are made of DNA.

1.1.2 The structure of DNA

The names of James Watson and Francis Crick are so closely linked with DNA that it is easy to forget that, when they began their collaboration in Cambridge, England in October 1951, the detailed structure of the

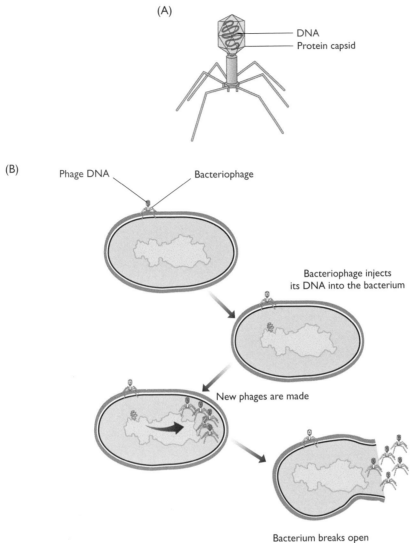

Figure 1.4   Bacteriophages are viruses that infect bacteria.

(A) The structure of a head-and-tail bacteriophage such as T2. The DNA genome of the phage is contained in the head part of the protein capsid. (B) The infection cycle. After injection into an *Escherichia coli* bacterium, the T2 phage genome directs synthesis of new phages. For T2, the infection cycle takes about 20 minutes at 37 °C and ends with lysis of the cell and release of 250–300 new phages. This is the **lytic infection cycle**. Some phages, such as λ, can also follow a **lysogenic infection cycle**, in which the phage genome becomes inserted into the bacterial chromosome and remains there, in quiescent form, for several generations of the bacterium (Section 4.2.1).

DNA polymer was already known. Their contribution was not to determine the structure of DNA *per se*, but to show that in living cells two DNA chains are intertwined to form the double helix. We will consider the two facets of DNA structure separately.

Nucleotides and polynucleotides

DNA is a linear, unbranched polymer in which the monomeric subunits are four chemically distinct nucleotides that can be linked together in any order in chains hundreds, thousands or even millions of units in length. Each nucleotide in a DNA polymer is made up of three components (*Figure 1.6*):

1. **2′-deoxyribose**, which is a **pentose**, a type of sugar composed of five carbon atoms. These five carbons are numbered 1′ (spoken as 'one-prime'), 2′, etc. The name '2′-deoxyribose' indicates that this particular sugar is a derivative of ribose, one in which the hydroxyl (–OH) group attached to the 2′-carbon of ribose has been replaced by a hydrogen (–H) group.

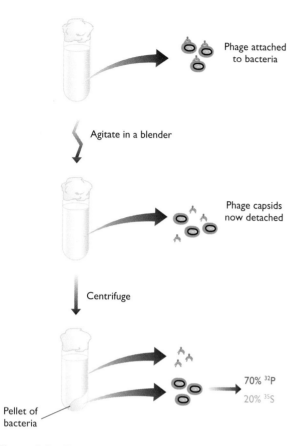

Phage attached to bacteria

Agitate in a blender

Phage capsids now detached

Centrifuge

Pellet of bacteria

70% ^{32}P
20% ^{35}S

Figure 1.5 The Hershey–Chase experiment.

The bacteriophages were labeled with ^{32}P and ^{35}S. A few minutes after infection, the culture was agitated to detach the empty phage capsids from the cell surface. The culture was then centrifuged and the radioactive content of the bacterial pellet determined. This pellet contained most of ^{32}P-labeled component of the phages (the DNA) but only 20% of the ^{35}S-labeled material (the phage protein). In a second experiment, Hershey and Chase showed that new phages produced at the end of an interrupted infection cycle contained less than 1% of the protein from the parent phages.

2. A **nitrogenous base**, one of **cytosine**, **thymine** (single-ring **pyrimidines**), **adenine** or **guanine** (double-ring **purines**). The base is attached to the 1′-carbon of the sugar by a **β-N-glycosidic bond** attached to nitrogen number 1 of the pyrimidine or number 9 of the purine.

3. A **phosphate group**, comprising one, two or three linked phosphate units attached to the 5′-carbon of the sugar. The phosphates are designated α, β and γ, with the α-phosphate being the one directly attached to the sugar.

A molecule made up of just the sugar and base is called a **nucleoside**; addition of the phosphates converts this to a nucleotide. Although cells contain nucleotides with one, two or three phosphate groups, only the nucleoside

triphosphates act as substrates for DNA synthesis. The full chemical names of the four nucleotides that polymerize to make DNA are:

- 2′-deoxyadenosine 5′-triphosphate
- 2′-deoxycytidine 5′-triphosphate
- 2′-deoxyguanosine 5′-triphosphate
- 2′-deoxythymidine 5′-triphosphate

The abbreviations of these four nucleotides are dATP, dCTP, dGTP and dTTP, respectively, or, when referring to a DNA sequence, A, C, G and T, respectively.

In a polynucleotide, individual nucleotides are linked together by **phosphodiester bonds** between their 5′- and 3′-carbons (*Figure 1.7*). From the structure of this linkage we can see that the polymerization reaction (*Figure 1.8*) involves removal of the two outer phosphates (the β- and γ-phosphates) from one nucleotide and replacement of the hydroxyl group attached to the 3′-carbon of the second nucleotide. Note that the two ends of the polynucleotide are chemically distinct, one having an unreacted triphosphate group attached to the 5′-carbon (the **5′** or **5′-P terminus**) and the other having an unreacted hydroxyl attached to the 3′-carbon (the **3′** or **3′-OH terminus**). This means that the polynucleotide has a chemical direction, expressed as either 5′→3′ (down in *Figure 1.8*) or 3′→5′ (up in *Figure 1.8*). An important consequence of the polarity of the phosphodiester bond is that the chemical reaction needed to extend a DNA polymer in the 5′→3′ direction is different to that needed to make a 3′→5′ extension. All natural **DNA polymerase** enzymes are only able to carry out 5′→3′ synthesis, which adds significant complications to the process by which double-stranded DNA is replicated (Section 13.2). The same limitation applies to RNA polymerases, the enzymes which make RNA copies of DNA molecules (Section 3.2.2).

RNA

Although our attention is firmly on DNA, the structure of RNA is so similar to that of DNA that it makes sense to introduce it here. RNA is also a polynucleotide but with two differences compared with DNA (*Figure 1.9*). First, the sugar in an RNA nucleotide is **ribose** and, second, RNA contains **uracil** instead of thymine. The four nucleotide substrates for synthesis of RNA are therefore:

- adenosine 5′-triphosphate
- cytidine 5′-triphosphate
- guanosine 5′-triphosphate
- uridine 5′-triphosphate

which are abbreviated to ATP, CTP, GTP and UTP, or A, C, G and U, respectively.

As with DNA, RNA polynucleotides contain 3′–5′ phosphodiester bonds, but these phosphodiester bonds are less stable than those in a DNA polynucleotide because of the indirect effect of the hydroxyl group at the 2′-position of the sugar. This may be one reason why the biological functions of RNA do not require the poly-

(A) A nucleotide

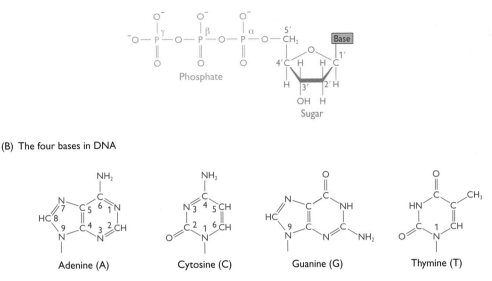

(B) The four bases in DNA

Figure 1.6 The structure of a nucleotide.

(A) The general structure of a deoxyribonucleotide, the type of nucleotide found in DNA. (B) The four bases that occur in deoxyribonucleotides.

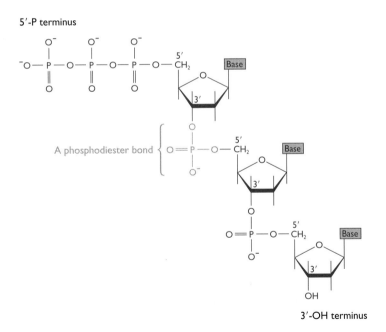

Figure 1.7 A short DNA polynucleotide showing the structure of the phosphodiester bond. Note that the two ends of the polynucleotide are chemically distinct.

nucleotide to be more than a few thousand nucleotides in length, at most. There are no RNA counterparts of the million-unit sized DNA molecules found in human chromosomes.

1.1.3 The double helix

In the years before 1950, various lines of evidence had shown that cellular DNA molecules are comprised of two or more polynucleotides assembled together in some

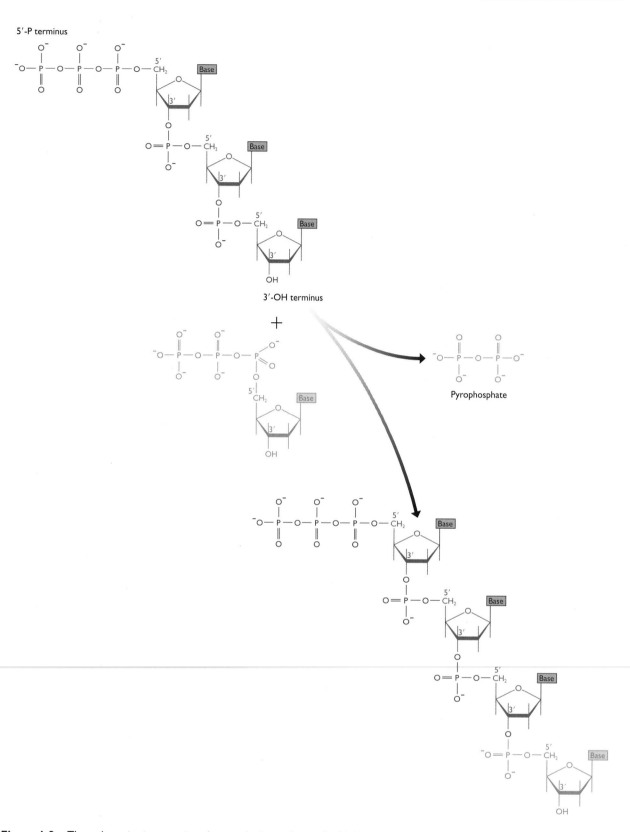

Figure 1.8 The polymerization reaction that results in synthesis of a DNA polynucleotide.

Synthesis occurs in the 5'→3' direction, with the new nucleotide being added to the 3'-carbon at the end of the existing polynucleotide. The β- and γ-phosphates of the nucleotide are removed as a pyrophosphate molecule.

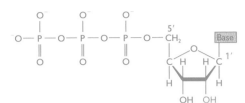

(A) A ribonucleotide

(B) Uracil

Figure 1.9 The chemical differences between DNA and RNA.

(A) RNA contains ribonucleotides, in which the sugar is ribose rather than 2'-deoxyribose. The difference is that a hydroxyl group rather than hydrogen atom is attached to the 2'-carbon. (B) RNA contains the pyrimidine called uracil instead of thymine.

way. The possibility that unraveling the nature of this assembly might provide insights into how genes work prompted Watson and Crick, among others, to try to solve the structure. According to Watson in his book *The Double Helix* (see Further Reading), their work was a desperate race against the famous American biochemist, Linus Pauling (Section 3.3.1), who initially proposed an incorrect triple helix model, giving Watson and Crick the time they needed to complete the double helix structure (Watson and Crick, 1953). It is now difficult to separate fact from fiction, especially regarding the part played by Rosalind Franklin, whose **X-ray diffraction** studies provided the bulk of the experimental data in support of the double helix and who was herself very close to solving the structure. The one thing that is clear is that the double helix, discovered by Watson and Crick on Saturday 7 March 1953, was the single most important break-through in biology during the 20th century.

The evidence that led to the double helix

Watson and Crick used four types of information to deduce the double helix structure:

■ *Biophysical data* of various kinds. The water content of DNA fibers was particularly important because it enabled the density of the DNA in a fiber to be estimated. The number of strands in the helix and the spacing between the nucleotides had to be compatible with the fiber density. Pauling's triple helix model was based on an incorrect density measure-

ment which suggested that the DNA molecule was more closely packed than it actually is.

■ *X-ray diffraction patterns* (Section 9.1.3), most of which were produced by Rosalind Franklin of Kings College, London, and which revealed the helical nature of the structure and indicated some of the key dimensions within the helix.

■ *The base ratios*, which had been discovered by Erwin Chargaff of Columbia University, New York. Chargaff carried out a lengthy series of chromatographic studies of DNA samples from various sources and showed that, although the values are different in different organisms, the amount of adenine is always the same as the amount of thymine, and the amount of guanine equals the amount of cytosine (*Figure 1.10*). These base ratios led to the

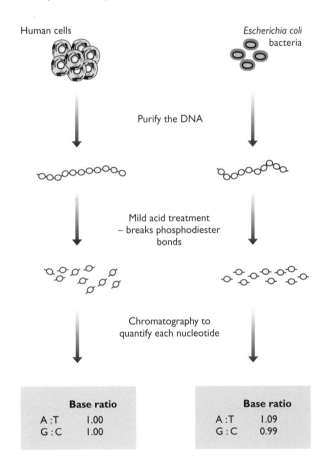

Figure 1.10 The base ratio experiments performed by Chargaff.

DNA was extracted from various organisms and treated with acid to hydrolyze the phosphodiester bonds and release the individual nucleotides. Each nucleotide was then quantified by chromatography. The data show some of the actual results obtained by Chargaff. These indicate that, within experimental error, the amount of adenine is the same as that of thymine, and the amount of guanine is the same as that of cytosine.

base-pairing rules, which were the key to the discovery of the double helix structure.

- *Model building*, which was the only major technique that Watson and Crick made use of themselves. Scale models of possible DNA structures enabled the relative positioning of the various atoms to be checked, to ensure that pairs of groups that formed bonds were not too far apart, and that other groups were not so close together as to interfere with one another.

The key features of the double helix

The double helix is right-handed, which means that if it were a spiral staircase and you were climbing upwards then the rail on the outside of the staircase would be on your right-hand side. The two strands run in opposite directions (*Figure 1.11A*). The helix is stabilized by two types of chemical interaction:

- **Base-pairing** between the two strands involves the formation of **hydrogen bonds** between an adenine on one strand and a thymine on the other strand, or between a cytosine and a guanine (*Figure 1.11B*). Hydrogen bonds are weak electrostatic attractions between an electronegative atom (such as oxygen or nitrogen) and a hydrogen atom attached to a second electronegative atom. Hydrogen bonds are longer than covalent bonds and are much weaker, typical bond energies being 1–10 kcal mol^{-1} at 25 °C, compared with up to 90 kcal mol^{-1} for a covalent bond. As well as their role in the DNA double helix, hydrogen bonds stabilize protein secondary structures. The two base-pair combinations – A base-paired with T, and G base-paired with C – explain the base ratios discovered by Chargaff. These are the only pairs that are permissible, partly because of the geometries of the nucleotide bases and the relative positions of the groups that are able to participate in hydrogen bonds, and partly because the pair must be between a purine and a pyrimidine; a purine–purine pair would be too big to fit within the helix, and a pyrimidine–pyrimidine pair would be too small.
- **Base-stacking**, sometimes called **π–π interactions**, involves hydrophobic interactions between adjacent base pairs and adds stability to the double helix once the strands have been brought together by base-pairing. These hydrophobic interactions arise because the hydrogen-bonded structure of water forces hydrophobic groups into the internal parts of a molecule.

Both base-pairing and base-stacking are important in holding the two polynucleotides together, but base-pairing has added significance because of its biological implications. The limitation that A can only base-pair with T, and G can only base-pair with C, means that DNA replication can result in perfect copies of a parent molecule through the simple expedient of using the sequences

---

> **Box 1.1: Base-pairing in RNA**
>
> RNA can also form base pairs, A pairing with U and G pairing with C. RNA can base-pair with a DNA molecule or with a second RNA molecule, but the predominant form of RNA base-pairing is intramolecular. This is seen with both ribosomal and transfer RNAs, two important types of RNA that we will encounter in later chapters. Base-pairing results in folded structures that enable these molecules to carry out their functions in expression of the genome (Sections 11.1.1 and 11.2.1). Within these RNAs, base-paired regions adopt helical conformations, but usually these extend for only a few tens of base pairs at most.

---

of the pre-existing strands to dictate the sequences of the new strands. This is **template-dependent DNA synthesis** and it is the system used by all cellular DNA polymerases (Section 4.1.1). Its counterpart, **template-dependent RNA synthesis**, is used by RNA polymerases to make RNA copies of genes, these copies preserving the biological information contained in the sequence of the genomic DNA molecule (Section 3.2.2). The only difference between DNA and RNA syntheses is that when RNA is made, the adenines in the DNA template do not specify thymines in the RNA copy. This is because RNA does not contain thymine; instead adenine pairs with uracil in DNA–RNA hybrids and in double-stranded RNA structures.

The double helix has structural flexibility

The double helix described by Watson and Crick, and shown in *Figure 1.11A*, is called the B-form of DNA. Its characteristic features lie in its dimensions: a helical diameter of 2.37 nm, a rise of 0.34 nm per base pair, and a pitch (i.e. distance taken up by a complete turn of the helix) of 3.4 nm, this corresponding to ten base pairs per turn. The DNA in living cells is thought to be predominantly in this B-form, but it is now clear that genomic DNA molecules are not entirely uniform in structure. This is mainly because each nucleotide in the helix has the flexibility to take up slightly different molecular shapes. To adopt these different conformations, the relative positions of the atoms in the nucleotide must change slightly. There are a number of possibilities but the most important conformational changes involve rotation around the β-N-glycosidic bond, changing the orientation of the base relative to the sugar, and rotation around the bond between the 3'- and 4'-carbons. Both rotations have a significant effect on the double helix: changing the base orientation influences the relative positioning of the two polynucleotides, and rotation around the 3'–4' bond affects the conformation of the sugar–phosphate backbone.

(A)

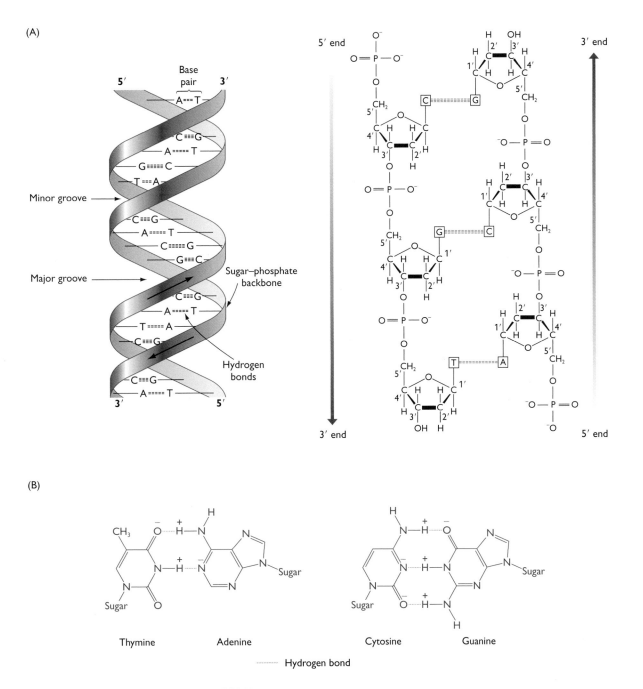

(B)

Thymine   Adenine   Cytosine   Guanine

·········· Hydrogen bond

Figure 1.11   The double helix structure of DNA.

(A) Two representations of the double helix. On the left the structure is shown with the sugar–phosphate 'backbones' of each polynucleotide drawn as a red ribbon with the base pairs in black. On the right the chemical structure for three base pairs is given. (B) A base-pairs with T, and G base-pairs with C. The bases are drawn in outline, with the hydrogen bonding indicated by dotted lines. Note that a G–C base pair has three hydrogen bonds whereas an A–T base pair has just two. The structures in part (A) are redrawn from Turner *et al.* (1997) (left) and Strachan and Read (1999) (right).

Rotations within individual nucleotides therefore lead to major changes in the overall structure of the helix. It has been recognized since the 1950s that changes in the dimensions of the double helix occur when fibers containing DNA molecules are exposed to different relative humidi-

ties. For example, the modified version of the double helix called the A-form (*Figure 1.12*) has a diameter of 2.55 nm, a rise of 0.29 nm per base pair and a pitch of 3.2 nm, corresponding to 11 base pairs per turn (*Table 1.1*). Other variations include B'-, C-, C'-, C''-, D-, E- and T-DNAs. All these

Table 1.1 Features of different conformations of the DNA double helix

| Feature | Conformation | | |
	B-DNA	A-DNA	Z-DNA
Type of helix	Right-handed	Right-handed	Left-handed
Helical diameter (nm)	2.37	2.55	1.84
Rise per base pair (nm)	0.34	0.29	0.37
Distance per complete turn (pitch) (nm)	3.4	3.2	4.5
Number of base pairs per complete turn	10	11	12
Topology of major groove	Wide, deep	Narrow, deep	Flat
Topology of minor groove	Narrow, shallow	Broad, shallow	Narrow, deep

---

Box 1.2: Units of length for DNA molecules

Because DNA is double-stranded, the lengths of molecules are described as so many **base pairs (bp)**. A **kilobase pair (kb)** is 10^3 bp and a **megabase pair (Mb)** is 10^6 bp. A **gigabase pair (Gb)** is 10^9 Mb. In summary:

  1 kb = 1000 bp
  1 Mb = 1000 kb = 1 000 000 bp
  1 Gb = 1000 Mb = 1 000 000 kb = 1 000 000 000 bp

Lengths of RNA molecules cannot be expressed in bp because most RNAs are single-stranded: their lengths must therefore be described as so many nucleotides.

---

attach to the double helix and regulate the activity of the genes contained within it. To carry out their function, each DNA-binding protein must attach at a specific position, near to the gene whose activity it must influence. This can be achieved, with a greater or lesser degree of ambiguity, by the protein reaching down into a groove, within which the DNA sequence can be 'read' without the helix being opened up by breaking the base pairs. A corollary of this is that a DNA-binding protein whose structure enables it to recognize a specific nucleotide sequence within, say, B-DNA might not be able to recognize that sequence if the DNA has taken up a different conformation. As we will see in Chapter 9, conformational variations along the length of a DNA molecule, together with other structural polymorphisms caused by the nucleotide sequence, could be important in determining the specificity of the interactions between the genome and its DNA-binding proteins.

1.2 The Human Genome

The critical feature of a DNA molecule is its nucleotide sequence. If the sequence of a DNA molecule is known then the genes that it contains can be identified and the activities of those genes can be studied in detail. Since the mid-1970s, molecular biologists have been able to obtain the sequences of longer and longer stretches of DNA, culminating in the 1990s with completion of the first complete sequences of entire genomes. The most important of these projects has been the one devoted to the human genome.

The **Human Genome Project** was conceived in 1984 and begun in earnest in 1990 with the primary aim of determining the nucleotide sequence of the entire human nuclear genome. The much smaller mitochondrial genome had been sequenced in the early 1980s (Anderson *et al.*, 1981). The project has been funded by governments and charities from across the world and has been the largest and most complex international collaboration ever attempted in any area of science. A second human genome project was set up by a private company – Celera Genomics of Maryland, USA – in 1998. Both projects completed a draft of the human genome sequence in 2001 and

are right-handed helices like the B-form. A more drastic reorganization is also possible, leading to the left-handed Z-DNA (*Figure 1.12*), a slimmer version of the double helix with a diameter of only 1.84 nm.

The bare dimensions of the various forms of the double helix do not reveal what are probably the most significant differences between them. These relate not to diameter and pitch, but to the extent to which the internal regions of the helix are accessible from the surface of the structure. As shown in *Figures 1.11* and *1.12*, the B-form of DNA does not have an entirely smooth surface; instead, two grooves spiral along the length of the helix. One of these grooves is relatively wide and deep and is called the **major groove**; the other is narrow and less deep and is called the **minor groove**. A-DNA also has two grooves (*Figure 1.12*), but with this conformation the major groove is even deeper, and the minor groove shallower and broader. Z-DNA is different again, with one groove virtually non-existent but the other very narrow and deep. In each form of DNA part of the internal surface of at least one of the grooves is formed by chemical groups attached to the nucleotide bases. In Chapter 9 we will see that expression of the biological information contained within a genome is mediated by DNA-binding proteins which

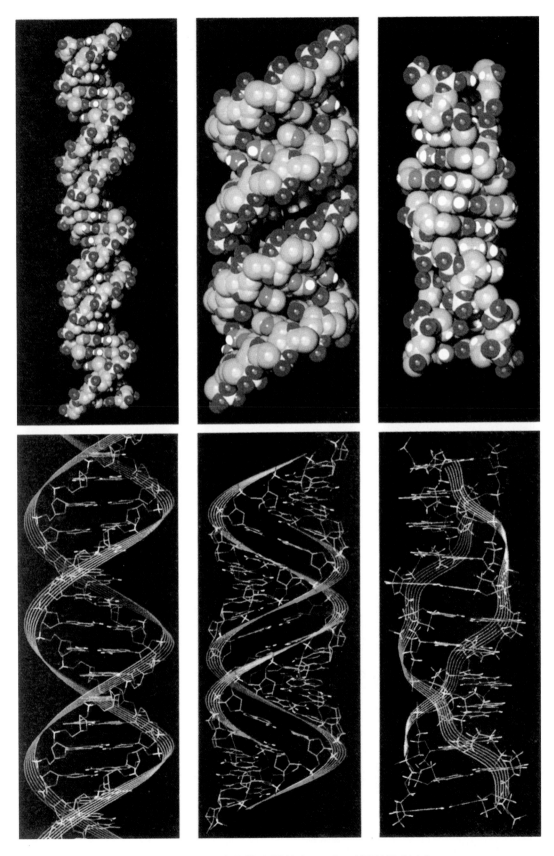

Figure 1.12   Computer-generated images of B-DNA (left), A-DNA (center) and Z-DNA (right).

Reprinted with permission from Kendrew A (ed.), *The Encyclopaedia of Molecular Biology*, Plate 1. Copyright 1994 Blackwell Science.

the results were published in the scientific journals *Nature* and *Science* in February of that year (IHGSC, 2001; Venter *et al.*, 2001). These drafts were not complete sequences, each representing only 83–84% of the entire genome, but their coverage was thought to include all of the most important parts of the genome, most of the remaining 16–17% being made up of sequences at the very ends of chromosomes (the **telomeres**) and around the **centromeres** (Section 2.2.1), where few, if any, genes are located (Bork and Copley, 2001).

Although only incomplete drafts, each of the genome projects has produced over 2.6 billion base pairs of sequence. This is such a large number that it is difficult to grasp the scale that it represents; an analogy is helpful. The typeface used for the text of this book enables approximately 60 nucleotides of DNA sequence to be written in a line 10 cm in length. If printed out in this format, the human genome sequence would stretch for 5000 km, the distance from Montreal to London, Los Angeles to Panama, Tokyo to Calcutta, Cape Town to Addis Ababa, or Auckland to Perth (*Figure 1.13*). The sequence would fill about 3000 books the size of this one. Understanding the sequence is clearly going to be an enormous task.

We should also bear in mind that although it is standard practice to refer to *the* human genome sequence, there are in fact many human genome sequences because every individual, except pairs of identical twins, have their own version. The differences between individual genomes are largely due to **single nucleotide polymorphisms (SNPs)**, positions in the genome where some individuals have one nucleotide (e.g. an A) and others have a different nucleotide (e.g. a G). Over 1.4 million SNPs have been identified, an average of one for every 2.0 kb of sequence (SNP Group, 2001). On average, every 2 kb also contains a **microsatellite** (also called a **short tandem repeat** or **STR**), which is a series of repeated nucleotides (e.g. CACACACA) in which the number of

repeats is variable in different individuals. Many of these SNPs and microsatellites have no effect on the function of the genome but many others do. For example, 60 000 SNPs lie within genes and at least some of these have an impact on the activities of these genes, leading to the variations that give each of us our own individual biological characteristics.

1.2.1 The content of the human nuclear genome

What do the DNA sequences reveal about the composition of the human nuclear genome? To begin we will examine a 50-kb segment of chromosome 7 (*Figure 1.14*), this segment forming part of the 'human β T-cell receptor locus', a much larger (685 kb) region of the genome that specifies proteins involved in the immune response (Rowen *et al.*, 1996). Our 50-kb segment contains the following genetic features:

- *One gene*. This gene is called TRY4 and it contains information for synthesis of the protein called trypsinogen, the inactive precursor of the digestive enzyme trypsin. TRY4 is one of a family of trypsinogen genes present in two clusters at either end of the β T-cell receptor locus. These genes have nothing to do with the immune response, they simply share this part of chromosome 7 with the β T-cell receptor locus. TRY4 is an example of a **discontinuous** gene, the information used in synthesis of the trypsinogen protein being split between five **exons**, separated by four non-coding **introns**.

- *Two gene segments*. These are V28 and V29-1, and each specifies a part of the β T-cell receptor protein after which the locus is named. V28 and V29-1 are not complete genes, only segments of a gene, and before being expressed they must be linked to other gene segments from elsewhere in the locus. This occurs in T lymphocytes and is an example of how a permanent change in the activity of the genome can arise during cellular differentiation (see Section 12.2.1). Like TRY4, both V28 and V29-1 are discontinuous.

- *One pseudogene*. A **pseudogene** is a non-functional copy of a gene, usually one whose nucleotide sequence has changed so that its biological information has become unreadable (see page 22). This particular pseudogene is called TRY5 and it is closely related to the functional members of the trypsinogen gene family.

- *52 genome-wide repeat sequences*. These are sequences that recur at many places in the genome. There are four main types of genome-wide repeat, called **LINEs** (long interspersed nuclear elements), **SINEs** (short interspersed nuclear elements), **LTR** (long terminal repeat) **elements** and **DNA transposons**. Examples of each type are seen in this short segment of the genome.

- *Two microsatellites*, which, as mentioned above, are sequences in which a short motif is repeated in tan-

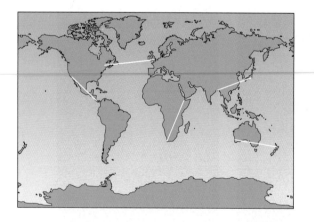

Figure 1.13   The immense length of the human genome.

The map illustrates the distance that would be covered by the human genome sequence if it were printed in the typeface used in this book.

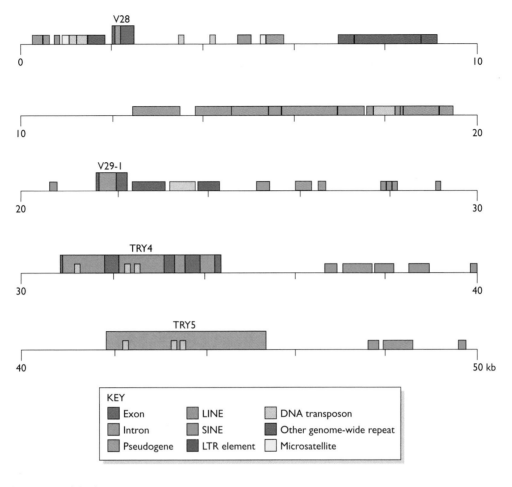

Figure 1.14 A segment of the human genome.

This map shows the location of genes, gene segments, pseudogenes, genome-wide repeats and microsatellites in a 50-kb segment of the human β T-cell receptor locus on chromosome 7. Redrawn from Rowen *et al.* (1996).

dem. One of the microsatellites seen here has the motif GA repeated 16 times, giving the sequence:

5′ –GAGAGAGAGAGAGAGA GAGAGAGAGAGAGAGA–3′
3′–CTCTCTCTCTCTCTCTCTCTCTCTCTCT–5′

The second microsatellite comprises six repeats of TATT.

■ Finally, approximately 50% of our 50-kb segment of the human genome is made up of stretches of non-genic, non-repetitive, single-copy DNA of no known function or significance.

Now we will look at these different components of the genome in greater detail.

Genes and related sequences

We look on the genes as the most important part of the human genome because these are the parts that contain biological information. Most genes specify one or more protein molecules, the 'expression' of these genes involving an RNA intermediate, called **messenger** or **mRNA**, which is transported from the nucleus to the cytoplasm where it directs synthesis of the protein coded by the gene (*Figure 1.15*). Other genes do not specify proteins, the end-products of their expression being **non-coding RNA**, which plays various roles in the cell (Section 3.2.1).

Each of the genes and gene segments present in the 50-kb sequence shown in *Figure 1.14* is discontinuous, the biological information being divided into a series of exons separated by non-coding introns. Most human genes are discontinuous, with an average of nine exons per gene, although some genes have many more than this. The record is held by the gene for a large muscle protein called titin, which has 178 exons and is also the longest known human gene at 80 780 bp. During gene expression, the initial RNA that is synthesized is a copy of the entire gene, including the introns as well as the exons. The process called **splicing** removes the introns from this **pre-mRNA** and joins the exons together to make the

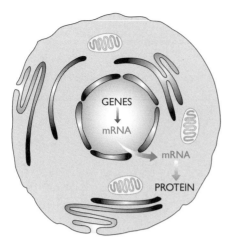

Figure 1.15 Messenger RNA (mRNA) is the intermediate between the genes and their protein products.

mRNA which eventually directs protein synthesis. At one time it was thought that splicing was a straightforward process, each exon being joined to its neighbor to produce a single mRNA from each discontinuous gene. Now it is known that many pre-mRNAs undergo **alternative** or **differential splicing**, giving rise to a series of mRNAs

containing different combinations of exons and each specifying a different protein (*Figure 1.16*). As well as the gene itself, the pre-mRNA transcribed from a gene also contains sequences from the regions preceding the first exon and following the last exon. These are called the **5'-untranslated region (5'-UTR)** and **3'-untranslated region (3'-UTR)**, respectively.

The various features of an 'average' human gene are shown in *Figure 1.17*. This diagram is useful for illustrating the components of a gene, but it should be remembered that many human genes do not conform with this 'average' structure. One type of variation is illustrated by V28 and V29-1 (see *Figure 1.14*). These are not intact genes but just gene segments, two of over 100 similar segments present at the β T-cell receptor locus. In individual T cells, the gene segments are linked together in various combinations to produce different functional receptor genes. This is not the same as splicing because the rearrangements do not involve the RNA transcribed from the genes, but the genes themselves. The final product is a gene that resembles the 'average' shown in *Figure 1.17*, but this gene is only assembled in the T cells. In other cells segments of the gene are scattered throughout the β T-cell receptor locus. There is also an α T-cell receptor locus, which contains 116 gene segments, and smaller γ and δ loci. Gene rearrangements also occur at the three loci that specify the various components of immunoglobulin proteins. We will examine these rearrangements in more

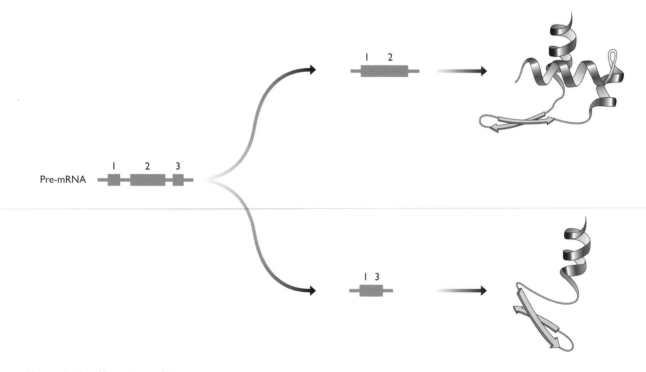

Figure 1.16 Alternative splicing.

Alternative splicing results in different combinations of exons becoming linked together, resulting in different proteins being synthesized from the same pre-mRNA.

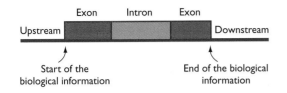
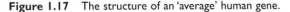

Figure 1.17 The structure of an 'average' human gene.

This gene has a two exons split by a single intron. For a protein-coding gene, the start of the biological information corresponds to the position of the initiation codon, and the end of the biological information is marked by the termination codon (Section 3.3.2). **'Upstream'** and **'downstream'** are two useful terms used to indicate the DNA sequences to either side of the gene.

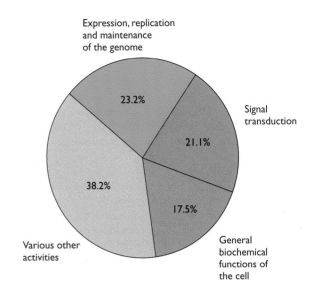

Figure 1.18 Categorization of the human gene catalog.

The pie chart shows a categorization of the identified human protein-coding genes. It omits approximately 13 000 genes whose functions are not yet known. The segment labeled 'various other activities' includes, among others, proteins involved in biochemical transport processes and protein folding, immunological proteins, and structural proteins. Based on *Figure 15* of Venter *et al.* (2001).

detail in Section 12.2.1 when we look at their impact on the regulation of genome expression. The point that we should bear in mind at the moment is that it is tempting to look on the gene segments at the T-cell receptor and immunoglobulin loci as 'unusual' because they do not conform to our standard representation of a gene as shown in *Figure 1.17*, but this is not a productive way to approach the study of genomes. By concentrating attention on a simplistic view of the 'average' gene we risk losing sight of the variations and embellishments that are critical to the overall activity of the genome.

The functions of human genes

The functions of about half of the 30 000–40 000 human genes are known or can be inferred with a reasonable degree of certainty. The vast majority code for proteins; less than 2500 specify the various types of non-coding RNA. Almost a quarter of the protein-coding genes are involved in expression, replication and maintenance of the genome (*Figure 1.18*) and another 20% specify components of the **signal transduction** pathways that regulate genome expression and other cellular activities in response to signals received from outside of the cell (Section 12.1). All of these genes can be looked on as having a function that is involved in one way or another with the activity of the genome. Enzymes responsible for the general biochemical functions of the cell account for another 17.5% of the known genes; the remainder are involved in activities such as transport of compounds into and out of cells, the folding of proteins into their correct three-dimensional structures, the immune response, and synthesis of structural proteins such as those found in the cytoskeleton and in muscles. It is possible that as the human gene catalog is made more complete the relative proportions of the genes in the three major categories in *Figure 1.18* will decrease. This is because these major categories represent the most studied areas of cell biology, which means that many of the relevant genes can be rec-

ognized because their protein products are known. Genes whose products have not yet been identified are more likely to be involved in the less well studied areas of cellular activity.

One thing that the gene catalog cannot tell us, and will not be able to tell us even when it is complete, is what makes a human being. The minimalist approach to molecular biology, whereby the study of individual genes or groups of genes is expected to lead eventually to a full biomolecular description of how a human being is constructed and functions, has been dealt a severe blow by the draft genome sequences. There are no amazing revelations about what makes humans different from apes. Even when the chimpanzee genome has been completely sequenced (which will not be for several years) it may still not be possible simply from genome comparisons to determine what makes us human (Baltimore, 2001). On the basis of gene number we are only three times more complex than a fruit fly and only twice as complex as the microscopic worm *Caenorhabditis elegans*. More detailed studies of how the human genome functions may reveal key features that underlie some of the special attributes of human beings, but genomics will never explain why a human was able to compose Mozart's 40th symphony, or indeed why it was composed by Mozart and not by an ordinary human.

Box 1.3: How many genes are there in the human genome?

The two genome projects differ in their estimate of the total number of genes in the human genome but the figure is almost certainly in the range 30 000–40 000 (Bork and Copley, 2001). The uncertainty arises because it can be quite difficult to recognize which sequences are genes and which are not, for reasons that we will explore in detail in Chapter 7. The numbers are much lower than was previously anticipated, with a 'best guess' of 80 000–100 000 still in vogue up to a few months before the draft sequences were completed (Pennisi, 2000). These early estimates were high because they were based on the supposition that, in most cases, a single gene spec-ifies a single mRNA and a single protein. According to this model, the number of genes in the human genome should be similar to the number of proteins in human cells, leading to the estimates of 80 000–100 000. The discovery that the number of genes is much lower than this indicates that differential splicing is more prevalent than was originally appreciated, and also suggests that the link between mRNA and protein might be more complex than previously thought. Processes such as **RNA editing** (Section 10.3.2), which enables a single mRNA to code for more than one protein, have been known about for some time but were considered to be rare. The small number of genes in the human genome might indicate that editing, as well as other 'unusual' phenomena, play a more important role in genome expression than we originally thought.

Pseudogenes and other evolutionary relics

The segment of chromosome 7 shown in *Figure 1.14* contains a single pseudogene, a non-functional copy of a gene. Pseudogenes are a type of evolutionary relic, an indication that the human genome is continually undergoing change. There are two main types of pseudogene:

- A **conventional pseudogene** is a gene that has been inactivated because its nucleotide sequence has changed by **mutation** (Section 14.1). Many mutations have only minor effects on the activity of a gene but some are more important and it quite possible for a single nucleotide change to result in a gene becoming completely non-functional. Once a pseudogene has become non-functional it will degrade through accumulation of more mutations and eventually will no longer be recognizable as a gene relic. TRY5 is an example of a conventional pseudogene.
- A **processed pseudogene** arises not by evolutionary decay but by an abnormal adjunct to gene expression. A processed pseudogene is derived from the mRNA copy of a gene by synthesis of a DNA copy which subsequently re-inserts into the genome (*Figure 1.19*). Because a processed pseudogene is a copy of an mRNA molecule, it does not contain any introns that were present in its parent gene. It also lacks the nucleotide sequences immediately upstream of the 5'-UTR of the parent gene, which is the region in which the signals used to switch on expression of the parent gene are located. The absence of these signals means that a processed pseudogene is inactive.

As well as pseudogenes, genomes also contain other evolutionary relics in the form of **truncated genes**, which lack a greater or lesser stretch from one end of the complete gene, and **gene fragments**, which are short isolated regions from within a gene (*Figure 1.20*). We will return to these relics when we look at multigene families in Chapter 2.

Genome-wide repeats and microsatellites

The draft sequences have shown that approximately 62% of the human genome comprises **intergenic regions**, the parts of the genome that lie between genes and which have no known function. These sequences used to be called **junk DNA** but the term is falling out of favor, partly because the number of surprises resulting from genome research over the last few years has meant that

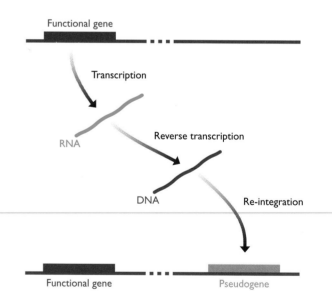

Figure 1.19 The origin of a processed pseudogene.

A processed pseudogene is thought to arise by integration into the genome of a copy of the mRNA transcribed from a functional gene. The process by which mRNA is copied into DNA is called reverse transcription and the product is called complementary DNA (cDNA). The cDNA may integrate into the same chromosome as its functional parent, or possibly into a different chromosome.

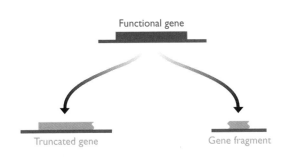

Figure 1.20 A truncated gene and a gene fragment.

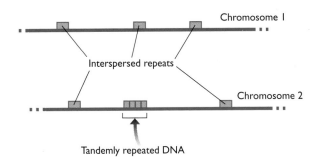

Figure 1.21 The two types of repetitive DNA: interspersed repeats and tandemly repeated DNA.

molecular biologists have become less confident in asserting that any part of the genome is unimportant simply because we do not currently know what its function might be. One thing that is clear is that the bulk of the intergenic DNA is made up of repeated sequences of one type or another. Because repeated sequences are important features of all genomes we will deal with them in detail during our general survey of genome anatomies in Chapter 2. Here we will limit ourselves to the key features of the human repeats.

Repetitive DNA can be divided into two categories (*Figure 1.21*): genome-wide or **interspersed repeats**, whose individual repeat units are distributed around the genome

in an apparently random fashion, and **tandemly repeated DNA**, whose repeat units are placed next to each other in an array. All four types of human genome-wide repeat – SINEs, LINEs, LTR elements and DNA transposons – are represented in *Figure 1.14*. An interesting feature of these genome-wide repeats is that each type appears to be derived from a **transposable element**, a mobile segment of DNA which is able to move around the genome from one place to another. Many of these elements leave copies of themselves when they move, which explains how they

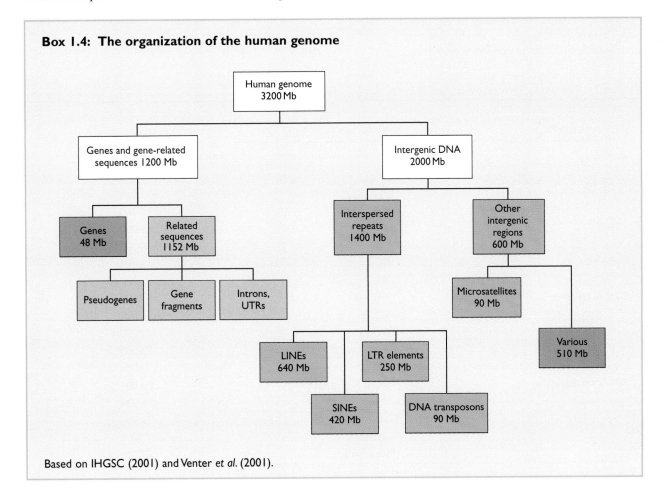

Box 1.4: The organization of the human genome

Based on IHGSC (2001) and Venter *et al.* (2001).

propagate and become common throughout the genome. There are two main classes of transposable element: those that transpose via an RNA intermediate and those that do not (Section 2.4.2). LINEs, SINEs and LTR elements are examples of the first class, and DNA transposons are examples of the second class. The four types of genome-wide repeat are distinguished because of their characteristic sequence features, but there are many variations and each type can be divided into a number of subcategories (*Table 1.2*). SINEs, for example, which are the most numerous genome-wide repeat, comprise three subtypes: Alu elements (approximately 1 090 000 copies in the genome), MIR (393 000 copies) and MIR3 (75 000 copies). Altogether, genome-wide repeats make up 44% of the draft genome sequences, some 1400 Mb of DNA.

The microsatellites shown in *Figure 1.14* are examples of tandemly repeated DNA. In a microsatellite the repeat unit is short – up to 13 bp in length – but other types of tandemly repeated sequence have longer units (Section 2.4.1). The commonest type of human microsatellite are dinucleotide repeats, with approximately 140 000 copies in the genome as whole (*Table 1.3*), about half of these being repeats of the motif 'CA'. Single-nucleotide repeats (e.g. AAAAA) are the next most common (about 120 000 in total). As with genome-wide repeats, it is not clear if microsatellites have a function. It is known that they arise through an error in the process responsible for copying of the genome during cell division (Section 14.1.1), and they might simply be unavoidable products of genome replication.

1.2.2 The human mitochondrial genome

The complete sequence of the human mitochondrial genome has been known for over 20 years (Anderson *et al.*, 1981). At just 16 569 bp, it is much smaller than the nuclear genome, and it contains just 37 genes. Thirteen of these genes code for proteins involved in the respiratory complex, the main biochemical component of the energy-generating mitochondria; the other 24 specify non-coding RNA molecules that are required for expression of the mitochondrial genome. The genes in this genome are much more closely packed than in the nuclear genome (*Figure 1.22*) and they do not contain introns. In many respects, the human mitochondrial genome is typical of the mitochondrial genomes of other animals. We will consider it in more detail when we look at organelle genomes in Chapter 2.

1.3 Why is the Human Genome Project Important?

The human genome has been the focus of biological research for the last decade and will continue to be the center of attention for many years to come. Why is all this activity being devoted to the human genome? There are many reasons.

First, the human gene catalog, containing a description of the sequence of every gene in the genome, will be immensely valuable, even if for many years the functions

Table 1.2   The types of genome-wide repeats in the human genome

Type of repeat	Subtype	Approximate number of copies in the human genome
SINEs		1 558 000
	Alu	1 090 000
	MIR	393 000
	MIR3	75 000
LINEs		868 000
	LINE-1	516 000
	LINE-2	315 000
	LINE-3	37 000
LTR elements		443 000
	ERV class I	112 000
	ERV(K) class II	8000
	ERV(L) class III	83 000
	MaLR	240 000
DNA transposons		294 000
	hAT	195 000
	Tc-1	75 000
	PiggyBac	2000
	Unclassified	22 000

Taken from IHGSC (2001). The numbers are approximate and are likely to be under-estimates (Li *et al.*, 2001).

Table 1.3 Microsatellites in the human genome

Length of repeat unit	Approximate number of copies in the human genome
1	120 000
2	140 000
3	37 500
4	105 000
5	56 000
6	49 000
7	27 000
8	35 500
9	27 500
10	27 500
11	28 000

From IHGSC (2001).

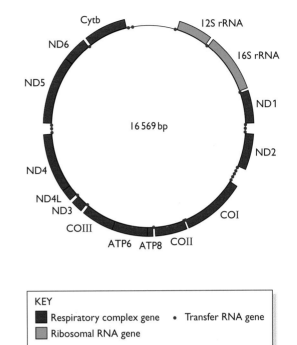

KEY

■ Respiratory complex gene   • Transfer RNA gene
■ Ribosomal RNA gene

Figure 1.22 The human mitochondrial genome.

The human mitochondrial genome is small and compact, with little wasted space, so much so that the ATP6 and ATP8 genes overlap. Abbreviations: ATP6, ATP8, genes for ATPase subunits 6 and 8; COI, COII, COIII, genes for cytochrome *c* oxidase subunits I, II and III; Cytb, gene for apocytochrome *b*; ND1–ND6, genes for NADH hydrogenase subunits 1–6. Ribosomal RNA and transfer RNA are two types of non-coding RNA (Section 3.2.1).

of some of the genes remain unknown. Not only will the catalog contain the sequences of the coding parts of every gene, it will also include the regulatory regions for these genes. Some of these genes are the ones that, when they function incorrectly, give rise to a genetic disease. The human gene catalog will provide rapid access to these genes, enabling the underlying basis to these diseases to be studied, hopefully leading to strategies for treatment and management.

While the catalog is being completed, attention will focus more and more on the **transcriptome** and **proteome** (Chapter 3), which are the keys to understanding how the information contained in the genome is utilized by the cell. The Human Genome Project, and the similar projects currently being carried out with other species' genomes, therefore opens the way to a comprehensive description of the molecular activities of human cells and the ways in which these activities are controlled. This is central to the continued development, not only of molecular biology and genetics, but also of those areas of biochemistry, cell biology and physiology now described as the **molecular life sciences**.

The genome projects will have additional benefits that at present can only be guessed at. We have seen that the human genome, in common with the genomes of many other organisms, contains extensive amounts of intergenic DNA. We think that most of the intergenic DNA has no function, but perhaps this is because we do not know enough about it. Could the intergenic DNA have a role,

but one that at present is too subtle for us to grasp? The first step in addressing this possibility is to obtain a complete description of the organization of the intergenic DNA in different genomes, so that common features, which might indicate a role for some or all of these sequences, can be identified.

There is one final reason for genome projects. The work stretches current technology to its limits. Genome analysis therefore represents the frontier of molecular biology, territory that was inaccessible just a few years ago and which still demands innovative approaches and a lot of sheer hard work. Scientists have always striven to achieve the almost impossible, and the motivation for many molecular biologists involved in genome projects is, quite simply, the challenge of the unknown.

References

Anderson S, Bankier AT, Barriel AG, et al. (1981) Sequence and organization of the human mitochondrial genome. *Nature*, **290**, 457–474.

Avery OT, MacLeod CM and McCarty M (1944) Studies on the chemical nature of the substance inducing transformation of pneumococcal types. *J. Exp. Med.*, **79**, 137–158.

Baltimore D (2001) Our genome unveiled. *Nature*, **409**, 814–816.

Bork P and Copley R (2001) Filling in the gaps. *Nature*, **409**, 818–820.

Griffith F (1928) The significance of pneumococcal types. *J. Hygiene*, **27**, 113–159.

Hershey AD and Chase M (1952) Independent functions of viral protein and nucleic acid in growth of bacteriophage. *J. Gen. Physiol.*, **36**, 39–56.

IHGSC (International Human Genome Sequencing Consortium) (2001) Initial sequencing and analysis of the human genome. *Nature*, **409**, 860–921.

Li W-H, Gu Z, Wang H and Nekrutenko A (2001) Evolutionary analyses of the human genome. *Nature*, **409**, 847–849.

Pennisi E (2000) And the gene number is . . .? *Science*, **288**, 1146–1147.

Rowen L, Koop BF and Hood L (1996) The complete 685-kilobase DNA sequence of the human β T cell receptor locus. *Science*, **272**, 1755–1762.

SNP Group (The International SNP Map Working Group) (2001) A map of human genome sequence variation containing 1.42 million single nucleotide polymorphisms. *Nature*, **409**, 928–933.

Strachan T and Read AP (1999) *Human Molecular Genetics,* 2nd edition. BIOS Scientific Publishers, Oxford.

Turner PC, McLennan AG, Bates AD and White MRH (1997) *Instant Notes in Molecular Biology*. BIOS Scientific Publishers, Oxford.

Venter JC, Adams MD, Myers EW, et al. (2001) The sequence of the human genome. *Science*, **291**, 1304–1351.

Watson JD and Crick FHC (1953) Molecular structure of nucleic acids: a structure for deoxyribose nucleic acid. *Nature*, **171**, 737–738.

Further Reading

Brock TD (1990) *The Emergence of Bacterial Genetics.* Cold Spring Harbor Laboratory Press, New York. — *A detailed history that puts into context the work on the transforming principle and the Hershey–Chase experiment.*

Judson HF (1979) *The Eighth Day of Creation.* Jonathan Cape, London. — *A highly readable account of the development of molecular biology up to the 1970s.*

Kay LE (1993) *The Molecular Vision of Life.* Oxford University Press, Oxford. — *Contains a particularly informative explanation of why genes were once thought to be made of protein.*

Lander ES and Weinberg RA (2000) Genomics: journey to the center of biology. *Science*, **287**, 1777–1782. — *A brief description of genetics and molecular biology from Mendel to the human genome sequence.*

McKusick VA (1989) The Human Genome Organisation: history, purposes and membership. *Genomics*, **5**, 385–387. — *Describes the original goals of the Human Genome Project.*

Orel V (1995) *Gregor Mendel: The First Geneticist.* Oxford University Press, Oxford. — *A fascinating biography.*

Watson JD (1968) *The Double Helix.* Atheneum, London. — *The most important discovery of 20th century biology, written as a soap opera.*

STUDY AIDS FOR CHAPTER 1

Key terms

Give short definitions of the following terms:

β-*N*-glycosidic bond
π–π interaction
3′ terminus
3′ untranslated region
5′ terminus
5′ untranslated region
Adenine
Alternative splicing
Bacteriophage
Base pair
Base-pairing
Base ratio
Base-stacking
Biological information
Capsid
Centromere
Chromosome theory
Conventional pseudogene
Cytochemistry
Cytosine
Deoxyribonuclease
Differential splicing
Diploid

Discontinuous gene
DNA
DNA polymerase
DNA transposon
Double helix
Downstream
Exon
Gamete
Gene
Gene fragment
Genome
Genome-wide repeat
Gigabase pair
Guanine
Haploid
Human Genome Project
Hydrogen bond
Intergenic region
Interspersed repeat
Intron
Junk DNA
Kilobase pair
LINE

LTR element
Lysogenic infection cycle
Lytic infection cycle
Major groove
Megabase pair
Messenger RNA
Microsatellite
Minor groove
Mitochondrial genome
Molecular biologist
Molecular life sciences
Mutation
Non-coding RNA
Nuclear genome
Nucleic acid
Nucleoside
Nucleotide
Pentose
Phage
Phosphodiester bond
Pre-mRNA
Processed pseudogene
Protease
Pseudogene
Purine
Pyrimidine

Ribose
RNA
RNA editing
Sex cell
Short tandem repeat
Signal transduction
SINE
Single nucleotide polymorphism
Somatic cell
Splicing
Tandemly repeated DNA
Telomere
Template-dependent DNA synthesis
Template-dependent RNA synthesis
Thymine
Transformation
Transforming principle
Transposable element
Truncated gene
Unit factor
Upstream
Uracil
X-ray diffraction

Self study questions

1. Give a brief description of the two components of the human genome.
2. DNA and genes were both discovered in the 1860s. Explain why the connection between two was not made until 80 years later.
3. Why did biologists originally think that protein is the genetic material?
4. Describe the two experiments that indicated that genes are made of DNA.
5. Draw a fully annotated diagram of the structure of a short DNA polynucleotide containing each of the four nucleotides. Indicate the changes that you would have to make if the drawing was of RNA not DNA.
6. Describe the evidence that led Watson and Crick to deduce that a cellular DNA molecule is a double helix.
7. Distinguish between base-pairing and base-stacking. What influences do these two types of interaction have on the structural flexibility of the double helix? Your answer should include a description of the conformations of the major and minor grooves in the various forms of the double helix.
8. Return to your diagram from Question 5. Show how your DNA polynucleotide could be extended by template-dependent DNA synthesis.
9. What differences and similarities would you expect to find if you compared your genome with those of your parents?
10. Make a photocopy of *Figure 1.14*. Annotate your photocopy as completely as you can by giving descriptions of the various components of the genomes that are present in this segment.
11. Draw and annotate the structure of an 'average' human gene. How do the gene segments V28 and V29-1 differ from this 'average'?
12. Outline the functional categories within the human gene catalog.
13. Distinguish between the two types of pseudogene.
14. Define the terms 'interspersed repetitive DNA' and 'tandemly repeated DNA' and give examples of both classes in the human genome.

Problem-based learning

1. To what extent were the results of the Avery and Hershey–Chase experiments accepted by the scientific community of the 1940s and 1950s?
2. The text (page 13) states that Watson and Crick discovered the double helix structure on Saturday 7 March 1953. Justify this statement.
3. If Watson and Crick had not existed then who would have discovered the double helix structure?
4. Is the 50-kb segment shown in *Figure 1.14* a good choice with which to illustrate the organization of the human genome?
5. Other than the example shown in *Figure 1.18*, in what ways might the human gene catalog be categorized? Would any other categorization be more informative than the functional one shown in *Figure 1.18*?

Genome Anatomies

Chapter Contents

2.1	*An Overview of Genome Anatomies*	**30**
	2.1.1 Genomes of eukaryotes	31
	2.1.2 Genomes of prokaryotes	33
2.2	*The Anatomy of the Eukaryotic Genome*	**35**
	2.2.1 Eukaryotic nuclear genomes	35
	2.2.2 Eukaryotic organelle genomes	46
2.3	*The Anatomy of the Prokaryotic Genome*	**49**
	2.3.1 The physical structure of the prokaryotic genome	50
	2.3.2 The genetic organization of the prokaryotic genome	54
2.4	*The Repetitive DNA Content of Genomes*	**59**
	2.4.1 Tandemly repeated DNA	59
	2.4.2 Interspersed genome-wide repeats	60

Learning outcomes

When you have read Chapter 2, you should be able to:

■ Draw diagrams illustrating the major differences between the genetic organizations of the genomes of humans, plants, insects, yeast and bacteria, and give an explanation for the C-value paradox

■ Describe the DNA–protein interactions that give rise to the chromatosome and the 30 nm chromatin fiber

■ State the functions of centromeres and telomeres and list their special structural features

■ Explain why chromosome banding patterns and the isochore model suggest that genes are not evenly distributed in eukaryotic chromosomes

■ Outline the differences between the gene contents of different eukaryotic genomes and explain, with examples, what is meant by 'multigene family'

■ Describe the physical features and gene contents of mitochondrial and chloroplast genomes and discuss the current hypothesis concerning the origins of organelle genomes

■ Describe the physical structure of the *Escherichia coli* genome and indicate the ways in which this structure is and is not typical of other prokaryotes

■ Define, with examples, the term 'operon'

■ Explain why prokaryotic genome sequences have complicated the species concept

■ Speculate on the content of the minimal prokaryotic genome and on the identity of distinctiveness genes

■ Define the term 'satellite DNA' and distinguish between satellite, minisatellite and microsatellite DNA

■ Give examples of the various types of RNA and DNA transposons, and outline their transposition pathways

THE HUMAN GENOME is by no means the only genome for which a complete or a draft sequence is available. By February 2001, when the draft human sequence was published, drafts were also available for the yeast *Saccharomyces cerevisiae*, the microscopic worm *Caenorhabditis elegans*, the fruit fly *Drosophila melanogaster*, the plant *Arabidopsis thaliana*, and one of the chromosomes of the malaria parasite *Plasmodium falciparum*. In addition, complete sequences had been obtained for over 30 microorganisms, including the bacteria *Escherichia coli* and *Mycobacterium tuberculosis* (*Table 2.1*). Work is progressing apace on the genomes of, among others, rice and the mouse, and plans are being made to start the chimpanzee genome project, the results of which will enable a direct comparison between the human genome and that of our closest non-human relative. The completed and ongoing projects are revealing a great deal about how genomes are organized, including a number of unexpected discoveries that have taken molecular biologists by surprise. In this chapter we will survey the information that has arisen from genome projects and merge this information with the knowledge that was acquired in the pre-genomic era of molecular biology.

2.1 An Overview of Genome Anatomies

Biologists recognize that the living world comprises two types of organism (*Figure 2.1*):

1. **Eukaryotes**, whose cells contain membrane-bound compartments, including a nucleus and organelles such as mitochondria and, in the case of plant cells, chloroplasts. Eukaryotes include animals, plants, fungi and protozoa.

2. **Prokaryotes**, whose cells lack extensive internal compartments. There are two very different groups of

Table 2.1 Examples of genomes for which a complete or draft sequence has been published

Species	Size of genome (Mb)	Approximate number of genes	References
Eukaryotes			
Arabidopsis thaliana (plant)	125	25 500	AGI (2000)
Caenorhabditis elegans (nematode worm)	97	19 000	CESC (1998)
Drosophila melanogaster (fruit fly)	180	13 600	Adams *et al.* (2000)
Homo sapiens (human)	3200	30 000–40 000	IHGSC (2001); Venter *et al.* (2001)
Saccharomyces cerevisiae (yeast)	12.1	5800	Goffeau *et al.* (1996)
Bacteria			
Escherichia coli K12	4.64	4400	Blattner *et al.* (1997)
Mycobacterium tuberculosis H37Rv	4.41	4000	Cole *et al.* (1998)
Mycoplasma genitalium	0.58	500	Fraser *et al.* (1995)
Pseudomonas aeruginosa PA01	6.26	5700	Stover *et al.* (2000)
Streptococcus pneumoniae	2.16	2300	Tettelin *et al.* (2001)
Vibrio cholerae El Tor N16961	4.03	4000	Heidelberg *et al.* (2000)
Yersinia pestis CO92	4.65	4100	Parkhill *et al.* (2001)
Archaea			
Archaeoglobus fulgidus	2.18	2500	Klenk *et al.* (1997)
Methanococcus jannaschii	1.66	1750	Bult *et al.* (1996)

For bacteria species, the strain designation (e.g. 'K12') is given if specified by the group who sequenced the genome. With many bacterial species, different strains have different genome sizes and gene contents (Section 2.3.2).

prokaryotes, distinguished from one another by characteristic genetic and biochemical features:

a. the **bacteria**, which include most of the commonly encountered prokaryotes such as the gram-negatives (e.g. *E. coli*), the gram-positives (e.g. *Bacillus subtilis*), the cyanobacteria (e.g. *Anabaena*) and many more;

b. the **archaea**, which are less well-studied, and have mostly been found in extreme environments such as hot springs, brine pools and anaerobic lake bottoms.

Eukaryotes and prokaryotes have quite different types of genome and we must therefore consider them separately.

2.1.1 Genomes of eukaryotes

Humans are fairly typical eukaryotes and the human genome is in many respects a good model for eukaryotic genomes in general. All of the eukaryotic nuclear genomes that have been studied are, like the human version, divided into two or more linear DNA molecules, each contained in a different chromosome; all eukaryotes also possess smaller, usually circular, mitochondrial genomes. The only general eukaryotic feature not illustrated by the human genome is the presence in plants and other photosynthetic organisms of a third genome, located in the chloroplasts.

Although the basic physical structures of all eukaryotic nuclear genomes are similar, one important feature is very different in different organisms. This is genome size, the smallest eukaryotic genomes being less than 10 Mb in length, and the largest over 100 000 Mb. As can be seen in *Table 2.2*, this size range coincides to a certain extent with the complexity of the organism, the simplest eukaryotes such as fungi having the smallest genomes, and higher eukaryotes such as vertebrates and flowering plants having the largest ones. This might appear to make sense as one would expect the complexity of an organism to be related to the number of genes in its genome – higher eukaryotes need larger genomes to accommodate the extra genes. However, the correlation is far from precise: if it was, then the nuclear genome of the yeast *S. cerevisiae*, which at 12 Mb is 0.004 times the size of the human nuclear genome, would be expected to contain $0.004 \times 35\,000$ genes, which is just 140. In fact the *S. cerevisiae* genome contains about 5800 genes.

For many years the lack of precise correlation between the complexity of an organism and the size of its genome was looked on as a bit of a puzzle, the so-called **C-value paradox**. In fact the answer is quite simple: space is saved in the genomes of less complex organisms because the genes are more closely packed together. The *S. cerevisiae* genome, the sequence of which was completed in 1996, illustrates this point, as we can see from the top two parts of *Figure 2.2*, where the 50-kb segment of the human

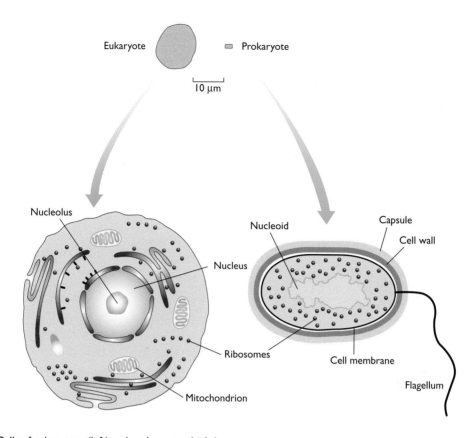

Figure 2.1 Cells of eukaryotes (left) and prokaryotes (right).

The top part of the figure shows a typical human cell and typical bacterium drawn to scale. The human cell is 10 μm in diameter and the bacterium is rod-shaped with dimensions of 1 × 2 μm. The lower drawings show the internal structures of eukaryotic and prokaryotic cells. Eukaryotic cells are characterized by their membrane-bound compartments, which are absent from prokaryotes. The bacterial DNA is contained in the structure called the nucleoid.

genome that we looked at in Chapter 1 is compared with a 50-kb segment of the yeast genome. The yeast genome segment, which comes from chromosome III (the first eukaryotic chromosome to be sequenced; Oliver *et al.*, 1992), has the following distinctive features:

- *It contains more genes than the human segment.* This region of yeast chromosome III contains 26 genes thought to code for proteins and two that code for **transfer RNAs (tRNAs)**, short non-coding RNA molecules involved in reading the genetic code during protein synthesis (Section 3.2.1).
- *Relatively few of the yeast genes are discontinuous.* In this segment of chromosome III none of the genes are discontinuous. In the entire yeast genome there are only 239 introns, compared with over 300 000 in the human genome.
- *There are fewer genome-wide repeats.* This part of chromosome III contains a single long terminal repeat (LTR) element, called *Ty2*, and four truncated LTR elements called delta sequences. These five genome-wide repeats make up 13.5% of the 50-kb segment, but this figure is not entirely typical of the

yeast genome as a whole. When all 16 yeast chromosomes are considered, the total amount of sequence taken up by genome-wide repeats is only 3.4% of the total. In humans, the genome-wide repeats make up 44% of the genome.

The picture that emerges is that the genetic organization of the yeast genome is much more economical than that of the human version. The genes themselves are more compact, having fewer introns, and the spaces between the genes are relatively short, with much less space taken up by genome-wide repeats and other non-coding sequences.

The hypothesis that more complex organisms have less compact genomes holds when other species are examined. The third part of *Figure 2.2* shows a 50-kb segment of the fruit-fly genome (Adams *et al.*, 2000). If we agree that a fruit fly is more complex than a yeast cell but less complex than a human then we would expect the organization of the fruit-fly genome to be intermediate between that of yeast and humans. This is what we see in *Figure 2.2C*, this 50-kb segment of the fruit-fly genome having 11 genes, more than in the human segment but fewer than in

Table 2.2 Sizes of eukaryotic genomes

Species	Genome size (Mb)
Fungi	
Saccharomyces cerevisiae	12.1
Aspergillus nidulans	25.4
Protozoa	
Tetrahymena pyriformis	190
Invertebrates	
Caenorhabditis elegans	97
Drosophila melanogaster	180
Bombyx mori (silkworm)	490
Strongylocentrotus purpuratus (sea urchin)	845
Locusta migratoria (locust)	5000
Vertebrates	
Takifugu rubripes (pufferfish)	400
Homo sapiens	3200
Mus musculus (mouse)	3300
Plants	
Arabidopsis thaliana (vetch)	125
Oryza sativa (rice)	430
Zea mays (maize)	2500
Pisum sativum (pea)	4800
Triticum aestivum (wheat)	16 000
Fritillaria assyriaca (fritillary)	120 000

Table 2.3 Compactness of the yeast, fruit-fly and human genomes

Feature	Yeast	Fruit fly	Human
Gene density (average number per Mb)	479	76	11
Introns per gene (average)	0.04	3	9
Amount of the genome that is taken up by genome-wide repeats	3.4%	12%	44%

coding for the alcohol dehydrogenase enzymes (San-Miguel *et al.*, 1996). This is the only gene in this 50-kb region, although there is a second one, of unknown function, approximately 100 kb beyond the right-hand end of the sequence shown here. Instead of genes, the dominant feature of this genome segment is the genome-wide repeats. The majority of these are of the LTR element type, which comprise virtually all of the non-coding part of the segment, and on their own are estimated to make up approximately 50% of the maize genome. It is becoming clear that one or more families of genome-wide repeats have undergone a massive proliferation in the genomes of certain species. This may provide an explanation for the most puzzling aspect of the C-value paradox, which is not the general increase in genome size that is seen in increasingly complex organisms, but the fact that similar organisms can differ greatly in genome size. A good example is provided by *Amoeba dubia* which, being a protozoan, might be expected to have a genome of 100–500 kb, similar to other protozoa such as *Tetrahymena pyriformis* (see *Table 2.2*). In fact the *Amoeba* genome is over 200 000 Mb. Similarly, we might guess that the genomes of crickets are similar in size to those of other insects, but these bugs have genomes of approximately 2000 Mb, 11 times that of the fruit fly.

2.1.2 Genomes of prokaryotes

Prokaryotic genomes are very different from eukaryotic ones. There is some overlap in size between the largest prokaryotic and smallest eukaryotic genomes, but on the whole prokaryotic genomes are much smaller. For example, the *E. coli* K12 genome is just 4639 kb, two-fifths the size of the yeast genome, and has only 4405 genes. The physical organization of the genome is also different in eukaryotes and prokaryotes. The traditional view has been that an entire prokaryotic genome is contained in a single circular DNA molecule. As well as this single 'chromosome', prokaryotes may also have additional genes on independent smaller, circular or linear DNA molecules called **plasmids** (*Figure 2.3*). Genes carried by plasmids are useful, coding for properties such as antibiotic resistance or the ability to utilize complex compounds such as toluene as a carbon source, but plasmids appear to be dispensable – a prokaryote can exist quite effectively

the yeast sequence. All of these genes are discontinuous, but seven have just one intron each. The picture is similar when the entire genome sequences of the three organisms are compared (*Table 2.3*). The gene density in the fruit-fly genome is intermediate between that of yeast and humans, and the average fruit-fly gene has many more introns than the average yeast gene but still three times fewer than the average human gene.

The comparison between the yeast, fruit-fly and human genomes also holds true when we consider the genome-wide repeats (see *Table 2.3*). These make up 3.4% of the yeast genome, about 12% of the fruit-fly genome, and 44% of the human genome. It is beginning to become clear that the genome-wide repeats play an intriguing role in dictating the compactness or otherwise of a genome. This is strikingly illustrated by the maize genome, which at 5000 Mb is larger than the human genome but still relatively small for a flowering plant. Only a few limited regions of the maize genome have been sequenced, but some remarkable results have been obtained, revealing a genome dominated by repetitive elements. *Figure 2.2D* shows a 50-kb segment of this genome, either side of one member of a family of genes

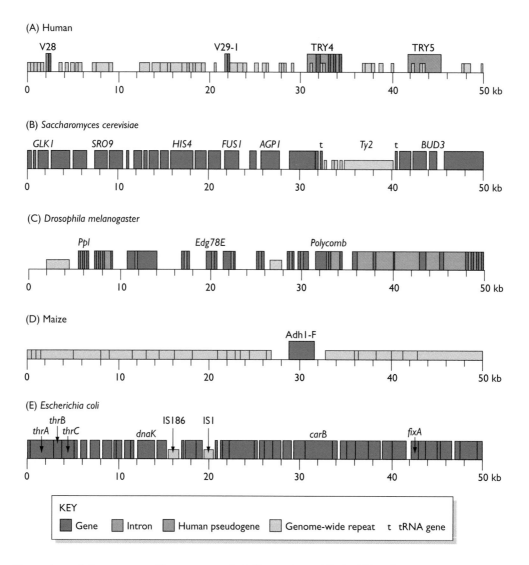

Figure 2.2 Comparison of the genomes of humans, yeast, fruit flies, maize and *Escherichia coli*.

(A) is the 50-kb segment of the human β T-cell receptor locus shown in *Figure 1.14*. This is compared with 50-kb segments from the genomes of (B) *Saccharomyces cerevisiae* (chromosome III; redrawn from Oliver *et al.*, 1992); (C) *Drosophila melanogaster* (redrawn from Adams *et al.*, 2000); (D) maize (redrawn from SanMiguel *et al.*, 1996) and (E) *E. coli* K12 (redrawn from Blattner *et al.*, 1997). See the text for more details.

without them. We now know that this traditional view of the prokaryotic genome has been biased by the extensive research on *E. coli*, which has been accompanied by the mistaken assumption that *E. coli* is a typical prokaryote. In fact, prokaryotes display a considerable diversity in genome organization, some having a unipartite genome, like *E. coli*, but others being more complex. *Borrelia burgdorferi* B31, for example, has a linear chromosome of 911 kb, carrying 853 genes, accompanied by 17 or 18 linear and circular molecules, which together contribute another 533 kb and at least 430 genes (Fraser *et al.*, 1997). Multipartite genomes are now known in many other bacteria and archaea.

In one respect, *E. coli* is fairly typical of other prokaryotes. After our discussion of eukaryotic gene organization, it will probably come as no surprise to learn that prokaryotic genomes are even more compact than those of yeast and other lower eukaryotes. We can see this fact illustrated in *Figure 2.2E*, which shows a 50-kb segment of the *E. coli* K12 genome. It is immediately obvious that there are more genes and less space between them, with 43 genes taking up 85.9% of the segment. Some genes have virtually no space between them: *thrA* and *thrB*, for example, are separated by a single nucleotide, and *thrC* begins at the nucleotide immediately following the last nucleotide of *thrB*. These three genes are an example of an

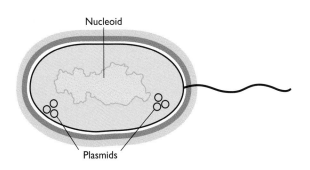

Figure 2.3 Plasmids are small circular DNA molecules that are found inside some prokaryotic cells.

operon, a group of genes involved in a single biochemical pathway (in this case, synthesis of the amino acid threonine) and expressed in conjunction with one another. Operons have been used as model systems for understanding how gene expression is regulated (Section 9.3.1). In general, prokaryotic genes are shorter than their eukaryotic counterparts, the average length of a bacterial gene being about two-thirds that of a eukaryotic gene, even after the introns have been removed from the latter (Zhang, 2000). Bacterial genes appear to be slightly longer than archaeal ones.

Two other features of prokaryotic genomes can be deduced from *Figure 2.2E*. First, there are no introns in the genes present in this segment of the *E. coli* genome. In fact *E. coli* has no discontinuous genes at all, and it is generally believed that this type of gene structure is virtually absent in prokaryotes, the few exceptions occurring mainly among the archaea. The second feature is the infrequency of repetitive sequences. Most prokaryotic genomes do not have anything equivalent to the high-copy-number genome-wide repeat families found in eukaryotic genomes. They do, however, possess certain sequences that might be repeated elsewhere in the genome, examples being the **insertion sequences** IS1 and IS186 that can be seen in the 50-kb segment shown in *Figure 2.2E*. These are further examples of transposable elements, sequences that have the ability to move around the genome and, in the case of insertion elements, to transfer from one organism to another, even sometimes between two different species (see page 64). The positions of the IS1 and IS186 elements shown in *Figure 2.2E* refer only to the particular *E. coli* isolate from which this sequence was obtained: if a different isolate is examined then the IS sequences could well be in different positions or might be entirely absent from the genome. Most other prokaryotic genomes have very few repeat sequences – there are virtually none in the 1.64 Mb genome of *Campylobacter jejuni* NCTC11168 (Parkhill *et al.*, 2000b) – but there are exceptions, notably the meningitis bacterium *Neisseria meningitidis* Z2491, which has over 3700 copies of 15 different types of repeat sequence, collectively making up almost 11% of the 2.18 Mb genome (Parkhill *et al.*, 2000a).

2.2 The Anatomy of the Eukaryotic Genome

We have already learnt that the human genome is split into two components: the nuclear genome and the mitochondrial genome (see *Figure 1.1*, page 5). This is the typical pattern for most eukaryotes, the bulk of the genome being contained in the chromosomes in the cell nucleus and a much smaller part located in the mitochondria and, in the case of photosynthetic organisms, in the chloroplasts. We will look first at the nuclear genome.

2.2.1 Eukaryotic nuclear genomes

The nuclear genome is split into a set of linear DNA molecules, each contained in a chromosome. No exceptions to this pattern are known: all eukaryotes that have been studied have at least two chromosomes and the DNA molecules are always linear. The only variability at this level of eukaryotic genome structure lies with chromosome number, which appears to be unrelated to the biological features of the organism. For example, yeast has 16 chromosomes, four times as many as the fruit fly. Nor is chromosome number linked to genome size: some salamanders have genomes 30 times bigger than the human version but split into half the number of chromosomes. These comparisons are interesting but at present do not tell us anything useful about the genomes themselves; they are more a reflection of the non-uniformity of the evolutionary events that have shaped genome architecture in different organisms.

Packaging of DNA into chromosomes

Chromosomes are much shorter than the DNA molecules that they contain: the average human chromosome has just under 5 cm of DNA. A highly organized packaging system is therefore needed to fit a DNA molecule into its chromosome. We must understand this packaging system before we start to think about how genomes function because the nature of the packaging has an influence on the processes involved in expression of individual genes (Section 8.2).

The important breakthroughs in understanding DNA packaging were made in the early 1970s by a combination of biochemical analysis and electron microscopy. It was already known that nuclear DNA is associated with DNA-binding proteins called **histones** but the exact nature of the association had not been delineated. In 1973–74 several groups carried out **nuclease protection experiments** on **chromatin** (DNA–histone complexes) that had been gently extracted from nuclei by methods designed to retain as much of the chromatin structure as possible. In a nuclease protection experiment the complex is treated with an enzyme that cuts the DNA at positions that are not 'protected' by attachment to a protein. The sizes of the resulting DNA fragments indicate the positioning of the protein complexes on the original DNA molecule (*Figure 2.4*). After limited nuclease treatment of purified chromatin, the bulk of the DNA fragments have lengths of approximately 200 bp and multiples thereof, suggesting a regular spacing of histone proteins along the DNA.

In 1974 these biochemical results were supplemented by electron micrographs of purified chromatin, which enabled the regular spacing inferred by the protection experiments to be visualized as beads of protein on the string of DNA (*Figure 2.5A*). Further biochemical analysis indicated that each bead, or **nucleosome**, contains eight histone protein molecules, these being two each of histones H2A, H2B, H3 and H4. Structural studies have shown that these eight proteins form a barrel-shaped **core octamer** with the DNA wound twice around the outside (*Figure 2.5B*). Between 140 and 150 bp of DNA (depending on the species) are associated with the nucleosome particle, and each nucleosome is separated by 50–70 bp of linker DNA, giving the repeat length of 190–220 bp previously shown by the nuclease protection experiments.

As well as the proteins of the core octamer, there is a group of additional histones, all closely related to one another and collectively called **linker histones**. In vertebrates these include histones H1a–e, H1°, H1t and H5. A single linker histone is attached to each nucleosome, to form the **chromatosome**, but the precise positioning of this linker histone is not known. Structural studies support the traditional model in which the linker histone acts as a clamp, preventing the coiled DNA from detaching from the nucleosome (*Figure 2.5C*; Zhou *et al.*, 1998; Travers, 1999). However, other results suggest that, at least in some organisms, the linker histone is not located on the extreme surface of the nucleosome–DNA assembly, as would be expected if it really were a clamp, but instead is inserted between the core octamer and the DNA (Pruss *et al.*, 1995; Pennisi, 1996).

The 'beads-on-a-string' structure shown in *Figure 2.5A* is thought to represent an unpacked form of chromatin that occurs only infrequently in living nuclei. Very gentle cell breakage techniques developed in the mid-1970s resulted in a more condensed version of the complex,

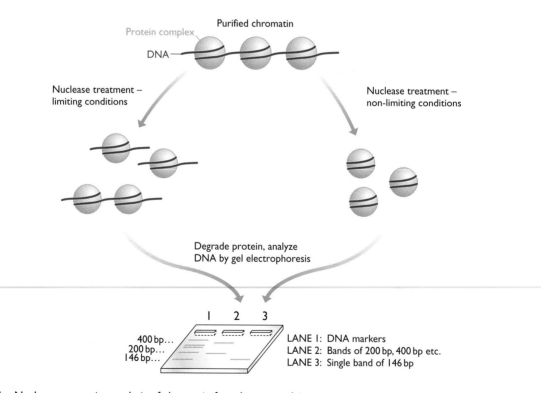

Figure 2.4 Nuclease protection analysis of chromatin from human nuclei.

Chromatin is gently purified from nuclei and treated with a nuclease enzyme. On the left, the nuclease treatment is carried out under limiting conditions so that the DNA is cut, on average, just once in each of the linker regions between the bound proteins. After removal of the protein, the DNA fragments are analyzed by **agarose gel electrophoresis** (see Technical Note 2.1) and found to be 200 bp in length, or multiples thereof. On the right, the nuclease treatment proceeds to completion, so all the DNA in the linker regions is digested. The remaining DNA fragments are all 146 bp in length. The results show that in this form of chromatin, protein complexes are spaced along the DNA at regular intervals, one for each 200 bp, with 146 bp of DNA closely attached to each protein complex.

Agarose gel electrophoresis

Separation of DNA and RNA molecules of different lengths

Gel electrophoresis is the standard method for separating DNA molecules of different lengths. It has many applications in size analysis of DNA fragments and can also be used to separate RNA molecules (see Technical Note 4.4, page 120).

Electrophoresis is the movement of charged molecules in an electric field: negatively charged molecules migrate towards the positive electrode, and positively charged molecule migrate towards the negative electrode. The technique was originally carried out in aqueous solution, but this is not particularly useful for DNA separations because the predominant factors influencing migration rate in solution phase are the shape of a molecule and its electric charge. Most DNA molecules are the same shape (linear) and although the charge of a DNA molecule is dependent on its length, the differences in charge are not sufficient to result in effective separation. The situation is different when electrophoresis is carried out in a gel, because now shape and charge are less important and molecular length is the critical determinant of migration rate. This is because the gel is a network of pores through which the DNA molecules have to travel to reach the positive electrode. Shorter molecules are less impeded by the pores than are longer molecules and so move through the gel more quickly. Molecules of different lengths therefore form bands in the gel.

Two types of gel are used in molecular biology: **agarose** gels, as described here, and **polyacrylamide** gels, which are covered in Technical Note 6.1 (page 165). Agarose is a polysaccharide that forms gels with pores ranging from 100 to 300 nm in diameter, the size depending on the concentration of agarose in the gel. Gel concentration therefore determines the range of DNA fragments that can be separated. For example, a 0.3% gel can be used for molecules between 5 and 50 kb, and a 5% gel for 100–500 bp molecules (Brown, 1998). The separation range is also affected by the electroendosmosis value (EEO) of the agarose, this

being a measure of the amount of bound sulfate and pyruvate anions. The greater the EEO, the slower the migration rate for a negatively charged molecule such as DNA.

An agarose gel is prepared by mixing the appropriate amount of agarose powder in a buffer solution, heating to dissolve the agarose, and then pouring the molten gel onto a Perspex plate with tape around the sides to prevent spillage. A comb is placed in the gel to form wells for the samples. The gel is allowed to set and the electrophoresis then carried out with the gel submerged under buffer. In order to follow the progress of the electrophoresis, one or two dyes of known migration rates are added to the DNA samples before loading. The bands of DNA are visualized by soaking the gel in ethidium bromide solution, this compound intercalating between DNA base pairs and fluorescing when activated with ultraviolet radiation.

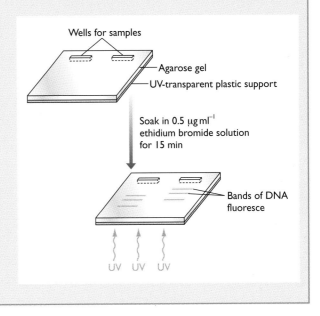

Wells for samples — Agarose gel — UV-transparent plastic support

Soak in 0.5 $\mu g\,ml^{-1}$ ethidium bromide solution for 15 min

Bands of DNA fluoresce

UV UV UV

---

called the **30 nm fiber** (it is approximately 30 nm in width). The exact way in which nucleosomes associate to form the 30 nm fiber is not known, but several models have been proposed, the most popular of which is the solenoid structure shown in *Figure 2.6*. The individual nucleosomes within the 30 nm fiber may be held together by interactions between the linker histones, or the attachments may involve the core histones, whose protein 'tails' extend outside the nucleosome (see *Figure 8.9*, page

230). The latter hypothesis is attractive because chemical modification of these tails results in the 30 nm fiber opening up, enabling genes contained within it to be activated (Section 8.2.1).

The special features of metaphase chromosomes

The 30 nm fiber is probably the major type of chromatin in the nucleus during interphase, the period between

(A)

(B)

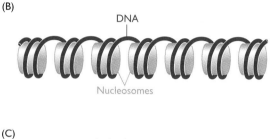

DNA

Nucleosomes

(C)

Linker histone

Figure 2.5   Nucleosomes.

(A) Electron micrograph of a purified chromatin strand showing the 'beads-on-a-string' structure. (Courtesy of Dr Barbara Hamkalo, University of California, Irvine.) (B) The model for the 'beads-on-a-string' structure in which each bead is a barrel-shaped nucleosome with the DNA wound twice around the outside. Each nucleosome is made up of eight proteins: a central tetramer of two histone H3 and two histone H4 subunits, plus a pair of H2A–H2B dimers, one above and one below the central tetramer (see *Figure 8.9*, page 230). (C) The precise position of the linker histone relative to the nucleosome is not known but, as shown here, the linker histone may act as a clamp, preventing the DNA from detaching from the outside of the nucleosome.

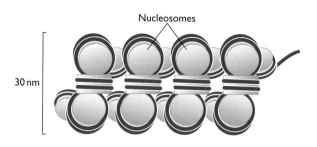

Nucleosomes

30 nm

Figure 2.6   The solenoid model for the 30 nm chromatin fiber.

In this model, the 'beads-on-a-string' structure of chromatin is condensed by winding the nucleosomes into a helix with six nucleosomes per turn. Higher levels of chromatin packaging are described in Section 8.1.2.

istic for a particular chromosome. This means that the set of chromosomes possessed by an organism can be represented as a **karyogram**, in which the banded appearance of each one is depicted. The human karyogram is shown in *Figure 2.8*.

Both the DNA in the centromere regions, and the proteins attached to it, have special features. The nucleotide sequence of centromeric DNA is best understood in the plant *Arabidopsis thaliana*, whose amenity to genetic analysis has enabled the positions of the centromeres on the DNA sequence to be located with some precision. Also, a special effort was made to sequence these centromeric regions, which are frequently excluded from draft genome sequences because of problems in obtaining an accurate reading through the highly repetitive structure that characterize these regions. *Arabidopsis* centromeres span 0.9–1.2 Mb of DNA and each one is made up largely of 180-bp repeat sequences. In humans the equivalent sequences are 171 bp and are called **alphoid DNA**. Before the *Arabidopsis* sequences were obtained it was thought that these repeat sequences were by far the principal component of centromeric DNA. However, *Arabidopsis* centromeres also contain multiple copies of genome-wide

nuclear divisions. When the nucleus divides, the DNA adopts a more compact form of packaging, resulting in the highly condensed **metaphase chromosomes** that can be seen with the light microscope and which have the appearance generally associated with the word 'chromosome' (*Figure 2.7*). The metaphase chromosomes form at a stage in the **cell cycle** after DNA replication has taken place and so each one contains two copies of its chromosomal DNA molecule. The two copies are held together at the **centromere**, which has a specific position within each chromosome. Individual chromosomes can therefore be recognized because of their size and the location of the centromere relative to the two ends. Further distinguishing features are revealed when chromosomes are stained. There are a number of different staining techniques (*Table 2.4*), each resulting in a banding pattern that is character-

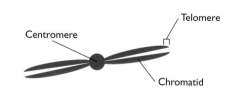

Centromere

Telomere

Chromatid

Figure 2.7   The typical appearance of a metaphase chromosome.

Metaphase chromosomes are formed after DNA replication has taken place, so each one is, in effect, two chromosomes linked together at the centromere. The arms are called the **chromatids**. A telomere is the extreme end of a chromatid.

Box 2.1: Unusual chromosome types

The karyograms of some organisms display unusual features not displayed by the human version. These include the following:

- **Minichromosomes** are relatively short in length but rich in genes. The chicken genome, for example, is split into 39 chromosomes: six **macrochromosomes** containing 66% of the DNA but only 25% of the genes, and 33 minichromosomes containing the remaining one-third of the genome and 75% of the genes. The gene density in the minichromosomes is therefore some six times greater than that in the macrochromosomes (McQueen et al., 1998).
- **B chromosomes** are additional chromosomes possessed by some individuals in a population, but not all. They are common in plants (Jones, 1995) and also known in fungi, insects and animals. B chromosomes appear to be fragmentary versions of normal chromosomes that result from unusual events during nuclear division. Some contain genes, often for rRNAs, but it is not clear if these genes are active. The presence of B chromosomes can affect the biological characteristics of the organism, particularly in plants where they are associated with reduced viability. It is presumed that B chromosomes are gradually lost from cell lineages as a result of irregularities in their inheritance pattern.
- **Holocentric chromosomes** do not have a single centromere but instead have multiple kinetochores spread along their length. *Caenorhabditis elegans* has holocentric chromosomes (Albertson and Thomson, 1982).

Table 2.4 Staining techniques used to produce chromosome banding patterns

Technique	Procedure	Banding pattern
G-banding	Mild proteolysis followed by staining with Giemsa	Dark bands are AT-rich Pale bands are GC-rich
R-banding	Heat denaturation followed by staining with Giemsa	Dark bands are GC-rich Pale bands are AT-rich
Q-banding	Stain with quinacrine	Dark bands are AT-rich Pale bands are GC-rich
C-banding	Denature with barium hydroxide and then stain with Giemsa	Dark bands contain constitutive heterochromatin (see Section 8.1.2)

---

repeats, along with a few genes, the latter at a density of 7–9 per 100 kb compared with 25 genes per 100 kb for the non-centromeric regions of *Arabidopsis* chromosomes (Copenhaver et al., 1999). The discovery that centromere DNA contains genes was a big surprise because it was thought that these regions were genetically inactive.

The special centromeric proteins in humans include at least seven that are not found elsewhere in the chromosome (Warburton, 2001). One of these proteins, CENP-A, is very similar to histone H3 and is thought to replace this histone in the centromeric nucleosomes. It is assumed that the small distinctions between CENP-A and H3 confer special properties on centromeric nucleosomes, but exactly what these properties might be and how they relate to the function of the centromere is not yet known. Part of the function of the centromere itself is revealed by the electron microscope, which shows that in a dividing cell a pair of plate-like **kinetochores** are present on the surface of the chromosome in the centromeric region. These structures act as the attachment points for the microtubules that radiate from the spindle pole bodies located at the nuclear surface and which draw the divided chromosomes into the daughter nuclei (*Figure 2.9*). Part of the kinetochore is made up of alphoid DNA plus CENP-A and other proteins, but its structure has not been described in detail (Vafa and Sullivan, 1997).

A second important part of the chromosome is the terminal region or **telomere**. Telomeres are important because they mark the ends of chromosomes and therefore enable the cell to distinguish a real end from an unnatural end caused by chromosome breakage – an essential requirement because the cell must repair the latter but not the former. Telomeric DNA is made up of hundreds of copies of a repeated motif, 5'–TTAGGG–3' in humans, with a short extension of the 3' terminus of the double-stranded DNA molecule (*Figure 2.10*). Two special proteins bind to the repeat sequences in human telomeres. These are called TRF1, which helps to regulate the length of the telomere, and TRF2, which maintains the single-strand extension. If TRF2 is inactivated then this extension is lost and the two polynucleotides fuse together in a covalent linkage (van Steensel et al., 1998). Other telomeric proteins are thought to form a linkage between the telomere and the periphery of the nucleus,

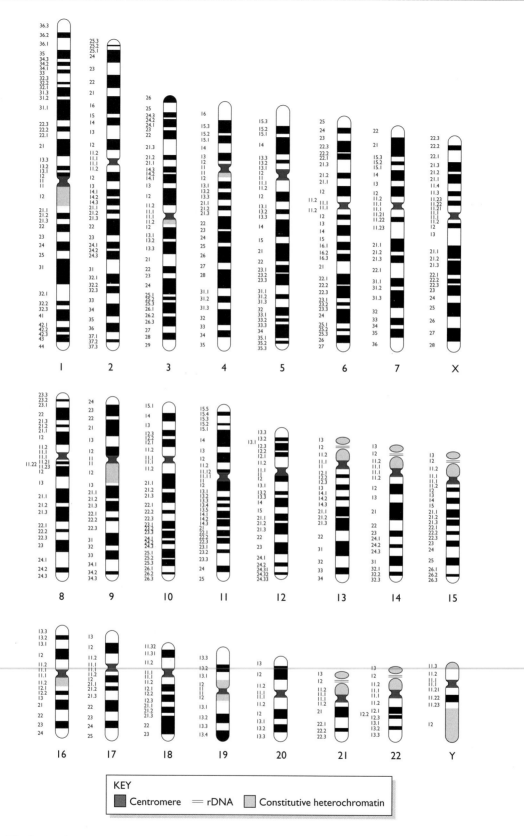

Figure 2.8 The human karyogram.

The chromosomes are shown with the G-banding pattern obtained after Giemsa staining. Chromosome numbers are given below each structure and the band numbers to the left. 'rDNA' is a region containing a cluster of repeat units for the ribosomal RNA genes, which specify a type of non-coding RNA (Section 3.2.1). 'Constitutive heterochromatin' is very compact chromatin which has few or no genes (Section 8.1.2). Redrawn from Strachan and Read (1999).

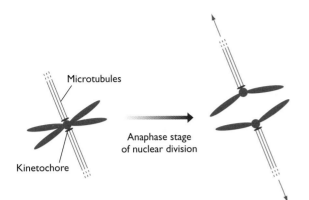

Figure 2.9 The role of the kinetochores during nuclear division.

During the anaphase period of nuclear division (see *Figures 5.14* and *5.15*, pages 137 and 138), individual chromosomes are drawn apart by the contraction of microtubules attached to the kinetochores.

Figure 2.10 Telomeres.

The sequence at the end of a human telomere. The length of the 3′ extension is different in each telomere. See Section 13.2.4 for more details about telomeric DNA.

the area in which the chromosome ends are localized (Tham and Zakian, 2000). Still others mediate the enzymatic activity that maintains the length of each telomere during DNA replication. We will return to this last activity in Section 13.2.4: it critical to the survival of the chromosome and may be a key to understanding cell senescence and death.

Where are the genes in a eukaryotic genome?

In the previous section we learnt that *Arabidopsis* centromeres contain genes but at a lesser density than that in the rest of the chromosomes. This alerts us to the fact that the genes are not arranged evenly along the length of a chromosome. In most organisms, genes appear to be distributed more at less at random, with substantial variations in gene density at different positions within a chromosome. The average gene density in *Arabidopsis* is 25 genes per 100 kb, but even outside of the centromeres and telomeres the density varies from 1 to 38 genes per 100 kb, as illustrated in *Figure 2.11* for the largest of the plant's five chromosomes. The same is true for human

chromosomes, where the density ranges from 0 to 64 genes per 100 kb.

The uneven gene distribution within human chromosomes was suspected for several years before the draft sequence was completed. There were two lines of evidence, one of which related to the banding patterns that are produced when chromosomes are stained. The dyes used in these procedures (see *Table 2.4*) bind to DNA molecules, but in most cases with preferences for certain base pairs. Giemsa, for example, has a greater affinity for DNA regions that are rich in A and T nucleotides. The dark G-bands in the human karyogram (see *Figure 2.8*) are therefore thought to be AT-rich regions of the genome. The base composition of the genome as a whole is 59.7% A + T so the dark G-bands must have AT contents substantially greater than 60%. Cytogeneticists therefore predicted that there would be fewer genes in dark G-bands because genes generally have AT contents of 45–50%. This prediction was confirmed when the draft genome sequence was compared with the human karyogram (IHGSC, 2001).

The second line of evidence pointing to uneven gene distribution derived from the **isochore** model of genome organization (Gardiner, 1996). According to this model, the genomes of vertebrates and plants (and possibly of other eukaryotes) are mosaics of segments of DNA, each at least 300 kb in length, with each segment having a uniform base composition that differs from that of the adjacent segments. Support for the isochore model comes from experiments in which genomic DNA is broken into fragments of approximately 100 kb, treated with dyes that bind specifically to AT- or GC-rich regions, and the pieces separated by density gradient centrifugation (Technical Note 2.2). When this experiment is carried out with human DNA, five fractions are seen, each representing a different isochore type with a distinctive base composition: two AT-rich isochores, called L1 and L2, and three GC-rich classes: H1, H2 and H3. The last of these, H3, is the least abundant in the human genome, making up only 3% of the total, but contains over 25% of the genes. This is a clear indication that genes are not distributed evenly through the human genome. The draft genome sequence suggests that the isochore theory over-simplifies what is, in reality, a much more complex pattern of variations in base composition along the length of each human chromosome (IHGSC, 2001). But even if it turns out to be a misconception, the isochore theory has played an important role in helping molecular biologists of the pre-sequence era to understand genome structure.

What genes are present in a eukaryotic genome?

There are various ways to categorize the genes in a eukaryotic genome. One possibility is to classify the genes according to their function, as shown in *Figure 1.18* (page 21) for the human genome. This system has the advantage that the fairly broad functional categories used in *Figure 1.18* can be further subdivided to produce

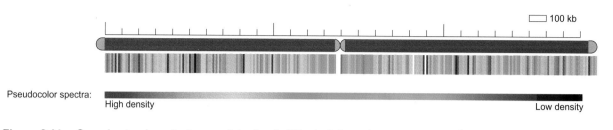

Pseudocolor spectra:

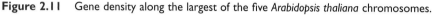

High density   Low density

Figure 2.11   Gene density along the largest of the five *Arabidopsis thaliana* chromosomes.

Chromosome 1, which is 29.1 Mb in length, is illustrated with the sequenced portions shown in red and the centromere and telomeres in blue. The gene map below the chromosome gives gene density in pseudocolor, from deep blue (low density) to red (high density). The density varies from 1 to 38 genes per 100 kb. Reprinted with permission from AGI (The Arabidopsis Genome Initiative), *Nature*, **408**, 797–815. Copyright 2000 Macmillan Magazines Limited.

a hierarchy of increasingly specific functional descriptions for smaller and smaller sets of genes. The weakness with this approach is that functions have not yet been assigned to many eukaryotic genes, so this type of classification leaves out a proportion of the total gene set. A more powerful method is to base the classification not on the functions of genes but on the structures of the proteins that they specify. A protein molecule is constructed from a series of **domains**, each of which has a particular biochemical function. Examples are the **zinc finger**, which is one of several domains that enable a protein to bind to a DNA molecule (Section 9.1.4), and the 'death domain', which is present in many proteins involved in **apoptosis**, the process of programmed cell death. Each domain has a characteristic amino acid sequence, perhaps not exactly the same sequence in every example of that domain, but close enough for the presence of a particular domain to be recognizable by examining the amino acid sequence of the protein. The amino acid sequence of a protein is specified by the nucleotide

---

Ultracentrifugation techniques

TECHNICAL 2.2 NOTE

Methodology for separation of cell components and large molecules

The development of high-speed centrifuges in the 1920s led to techniques for separating organelles and other fractions from disrupted cells. The first technique to be used was **differential centrifugation**, in which pellets of successively lighter cell components are collected by centrifuging cell extracts at different speeds. Intact nuclei, for example, are relatively large and can be collected from a cell extract by centrifugation at 1000 *g* for 10 minutes; mitochondria, being lighter, require centrifugation at 20 000 *g* for 20 minutes. Different cell components can be obtained in fairly pure form by careful manipulation of the centrifugation parameters.

   Density gradient centrifugation was used for the first time in 1951. In this procedure, the cell fraction is not centrifuged in a normal aqueous solution. Instead, a sucrose solution is layered into the tube in such a way that a density gradient is formed, the solution being more concentrated and hence denser towards the bottom of the tube. The cell fraction is placed on the top of the gradient and the tube centrifuged at a very high speed: at least 500 000 *g* for several hours. Under these conditions, the rate of migration of a cell component through the gradient depends on its **sedimentation coefficient**, which in turn depends on its molecular mass and shape. For example, eukaryotic ribosomes have a sedimentation coefficient of 80S (S stands for Svedberg units, Svedberg being the Swedish scientist who pioneered the biological applications of ultracentrifugation), whereas bacterial ribosomes, being smaller, have a sedimentation coefficient of 70S.

   In a second type of density gradient centrifugation a solution such as 8 M cesium chloride is used, which is substantially denser than the sucrose solution used to measure S values. The starting solution is uniform, the gradient being established during the centrifugation. Cellular components migrate down through the centrifuge tube but molecules such as DNA and proteins do not reach the bottom; instead each one comes to rest at a position where the density of the matrix equals its own **buoyant density** (see *Figure 2.24*, page 60). This technique has many applications in molecular biology, being able to separate DNA fragments of different base compositions and DNA molecules with different conformations (e.g. supercoiled, circular and linear DNA). It can also distinguish between normal DNA and DNA labeled with a heavy isotope of nitrogen (Section 13.1.1).

sequence of its gene, so the domains present in a protein can be determined from the nucleotide sequence of the gene that codes for that protein. The genes in a genome can therefore be categorized according to the protein domains that they specify. This method has the advantage that it can be applied to genes whose functions are not known and hence can encompass a larger proportion of the set of genes in a genome.

Classification schemes based on gene function suggest that all eukaryotes possess the same basic set of genes, but that more complex species have a greater number of genes in each category. For example, humans have the greatest number of genes in all but one of the categories used in *Figure 2.12*, the exception being 'metabolism' where *Arabidopsis* comes out on top as a result of its photosynthetic capability, which requires a large set of genes not present in the other four genomes included in this comparison. This functional classification reveals other interesting features, notably that *C. elegans* has a relatively high number of genes whose functions are involved in cell–cell signaling, which is surprising given that this organism has just 959 cells. Humans, who have 10^{13} cells, have only 250 more genes for cell–cell signaling. In general, this type of analysis emphasizes the similarities between genomes, but does not reveal the genetic basis of the vastly different types of biological information con-

tained in the genomes of, for example, fruit flies and humans. The domain approach holds more promise in this respect because it shows that the human genome specifies a number of protein domains that are absent from the genomes of other organisms, these domains including several involved in activities such as cell adhesion, electric couplings, and growth of nerve cells (*Table 2.5*). These functions are interesting because they are ones that we look on as conferring the distinctive features of vertebrates compared with other types of eukaryote.

Is it possible to identify a set of genes that are present in vertebrates but not in other eukaryotes? This analysis can only be done in an approximate way at present because only a few genome sequences are available. It currently appears that approximately one-fifth to one-quarter of the genes in the human genome are unique to vertebrates, and a further quarter are found only in vertebrates and other animals (*Figure 2.13*).

Families of genes

Since the earliest days of DNA sequencing it has been known that **multigene families** – groups of genes of identical or similar sequence – are common features of many genomes. For example, every eukaryote that has been studied (as well as all but the simplest bacteria) has

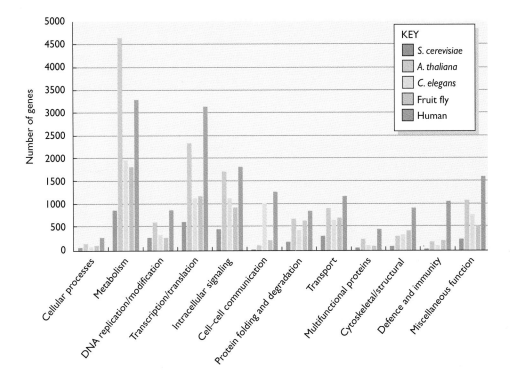

Figure 2.12   Comparison of the gene catalogs of *Saccharomyces cerevisiae*, *Arabidopsis thaliana*, *Caenorhabditis elegans*, fruit fly and humans.

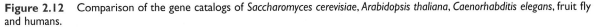

Genes are categorized according to their function, as deduced from the protein domains specified by each gene. Redrawn from IHGSC (2001).

Table 2.5 Examples of protein domains specified by different genomes

Domain	Function	Number of genes in the genome containing the domain				
		Human	Fruit fly	*Caenorhabditis*	*Arabidopsis*	Yeast
Zinc finger, C_2H_2 type	DNA binding	564	234	68	21	34
Zinc finger, GATA type	DNA binding	11	5	8	26	9
Homeobox	Gene regulation during development	160	100	82	66	6
Death	Programmed cell death	16	5	7	0	0
Connexin	Electrical coupling between cells	14	0	0	0	0
Ephrin	Nerve cell growth	7	2	4	0	0

Taken from Venter *et al.* (2001). For more information on zinc fingers and the homeobox domain see Section 9.1.4.

multiple copies of the genes for the non-coding **ribosomal RNAs** (**rRNAs**; Section 3.2.1). This is illustrated by the human genome, which contains approximately 2000 genes for the 5S rRNA (so-called because it has a sedimentation coefficient of 5S; see Technical Note 2.2), all located in a single cluster on chromosome 1. There are also about 280 copies of a repeat unit containing the 28S, 5.8S and 18S rRNA genes, grouped into five clusters of 50–70 repeats, one on each of chromosomes 13, 14, 15, 21 and 22 (see *Figure 2.8*). Ribosomal RNAs are components of the protein-synthesizing particles called **ribosomes**, and it is presumed that their genes are present in multiple copies because there is a heavy demand for rRNA synthesis during cell division, when several tens of thousands of new ribosomes must be assembled.

The rRNA genes are examples of 'simple' or 'classical' multigene families, in which all the members have identical or nearly identical sequences. These families are believed to have arisen by gene duplication, with the sequences of the individual members kept identical by an evolutionary process that, as yet, has not been fully described (Section 15.2.1). Other multigene families, more common in higher eukaryotes than in lower eukaryotes, are called 'complex' because the individual members, although similar in sequence, are sufficiently different for the gene products to have distinctive properties. One of the best examples of this type of multigene family are the mammalian globin genes. The globins are the blood proteins that combine to make hemoglobin, each molecule of hemoglobin being made up of two α-type and two β-type globins. In humans the α-type globins are coded by a small multigene family on chromosome 16 and the β-type globins by a second family on chromosome 11 (*Figure 2.14*). These genes were among the first to be sequenced,

---

Box 2.2: Two examples of unusual gene organization

Overlapping genes

These are occasionally found in small compact genomes such as those of viruses. Usually the amino acid sequences of the proteins coded by a pair of overlapping genes are not similar because the mRNAs are translated from different reading frames, as illustrated by the *D* and *E* genes of the *Escherichia coli* bacteriophage φX174.

```
                          met  val  ...   Gene E
                         ┌──┐ ┌──┐
        DNA sequence  ...GTTTATGGTA...
                        └─┘└─┘└─┘└─┘
                         val  tyr  gly   ...   Gene D
```

Overlapping genes are very rare in the nuclear genomes of higher organisms although there are examples in the compact mitochondrial genomes of some animals, including humans (see *Figure 1.22*, page 25).

Genes-within-genes

These are relatively common features of nuclear genomes, one gene being contained within an intron of a second gene. An example in the human genome is the neurofibromatosis type 1 gene, which has three short genes (called *OGMP*, *EVI2A* and *EVI2B*) within one of its introns. Each of these internal genes is also split into exons and introns.

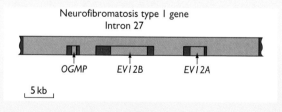

Neurofibromatosis type 1 gene
Intron 27

OGMP   EVI2B   EVI2A

5 kb

Recently, it has been discovered that many small nucleolar RNAs (snoRNAs) — non-coding RNAs that are involved in chemical modification of other RNAs (Section 10.3.1) — are specified by genes within introns.

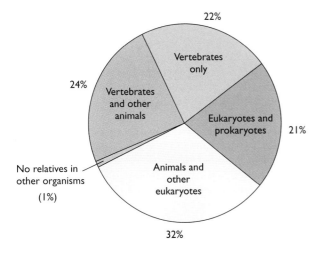

Figure 2.13 Relationship between the human gene catalog and the catalogs of other groups of organism.

The pie chart categorizes the human gene catalog according to the distribution of individual genes in other organisms. The chart shows, for example, that 22% of the human gene catalog is made up of genes that are specific to vertebrates, and that another 24% comprises genes specific to vertebrates and other animals. Genes are categorized according to their function, as deduced from the protein domains specified by each gene. Redrawn from IHGSC (2001).

back in the late 1970s (Fritsch *et al.*, 1980). The sequence data showed that the genes in each family are similar to one another, but by no means identical. In fact the nucleotide sequences of the two most different genes in the β-type cluster, coding for the β- and ε-globins, display only 79.1% identity. Although this is similar enough

for both proteins to be β-type globins, it is sufficiently different for them to have distinctive biochemical properties. Similar variations are seen in the α-cluster.

Why are the members of the globin gene families so different from one another? The answer was revealed when the expression patterns of the individual genes were studied. It was discovered that the genes are expressed at different stages in human development: for example, in the β-type cluster ε is expressed in the early embryo, G_γ and A_γ (whose protein products differ by just one amino acid) in the fetus, and δ and β in the adult (*Figure 2.14*). The different biochemical properties of the resulting globin proteins are thought to reflect slight changes in the physiological role that hemoglobin plays during the course of human development.

In some multigene families, the individual members are clustered, as with the globin genes, but in others the genes are dispersed around the genome. An example of a dispersed family is the five human genes for aldolase, an enzyme involved in energy generation, which are located on chromosomes 3, 9, 10, 16 and 17. The important point is that, even though dispersed, the members of the multigene family have sequence similarities that point to a common evolutionary origin. When these sequence comparisons are made it is sometimes possible to see relationships not only within a single gene family but also between different families. All of the genes in the α- and β-globin families, for example, have some sequence similarity and are thought to have evolved from a single ancestral globin gene. We therefore refer to these two multigene families as comprising a single globin **gene superfamily**, and from the similarities between the individual genes we can chart the duplication events that have given rise to the series of genes that we see today (Section 15.2.1).

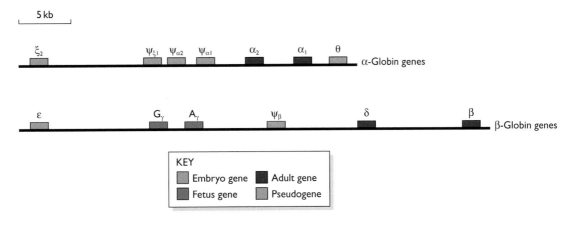

Figure 2.14 The human α- and β-globin gene clusters.

The α-globin cluster is located on chromosome 16 and the β-cluster on chromosome 11. Both clusters contain genes that are expressed at different developmental stages and each includes at least one pseudogene. Note that expression of the α-type gene ξ_2 begins in the embryo and continues during the fetal stage; there is no fetal-specific α-type globin. The θ pseudogene is expressed but its protein product is inactive. None of the other pseudogenes is expressed. For more information on the developmental regulation of the β-globin genes, see Section 8.1.2.

2.2.2 Eukaryotic organelle genomes

Now we move out of the nucleus to examine the genomes present in the mitochondria and chloroplasts of eukaryotic cells. The possibility that some genes might be located outside of the nucleus – **extrachromosomal genes** as they were initially called – was first raised in the 1950s as a means of explaining the unusual inheritance patterns of certain genes in the fungus *Neurospora crassa*, the yeast *S. cerevisiae* and the photosynthetic alga *Chlamydomonas reinhardtii*. Electron microscopy and biochemical studies at about the same time provided hints that DNA molecules might be present in mitochondria and chloroplasts. Eventually, in the early 1960s, these various lines of evidence were brought together and the existence of mitochondrial and chloroplast genomes, independent of and distinct from the nuclear genome, was accepted.

Physical features of organelle genomes

Almost all eukaryotes have mitochondrial genomes, and all photosynthetic eukaryotes have chloroplast genomes. Initially, it was thought that virtually all organelle genomes were circular DNA molecules. Electron microscopy studies had shown both circular and linear DNA in some organelles, but it was assumed that the linear molecules were simply fragments of circular genomes that had become broken during preparation for electron microscopy. We still believe that most mitochondrial and chloroplast genomes are circular, but we now recognize that there is a great deal of variability in different organisms. In many eukaryotes the circular genomes coexist in the organelles with linear versions and, in the case of chloroplasts, with smaller circles that contain subcomponents of the genome as a whole. The latter pattern reaches its extreme in the marine algae called dinoflagellates, whose chloroplast genomes are split into many small circles, each containing just a single gene (Zhang *et al.*, 1999). We also now realize that the mitochondrial genomes of some microbial eukaryotes (e.g. *Paramecium*, *Chlamydomonas* and several yeasts) are always linear (Nosek *et al.*, 1998).

Copy numbers for organelle genomes are not particularly well understood. Each human mitochondrion contains about 10 identical molecules, which means that there are about 8000 per cell, but in *S. cerevisiae* the total number is probably smaller (less than 6500) even though there may be over 100 genomes per mitochondrion. Photosynthetic microorganisms such as *Chlamydomonas* have approximately 1000 chloroplast genomes per cell, about one-fifth the number present in a higher plant cell. One mystery, which dates back to the 1950s and has never been satisfactorily solved, is that when organelle genes are studied in genetic crosses the results suggest that there is just one copy of a mitochondrial or chloroplast genome per cell. This is clearly not the case but indicates that our understanding of the transmission of organelle genomes from parent to offspring is less than perfect.

Mitochondrial genome sizes are variable (*Table 2.6*) and are unrelated to the complexity of the organism. Most multicellular animals have small mitochondrial genomes with a compact genetic organization, the genes being close together with little space between them. The human mitochondrial genome (see *Figure 1.22*, page 25), at 16 569 bp, is typical of this type. Most lower eukaryotes such as *S. cerevisiae* (*Figure 2.15*), as well as flowering plants, have larger and less compact mitochondrial genomes, with a number of the genes containing introns. Chloroplast genomes have less variable sizes (*Table 2.6*) and most have a structure similar to that shown in *Figure 2.16* for the rice chloroplast genome.

The genetic content of organelle genomes

Organelle genomes are much smaller than their nuclear counterparts and we therefore anticipate that their gene contents are much more limited, which is indeed the case. Again, mitochondrial genomes display the greater variability, gene contents ranging from five for the malaria parasite *P. falciparum* to 92 for the protozoan *Reclinomonas americana* (*Table 2.7*; Lang *et al.*, 1997; Palmer, 1997a). All mitochondrial genomes contain genes for the non-coding rRNAs and at least some of the protein components of the respiratory chain, the latter being the main biochemical feature of the mitochondrion. The more gene-rich genomes also code for tRNAs, ribosomal proteins, and proteins involved in transcription, translation and transport of other proteins into the mitochondrion from the surrounding cytoplasm (*Table 2.7*). Most chloroplast genomes appear to possess the same set of 200 or so genes, again coding for rRNAs and tRNAs, as well as ribosomal proteins and proteins involved in photosynthesis (see *Figure 2.16*).

A general feature of organelle genomes emerges from *Table 2.7*. These genomes specify some of the proteins found in the organelle, but not all of them. The other proteins are coded by nuclear genes, synthesized in the cytoplasm, and transported into the organelle. If the cell has mechanisms for transporting proteins into mitochondria and chloroplasts, then why not have all the organelle proteins specified by the nuclear genome? We do not yet have a convincing answer to this question, although it has been suggested that at least some of the proteins coded by organelle genomes are extremely hydrophobic and cannot be transported through the membranes that surround mitochondria and chloroplasts, and so simply cannot be moved into the organelle from the cytoplasm (Palmer, 1997b). The only way in which the cell can get them into the organelle is to make them there in the first place.

The origins of organelle genomes

The discovery of organelle genomes led to many speculations about their origins. Today most biologists accept that the **endosymbiont theory** is correct, at least in outline, even

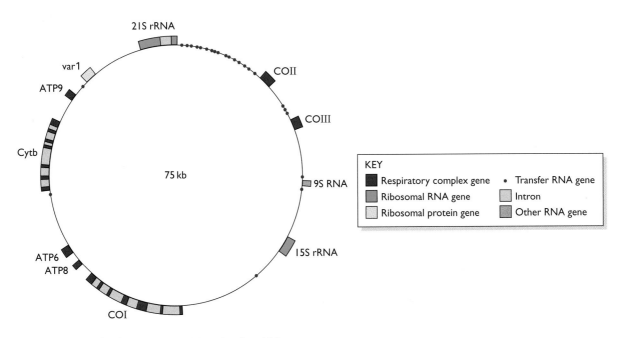

Figure 2.15 The *Saccharomyces cerevisiae* mitochondrial genome.

Because of their relatively small sizes, many mitochondrial genomes have been completely sequenced. In the yeast genome, the genes are more spaced out than in the human mitochondrial genome (*Figure 1.22*, page 25) and some of the genes have introns. This type of organization is typical of many lower eukaryotes and plants. The yeast genome contains five additional open reading frames (not shown on this map) that have not yet been shown to code for functional gene products, and there are also several genes located within the introns of the discontinuous genes. Most of the latter code for maturase proteins involved in splicing the introns from the transcripts of these genes (Section 10.2.3). Abbreviations: ATP6, ATP8, ATP9, genes for ATPase subunits 6, 8 and 9, respectively; COI, COII, COIII, genes for cytochrome c oxidase subunits I, II and III, respectively; Cytb, gene for apocytochrome *b*; var1, gene for a ribosome-associated protein. Ribosomal RNA and transfer RNA are two types of non-coding RNA (Section 3.2.1). The 9S RNA gene specifies the RNA component of the enzyme ribonuclease P (Section 10.2.2).

though it was considered quite unorthodox when first proposed in the 1960s. The endosymbiont theory is based on the observation that the gene expression processes occurring in organelles are similar in many respects to equivalent processes in bacteria. In addition, when nucleotide sequences are compared organelle genes are found to be more similar to equivalent genes from bacteria than they are to eukaryotic nuclear genes. The endosymbiont theory therefore holds that mitochondria and chloroplasts are the relics of free-living bacteria that formed a symbiotic association with the precursor of the eukaryotic cell, way back at the very earliest stages of evolution.

Support for the endosymbiont theory has come from the discovery of organisms which appear to exhibit stages of endosymbiosis that are less advanced than seen with mitochondria and chloroplasts. For example, an early stage in endosymbiosis is displayed by the protozoan *Cyanophora paradoxa*, whose photosynthetic structures, called **cyanelles**, are different from chloroplasts and instead resemble ingested cyanobacteria. Similarly, the *Rickettsia*, which live inside eukaryotic cells, might be modern versions of the bacteria that gave rise to mitochondria (Andersson *et al.*, 1998). It has also been sug-

gested that the hydrogenosomes of trichomonads (unicellular microbes, many of which are parasites), some of which have a genome but most of which do not, represent an advanced type of mitochondrial endosymbiosis (Palmer, 1997b; Akhmanova *et al.*, 1998).

If mitochondria and chloroplasts were once free-living bacteria, then since the endosymbiosis was set up there must have been a transfer of genes from the organelle into the nucleus. We do not understand how this occurred, or indeed whether there was a mass transfer of many genes at once, or a gradual trickle from one site to the other. But we do know that DNA transfer from organelle to nucleus, and indeed between organelles, still occurs. This was discovered in the early 1980s, when the first partial sequences of chloroplast genomes were obtained. It was found that in some plants the chloroplast genome contains segments of DNA, often including entire genes, that are copies of parts of the mitochondrial genome. The implication is that this so-called **promiscuous DNA** has been transferred from one organelle to the other. We now know that this is not the only type of transfer that can occur. The *Arabidopsis* mitochondrial genome contains various segments of nuclear DNA as well as 16 fragments

Table 2.6   Sizes of mitochondrial and chloroplast genomes

Species	Type of organism	Genome size (kb)
Mitochondrial genomes		
Plasmodium falciparum	Protozoan (malaria parasite)	6
Chlamydomonas reinhardtii	Green alga	16
Mus musculus	Vertebrate (mouse)	16
Homo sapiens	Vertebrate (human)	17
Metridium senile	Invertebrate (sea anemone)	17
Drosophila melanogaster	Invertebrate (fruit fly)	19
Chondrus crispus	Red alga	26
Aspergillus nidulans	Ascomycete fungus	33
Reclinomonas americana	Protozoa	69
Saccharomyces cerevisiae	Yeast	75
Suillus grisellus	Basidiomycete fungus	121
Brassica oleracea	Flowering plant (cabbage)	160
Arabidopsis thaliana	Flowering plant (vetch)	367
Zea mays	Flowering plant (maize)	570
Cucumis melo	Flowering plant (melon)	2500
Chloroplast genomes		
Pisum sativum	Flowering plant (pea)	120
Marchantia polymorpha	Liverwort	121
Oryza sativa	Flowering plant (rice)	136
Nicotiana tabacum	Flowering plant (tobacco)	156
Chlamydomonas reinhardtii	Green alga	195

Table 2.7   Features of mitochondrial genomes

Feature	*Plasmodium falciparum*	*Chlamydomonas reinhardtii*	*Homo sapiens*	*Saccharomyces cerevisiae*	*Arabidopsis thaliana*	*Reclinomonas americana*
Total number of genes	5	12	37	35	52	92
Types of genes						
Protein-coding genes	5	7	13	8	27	62
Respiratory complex	0	7	13	7	17	24
Ribosomal proteins	0	0	0	1	7	27
Transport proteins	0	0	0	0	3	6
RNA polymerase	0	0	0	0	0	4
Translation factor	0	0	0	0	0	1
Non-coding RNA genes	0	5	24	26	25	30
Ribosomal RNA genes	0	2	2	2	3	3
Transfer RNA genes	0	3	22	24	22	26
Other RNA genes	0	0	0	1	0	1
Number of introns	0	1	0	8	23	1
Genome size (kb)	6	16	17	75	367	69

Based on Palmer (1997a).

Figure 2.16 The rice chloroplast genome.

Only those genes with known functions are shown. A number of the genes contain introns which are not indicated on this map. These discontinuous genes include several of those for tRNAs, which is why the tRNA genes are of different lengths even though the tRNAs that they specify are all of similar size.

of the chloroplast genome, including six tRNA genes that have retained their activity after transfer to the mitochondrion. The nuclear genome of this plant includes several short segments of the chloroplast and mitochondrial genomes as well as a 270-kb piece of mitochondrial DNA located within the centromeric region of chromosome 2 (Copenhaver *et al.*, 1999; AGI, 2000). The transfer of mitochondrial DNA to vertebrate nuclear genomes has also been documented.

2.3 The Anatomy of the Prokaryotic Genome

Because of the relatively small sizes of prokaryotic genomes, many complete genome sequences for various bacteria and archaea have been published over the last few years. As a result, we are beginning to understand a great deal about the anatomies of prokaryotic genomes, and in many respects we know more about these organisms than we do about eukaryotes.

2.3.1 The physical structure of the prokaryotic genome

Most prokaryotic genomes are less than 5 Mb in size, although a few are substantially larger than this: *B. megaterium*, for example, has a huge genome of 30 Mb. The traditional view has been that in a typical prokaryote the genome is contained in a single circular DNA molecule, localized within the **nucleoid** – the lightly staining region of the otherwise featureless prokaryotic cell (see *Figure 2.1*, page 32). This is certainly true for *E. coli* and many of the other commonly studied bacteria. However, as we will see, our growing knowledge of prokaryotic genomes is leading us to question several of the preconceptions that became established during the pre-genome era of microbiology. These preconceptions relate both to the physical structure of the prokaryotic genome and its genetic organization.

The traditional view of the bacterial 'chromosome'

As with eukaryotic chromosomes, a prokaryotic genome has to squeeze into a relatively tiny space (the circular *E. coli* chromosome has a circumference of 1.6 mm whereas an *E. coli* cell is just 1.0×2.0 µm) and, as with eukaryotes, this is achieved with the help of DNA-binding proteins that package the genome in an organized fashion. The resulting structure has no substantial similarities with a eukaryotic chromosome, but we still use 'bacterial chromosome' as a convenient term to describe it.

Most of what we know about the organization of DNA in the nucleoid comes from studies of *E. coli*. The first feature to be recognized was that the circular *E. coli* genome is **supercoiled**. Supercoiling occurs when additional turns are introduced into the DNA double helix (positive supercoiling) or if turns are removed (negative supercoiling). With a linear molecule, the torsional stress introduced by over- or under-winding is immediately released by rotation of the ends of the DNA molecule, but a circular molecule, having no ends, cannot reduce the strain in this way. Instead the circular molecule responds by winding around itself to form a more compact structure (*Figure 2.17*). Supercoiling is therefore an ideal way to package a circular molecule into a small space. Evidence that supercoiling is involved in packaging the circular *E. coli* genome was first obtained in the 1970s from examination of isolated nucleoids, and subsequently confirmed as a feature of DNA in living cells in 1981. In *E. coli*, the supercoiling is thought to be generated and controlled by two enzymes, DNA gyrase and DNA topoisomerase I, which we will look at in more detail in Section 13.1.2 when we examine the roles of these enzymes in DNA replication.

Studies of isolated nucleoids and of living cells have shown that the *E. coli* DNA molecule does not have unlimited freedom to rotate once a break is introduced. The most likely explanation is that the bacterial DNA is attached to proteins that restrict its ability to relax, so that

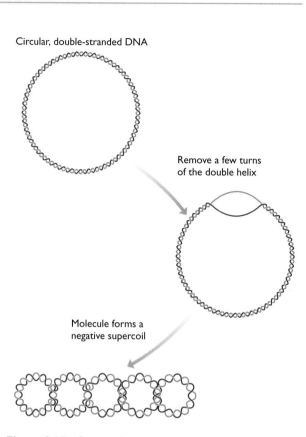

Circular, double-stranded DNA

Remove a few turns of the double helix

Molecule forms a negative supercoil

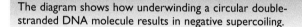

Figure 2.17 Supercoiling.

The diagram shows how underwinding a circular double-stranded DNA molecule results in negative supercoiling.

rotation at a break site results in loss of supercoiling from only a small segment of the molecule (*Figure 2.18*). The current model has the *E. coli* DNA attached to a protein core from which 40–50 supercoiled loops radiate out into the cell. Each loop contains approximately 100 kb of supercoiled DNA, the amount of DNA that becomes unwound after a single break.

The protein component of the nucleoid includes DNA gyrase and DNA topoisomerase I, the two enzymes that are primarily responsible for maintaining the supercoiled state, as well as a set of at least four proteins believed to have a more specific role in packaging the bacterial DNA. The most abundant of these packaging proteins is HU, which is structurally very different to eukaryotic histones but acts in a similar way, forming a tetramer around which approximately 60 bp of DNA becomes wound. There are some 60 000 HU proteins per *E. coli* cell, enough to cover about one-fifth of the DNA molecule, but it is not known if the tetramers are evenly spaced along the DNA or restricted to the core region of the nucleoid.

Complications on the E. coli theme

In recent years it has become clear that the straightforward view of prokaryotic genome anatomy developed

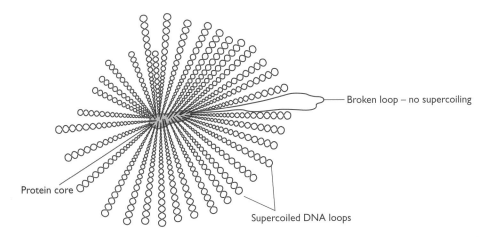

Figure 2.18 A model for the structure of the *Escherichia coli* nucleoid.

Between 40 and 50 supercoiled loops of DNA radiate from the central protein core. One of the loops is shown in circular form, indicating that a break has occurred in this segment of DNA, resulting in a loss of the supercoiling.

from studies of *E. coli* is an over-simplification. Although the majority of bacterial and archaeal chromosomes are indeed circular, an increasing number of linear ones are being found. The first of these, for *Borrelia burgdorferi*, the organism that causes Lyme disease, was described in 1989 (Ferdows and Barbour, 1989) and during the following years similar discoveries were made for *Streptomyces* and other bacteria (Chen, 1996).

A second complication concerns the precise status of plasmids with regard to the prokaryotic genome. A plasmid is a small piece of DNA, often, but not always circular, that coexists with the main chromosome in a bacterial cell (see *Figure 2.3*, page 35). Some types of plasmid are able to integrate into the main genome, but others are thought to be permanently independent. Plasmids carry genes that are not usually present in the main chromosome, but in many cases these genes are non-essential to the bacterium, coding for characteristics such as antibiotic resistance, which the bacterium does not need if the envi-

ronmental conditions are amenable (*Table 2.8*). As well as this apparent dispensability, many plasmids are able to transfer from one cell to another, and the same plasmids are sometimes found in bacteria that belong to different species. These various features of plasmids suggest that they are independent entities and that in most cases the plasmid content of a prokaryotic cell should not be included in the definition of its genome.

With a bacterium such as *E. coli* K12, which has a 4.6-Mb chromosome and can harbor various combinations of plasmids, none of which is more than a few kb in size and all of which are dispensable, it is acceptable to define the main chromosome as the 'genome'. With other prokaryotes it is not so easy (*Table 2.9*). *Vibrio cholerae*, the pathogenic bacterium that causes cholera, has two circular DNA molecules, one of 2.96 Mb and the other of 1.07 Mb, with 71% of the organism's 3885 genes on the larger of these (Heidelberg *et al.*, 2000). It would appear obvious that these

Table 2.8 Features of typical plasmids

Type of plasmid	Gene functions	Examples
Resistance	Antibiotic resistance	Rbk of *Escherichia coli* and other bacteria
Fertility	Conjugation and DNA transfer between bacteria	F of *E. coli*
Killer	Synthesis of toxins that kill other bacteria	Col of *E. coli*, for colicin production
Degradative	Enzymes for metabolism of unusual molecules	TOL of *Pseudomonas putida*, for toluene metabolism
Virulence	Pathogenicity	Ti of *Agrobacterium tumefaciens*, conferring the ability to cause crown gall disease on dicotyledonous plants

Supercoiled domains in the *Escherichia coli* nucleoid

Studies with extracted nucleoids suggested that the *E. coli* genome is organized into a series of supercoiled domains. Was this model correct for living bacteria?

Early attempts to understand the organization of bacterial DNA made use of isolated nucleoids from *E. coli* cell extracts. These experiments showed that *E. coli* DNA is supercoiled, and also suggested that relaxation of supercoils after strand breakage is limited to the single region in which the break occurs, more extensive relaxation being prevented by the presence of proteins or other structures that bind to the DNA and separate the molecule into independent domains. This model was considered unproven because domains were observed only when the cell extracts contained contaminating RNA molecules. If these RNA molecules were degraded by treating with ribonuclease then the DNA molecule lost all its supercoiling. The supercoiled domains might therefore be an artefact of the extraction procedure. To resolve this problem it was necessary to devise a way of directly examining the structure of the nucleoid in living cells.

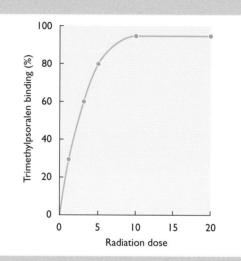

The experiment

One approach was to exploit the ability of trimethylpsoralen to distinguish between supercoiled and relaxed DNA. When photoactivated by a pulse of light at 360 nm, this compound binds to double-stranded DNA at a rate that is directly proportional to the degree of torsional stress possessed by the molecule. The degree of supercoiling can therefore be assayed by measuring the amount of trimethylpsoralen that binds to the molecule in unit time. The technique was devised by Richard Sinden and David Pettijohn of the University of Colorado Health Sciences Center, and used by them in 1981 to probe the structure of the nucleoids in living *E. coli* bacteria. Cells were exposed to varying doses of radiation to introduce single-strand breaks into the DNA molecule, and then suspended in a buffer containing radioactively labeled trimethylpsoralen. After photoactivation, DNA was extracted from each batch of cells and the amount of bound trimethylpsoralen measured by radioactive counting. Typical results are shown in the graph.

The conclusion

The amount of trimethylpsoralen bound by the nucleoids of non-irradiated cells confirmed that DNA is supercoiled in living *E. coli*. The results also supported the domain model. If the *E. coli* nucleoid is not organized into domains then one break in the DNA molecule would lead to complete loss of supercoiling: irradiation would therefore have an all-or-nothing effect on trimethylpsoralen binding. This was not

observed. Instead, the results showed that supercoiling is reduced in a gradual, incremental fashion as the irradiation dose increases. This is the response predicted by the domain model, in which the overall supercoiling of the molecule is gradually relaxed as greater doses of radiation cause breaks within an increasing number of domains. The trimethylpsoralen method therefore confirms that the domain model is correct.

Can these results also be used to estimate the number of domains in the *E. coli* nucleoid? The number of single-strand breaks caused by a given dose of radiation can be determined by purifying the DNA, denaturing with alkali to separate the strands, and measuring the average length of the polynucleotides that remain. The number of breaks needed to produce a complete loss of supercoiling can therefore be determined, but this figure is not the same as the number of domains. Radiation 'hits' occur at random, so some domains may accumulate two or more breaks before all of the others have received their first hits. However, 'randomness' is quantifiable statistically, so the total number of domains can be calculated from the numbers of strand breaks needed to produce different degrees of reduction in the overall amounts of supercoiling. Using this approach, Sinden and Pettijohn calculated that the *E. coli* nucleoid contains 43 ± 10 domains.

Reference

Sinden RR and Pettijohn DE (1981) Chromosomes in living *Escherichia coli* cells are segregated into domains of supercoiling. *Proc. Natl Acad. Sci. USA*, **78**, 224–228.

Table 2.9   Examples of genome organization in prokaryotes

Species	Genome organization		
	DNA molecules	**Size (Mb)**	**Number of genes**
Escherichia coli K-12	One circular molecule	4.639	4397
Vibrio cholerae El Tor N16961	Two circular molecules		
	Main chromosome	2.961	2770
	Megaplasmid	1.073	1115
Deinococcus radiodurans R1	Four circular molecules		
	Chromosome 1	2.649	2633
	Chromosome 2	0.412	369
	Megaplasmid	0.177	145
	Plasmid	0.046	40
Borrelia burgdorferi B31	seven or eight circular molecules, 11 linear molecules		
	Linear chromosome	0.911	853
	Circular plasmid cp9	0.009	12
	Circular plasmid cp26	0.026	29
	Circular plasmid cp32*	0.032	Not known
	Linear plasmid lp17	0.017	25
	Linear plasmid lp25	0.024	32
	Linear plasmid lp28-1	0.027	32
	Linear plasmid lp28-2	0.030	34
	Linear plasmid lp28-3	0.029	41
	Linear plasmid lp28-4	0.027	43
	Linear plasmid lp36	0.037	54
	Linear plasmid lp38	0.039	52
	Linear plasmid lp54	0.054	76
	Linear plasmid lp56	0.056	Not known

*There are 5 or 6 similar versions of plasmid cp32 per bacterium.

two DNA molecules together constitute the *Vibrio* genome, but closer examination reveals that most of the genes for the central cellular activities such as genome expression and energy generation, as well as the genes that confer pathogenicity, are located on the larger molecule. The smaller molecule contains many essential genes but also has certain features that are considered characteristic of plasmids, notably an **integron**, a set of genes and other DNA sequences that enable plasmids to capture genes from bacteriophages and other plasmids. It therefore appears possible that the smaller genome is a 'megaplasmid' that was acquired by the ancestor to *Vibrio* at some period in the bacterium's evolutionary past. *Deinococcus radiodurans* R1, whose genome is of particular interest because it contains many genes that help this bacterium resist the harmful effects of radiation, is constructed on similar lines, with essential genes distributed among two circular chromosomes and two plasmids (White *et al.*, 1999). However, the *Vibrio* and *Deinococcus* genomes are relatively non-complex compared with *Borrelia*

burgdorferi B31, whose linear chromosome of 911 kb, carrying 853 genes, is accompanied by 17 or 18 linear and circular plasmids which together contribute another 533 kb and at least 430 genes (Fraser *et al.*, 1997). Although the functions of most of these genes are unknown, those that have been identified include several that would not normally be considered dispensable, such as genes for membrane proteins and purine biosynthesis. The implication is that at least some of the *Borrelia* plasmids are essential components of the genome, leading to the possibility that some prokaryotes have highly multipartite genomes, comprising a number of separate DNA molecules, more akin to what we see in the eukaryotic nucleus rather than the 'typical' prokaryotic arrangement. This interpretation of the *Borrelia* genome is still controversial, and is complicated by the fact that the related bacterium *Treponema pallidum*, whose genome is a single circular DNA molecule of 1138 kb containing 1041 genes (Fraser *et al.*, 1998), does not contain any of the genes present on the *Borrelia* plasmids.

The final complication regarding the physical structures of prokaryotic genomes concerns differences between the packaging systems for bacterial and archaeal DNA molecules. One reason why the archaea are looked upon as a distinct group of organisms, different from the bacteria, is that archaea do not possess packaging proteins such as HU but instead have proteins that are much more similar to histones. Currently we have no information on the structure of the archaeal nucleoid, but the assumption is that these histone-like proteins play a central role in DNA packaging.

2.3.2 The genetic organization of the prokaryotic genome

We have already learnt that bacterial genomes have compact genetic organizations with very little space between genes (see *Figure 2.2*, page 34). To re-emphasize this point, the complete circular gene map of the *E. coli* K12 genome is shown in *Figure 2.19*. There *is* non-coding DNA in the *E. coli* genome, but it accounts for only 11% of the total and it is distributed around the genome in small segments that do not show up when the map is drawn at this scale. In this regard, *E. coli* is typical of all prokaryotes whose genomes have so far been sequenced – prokaryotic genomes have very little wasted space. There are theories

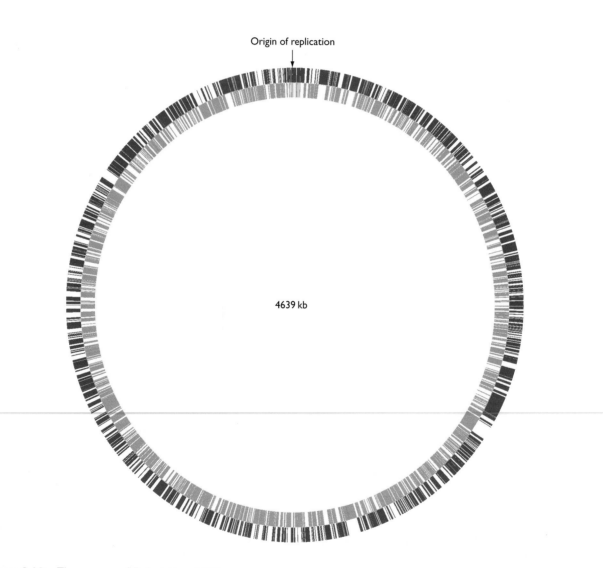

Origin of replication

4639 kb

Figure 2.19 The genome of *Escherichia coli* K12.

The map is shown with the origin of replication (Section 13.2.1) positioned at the top. Genes on the outside of the circle are transcribed in the clockwise direction and those on the inside are transcribed in the anticlockwise direction. Image supplied courtesy of Dr FR Blattner, Laboratory of Genetics, University of Wisconsin-Madison. Reproduced with permission.

that this compact organization is beneficial to prokaryotes, for example by enabling the genome to be replicated relatively quickly, but these ideas have never been supported by hard experimental evidence.

Operons are characteristic features of prokaryotic genomes

One characteristic feature of prokaryotic genomes illustrated by *E. coli* is the presence of operons. In the years before genome sequences, it was thought that we understood operons very well; now we are not so sure.

An operon is a group of genes that are located adjacent to one another in the genome, with perhaps just one or two nucleotides between the end of one gene and the start of the next. All the genes in an operon are expressed as a single unit. This type of arrangement is common in prokaryotic genomes. A typical *E. coli* example is the **lactose operon**, the first operon to be discovered (Jacob and Monod, 1961), which contains three genes involved in conversion of the disaccharide sugar lactose into its monosaccharide units – glucose and galactose (*Figure 2.20A*). The monosaccharides are substrates for the energy-generating glycolytic pathway, so the function of the genes in the lactose operon is to convert lactose into

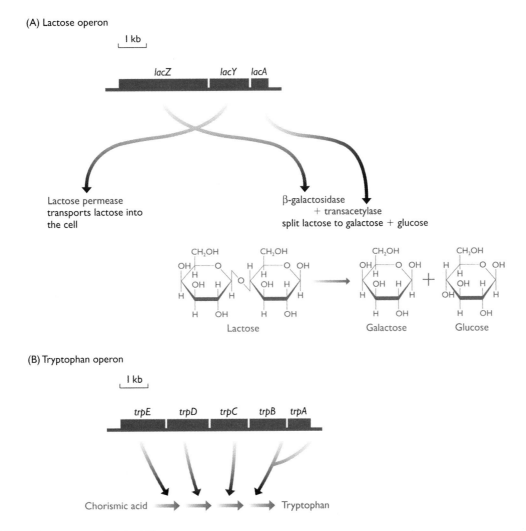

Figure 2.20 Two operons of *Escherichia coli.*

(A) The lactose operon. The three genes are called *lacZ*, *lacY* and *lacA*, the first two separated by 52 bp and the second two by 64 bp. All three genes are expressed together, *lacY* coding for the lactose permease which transports lactose into the cell, and *lacZ* and *lacA* coding for enzymes that split lactose into its component sugars – galactose and glucose. (B) The tryptophan operon, which contains five genes coding for enzymes involved in the multistep biochemical pathway that converts chorismic acid into the amino acid tryptophan. The genes in the tryptophan operon are closer together than those in the lactose operon: *trpE* and *trpD* overlap by 1 bp, as do *trpB* and *trpA*; *trpD* and *trpC* are separated by 4 bp, and *trpC* and *trpB* by 12 bp. For more details on the regulation of these operons, see Sections 9.3.1 and 12.1.1.

a form that can be utilized by *E. coli* as an energy source. Lactose is not a common component of *E. coli*'s natural environment, so most of the time the operon is not expressed and the enzymes for lactose utilization are not made by the bacterium. When lactose becomes available, it switches on the operon; all three genes are expressed together, resulting in coordinated synthesis of the lactose-utilizing enzymes. This is the classic example of gene regulation in bacteria, and is examined in detail in Section 9.3.1.

Altogether there are almost 600 operons in the *E. coli* K12 genome, each containing two or more genes, and a similar number are present in *Bacillus subtilis*. In most cases the genes in an operon are functionally related, coding for a set of proteins that are involved in a single biochemical activity such as utilization of a sugar as an energy source or synthesis of an amino acid. An example of the latter is the tryptophan operon of *E. coli* (*Figure 2.20B*). Microbial geneticists are attracted to the simplicity of this system whereby a bacterium is able to control its various biochemical activities by regulating the expression of groups of related genes linked together in operons. This may be a correct interpretation of the function of operons in *E. coli*, *Bacillus subtilis* and many other prokaryotes, but in at least some species the picture is less straightforward. Both the archaeon *Methanococcus jannaschii* and the bacterium *Aquifex aeolicus* have operons, but the genes in an individual operon rarely have any biochemical relationship. For example, one of the operons in the *A. aeolicus* genome contains six linked genes, these genes coding for two proteins involved in DNA recombination (Section 14.3), an enzyme used in protein synthesis, a protein required for motility, an enzyme involved in nucleotide synthesis, and an enzyme for lipid synthesis (*Figure 2.21*; Deckert *et al.*, 1998). This is typical of the operon structure in the *A. aeolicus* and *M. jannaschii* genomes. In other words, the notion that expression of an operon leads to the coordinated synthesis of enzymes required for a single biochemical pathway does not hold for these species.

Genome projects have therefore confused our understanding of operons. It is certainly too early to abandon

the belief that operons play a central role in biochemical regulation in many bacteria, but we need to explain the unexpected features of the operons in *A. aeolicus* and *M. jannaschii*. It has been pointed out that both *A. aeolicus* and *M. jannaschii* are autotrophs, which means that, unlike many prokaryotes, they are able to synthesize organic compounds from carbon dioxide (Deckert *et al.*, 1998), but how this similarity between the species might be used to explain their operon structures is not clear.

Prokaryotic genomes and the species concept

Genome projects have also confused our understanding of what constitutes a 'species' in the prokaryotic world. This has always been a problem in microbiology because the standard biological definitions of species have been difficult to apply to microorganisms. The early taxonomists such as Linneaus described species in morphological terms, all members of one species having the same or very similar structural features. This form of classification was in vogue until the early 20th century and was first applied to microorganisms in the 1880s by Robert Koch and others, who used staining and biochemical tests to distinguish between bacterial species. However, it was recognized that this type of classification was imprecise because many of the resulting species were made up of a variety of types with quite different properties. An example is provided by *E. coli* which, like many bacterial species, includes strains with distinctive pathogenic characteristics, ranging from harmless through to lethal. During the 20th century, biologists redefined the species concept in evolutionary terms and we now look on a species as a group of organisms that can interbreed with one another. If anything, this is more problematic with microorganisms because there are a variety of methods by which genes can be exchanged between prokaryotes that, according to their biochemical and physiological properties, are different species (Box 2.3). The barrier to **gene flow** that is central to the species concept therefore does not hold with prokaryotes.

Genome sequencing has emphasized the difficulties in applying the species concept to prokaryotes. It has become clear that different strains of a single species can have very different genome sequences, and may even have individual sets of strain-specific genes. This was first shown by a comparison between two strains of *Helicobacter pylori*, which causes gastric ulcers and other diseases of the human digestive tract. The two strains were isolated in the UK and USA and have genomes of 1.67 Mb and 1.64 Mb, respectively. The larger genome contains 1552 genes and the smaller one 1495 genes, 1406 of these genes being present in both strains. In other words, some 6–7% of the gene content of each strain is unique to that strain (Alm *et al.*, 1999). A much more extreme distinction between strains was revealed when the sequence of the common laboratory strain of *E. coli*, K12, was compared with that of one of the most pathogenic strains, O157:H7 (Perna *et al.*,

gatC   recA   pilU   cmk   pgsA   recJ

Figure 2.21 A typical operon in the genome of *Aquifex aeolicus*.

The genes code for the following proteins: *gatC*, glutamyl-tRNA aminotransferase subunit C, which plays a role in protein synthesis (Section 11.2); *recA*, recombination protein RecA; *pilU*, twitching mobility protein; *cmk*, cytidylate kinase, required for synthesis of cytidine nucleotides; *pgsA*, phosphotidylglycerophosphate synthase, an enzyme involved in lipid biosynthesis; *recJ*, single-strand-specific endonuclease RecJ, which is another recombination protein (Section 14.3).

Box 2.3: Mechanisms for gene flow between prokaryotes

There are three ways in which genes can be passed from one bacterial species to another:

- During **conjugation** two bacteria come into physical contact and one bacterium (the donor) transfers DNA to the second bacterium (the recipient). The transferred DNA can be a copy of some or possibly all of the donor cell's chromosome, or a segment of chromosomal DNA integrated in a plasmid. The latter is called **episome transfer**.

- **Transduction** involves transfer of a small segment of DNA from donor to recipient via a bacteriophage.

- During **transformation** the recipient cell takes up from its environment a fragment of DNA released from a donor cell.

The relative importance of these processes in gene flow between species is not known. Each mechanism was first recognized as a means by which DNA can be transferred between *Escherichia coli* strains, and the extent to which the equivalent forms of transfer can occur between species has not yet been studied. The three processes have been used experimentally to map the positions of genes on bacterial genomes (Section 5.2.4).

2001). The lengths of the two genomes are significantly different – 4.64 Mb for K12 and 5.53 Mb for O157:H7 – with the extra DNA in the pathogenic strain scattered around the genome at almost 200 separate positions. These 'O-islands' contain 1387 genes not present in *E. coli* K12, many of these genes coding for toxins and other proteins that are clearly involved in the pathogenic properties of O157:H7. But it is not simply a case of O157:H7 containing extra genes that make it pathogenic. K12 also has 234 segments of its own unique DNA, and although these 'K-islands' are, on average, smaller than the O-islands, they still contain 528 genes that are absent from O157:H7. The situation therefore is that *E. coli* O157:H7 and *E. coli* K12 each has a set of strain-specific genes, which make up 26% and 12% of the gene catalogs, respectively. This is substantially more variation than can be tolerated by the species concept as applied to higher organisms, and is difficult to reconcile with any definition of species yet devised for microorganisms.

The difficulties become even more acute when other bacterial and archaeal genomes are examined. Because of the ease with which genes can flow between different prokaryotic species (Box 2.3), it was anticipated that the same genes would occasionally be found in different species, but the extent of **lateral gene transfer** revealed by sequencing has taken everybody by surprise (Ochman *et al.*, 2000). Most genomes contain a few hundred kb of DNA acquired directly from a different species, and in some cases the figure is higher: 12.8% of the *E. coli* K12 genome, corresponding to 0.59 Mb, has been obtained in this way (*Figure 2.22*). A second surprise is that transfer

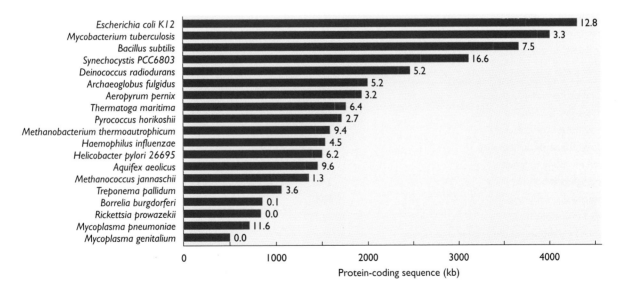

Figure 2.22   The impact of lateral gene transfer on the content of prokaryotic genomes.

The chart shows the DNA that is unique to a particular species in blue and the DNA that has been acquired by lateral gene transfer in red. The number at the end of each bar indicates the percentage of the genome that derives from lateral transfer. Note that intergenic regions are omitted from this analysis. Redrawn from Ochman *et al.* (2000).

has occurred between very different species, even between bacteria and archaea. For example, the thermophilic bacterium *Thermatoga maritima* has 1877 genes, 451 of which appear to have been obtained from archaeons (Nelson *et al.*, 1999). Transfer in the other direction, from bacteria to archaea, is equally prevalent. The picture that is emerging is one in which prokaryotes living in similar ecological niches exchange genes with one another in order to increase their individual fitness for survival in their particular environment. Many of the *Thermatoga* genes that have been obtained from archaeons have probably helped this bacterium acquire its ability to tolerate high temperatures.

Lateral gene transfer has clearly played an important role in the evolution of prokaryotes. Unlike higher organisms, the evolutionary histories of bacteria and archaea cannot be described as a simple branching pattern, but instead have to incorporate the horizontal flow of genes between species (*Figure 2.23*). As well as its impact on our understanding of evolution, lateral gene transfer also has implications for the way in which evolutionary relationships are inferred by **molecular phylogenetics** (Chapter 16). With higher organisms, comparisons between the sequences of equivalent genes in different species can be used to reconstruct the evolutionary relationships between those species. This type of analysis assumes that evolution has followed the simple branching pattern illustrated in *Figure 2.23A*, and cannot, therefore, be used to infer relationships between prokaryotes if there is a chance that the genes being analyzed have been transferred laterally between any of

the species being studied. But the analysis was used for many years before the extent of lateral gene transfer was recognized and microbiologists are now faced with the need to reassess the validity of the evolutionary schemes that were established in the pre-genome era (Doolittle, 1999).

Speculation on the minimal genome content and the identity of distinctiveness genes

Even though a number of prokaryotic genome sequences have now been published, it is not yet possible to describe a complete catalog of the gene content of any one species, for the simple reason that the functions of many of the genes are unknown. Over 1500 of the *E. coli* K12 genes, for example, have not yet been assigned a function. Despite the incompleteness of the information, it is still interesting to examine the roles of the genes whose functions are known, and to appreciate the number of different genes involved in each of the various biochemical activities that a bacterium such as *E. coli* is able to carry out (*Table 2.10*).

Gene catalogs are even more interesting when comparisons are made between different species. We see, for example, that whereas 243 of the identified genes in the *E. coli* genome are involved in energy metabolism, *Haemophilus influenzae* has only 112 genes in this category and *Mycoplasma genitalium* just 31. These comparisons have led to speculation about the smallest number of genes needed to specify a free-living cell. Theoretical considerations initially led to the suggestion that 256 genes

Table 2.10 Partial gene catalogs for *Escherichia coli* K12, *Haeomphilus influenzae* Rd and *Mycoplasma genitalium*

Category	Number of genes in		
	E. coli K12	*H. influenzae* Rd	*M. genitalium*
Total protein-coding genes	4288	1727	470
Biosynthesis of amino acids	131	68	1
Biosynthesis of cofactors	103	54	5
Biosynthesis of nucleotides	58	53	19
Cell envelope proteins	237	84	17
Energy metabolism	243	112	31
Intermediary metabolism	188	30	6
Lipid metabolism	48	25	6
DNA replication, recombination and repair	115	87	32
Protein folding	9	6	7
Regulatory proteins	178	64	7
Transcription	55	27	12
Translation	182	141	101
Uptake of molecules from the environment	427	123	34

Taken from Fraser *et al.* (1995) and Blattner *et al.* (1997). The numbers refer only to genes whose functions are known and are therefore approximate, because each genome also contains many genes whose functions have not yet been identified.

(A) No lateral gene transfer between species

Ancestral species

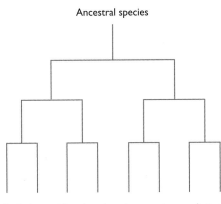

Evolutionary histories of modern species are distinct

(B) Lateral gene transfer occurs between species

Ancestral species

Lateral gene transfer

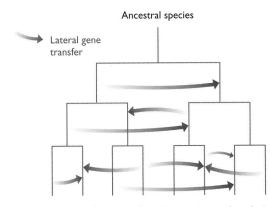

Evolutionary histories of modern species are less distinct

Figure 2.23 Lateral gene transfer obscures the evolutionary relationships between species.

In (A) a group of eight modern species has evolved from an ancestor without lateral gene transfer. The evolutionary relationships between the species can be inferred by comparisons between their DNA sequences, using the molecular phylogenetics techniques described in Chapter 16. In (B) extensive lateral gene transfer has occurred. The evolutionary histories of the modern species cannot now be inferred by standard molecular phylogenetics because one or more of the species may have acquired the sequences that are being compared by lateral gene transfer rather than by inheritance from a direct ancestor.

are the minimum required (Mushegian and Koonin, 1996), but experiments in which increasing numbers of *Mycoplasma* genes have been mutated suggest that 265–350 are needed (Hutchison *et al.*, 1999). There has been similar interest in searching for 'distinctiveness' genes – ones that distinguish one species from another. Of the 470 genes in the *M. genitalium* genome, 350 are also present in the distantly related bacterium *Bacillus subtilis*

(Doolittle, 1997), which suggests that the biochemical and structural features that distinguish a *Mycoplasma* from a *Bacillus* are encoded in the 120 or so genes that are unique to the former. Unfortunately, the identities of these supposed distinctiveness genes do not provide any obvious clues about what makes a bacterium a *Mycoplasma* rather than anything else.

2.4 The Repetitive DNA Content of Genomes

Repetitive DNA is the one aspect of genome structure that we have not examined in detail. *Figure 2.2* (page 34) showed us that repetitive DNA is found in all organisms and that in some, including humans, it makes up a substantial fraction of the entire genome. There are various types of repetitive DNA, and several classification systems have been devised. The scheme that we will use begins by dividing the repeats into those that are clustered into tandem arrays and those that are dispersed around the genome.

2.4.1 Tandemly repeated DNA

Tandemly repeated DNA is a common feature of eukaryotic genomes but is found much less frequently in prokaryotes. This type of repeat is also called **satellite DNA** because DNA fragments containing tandemly repeated sequences form 'satellite' bands when genomic DNA is fractionated by density gradient centrifugation (see Technical Note 2.2). For example, when broken into fragments 50–100 kb in length, human DNA forms a main band (buoyant density 1.701 g cm^{-3}) and three satellite bands (1.687, 1.693 and 1.697 g cm^{-3}). The main band contains DNA fragments made up mostly of single-copy sequences with GC compositions close to 40.3%, the average value for the human genome. The satellite bands contain fragments of repetitive DNA, and hence have GC contents and buoyant densities that are atypical of the genome as a whole (*Figure 2.24*).

Satellite DNA is found at centromeres and elsewhere in eukaryotic chromosomes

The satellite bands in density gradients of eukaryotic DNA are made up of fragments composed of long series of tandem repeats, possibly hundreds of kb in length. A single genome can contain several different types of satellite DNA, each with a different repeat unit, these units being anything from < 5 to > 200 bp. The three satellite bands in human DNA include at least four different repeat types.

We have already encountered one type of human satellite DNA, the alphoid DNA repeats found in the centromere regions of chromosomes (see page 38). Although some satellite DNA is scattered around the

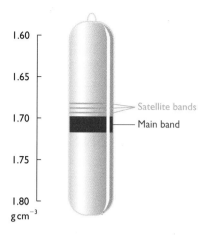

Figure 2.24 Satellite DNA from the human genome.

Human DNA has an average GC content of 40.3% and average buoyant density of 1.701 g cm^{-3}. Fragments made up mainly of single-copy DNA have a GC content close to this average and are contained in the main band in the density gradient. The satellite bands at 1.687, 1.693 and 1.697 g cm^{-3} consist of fragments containing repetitive DNA. The GC contents of these fragments depend on their repeat motif sequences and are different from the genome average, meaning that these fragments have different buoyant densities to single-copy DNA and migrate to different positions in the density gradient.

genome, most is located in the centromeres, where it may play a structural role, possibly as binding sites for one or more of the special centromeric proteins (see page 39). Alternatively, the repetitive DNA content of the centromere might be a reflection of the fact that this is the last region of the chromosome to be replicated. In order to delay its replication until the very end of the cell cycle, the centromere DNA must lack sequences that can act as origins of replication. The repetitive nature of centromeric DNA may be a means of ensuring that such origins are absent (Csink and Henikoff, 1998).

Minisatellites and microsatellites

Although not appearing in satellite bands on density gradients, two other types of tandemly repeated DNA are also classed as 'satellite' DNA. These are **minisatellites** and **microsatellites**. Minisatellites form clusters up to 20 kb in length, with repeat units up to 25 bp; microsatellite clusters are shorter, usually < 150 bp, and the repeat unit is usually 13 bp or less.

Minisatellite DNA is a second type of repetitive DNA that we are already familiar with because of its association with structural features of chromosomes. Telomeric DNA, which in humans comprises hundreds of copies of the motif 5'–TTAGGG–3' (see *Figure 2.10*, page 41), is an

example of a minisatellite. We know a certain amount about how telomeric DNA is formed, and we know that it has an important function in DNA replication (Section 13.2.4). In addition to telomeric minisatellites, some eukaryotic genomes contain various other clusters of minisatellite DNA, many, although not all, near the ends of chromosomes. The functions of these other minisatellite sequences have not been identified.

The function of microsatellites is equally mysterious. The typical microsatellite consists of a 1-, 2-, 3- or 4-bp unit repeated 10–20 times, as illustrated by the microsatellites in the human β T-cell receptor locus (Section 1.2.1). Although each microsatellite is relatively short, there are many of them in the genome (see *Table 1.3*; page 25). In humans, for example, microsatellites with a CA repeat, such as:

$$5'-CACACACACACACAC-3'$$
$$3'-GTGTGTGTGTGTGTG-5'$$

make up 0.25% of the genome, 8 Mb in all. Single base-pair repeats such as:

$$5'-AAAAAAAAAAAAAAA-3'$$
$$3'-TTTTTTTTTTTTTTT-5'$$

make up another 0.15%.

Although their function, if any, is unknown, microsatellites have proved very useful to geneticists. Many microsatellites are variable, meaning that the number of repeat units in the array is different in different members of a species. This is because 'slippage' sometimes occurs when a microsatellite is copied during DNA replication, leading to insertion or, less frequently, deletion of one or more of the repeat units (see *Figure 14.5*; page 423). No two humans alive today have exactly the same combination of microsatellite length variants: if enough microsatellites are examined then a unique **genetic profile** can be established for every person. The only exceptions are genetically identical twins. Genetic profiling is well known as a tool in forensic science (*Figure 2.25*), but identification of criminals is a fairly trivial application of microsatellite variability. More sophisticated methodology makes use of the fact that a person's genetic profile is inherited partly from the mother and partly from the father. This means that microsatellites can be used to establish kinship relationships and population affinities, not only for humans but also for other animals, and for plants.

2.4.2 Interspersed genome-wide repeats

Tandemly repeated DNA sequences are thought to have arisen by expansion of a progenitor sequence, either by replication slippage, as described for microsatellites, or by DNA recombination processes (Section 14.3). Both of these events are likely to result in a series of linked repeats, rather than individual repeat units scattered

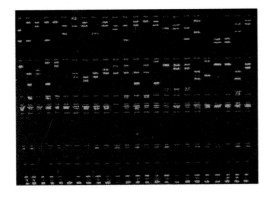

Figure 2.25 The use of microsatellite analysis in genetic profiling.

In this example, microsatellites located on the short arm of chromosome 6 have been amplified by the polymerase chain reaction (PCR; Section 4.3). The PCR products are labeled with a blue or green fluorescent marker and run in a polyacrylamide gel (see Technical Note 6.1, page 165), each lane showing the genetic profile of a different individual. No two individuals have the same genetic profile because each person has a different set of microsatellite length variants, the variants giving rise to bands of different sizes after PCR. The red bands are DNA size markers. Image supplied courtesy of PE Biosystems, Warrington, UK, and reproduced with permission.

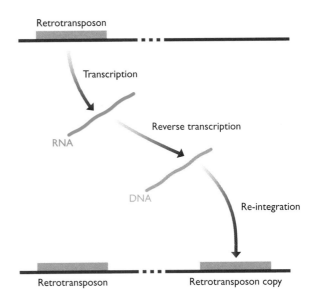

Figure 2.26 Retrotransposition.

Compare with *Figure 1.19* (page 22), and note that the events are essentially the same as those that result in a processed pseudogene.

around the genome. Interspersed repeats must therefore have arisen by a different mechanism, one that can result in a copy of a repeat unit appearing in the genome at a position distant from the location of the original sequence. The most frequent way in which this occurs is by **transposition**, and most interspersed repeats have inherent transpositional activity.

Transposition via an RNA intermediate

The precise mechanics of transposition need not worry us until we deal with recombination and related rearrangements to the genome in Section 14.3. All that we need to know at this point is that there are two alternative modes of transposition, one that involves an RNA intermediate and one that does not. The version that involves an RNA intermediate is called **retrotransposition**. The basic mechanism involves three steps (*Figure 2.26*):

1. An RNA copy of the **transposon** is synthesized by the normal process of transcription.
2. The RNA transcript is copied into DNA, which initially exists as an independent molecule outside of the genome. This conversion of RNA to DNA, the reverse of the normal transcription process, requires a special enzyme called **reverse transcriptase**. Often the reverse transcriptase is coded by a gene within

the transposon and is translated from the RNA copy synthesized in step 1.
3. The DNA copy of the transposon integrates into the genome, possibly back into the same chromosome occupied by the original unit, or possibly into a different chromosome. The end result is that there are now two copies of the transposon, at different points in the genome.

RNA transposons or **retroelements** are features of eukaryotic genomes but have not so far been discovered in prokaryotes. They have attracted a great deal of attention because there are clear similarities between some types of retroelement and the free-living viruses called **retroviruses**, which include both benign forms and also virulent types such as the human immunodeficiency viruses that cause AIDS. These structural relationships are illustrated in *Figure 2.27* and can be summarized as follows:

- **Retroviruses** are viruses whose genomes are made of RNA and have the genetic organization shown in *Figure 2.27A*. They infect many types of vertebrate. Once inside a cell, the RNA genome is copied into DNA by the reverse transcriptase specified by the viral *pol* gene, and the DNA copy integrates into the host genome. New viruses can be produced by copying the integrated DNA into RNA and packaging this into virus coat proteins, the latter coded by the *env* genes on the virus genome.

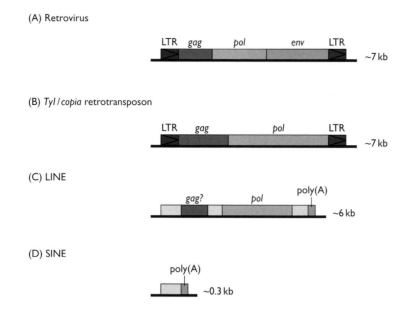

Figure 2.27 Retroelements.

A comparison of the structures of four types of retroelement. Retroviruses and retrotransposons are LTR elements that possess long terminal repeats at each end. The *gag* gene codes for a series of proteins located in the virus core; *pol* codes for the reverse transcriptase and other enzymes involved in replication of the element; *env* codes for coat proteins. LINEs and SINEs are non-LTR retroelements or retroposons. Both have a poly(A) region (a long series of A nucleotides) at one end.

■ **Endogenous retroviruses (ERVs)** are retroviral genomes integrated into vertebrate chromosomes. Some are still active and might, at some stage in a cell's lifetime, direct synthesis of exogenous viruses, but most are decayed relics that no longer have the capacity to form viruses (Patience *et al.*, 1997). These inactive sequences are genome-wide repeats but they are not capable of additional proliferation.

■ **Retrotransposons** have sequences similar to ERVs but are features of non-vertebrate eukaryotic genomes (i.e. plants, fungi, invertebrates and microbial eukaryotes) rather than vertebrates. Retrotransposons have very high copy numbers in some genomes, with many different types present. Most of the genome-wide repeats in maize are retrotransposons and in this plant these elements probably make up almost half the genome (see *Figure 2.2*, page 34). There are two types of retrotransposon: the *Ty3/gypsy* family (*Ty3* and *gypsy* are examples of this class in yeast and fruit fly, respectively), whose members possess the same set of genes as an ERV, and the *Ty1/copia* family, members of which lack the *env* gene (*Figure 2.27B*). Both types are able to transpose (via the mechanism depicted in *Figure 2.26*) but the absence of the *env* gene means that the

Ty1/copia group cannot form infectious virus particles. In fact, despite the presence of *env* in the *Ty3/gypsy* genome, it has only recently been recognized that some of these elements can form viruses and hence should be looked upon as non-vertebrate retroviruses (Song *et al.*, 1994; Peterson-Burch *et al.*, 2000). Although technically they are interspersed elements, retrotransposons are sometimes found in clusters in a genome sequence as a result of the presence of preferred integration sites for transposing elements.

The three types of retroelement described so far are **LTR elements**, as they have long terminal repeats at either end which play a role in the transposition process (see *Figure 2.27*). Other retroelements do not have LTRs. These are called **retroposons** and in mammals include the following:

■ **LINEs (long interspersed nuclear elements)** contain a reverse-transcriptase-like gene probably involved in the retrotransposition process (*Figure 2.27C*). An example is the human element LINE-1, which is 6.1 kb and has a copy number of 516 000 in the human genome (see *Table 1.2*, page 24).

■ **SINEs (short interspersed nuclear elements)** do not have a reverse transcriptase gene but can still transpose, probably by 'borrowing' reverse transcriptase enzymes that have been synthesized by other retroelements (*Figure 2.27D*). The commonest SINE in the human genome is **Alu**, which has a copy number of over 1 million (*Table 1.2*). Alu seems to be derived from the gene for the 7SL RNA, a non-coding RNA involved in movement of proteins around the cell. The first Alu element may have arisen by the accidental reverse transcription of a 7SL RNA molecule and integration of the DNA copy into the human genome. Some Alu elements are actively copied into RNA, providing the opportunity for proliferation of the element.

DNA transposons

Not all transposons require an RNA intermediate. Many are able to transpose in a more direct DNA to DNA manner. With these elements we are aware of two distinct transposition mechanisms (*Figure 2.28*), one involving

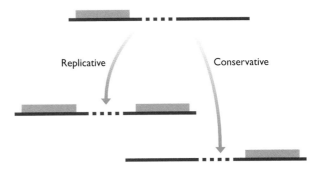

Figure 2.28   Two mechanisms of transposition used by DNA transposons.

For more details see Section 14.3.3.

direct interaction between the donor transposon and the target site, resulting in copying of the donor element (**replicative transposition**), and the second involving excision of the element and re-integration at a new site

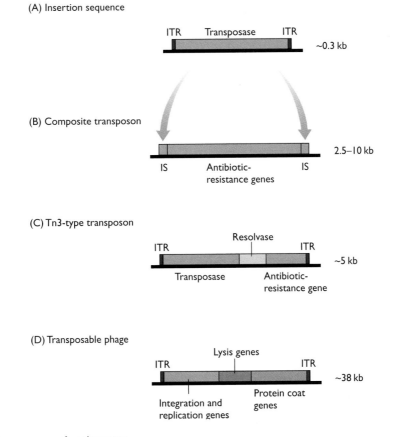

Figure 2.29   DNA transposons of prokaryotes.

Four types are shown. Insertion sequences, Tn3-type transposons and transposable phages are flanked by short (< 50 bp) inverted terminal repeat (ITR) sequences. The resolvase gene of the Tn3-type transposon codes for a protein involved in the transposition process.

(**conservative transposition**). Both mechanisms require enzymes which are usually coded by genes within the transposon. The molecular events that occur during these two types of transposition are described in Section 14.3.3.

In eukaryotes, DNA transposons are less common than retrotransposons (see *Table 1.2*, page 24), but they have a special place in genetics because a family of plant DNA transposons – the *Ac/Ds* elements of maize – were the first transposable elements to be discovered, by Barbara McClintock in the 1950s. Her conclusions – that some genes are mobile and can move from one position to another in a chromosome – were based on exquisite genetic experiments, the molecular basis of transposition not being understood until the late 1970s.

DNA transposons are a much more important component of prokaryotic genome anatomies than the RNA transposons. The insertion sequences, IS1 and IS186, present in the 50-kb segment of *E. coli* DNA that we examined in Section 2.1.2 (see *Figure 2.2*, page 34), are examples of DNA transposons, and a single *E. coli* genome may contain as many as 20 of these of various types. Most of the sequence of an IS is taken up by one or two genes that specify the **transposase** enzyme that catalyzes its transposition (*Figure 2.29A*). IS elements can transpose either replicatively or conservatively. Other kinds of DNA transposon known in *E. coli*, and fairly typical of prokaryotes in general, are as follows:

- **Composite transposons** are basically a pair of IS elements flanking a segment of DNA, usually containing one or more genes, often ones coding for antibiotic resistance (Figure 2.29B). The transposition of a composite transposon is catalyzed by the transposase coded by one or both of the IS elements. Composite transposons use the conservative mechanism of transposition.

- **Tn3-type transposons** have their own transposase gene and so do not require flanking IS elements in order to transpose (*Figure 2.29C*). Tn3 elements transpose replicatively.

- **Transposable phages** are bacterial viruses which transpose replicatively as part of their normal infection cycle (*Figure 2.29D*).

References

Adams MA, Celniker SE, Holt RA, et al. (2000) The genome sequence of *Drosophila melanogaster. Science*, **287**, 2185–2195.

AGI (The Arabidopsis Genome Initiative) (2000) Analysis of the genome sequence of the flowering plant *Arabidopsis thaliana. Nature*, **408**, 796–815.

Akhmanova A, Voncken F, van Allen T, et al. (1998) A hydrogenosome with a genome. *Nature*, **396**, 527–528.

Albertson DG and Thomson JN (1982) The kinetochores of *Caenorhabditis elegans. Chromosoma*, **86**, 409–428.

Alm RA, Ling L-SL, Moir DT, et al. (1999) Genomic-sequence comparison of two unrelated isolates of the human gastric pathogen *Helicobacter pylori. Nature*, **397**, 176–180.

Andersson SGE, Zomorodipour A, Andersson JO, et al. (1998) The genome sequence of *Rickettsia prowazekii* and the origin of mitochondria. *Nature*, **396**, 133–140.

Blattner FR, Plunkett G, Bloch CA, et al. (1997) The complete genome sequence of *Escherichia coli* K-12. *Science*, **277**, 1453–1462.

Brown TA (1998) *Molecular Biology Labfax*, 2nd edition, Vol. 1. Academic Press, London.

Bult CJ, White O, Olsen GJ, et al. (1996) Complete genome sequence of the methanogenic archaeon *Methanococcus jannaschii. Science*, **273**, 1058–1073.

CESC (The C. elegans Sequencing Consortium) (1998) Genome sequence of the nematode *C. elegans*: a platform for investigating biology. *Science*, **282**, 2012–2018.

Chen CW (1996) Complications and implications of linear bacterial chromosomes. *Trends Genet.* **12**, 192–196.

Cole ST, Brosch R, Parkhill J, et al. (1998) Deciphering the biology of *Mycobacterium tuberculosis* from the complete genome sequence. *Nature*, **393**, 537–544.

Copenhaver GP, Nickel K, Kuromori T, et al. (1999) Genetic definition and sequence analysis of *Arabidopsis* centromeres. *Science*, **286**, 2468–2474.

Csink AK and Henikoff S (1998) Something from nothing: the evolution and utility of satellite repeats. *Trends Genet.*, **14**, 200–204.

Deckert G, Warren PV, Gaasterland T, et al. (1998) The complete genome of the hyperthermophile bacterium *Aquifex aeolicus. Nature*, **392**, 353–358.

Doolittle RF (1997) Microbial genomes opened up. *Nature*, **392**, 339–342.

Doolittle WF (1999) Lateral genomics. *Trends Genet.*, **15**, M5–M8.

Ferdows MS and Barbour AG (1989) Megabase-sized linear DNA in the bacterium *Borrelia burgdorferi*, the Lyme disease agent. *Proc. Natl Acad. Sci. USA*, **86**, 5969–5973.

Fraser CM, Gocayne JD, White O, et al. (1995) The minimal gene complement of *Mycoplasma genitalium. Science*, **270**, 397–403.

Fraser CM, Casjens S, Huang WM, et al. (1997) Genomic sequence of a Lyme disease spirochaete, *Borrelia burgdorferi. Nature*, **390**, 580–586.

Fraser CM, Norris SJ, Weinstock GM, et al. (1998) Complete genome sequence of *Treponema pallidum*, the syphilis spirochete. *Science*, **281**, 375–388.

Fritsch EF, Lawn RM and Maniatis T (1980) Molecular cloning and characterization of the human α-like globin gene cluster. *Cell*, **19**, 959–972.

Gardiner K (1996) Base composition and gene distribution: critical patterns in mammalian genome organization. *Trends Genet.*, **12**, 519–524.

Goffeau A, Barrell BG, Bussey H, et al. (1996) Life with 6000 genes. *Science*, **274**, 562–567.

Heidelberg JF, Eisen JA, Nelson WC, et al. (2000) DNA sequence of both chromosomes of the cholera pathogen *Vibrio cholerae. Nature*, **406**, 477–483.

Hutchison CA, Peterson SN, Gill SR, et al. (1999) Global transposon mutagenesis and a minimal Mycoplasma genome. *Science*, **286**, 2165–2169.

IHGSC (International Human Genome Sequencing Consortium) (2001) Initial sequencing and analysis of the human genome. *Nature*, **409**, 860–921.

Jacob F and Monod J (1961) Genetic regulatory mechanisms in the synthesis of proteins. *J. Mol. Biol.*, **3**, 318–356.

Jones RN (1995) B chromosomes in plants. *New Phytol.*, **131**, 411–434.

Klenk H-P, Clayton RA, Tomb J-F, et al. (1997) The complete genome sequence of the hyperthermophilic, sulphate-reducing archaeon *Archaeoglobus fulgidus*. *Nature*, **390**, 364–370.

Lang BF, Burger G, O'Kelly CJ, et al. (1997) An ancestral mitochondrial DNA resembling a eubacterial genome in miniature. *Nature*, **387**, 493–497.

McQueen HA, Siriaco G and Bird AP (1998) Chicken minichromosomes are hyperacetylated, early replicating, and gene rich. *Genome Res.*, **8**, 621–630.

Mushegian AR and Koonin EV (1996) A minimal gene set for cellular life derived by comparison of complete bacterial genomes. *Proc. Natl Acad. Sci. USA*, **93**, 10 268–10 273.

Nelson KE, Clayton RA, Gill SR, et al. (1999) Evidence for lateral gene transfer between Archaea and Bacteria from genome sequence of *Thermatoga maritima*. *Nature*, **399**, 323–329.

Nosek J, Tomáska L, Fukuhara H, Suyama Y and Kovác L (1998) Linear mitochondrial genomes: 30 years down the line. *Trends Genet.*, **14**, 184–188.

Ochman H, Lawrence JG and Groisman EA (2000) Lateral gene transfer and the nature of bacterial innovation. *Nature*, **405**, 299–304.

Oliver SG, van der Aart QJM, Agostini-Carbone ML, et al. (1992) The complete DNA sequence of yeast chromosome III. *Nature*, **357**, 38–46.

Palmer JD (1997a) The mitochondrion that time forgot. *Nature*, **387**, 454–455.

Palmer JD (1997b) Organelle genomes: going, going, gone! *Science*, **275**, 790–791.

Parkhill J, Achtman M, James KD, et al. (2000a) Complete genome sequence of a serogroup A strain of *Neisseria meningitidis* Z2491. *Nature*, **404**, 502–506.

Parkhill J, Wren BW, Mungall K, et al. (2000b) The genome sequence of the food-borne pathogen *Campylobacter jejuni* reveals hypervariable sequences. *Nature*, **403**, 665–668.

Parkhill J, Wren BW, Thomson NR, et al. (2001) Genome sequence of *Yersinia pestis*, the causative agent of plague. *Nature*, **413**, 523–527.

Patience C, Wilkinson DA and Weiss RA (1997) Our retroviral heritage. *Trends Genet.*, **13**, 116–120.

Pennisi E (1996) Linker histones, DNA's protein custodians, gain new respect. *Science*, **274**, 503–504.

Perna NT, Plunkett G, Burland V, et al. (2001) Genome sequence of enterohaemorrhagic *Escherichia coli* O157:H7. *Nature*, **409**, 529–532.

Peterson-Burch BD, Wright DA, Laten HM and Voytas DF (2000) Retroviruses in plants? *Trends Genet.*, **16**, 151–152.

Pruss D, Hayes JJ and Wolffe AP (1995) Nucleosomal anatomy – where are the histones? *Bioessays*, **17**, 161–170.

SanMiguel P, Tikhonov A, Jin Y-K, et al. (1996) Nested retrotransposons in the intergenic regions of the maize genome. *Science*, **274**, 765–768.

Song SU, Gerasimova T, Kurkulos M, Boeke JD and Corces VG (1994) An env-like protein encoded by a *Drosophila* retroelement: evidence that *gypsy* is an infectious retrovirus. *Genes Dev.*, **8**, 2046–2057.

Stover CK, Pham XQ, Erwin AL, et al. (2000) Complete genome sequence of *Pseudomonas aeruginosa* PA01, an opportunistic pathogen. *Nature*, **406**, 959–964.

Strachan T and Read AP (1999) *Human Molecular Genetics*, 2nd edition. BIOS Scientific Publishers, Oxford.

Tettelin H, Nelson KE, Paulsen IT, et al. (2001) Complete genome sequence of a virulent isolate of *Streptococcus pneumoniae*. *Science*, **293**, 498–506.

Tham W-H and Zakian VA (2000) Telomeric tethers. *Nature*, **403**, 34–35.

Travers A (1999) The location of the linker histone on the nucleosome. *Trends Biochem. Sci.*, **24**, 4–7.

Vafa O and Sullivan KF (1997) Chromatin containing CENP-A and alpha-satellite DNA is a major component of the inner kinetochore plate. *Curr. Biol.*, **7**, 897–900.

van Steensel B, Smogorzewska A and de Lange T (1998) TRF2 protects human telomeres from end-to-end fusions. *Cell*, **92**, 401–413.

Venter JC, Adams MD, Myers EW, et al. (2001) The sequence of the human genome. *Science*, **291**, 1304–1351.

Warburton PE (2001) Epigenetic analysis of kinetochore assembly on variant human centromeres. *Trends Genet.*, **17**, 243–247.

White O, Eisen JA, Heidelberg JF, et al. (1999) Genome sequence of the radioresistant bacterium *Deinococcus radiodurans* R1. *Science*, **286**, 1571–1577.

Zhang J (2000) Protein-length distributions for the three domains of life. *Trends Genet.*, **16**, 107–109.

Zhang Z, Green BR and Cavalier-Smith T (1999) Single gene circles in dinoflagellate genomes. *Nature*, **400**, 155–159.

Zhou Y-B, Gerchman SE, Ramakrishnan V, Travers A and Muyldermans S (1998) Position and orientation of the globular domain of linker histone H5 on the nucleosome. *Nature*, **395**, 402–405.

Further Reading

Comfort NC (2001) *The Tangled Field: Barbara McClintock's Search for the Patterns of Genetic Control.* Harvard University Press, Cambridge, MA. — *A biography of the geneticist who discovered transposable elements; for a highly condensed version, see* Trends Genet., **17**, 475–478.

Drlica K and Riley M (1990) *The Bacterial Chromosome.* American Society for Microbiology, Washington, DC. — *A source of information on all aspects of bacterial DNA.*

Federoff NV (1983) Controlling elements in maize. In: J Shapiro, ed. *Mobile Genetic Elements.* Academic Press, New York, pp. 1–63. — *Includes a summary of the experiments carried out by McClintock that led to the discovery of transposons.*

Gill P, Ivanov PL, Kimpton C, et al. (1994) Identification of the remains of the Romanov family by DNA analysis. *Nature Genet.*, **6**, 130–135. — *An example of the use of genetic profiling based on microsatellite variability.*

Koonin EV (2000) How many genes can make a cell: the minimal-gene-set concept. *Ann. Rev. Genomics Hum. Genet.*, **1**, 99–116. — *Describes the experimental and theoretical work that is being done on the minimal genome.*

Kumar A and Bennetzen JL (1999) Plant retrotransposons. *Ann. Rev. Genet.*, **33**, 479–532. — *Detailed review of this subject.*

Lang BV, Gray MW and Burger G (1999) Mitochondrial genome evolution and the origin of eukaryotes. *Ann. Rev. Genet.*, **33**, 351–397. — *Covers all of the essential material regarding mitochondrial genomes.*

Margulis L (1970) *Origin of Eukaryotic Cells.* Yale University Press, New Haven. — *The first description of the endosymbiont theory for the origin of mitochondria and chloroplasts.*

Petrov DA (2001) Evolution of genome size: new approaches to an old problem. *Trends Genet.*, **17**, 23–28. — *Reviews the C-value paradox and the genetic processes that might result in differences in genome size.*

Ramakrishnan V (1997) Histone H1 and chromatin higher-order structure. *Crit. Rev. Eukaryotic Gene Expression*, **7**, 215–230. — *Detailed descriptions of models for the 30 nm chromatin fiber.*

Schueler MG, Higgins AW, Rudd MK, Gustashaw K and Willard HW (2001) Genomic and genetic definition of a functional human centromere. *Science*, **294**, 109–115. — *Details of the sequence features of human centromeres.*

STUDY AIDS FOR CHAPTER 2

Key terms

Give short definitions of the following terms:

30 nm fiber	Linker histone
Agarose gel	LTR element
Alphoid DNA	Macrochromosome
Alu element	Metaphase chromosome
Apoptosis	Microsatellite
Archaea	Minichromosome
B chromosome	Minisatellite
Bacteria	Molecular phylogenetics
Buoyant density	Multigene family
Cell cycle	Nuclease protection
Centromere	experiment
Chromatid	Nucleoid
Chromatin	Nucleosome
Chromatosome	Operon
Composite transposon	Overlapping genes
Conjugation	Plasmid
Conservative transposition	Polyacrylamide gel
Core octamer	Prokaryote
C-value paradox	Promiscuous DNA
Cyanelle	Replicative transposition
Density gradient centrifugation	Retroelement
Differential centrifugation	Retroposon
Domain	Retrotransposition
Endogenous retrovirus	Retrotransposon
Endosymbiont theory	Retrovirus
Episome transfer	Reverse transcriptase
Eukaryote	Ribosomal RNA
Extrachromosomal gene	Ribosome
Gene flow	Satellite DNA
Gene superfamily	Sedimentation coefficient
Genes-within-genes	SINE
Genetic profile	Supercoiling
Histone	Telomere
Holocentric chromosome	Tn3-type transposon
Insertion sequence	Transduction
Integron	Transfer RNA
Isochore	Transformation
Karyogram	Transposable phage
Kinetochore	Transposase
Lactose operon	Transposition
Lateral gene transfer	Transposon
LINE	Zinc finger

Self study questions

1. Briefly describe the main types of living organism recognized by biologists.
2. What is meant by the 'C-value paradox'? What is the underlying reason for this paradox?
3. Compare the genetic organizations of the human, yeast, fruit fly and maize genomes.
4. How does the genetic organization of the *Escherichia coli* genome differ from that of a eukaryote?
5. Explain how nuclease protection experiments can be used to study chromatin structure.
6. Write a short essay describing how DNA is packaged into chromatosomes and into the 30 nm fiber.
7. Draw a typical metaphase chromosome and label the centromere and telomeres. What are the functions and key features of the centromere and telomeres?
8. Explain why chromosome banding patterns and the isochore model suggest that genes are not evenly distributed in a eukaryotic chromosome.
9. Why is it informative to categorize eukaryotic genes on the basis of the protein domains that they encode?
10. Using examples from the human genome, explain what is meant by the term 'multigene family'.
11. Compare and contrast the physical structures and genetic organizations of the mitochondrial genomes of different types of eukaryote.
12. List the key features of the typical chloroplast genome.
13. Discuss the evidence for the endosymbiont theory of organelle origins.
14. Define the term 'nucleoid'. Give details of an experiment that has provided information on the packaging of DNA in the *Escherichia coli* nucleoid.
15. To what extent is the physical structure of the *Escherichia coli* genome typical of that of prokaryotes in general?
16. Describe the structure and function of the lactose operon of *Escherichia coli*. Is the lactose operon typical of operons in other prokaryotes?
17. What is lateral gene transfer and how has it affected our interpretation of the species concept as applied to prokaryotes?
18. Outline the speculations that have been made about the content of the minimal prokaryotic genome and the identity of distinctiveness genes.
19. Describe how centrifugation techniques can be used to study the repetitive DNA content of a genome.
20. Use examples from the human genome to distinguish between satellite, minisatellite and microsatellite DNA.
21. How are microsatellites used in forensic science?
22. Draw a diagram to illustrate the mechanism of retrotransposition. Annotate your diagram to show the differences

between the retrotransposition processes of retroviruses and retrotransposons.

23. List the similarities and differences between LINEs and SINEs.
24. Distinguish between the two transposition mechanisms for a DNA transposon. Give examples of DNA transposons stating which mechanism each follows.

Problem-based learning

1. To what extent is it possible to describe the 'typical' features of a eukaryotic genome?
2. What impact is DNA packaging likely to have on the expression of individual genes?
3. A great deal is now known about the special features of centromeric DNA in *Arabidopsis thaliana*. How different is human centromeric DNA?
4. Defend or attack the isochore model.
5. Why do organelle genomes exist?
6. Should the traditional view of the prokaryotic genome as a single circular DNA molecule be abandoned? If so, what new definition of 'prokaryotic genome' should be adopted?
7. Can the concept of bacterial species survive genome sequencing?
8. Most eukaryotic genomes have substantial repetitive DNA contents. Devise hypotheses for the origins of this repetitive DNA. Could your hypotheses be tested?

Transcriptomes and Proteomes

Chapter Contents

3.1	*Genome Expression in Outline*	70
3.2	*The RNA Content of the Cell*	72
	3.2.1 Coding and non-coding RNA	72
	3.2.2 Synthesis of RNA	74
	3.2.3 The transcriptome	78
3.3	*The Protein Content of the Cell*	80
	3.3.1 Protein structure	80
	3.3.2 The link between the transcriptome and the proteome	83
	3.3.3 The link between the proteome and the biochemistry of the cell	88

Learning outcomes

When you have read Chapter 3, you should be able to:

- Define the terms 'transcriptome' and 'proteome'

- Draw a diagram illustrating the modern interpretation of the genome expression pathway, indicating the main points at which genome expression is regulated

- Distinguish between coding and non-coding RNA and give examples of each type

- Outline the process by which RNA is synthesized in the cell

- List the major types of RNA processing events that occur in living cells

- Describe how transcriptomes are studied and discuss the applications of this type of research

- Give a detailed description of the various levels of protein structure

- Explain why amino acids underlie protein diversity

- Outline how the meaning of each codon in the genetic code was elucidated

- Describe the key features of the genetic code

- Explain why the function of a protein is dependent on its amino acid sequence

- List the major roles of proteins in living organisms and relate this diversity to the function of the genome

THE GENOME is a store of biological information but on its own it is unable to release that information to the cell. Utilization of the biological information requires the co-ordinated activity of enzymes and other proteins, which participate in a complex series of biochemical reactions referred to as **genome expression**. The details of genome expression are described in Part 3. Before reaching this detailed discussion, an overview of the key events involved in genome expression will be valuable, in order to establish a foundation of knowledge onto which the more comprehensive understanding can subsequently be built. This chapter provides that overview.

3.1 Genome Expression in Outline

The initial product of genome expression is the **transcriptome**, a collection of RNA molecules derived from those protein-coding genes whose biological information is required by the cell at a particular time (*Figure 3.1*). These RNA molecules direct synthesis of the final product of genome expression, the **proteome**, the cell's repertoire of proteins, which specifies the nature of the biochemical reactions that the cell is able to carry out.

The transcriptome is constructed by the process called **transcription**, in which individual genes are copied into RNA molecules. Construction of the proteome involves **translation** of these RNA molecules into protein. Transcription and translation are important terms but it is unfortunate that the expression of individual genes is sometimes described simply as the two-step process 'DNA makes RNA makes protein' (*Figure 3.2A*). This is an inadequate over-simplification of the much more complex series of events involved in synthesis and maintenance of the transcriptome and proteome of even the simplest type of cell. In reality, genome expression comprises the following steps (*Figure 3.2B*):

- **Accessing the genome.** This involves various processes that influence chromatin structure and nucleosome positioning in the parts of the genome that contain active genes, ensuring that these genes are accessible and are not buried deep within highly packaged parts of the chromosomes.

- **Assembly of the transcription initiation complex,** which comprises the set of proteins that work together to copy genes into RNA. Assembly of initiation complexes is a highly targeted process because these complexes must be constructed at precise positions in the genome, adjacent to active genes, and nowhere else.

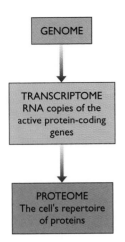

Figure 3.1 The genome, transcriptome and proteome.

- **Synthesis of RNA,** during which the gene is transcribed into an RNA copy.
- **Processing of RNA** involves a series of alterations that are made to the sequence of the RNA molecule and to its chemical structure, and which must occur before the RNA molecules can be translated into protein or, in the case of non-coding RNA, before they can carry out their other functions in the cell.
- **RNA degradation** is the controlled turnover of RNA molecules. Degradation is not simply a means of getting rid of unwanted RNAs: it plays an active role in determining the make up of the transcriptome and hence is an integral step in genome expression.

- **Assembly of the translation initiation complex** occurs near the 5′ termini of coding RNA molecules, and is a prerequisite for translation of these molecules.
- **Protein synthesis** is the synthesis of a protein by translation of an RNA molecule.
- **Protein folding and protein processing** may occur together. Folding results in the protein taking up its correct three-dimensional configuration. Processing involves modification of the protein by addition of chemical groups and, for some proteins, removal of one or more segments of the protein.
- **Protein degradation** has an important influence on the composition of the proteome and, like RNA degradation, is an integral component of genome expression.

Genome expression is clearly much more complicated than 'DNA makes RNA makes protein'. A particular weakness of this two-step interpretation is that it draws attention away from the points in the expression pathway at which the flow of information from genome to proteome can be regulated. Control mechanisms exist for regulating every one of the steps shown in *Figure 3.2B*, enabling the composition of the transcriptome and proteome to be altered in a rapid and precise manner, and allowing the cell to adjust its biochemical capabilities in response to changes in the extracellular environment and to signals received from other cells. As we will see in Chapter 12, these regulatory events underlie not only the functioning of individual cells but also the processes of differentiation and development.

Box 3.1: Cross-references to Part 3 of *Genomes*

The stages of genome expression described in the Chapter 3 are covered in more detail in the following chapters of Part 3.

Stage of genome expression	Cross reference to Part 3
Accessing the genome	Chapter 8
Assembly of the transcription initiation complex	Chapter 9
Synthesis of RNA	Sections 10.1 and 10.2
Processing of RNA	Sections 10.1–10.3
RNA degradation	Section 10.4
Assembly of the translation initiation complex	Section 11.2.2
Protein synthesis	Sections 11.2.3 and 11.2.4
Protein folding and protein processing	Section 11.3
Protein degradation	Section 11.4

The processes responsible for regulation of the individual stages in genome expression are described in the Sections listed above, and a more general consideration of genome regulation in the context of differentiation and development is contained in Chapter 12.

(A)

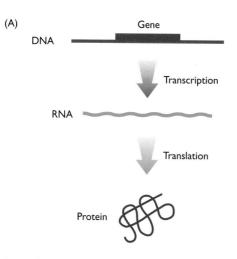

(B)

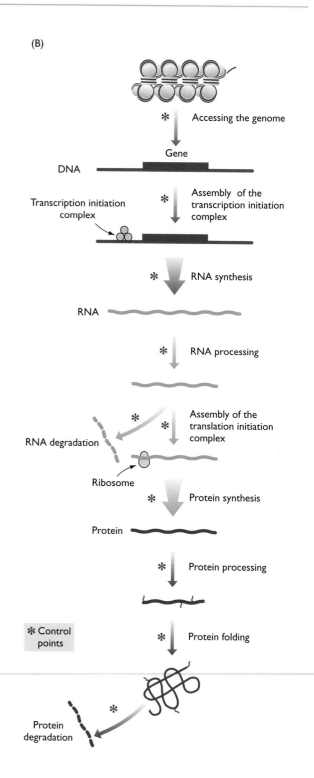

Figure 3.2 Two views of genome expression.

(A) shows the old-fashioned depiction of gene expression, summarized as 'DNA makes RNA makes protein', the two steps being called transcription and translation. (B) gives a more accurate outline of the events involved in genome expression, especially in higher organisms. Note that these schemes apply only to protein-coding genes. Those genes that give rise to non-coding RNAs are transcribed and processed as shown but the RNAs are not translated.

3.2 The RNA Content of the Cell

A typical bacterium contains 0.05–0.10 pg of RNA, making up about 6% of its total weight. A mammalian cell, being much larger, contains more RNA, 20–30 pg in all, but this represents only 1% of the cell as a whole (Alberts *et al.*, 1994). It is important to appreciate that not all of this RNA constitutes the transcriptome. The latter is just the coding RNA – those molecules that have been transcribed from protein-coding genes and which are therefore capable of being translated into protein. Most of the cellular RNA does not fall into this category because it is non-coding. An understanding of the distinctive features of coding and non-coding RNA is therefore essential before we continue with our overview of genome expression.

3.2.1 Coding and non-coding RNA

The best way to understand the RNA content of a cell is to divide it into categories and subcategories depending on function. There are several ways of doing this, the most informative scheme being the one shown in *Figure 3.3*. The primary division is between coding RNA and non-coding RNA. The coding RNA comprises the transcriptome and is made up of just one class of molecule:

- **Messenger RNAs (mRNAs)**, which are transcripts of protein-coding genes and hence are translated into protein in the latter stages of genome expression.

Messenger RNAs rarely make up more than 4% of the total RNA and are short-lived, being degraded soon after synthesis. Bacterial mRNAs have half-lives of no more than a few minutes and in eukaryotes most mRNAs are degraded a few hours after synthesis. This rapid turnover means that the composition of the transcriptome is not fixed and can quickly be restructured by changing the rate of synthesis of individual mRNAs.

The second type of RNA is non-coding. This is more diverse than the coding RNA and comprises transcripts with a number of different functions, all of which are performed by the RNA molecules themselves. In both prokaryotes and eukaryotes the two main types of non-coding RNA are:

- **Ribosomal RNAs (rRNAs)**, which are the most abundant RNAs in the cell, making up over 80% of the total in actively dividing bacteria. These molecules are components of ribosomes, the structures on which protein synthesis takes place (Section 11.2).
- **Transfer RNAs (tRNAs)** are small molecules that are also involved in protein synthesis, carrying amino acids to the ribosome and ensuring that these are linked together in the order specified by the nucleotide sequence of the mRNA that is being translated (Section 11.1).

Ribosomal and tRNAs are present in the cells of all species. The other non-coding RNA types are more limited in their distribution (see *Figure 3.3*). Eukaryotes, for example, have a variety of short non-coding RNAs that are usually divided into three categories, the names indicating their primary locations in the cell:

- **Small nuclear RNA (snRNA**; also called **U-RNA** because these molecules are rich in uridine nucleotides), which is involved in mRNA processing (Section 10.1.3);
- **Small nucleolar RNA (snoRNA)**, which plays a central role in the processing of rRNA molecules (Section 10.3.1);
- **Small cytoplasmic RNA (scRNA)**, a diverse group including molecules with a range of functions, some understood and others still mysterious.

Bacteria and archaea also contain non-coding RNAs other than rRNA and tRNA but these molecules do not make up a substantial fraction of the total RNA. In bacteria they include one interesting RNA type, apparently present in most if not all species, called **transfer-messenger RNA (tmRNA)**, which looks like a tRNA attached to an mRNA, and which adds short peptide tags onto proteins that have been synthesized incorrectly, labeling them for immediate degradation (Muto *et al.*, 1998).

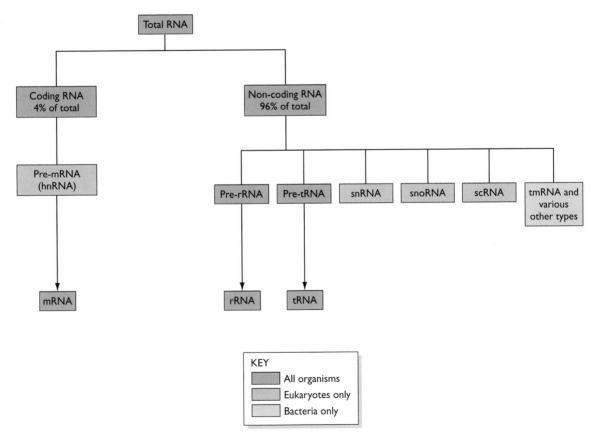

Figure 3.3 The RNA content of a cell.

This scheme shows the types of RNA present in all organisms (eukaryotes, bacteria and archaea) and those categories found only in eukaryotic or bacterial cells. The non-coding RNAs of archaea have not yet been fully characterized and it is not clear which types are present in addition to rRNA and tRNA. For abbreviations, see the text.

Box 3.2: Non-coding RNA specified by the human genome

The human genome has genes for the following types of non-coding RNA.

Type of non-coding RNA	Function	See Section
rRNA	Protein synthesis	11.2
tRNA	Protein synthesis	11.1
snRNA	mRNA processing	10.1.3
snoRNA	rRNA processing	10.3.1
7SL RNA	Type of scRNA; involved in transport of proteins in the cell	–
7SK RNA	Type of scRNA; function unknown	–
RNA component of RNase P RNA	Processing of tRNA molecules	10.2.2
RNA component of RNase MRP	Processing of rRNA molecules	10.2.2
RNA component of telomerase RNA	Synthesis of telomeres	13.2.4
hyRNA	Components of Ro RNA, function unknown	–
Vault RNAs	Unknown	–
H19	Unknown	–
Xist	X chromosome inactivation	8.2.2

Based on IHGSC (2001).

3.2.2 Synthesis of RNA

The enzymes responsible for transcription of DNA into RNA are called **DNA-dependent RNA polymerases**. The name indicates that the enzymatic reaction that they catalyze results in polymerization of RNA from ribonucleotides, and occurs in a DNA-dependent manner, meaning that the sequence of nucleotides in a DNA template dictates the sequence of nucleotides in the RNA that is made (*Figure 3.4*). It is permissible to shorten the enzyme name to **RNA polymerase**, as the context in which the name is used means that there is rarely confusion with the **RNA-dependent RNA polymerases** that are involved in replication and expression of some virus genomes. Note that there are also **template-independent**

RNA polymerases, including one – **poly(A) polymerase** – with an important role in genome expression (Section 10.1.2). The chemical basis of the template-dependent synthesis of RNA is illustrated in *Figure 3.5*. Ribonucleotides are added one after another to the growing 3′ end of the RNA transcript, the identity of each nucleotide being specified by the base-pairing rules: A base-pairs with T or U; G base-pairs with C. During each nucleotide addition, the β- and γ-phosphates are removed from the incoming nucleotide, and the hydroxyl group is removed from the 3′-carbon of the nucleotide at the end of the chain, precisely the same as for DNA polymerization (see *Figure 1.8*, page 12).

RNA polymerase is the central component of the transcription initiation complex. Every time a gene is transcribed, a new complex has to be assembled immediately upstream of the gene. The initiation complexes are constructed at the appropriate positions, and not at random points within the genome, because their target sites are marked by specific nucleotide sequences called **promoters**, which are only found upstream of genes. In bacteria, promoters are directly recognized by the RNA polymerase enzyme, but in eukaryotes and archaea an intermediary DNA-binding protein is required, which attaches to the DNA and forms a platform to which the RNA polymerase binds (*Figure 3.6*). Promoters are described in detail in Section 9.2.2, which considers assembly of transcription initiation complexes.

Processing of precursor RNA

As well as the mature RNAs described above, cells also contain precursor molecules (see *Figure 3.3*). Many RNAs, especially in eukaryotes, are initially synthesized as precursor or **pre-RNA**, which has to be processed before

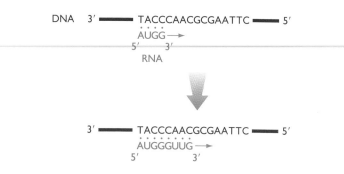

Figure 3.4   Template-dependent RNA synthesis.

The RNA transcript is synthesized in the 5′→3′ direction, reading the DNA in the 3′→5′ direction, with the sequence of the transcript determined by base-pairing to the DNA template.

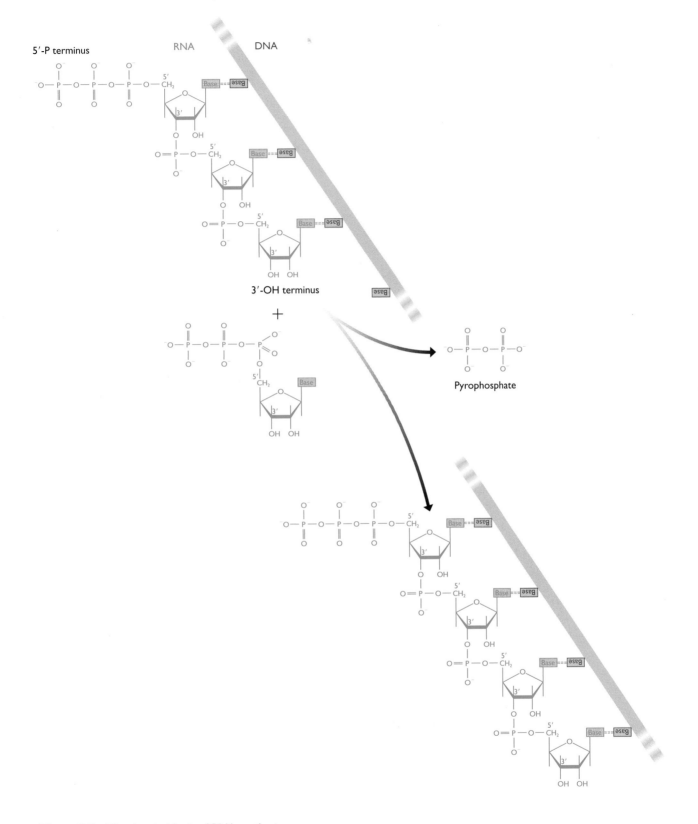

Figure 3.5 The chemical basis of RNA synthesis.

Compare this reaction with polymerization of DNA, as illustrated in *Figure 1.8*, page 12.

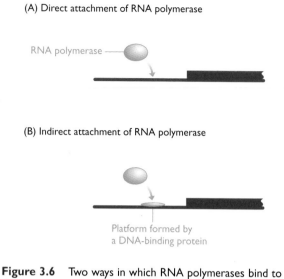

(A) Direct attachment of RNA polymerase

RNA polymerase

(B) Indirect attachment of RNA polymerase

Platform formed by
a DNA-binding protein

Figure 3.6 Two ways in which RNA polymerases bind to their promoters.

(A) shows the direct recognition of the promoter by the RNA polymerase, as occurs in bacteria. (B) shows recognition of the promoter by a DNA-binding protein which forms a platform onto which the RNA polymerase binds. This indirect mechanism occurs with eukaryotic and archaeal RNA polymerases.

it can carry out its function. The various processing events, all of which are described in detail in Chapter 10, include the following (*Figure 3.7*):

- **End-modifications** occur during the synthesis of eukaryotic and archaeal mRNAs, most of which have a single, unusual nucleotide called a **cap** attached at the 5′ end and a **poly(A) tail** attached to the 3′ end. The cap and poly(A) tails are involved in assembly of the translation initiation complex on these mRNAs (Section 10.2.2).

- **Splicing** is the removal of introns from a precursor RNA. Many eukaryotic protein-coding genes contain introns and these are copied when the gene is transcribed. The introns are removed from the **pre-mRNA** by cutting and joining reactions. Unspliced pre-mRNA forms the nuclear RNA fraction called **heterogenous nuclear RNA (hnRNA)**. Some eukaryotic pre-rRNAs and pre-tRNAs also contain introns, as do some archaeal transcripts, but they are extremely rare in bacteria.

- **Cutting events** are particularly important in the processing of rRNA and tRNA, many of which are initially synthesized from transcription units that specify more than one molecule. The **pre-rRNAs** and **pre-tRNAs** must therefore be cut into pieces to produce the mature RNAs. This type of processing occurs in both prokaryotes and eukaryotes.

- **Chemical modifications** are made to rRNAs, tRNAs and mRNAs. The rRNAs and tRNAs of all organisms are modified by addition of new chemical groups, these groups being added to specific nucleotides within each RNA. Chemical modification of mRNA, called **RNA editing**, is seen in a diverse group of eukaryotes.

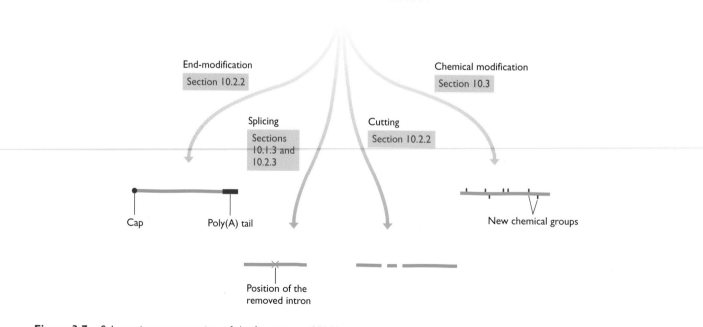

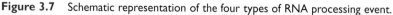

Pre-RNA

End-modification
Section 10.2.2

Chemical modification
Section 10.3

Splicing
Sections 10.1.3 and 10.2.3

Cutting
Section 10.2.2

Cap   Poly(A) tail

New chemical groups

Position of the removed intron

Figure 3.7 Schematic representation of the four types of RNA processing event.

Not all events occur in all organisms – see the text for details.

Processing of mRNA has an important influence on the composition of the transcriptome. RNA editing, for example, can result in a single pre-mRNA being converted into two different mRNAs coding for quite distinct proteins (*Figure 3.8A*). This does not appear to be particularly common, but **alternative splicing**, in which one pre-mRNA gives rise to two or more mRNAs by assembly of different combinations of exons (*Figure 3.8B*), is fairly widespread. The mRNAs resulting from both editing and alternative splicing often display tissue specificity, the processing events increasing the coding capabilities of the genome without the requirement for an increased gene number.

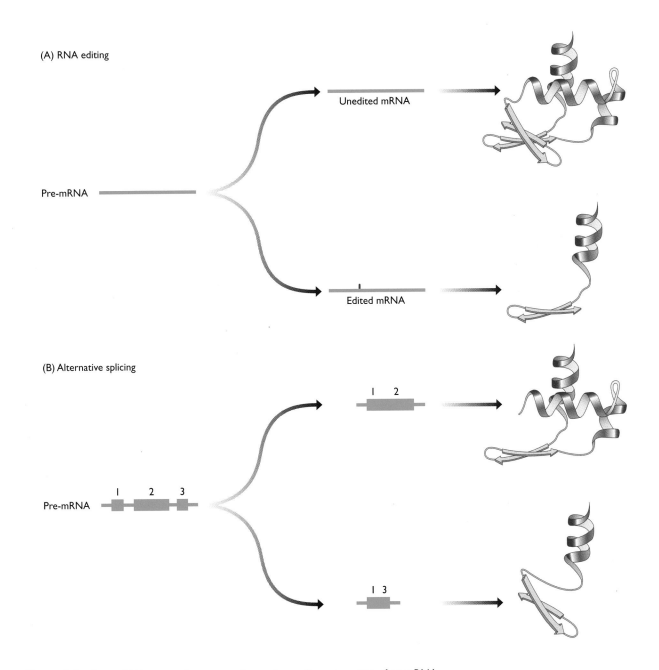

Figure 3.8   Some RNA processing events change the coding properties of an mRNA.

(A) RNA editing can change the sequence of an mRNA, resulting in synthesis of a different protein. An example occurs with the human mRNA for apolipoprotein B, as shown in *Figure 10.29*, page 306. (B) Alternative splicing results in different combinations of exons becoming linked together, again resulting in different proteins being synthesized from the same pre-mRNA. *Figure 10.20*, page 293 shows how alternative splicing underlies sex determination in *Drosophila*.

3.2.3 The transcriptome

Although the transcriptome makes up less than 4% of the total cell RNA, it is the most significant component because it contains the coding RNAs that specify the composition of the proteome and hence determine the biochemical capacity of the cell. One important point to note is that the transcriptome is never synthesized *de novo*. Every cell receives part of its parent's transcriptome when it is first brought into existence by cell division, and maintains a transcriptome throughout its lifetime. Even quiescent cells in bacterial spores or in the seeds of plants have a transcriptome, although expression of that transcriptome into protein may be completely switched off. Transcription does not therefore result in *synthesis* of the transcriptome but instead *maintains* the transcriptome by replacing mRNAs that have been degraded, and brings about *changes* to the composition of the transcriptome via the switching on and off of different sets of genes.

Even in the simplest organisms such as bacteria and yeast, many genes are active at any one time. Transcriptomes are therefore complex, containing copies of hundreds, if not thousands, of different mRNAs. Usually, each mRNA makes up only a small fraction of the whole, with the most common type rarely contributing more than 1% of the total mRNA. Exceptions are those cells that have highly specialized biochemistries, which are reflected by transcriptomes in which one or a few mRNAs predominate. Developing wheat seeds are an example: these synthesize large amounts of the gliadin proteins, which accumulate in the dormant grain and provide a source of amino acids for the germinating seedling. Within the developing seeds, the gliadin mRNAs can make up as much as 30% of the transcriptomes of certain cells.

Perhaps surprisingly, it is relatively easy to determine the composition of a transcriptome, and to make comparisons between different transcriptomes, using the **microarray** and **DNA chip** technologies described in Section 7.3.1. To illustrate the types of analysis that are possible we will examine some of the recent research on the yeast and human transcriptomes.

Studies of the yeast transcriptome

With less than 6000 genes, the yeast *Saccharomyces cerevisiae* is ideally suited for transcriptome studies, and many of the pioneering projects have been carried out with this organism. One of the first discoveries was that, although mRNAs are being degraded and re-synthesized all the time, the composition of the yeast transcriptome undergoes very little change if the biochemical features of the environment remain constant (DeRisi *et al.*, 1997). When yeast is grown in a glucose-rich medium, which allows the cells to divide at their maximum rate, the transcriptome is almost completely stable, only 19 mRNAs displaying a greater than two-fold change in abundance over a period of 2 hours (*Figure 3.9*). Significant alterations to the transcriptome are seen only when the glucose in the growth medium becomes depleted, forcing the cells to switch from aerobic to anaerobic respiration. During this switch, the levels of over 700 mRNAs increase by a factor of two or more, and another 1000 mRNAs decline to less than half their original amount. The changing environment clearly results in a restructuring of the transcriptome to meet the new biochemical demands of the cell.

The yeast transcriptome also undergoes restructuring during cellular differentiation. This has been established by studying sporulation (spore formation), which is induced by starvation and other stressful environmental conditions. The sporulation pathway can be divided into four stages – early, middle, mid-late and late – on the basis of the morphological and biochemical events that occur (*Figure 3.10*). Previous studies have shown, not unexpectedly, that each stage is characterized by expression of a different set of genes. Transcriptome studies have added to our understanding of the sporulation process in several ways (Chu *et al.*, 1998). Most significantly, the changes that occur to the composition of the transcriptome indicate that the early stage of sporulation can be subdivided into three distinct phases, called early (I), early (II) and early-middle. The levels of over 250 mRNAs increase significantly during early sporulation, and another 158 mRNAs increase specifically during the middle stage. A further 61 mRNAs increase in abundance during the mid-late period and five more during the late phase. There are also 600 mRNAs that decrease in abundance during sporulation, these presumably coding for proteins that are needed during vegetative growth but whose synthesis must be switched off when spores are being formed.

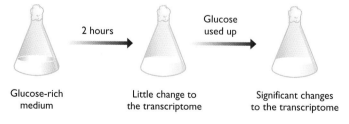

Glucose-rich medium → 2 hours → Little change to the transcriptome → Glucose used up → Significant changes to the transcriptome

Figure 3.9 The yeast transcriptome is stable during growth of cells in a glucose-rich medium, but undergoes significant changes when the glucose is used up and the cells switch from aerobic to anaerobic respiration.

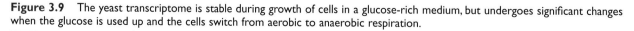

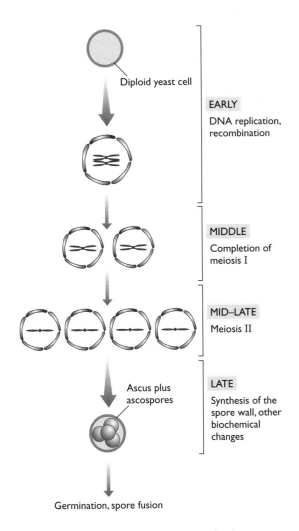

Figure 3.10   The sporulation pathway of *Saccharomyces cerevisiae*.

The middle three drawings show the nuclear divisions that occur during sporulation. See *Figure 5.15*, page 138 for details of the events involved in meiosis I and meiosis II.

This work on yeast sporulation is important for two reasons. First, by describing the changes in genome expression that occur during sporulation, the transcriptome analyses open the way to studies of the interactions between the genome and the environmental signals that trigger sporulation. Studies of this type, in a relatively simple organism such as yeast, act as an important model for the more complex developmental processes that operate in higher eukaryotes, including humans. Secondly, several of the mRNAs whose levels change significantly during sporulation are transcripts of genes whose functions were previously unknown. Transcriptome studies therefore help to annotate a genome sequence, aiding identification of genes whose roles in the genome have not been determined by other methods. We will return to this issue in Chapter 7.

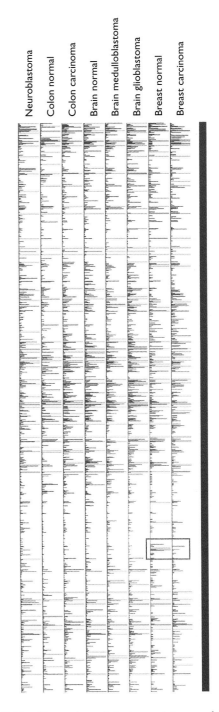
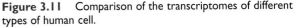

Figure 3.11   Comparison of the transcriptomes of different types of human cell.

The diagram shows human chromosome 11 aligned vertically. The bar charts indicate the expression levels in different cell types of the genes on this chromosome. The lengths of the blue bars are proportional to the extent of gene expression, and the red bars indicate genes whose expression levels are higher than can be illustrated on this scale. The box highlights significant differences between the transcriptomes of normal and cancerous breast tissue cells. Reprinted with permission from Caron *et al. Science*, **291**, 1289–1292. Copyright 2001 American Association for the Advancement of Science.

The human transcriptome

With six times as many genes, the human transcriptome is substantially more complex than that of yeast, and studies of its composition are still in their infancy. Some interesting results have, nonetheless, been obtained. For example, the transcriptomes of eight different cell types have been mapped onto the draft human genome sequence (Caron *et al.*, 2001), resulting in global views of the pattern of gene expression along entire chromosomes (*Figure 3.11*). As well as providing a 'blueprint' for the transcriptomes of each of the eight cell types, this work has underlined the extent to which the transcriptome of a cancerous tissue differs from that of the normal version. Transcriptome restructuring as a result of cancer was first discovered several years ago, when it was shown that 289 mRNAs are present in significantly different amounts in the transcriptomes of normal colon epithelial cells compared with cancerous colon cells, and that about half of these mRNAs also display an altered abundance in pancreatic cancer cells (Zhang *et al.*, 1997). It is hoped that by understanding the differences between the transcriptomes of normal and cancerous cells it will be possible to devise new ways of treating the cancers.

Transcriptome studies also have applications in cancer diagnosis. The initial breakthrough in this respect came in 1999 when it was shown that the transcriptome of acute lymphoblastic leukemia cells is different from that of acute myeloid leukemia cells (Golub *et al.*, 1999). Twenty-seven lymphoblastic and eleven myeloid cancers were studied and, although all the transcriptomes were slightly different, the distinctions between the two types were sufficient for unambiguous identifications to be made. The significance of this work lies with the improved remission rates that are achievable if a cancer is identified accurately at an early stage, before clear morphological indicators are seen. This is not relevant with these two types of leukemia because these can be distinguished by non-genetic means, but it is important with other cancers such as non-Hodgkin lymphoma. The commonest version of this disease is called diffuse large B-cell lymphoma, and for many years it was thought that all tumors of this type were the same. Transcriptome studies changed this view and showed that B-cell lymphoma can be divided into two distinct subtypes (Alizadeh *et al.*, 2000). The distinctions between the transcriptomes of the two subtypes enable each one to be related to a different class of B cells, stimulating and directing the search for specific treatments that are tailored for each lymphoma.

3.3 The Protein Content of the Cell

The proteome is the final product of genome expression and comprises all the proteins present in a cell at a particular time. A 'typical' mammalian cell, for example a liver hepatocyte, is thought to contain 10 000–20 000 different proteins, about 8×10^9 individual molecules in all, representing approximately 0.5 ng of protein or 18–20% of the total cell weight (Alberts *et al.*, 1994; Lodish *et al.*, 2000). The copy numbers of individual proteins vary enormously, from less than 20 000 molecules per cell for the rarest types to 100 million copies for the commonest ones. Any protein that is present at a copy number of greater than 50 000 per cell is considered to be relatively abundant, and in the average mammalian cell some 2000 proteins fall into this category. When the proteomes of different types of mammalian cell are examined, very few differences are seen among these abundant proteins, suggesting that most of them are **housekeeping** proteins which perform general biochemical activities that occur in all cells. The proteins that provide the cell with its specialized function are often quite rare, although there are exceptions such as the vast amounts of hemoglobin that are present only in red blood cells (Alberts *et al.*, 1994).

The proteome can be looked upon as the central link between the genome and the cell: it is, on the one hand, the culmination of genome expression and, on the other hand, the starting point for the biochemical activities that constitute cellular life (*Figure 3.12*). In order to comprehend how the proteome makes this connection we must first understand the structure of proteins.

3.3.1 Protein structure

A protein, like a DNA molecule, is a linear unbranched polymer. In proteins the monomeric subunits are called **amino acids** (*Figure 3.13*) and the resulting polymers, or **polypeptides**, are rarely more than 2000 units in length. As with DNA, the key features of protein structure were determined in the first half of the 20th century, this phase of protein biochemistry culminating in the 1940s and early 1950s with the elucidation by Pauling and Corey of the major conformations or **secondary structures** taken up by polypeptides (Pauling *et al.*, 1951; Pauling and Corey, 1953). In recent years, interest has focused on how these secondary structures combine to produce the complex three-dimensional shapes of proteins.

The four levels of protein structure

Proteins are traditionally looked upon as having four distinct levels of structure. These levels are hierarchical, the protein being built up stage by stage, with each level of structure depending on the one below it:

Figure 3.12   The central role of the proteome.

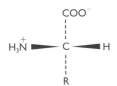

Figure 3.13   The general structure of an amino acid.

All amino acids have the same general structure, comprising a central α-carbon attached to a hydrogen atom, a carboxyl group, an amino group and an R group. The R group is different for each amino acid (see *Figure 3.17*).

1. *The primary structure* of the protein is formed by joining amino acids into a polypeptide. The amino acids are linked by **peptide bonds** which are formed by a condensation reaction between the carboxyl group of one amino acid and the amino group of a second amino acid (*Figure 3.14*). In passing, note that, as with a polynucleotide, the two ends of the polypeptide are chemically distinct: one has a free amino group and is called the **amino, NH₂-, or N terminus**; the other has a free carboxyl group and is called the **carboxyl, COOH-, or C terminus**. The direction of the polypeptide can therefore be expressed as either N→C (left to right in *Figure 3.14*) or C→N (right to left in *Figure 3.14*).

2. *The secondary structure* refers to the different conformations that can be taken up by the polypeptide. The two main types of secondary structure are the **α-helix** and **β-sheet** (*Figure 3.15*), both of which are stabilized by hydrogen bonds that form between different amino acids in the polypeptide. Most polypeptides are long enough to be folded into a series of secondary structures, one after another along the molecule.

3. *The tertiary structure* results from folding the secondary structural components of the polypeptide into a three-dimensional configuration (*Figure 3.16*). The tertiary structure is stabilized by various chemical forces, notably hydrogen bonding between individual amino acids, and hydrophobic forces, which dictate that amino acids with non-polar (i.e. 'water-hating') side-groups must be shielded from water by embedding within the internal regions of the protein. There may also be covalent linkages called **disulfide bridges** between cysteine amino acids at various places in the polypeptide.

4. *The quaternary structure* involves the association of two or more polypeptides, each folded into its tertiary structure, into a multi-subunit protein. Not all proteins form quaternary structures, but it is a feature of many proteins with complex functions, including several involved in genome expression. Some quaternary structures are held together by disulfide bridges between different polypeptides, but many proteins comprise looser associations of subunits stabilized by hydrogen bonding and

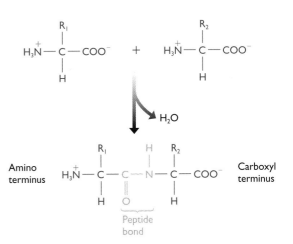

Figure 3.14   In polypeptides, amino acids are linked by peptide bonds.

The drawing shows the chemical reaction that results in two amino acids becoming linked together by a peptide bond. The reaction is called a condensation because it results in elimination of water.

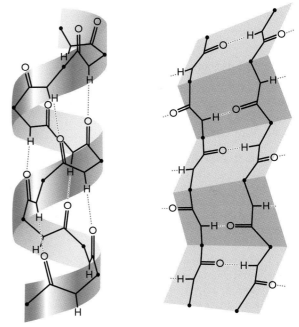

Figure 3.15   The two main secondary structural units found in proteins: (A) the α-helix, and (B) the β-sheet.

The polypeptide chains are shown in outline with the positions of the α-carbons indicated by small dots. The R groups have been omitted for clarity. Each structure is stabilized by hydrogen bonds between the C=O and N–H groups of different peptide bonds. The β-sheet conformation that is shown is anti-parallel, the two chains running in opposite directions. Parallel β-sheets also occur.

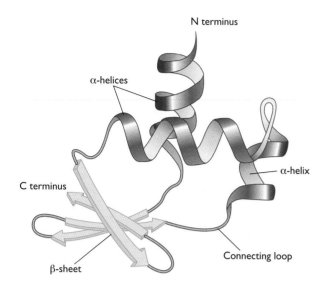

Figure 3.16 The tertiary structure of a protein.

This imaginary protein structure comprises three α-helices, shown as coils, and a four-stranded β-sheet, indicated by the arrows. Redrawn from Turner *et al.* (1997).

hydrophobic effects, and can revert to their component polypeptides, or change their subunit composition, according to the functional requirements.

Amino acid diversity underlies protein diversity

Proteins are functionally diverse because the amino acids from which proteins are made are themselves chemically diverse. Different sequences of amino acids therefore result in different combinations of chemical reactivities, these combinations dictating not only the overall structure of the resulting protein but also the positioning on the surface of the structure of reactive groups that determine the chemical properties of the protein.

Amino acid diversity derives from the R group because this part is different in each amino acid and varies greatly in structure. Proteins are made up from a set of 20 amino acids (*Figure 3.17*; *Table 3.1*). Some of these have R groups that are small, relatively simple structures such as a single hydrogen atom (in the amino acid called glycine) or a methyl group (alanine). Others are large complex aromatic side chains (phenylalanine, tryptophan and tyrosine). Most are uncharged, but two are negatively charged (aspartic acid and glutamic acid) and three are positively charged (arginine, histidine and lysine). Some are polar (e.g. glycine, serine and threonine), others are non-polar (e.g. alanine, leucine and valine).

The 20 amino acids shown in *Figure 3.17* are the ones that are conventionally looked on as being specified by the genetic code (Section 3.3.2). They are therefore the amino acids that are linked together when polypeptides are assembled during the protein-synthesis phase of genome expression. However, these 20 amino acids do not on their own represent the limit of the chemical diversity of proteins. The diversity is even greater because of two factors:

■ At least one additional amino acid – selenocysteine (*Figure 3.18*) – can be inserted into a polypeptide chain during protein synthesis, its insertion directed by a modified reading of the genetic code (Section 3.3.2).

Box 3.3: Non-covalent bonds in proteins

The secondary and higher levels of protein structure are stabilized by various types of non-covalent bonds and interactions:

■ **Hydrogen bonds** are weak electrostatic attractions between an electronegative atom (such as oxygen or nitrogen) and a hydrogen atom attached to a second electronegative atom. Hydrogen bonds are longer and much weaker than covalent bonds, typical bond energies being 1–10 kcal mol^{-1} at 25 °C, compared with up to 90 kcal mol^{-1} for a covalent bond. Hydrogen bonds stabilize protein secondary structures.

■ **van der Waals forces** can be attractive or repulsive. Attractive forces result from fluctuations between the electron charge densities of adjacent atoms and have energies of 0.1–0.2 kcal mol^{-1}. They promote formation of secondary structures in proteins. Repulsive van der Waals forces occur as a result of repulsion between electrons when two atoms come too close together. These repulsive forces place limitations on the degree of packing that can be achieved within a protein.

■ **Electrostatic interactions** or ionic bonds form between charged groups. In aqueous environments they are relatively weak – approximately 3 kcal mol^{-1} – because of the shielding effect of water. Electrostatic interactions occur between the R groups of charged amino acids within proteins and are also important on the protein surface, in particular in attachment of a protein to DNA (Section 9.1.5).

■ **Hydrophobic effects** arise because the hydrogen-bonded structure of water forces hydrophobic groups into the internal parts of a protein. Hydrophobic effects are not true bonds but they are the main determinants of protein tertiary structure.

(A) Non-polar R groups

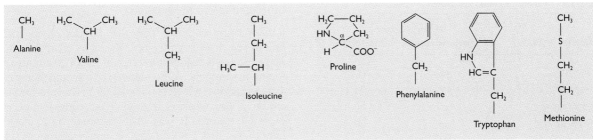

(B) Polar R groups

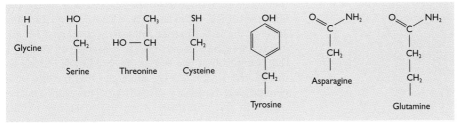

(C) Negatively charged R groups

(D) Positively charged R groups

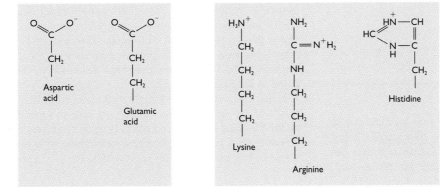

Figure 3.17   Amino acid R groups.

These 20 amino acids are the ones that are conventionally looked upon as being specified by the genetic code (Section 3.3.2). The classification into non-polar, polar etc. is as described in Lehninger (1970).

■ During protein processing, some amino acids are modified by the addition of new chemical groups, for example by acetylation or phosphorylation, or by attachment of large side chains made up of sugar units (Section 11.3.3).

Proteins therefore have an immense amount of chemical variability, some of this directly specified by the genome, the remainder arising by protein processing.

3.3.2   The link between the transcriptome and the proteome

The flow of information from DNA to RNA by transcription does not provide any conceptual difficulty. DNA and RNA polynucleotides have very similar struc-

tures and we can easily understand how an RNA copy of a gene can be made by template-dependent synthesis using the base-pairing rules with which we are familiar. The second phase of genome expression, during which the mRNA molecules of the transcriptome direct synthesis of proteins, is less easy to understand simply by considering the structures of the molecules that are involved. In the early 1950s, shortly after the double helix structure of DNA had been discovered, several molecular biologists attempted to devise ways in which amino acids could attach to mRNAs in an ordered fashion, but in all of these schemes at least some of the bonds had to be shorter or longer than was possible according to the laws of physical chemistry, and each idea was quietly dropped. Eventually, in 1957, Francis Crick cut a way through the confusion by predicting the existence of

Table 3.1 Amino acid abbreviations

Amino acid	Abbreviation	
	Three-letter	One-letter
Alanine	Ala	A
Arginine	Arg	R
Asparagine	Asn	N
Aspartic acid	Asp	D
Cysteine	Cys	C
Glutamic acid	Glu	E
Glutamine	Gln	Q
Glycine	Gly	G
Histidine	His	H
Isoleucine	Ile	I
Leucine	Leu	L
Lysine	Lys	K
Methionine	Met	M
Phenylalanine	Phe	F
Proline	Pro	P
Serine	Ser	S
Threonine	Thr	T
Tryptophan	Trp	W
Tyrosine	Tyr	Y
Valine	Val	V

an adaptor molecule (Crick, 1990) that would form a bridge between the mRNA and the polypeptide being synthesized. Soon afterwards it was realized that the non-coding tRNAs were these adaptor molecules, and, once this fact had been established, a detailed understanding of the mechanism by which proteins are synthesized was quickly built up. We will examine this process in Section 11.1.

The other aspect of protein synthesis that interested molecular biologists in the 1950s and 1960s was the **informational problem**. This refers to the second important component of the link between the transcriptome and proteome: the **genetic code** which specifies how the nucleotide sequence of an mRNA is translated into the amino acid sequence of a protein.

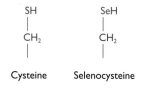

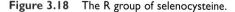

Figure 3.18 The R group of selenocysteine.

Selenocysteine is the same as cysteine but with the sulfur replaced with a selenium atom.

The genetic code specifies how an mRNA sequence is translated into a polypeptide

It was recognized in the 1950s that a triplet genetic code – one in which each codeword or **codon** comprises three nucleotides – is required to account for all 20 amino acids found in proteins. A two-letter code would have only $4^2 = 16$ codons, which is not enough to account for all 20 amino acids, whereas a three-letter code would give $4^3 = 64$ codons. It was also assumed, as a working hypothesis, that mRNAs contain non-overlapping series of codons, and that these are colinear with the polypeptides that they encode (*Figure 3.19*). Although experimental proof of these assumptions was difficult to obtain, all three turned out to be correct. There are qualifications, such as the lack of strict colinearity displayed by many eukaryotic genes because of the presence of introns, but this complication was not appreciated until introns were discovered in 1977, long after the main work on the genetic code had been carried out. This work was completed in the mid-1960s when the meanings of all 64 codons were determined, partly by analysis of polypeptides resulting from translation of artificial mRNAs of known or predictable sequence in cell-free protein-synthesizing systems, and partly by determining which amino acids associated with which RNA sequences in an assay based on purified ribosomes, the protein–RNA complexes that carry out protein synthesis in the cell (Section 11.2). These experiments are described in more detail in Research Briefing 3.1.

When the work on the genetic code was completed it was realized that the 64 codons fall into groups, the members of each group coding for the same amino acid (*Figure 3.20*). Only tryptophan and methionine have just a single codon each: all others are coded by two, three, four

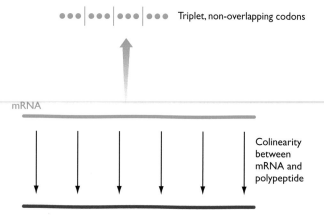

Figure 3.19 Early assumptions about the genetic code.

It was assumed that the genetic code is triplet, that codons do not overlap, and that there is a colinear relationship between the sequences of an mRNA and the polypeptide it encodes.

RESEARCH 3.1 BRIEFING

Elucidation of the genetic code

Assigning amino acids to triplet codons was the main objective of molecular biology during the first half of the 1960s.

Once it had been accepted that the genetic code is triplet and non-overlapping, and that genes display colinearity with the proteins that they specify, attention turned to elucidation of the code, with the objective of assigning amino acids to individual codons. Two types of experiment enabled the code of *Escherichia coli* to be completely worked out during the years 1961–66.

Cell-free translation of artificial RNAs

The first type of experiment was pioneered by Marshall Nirenberg and Heinrich Matthaei at the National Institutes of Health, Maryland, and made use of an extract, prepared from *E. coli* cells, that contained all the components needed to carry out translation except for mRNA. This **cell-free protein synthesizing system** therefore only made protein when RNA was added. If the sequence of the added RNA was known or could be predicted then the composition of the proteins that were made could be used to assign codons to amino acids.

The system was first used with the simplest artificial RNAs, those that contain just a single nucleotide, such as the **homopolymer** poly(U), whose sequence is 5′–UUU ... UUU–3′. When added into the cell-free system, poly(U) directed synthesis of polyphenylalanine, showing that the codon 5′–UUU–3′ codes for phenylalanine. Equivalent experiments enabled 5′–AAA–3′ to be assigned to lysine and 5′–CCC–3′ to proline. For unexplained reasons poly(G) gave no protein product.

Reliable methods for sequencing RNA were not developed until the late 1960s. This meant that when the cell-free experiments were performed it was not possible to determine the exact sequences of artificial **heteropolymers** – RNAs containing more than one nucleotide. These could still, however, be used in the cell-free system because their codon compositions could be deduced statistically from the identity and relative amounts of the nucleotides used in the reaction mixture from which each heteropolymer was made. Random heteropolymers of different compositions enabled amino acids to be assigned to over half the codons in the genetic code. A few more codons were assigned when the technique for synthesis of artificial RNAs was refined so that ordered heteropolymers could be made. These RNAs are polymerized from dinucleotides such as 5′–GC–3′ and so have predictable sequences, the ordered

heteropolymer poly(GC) having the sequence 5′–GCGC ... GCGC–3′ and therefore containing just two codons, 5′–GCG–3′ and 5′–CGC–3′.

The triplet binding assay

The genetic code could not be completed with the standard cell-free protein synthesizing system because it was simply not possible to devise random and ordered heteropolymers that enabled every codon to be assigned unambiguously. A new approach was therefore needed. This was the triplet binding assay, devised by Nirenberg and Philip Leder in 1964 and based on their discovery that a ribosome will attach to an RNA triplet if the appropriate aminoacyl-tRNA (a tRNA linked to its amino acid; see Section 11.1.1) is also present. The code could therefore be completed (and all previously assigned codons checked) by synthesizing triplets of all possible sequences and testing each one individually with different aminoacyl-tRNAs.

The final remaining ambiguity concerned the termination codons, which could not be directly identified by the triplet binding assay as it could not be proven that the inability of 5′–UAA–3′, 5′–UAG–3′ and 5′–UGA–3′ to bind any aminoacyl-tRNA was not simply the result of deficiencies in this assay system. Confirmation that these are the termination codons was provided by Sydney Brenner and co-workers at Cambridge, UK, through genetic analysis of **suppressor mutations**, which result in one or other of the termination codons being recognized as a codon for an amino acid.

References

Brenner S, Stretton AOW and Kaplan S (1965) Genetic code: the 'nonsense' triplets for chain termination and their suppression. *Nature*, **206**, 994–998.
Crick FHC (1962) The genetic code. *Sci. Am.*, **207**(4), 66–74.
Crick FHC (1966) The genetic code III. *Sci Am.*, **215**(4), 55–62.
Nirenberg MW (1963) The genetic code II. *Sci.Am.*, **208**(3), 80–94.
Nirenberg MW and Matthaei H (1961) The dependence of cell-free protein synthesis in *E. coli* upon naturally occurring or synthetic polyribonucleic acids. *Proc. Natl Acad. Sci. USA*, **47**, 1588–1602.
Nirenberg MW and Leder P (1964) RNA codewords and protein synthesis. *Science*, **145**, 1399–1407.

or six codons. This feature of the code is called **degeneracy**. The code also has four **punctuation codons**, which indicate the points within an mRNA where translation of the nucleotide sequence should start and finish (*Figure 3.21*). The **initiation codon** is usually 5'–AUG–3', which also specifies methionine (so most newly synthesized polypeptides start with methionine), although with a few mRNAs other codons such as 5'–GUG–3' and 5'–UUG–3' are used. The three **termination codons** are 5'–UAG–3', 5'–UAA–3' and 5'–UGA–3'; these are sometimes called amber, opal and ochre, respectively, these being the whimsical names given to the original *Escherichia coli* mutants whose analysis led to their discovery.

The genetic code is not universal

It was originally thought that the genetic code must be the same in all organisms. The argument was that, once established, it would be impossible for the code to change because giving a new meaning to any single codon would result in widespread disruption of the amino acid sequences of proteins. This reasoning seems sound, so it is surprising that, in reality, the genetic code is not universal. The code shown in *Figure 3.20* holds for the vast majority of genes in the vast majority of organisms, but deviations are widespread. In particular, mitochondrial genomes often use a non-standard code (*Table 3.2*). This was first discovered in 1979 by Frederick Sanger's group in Cambridge, UK, who found that several human mitochondrial mRNAs contain the sequence 5'–UGA–3', which normally codes for termination, at internal positions where protein synthesis was not expected to stop. Comparisons with the amino acid sequences of the proteins coded by these mRNAs showed that 5'–UGA–3'

is a tryptophan codon in human mitochondria, and that this is just one of four code deviations in this particular genetic system. Mitochondrial genes in other organisms also display code deviations, although at least one of these – the use of 5'–CGG–3' as a tryptophan codon in plant mitochondria – is probably corrected by RNA editing (Section 10.3.2) before translation occurs (Covello and Gray, 1989; Gualberto *et al.*, 1989).

Non-standard codes are also known in the nuclear genomes of lower eukaryotes. Often a modification is restricted to just a small group of organisms and frequently it involves reassignment of the termination codons (*Table 3.2*). Modifications are less common among prokaryotes but one example is known in *Mycoplasma* species. A more important type of code variation in nuclear genomes is **context-dependent codon reassignment**, which occurs when the protein to be synthesized contains selenocysteine. This applies to many organisms, both prokaryotes and eukaryotes, including humans,

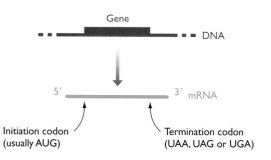

Figure 3.21 The positions of the punctuation codons in an mRNA.

UUU	phe	UCU		UAU	tyr	UGU	cys
UUC		UCC	ser	UAC		UGC	
UUA	leu	UCA		UAA	stop	UGA	stop
UUG		UCG		UAG		UGG	trp
CUU		CCU		CAU	his	CGU	
CUC	leu	CCC	pro	CAC		CGC	arg
CUA		CCA		CAA	gln	CGA	
CUG		CCG		CAG		CGG	
AUU		ACU		AAU	asn	AGU	ser
AUC	ile	ACC	thr	AAC		AGC	
AUA		ACA		AAA	lys	AGA	arg
AUG	met	ACG		AAG		AGG	
GUU		GCU		GAU	asp	GGU	
GUC	val	GCC	ala	GAC		GGC	gly
GUA		GCA		GAA	glu	GGA	
GUG		GCG		GAG		GGG	

Figure 3.20 The genetic code.

See *Table 3.1* for the three-letter abbreviations of the amino acids.

Box 3.4: The origin and evolution of the genetic code

The evolution of the genetic code has provoked argument ever since DNA was established as the genetic material back in the 1950s. Many geneticists support the 'frozen accident' theory, which suggests that codons were randomly allocated to amino acids during the earliest stages of evolution, the code subsequently becoming 'frozen' because any changes would result in widespread disruption of the amino acid sequences of proteins. Various lines of evidence suggest that the code might have evolved in a less random manner (Knight et al., 1999; Szathmáry, 1999). First, controversial experimental results suggest that at least some amino acids bind directly to RNAs containing the appropriate codons, this occurring in the absence of the tRNA that mediates the interaction in present-day cells. These results suggest

that the code might have originated from natural interactions between RNA molecules and amino acids. If correct, the implication is that there is some kind of chemical relationship between an amino acid and its codon(s). Second, the deviations from the standard code listed in Table 3.2 indicate that the same codon reallocations have occurred more than once. For example, three unrelated groups of protozoa use 5'–UAA–3' and 5'–UAG–3' as glutamine rather than termination codons, suggesting that these codons were reallocated in the same way on at least three separate occasions. Similarly, the use of 5'–UGA–3' as a tryptophan codon must have evolved twice, once in mitochondria and once in the ancestors of Mycoplasma. If the relationship between codon and amino acid is entirely random, as suggested by the 'frozen accident' theory, then we would not expect to see the same codon reallocations recurring on different occasions.

Table 3.2 Examples of deviations from the standard genetic code

Organism	Codon	Should code for	Actually codes for
Mitochondrial genomes			
Mammals	UGA	Stop	Trp
	AGA, AGG	Arg	Stop
	AUA	Ile	Met
Drosophila	UGA	Stop	Trp
	AGA	Arg	Ser
	AUA	Ile	Met
Saccharomyces cerevisiae	UGA	Stop	Trp
	CUN	Leu	Thr
	AUA	Ile	Met
Fungi	UGA	Stop	Trp
Maize	CGG	Arg	Trp
Nuclear and prokaryotic genomes			
Several protozoa	UAA, UAG	Stop	Gln
Candida cylindracea	CUG	Leu	Ser
Micrococcus sp.	AGA	Arg	Stop
	AUA	Ile	Stop
Euplotes sp.	UGA	Stop	Cys
Mycoplasma sp.	UGA	Stop	Trp
	CGG	Arg	Stop
Context-dependent codon reassignments			
Various	UGA	Stop	Selenocysteine

Abbreviation: N, any nucleotide.

because selenoproteins are widespread (see *Table 3.3*). Selenocysteine is coded by 5′–UGA–3′, which therefore has two meanings because it is still used as a termination codon in the organisms concerned (*Table 3.2*). The 5′–UGA–3′ codons that specify selenocysteine are distinguished from those that are true termination codons by the presence of a hairpin loop structure in the mRNA, positioned just downstream of the selenocysteine codon in prokaryotes and in the 3′ untranslated region (i.e. the part of the mRNA after the termination codon) in eukaryotes. Recognition of the codon requires interaction between the hairpin and a special protein that is involved in translation of these mRNAs (Low and Berry, 1996).

3.3.3 The link between the proteome and the biochemistry of the cell

How does the proteome convert the biological information that it has received from the genome into the biochemical capabilities of the cell? Two fundamental aspects of protein chemistry enable this final step in genome expression to be achieved. The first of these is the hierarchical nature of the four levels of protein structure, which provides a direct link between the amino acid sequence of a protein and its chemical properties. The second aspect is the multiplicity of the chemical properties that can be displayed by different proteins, this variability enabling proteins to carry out a huge range of different biochemical activities.

The amino acid sequence of a protein determines its function

The links between the different levels of protein structure are most clearly understood at the primary-to-secondary level. Because of the chemical properties of their R groups, certain amino acids are more commonly found in

α-helices while others have a predisposition for β-sheets. A secondary structure therefore forms around a group of amino acids that favor that particular secondary structure and initiate its formation. It then extends to include adjacent amino acids that either favor the structure or have no strong disinclination towards it, and finally terminates when one or more blocking amino acids, which cannot participate in that particular type of structure, are reached (*Figure 3.22*). This process of nucleation, extension and delimitation is repeated along the polypeptide until each part of the chain has adopted its preferred secondary structure.

By identifying which amino acids are most frequently located in which secondary structures, and by studying the structures taken up by small polypeptides of known sequence, biochemists have been able to deduce rules for this level of protein folding, and, to a certain extent, can predict which secondary structures will be adopted by a polypeptide simply by examining its primary sequence (Barton, 1995). It is less easy to predict the outcomes of the next two stages of protein folding, which result in the secondary structural units becoming arranged into the tertiary structure, and tertiary units associating to form quaternary multi-subunit structures. The tertiary structures of most proteins are made up of two or more structural **domains**, possibly with little interaction between them; these domains are thought to fold independently of one another. Understanding how this occurs is complicated by the fact that many domains include secondary structural units from quite different regions of a polypeptide. But the difficulty in identifying rules for predicting tertiary and quaternary structures does not detract from the fact that these higher levels of structure are determined by the amino acid sequence of the polypeptide. This is illustrated by the ability of proteins that have been unfolded in the test tube, for example by treatment with urea, to refold spontaneously into

Table 3.3 Examples of proteins that contain selenocysteine

Protein	Organism
Prokaryotic enzymes	
Formate dehydrogenase	*Clostridium thermoaceticum, Clostridium thermoautotrophicum, Enterobacter aerogenes, Escherichia coli, Methanococcus vaniellii*
Glycine reductase	*Clostridium purinolyticum, Clostridium sticklandii*
NiFeSe hydrogenase	*Desulfomicrobium baculatum, Methanococcus voltae*
Eukaryotic enzymes	
Glutathione peroxidase	Human, cow, rat, mouse
Selenoprotein P	Human, cow, rat
Selenoprotein W	Rat
Type 1 deiodinase	Human, rat, mouse, dog
Type 2 deiodinase	Frog
Type 3 deiodinase	Human, rat, frog

See Low and Berry (1996).

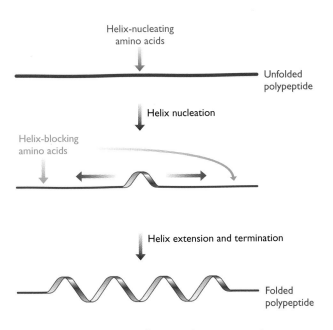

Helix-nucleating amino acids

Unfolded polypeptide

Helix nucleation

Helix-blocking amino acids

Helix extension and termination

Folded polypeptide

Figure 3.22   Formation of a secondary structure in a polypeptide.

An α-helix is shown nucleating at a position containing amino acids that favor helix formation, and extending in either direction until groups of amino acids that block helix formation are reached.

their correct structures when the treatment is reversed (Section 11.3.1).

Spontaneous refolding does not occur with all proteins and is particularly difficult to achieve with larger ones, the problem appearing to be that the protein can adopt alternative partially folded structures at various stages of the folding process, only one of which leads to the correctly folded tertiary configuration. If the protein makes the wrong 'choice' it can end up at a dead end in which it is partially folded in an incorrect manner but from which it cannot escape. In cells, proteins called **molecular chaperones** aid the folding of other proteins, probably by reducing the likelihood that the protein being folded adopts the wrong intermediate structure (Section 11.3.1). The existence of alternative folding pathways, and of proteins that aid the folding of other proteins, complicates the premise that a protein's folded structure is dictated by its amino acid sequence, but the premise still holds. The role of molecular chaperones is not to impose a new structure on a protein, but merely to increase the efficiency of the protein's natural, sequence-directed, folding pathway.

The multiplicity of protein function

The biological information encoded by the genome finds its final expression in a protein whose biological pro-

perties are determined by the spatial arrangement of chemical groups on its surface and within its folded structure. By specifying proteins of different types, the genome is able to construct and maintain a proteome whose overall biological properties form the underlying basis of life. The proteome can play this role because of the huge diversity of protein structures that can be formed, the diversity enabling proteins to carry out a variety of biological functions. These functions include the following:

- *Biochemical catalysis* is the role of the special type of proteins called enzymes. The central metabolic pathways, which provide the cell with energy, are catalyzed by enzymes, as are the biosynthetic processes that result in construction of nucleic acids, proteins, carbohydrates and lipids. Biochemical catalysis also drives genome expression through the activities of enzymes such as RNA polymerase.
- *Structure*, which at the cellular level is determined by the proteins that make up the cytoskeleton, is also the primary function of some extracellular proteins. An example is collagen, which is an important component of bones and tendons.
- *Movement* is conferred by contractile proteins, of which actin and myosin in cytoskeletal fibers are the best known examples.
- *Transport* of materials around the body is an important protein activity: for example, hemoglobin transports oxygen in the bloodstream, and serum albumin transports fatty acids.
- *Regulation* of cellular processes is mediated by signaling proteins such as STATs (Section 12.1.2) and by proteins such as **activators** that bind to the genome and influence the expression levels of individual genes and groups of genes (Section 9.3.2). The activities of groups of cells are regulated and coordinated by extracellular hormones and cytokines, many of which are proteins (e.g. insulin, the hormone that controls blood sugar levels, and the interleukins, a group of cytokines that regulate cell division and differentiation).
- *Protection* of the body and of individual cells is the function of a range of proteins, including the antibodies, and those proteins involved in the blood clotting response.
- *Storage* functions are performed by proteins such as ferritin, which acts as an iron store in the liver, and the gliadins, which store amino acids in dormant wheat seeds.

This multiplicity of protein function provides the proteome with its ability to convert the blueprint contained in the genome into the essential features of the life process.

References

Alberts B, Bray D, Lewis J, Raff M, Roberts K and Watson JD (1994) *Molecular Biology of the Cell*, 3rd edition. Garland Publishing, New York.

Alizadeh AA, Eisen MB, Davis RE, et al. (2000) Distinct types of diffuse large B-cell lymphoma identified by gene expression profiling. *Nature*, **403**, 503–511.

Barton GJ (1995) Protein secondary structure prediction. *Curr. Opin. Struct. Biol.*, **5**, 372–376.

Caron H, van Schail B, van der Mee S, et al. (2001) The human transcriptome map: clustering of highly expressed genes in chromosomal domains. *Science*, **291**, 1289–1292.

Chu S, DeRisi J, Eisen M, et al. (1988) The transcriptional program of sporulation in budding yeast. *Science*, **282**, 699–705.

Covello PS and Gray MW (1989) RNA editing in plant mitochondria. *Nature*, **341**, 662–666.

Crick FHC (1990) *What Mad Pursuit: A Personal View of Scientific Discovery*. Penguin Books, London.

DeRisi JL, Iyer VR and Brown PO (1997) Exploring the metabolic and genetic control of gene expression on a genomic scale. *Science*, **278**, 680–686.

Golub TR, Slonim DK, Tamayo P, et al. (1999) Molecular classification of cancer: class discovery and class prediction by gene expression monitoring. *Science*, **286**, 531–537.

Gualberto JM, Lamattinia L, Bonnard G, Weil J-H and Grienenberger J-M (1989) RNA editing in wheat mitochondria results in the conservation of protein sequences. *Nature*, **341**, 660–662.

IHGSC (International Human Genome Sequencing Consortium) (2001) Initial sequencing and analysis of the human genome. *Nature*, **409**, 860–921.

Knight RD, Freeland SJ and Landweber LF (1999) Selection, history and chemistry: the three faces of the genetic code. *Trends Biochem. Sci.*, **24**, 241–247.

Lehninger A (1970) *Biochemistry*. Worth, New York.

Lodish H, Berk A, Zipursky AL, Matsudaira P, Baltimore D and Darnell J (2000) *Molecular Cell Biology*, 4th edition. W. H. Freeman, New York.

Low SC and Berry MJ (1996) Knowing when not to stop: selenocysteine incorporation in eukaryotes. *Trends Biochem. Sci.*, **21**, 203–208.

Muto A, Ushida C and Himeno H (1998) A bacterial RNA that functions as both a tRNA and an mRNA. *Trends Biochem. Sci.*, 23, 25–29.

Pauling L and Corey RB (1953) Configurations of polypeptide chains with favored orientations around single bonds: two new pleated sheets. *Proc. Natl Acad. Sci. USA*, **37**, 729–740.

Pauling L, Corey RB and Branson HR (1951) The structure of proteins: two hydrogen-bonded helical configurations of the polypeptide chain. *Proc. Natl Acad. Sci. USA*, **27**, 205–211.

Szathmáry E (1999) The origin of the genetic code: amino acids as cofactors in an RNA world. *Trends Genet.*, 15, 223–229.

Turner PC, McLennan AG, Bates AD and White MRH (1997) *Instant Notes in Molecular Biology*. BIOS Scientific Publishers, Oxford.

Zhang L, Zhou W, Velculescu VE, et al. (1997) Gene expression in normal and cancer cells. *Science*, **276**, 1268–1272.

Further Reading

Branden C and Tooze J (1991) *Introduction to Protein Structure*. Garland, New York. — *The best comprehensive text on this subject.*

Judson HF (1979) *The Eighth Day of Creation*. Jonathan Cape, London. — *Includes a great deal of fascinating information on the experiments that led to elucidation of the genetic code.*

Perou CM, Brown PO and Botstein D (2000) Tumor classification using gene expression patterns from DNA microarrays. In: *New Technologies for Life Sciences: A Trends Guide*. Elsevier Science, London. pp. 67–76 — *A review of the work that has recently been done on the transcriptomes of normal and cancerous cells.*

STUDY AIDS FOR CHAPTER 3

Key terms

Give short definitions of the following terms:

α-helix
β-sheet
Alternative splicing
Amino acid
C terminus
Cap
Cell-free protein synthesizing system
Coding RNA
Codon
Context-dependent codon reassignment
Degeneracy
Disulfide bridge
DNA-dependent RNA polymerase

Domain
Electrostatic interactions
End-modification
Genetic code
Genome expression
Heterogenous nuclear RNA
Heteropolymer
Homopolymer
Housekeeping protein
Hydrogen bond
Hydrophobic effect
Informational problem
Initiation codon
Messenger RNA
Molecular chaperone
N terminus

Non-coding RNA
Peptide bond
Poly(A) polymerase
Poly(A) tail
Polypeptide
Pre-mRNA
Pre-RNA
Pre-rRNA
Pre-tRNA
Primary structure
Promoter
Proteome
Punctuation codon
Quaternary structure
Ribosomal RNA
Ribosome
RNA-dependent RNA polymerase
RNA editing

Secondary structure
Small cytoplasmic RNA
Small nuclear RNA
Small nucleolar RNA
Splicing
Suppressor mutation
Template-independent RNA polymerase
Termination codon
Tertiary structure
Transcription
Transcriptome
Transfer RNA
Transfer-messenger RNA
Translation
Triplet binding assay
U-RNA
van der Waals forces

Self study questions

1. Draw a diagram illustrating the steps involved in genome expression. Indicate the points at which the flow of information from genome to proteome can be regulated.

2. Distinguish between 'coding' and 'non-coding' RNA. List the types of coding and non-coding RNA found in prokaryotic and eukaryotic cells and briefly describe the functions of each type.

3. Outline how RNA is synthesized in living cells.

4. Briefly describe the types of RNA processing event that are important in genome expression.

5. Describe, with examples, how transcriptomes are studied.

6. Describe the four levels of protein structure. Your answer should include an indication of the types of chemical interaction that are important at each structural level.

7. Explain how the diversity of amino acid structure is related to the functional diversity of proteins.

8. Outline the experiments that enabled the meaning of each codon in the genetic code to be elucidated.

9. List the key features of the genetic code. What variations of the standard genetic code are known?

10. Explain how protein folding enables the amino acid sequence of a protein to determine the function of that protein.

11. Give examples of the various roles that proteins play in living organisms.

Problem-based learning

1. The transcriptome and proteome are looked on as, respectively, an intermediate and the end-product of genome expression. Evaluate the strengths and limitations of these terms for our understanding of genome expression.

2. To what extent can transcriptome studies identify the functions of genes?

3. The 20 amino acids shown in *Figure 3.17* are not the only amino acids found in living cells. Devise a hypothesis to explain why these 20 amino acids are the only ones that are specified by the genetic code. Can your hypothesis be tested?

4. What experiments were carried out in order to establish that the genetic code is a triplet and non-overlapping code and that proteins are colinear with their genes?

5. 'The role of molecular chaperones is not to impose a new structure on a protein, but merely to increase the efficiency of the protein's natural, sequence-directed, folding pathway' (page 89). Defend this statement.

Studying Genomes

Virtually everything we know about genomes and genome expression has been discovered by scientific research; theoretical studies have played very little role in this or any other area of molecular and cell biology. It is possible to learn 'facts' about genomes without knowing very much about how those facts were obtained, but in order to gain a real understanding of the subject we must first examine in detail the techniques and scientific approaches that have been used to study genomes. Armed with this technical background, you will be able to tackle the remainder of the book with the confidence that comes from a thorough appreciation of how molecular biology research is carried out.

Part 2 of *Genomes* takes a staged approach to genetic and genomic research methods. First, in Chapter 4, we examine the techniques, centered on the polymerase chain reaction and DNA cloning, that are used to study DNA molecules. These techniques are very effective with short segments of DNA, including individual genes, enabling a wealth of information to be obtained at this level.

Chapter 5 moves on to the methods developed to construct maps of genomes, and describes how the techniques of genetic mapping, first developed almost a century ago, have been supplemented with complementary methods for direct physical mapping of genomes.

Chapter 6 makes the link between mapping and sequencing, and shows that although a map can be a valuable aid to assembly of a long DNA sequence, mapping is not always a prerequisite to genome sequencing.

Finally, in Chapter 7, we look at the various approaches that are used to understand a genome sequence. As you read this chapter you will begin to appreciate that locating the genes in a genome sequence, and identifying their functions, is one of the major research challenges of modern biology.

Chapter 4

Studying DNA

Chapter 5

Mapping Genomes

Chapter 6

Sequencing Genomes

Chapter 7

Understanding a Genome Sequence

Studying DNA

Chapter Contents

4.1   *Enzymes for DNA Manipulation*   97

   4.1.1   **DNA polymerases**   98

   4.1.2   **Nucleases**   102

   4.1.3   **DNA ligases**   105

   4.1.4   **End-modification enzymes**   107

4.2   *DNA Cloning*   108

   4.2.1   **Cloning vectors and the way they are used**   109

4.3   *The Polymerase Chain Reaction (PCR)*   119

   4.3.1   **Carrying out a PCR**   120

   4.3.2   **The applications of PCR**   121

Learning outcomes

When you have read Chapter 4, you should be able to:

- Give outline descriptions of the events involved in DNA cloning and the polymerase chain reaction (PCR), and state the applications and limitations of these techniques

- Describe the activities and main applications of the different types of enzyme used in recombinant DNA research

- Identify the important features of DNA polymerases and distinguish between the various DNA polymerases used in genomics research

- Describe, with examples, the way that restriction endonucleases cut DNA and explain how the results of a restriction digest are examined

- Distinguish between blunt- and sticky-end ligation and explain how the efficiency of blunt-end ligation can be increased

- Give details of the key features of plasmid cloning vectors and describe how these vectors are used in cloning experiments, using pBR322 and pUC8 as examples

- Describe how bacteriophage λ vectors are used to clone DNA

- Give examples of vectors used to clone long pieces of DNA, and evaluate the strengths and weaknesses of each type

- Summarize how DNA is cloned in yeast, animals and plants

- Describe how a PCR is performed, paying particular attention to the importance of the primers and the temperatures used during the thermal cycling

THE TOOLKIT OF TECHNIQUES used by molecular biologists to study DNA molecules was assembled during the 1970s and 1980s. Before then, the only way in which individual genes could be studied was by classical genetics, using the procedures that we will examine in Chapter 5. Classical genetics is a powerful approach to gene analysis and many of the fundamental discoveries in molecular biology were made in this way. The **operon theory** proposed by Jacob and Monod in 1961 (Section 9.3.1), which describes how the expression of some bacterial genes is regulated, was perhaps the most heroic achievement of this era of genetics. But the classical approach is limited because it does not involve the direct examination of genes, information on gene structure and activity being inferred from the biological characteristics of the organism being studied. By the late 1960s these indirect methods had become insufficient for answering the more detailed questions that molecular biologists had begun to ask about the expression pathways of individual genes. These questions could only be addressed by examining directly the segments of DNA containing the genes of interest.

This was not possible using the current technology, so a new set of techniques had to be invented.

The development of these new techniques was stimulated by breakthroughs in biochemical research which, in the early 1970s, provided molecular biologists with enzymes that could be used to manipulate DNA molecules in the test tube. These enzymes occur naturally in living cells and are involved in processes such as DNA replication, repair and recombination (see Chapters 13 and 14). In order to determine the functions of these enzymes, many of them were purified and the reactions that they catalyze studied in the test tube. Molecular biologists then adopted the pure enzymes as tools for manipulating DNA molecules in pre-determined ways, using them to make copies of DNA molecules, to cut DNA molecules into shorter fragments, and to join them together again in combinations that do not exist in nature (*Figure 4.1*). These manipulations, which are described in Section 4.1, form the basis of **recombinant DNA technology**, in which new or 'recombinant' DNA molecules are constructed in the test tube from pieces of naturally occurring chromosomes and plasmids. Recombinant

DNA methodology led to the development of **DNA** or **gene cloning**, in which short DNA fragments, possibly containing a single gene, are inserted into a plasmid or virus chromosome and then replicated in a bacterial or eukaryotic host (*Figure 4.2*). We will examine exactly how gene cloning is performed, and the reasons why this technique resulted in a revolution in molecular biology, in Section 4.2.

Gene cloning was well established by the end of the 1970s. The next major technical breakthrough came some 5 years later when the **polymerase chain reaction (PCR)** was invented (Mullis, 1990). PCR is not a complicated technique – all that it achieves is the repeated copying of a short segment of a DNA molecule (*Figure 4.3*) – but it has become immensely important in many areas of bio-logical research, not least the study of genomes. PCR is covered in detail in Section 4.3.

4.1 Enzymes for DNA Manipulation

Recombinant DNA technology was one of the main factors that contributed to the rapid advance in knowledge concerning gene expression that occurred during the 1970s and 1980s. The basis of recombinant DNA technology is the ability to manipulate DNA molecules in the test tube. This, in turn, depends on the availability of purified enzymes whose activities are known and can be controlled, and which can therefore be used to make specified changes to the DNA molecules that are being manipulated. The enzymes available to the molecular biologist fall into four broad categories:

- **DNA polymerases** (Section 4.1.1), which are enzymes that synthesize new polynucleotides complementary to an existing DNA or RNA template (*Figure 4.4A*);
- **Nucleases** (Section 4.1.2), which degrade DNA molecules by breaking the phosphodiester bonds that link one nucleotide to the next (*Figure 4.4B*);
- **Ligases** (Section 4.1.3), which join DNA molecules together by synthesizing phosphodiester bonds between nucleotides at the ends of two different molecules, or at the two ends of a single molecule (*Figure 4.4C*);
- **End-modification enzymes** (Section 4.1.4), which make changes to the ends of DNA molecules, adding an important dimension to the design of ligation experiments, and providing one means of labeling DNA molecules with radioactive and other markers (Technical Note 4.1).

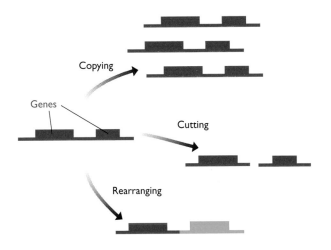

Figure 4.1 Examples of the manipulations that can be carried out with DNA molecules.

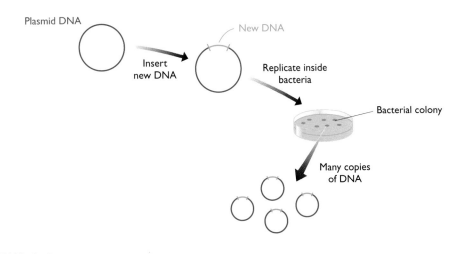

Figure 4.2 DNA cloning.

In this example, the fragment of DNA to be cloned is inserted into a plasmid vector which is subsequently replicated inside a bacterial host.

DNA labeling

Attachment of radioactive, fluorescent or other types of marker to DNA molecules.

DNA labeling is a central part of many molecular biology procedures, including Southern hybridization (Section 4.1.2), fluorescent *in situ* hybridization (FISH; Section 5.3.2) and DNA sequencing (Section 6.1). It enables the location of a particular DNA molecule – on a nitrocellulose or nylon membrane, in a chromosome or in a gel – to be determined by detecting the signal emitted by the marker. Labeled RNA molecules are also used in some applications (Technical Note 4.4, page 120).

Radioactive markers are frequently used for labeling DNA molecules. Nucleotides can be synthesized in which one of the phosphorus atoms is replaced with ^{32}P or ^{33}P, one of the oxygen atoms in the phosphate group is replaced with ^{35}S, or one or more of the hydrogen atoms is replaced with ^{3}H (see *Figure 1.6*, page 11). Radioactive nucleotides still act as substrates for DNA polymerases and so are incorporated into a DNA molecule by any strand-synthesis reaction catalyzed by a DNA polymerase. Labeled nucleotides or individual phosphate groups can also be attached to one or both ends of a DNA molecule by the reactions catalyzed by T4 polynucleotide kinase or terminal deoxynucleotidyl transferase (Section 4.1.4). The radioactive signal can be detected by scintillation counting, but for most molecular biology applications positional information is needed, so detection is by exposure of an X-ray-sensitive film (**autoradiography**) or a radiation-sensitive phosphorescent screen (**phosphorimaging**). The choice between the various radioactive labels depends on the requirements of the procedure. High sensitivity is possible with ^{32}P because this isotope has a high emission energy, but sensitivity is accompanied by low resolution because of signal scattering. Low-emission isotopes such as ^{35}S or ^{3}H give less sensitivity but greater resolution.

Health and environmental issues have meant that radioactive markers have become less popular in recent years and for many procedures they are now largely superseded by non-radioactive alternatives. The most useful of these are fluorescent markers, which are central components of techniques such as FISH (Section 5.3.2) and automated DNA sequencing (Section 6.1.1). Fluorescent labels with various emission wavelengths (i.e. of different colors) are incorporated into nucleotides or attached directly to DNA molecules, and are detected with a suitable film, by fluorescence microscopy, or with a fluorescence detector. Other types of non-radioactive labeling make use of chemiluminescent emissions, but these have the disadvantage that the signal is not generated directly by the label, but instead must be 'developed' by treatment of the labeled molecule with chemicals. A popular method involves labeling the DNA with the enzyme alkaline phosphatase, which is detected by applying dioxetane, which the enzyme dephosphorylates to produce the chemiluminescence.

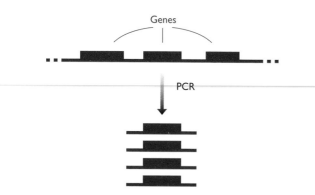

Figure 4.3 The polymerase chain reaction (PCR) is used to make copies of a selected segment of a DNA molecule.

In this example, a single gene is copied.

4.1.1 DNA polymerases

Many of the techniques used to study DNA depend on the synthesis of copies of all or part of existing DNA or RNA molecules. This is an essential requirement for PCR (Section 4.3), DNA sequencing (Section 6.1), DNA labeling (Technical Note 4.1) and many other procedures that are central to molecular biology research. An enzyme that synthesizes DNA is called a **DNA polymerase** and one that copies an existing DNA or RNA molecule is called a **template-dependent DNA polymerase**.

The mode of action of a template-dependent DNA polymerase

A template-dependent DNA polymerase makes a new DNA polynucleotide whose sequence is dictated, via the base-pairing rules, by the sequence of nucleotides in the DNA molecule that is being copied (*Figure 4.5*). The mode of action is very similar to that of a template-dependent RNA polymerase (Section 3.2.2), the new polynucleotide

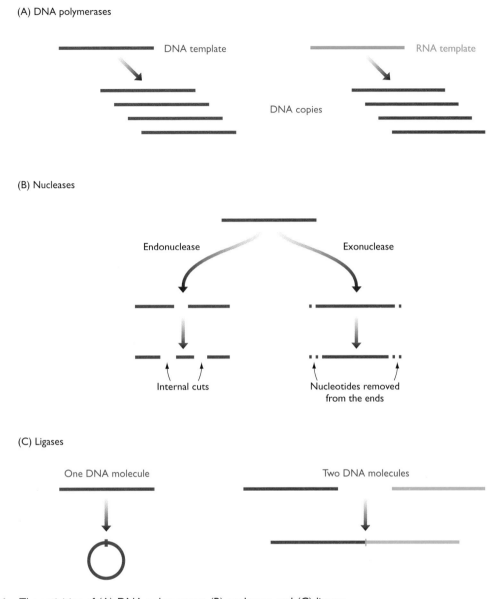

Figure 4.4 The activities of (A) DNA polymerases, (B) nucleases, and (C) ligases.

In (A), the activity of a DNA-dependent DNA polymerase is shown on the left and that of an RNA-dependent DNA polymerase on the right. In (B), the activities of endonucleases and exonucleases are shown. In (C) the red DNA molecule is ligated to itself on the left, and to a second molecule on the right.

being synthesized in the 5'→3' direction: DNA polymerases that make DNA in the other direction are unknown in nature.

One important difference between template-dependent DNA synthesis and the equivalent process for synthesis of RNA is that a DNA polymerase is unable to use an entirely single-stranded molecule as the template. In order to initiate DNA synthesis there must be a short double-stranded region to provide a 3' end onto which the enzyme will add new nucleotides (*Figure 4.6A*). The way in which this requirement is met in living cells when the genome is replicated is described in Chapter 13. In the test tube, a DNA copying reaction is initiated by attaching to the template a short synthetic **oligonucleotide**, usually about 20 nucleotides in length, which acts as a **primer** for DNA synthesis. At first glance, the need for a primer might appear to be an undesired complication in the use of DNA polymerases in recombinant DNA technology, but nothing could be further from the truth. Because annealing of the primer to the template depends on complementary base-pairing, the position within the template molecule at which DNA copying is initiated can

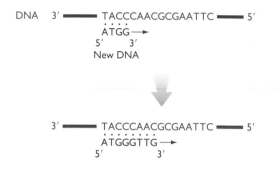

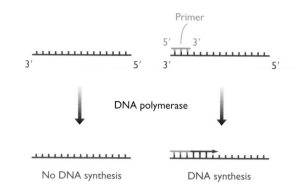

(A) DNA synthesis requires a primer

Figure 4.5 The activity of a DNA-dependent DNA polymerase.

New nucleotides are added on to the 3' end of the growing polynucleotide, the sequence of this new polynucleotide being determined by the sequence of the template DNA.

(B) The primer determines which part of a DNA molecule is copied

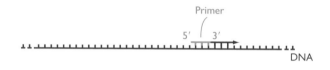

Figure 4.6 The role of the primer in template-dependent DNA synthesis.

(A) A DNA polymerase requires a primer in order to initiate the synthesis of a new polynucleotide. (B) The sequence of this oligonucleotide determines the position at which it attaches to the template DNA and hence specifies the region of the template that will be copied. When a DNA polymerase is used to make new DNA *in vitro*, the primer is usually a short oligonucleotide made by chemical synthesis. For details of how DNA synthesis is primed *in vivo*, see *Figure 13.12*, page 396.

be specified by synthesizing a primer with the appropriate nucleotide sequence (*Figure 4.6B*). A short specific segment of a much longer template molecule can therefore be copied, which is much more valuable than the random copying that would occur if DNA synthesis did not need to be primed. You will fully appreciate the importance of priming when we deal with PCR in Section 4.3.

A second general feature of template-dependent DNA polymerases is that many of these enzymes are multi-functional, being able to degrade DNA molecules as well as synthesize them. This is a reflection of the way in which DNA polymerases act in the cell during genome replication (Section 13.2.2). As well as their 5'→3' DNA synthesis capability, DNA polymerases can also have one or both of the following exonuclease activities (*Figure 4.7*):

- A **3'→5' exonuclease** activity enables the enzyme to remove nucleotides from the 3' end of the strand that it has just synthesized. This is called the **proof-reading** activity because it allows the polymerase to correct errors by removing a nucleotide that has been inserted incorrectly.
- A **5'→3' exonuclease** activity is less common, but is possessed by some DNA polymerases whose natural function in genome replication requires that they must be able to remove at least part of a poly-nucleotide that is already attached to the template strand that the polymerase is copying.

The types of DNA polymerases used in research

Several of the template-dependent DNA polymerases that are used in molecular biology research (*Table 4.1*) are versions of the *Escherichia coli* DNA polymerase I enzyme, which plays a central role in replication of this bacterium's genome (Section 13.2.2). This enzyme, sometimes called the **Kornberg polymerase**, after its discoverer Arthur Kornberg (Kornberg, 1960), has both the 3'→5' and 5'→3' exonuclease activities, which limits

it usefulness in DNA manipulation. Its main application is in DNA labeling, as described in Technical Note 4.1.

Of the two exonuclease activities, it is the 5'→3' version that causes most problems when a DNA polymerase is used to manipulate molecules in the test tube. This is because an enzyme that possesses this activity is able to remove nucleotides from the 5' ends of polynucleotides that have just been synthesized (*Figure 4.8*). It is unlikely that the polynucleotides will be completely degraded, because the polymerase function is usually much more active than the exonuclease, but some techniques will not work if the 5' ends of the new polynucleotides are short-ened in any way. In particular, DNA sequencing is based on synthesis of new polynucleotides, all of which share exactly the same 5' end, marked by the primer used to initiate the sequencing reactions. If any nibbling of the 5' ends occurs, then it is impossible to determine the correct DNA sequence. When DNA sequencing was first intro-duced in the late 1970s, it made use of a modified version of the Kornberg enzyme called the **Klenow polymerase**. The Klenow polymerase was initially prepared by cutting the natural *E. coli* DNA polymerase I enzyme into two segments with a protease. One of these segments retained the polymerase and 3'→5' exonuclease activities, but

(A) 5′→3′ DNA synthesis

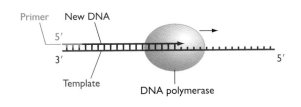

(B) 3′→5′ exonuclease activity

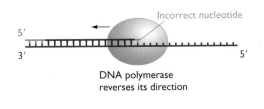

DNA polymerase
reverses its direction

(C) 5′→3′ exonuclease activity

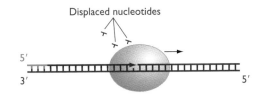

Figure 4.7 The DNA synthesis and exonuclease activities of DNA polymerases.

All DNA polymerases can make DNA and many also have one or both of the exonuclease activities.

lacked the 5′→3′ exonuclease of the untreated enzyme. The enzyme is still often called the Klenow *fragment* in memory of this old method of preparation, but nowadays it is almost always prepared from *E. coli* cells whose polymerase gene has been engineered so that the resulting enzyme has the desired properties. But in fact the Klenow polymerase is now rarely used in sequencing and has its major application in DNA labeling (see Technical Note

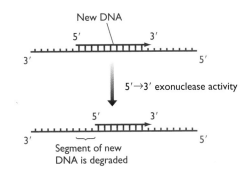

Figure 4.8 The 5′→3′ exonuclease activity of a DNA polymerase can degrade the 5′ end of a polynucleotide that has just been synthesized.

4.1). This is because an enzyme called **Sequenase** (see *Table 4.1*), which has superior properties as far a sequencing is concerned, was developed in the 1980s. We will return to the features of Sequenase, and why they make the enzyme ideal for sequencing, in Box 6.1 (page 166).

The *E. coli* DNA polymerase I enzyme has an optimum reaction temperature of 37 °C, this being the usual temperature of the natural environment of the bacterium, inside the intestines of mammals such as humans. Test-tube reactions with either the Kornberg or Klenow polymerases, and with Sequenase, are therefore incubated at 37 °C, and terminated by raising the temperature to 75 °C or above, which causes the protein to unfold or **denature**, destroying its activity. This regimen is perfectly adequate for most molecular biology techniques but, for reasons that will become clear in Section 4.3, PCR requires a **thermostable DNA polymerase** – one that is able to function at temperatures much higher than 37 °C. Suitable enzymes can be obtained from bacteria such as *Thermus aquaticus*, which live in hot springs at temperatures up to 95 °C, and whose DNA polymerase I enzyme has an optimum working temperature of 72 °C. The biochemical basis of protein thermostability is not fully understood, but probably centers on structural features that reduce the amount of protein unfolding that occurs at elevated temperatures.

One additional type of DNA polymerase is important in molecular biology research. This is **reverse transcriptase**, which is an **RNA-dependent DNA polymerase** and so

Table 4.1 Features of template-dependent DNA polymerases used in molecular biology research

Polymerase	Description	Main use	Cross reference
DNA polymerase I	Unmodified *E. coli* enzyme	DNA labeling	Technical Note 4.1 (page 98)
Klenow polymerase	Modified version of *E. coli* DNA polymerase I	DNA labeling	Technical Note 4.1 (page 98)
Sequenase	Modified version of phage T7 DNA polymerase I	DNA sequencing	Box 6.1 (page 166)
Taq polymerase	*Thermus aquaticus* DNA polymerase I	PCR	Section 4.3
Reverse transcriptase	RNA-dependent DNA polymerase, obtained from various retroviruses	cDNA synthesis	Section 5.3.3

makes DNA copies of RNA rather than DNA templates. Reverse transcriptases are involved in the replication cycles of retroviruses (Section 2.4.2), including the human immunodeficiency viruses that cause AIDS, these having RNA genomes that are copied into DNA after infection of the host. In the test tube, a reverse transcriptase can be used to make DNA copies of mRNA molecules. These copies are called **complementary DNAs (cDNAs)**. Their synthesis is important in some types of gene cloning and in techniques used to map the regions of a genome that specify particular mRNAs (Section 7.1.2).

4.1.2 Nucleases

A range of nucleases have found applications in recombinant DNA technology (*Table 4.2*). Some nucleases have a broad range of activities but most are either **exonucleases**, removing nucleotides from the ends of DNA and/or RNA molecules, or **endonucleases**, making cuts at internal phosphodiester bonds. Some nucleases are specific for DNA and some for RNA; some work only on double-stranded DNA and others only on single-stranded DNA, and some are not fussy what they work on. We will encounter various examples of nucleases in later chapters when we deal with the techniques in which they used. Only one type of nuclease will be considered in detail here: the **restriction endonucleases**, which play a central role in all aspects of recombinant DNA technology.

Restriction endonucleases enable DNA molecules to be cut at defined positions

A restriction endonuclease is an enzyme that binds to a DNA molecule at a specific sequence and makes a double-stranded cut at or near that sequence. Because of the sequence specificity, the positions of cuts within a DNA molecule can be predicted, assuming that the DNA sequence is known, enabling defined segments to be excised from a larger molecule. This ability underlies gene cloning and all other aspects of recombinant DNA technology in which DNA fragments of known sequence are required.

There are three types of restriction endonuclease. With Types I and III there is no strict control over the position of the cut relative to the specific sequence in the DNA molecule that is recognized by the enzyme. These

enzymes are therefore less useful because the sequences of the resulting fragments are not precisely known. Type II enzymes do not suffer from this disadvantage because the cut is always at the same place, either within the recognition sequence or very close to it (*Figure 4.9*). For example, the Type II enzyme called *Eco*RI (isolated from *E. coli*) cuts DNA only at the hexanucleotide 5′–GAATTC–3′. Digestion of DNA with a Type II enzyme therefore gives a reproducible set of fragments whose sequences are predictable if the sequence of the target DNA molecule is known. Over 2500 Type II enzymes have been isolated and more than 300 are available for use in the laboratory (Brown, 1998). Many enzymes have hexanucleotide target sites, but others recognize shorter or longer sequences (*Table 4.3*). There are also examples of enzymes with degenerate recognition sequences, meaning that they cut DNA at any of a family of related sites. *Hin*fI (from *Haemophilus influenzae*), for example, recognizes 5′–GANTC–3′, where 'N' is any nucleotide, and so

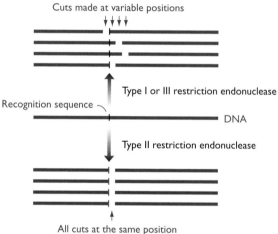

Figure 4.9 Cuts produced by restriction endonucleases.

In the top part of the diagram, the DNA is cut by a Type I or Type III restriction endonuclease. The cuts are made in slightly different positions relative to the recognition sequence, so the resulting fragments have different lengths. In the lower part, a Type II enzyme is used. Each molecule is cut at exactly the same position to give exactly the same pair of fragments.

Table 4.2 Features of important nucleases used in molecular biology research

Nuclease	Description	Main use	Cross reference
Restriction endonucleases	Sequence-specific DNA endonucleases, from many sources	Many applications	Section 4.1.2
S1 nuclease	Endonuclease specific for single-stranded DNA and RNA, from the fungus *Aspergillus oryzae*	Transcript mapping	Section 7.1.2
Deoxyribonuclease I	Endonuclease specific for double-stranded DNA and RNA, from *Escherichia coli*	Nuclease footprinting	Section 9.1.1

Table 4.3   Some examples of restriction endonucleases

Enzyme	Recognition sequence	Type of ends	End sequences
AluI	5′–AGCT–3′ 3′–TCGA–5′	Blunt	5′–AG CT–3′ 3′–TC GA–5′
Sau3AI	5′–GATC–3′ 3′–CTAG–5′	Sticky, 5′ overhang	5′– GATC–3′ 3′–CTAG –5′
HinfI	5′–GANTC–3′ 3′–CTNAG–5′	Sticky, 5′ overhang	5′–G ANTC–3′ 3′–CTNA G–5′
BamHI	5′–GGATCC–3′ 3′–CCTAGG–5′	Sticky, 5′ overhang	5′–G GATCC–3′ 3′–CCTAG G–5′
BsrBI	5′–CCGCTC–3′ 3′–GGCGAG–5′	Blunt	5′– NNNCCGCTC–3′ 3′– NNNGGCGAG–5′
EcoRI	5′–GAATTC–3′ 3′–CTTAAG–5′	Sticky, 5′ overhang	5′–G AATTC–3′ 3′–CTTAA G–5′
PstI	5′–CTGCAG–3′ 3′–GACGTC–5′	Sticky, 3′ overhang	5′–CTGCA G–3′ 3′–G ACGTC–5′
NotI	5′–GCGGCCGC–3′ 3′–CGCCGGCG–5′	Sticky, 5′ overhang	5′–GC GGCCGC–3′ 3′–CGCCGG CG–5′
BglI	5′–GCCNNNNNGGC–3′ 3′–CGGNNNNNCCG–5′	Sticky, 3′ overhang	5′–GCCNNNN NGGC–3′ 3′–CGGN NNNNCCG–5′

N = any nucleotide.

Note that most, but not all, recognition sequences have inverted symmetry: when read in the 5′→3′ direction, the sequence is the same in both strands.

cuts at 5′–GAATC–3′, 5′–GATTC–3′, 5′–GAGTC–3′ and 5′–GACTC–3′. Most enzymes cut within the recognition sequence, but a few, such a *BsrBI*, cut at a specified position outside of this sequence.

Restriction enzymes cut DNA in two different ways. Many make a simple double-stranded cut, giving a **blunt** or **flush end**; others cut the two DNA strands at different positions, usually two or four nucleotides apart, so that the resulting DNA fragments have short single-stranded overhangs at each end. These are called **sticky** or **cohesive ends** because base-pairing between them can stick the DNA molecule back together again (*Figure 4.10A*). Some sticky-end cutters give 5′ overhangs (e.g. *Sau3AI*, *Hinf*I) whereas others leave 3′ overhangs (e.g. *Pst*I) (*Figure 4.10B*). One feature that is particularly important in recombinant DNA technology is that some pairs of restriction enzymes have different recognition sequences but give the same sticky ends, examples being *Sau3AI* and *Bam*HI, which both give a 5′–GATC–3′ sticky end even though *Sau3AI* has a 4-bp recognition sequence and *Bam*HI recognizes a 6-bp sequence (*Figure 4.10C*).

Examining the results of a restriction digest

After treatment with a restriction endonuclease, the resulting DNA fragments can be examined by agarose gel electrophoresis (see Technical Note 2.1, page 37) to determine their sizes. Depending on the concentration of agarose in the gel, fragments between 100 bp and 50 kb can be separated into sharp bands after electrophoresis (*Figure 4.11*). Fragments less than 150 bp can be separated in a 4% or 5% agarose gel, making it possible to distinguish bands representing molecules that differ in size by just a single nucleotide. With larger fragments, however, it is not always possible to separate molecules of similar size, even in gels of lower agarose concentration. If the starting DNA is long, and so gives rise to many fragments after digestion with a restriction enzyme, then the gel may simply show a smear of DNA because there are fragments of every possible length that all merge together. This is the usual result when genomic DNA is restricted.

If the sequence of the starting DNA is known then the sequences, and hence the sizes, of the fragments resulting from treatment with a particular restriction enzyme can be predicted. The band for a desired fragment (for example, one containing a gene) can then be identified, cut out of the gel, and the DNA purified. Even if its size is unknown, a fragment containing a gene or another segment of DNA of interest can be identified by the technique called **Southern hybridization**, providing that some of the sequence within the fragment is known or can be predicted. The first step is to transfer the restriction fragments from the agarose gel to a nitrocellulose or

(A) Blunt and sticky ends

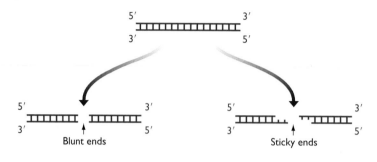

(B) 5′ and 3′ overhangs

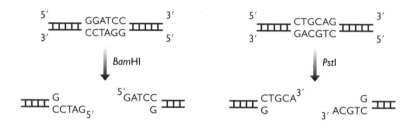

(C) The same sticky end produced by different enzymes

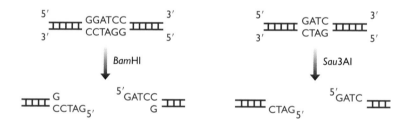

Figure 4.10 The results of digestion of DNA with different restriction endonucleases.

(A) Blunt ends and sticky ends. (B) Different types of sticky end: the 5′ overhangs produced by *BamHI* and the 3′ overhangs produced by *PstI*. (C) The same sticky ends produced by two different restriction endonucleases: a 5′ overhang with the sequence 5′–GATC–3′ is produced by both *BamHI* (recognizes 5′–GGATCC–3′) and *Sau3AI* (recognizes 5′–GATC–3′).

nylon membrane. This is done by placing the membrane on the gel and allowing buffer to soak through, taking the DNA from the gel to the membrane, where it becomes bound (*Figure 4.12A*). This process results in the DNA bands becoming immobilized in the same relative positions on the surface of the membrane.

The next step is to prepare a **hybridization probe**, which is a labeled DNA molecule whose sequence is complementary to the target DNA that we wish to detect. The probe could, for example, be a synthetic oligonucleotide whose sequence matches part of an interesting gene. Because the probe and target DNAs are complementary, they can base-pair or **hybridize**, the position of the hybridized probe on the membrane being identified by

detecting the signal given out by a label attached to the probe. To carry out the hybridization, the membrane is placed in a glass bottle with the labeled probe and some buffer, and the bottle gently rotated for several hours so that the probe has plenty of opportunity to hybridize to its target DNA. The membrane is then washed to remove any probe that has not become hybridized, and the signal from the label is detected (see Technical Note 4.1, page 98). In the example shown in *Figure 4.12B* the probe is radioactively labeled and the signal is detected by **autoradiography**. The band that is seen on the autoradiograph is the one that corresponds to the restriction fragment that hybridizes to the probe and which therefore contains the gene that we are searching for.

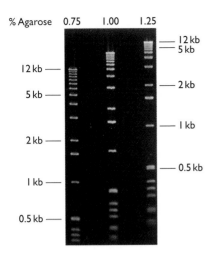

Figure 4.11 Separation of DNA molecules by agarose gel electrophoresis.

The range of fragment sizes that can be resolved depends on the concentration of agarose in the gel. Electrophoresis has been performed with three different concentrations of agarose. The labels indicate the sizes of bands in the left and right lanes. Photo courtesy of BioWhittaker Molecular Applications.

4.1.3 DNA ligases

DNA fragments that have been generated by treatment with a restriction endonuclease can be joined back together again, or attached to a new partner, by a DNA ligase. The reaction requires energy, which is provided by adding either ATP or NAD to the reaction mixture, depending on the type of ligase that is being used.

The most widely used DNA ligase is obtained from *E. coli* cells infected with T4 bacteriophage. It is involved in replication of the phage DNA and is encoded by the T4 genome. Its natural role is to synthesize phosphodiester bonds between unlinked nucleotides present in one polynucleotide of a double-stranded molecule (*Figure 4.13A*). In order to join together two restriction fragments, the ligase has to synthesize two phosphodiester bonds, one in each strand (*Figure 4.13B*). This is by no means beyond the capabilities of the enzyme, but the reaction can occur only if the ends to be joined come close enough to one another by chance – the ligase is not able to catch hold of them and bring them together. If the two molecules have complementary sticky ends, and the ends come together by random diffusion events in the ligation mixture, then transient base pairs might form between the two over-hangs. These base pairs are not particularly stable but they may persist for sufficient time for a ligase enzyme to attach to the junction and synthesize phosphodiester bonds to fuse the ends together (*Figure 4.13C*). If the molecules are blunt ended, then they cannot base-pair to one another, not even temporarily, and ligation is a much less efficient process, even when the DNA concentration is high and pairs of ends are in relatively close proximity.

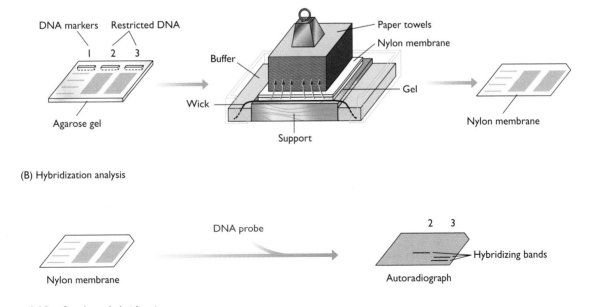

Figure 4.12 Southern hybridization.

(A) Transfer of DNA from the gel to the membrane. (B) The membrane is probed with a radioactively labeled DNA molecule. On the resulting autoradiograph, one hybridizing band is seen in lane 2, and two in lane 3.

(A) The role of DNA ligase *in vivo*

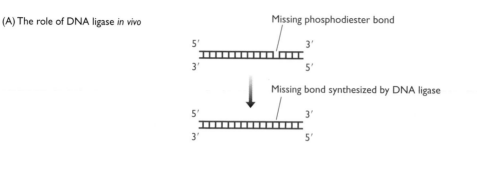

(B) Ligation *in vitro*

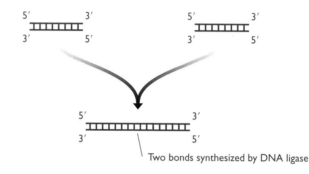

(C) Sticky-end ligation is more efficient

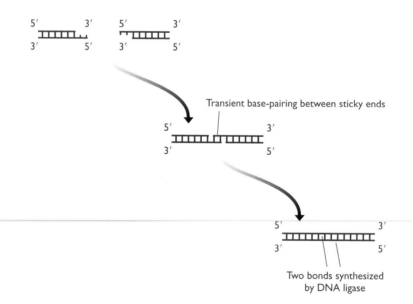

Figure 4.13 Ligation of DNA molecules with DNA ligase.

(A) In living cells, DNA ligase synthesizes a missing phosphodiester bond in one strand of a double-stranded DNA molecule. (B) To link two DNA molecules *in vitro*, DNA ligase must make two phosphodiester bonds, one in each strand. (C) Ligation *in vitro* is more efficient when the molecules have compatible sticky ends, because transient base-pairing between these ends holds the molecules together and so increases the opportunity for DNA ligase to attach and synthesize the new phosphodiester bonds. For the role of DNA ligase during DNA replication *in vivo*, see *Figures 13.17* and *13.19*.

The greater efficiency of sticky-end ligation has stimulated the development of methods for converting blunt ends into sticky ends. In one method, short double-stranded molecules called **linkers** or **adaptors** are attached to the blunt ends. Linkers and adaptors work in slightly different ways but both contain a recognition sequence for a restriction endonuclease and so produce a sticky end after treatment with the appropriate enzyme (*Figure 4.14*). Another way to create a sticky end is by

homopolymer **tailing**, in which nucleotides are added one after the other to the 3' terminus at a blunt end (*Figure 4.15*). The enzyme involved is called **terminal deoxynucleotidyl transferase**, which we will meet in the next section. If the reaction contains the DNA, enzyme, and only one of the four nucleotides, then the new stretch of single-stranded DNA that is made consists entirely of just that single nucleotide. It could, for example, be a poly(G) tail, which would enable the molecule to base-pair to other molecules that carry poly(C) tails, created in the same way but with dCTP rather than dGTP in the reaction mixture.

4.1.4 End-modification enzymes

Terminal deoxynucleotidyl transferase (see *Figure 4.15*), obtained from calf thymus tissue, is one example of an end-modification enzyme. It is, in fact, a **template-independent DNA polymerase**, because it is able to synthesize a new DNA polynucleotide without base-pairing of the incoming nucleotides to an existing strand of DNA or RNA. Its main role in recombinant DNA technology is in homopolymer tailing, as described above.

Two other end-modification enzymes are also frequently used. These are **alkaline phosphatase** and **T4 polynucleotide kinase**, which act in complementary ways. Alkaline phosphatase, which is obtained from various sources, including *E. coli* and calf intestinal tissue, removes phosphate groups from the 5' ends of DNA molecules, which prevents these molecules from being ligated to one another. Two ends carrying 5' phosphates can be ligated to one another, and a phosphatased end can ligate to a non-phosphatased end, but a link cannot be formed between a pair of ends if neither carries a

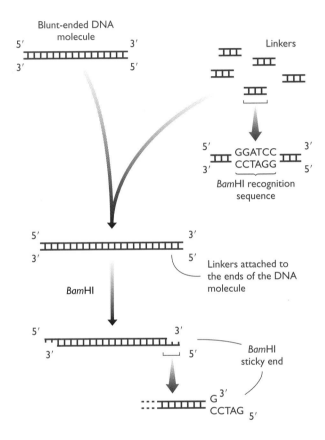

Figure 4.14   Linkers are used to place sticky ends on to a blunt-ended molecule.

In this example, each linker contains the recognition sequence for the *Bam*HI restriction endonuclease. DNA ligase attaches the linkers to the ends of the blunt-ended molecule in a reaction that is made relatively efficient because the linkers are present at a high concentration. The restriction enzyme is then added to cleave the linkers and produce the sticky ends. Note that during the ligation the linkers ligate to one another, so a series of linkers (a **concatamer**) is attached to each end of the blunt molecule. When the restriction enzyme is added, these linker concatamers are cut into segments, with half of the innermost linker left attached to the DNA molecule. Adaptors are similar to linkers but each one has one blunt end and one sticky end. The blunt-ended DNA is therefore given sticky ends simply by ligating it to the adaptors; there is no need to carry out the restriction step. For more details, see Brown (2001).

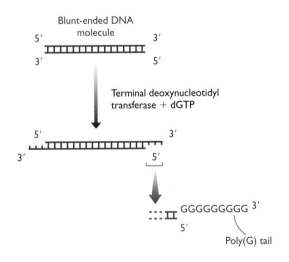

Figure 4.15   Homopolymer tailing.

In this example, a poly(G) tail is synthesized at each end of a blunt-ended DNA molecule. Tails comprising other nucleotides are synthesized by including the appropriate dNTP in the reaction mixture.

5′ phosphate. Judicious use of alkaline phosphatase can therefore direct the action of a DNA ligase in a predetermined way so that only desired ligation products are obtained. T4 polynucleotide kinase, obtained from *E. coli* cells infected with T4 phage, performs the reverse reaction to alkaline phosphatase, adding phosphates to 5′ ends. Like alkaline phosphatase, the enzyme is used during complicated ligation experiments, but its main application is in the end-labeling of DNA molecules (see Technical Note 4.1, page 98).

4.2 DNA Cloning

DNA cloning is a logical extension of the ability to manipulate DNA molecules with restriction endonucleases and ligases. Imagine that an animal gene has been obtained as a single restriction fragment after digestion of a larger

molecule with the restriction enzyme *Bam*HI, which leaves 5′–GATC–3′ sticky ends (*Figure 4.16*). Imagine also that a small *E. coli* plasmid has been purified and treated with *Bam*HI, which cuts the plasmid in a single position. The circular plasmid has therefore been converted into a linear molecule, again with 5′–GATC–3′ sticky ends. Mix the two DNA molecules together and add DNA ligase. Various recombinant ligation products will be obtained, one of which comprises the circularized plasmid with the animal gene inserted into the position originally taken by the *Bam*HI restriction site. If the recombinant plasmid is now re-introduced into *E. coli*, and the inserted gene has not disrupted its replicative ability, then the plasmid plus inserted gene will be replicated and copies passed to the daughter bacteria after cell division. More rounds of plasmid replication and cell division will result in a colony of recombinant *E. coli* bacteria, each bacterium containing multiple copies of the animal gene. This series

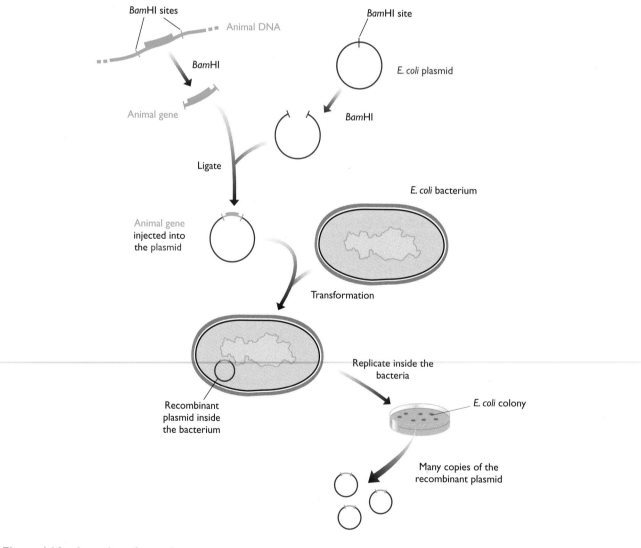

Figure 4.16 An outline of gene cloning.

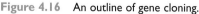

See the text for details.

of events, as illustrated in *Figure 4.16*, constitutes the process called DNA or gene cloning.

4.2.1 Cloning vectors and the way they are used

In the experiment shown in *Figure 4.16*, the plasmid acts as a **cloning vector**, providing the replicative ability that enables the cloned gene to be propagated inside the host cell. Plasmids replicate efficiently in bacterial hosts because each plasmid possesses an **origin of replication** which is recognized by the DNA polymerases and other proteins that normally replicate the bacterium's chromosomes (Section 13.2.1). The host cell's replicative machinery therefore propagates the plasmid, plus any new genes that have been inserted into it. Bacteriophage genomes can also be used as cloning vectors because they too possess origins of replication that enable them to be propagated inside bacteria, either by the host enzymes or by DNA polymerases and other proteins specified by phage genes. The next two sections describe how plasmid and phage vectors are used to clone DNA in *E. coli*.

Plasmids are uncommon in eukaryotes, although *Saccharomyces cerevisiae* possesses one that is sometimes used for cloning purposes; most eukaryotic vectors are therefore based on virus genomes. Alternatively, with a eukaryotic host the replication requirement can be bypassed by performing the experiment in such a way that the DNA to be cloned becomes inserted into one of the host chromosomes. These approaches to cloning in eukaryotic cells are described later in the chapter.

Vectors based on E. coli *plasmids*

The easiest way to understand how a cloning vector is used is to start with the simplest *E. coli* plasmid vectors, which illustrate all of the basic principles of DNA cloning. We will then be able to turn our attention to the special features of phage vectors and vectors used with eukaryotes.

One of the first plasmid vectors to be developed was pBR322 (Bolivar *et al.*, 1977), which was constructed by ligating restriction fragments from three naturally occurring *E. coli* plasmids: R1, R6.5 and pMB1. The pBR322 plasmid is small (just 4363 bp) and, as well as the origin of replication, it carries genes coding for enzymes that enable the host bacterium to withstand the growth-inhibitory effects of two antibiotics: ampicillin and tetracycline (*Figure 4.17*). This means that cells containing a pBR322 plasmid can be distinguished from those that do not by plating the bacteria onto agar medium containing ampicillin and/or tetracycline. Normal *E. coli* cells are sensitive to these antibiotics and cannot grow when either of the two antibiotics is present. Ampicillin and tetracycline resistance are therefore **selectable markers** for pBR322.

The manipulations shown in *Figure 4.16*, resulting in construction of a recombinant plasmid, are carried out in the test tube with purified DNA. Pure pBR322 DNA can

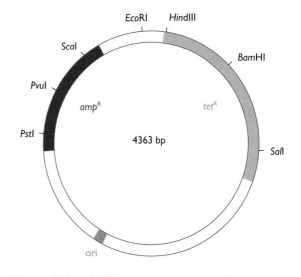

Figure 4.17 pBR322.

The map shows the positions of the ampicillin-resistance gene (*amp^R*), the tetracycline-resistance gene (*tet^R*), the origin of replication (ori) and the recognition sequences for seven restriction endonucleases.

be obtained quite easily from extracts of bacterial cells (Technical Note 4.2), but how can the manipulated plasmids be re-introduced into the bacteria? The answer is to make use of the natural processes for **transformation** of bacteria, which result in the uptake of 'naked' DNA by a bacterial cell. This is the process studied by Avery and his colleagues in the experiments which showed that bacterial genes are made of DNA (Section 1.1.1). Transformation is not a particularly efficient process in many bacteria, including *E. coli*, but the rate of uptake can be enhanced significantly by suspending the cells in calcium chloride before adding the DNA, and then briefly incubating the mixture at 42 °C. Even after this enhancement, only a very small proportion of the cells take up a plasmid. This is why the antibiotic-resistance markers are so important – they allow the small number of transformants to be selected from the large background of non-transformed cells.

The map of pBR322 shown in *Figure 4.17* indicates the positions of the recognition sequences for seven restriction enzymes, each of which cuts the plasmid at just one location. Note that six of these sites lie within one or other of the genes for antibiotic resistance. This means that if a new fragment of DNA is ligated into one of these six sites, then the antibiotic-resistance properties of the plasmid become altered – the plasmid loses the ability to confer either ampicillin or tetracycline resistance on the host cells. This is called **insertional inactivation** of the selectable marker and is the key to distinguishing a **recombinant** plasmid – one that contains an inserted piece of DNA – from a non-recombinant plasmid that has no new DNA. Identifying recombinants is important because the manipulations illustrated in *Figure 4.16* result in a variety of ligation products, including plasmids that have recir-

DNA purification

TECHNICAL 4.2 NOTE

Techniques for the preparation of pure samples of DNA from living cells play a central role in molecular biology research.

The first step in DNA purification is to break open the cells from which the DNA will be obtained. With some types of material this step is easy: cultured animal cells, for example, are broken open simply by adding a detergent such as sodium dodecyl sulfate (SDS), which disrupts the cell membranes, releasing the cell contents. Other types of cell have strong walls and so demand a harsher treatment. Plant cells are usually frozen and then ground in a mortar and pestle, this being the only effective way of breaking their cellulose walls. Bacteria such as *Escherichia coli* can be lysed by a combination of enzymatic and chemical treatment. The enzyme is **lysozyme**, obtained from egg white, which breaks down the polymeric compounds in the bacterial cell wall; the chemical is ethylenediamine tetra-acetate (EDTA), which chelates magnesium ions, further reducing the integrity of the cell wall. Disruption of the cell membrane by adding a detergent then causes the cells to burst.

Once the cells have been broken, two different methods can be used to purify the DNA from the resulting extract. The first involves degrading or removing all the cellular components other than the DNA, an approach that works best if the cells do not contain large amounts of lipid or carbohydrate. The extract is first centrifuged at low speed to remove debris such as pieces of cell wall, which form a pellet at the bottom of the tube. The supernatant is transferred to another test tube and mixed with phenol, which causes the protein to precipitate at the interface between the organic and aqueous layers. The aqueous layer, which contains the dissolved nucleic acids, is collected and a ribonuclease enzyme added, which breaks the RNA into a mixture of nucleotides and short oligonucleotides. The DNA polynucleotides, which remain intact, can now be precipitated by adding ethanol, pelleted by centrifugation, and resuspended in an appropriate volume of buffer.

In the second method for DNA purification, rather than degrading everything other than DNA, the DNA itself is selectively removed from the extract. One way of doing this is by adding the detergent cetyltrimethylammonium bromide (CTAB), which forms an insoluble complex with nucleic acids. The precipitate is collected by centrifugation and resuspended in a high-salt solution, which causes the complex to break down, releasing the nucleic acids. Ribonuclease treatment followed by ethanol precipitation yields pure DNA from this mixture. Another popular technique makes use of the tight binding between DNA and silica particles that occurs in the presence of a denaturing chemical such as guanidinium thiocyanate. The silica particles and guanidinium thiocyanate can be added directly to the extract and the DNA collected by centrifugation. Alternatively, the silica can be placed in a chromatography column and the extract, plus guanidinium thiocyanate, passed through. The DNA binds to the silica particles in the column and is subsequently recovered by washing away the guanidinium thiocyanate with water, the DNA now detaching from the silica and eluting from the column.

The two approaches described above purify all the DNA in a cell. Special methods are needed if the aim is to obtain just plasmid DNA (for example, recombinant cloning vectors) from bacterial cells. One popular method makes use of the fact that, although both plasmids and the bacterial chromosome are made up of supercoiled DNA, lysis of the bacterial cell inevitably leads to a certain amount of disruption of the nucleoid, leading to breakage of the loops of supercoiled chromosomal DNA (Section 2.3.1). A cell extract therefore contains supercoiled plasmid DNA and *non-supercoiled* chromosomal DNA, and the plasmids can be purified by a method that distinguishes between DNA molecules with these different conformations. One technique involves adding sodium hydroxide until the pH of the cell extract reaches 12.0–12.5, which causes the base pairs in non-supercoiled DNA to break. The resulting single strands tangle up into an insoluble network that can be removed by centrifugation, leaving the supercoiled plasmids in the supernatant.

cularized without insertion of new DNA. To identify recombinants, the resistance properties of colonies are tested by transferring cells from agar containing one antibiotic onto agar containing the second antibiotic. For example, if the *Bam*HI site has been used then recombinants will be ampicillin resistant but tetracycline sensitive, because the *Bam*HI site lies within the region that specifies resistance to tetracycline. After transformation, cells are plated onto ampicillin agar (*Figure 4.18*). All cells that contain a pBR322 plasmid, whether recombinant or not, divide and produce a colony. The colonies are then transferred onto tetracycline agar by **replica plating**, which results in the colonies on the second plate retaining the relative positions that they had on the first plate. Some colonies do not grow on the tetracycline agar because their cells contain recombinant pBR322 molecules with a disrupted tetracycline-resistance gene. These are the colonies we are looking for because they contain the cloned gene, so we return to the ampicillin plate, from which samples of the cells can be recovered.

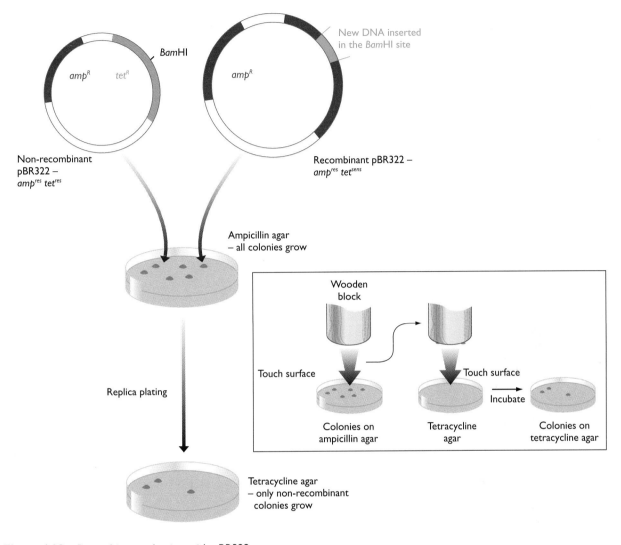

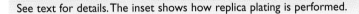

Figure 4.18   Recombinant selection with pBR322.

See text for details. The inset shows how replica plating is performed.

Replica plating is not a difficult technique but it takes time. It would be much better if recombinants could be distinguished from non-recombinants simply by plating onto a single agar medium. This is possible with most of the modern plasmid cloning vectors, including pUC8 (*Figure 4.19*; Vieira and Messing, 1982). This vector carries the ampicillin-resistance gene from pBR322, along with a second gene, called *lacZ'*, which is part of the *E. coli* gene for the enzyme β-galactosidase. The remainder of the *lacZ* gene is located in the chromosome of the special *E. coli* strain that is used when cloning genes with pUC8. The proteins specified by the gene segments on the plasmid and on the chromosome are able to combine to produce a functional β-galactosidase enzyme. The presence of functional β-galactosidase molecules in the cells can be checked by a histochemical test with a compound called X-gal (5-bromo-4-chloro-3-indolyl-β-D-galactopyranoside), which the enzyme converts into a blue product. The *lacZ'* gene contains a cluster of unique restriction sites; insertion of new DNA into any one of these sites results in insertional inactivation of the gene and hence loss of β-galactosidase activity. Recombinants and non-recombinants can therefore be distinguished simply by plating the transformed cells onto agar containing ampicillin and X-gal (*Figure 4.19*). All colonies that grow on this medium are made up of transformed cells because only transformants are ampicillin resistant. Some colonies are blue and some are white. Those that are blue contain cells with functional β-galactosidase enzymes and hence with undisrupted *lacZ'* genes; these colonies are therefore non-recombinants. The white colonies comprise cells without β-galactosidase activity and hence with disrupted *lacZ'* genes; these are recombinants.

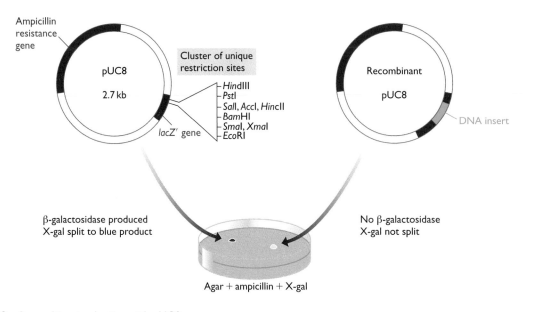

Figure 4.19 Recombinant selection with pUC8.

See the text for details.

Cloning vectors based on E. coli *bacteriophage genomes*

E. coli bacteriophages were developed as cloning vectors back in the earliest days of the recombinant DNA revolution. The main reason for seeking a different type of vector was the inability of plasmids such as pBR322 and pUC8 to handle DNA fragments greater than about 10 kb in size, larger inserts undergoing rearrangements or interfering with the plasmid replication system in such a way that the recombinant DNA molecules become lost from the host cells. The first attempts to develop vectors able to handle larger fragments of DNA centered on bacteriophage λ. The infection cycle of λ is similar to that of the T2 phages studied by Hershey and Chase in the experiments that alerted molecular biologists to the fact that genes are made of DNA (Section 1.1.1), but there is one important difference. As well as following the **lytic infection cycle** (see *Figure 1.4B*, page 9), the λ genome is able to integrate into the bacterial chromosome, where it can remain quiescent for many generations, being replicated along with the host chromosome every time the cell divides. This is called the **lysogenic infection cycle** (*Figure 4.20*).

The λ genome is 48.5 kb, of which some 15 kb or so is 'optional' in that it contains genes that are only needed for integration of the phage DNA into the *E. coli* chromosome (*Figure 4.21A*). These segments can therefore be deleted without impairing the ability of the phage to infect bacteria and direct synthesis of new λ particles by the lytic cycle. Two types of vector have been developed (*Figure 4.21B*):

■ **Insertion vectors**, in which part or all of the optional DNA has been removed and a unique restriction site introduced at some position within the trimmed down genome;

■ **Replacement vectors**, in which the optional DNA is contained within a **stuffer fragment**, flanked by a pair of restriction sites, that is replaced when the DNA to be cloned is ligated into the vector.

The λ genome is linear, but the two natural ends of the molecule have 12-nucleotide single-stranded overhangs, called *cos* sites, which have complementary sequences and so can base-pair to one another. A λ cloning vector can therefore be obtained as a circular molecule which can be manipulated in the test tube in the same way as a plasmid, and re-introduced into *E. coli* by **transfection**, the term used for uptake of naked phage DNA. Alternatively, a more efficient uptake system called *in vitro* **packaging** can be utilized (Hohn and Murray, 1977). This procedure starts with the linear version of the cloning vector, the initial restriction cutting the molecule into two segments, the left and right arms, each with a *cos* site at one end. The ligation is carried out with carefully measured quantities of each arm and the DNA to be cloned, the aim being to produce concatamers in which the different fragments are linked together in the order left arm–new DNA–right arm, as shown in *Figure 4.22*. The concatamers are then added to an *in vitro* packaging mix, which contains all the proteins needed to make a λ phage particle. These proteins form phage particles spontaneously, and will place inside the particles any DNA fragment that is between 37 and 52 kb in length and is flanked by *cos* sites. The packaging mix therefore cuts left arm–new DNA–right arm combinations of 37–52 kb out of the concatamers and constructs λ phages around them. The phages are then mixed with *E. coli* cells, and the natural infection process transports the vector plus new DNA into the bacteria.

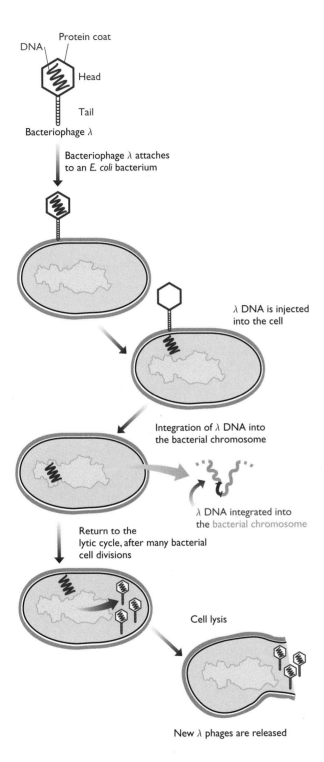

Protein coat
DNA
Head
Tail
Bacteriophage λ

Bacteriophage λ attaches
to an *E. coli* bacterium

λ DNA is injected
into the cell

Integration of λ DNA into
the bacterial chromosome

λ DNA integrated into
the bacterial chromosome

Return to the
lytic cycle, after many bacterial
cell divisions

Cell lysis

New λ phages are released

Figure 4.20 The lysogenic infection cycle of bacteriophage λ.

Compare with the lytic infection cycle of T2
bacteriophage, shown in *Figure 1.4B* (page 9). The special
feature of the lysogenic cycle is the insertion of the phage
genome into the bacterium's chromosomal DNA, where it
can remain quiescent for many generations.

After infection, the cells are spread onto an agar plate.
The objective is not to obtain individual colonies but to
produce an even layer of bacteria across the entire surface
of the agar. Bacteria that were infected with the packaged
cloning vector die within about 20 minutes because the λ
genes contained in the arms of the vector direct replication of the DNA and synthesis of new phages by the lytic
cycle, each of these new phages containing its own copy
of the vector plus cloned DNA. Death and lysis of the
bacterium releases these phages into the surrounding
medium, where they infect new cells and begin another
round of phage replication and lysis. The end result is a
zone of clearing, called a **plaque**, which is visible on the
lawn of bacteria that grows on the agar plate (*Figure 4.23*).
With some λ vectors, all plaques are made up of
recombinant phages because ligation of the two arms
without insertion of new DNA results in a molecule that
is too short to be packaged. With other vectors it is
necessary to distinguish recombinant plaques from non-
recombinant ones. Various methods are used, including
the β-galactosidase system described above for the
plasmid vector pUC8 (see *Figure 4.19*), which is also
applicable to those λ vectors that carry a fragment of the
lacZ gene into which the DNA to be cloned is inserted.

Vectors for longer pieces of DNA

The λ phage particle can accommodate up to 52 kb of
DNA, so if the genome has 15 kb removed then up to
18 kb of new DNA can be cloned. This limit is higher than
that for plasmid vectors, but is still very small compared
with the sizes of intact genomes. The comparison is
important because a **clone library** – a collection of clones
whose inserts cover an entire genome – is the starting
point for a project aimed at determining the sequence of
that genome (Chapter 6). If a λ vector is used with human
DNA, then over half a million clones are needed for there
to be a 95% chance of any particular part of the genome
being present in the library (*Table 4.4*). It is possible to
prepare a library comprising half a million clones, especially if automated techniques are used, but such a large
collection is far from ideal. It would be much better to
reduce the number of clones by using a vector that is able
to handle fragments of DNA longer than 18 kb. Many of
the developments in cloning technology over the last 20
years have been aimed at finding ways of doing this.

One possibility is to use a **cosmid** – a plasmid that
carries a λ *cos* site (*Figure 4.24*). Concatamers of cosmid
molecules, linked at their *cos* sites, act as substrates for *in
vitro* packaging because the *cos* site is the only sequence
that a DNA molecule needs in order to be recognized as a
'λ genome' by the proteins that package DNA into λ
phage particles. Particles containing cosmid DNA are as
infective as real λ phages, but once inside the cell the
cosmid cannot direct synthesis of new phage particles and
instead replicates as a plasmid. Recombinant DNA is
therefore obtained from colonies rather than plaques. As
with other types of λ vector, the upper limit for the length
of the cloned DNA is set by the space available within the

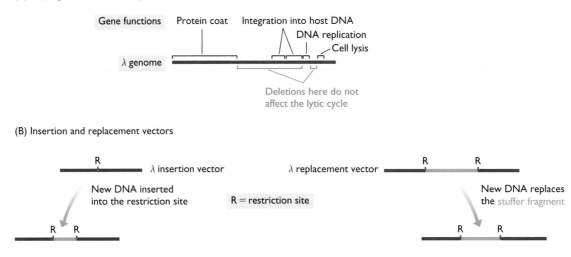

Figure 4.21 Cloning vectors based on bacteriophage λ.

(A) In the λ genome, the genes are arranged into functional groups. For example, the region marked as 'protein coat' comprises 21 genes coding for proteins that are either components of the phage capsid or are required for capsid assembly, and 'cell lysis' comprises four genes involved in lysis of the bacterium at the end of the lytic phase of the infection cycle. The regions of the genome that can be deleted without impairing the ability of the phage to follow the lytic cycle are indicated in green. (B) The differences between a λ insertion vector and a λ replacement vector.

λ phage particle. A cosmid can be 8 kb or less in size, so up to 44 kb of new DNA can be inserted before the packaging limit of the λ phage particle is reached. This reduces the size of the human genomic library to about a quarter of a million clones, which is an improvement compared with a λ library, but is still a massive number of clones to have to work with.

The first major breakthrough in attempts to clone DNA fragments much longer than 50 kb came with the invention of **yeast artificial chromosomes** or **YACs** (Burke *et al.*, 1987). These vectors are propagated in *S. cerevisiae* rather than in *E. coli* and are based on chromosomes, rather than on plasmids or viruses. The first YACs were constructed after studies of natural chromosomes had shown that, in addition to the genes that it carries, each chromosome has three important components:

- The **centromere**, which plays a critical role during nuclear division (see *Figure 2.7*, page 38);
- The **telomeres**, the special sequences which mark the ends of chromosomal DNA molecules (see *Figure 2.10*, page 41);
- One or more **origins of replication**, which initiate synthesis of new DNA when the chromosome divides (Section 13.2.1).

In a YAC, the DNA sequences that underlie these chromosomal components are linked together with one or more

Table 4.4 Sizes of human genomic libraries prepared in different types of cloning vector

Type of vector	Insert size (kb)	Number of clones*	
		P = 95%	P = 99%
λ replacement	18	532 500	820 000
Cosmid, fosmid	40	240 000	370 000
PI	125	77 000	118 000
BAC, PAC	300	32 000	50 000
YAC	600	16 000	24 500
Mega-YAC	1400	6850	10 500

*Calculated from the equation:

$$N = \frac{\ln(1-P)}{\ln(1-\frac{a}{b})}$$

where *N* is the number of clones required, *P* is the probability that any given segment of the genome is present in the library, *a* is the average size of the DNA fragments inserted into the vector, and *b* is the size of the genome.

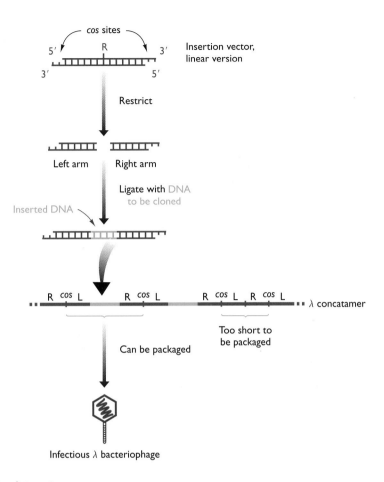

Figure 4.22   Cloning with a λ insertion vector.

The linear form of the vector is shown at the top of the diagram. Treatment with the appropriate restriction endonuclease produces the left and right arms, both of which have one blunt end and one end with the 12-nucleotide overhang of the *cos* site. The DNA to be cloned is blunt ended and so is inserted between the two arms during the ligation step. These arms also ligate to one another via their *cos* sites, forming a concatamer. Some parts of the concatamer comprise left arm–insert DNA–right arm and, assuming this combination is 37–52 kb in length, will be enclosed inside the capsid by the *in vitro* packaging mix. Parts of the concatamer made up of left arm ligated directly to right arm, without new DNA, are too short to be packaged.

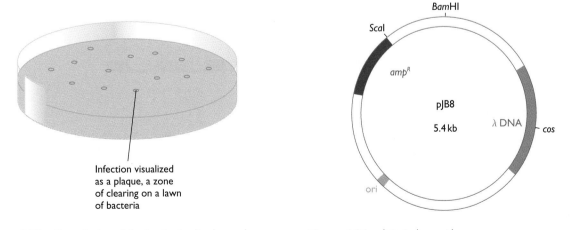

Figure 4.23   Bacteriophage infection is visualized as a plaque on a lawn of bacteria.

Figure 4.24   A typical cosmid.

pJB8 is 5.4 kb in size and carries the ampicillin-resistance gene (*amp^R*), a segment of λ DNA containing the *cos* site, and an *Escherichia coli* origin of replication (ori).

selectable markers and at least one restriction site into which new DNA can be inserted (*Figure 4.25*). All of these components can be contained in a DNA molecule of 10–15 kb. Natural yeast chromosomes range from 230 kb to over 1700 kb, so YACs have the potential to clone Mb-sized DNA fragments. This potential has been realized,

standard YACs being able to clone 600 kb fragments, with special types able to handle DNA up to 1400 kb in length. Currently this is the highest capacity of any type of cloning vector, and several genome projects have made extensive use of YACs. Unfortunately, with some types of YAC there have been problems with insert stability, the cloned DNA

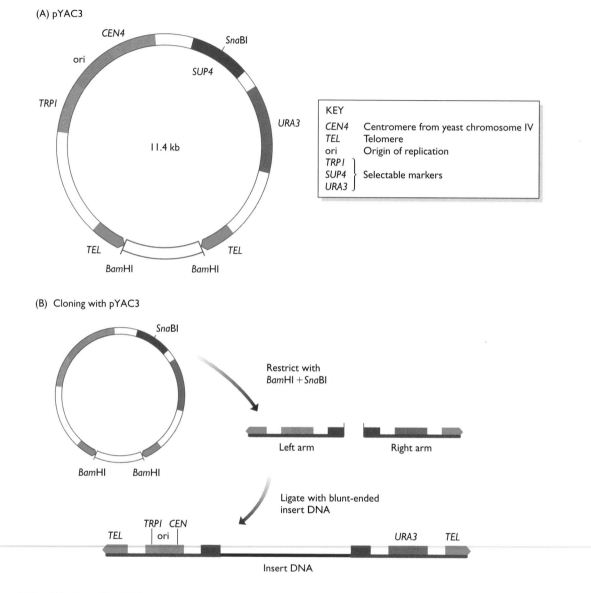

Figure 4.25 Working with a YAC.

(A) The cloning vector pYAC3. (B) To clone with pYAC3, the circular vector is digested with *Bam*HI and *Sna*BI. *Bam*HI restriction removes the stuffer fragment held between the two telomeres in the circular molecule. *Sna*BI cuts within the *SUP4* gene and provides the site into which new DNA will be inserted. Ligation of the two vector arms with new DNA produces the structure shown at the bottom. This structure carries functional copies of the *TRP1* and *URA3* selectable markers. The host strain has inactivated copies of these genes, which means that it requires tryptophan and uracil as nutrients. After transformation, cells are plated onto a minimal medium, lacking tryptophan and uracil. Only cells that contain the vector, and so can synthesize tryptophan and uracil, are able to survive on this medium and produce colonies. Note that if a vector comprises two right arms, or two left arms, then it will not give rise to colonies because the transformed cells will still require one of the nutrients. The presence of insert DNA in the cloned vector molecules is checked by testing for inactivation of *SUP4*. This is done by a color test: on the appropriate medium, colonies containing recombinant vectors (i.e. with an insert) are white; non-recombinants (vector but no insert) are red.

Working with a clone library

Clone collections used as a source of genes and other DNA segments.

Clone libraries have been used in molecular biology research for many years and their importance extends well beyond their role as the starting point for a genome sequencing project. Since the 1970s, clone libraries have been prepared from different organisms as a means of obtaining individual genes and other DNA segments for further study by sequencing and other recombinant DNA techniques.

Clone libraries can be prepared from either genomic DNA or cDNA (Section 7.1.2), using a plasmid or bacteriophage vector. The clones are usually stored as bacterial colonies or plaques on 23×23 cm agar plates, with 100 000–150 000 clones per plate. A complete human library can therefore be contained in just 1–8 plates, depending on the type of cloning vector that has been used (see Table 4.4). Three methods can identify the clone that contains the gene or other piece of DNA that is being sought:

- **Hybridization analysis** can be performed with an oligonucleotide or other DNA molecule that is known to hybridize to the sequence of interest. To do this, a nylon or nitrocellulose membrane is placed on the surface of the agar dish and then carefully removed to 'lift off' the colonies or plaques. Treatment with alkali and protease degrades the cellular material, leaving behind the DNA from each clone, which is then bound tightly to the surface of the membrane by heating or ultraviolet irradiation. The labeled probe is now applied to the membrane, in the same way as in Southern hybridization (Figure 4.12), and the

position at which the probe attaches determined by the appropriate detection method. The position of the hybridization signal on the membrane corresponds to the location of the clone of interest on the agar plate.
- **PCR** (Section 4.3) can be used to screen clones for the sequence of interest. This cannot be done in situ, so individual clones must be transferred to the wells of microtiter trays. The PCR approach to clone identification is therefore relatively cumbersome because only a few hundred clones can be accommodated in a single tray. PCRs using primers specific for the sequence of interest are performed with each clone in turn, possibly using a combinatorial approach, which reduces the number of PCRs that are needed in order to identify the one that gives a positive result (see Figure 6.14, page 178).
- **Immunological techniques** can be used if the sequence being sought is a gene that is expressed in the cell in which the clone library has been prepared. If gene expression is occurring then the protein product will be made, and this can be detected by screening the library with a labeled antibody that binds only to that protein. As in hybridization analysis, the clones are first transferred onto a membrane and then treated to break down the cells and bind the protein to the membrane surface. Exposure of the membrane to the labeled antibody then reveals the position of the clone containing the gene of interest.

becoming rearranged into new sequence combinations (Anderson, 1993). For this reason there is also great interest in other types of vectors, ones that cannot clone such large pieces of DNA but which suffer less from instability problems. These vectors include the following:

- **Bacteriophage P1 vectors** (Sternberg, 1990) are very similar to λ vectors, being based on a deleted version of a natural phage genome, the capacity of the cloning vector being determined by the size of the deletion and the space within the phage particle. The P1 genome is larger than the λ genome, and the phage particle is bigger, so a P1 vector can clone larger fragments of DNA than a λ vector, up to 125 kb using current technology.
- **Bacterial artificial chromosomes** or **BACs** (Shizuya et al., 1992) are based on the naturally occurring F plasmid of E. coli. Unlike the plasmids used to construct the early cloning vectors, the F plasmid is

relatively large and vectors based on it have a higher capacity for accepting inserted DNA. BACs can be used to clone fragments of 300 kb and longer.
- **P1-derived artificial chromosomes** or **PACs** (Ioannou et al., 1994) combine features of P1 vectors and BACs and have a capacity of up to 300 kb.
- **Fosmids** (Kim et al., 1992) contain the F plasmid origin of replication and a λ cos site. They are similar to cosmids but have a lower copy number in E. coli, which means that they are less prone to instability problems.

The sizes of human genome libraries prepared in these various types of vector are given in Table 4.4.

Cloning in organisms other than E. coli

Cloning is not merely an aid to DNA sequencing: it also provides a means of studying the mode of expression of a gene and the way in which expression is regulated, of carrying out genetic engineering experiments aimed at modifying the biological characteristics of the host organism, and of synthesizing important animal proteins, such as pharmaceuticals, in a new host cell from which the proteins can be obtained in larger quantities than is possible by conventional purification from animal tissue. These multifarious applications demand that genes must frequently be cloned in organisms other than *E. coli*.

Cloning vectors based on plasmids or phages have been developed for most of the well studied species of bacteria such as *Bacillus*, *Streptomyces* and *Pseudomonas*,

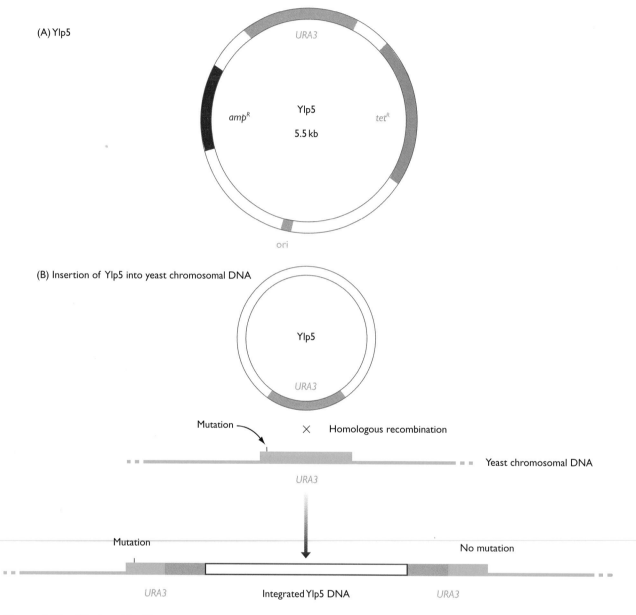

Figure 4.26 Cloning with a YIp.

(A) YIp5, a typical yeast integrative plasmid. The plasmid contains the ampicillin-resistance gene (*amp^R*), the tetracycline-resistance gene (*tet^R*), the yeast gene *URA3*, and an *Escherichia coli* origin of replication (ori). The presence of the *E. coli* ori means that recombinant YIp5 molecules can be constructed in *E. coli* before their transfer into yeast cells. YIp5 is therefore a **shuttle vector** – it can be shuttled between two species. (B) YIp5 has no origin of replication that can function inside yeast cells, but can survive if it integrates into the yeast chromosomal DNA by homologous recombination between the plasmid and chromosomal copies of the *URA3* gene. The chromosomal gene carries a small mutation which means that it is non-functional and the host cells are *ura3$^-$*. One of the pair of *URA3* genes that are formed after integration of the plasmid DNA is mutated, but the other is not. Recombinant cells are therefore *ura$^+$* and can be selected by plating on to minimal medium, which does not contain uracil.

these vectors being used in exactly the same way as the *E. coli* analogs. Plasmid vectors are also available for yeasts and fungi. Some of these carry the origin of replication from the **2 μm circle**, a plasmid present in many strains of *S. cerevisiae*, but other plasmid vectors only have an *E. coli* origin. An example is YIp5, an *S. cerevisiae* vector that is simply a pBR322 plasmid that contains a copy of the yeast gene called *URA3* (*Figure 4.26A*). What is the logic behind the construction of YIp5? When used in a cloning experiment, the vector is initially used with *E. coli* as the host, up to the stage where the desired recombinant molecule has been constructed by restriction and ligation. The recombinant vector is then purified from *E. coli* and transferred into *S. cerevisiae*, usually by mixing the DNA with **protoplasts** – yeast cells whose walls have been removed by enzyme treatment. Without an origin of replication the vector is unable to propagate independently inside yeast cells, but it can survive if it becomes integrated into one of the yeast chromosomes, which can occur by **homologous recombination** (Section 7.2.2) between the *URA3* gene carried by the vector and the chromosomal copy of this gene (*Figure 4.26B*). 'YIp' in fact stands for 'yeast integrative plasmid'. Once integrated the YIp, plus any DNA that has been inserted into it, replicates along with the host chromosomes.

Integration into chromosomal DNA is also a feature of many of the cloning systems used with animals and plants, and forms the basis of the construction of **knockout mice**, which are used to determine the functions of previously unknown genes that are discovered in the human genome (Section 7.2.2). The vectors are animal equivalents of YIps. Adenoviruses and retroviruses are used to clone genes in animals when the objective is to treat a genetic disease or a cancer by **gene therapy** (Lemoine and Cooper, 1998). A similar range of vectors has been developed for cloning genes in plants. Plasmids can be introduced into plant embryos by bombardment with DNA-coated microprojectiles, a process called **biolistics** (Klein *et al.*, 1987), integration of the plasmid DNA into the plant chromosomes, followed by growth of the embryo, resulting in a plant that contains the cloned DNA in most or all of its cells. Some success has also been achieved with plant vectors based on the genomes of caulimoviruses and geminiviruses (Timmermans *et al.*, 1994; Viaplana *et al.*, 2001), but the most interesting types of plant cloning vector are those derived from the **Ti plasmid** (Hansen and Wright, 1999). This is a large bacterial plasmid found in the soil microorganism *Agrobacterium tumefaciens*, part of which, the **T-DNA**, becomes integrated into a plant chromosome when the bacterium infects a plant stem and causes crown root disease. The T-DNA carries a number of genes that are expressed inside the plant cells and induce the various physiological changes that characterize the disease. Vectors such as pBIN19 (*Figure 4.27*) have been designed to make use of this natural genetic engineering system (Bevan, 1984). The recombinant vector is introduced into *A. tumefaciens* cells, which are allowed to infect a cell suspension or

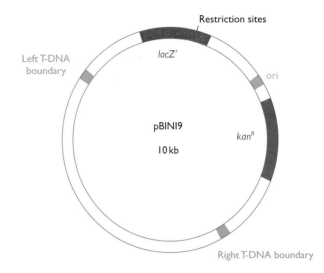

Figure 4.27   The plant cloning vector pBIN19.

pBIN19 carries the *lacZ'* gene (see *Figure 4.19*), the kanamycin-resistance gene (*kan^R*), an *Escherichia coli* origin of replication (ori), and the two boundary sequences from the T-DNA region of the Ti plasmid. These two boundary sequences recombine with plant chromosomal DNA, inserting the segment of DNA between them into the plant DNA. The orientation of the boundary sequences in pBIN19 means that the *lacZ'* and *kan^R* genes, as well as any new DNA ligated into the restriction sites within *lacZ'*, are transferred to the plant DNA. Recombinant plant cells are selected by plating onto kanamycin agar, and then regenerated into whole plants. Note that pBIN19 is another example of a shuttle vector, recombinant molecules being constructed in *E. coli*, using the *lacZ'* selection system, before transfer to *Agrobacterium tumefaciens* and thence to the plant. For more details, see Brown (2001).

plant callus culture, from which mature transformed plants can be regenerated.

4.3 The Polymerase Chain Reaction (PCR)

In essence, DNA cloning results in the purification of a single fragment of DNA from a complex mixture of DNA molecules. Cloning is a powerful technique and its impact on our understanding of genes and genomes has been immeasurable. Cloning does, however, have one major disadvantage: it is a time-consuming and, in parts, difficult procedure. It takes several days to perform the manipulations needed to insert DNA fragments into a cloning vector and then introduce the ligated molecules into the host cells and select recombinants. If the experimental strategy involves generation of a large clone library, followed by screening of the library to identify a clone that contains a gene of interest (see Technical Note 4.3), then several more weeks or even months might be needed to complete the project.

Techniques for studying RNA

Many of the techniques devised for studying DNA molecules can be adapted for use with RNA.

■ **Agarose gel electrophoresis** of RNA is carried out after denaturation of the RNA so that the migration rate of each molecule is dependent entirely on its length, and is not influenced by the intramolecular base pairs that can form in many RNAs (e.g. *Figure 11.11*, page 323). The denaturant, usually formaldehyde or glyoxal, is added to the sample before it is loaded onto the gel.

■ **Northern hybridization** refers to the procedure whereby an RNA gel is blotted onto a nylon membrane and hybridized to a labeled probe (see *Figure 7.4*, page 192). This is equivalent to Southern hybridization and is done in a similar way.

■ **Labeled RNA molecules** are usually prepared by copying a DNA template into RNA in the presence of a labeled ribonucleotide. The RNA polymerase enzymes of SP6, T3 or T7 bacteriophages are used because they can produce up to 30 μg of labeled RNA from 1 μg of DNA in 30 minutes. RNA can also be end-labeled by treatment with purified poly(A) polymerase (Section 10.1.2).

■ **PCR** of RNA molecules requires a modification to the first step of the normal reaction. *Taq* polymerase cannot copy an RNA molecule so the first step is catalyzed by a reverse transcriptase, which makes a DNA copy of the RNA template. This DNA copy is then amplified by *Taq* polymerase. The technique is called **reverse transcriptase-PCR** or **RT-PCR**. The discovery of thermostable enzymes that make DNA copies of both RNA and DNA templates (e.g. the *Tth*

DNA polymerase from the bacterium *Thermus thermophilus*) raises the possibility of carrying out RT-PCR in a single reaction with just one enzyme.

■ **RNA sequencing** methods exist but are difficult to perform and are applicable only to small molecules. The methods are similar to chemical degradation sequencing of DNA (Box 6.2, page 170) but employ sequence-specific endonucleases rather than chemicals to generate the cleaved molecules. In practice, the sequence of an RNA molecule is usually obtained by converting it into cDNA (see *Figure 5.32*, page 155) and sequencing by the chain termination method (Section 6.1.1).

■ **Specialist methods** have been developed for mapping the positions of RNA molecules on to DNA sequences, for example to determine the start and end points of transcription and to locate the positions of introns in a DNA sequence. These methods are described in Section 7.1.2.

The only major deficiency in the RNA toolkit is the absence of enzymes with the degree of sequence specificity displayed by the restriction endonucleases that are so important in DNA manipulations. Other than this, the only drawback with RNA work is the ease with which RNAs are degraded by ribonucleases that are released when cells are disrupted (as during RNA extraction), and which are also present on the hands of laboratory workers and which tend to contaminate glassware and solutions. This means that rigorous laboratory procedures (e.g. cleaning of glassware with chemicals that destroy ribonucleases) have to be adopted in order to keep RNA molecules intact.

PCR complements DNA cloning in that it enables the same result to be achieved – purification of a specified DNA fragment – but in a much shorter time, perhaps just a few hours (Saiki *et al.*, 1988). PCR is complementary to, not a replacement for, cloning because it has its own limitations, the most important of which is the need to know the sequence of at least part of the fragment that is to be purified. Despite this constraint, PCR has acquired central importance in many areas of molecular biology research. We will examine the technique first and then survey its applications.

4.3.1 Carrying out a PCR

PCR results in the repeated copying of a selected region of a DNA molecule (see *Figure 4.3*, page 98). Unlike cloning,

PCR is a test-tube reaction and does not involve the use of living cells: the copying is carried out not by cellular enzymes but by the purified, thermostable DNA polymerase of *T. aquaticus* (Section 4.1.1). The reason why a thermostable enzyme is needed will become clear when we look in more detail at the events that occur during PCR.

To carry out a PCR experiment, the target DNA is mixed with *Taq* DNA polymerase, a pair of oligonucleotide primers, and a supply of nucleotides. The amount of target DNA can be very small because PCR is extremely sensitive and will work with just a single starting molecule. The primers are needed to initiate the DNA synthesis reactions that will be carried out by the *Taq* polymerase (see *Figure 4.6*, page 100). They must attach to the target DNA at either side of the segment that is to be copied; the sequences of these attachment sites must therefore be known so that primers of the appropriate sequences can be synthesized.

The reaction is started by heating the mixture to 94 °C. At this temperature the hydrogen bonds that hold together the two polynucleotides of the double helix are broken, so the target DNA becomes denatured into single-stranded molecules (*Figure 4.28*). The temperature is then reduced to 50–60 °C, which results in some rejoining of the single strands of the target DNA, but also allows the primers to attach to their annealing positions. DNA synthesis can now begin, so the temperature is raised to 72 °C, the optimum for *Taq* polymerase. In this first stage of the PCR, a set of 'long products' is synthesized from each strand of the target DNA. These polynucleotides have identical 5′ but random 3′ ends, the latter representing positions where DNA synthesis terminates by chance. When the cycle of denaturation–annealing–synthesis is repeated, the long products act as templates for new DNA synthesis, giving rise to 'short products' whose 5′ and 3′ ends are both set by the primer annealing positions (*Figure 4.29*). In subsequent cycles, the number of short products accumulates in an exponential fashion (doubling during each cycle) until one of the components of the reaction becomes depleted. This means that after 30 cycles, there will be over 250 million short products

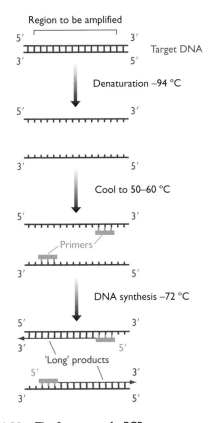

Figure 4.28   The first stage of a PCR.

See the text for details.

derived from each starting molecule. In real terms, this equates to several micrograms of PCR product from a few nanograms or less of target DNA.

The results of a PCR can be determined in various ways. Usually, the products are analyzed by agarose gel electrophoresis, which will reveal a single band if the PCR has worked as expected and has amplified a single segment of the target DNA (*Figure 4.30*). Alternatively, the sequence of the product can be determined, using techniques described in Section 6.1.1.

4.3.2 The applications of PCR

PCR is such a straightforward procedure that it is sometimes difficult to understand how it can have become so important in modern research. First we will deal with its limitations. In order to synthesize primers that will anneal at the correct positions, the sequences of the boundary regions of the DNA to be amplified must be known. This means that PCR cannot be used to purify fragments of genes or other parts of a genome that have never been studied before. A second constraint is the length of DNA that can be copied. Regions of up to 5 kb can be amplified without too much difficulty, and longer amplifications – up to 40 kb – are possible using modifications of the standard technique. However, the >100 kb fragments that are needed for genome sequencing projects are unattainable by PCR.

What are the strengths of PCR? Primary among these is the ease with which products representing a single segment of the genome can be obtained from a number of different DNA samples. We will encounter one important example of this in the next chapter when we look at how DNA markers are typed in genetic mapping projects (Section 5.2.2). PCR is used in a similar way to screen human DNA samples for mutations associated with genetic diseases such as thalassemia and cystic fibrosis. It also forms the basis of genetic profiling, in which variations in microsatellite length are typed (see *Figure 2.25*, page 61).

A second important feature of PCR is its ability to work with minuscule amounts of starting DNA. This means that PCR can be used to obtain sequences from the trace amounts of DNA that are present in hairs, bloodstains and other forensic specimens, and from bones and other remains preserved at archaeological sites. In clinical diagnosis, PCR is able to detect the presence of viral DNA well before the virus has reached the levels needed to initiate a disease response. This is particularly important in the early identification of viral-induced cancers because it means that treatment programs can be initiated before the cancer becomes established.

The above are just a few of the applications of PCR. The technique is now a major component of the molecular biologist's toolkit and we will discover many more examples of its use in the study of genomes as we progress through the remaining chapters of this book.

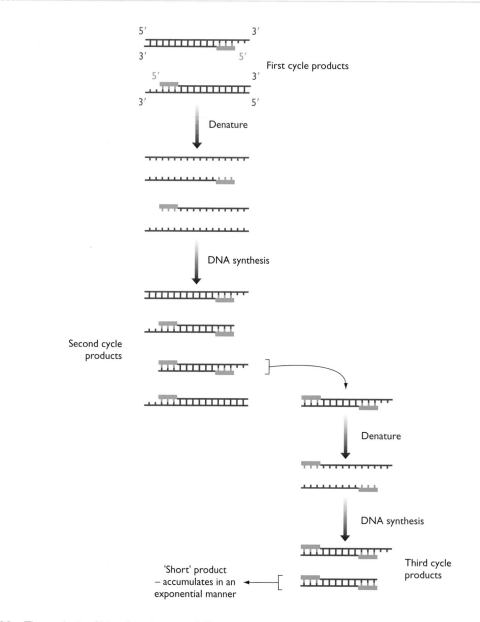

Figure 4.29   The synthesis of 'short' products in a PCR.

The first cycle products from *Figure 4.28* are shown at the top. The next cycle of denaturation–annealing–synthesis leads to four products, two of which are identical to the first cycle products and two of which are made entirely of new DNA. During the third cycle, the latter give rise to 'short' products which, in subsequent cycles, accumulate in an exponential fashion.

Figure 4.30   Analysing the results of a PCR by agarose gel electrophoresis.

The PCR has been carried out in a microfuge tube. A sample is loaded into lane 2 of an agarose gel. Lane 1 contains DNA size markers, and lane 3 contains a sample of a PCR done by a colleague. After electrophoresis, the gel is stained with ethidium bromide (see Technical Note 2.1, page 37). Lane 2 contains a single band of the expected size, showing that the PCR has been successful. In lane 3 there is no band – this PCR has not worked.

References

Anderson C (1993) Genome shortcut leads to problems. *Science,* **259**, 1684–1687.

Bevan M (1984) Binary *Agrobacterium* vectors for plant transformation. *Nucleic Acids Res.,* **12**, 8711–8721.

Bolivar F, Rodriguez RL, Greene PJ, et al. (1977) Construction and characterisation of new cloning vectors. II. A multi-purpose cloning system. *Gene,* **2**, 95–113.

Brown TA (1998) *Molecular Biology Labfax,* 2nd edition, Vol. I. Academic Press, London.

Brown TA (2001) *Gene Cloning and DNA Analysis: An Introduction,* 4th edition. Blackwell Scientific Publishers, Oxford.

Burke DT, Carle GF and Olson MV (1987) Cloning of large segments of exogenous DNA into yeast by means of artificial chromosome vectors. *Science,* **236**, 806–812.

Hansen G and Wright MS (1999) Recent advances in the transformation of plants. *Trends Plant Sci.,* **4**, 226–231.

Hohn B and Murray K (1977) Packaging recombinant DNA molecules into bacteriophage particles *in vitro. Proc. Natl Acad. Sci. USA,* **74**, 3259–3263.

Ioannou PA, Amemiya CT, Garnes J, et al. (1994) P1-derived vector for the propagation of large human DNA fragments. *Nature Genet.,* **6**, 84–89.

Kim U-J, Shizuya H, de Jong PJ, Birren B and Simon MI (1992) Stable propagation of cosmid and human DNA inserts in an F factor based vector. *Nucleic Acids Res.,* **20**, 1083–1085.

Klein TM, Wolf ED, Wu R and Sanford JC (1987) High velocity microprojectiles for delivering nucleic acids into living cells. *Nature,* **327**, 70–73.

Kornberg A (1960) Biologic synthesis of deoxyribonucleic acid. *Science,* **131**, 1503–1508.

Lemoine N and Cooper D (1998) *Gene Therapy.* BIOS Scientific Publishers, Oxford.

Mullis KB (1990) The unusual origins of the polymerase chain reaction. *Sci. Am.,* **262**(4), 56–65.

Saiki RK, Gelfand DH, Stoffel S, et al. (1988) Primer-directed enzymatic amplification of DNA with a thermostable DNA polymerase. *Science,* **239**, 487–491.

Shizuya H, Birren B, Kim UJ, et al. (1992) Cloning and stable maintenance of 300-kilobase-pair fragments of human DNA in *Escherichia coli* using an F-factor-based vector. *Proc. Natl Acad. Sci. USA,* **89**, 8794–8797.

Sternberg N (1990) Bacteriophage P1 cloning system for the isolation, amplification, and recovery of DNA fragments as large as 100 kilobase pairs. *Proc. Natl Acad. Sci. USA,* **87**, 103–107.

Timmermans MCP, Das OP and Messing J (1994). Geminiviruses and their uses as extrachromosomal replicons. *Ann. Rev. Plant Physiol. Plant Mol. Biol.,* **45**, 79–112.

Viaplana R, Turner DS and Covey SN (2001) Transient expression of a GUS reporter gene from cauliflower mosaic virus replacement vectors in the presence and absence of helper virus. *J. Gen. Virol.,* **82**, 59–65.

Vieira J and Messing J (1982) The pUC plasmids, an M13mp7-derived system for insertion mutagenesis and sequencing with synthetic universal primers. *Gene,* **19**, 259–268.

Further Reading

Blackman K (2001) The advent of genetic engineering. *Trends Biochem. Sci.,* **26**, 268–270. — *A personal account of the early days of gene cloning.*

Dale JW (1998) *Molecular Genetics of Bacteria,* 3rd edition. Wiley, Chichester. — *Provides a detailed description of plasmids and bacteriophages.*

Monaco AP and Larin Z (1994) YACs, BACs, PACs and MACs – artificial chromosomes as research tools. *Trends Biotechnol.,* **12**, 280–286. — *A good review of high-capacity cloning vectors.*

Southern EM (2000) Blotting at 25. *Trends Biochem. Sci.,* **25**, 585–588. — *The origins of Southern hybridization.*

Watson JD, Gilman M, Witkowski J and Zoller M (1992) *Recombinant DNA,* 2nd edition. W.H. Freeman, New York. — *Detailed descriptions of basic recombinant DNA methodology.*

STUDY AIDS FOR CHAPTER 4

Key terms

Give short definitions of the following terms:

2 μm circle
Adaptor
Alkaline phosphatase
Autoradiography
Bacterial artificial chromosome (BAC)
Bacteriophage P1 vector
Biolistics
Blunt end
Clone library
Cloning vector
Cohesive end

Complementary DNA (cDNA)
Concatamer
cos site
Cosmid
Denaturation of protein
DNA cloning
DNA polymerase
End-modification enzyme
Endonuclease
Exonuclease
Flush end

Fosmid
Gene cloning
Gene therapy
Homologous recombination
Homopolymer tailing
Hybridization
Hybridization probe
In vitro packaging
Insertion vector
Insertional inactivation
Klenow polymerase
Knockout mice
Kornberg polymerase
Ligase

Linker
Lysogenic infection cycle
Lysozyme
Lytic infection cycle
Northern hybridization
Nuclease
Oligonucleotide
Origin of replication
P1-derived artificial chromosome (PAC)
Phosphorimaging
Plaque
Polymerase chain reaction (PCR)

Primer
Proofreading
Protoplast
Recombinant
Recombinant DNA
  technology
Replacement vector
Replica plating
Restriction endonuclease
Reverse transcriptase
Reverse transcriptase-PCR
RNA-dependent DNA
  polymerase
Selectable marker
Sequenase
Shuttle vector
Southern hybridization

Sticky end
Stuffer fragment
T4 polynucleotide kinase
T-DNA
Template-dependent DNA
  polymerase
Template-independent DNA
  polymerase
Terminal deoxynucleotidyl
  transferase
Thermostable
Ti plasmid
Transfection
Transformation
Yeast artificial chromosome
  (YAC)

Self study questions

1. Draw diagrams that outline the events that occur during (a) DNA cloning, and (b) PCR. What are the limitations of each of these two techniques?
2. List the types of enzyme used in recombinant DNA research.
3. Distinguish between the two types of exonuclease activity that can be possessed by a DNA polymerase, and explain how these activities influence the potential applications of individual DNA polymerases in recombinant DNA research.
4. Using examples, describe the various types of end produced after digestion of DNA with a restriction endonuclease.
5. How are agarose gel electrophoresis and Southern hybridization used to examine the results of a restriction digest?
6. Explain why the efficiency of blunt-end ligation is less than that of sticky-end ligation. What steps can be taken to improve the efficiency of blunt-end ligation?

7. Draw diagrams of (a) pBR322, and (b) pUC8. Explain how the differences between these two vectors influence the ways in which they are used to clone DNA fragments.
8. Distinguish between the lytic and lysogenic infection cycles for a bacteriophage.
9. Write a short description of the way in which a bacteriophage λ vector is used to clone DNA. How does a cosmid differ from a standard λ vector?
10. Draw a diagram showing a typical YAC. Indicate the key features and explain how a YAC is used to clone DNA.
11. What problems might arise when a YAC is used to clone a large fragment of DNA? To what extent can these problems be solved by the use of other types of high-capacity cloning vector?
12. How is DNA cloned in organisms other than *Escherichia coli?*
13. Describe how a PCR is carried out, paying particular attention to the role of the primers and the temperatures used during the thermal cycling.

Problem-based learning

1. Soon after the first gene cloning experiments were carried out in the early 1970s, a number of scientists argued that there should be a temporary moratorium on this type of research. What was the basis of these scientists' fears and to what extent were these fears justified?
2. What would be the features of an ideal cloning vector? To what extent are these requirements met by any of the existing cloning vectors?
3. The specificity of the primers is a critical feature of a successful PCR. If the primers anneal at more than one position in the target DNA then products additional to the one being sought will be synthesized. Explore the factors that determine primer specificity and evaluate the influence of the annealing temperature on the outcome of a PCR.

5

Mapping Genomes

Chapter Contents

5.1   Genetic and Physical Maps   128

5.2   Genetic Mapping   128

   5.2.1   Genes were the first markers to
be used   129

   5.2.2   DNA markers for genetic
mapping   129

   5.2.3   Linkage analysis is the basis of
genetic mapping   134

   5.2.4   Linkage analysis with different
types of organism   140

5.3   Physical Mapping   145

   5.3.1   Restriction mapping   145

   5.3.2   Fluorescent *in situ* hybridization
(FISH)   152

   5.3.3   Sequence tagged site (STS)
mapping   153

Learning outcomes

When you have read Chapter 5, you should be able to:

- ■ Explain why a map is an important aid to genome sequencing

- ■ Distinguish between the terms 'genetic map' and 'physical map'

- ■ Describe the different types of marker used to construct genetic maps, and state how each type of marker is scored

- ■ Summarize the principles of inheritance as discovered by Mendel, and show how subsequent genetic research led to the development of linkage analysis

- ■ Explain how linkage analysis is used to construct genetic maps, giving details of how the analysis is carried out in various types of organism, including humans and bacteria

- ■ State the limitations of genetic mapping

- ■ Evaluate the strengths and weaknesses of the various methods used to construct physical maps of genomes

- ■ Describe how restriction mapping is carried out

- ■ Describe how fluorescent *in situ* hybridization (FISH) is used to construct a physical map, including the modifications used to increase the sensitivity of this technique

- ■ Explain the basis of sequence tagged site (STS) mapping, and list the various DNA sequences that can be used as STSs

- ■ Describe how radiation hybrids and clone libraries are used in STS mapping

---

THE NEXT TWO CHAPTERS describe the techniques and strategies used to obtain genome sequences. DNA sequencing is obviously paramount among these techniques, but sequencing has one major limitation: even with the most sophisticated technology it is rarely possible to obtain a sequence of more than about 750 bp in a single experiment. This means that the sequence of a long DNA molecule has to be constructed from a series of shorter sequences. This is done by breaking the molecule into fragments, determining the sequence of each one, and using a computer to search for overlaps and build up the master sequence (*Figure 5.1*). This **shotgun method** is the standard approach for sequencing small prokaryotic genomes, but is much more difficult with larger genomes because the required data analysis becomes disproportionately more complex as the number of fragments increases (for n fragments the number of possible overlaps is given by $2n^2 - 2n$). A second problem with the shotgun method is that it can lead to errors when repetitive regions of a genome are analyzed. When a repetitive sequence is broken into fragments, many of the resulting pieces contain the same, or very

similar, sequence motifs. It would be very easy to reassemble these sequences so that a portion of a repetitive region is left out, or even to connect together two quite separate pieces of the same or different chromosomes (*Figure 5.2*).

The difficulties in applying the shotgun method to a large molecule that has a significant repetitive DNA content means that this approach cannot be used on its own to sequence a eukaryotic genome. Instead, a genome **map** must first be generated. A genome map provides a guide for the sequencing experiments by showing the positions of genes and other distinctive features. Once a genome map is available, the sequencing phase of the project can proceed in either of two ways (*Figure 5.3*):

- ■ By the **whole-genome shotgun** method (Section 6.2.3), which takes the same approach as the standard shotgun procedure but uses the distinctive features on the genome map as landmarks to aid assembly of the master sequence from the huge numbers of short sequences that are obtained. Reference to the map also ensures that regions

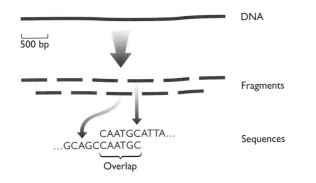

DNA

500 bp

Fragments

Sequences

CAATGCATTA...
...GCAGCCAATGC

Overlap

Figure 5.1 The shotgun approach to sequence assembly.

The DNA molecule is broken into small fragments, each of which is sequenced. The master sequence is assembled by searching for overlaps between the sequences of individual fragments. In practice, an overlap of several tens of base pairs would be needed to establish that two sequences should be linked together.

containing repetitive DNA are assembled correctly. The whole-genome shotgun approach is a rapid way of obtaining a eukaryotic genome sequence, but there are still doubts about the degree of accuracy that can be achieved.

- By the **clone contig** approach (Section 6.2.2). In this method the genome is broken into manageable segments, each a few hundred kb or a few Mb in length, which are short enough to be sequenced accurately by the shotgun method. Once the sequence of a segment has been completed, it is positioned at its correct location on the map. This step-by-step approach takes longer than whole-genome shotgun sequencing, but is thought to produce a more accurate and error-free sequence.

With both approaches, the map provides the framework for carrying out the sequencing phase of the project. If the map indicates the positions of genes, then it can also be used to direct the initial part of a clone contig project

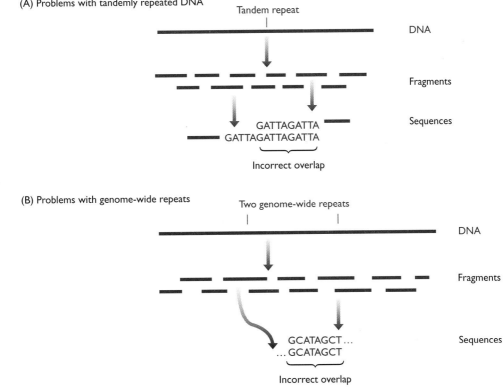

(A) Problems with tandemly repeated DNA

Tandem repeat

DNA

Fragments

Sequences

GATTAGATTA
GATTAGATTAGATTA

Incorrect overlap

(B) Problems with genome-wide repeats

Two genome-wide repeats

DNA

Fragments

Sequences

GCATAGCT...
...GCATAGCT

Incorrect overlap

Figure 5.2 Problems with the shotgun approach.

(A) The DNA molecule contains a tandemly repeated element made up of many copies of the sequence GATTA. When the sequences are examined, an overlap is identified between two fragments, but these are from either end of the tandem repeat. If the error is not recognized then the internal region of the tandem repeat will be omitted from the master sequence. (B) In the second example, the DNA molecule contains two copies of a genome-wide repeat. When the sequences are examined, two fragments appear to overlap, but one fragment contains the left-hand part of one repeat and the other fragment has the right-hand part of the second repeat. In this case, failure to recognize the error would lead to the segment of DNA between the two genome-wide repeats being left out of the master sequence. If the two genome-wide repeats were on different chromosomes, then the sequences of these chromosomes might mistakenly be linked together.

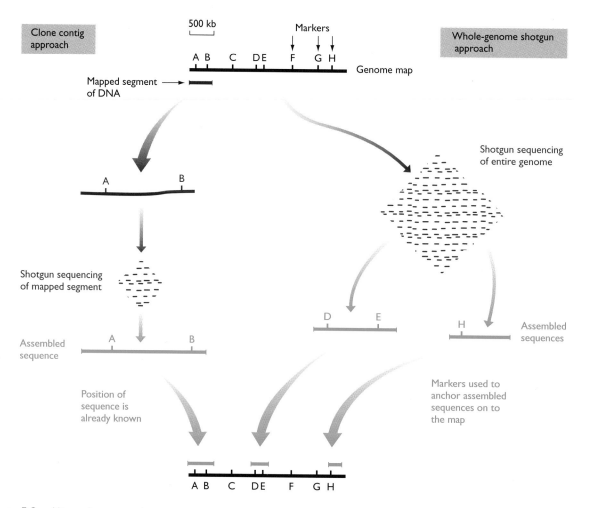

Figure 5.3 Alternative approaches to genome sequencing.

A genome consisting of a linear DNA molecule of 2.5 Mb has been mapped, and the positions of eight markers (A–H) are known. On the left, the clone contig approach starts with a segment of DNA whose position on the genome map has been identified because it contains markers A and B. The segment is sequenced by the shotgun method and the master sequence placed at its known position on the map. On the right, the whole-genome shotgun approach involves random sequencing of the entire genome. This results in pieces of contiguous sequence, possibly hundreds of kb in length. If a contiguous sequence contains a marker, then it can be positioned on the genome map. Note that, with either method, the more markers there are on the genome map the better. For more details of these sequencing strategies, see Section 6.2.

to the interesting regions of a genome, so that the sequences of important genes are obtained as quickly as possible.

5.1 Genetic and Physical Maps

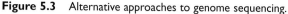

The convention is to divide genome mapping methods into two categories.

■ **Genetic mapping** is based on the use of genetic techniques to construct maps showing the positions of genes and other sequence features on a genome. Genetic techniques include cross-breeding experi-

ments or, in the case of humans, the examination of family histories (pedigrees). Genetic mapping is described in Section 5.2.

■ **Physical mapping** uses molecular biology techniques to examine DNA molecules directly in order to construct maps showing the positions of sequence features, including genes. Physical mapping is described in Section 5.3.

5.2 Genetic Mapping

As with any type of map, a genetic map must show the positions of distinctive features. In a geographic map these **markers** are recognizable components of the landscape, such as rivers, roads and buildings. What markers can we use in a genetic landscape?

5.2.1 Genes were the first markers to be used

The first genetic maps, constructed in the early decades of the 20th century for organisms such as the fruit fly, used genes as markers. This was many years before it was understood that genes are segments of DNA molecules. Instead, genes were looked upon as abstract entities responsible for the transmission of heritable characteristics from parent to offspring. To be useful in genetic analysis, a heritable characteristic has to exist in at least two alternative forms or **phenotypes**, an example being tall or short stems in the pea plants originally studied by Mendel. Each phenotype is specified by a different **allele** of the corresponding gene. To begin with, the only genes that could be studied were those specifying phenotypes that were distinguishable by visual examination. So, for example, the first fruit-fly maps showed the positions of genes for body color, eye color, wing shape and suchlike, all of these phenotypes being visible simply by looking at the flies with a low-power microscope or the naked eye. This approach was fine in the early days but geneticists soon realized that there were only a limited number of visual phenotypes whose inheritance could be studied, and in many cases their analysis was complicated because a single phenotype could be affected by more than one gene. For example, by 1922 over 50 genes had been mapped onto the four fruit-fly chromosomes, but nine of these were for eye color; in later research, geneticists studying fruit flies had to learn to distinguish between fly eyes that were colored red, light red, vermilion, garnet, carnation, cinnabar, ruby, sepia, scarlet, pink, cardinal, claret, purple or brown. To make gene maps more comprehensive it would be necessary to find characteristics that were more distinctive and less complex than visual ones.

The answer was to use biochemistry to distinguish phenotypes. This has been particularly important with two types of organisms – microbes and humans. Microbes, such as bacteria and yeast, have very few visual characteristics so gene mapping with these organisms has to rely on biochemical phenotypes such as those listed in *Table 5.1*. With humans it is possible to use visual

characteristics, but since the 1920s studies of human genetic variation have been based largely on biochemical phenotypes that can be scored by blood typing. These phenotypes include not only the standard blood groups such as the ABO series (Yamamoto *et al.*, 1990), but also variants of blood serum proteins and of immunological proteins such as the human leukocyte antigens (the HLA system). A big advantage of these markers is that many of the relevant genes have **multiple alleles**. For example, the gene called *HLA-DRB1* has at least 290 alleles and *HLA-B* has over 400. This is relevant because of the way in which gene mapping is carried out with humans (Section 5.2.4). Rather than setting up many breeding experiments, which is the procedure with experimental organisms such as fruit flies or mice, data on inheritance of human genes have to be gleaned by examining the phenotypes displayed by members of a single family. If all the family members have the same allele for the gene being studied then no useful information can be obtained. It is therefore necessary for the relevant marriages to have occurred, by chance, between individuals with different alleles. This is much more likely if the gene being studied has 290 rather than two alleles.

5.2.2 DNA markers for genetic mapping

Genes are very useful markers but they are by no means ideal. One problem, especially with larger genomes such as those of vertebrates and flowering plants, is that a map based entirely on genes is not very detailed. This would be true even if every gene could be mapped because, as we saw in Chapter 2, in most eukaryotic genomes the genes are widely spaced out with large gaps between them (see *Figure 2.2*, page 34). The problem is made worse by the fact that only a fraction of the total number of genes exist in allelic forms that can be distinguished conveniently. Gene maps are therefore not very comprehensive. We need other types of marker.

Mapped features that are not genes are called **DNA markers**. As with gene markers, a DNA marker must have at least two alleles to be useful. There are three types of DNA sequence feature that satisfy this requirement: restriction fragment length polymorphisms (RFLPs), simple sequence length polymorphisms (SSLPs), and single nucleotide polymorphisms (SNPs).

Table 5.1 Typical biochemical markers used for genetic analysis of *Saccharomyces cerevisiae*

Marker	Phenotype	Method by which cells carrying the marker are identified
ADE2	Requires adenine	Grows only when adenine is present in the medium
CAN1	Resistant to canavanine	Grows in the presence of canavanine
CUP1	Resistant to copper	Grows in the presence of copper
CYH1	Resistant to cycloheximide	Grows in the presence of cycloheximide
LEU2	Requires leucine	Grows only when leucine is present in the medium
SUC2	Able to ferment sucrose	Grows if sucrose is the only carbohydrate in the medium
URA3	Requires uracil	Grows only when uracil is present in the medium

Restriction fragment length polymorphisms (RFLPs)

RFLPs were the first type of DNA marker to be studied. Recall that restriction enzymes cut DNA molecules at specific recognition sequences (Section 4.1.2). This sequence specificity means that treatment of a DNA molecule with a restriction enzyme should always produce the same set of fragments. This is not always the case with genomic DNA molecules because some restriction sites are polymorphic, existing as two alleles, one allele displaying the correct sequence for the restriction site and therefore being cut when the DNA is treated with the enzyme, and the second allele having a sequence alteration so the restriction site is no longer recognized. The result of the sequence alteration is that the two adjacent restriction fragments remain linked together after treatment with the enzyme, leading to a length polymorphism (*Figure 5.4*). This is an RFLP and its position on a genome map can be worked out by following the inheritance of its alleles, just as is done when genes are used as markers. There are thought to be about 10^5 RFLPs in the human genome, but of course for each RFLP there can only be two alleles (with and without the site). The value of RFLPs in human gene mapping is therefore limited by the high possibility that the RFLP being studied shows no variability among the members of an interesting family.

In order to score an RFLP, it is necessary to determine the size of just one or two individual restriction fragments against a background of many irrelevant fragments. This is not a trivial problem: an enzyme such as *Eco*RI, with a 6-bp recognition sequence, should cut approximately once every $4^6 = 4096$ bp and so would give almost 800 000 fragments when used with human DNA. After separation by agarose gel electrophoresis (see Technical Note 2.1, page 37), these 800 000 fragments produce a smear and the RFLP cannot be distinguished. Southern hybridization, using a probe that spans the polymorphic restriction site, provides one way of visualizing the RFLP

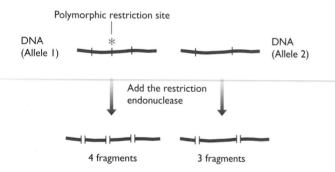

Figure 5.4 A restriction fragment length polymorphism (RFLP).

The DNA molecule on the left has a polymorphic restriction site (marked with the asterisk) that is not present in the molecule on the right. The RFLP is revealed after treatment with the restriction enzyme because one of the molecules is cut into four fragments whereas the other is cut into three fragments.

(*Figure 5.5A*), but nowadays PCR is more frequently used. The primers for the PCR are designed so that they anneal either side of the polymorphic site, and the RFLP is typed by treating the amplified fragment with the restriction enzyme and then running a sample in an agarose gel (*Figure 5.5B*).

Simple sequence length polymorphisms (SSLPs)

SSLPs are arrays of repeat sequences that display length variations, different alleles containing different numbers of repeat units (*Figure 5.6A*). Unlike RFLPs, SSLPs can be multi-allelic as each SSLP can have a number of different length variants. There are two types of SSLP, both of which were described in Section 2.4.1:

- **Minisatellites**, also known as **variable number of tandem repeats** (**VNTRs**), in which the repeat unit is up to 25 bp in length;
- **Microsatellites** or **simple tandem repeats** (**STRs**), whose repeats are shorter, usually dinucleotide or tetranucleotide units.

Microsatellites are more popular than minisatellites as DNA markers, for two reasons. First, minisatellites are not spread evenly around the genome but tend to be found more frequently in the telomeric regions at the ends of chromosomes. In geographic terms, this is equivalent to trying to use a map of lighthouses to find one's way around the middle of an island. Microsatellites are more conveniently spaced throughout the genome. Second, the quickest way to type a length polymorphism is by PCR (*Figure 5.6B*), but PCR typing is much quicker and more accurate with sequences less than 300 bp in length. Most minisatellite alleles are longer than this because the repeat units are relatively large and there tend to be many of them in a single array, so PCR products of several kb are needed to type them. Typical microsatellites consist of 10–30 copies of a repeat that is usually no longer than 4 bp in length, and so are much more amenable to analysis by PCR. There are 6.5×10^5 microsatellites in the human genome (see *Table 1.3*, page 25).

Single nucleotide polymorphisms (SNPs)

These are positions in a genome where some individuals have one nucleotide (e.g. a G) and others have a different nucleotide (e.g. a C) (*Figure 5.7*). There are vast numbers of SNPs in every genome, some of which also give rise to RFLPs, but many of which do not because the sequence in which they lie is not recognized by any restriction enzyme. In the human genome there are at least 1.42 million SNPs, only 100 000 of which result in an RFLP (SNP Group, 2001).

Although each SNP could, potentially, have four alleles (because there are four nucleotides), most exist in just two forms, so these markers suffer from the same drawback as RFLPs with regard to human genetic mapping: there is a high possibility that a SNP does not display any variability in the family that is being studied. The advantages of

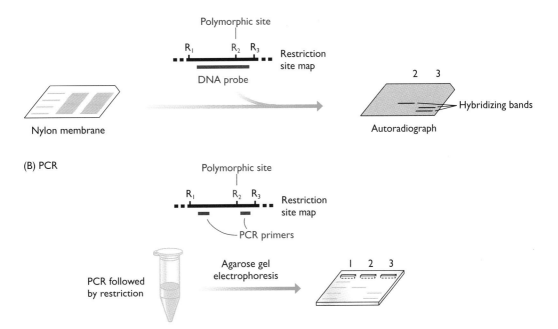

(A) Southern hybridization

(B) PCR

Figure 5.5   Two methods for scoring an RFLP.

(A) RFLPs can be scored by Southern hybridization. The DNA is digested with the appropriate restriction enzyme and separated in an agarose gel. The smear of restriction fragments is transferred to a nylon membrane and probed with a piece of DNA that spans the polymorphic restriction site. If the site is absent then a single restriction fragment is detected (lane 2); if the site is present then two fragments are detected (lane 3). (B) The RFLP can also be typed by PCR, using primers that anneal either side of the polymorphic restriction site. After the PCR, the products are treated with the appropriate restriction enzyme and then analyzed by agarose gel electrophoresis. If the site is absent then one band is seen on the agarose gel; if the site is present then two bands are seen.

Box 5.1: Why do SNPs have only two alleles?

Any of the four nucleotides could be present at any position in the genome, so it might be imagined that each single nucleotide polymorphism (SNP) should have four alleles. Theoretically this is possible but in practice most SNPs exist as just two variants. This is because of the way in which SNPs arise and spread in a population. An SNP originates when a **point mutation** (Section 14.1.1) occurs in a genome, converting one nucleotide into another. If the mutation is in the reproductive cells of an individual, then one or more of the offspring might inherit the mutation and, after many generations, the SNP may eventually become established in the population. But there are just two alleles – the original sequence and the mutated version. For a third allele to arise, a new mutation must occur at the same position in the genome in another individual, and this individual and his or her offspring must reproduce in such a way that the new allele becomes established. This scenario is not impossible but it is unlikely; consequently the vast majority of SNPs are biallelic. For more on the way in which alleles spread in populations, see Box 16.3 (page 497).

SNPs are their abundant numbers and the fact that they can be typed by methods that do not involve gel electrophoresis. This is important because gel electrophoresis has proved difficult to automate so any detection method that uses it will be relatively slow and labor-intensive. SNP detection is more rapid because it is based on **oligonucleotide hybridization analysis**. An oligonucleotide is a short single-stranded DNA molecule, usually less than 50 nucleotides in length, that is synthesized in the test tube. If the conditions are just right, then an oligonucleotide will hybridize with another DNA molecule only if the oligonucleotide forms a completely base-paired structure with the second molecule. If there is a single mismatch – a single position within the oligonucleotide that does not form a base pair – then hybridization does not occur (*Figure 5.8*). Oligonucleotide hybridization can therefore discriminate between the two alleles of an SNP. Various screening strategies have been

(A) Two variants of an SSLP

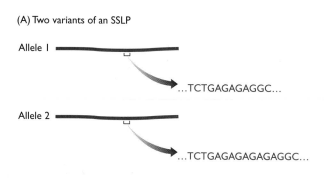

...TCTGAGAGAGGC...

...TCTGAGAGAGAGAGGC...

(B) Typing an SSLP by PCR

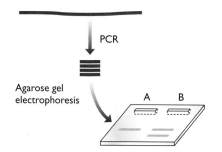

PCR

Agarose gel electrophoresis

A   B

Figure 5.6 SSLPs and how they are typed.

(A) Two alleles of a microsatellite SSLP. In allele 1 the motif 'GA' is repeated three times, and in allele 2 it is repeated five times. (B) How the SSLP could be typed by PCR. The region surrounding the SSLP is amplified and the products loaded into lane A of the agarose gel. Lane B contains DNA markers that show the sizes of the bands given after PCR of the two alleles. The band in lane A is the same size as the larger of the two DNA markers, showing that the DNA that was tested contained allele 2.

devised (Mir and Southern, 2000), including **DNA chip** technology (Technical Note 5.1) and **solution hybridization techniques.**

■ A **DNA chip** is a wafer of glass or silicon, 2.0 cm² or less in area, carrying many different oligonucleotides in a high-density array. The DNA to be tested is labeled with a fluorescent marker and pipetted onto the surface of the chip. Hybridization is detected by examining the chip with a fluorescence microscope, the positions at which the fluorescent signal is emitted indicating which oligonucleotides have hybridized with the test DNA. Many SNPs can therefore be scored in a single experiment (Wang *et al.*, 1998; Gerhold *et al.*, 1999).

■ **Solution hybridization techniques** are carried out in the wells of a microtiter tray, each well containing a different oligonucleotide, and use a detection system that can discriminate between unhybridized single-stranded DNA and the double-stranded product that results when an oligonucleotide hybridizes to the test DNA. Several systems have been developed, one of which makes use of a pair of

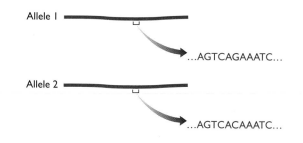

...AGTCAGAAATC...

...AGTCACAAATC...

Figure 5.7 A single nucleotide polymorphism (SNP).

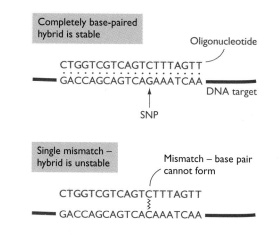

Completely base-paired hybrid is stable

Oligonucleotide

CTGGTCGTCAGTCTTTAGTT
GACCAGCAGTCAGAAATCAA

DNA target

SNP

Single mismatch – hybrid is unstable

Mismatch – base pair cannot form

CTGGTCGTCAGTCTTTAGTT
GACCAGCAGTCACAAATCAA

Figure 5.8 Oligonucleotide hybridization is very specific.

Under highly stringent hybridization conditions, a stable hybrid occurs only if the oligonucleotide is able to form a completely base-paired structure with the target DNA. If there is a single mismatch then the hybrid does not form. To achieve this level of stringency, the incubation temperature must be just below the **melting temperature** or T_m of the oligonucleotide. At temperatures above the T_m, even the fully base-paired hybrid is unstable. At more than 5 °C below the T_m, mismatched hybrids might be stable. The T_m for the oligonucleotide shown in the figure would be about 58 °C. The T_m in °C is calculated from the formula
$T_m = (4 \times$ number of G and C nucleotides) $+$ $(2 \times$ number of A and T nucleotides). This formula gives a rough indication of the T_m for oligonucleotides of 15–30 nucleotides in length.

labels comprising a fluorescent dye and a compound that quenches the fluorescent signal when brought into close proximity with the dye. The dye is attached to one end of an oligonucleotide and the quenching compound to the other end. Normally there is no fluorescence because the oligonucleotide is designed in such a way that the two ends base-pair to one another, placing the quencher next to the dye (*Figure 5.9*). Hybridization between oligonucleotide and test DNA disrupts this base pairing, moving the quencher away from the dye and enabling the fluorescent signal to be generated (Tyagi *et al.*, 1998).

DNA microarrays and chips

High-density arrays of DNA molecules for parallel hybridization analyses.

DNA microarrays and chips are designed to allow many hybridization experiments to be performed in parallel. Their main applications have been in the screening of polymorphisms such as SNPs (Section 5.2.2) and comparing the RNA populations of different cells (Section 7.3.1). They also have potential in novel DNA sequencing methodologies (Section 6.1.2).

Although the terminology is inexact, microarrays and chips are, strictly speaking, two distinct types of matrix (Gerhold et al., 1999). In both architectures, a large number of DNA probes, each one with a different sequence, are immobilized at defined positions on a solid surface. The probes can be synthetic oligonucleotides or other short DNA molecules such as cDNAs. In the earliest technology, the oligonucleotides or cDNAs were spotted onto a glass microscope slide or a piece of nylon membrane to form a **microarray**. With this approach, only a relatively low density can be achieved – typically 6400 spots as an 80 × 80 array in an area of 18 mm × 18 mm, which is sufficient for examining RNA populations but less applicable to the high throughput analyses needed to type SNPs.

To prepare really high density arrays, oligonucleotides are synthesized in situ on the surface of a wafer of glass or silicon, resulting in a **DNA chip**. The normal method for oligonucleotide synthesis involves adding nucleotides one by one to the growing end of an oligonucleotide, the sequence determined by the order in which the dNTP substrates are added to the reaction mixture. If used for synthesis on a chip, this method would result in every oligonucleotide having the same sequence. Instead, modified dNTP substrates are used, ones that have to be light-activated before they will attach to the end of a growing oligonucleotide. The dNTPs are added one after another to the chip surface, **photolithography** being used to direct pulses of light onto individual positions in the array and hence to determine which of the growing oligonucleotides will be extended by addition of the particular dNTP added at each step. A density of up to 250 000 oligonucleotides per cm² is possible, so if used for SNP screening, 125 000 polymorphisms can be typed in a single

experiment, presuming there are oligonucleotides for both alleles of each SNP.

Chips and microarrays are not complicated to use (see figure below). The array is incubated with labeled target DNA to allow hybridization to take place. Which oligonucleotides have hybridized to the target DNA is determined by scanning the surface of the array and recording the positions at which the signal emitted by the label is detectable. Radioactive labels can be used with a low-density microarray, signals being detected electronically by **phosphorimaging**. This does not provide adequate resolution for a high-density chip, so with these it is necessary to use a fluorescent label. The fluorescent signal is detected by laser scanning or, more routinely, by fluorescent confocal microscopy.

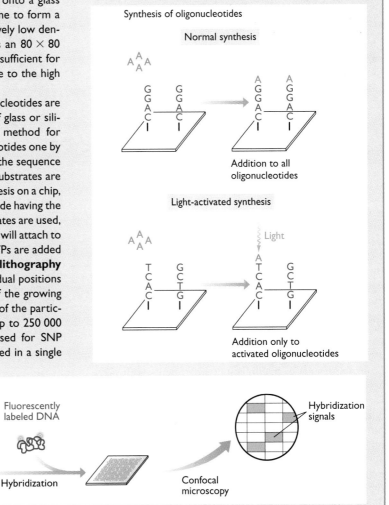

Synthesis of oligonucleotides

Normal synthesis

Addition to all oligonucleotides

Light-activated synthesis

Addition only to activated oligonucleotides

Using a DNA chip

Fluorescently labeled DNA

DNA chip

Hybridization

Confocal microscopy

Hybridization signals

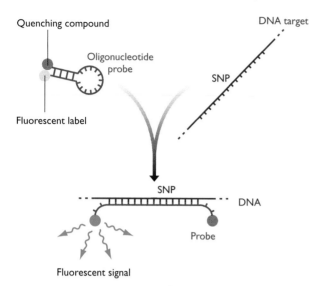

Figure 5.9   One way of detecting an SNP by solution hybridization.

> The oligonucleotide probe has two end-labels. One of these is a fluorescent dye and the other is a quenching compound. The two ends of the oligonucleotide base-pair to one another, so the fluorescent signal is quenched. When the probe hybridizes to its target DNA, the ends of the molecule become separated, enabling the fluorescent dye to emit its signal. The two labels are called 'molecular beacons'.

5.2.3 Linkage analysis is the basis of genetic mapping

Now that we have assembled a set of markers with which to construct a genetic map we can move on to look at the mapping techniques themselves. These techniques are all based on **genetic linkage**, which in turn derives from the seminal discoveries in genetics made in the mid 19th century by Gregor Mendel.

The principles of inheritance and the discovery of linkage

Genetic mapping is based on the principles of inheritance as first described by Gregor Mendel in 1865 (Orel, 1995). From the results of his breeding experiments with peas, Mendel concluded that each pea plant possesses two alleles for each gene, but displays only one phenotype. This is easy to understand if the plant is pure-breeding, or **homozygous**, for a particular characteristic, as it then possesses two identical alleles and displays the appropriate phenotype (*Figure 5.10A*). However, Mendel showed that if two pure-breeding plants with different phenotypes are crossed then all the progeny (the F_1 generation) display the same phenotype. These F_1 plants must be **heterozygous**, meaning that they possess two different alle-

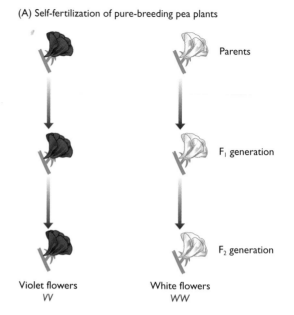

(A) Self-fertilization of pure-breeding pea plants

Violet flowers   White flowers
VV WW

(B) Cross-fertilization of two pure-breeding types

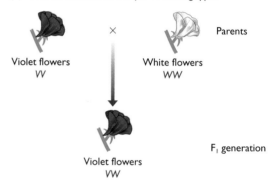

Violet flowers   ×   White flowers
VV WW

Violet flowers
VW

Figure 5.10   Homozygosity and heterozygosity.

> Mendel studied seven pairs of contrasting characteristics in his pea plants, one of which was violet and white flower color, as shown here. (A) Pure-breeding plants always give rise to flowers with the parental color. These plants are homozygotes, each possessing a pair of identical alleles, denoted here by VV for violet flowers and WW for white flowers. (B) When two pure-breeding plants are crossed, only one of the phenotypes is seen in the F_1 generation. Mendel deduced that the genotype of the F_1 plants was VW, so V was the dominant allele and W was the recessive allele.

les, one for each phenotype, one allele inherited from the mother and one from the father. Mendel postulated that in this heterozygous condition one allele overrides the effects of the other allele; he therefore described the phenotype expressed in the F_1 plants as being **dominant** over the second, **recessive** phenotype (*Figure 5.10B*). This is the perfectly correct interpretation of the interaction between the pairs of alleles studied by Mendel, but we now appre-

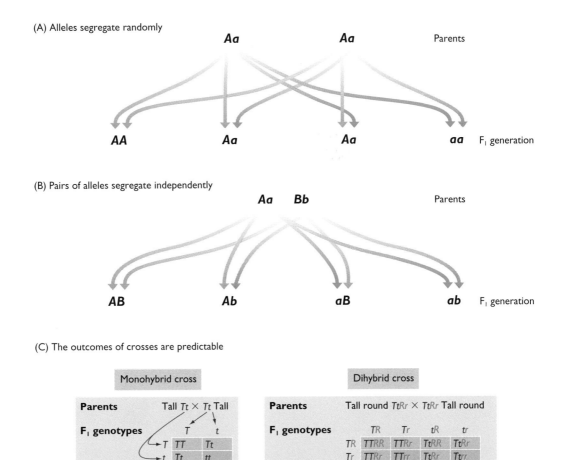

Figure 5.11 Mendel's Laws enable the outcome of genetic crosses to be predicted.

(A) Mendel's First Law states that alleles segregate randomly. The example shows inheritance of alleles *A* and *a* in a cross involving two heterozygous parents. Each member of the F_1 generation has an equal chance of inheriting *A* or *a* from either of the parents. (B) The Second Law says that pairs of alleles segregate independently. One parent is shown – a plant that is heterozygous for both genes A and B. Each member of the F_1 generation has an equal chance of inheriting either allele of either gene from this parent. (C) Two crosses with their predicted outcomes. In a monohybrid cross, the alleles of a single gene are followed, in this case allele *T* for tall pea plants and allele *t* for short pea plants. *T* is dominant and *t* is recessive. The grid shows the predicted genotypes and phenotypes of the F_1 generation. When Mendel carried out this cross he obtained 787 tall pea plants and 277 short plants, a ratio of 2.84 : 1. In the dihybrid cross, two genes are followed. The second gene determines the shape of the peas, the alleles being *R* (round, the dominant allele) and *r* (wrinkled, which is recessive). Again, the predicted genotypes and phenotypes are shown.

ciate that this simple dominant–recessive rule can be complicated by situations that he did not encounter. One of these is **incomplete dominance**, where the heterozygous phenotype is intermediate between the two homozygous forms. An example is when red carnations are crossed with white ones, the F_1 heterozygotes being pink. Another complication is **codominance**, when both alleles are detectable in the heterozygote. Codominance is the typical situation for DNA markers.

As well as discovering dominance and recessiveness,

Mendel carried out additional crosses that enabled him to establish two Laws of Genetics. The First Law states that *alleles segregate randomly*. In other words, if the parent's alleles are *A* and *a*, then a member of the F_1 generation has the same chance of inheriting *A* as it has of inheriting *a* (*Figure 5.11A*). The Second Law is that *pairs of alleles segregate independently*, so that inheritance of the alleles of gene A is independent of inheritance of the alleles of gene B (*Figure 5.11B*). Because of these laws, the outcomes of genetic crosses are predictable (*Figure 5.11C*).

When Mendel's work was rediscovered in 1900, his Second Law worried the early geneticists because it was soon established that genes reside on chromosomes, and it was realized that all organisms have many more genes than chromosomes. Chromosomes are inherited as intact units, so it was reasoned that the alleles of some pairs of genes will be inherited together because they are on the same chromosome (*Figure 5.12*). This is the principle of genetic linkage, and it was quickly shown to be correct, although the results did not turn out exactly as expected. The complete linkage that had been anticipated between many pairs of genes failed to materialize. Pairs of genes were either inherited independently, as expected for genes in different chromosomes, or, if they showed linkage, then it was only **partial linkage**: sometimes they

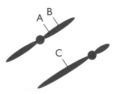

Figure 5.12 Genes on the same chromosome should display linkage.

Genes A and B are on the same chromosome and so should be inherited together. Mendel's Second Law should therefore not apply to the inheritance of A and B, but holds for the inheritance of A and C, or B and C. Mendel did not discover linkage because the seven genes that he studied were each on a different pea chromosome.

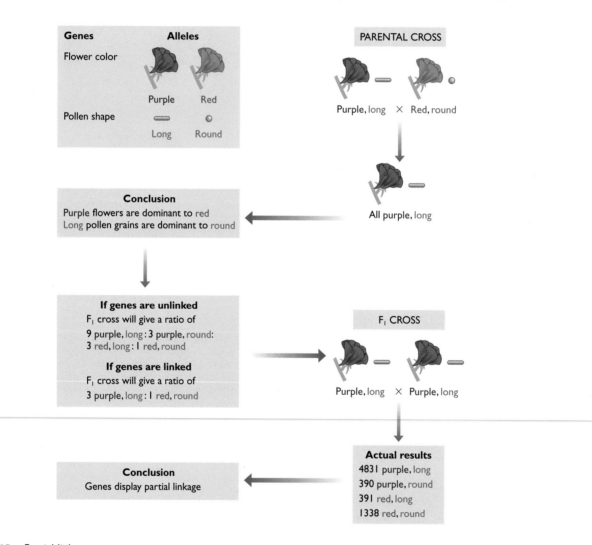

Figure 5.13 Partial linkage.

Partial linkage was discovered in the early 20th century. The cross shown here was carried out by Bateson, Saunders and Punnett in 1905 with sweet peas. The parental cross gives the typical dihybrid result (see *Figure 5.11C*), with all the F_1 plants displaying the same phenotype, indicating that the dominant alleles are purple flowers and long pollen grains. The F_1 cross gives unexpected results as the progeny show neither a 9 : 3 : 3 : 1 ratio (expected for genes on different chromosomes) nor a 3 : 1 ratio (expected if the genes are completely linked). An unusual ratio is typical of partial linkage.

were inherited together and sometimes they were not (*Figure 5.13*). The resolution of this contradiction between theory and observation was the critical step in the development of genetic mapping techniques.

Partial linkage is explained by the behavior of chromosomes during meiosis

The critical breakthrough was achieved by Thomas Hunt Morgan, who made the conceptual leap between partial linkage and the behavior of chromosomes when the nucleus of a cell divides. Cytologists in the late 19th century had distinguished two types of nuclear division: **mitosis** and **meiosis**. Mitosis is more common, being the process by which the diploid nucleus of a somatic cell divides to produce two daughter nuclei, both of which are diploid (*Figure 5.14*). Approximately 10^{17} mitoses are needed to produce all the cells required during a human lifetime. Before mitosis begins, each chromosome in the nucleus is replicated, but the resulting daughter chromosomes do not immediately break away from one another. To begin with they remain attached at their centromeres

and by **cohesin** proteins which act as 'molecular glue' holding together the arms of the replicated chromosomes (see *Figure 13.23*, page 405). The daughters do not separate until later in mitosis when the chromosomes are distributed between the two new nuclei. Obviously it is important that each of the new nuclei receives a complete set of chromosomes, and most of the intricacies of mitosis appear to be devoted to achieving this end.

Mitosis illustrates the basic events occurring during nuclear division but is not directly relevant to genetic mapping. Instead, it is the distinctive features of meiosis that interest us. Meiosis occurs only in reproductive cells, and results in a diploid cell giving rise to four haploid **gametes**, each of which can subsequently fuse with a gamete of the opposite sex during sexual reproduction. The fact that meiosis results in four haploid cells whereas mitosis gives rise to two diploid cells is easy to explain: meiosis involves two nuclear divisions, one after the other, whereas mitosis is just a single nuclear division. This is an important distinction, but the critical difference between mitosis and meiosis is more subtle. Recall that in a diploid cell there are two separate copies

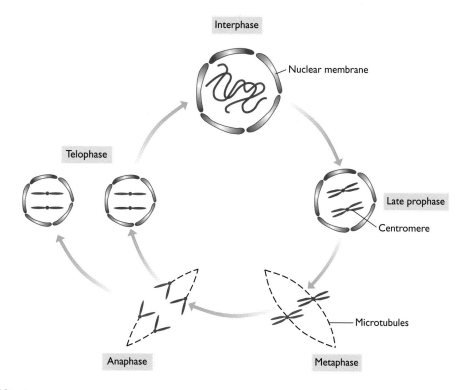

Figure 5.14 Mitosis.

During interphase (the period between nuclear divisions) the chromosomes are in their extended form (Section 2.2.1). At the start of mitosis the chromosomes condense and by late prophase have formed structures that are visible with the light microscope. Each chromosome has already undergone DNA replication but the two daughter chromosomes are held together by the centromere. During metaphase the nuclear membrane breaks down (in most eukaryotes) and the chromosomes line up in the center of the cell. Microtubules now draw the daughter chromosomes towards either end of the cell. In telophase, nuclear membranes re-form around each collection of daughter chromosomes. The result is that the parent nucleus has given rise to two identical daughter nuclei. For simplicity, just one pair of homologous chromosomes is shown; one member of the pair is red, the other is blue.

of each chromosome (Chapter 1). We refer to these as pairs of **homologous chromosomes**. During mitosis, homologous chromosomes remain separate from one another, each member of the pair replicating and being passed to a daughter nucleus independently of its homolog. In meiosis, however, the pairs of homologous chromosomes are by no means independent. During meiosis I, each chromosome lines up with its homolog to form a **bivalent** (*Figure 5.15*). This occurs after each chromosome has replicated, but before the replicated structures split, so the bivalent in fact contains four chromosome copies, each of which is destined to find its way into one of the four gametes that will be produced at the end of the meiosis. Within the bivalent, the chromosome arms (the **chromatids**) can undergo physical breakage and exchange of segments of DNA. The process is called **crossing-over** or **recombination** and was discovered by the Belgian cytologist Janssens in

1909. This was just 2 years before Morgan started to think about partial linkage.

How did the discovery of crossing-over help Morgan explain partial linkage? To understand this we need to think about the effect that crossing-over can have on the inheritance of genes. Let us consider two genes, each of which has two alleles. We will call the first gene A and its alleles *A* and *a*, and the second gene B with alleles *B* and *b*. Imagine that the two genes are located on chromosome number 2 of *Drosophila melanogaster*, the species of fruit fly studied by Morgan. We are going to follow the meiosis of a diploid nucleus in which one copy of chromosome 2 has alleles *A* and *B*, and the second has *a* and *b*. This situation is illustrated in *Figure 5.16*. Consider the two alternative scenarios:

1. ***A crossover does not occur between genes A and B.*** If this is what happens then two of the resulting

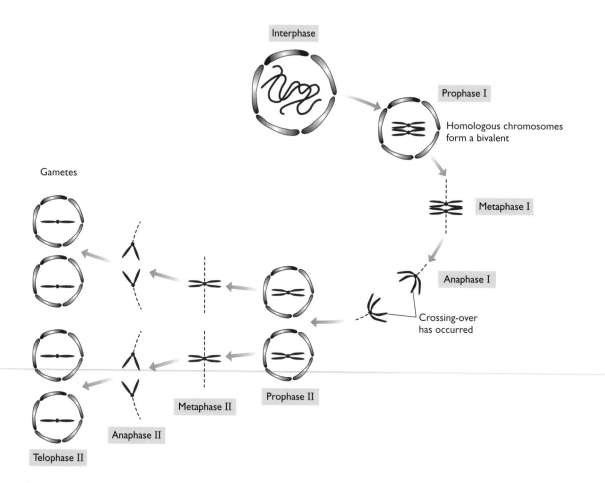

Figure 5.15 Meiosis.

The events involving one pair of homologous chromosomes are shown; one member of the pair is red, the other is blue. At the start of meiosis the chromosomes condense and each homologous pair lines up to form a bivalent. Within the bivalent, crossing-over might occur, involving breakage of chromosome arms and exchange of DNA. Meiosis then proceeds by a pair of mitotic nuclear divisions that result initially in two nuclei, each with two copies of each chromosome still attached at their centromeres, and finally in four nuclei, each with a single copy of each chromosome. These final products of meiosis, the gametes, are therefore haploid. The molecular basis of recombination is described in Section 14.3.

gametes will contain chromosome copies with alleles *A* and *B*, and the other two will contain *a* and *b*. In other words, two of the gametes have the **genotype** *AB* and two have the genotype *ab*.

2. ***A crossover does occur between genes A and B.*** This leads to segments of DNA containing gene B being exchanged between homologous chromosomes. The eventual result is that each gamete has a different genotype: 1 *AB*, 1 *aB*, 1 *Ab*, 1 *ab*.

Now think about what would happen if we looked at the results of meiosis in a hundred identical cells. If crossovers never occur then the resulting gametes will have the following genotypes:

200 *AB*

200 *ab*

This is complete linkage: genes A and B behave as a single unit during meiosis. But if (as is more likely) crossovers occur between A and B in some of the nuclei, then the allele pairs will not be inherited as single units. Let us say that crossovers occur during 40 of the 100 meioses. The following gametes will result:

160 *AB*

160 *ab*

40 *Ab*

40 *aB*

The linkage is not complete, it is only partial. As well as the two **parental** genotypes (*AB*, *ab*) we see gametes with **recombinant** genotypes (*Ab*, *aB*).

From partial linkage to genetic mapping

Once Morgan had understood how partial linkage could be explained by crossing-over during meiosis he was able to devise a way of mapping the relative positions of genes on a chromosome. In fact the most important work was done not by Morgan himself, but by an undergraduate in his laboratory, Arthur Sturtevant (Sturtevant, 1913). Sturtevant assumed that crossing-over was a random event, there being an equal chance of it occurring at any position along a pair of lined-up chromatids. If this assumption is correct then two genes that are close together will be separated by crossovers less frequently than two genes that are more distant from one another. Furthermore, the frequency with which the genes are unlinked by crossovers will be directly proportional to how far apart they are on their chromosome. The **recombination frequency** is therefore a measure of the distance between two genes. If you work out the recombination frequencies for different pairs of genes, you can construct a map of their relative positions on the chromosome (*Figure 5.17*).

It turns out that Sturtevant's assumption about the randomness of crossovers was not entirely justified. Comparisons between genetic maps and the actual positions of genes on DNA molecules, as revealed by physical mapping and DNA sequencing, have shown that some

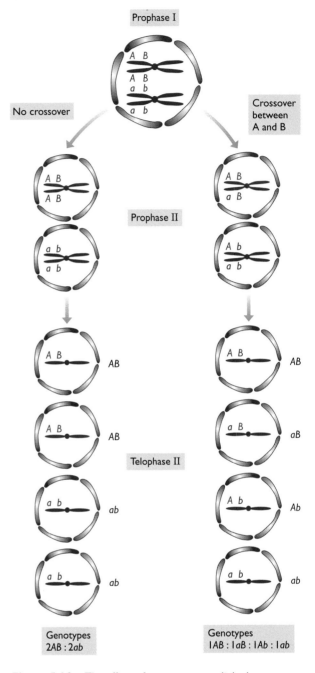

Figure 5.16 The effect of a crossover on linked genes.

The drawing shows a pair of homologous chromosomes, one red and the other blue. A and B are linked genes with alleles *A*, *a*, *B* and *b*. On the left is a meiosis with no crossover between A and B: two of the resulting gametes have the genotype *AB* and the other two are *ab*. On the right, a crossover occurs between A and B: the four gametes display all of the possible genotypes: *AB*, *aB*, *Ab* and *ab*.

regions of chromosomes, called **recombination hotspots**, are more likely to be involved in crossovers than others. This means that a genetic map distance does not necessarily indicate the physical distance between two markers

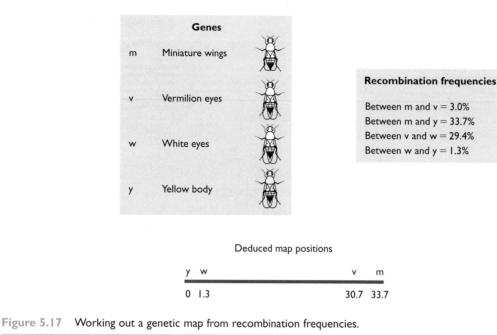

Figure 5.17 Working out a genetic map from recombination frequencies.

The example is taken from the original experiments carried out with fruit flies by Arthur Sturtevant. All four genes are on the X chromosome of the fruit fly. Recombination frequencies between the genes are shown, along with their deduced map positions.

(see *Figure 5.22*, page 147). Also, we now realize that a single chromatid can participate in more than one crossover at the same time, but that there are limitations on how close together these crossovers can be, leading to more inaccuracies in the mapping procedure. Despite these qualifications, **linkage analysis** usually makes correct deductions about gene order, and distance estimates are sufficiently accurate to generate genetic maps that are of value as frameworks for genome sequencing projects.

5.2.4 Linkage analysis with different types of organism

To see how linkage analysis is actually carried out, we need to consider three quite different situations:

- Linkage analysis with species such as fruit flies and mice, with which we can carry out planned breeding experiments;
- Linkage analysis with humans, with whom we cannot carry out planned experiments but instead make use of family pedigrees;
- Linkage analysis with bacteria, which do not undergo meiosis.

Linkage analysis when planned breeding experiments are possible

The first type of linkage analysis is the modern counterpart of the method developed by Morgan and his colleagues. The method is based on analysis of the progeny of experimental crosses set up between parents of known

genotypes and is, at least in theory, applicable to all eukaryotes. Ethical considerations preclude this approach in humans, and practical problems such as the length of the gestation period and the time taken for the newborn to reach maturity (and hence to participate in subsequent crosses) limit the effectiveness of the method with some animals and plants.

If we return to *Figure 5.16* we see that the key to gene mapping is being able to determine the genotypes of the gametes resulting from meiosis. In a few situations this is possible by directly examining the gametes. For example, the gametes produced by some microbial eukaryotes, including the yeast *Saccharomyces cerevisiae*, can be grown into colonies of haploid cells, whose genotypes can be determined by biochemical tests. Direct genotyping of gametes is also possible with higher eukaryotes if DNA markers are used, as PCR can be carried out with the DNA from individual spermatozoa, enabling RFLPs, SSLPs and SNPs to be typed. Unfortunately, sperm typing is laborious. Routine linkage analysis with higher eukaryotes is therefore carried out not by examining the gametes directly but by determining the genotypes of the diploid progeny that result from fusion of two gametes, one from each of a pair of parents. In other words, a genetic cross is performed.

The complication with a genetic cross is that the resulting diploid progeny are the product not of one meiosis but of two (one in each parent), and in most organisms crossover events are equally likely to occur during production of the male and female gametes. Somehow we have to be able to disentangle from the genotypes of the diploid progeny the crossover events that occurred in each of these two meioses. This means that the cross has to be set up with care. The standard procedure is to use a **test cross**. This is illustrated in *Figure 5.18*, Scenario 1, where we have set up a test cross to map the two genes we met earlier: gene A (alleles *A* and *a*) and gene B (alleles *B* and *b*), both on chromosome 2 of the fruit fly. The critical feature of a test cross is the genotypes of the two parents:

- One parent is a **double heterozygote**. This means that all four alleles are present in this parent: its genotype is *AB/ab*. This notation indicates that one pair of the homologous chromosomes has alleles *A* and *B*, and the other has *a* and *b*. Double heterozygotes can be obtained by crossing two pure-breeding strains, for example *AB/AB × ab/ab*.

- The second parent is a pure-breeding **double homozygote**. In this parent both homologous copies of chromosome 2 are the same: in the example shown in Scenario 1 both have alleles *a* and *b* and the genotype of the parent is *ab/ab*.

The double heterozygote has the same genotype as the cell whose meiosis we followed in *Figure 5.16*. Our objective is therefore to infer the genotypes of the gametes produced by this parent and to calculate the fraction that are recombinants. Note that all the gametes produced by the second parent (the double homozygote) will have the genotype *ab* regardless of whether they are parental or recombinant gametes. Alleles *a* and *b* are both recessive, so meiosis in this parent is, in effect, invisible when the genotypes of the progeny are examined. This means that, as shown in Scenario 1 in *Figure 5.18*, the genotypes of the diploid progeny can be unambiguously converted into the genotypes of the gametes from the double heterozygous parent. The test cross therefore enables us to make a direct examination of a single meiosis and hence to calculate a recombination frequency and map distance for the two genes being studied.

Just one additional point needs to be considered. If, as in Scenario 1 in *Figure 5.18*, gene markers displaying dominance and recessiveness are used, then the double

---

Box 5.2: Multipoint crosses

The power of linkage analysis is enhanced if more than two markers are followed in a single cross. Not only does this generate recombination frequencies more quickly, but it also enables the relative order of markers on a chromosome to be determined by simple inspection of the data. This is because two recombination events are required to unlink the central marker from the two outer markers in a series of three. Either of the two outer markers can be unlinked by just a single recombination.

A double recombination is less likely than a single one, so unlinking of the central marker will occur relatively

infrequently. An example of the data obtained from a three-point cross is shown in the table. A cross has been set up between a triple heterozygote (*ABC/abc*) and a triple homozygote (*abc/abc*). The most frequent progeny are those with one of the two parental genotypes, resulting from an absence of recombination events in the region containing the markers A, B and C. Two other classes of progeny are relatively frequent (51 and 63 progeny in the example shown). Each of these is presumed to arise from a single recombination. Inspection of their genotypes shows that in the first of these two classes, marker A has become unlinked from B and C, and in the second class, marker B has become unlinked from A and C. The implication is that A and B are the outer markers. This is confirmed by the number of progeny in which marker C has become unlinked from A and B. There are only two of these, showing that a double recombination is needed to produce this genotype. Marker C is therefore between A and B.

Single crossover

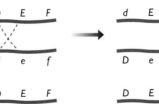

Double crossover

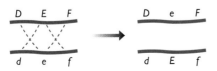

Genotypes of progeny	Number of progeny	Inferred recombination events
ABC/abc abc/abc	987	None (parental genotypes)
aBC/abc Abc/abc	51	One, between A and B + C
AbC/abc aBc/abc	63	One, between B and A + C
ABc/abc abC/abc	2	Two, one between C and A and one between C and B

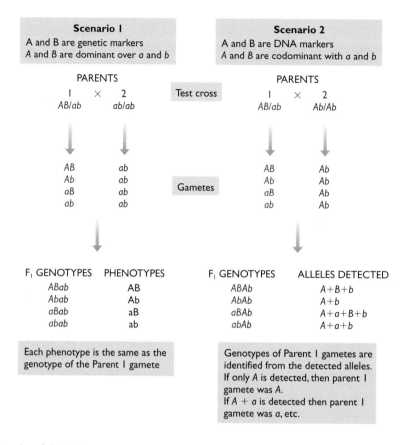

Figure 5.18 Two examples of the test cross.

In Scenario 1, A and B are genetic markers with alleles A, a, B and b. The resulting progeny are scored by examining their phenotypes. Because the double homozygous parent (Parent 2) has both recessive alleles – a and b – it effectively makes no contribution to the phenotypes of the progeny. The phenotype of each individual in the F_1 generation is therefore the same as the genotype of the gamete from Parent 1 that gave rise to that individual. In Scenario 2, A and B are DNA markers whose allele pairs are codominant. In this particular example, the double homozygous parent has the genotype Ab/Ab. The alleles present in each F_1 individual are directly detected, for example by PCR. These allele combinations enable the genotype of the Parent 1 gamete that gave rise to each individual to be deduced.

homozygous parent must have alleles for the two recessive phenotypes; however, if codominant DNA markers are used, then the double homozygous parent can have any combination of homozygous alleles (i.e. *AB/AB*, *Ab/Ab*, *aB/aB* and *ab/ab*). Scenario 2 in *Figure 5.18* shows the reason for this.

Gene mapping by human pedigree analysis

With humans it is of course impossible to pre-select the genotypes of parents and set up crosses designed specifically for mapping purposes. Instead, data for the calculation of recombination frequencies have to be obtained by examining the genotypes of the members of successive generations of existing families. This means that only limited data are available, and their interpretation is often difficult because a human marriage rarely results in a convenient test cross, and often the genotypes of one or more family members are unobtainable because those individuals are dead or unwilling to cooperate.

The problems are illustrated by *Figure 5.19*. In this example we are studying a genetic disease present in a family of two parents and six children. Genetic diseases are frequently used as gene markers in humans, the disease state being one allele and the healthy state being a second allele. The pedigree in *Figure 5.19A* shows us that the mother is affected by the disease, as are four of her children. We know from family accounts that the maternal grandmother also suffered from this disease, but both she and her husband – the maternal grandfather – are now dead. We can include them in the pedigree, with slashes indicating that they are dead, but we cannot obtain any further information on their genotypes. Our aim is to map the position of the gene for the genetic disease. For this purpose we are studying its linkage to a microsatellite marker M, four alleles of which – M_1, M_2, M_3 and M_4 – are present in the living family members. The question is, how many of the children are recombinants?

If we look at the genotypes of the six children we see that numbers 1, 3 and 4 have the disease allele and the

(A) The pedigree

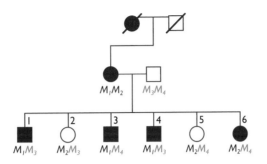

(B) Possible interpretations of the pedigree

MOTHER'S CHROMOSOMES

		Hypothesis 1	**Hypothesis 2**
		Disease M_1	Healthy M_1
		Healthy M_2	Disease M_2
CHILD 1	Disease M_1	Parental	Recombinant
CHILD 2	Healthy M_2	Parental	Recombinant
CHILD 3	Disease M_1	Parental	Recombinant
CHILD 4	Disease M_1	Parental	Recombinant
CHILD 5	Healthy M_2	Parental	Recombinant
CHILD 6	Disease M_2	Recombinant	Parental
Recombination frequency		$1/6 = 16.7\%$	$5/6 = 83.3\%$

(C) Resurrection of the maternal grandmother

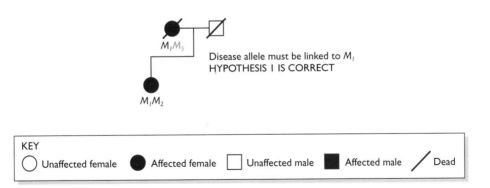

Disease allele must be linked to M_1
HYPOTHESIS 1 IS CORRECT

KEY
○ Unaffected female   ● Affected female   ☐ Unaffected male   ■ Affected male   ╱ Dead

Figure 5.19   An example of human pedigree analysis.

(A) The pedigree shows inheritance of a genetic disease in a family of two living parents and six children, with information about the maternal grandparents available from family records. The disease allele (closed symbols) is dominant over the healthy allele (open symbols). The objective is to determine the degree of linkage between the disease gene and the microsatellite M by typing the alleles for this microsatellite (M_1, M_2, etc.) in living members of the family. (B) The pedigree can be interpreted in two different ways: Hypothesis 1 gives a low recombination frequency and indicates that the disease gene is tightly linked to microsatellite M; Hypothesis 2 suggests that the gene and microsatellite are much less closely linked. In (C), the issue is resolved by the reappearance of the maternal grandmother, whose microsatellite genotype is consistent only with Hypothesis 1. See the text for more details.

microsatellite allele M_1. Numbers 2 and 5 have the healthy allele and M_2. We can therefore construct two alternative hypotheses. The first is that the two copies of the relevant homologous chromosomes in the mother have the genotypes *Disease-M$_1$* and *Healthy-M$_2$*; therefore children 1, 2, 3, 4 and 5 have parental genotypes and child 6 is the one and only recombinant (*Figure 5.19B*). This would suggest that the disease gene and the microsatellite are relatively closely linked and that crossovers between them occur infrequently. The alternative hypothesis is that the mother's chromosomes have the genotypes *Healthy-M$_1$* and *Disease-M$_2$*; this would mean that children 1–5 are recombinants, and child 6 has the parental genotype. This would mean that the gene and microsatellite are relatively far apart on the chromosome. We cannot determine which of these hypotheses is correct: the data are frustratingly ambiguous.

The most satisfying solution to the problem posed by the pedigree in *Figure 5.19* would be to know the genotype of the grandmother. Let us pretend that this is a soap opera family and that the grandmother is not really dead. To everyone's surprise she reappears just in time to save the declining audience ratings. Her genotype for microsatellite M turns out to be $M_1 M_5$ (*Figure 5.19C*). This tells us that the disease allele is on the same chromosome as M_1. We can therefore conclude with certainty that Hypothesis 1 is correct and that only child 6 is a recombinant.

Resurrection of key individuals is not usually an option open to real-life geneticists, although DNA can be obtained from old pathology specimens such as slides and Guthrie cards. Imperfect pedigrees are analyzed statistically, using a measure called the **lod score** (Morton, 1955). This stands for <u>lo</u>garithm of the <u>o</u>dds that the genes are linked and is used primarily to determine if the two markers being studied lie on the same chromosome, in other words if the genes are linked or not. If the lod analysis establishes linkage then it can also provide a measure of the most likely recombination frequency. Ideally the available data will derive from more than one pedigree, increasing the confidence in the result. The analysis is less ambiguous for families with larger numbers of children, and, as we saw in *Figure 5.19*, it is important that the members of at least three generations can be genotyped. For this reason, family collections have been established, such as the one maintained by the Centre d'Études du Polymorphisme Humaine (CEPH) in Paris (Dausset *et al.*, 1990). The CEPH collection contains cultured cell lines from families in which all four grandparents as well as at least eight second-generation children could be sampled. This collection is available for DNA marker mapping by any researcher who agrees to submit the resulting data to the central CEPH database.

Genetic mapping in bacteria

The final type of genetic mapping that we must consider is the strategy used with bacteria. The main difficulty that geneticists faced when trying to develop genetic mapping

techniques for bacteria is that these organisms are normally haploid, and so do not undergo meiosis. Some other way therefore had to be devised to induce crossovers between homologous segments of bacterial DNA. The answer was to make use of three natural methods that exist for transferring pieces of DNA from one bacterium to another (*Figure 5.20*):

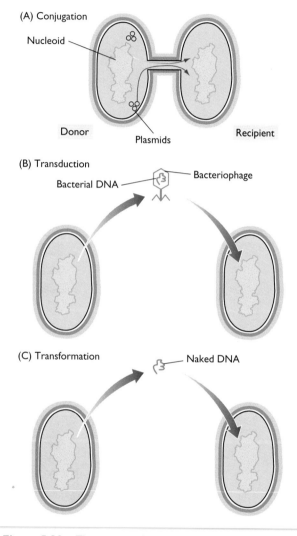

Figure 5.20 Three ways of achieving DNA transfer between bacteria.

(A) Conjugation can result in transfer of chromosomal or plasmid DNA from the donor bacterium to the recipient. Conjugation involves physical contact between the two bacteria, with transfer thought to occur through a narrow tube called the **pilus**. (B) Transduction is the transfer of a small segment of the donor cell's DNA via a bacteriophage. (C) Transformation is similar to transduction but 'naked' DNA is transferred. The events illustrated in (B) and (C) are often accompanied by death of the donor cell. In (B), death occurs when the bacteriophages emerge from the donor cell; in (C), release of DNA from the donor cell is usually a consequence of the cell's death through natural causes.

- In **conjugation** two bacteria come into physical contact and one bacterium (the donor) transfers DNA to the second bacterium (the recipient). The transferred DNA can be a copy of some or possibly all of the donor cell's chromosome, or it could be a segment of chromosomal DNA – up to 1 Mb in length – integrated in a plasmid (Section 2.1.2). The latter is called **episome transfer**.
- **Transduction** involves transfer of a small segment of DNA – up to 50 kb or so – from donor to recipient via a bacteriophage.
- In **transformation** the recipient cell takes up from its environment a fragment of DNA, rarely longer than 50 kb, released from a donor cell.

After transfer, a double crossover must occur so that the DNA from the donor bacterium is integrated into the recipient cell's chromosome (*Figure 5.21A*). If this does not occur then the transferred DNA is lost when the recipient cell divides. The only exception is after episome transfer, plasmids being able to propagate independently of the host chromosome.

Biochemical markers are invariably used, the dominant or **wild-type** phenotype being possession of a biochemical characteristic (e.g. ability to synthesize tryptophan) and the recessive phenotype being the complementary characteristic (e.g. inability to synthesize tryptophan). The gene transfer is usually set up between a donor strain that possesses the wild-type alleles and a recipient with the recessive alleles, transfer into the recipient strain being monitored by looking for acquisition of the biochemical function(s) specified by the genes being studied. The precise details of the mapping procedure depend on the type of gene transfer that is being used. In conjugation mapping the donor DNA is transferred as a continuous thread into the recipient, and gene positions are mapped by timing the entry of the wild-type alleles into the recipient (*Figure 5.21B*). Transduction and transformation mapping enable genes that are relatively close together to be mapped, because the transferred DNA segment is short (< 50 kb), so the probability of two genes being transferred together depends on how close together they are on the bacterial chromosome (*Figure 5.21C*).

5.3 Physical Mapping

A map generated by genetic techniques is rarely sufficient for directing the sequencing phase of a genome project. This is for two reasons:

- *The resolution of a genetic map depends on the number of crossovers that have been scored.* This is not a major problem for microorganisms because these can be obtained in huge numbers, enabling many crossovers to be studied, resulting in a highly detailed genetic map in which the markers are just a few kb apart. For example, when the *Escherichia coli* genome sequencing project began in 1990, the latest

genetic map for this organism comprised over 1400 markers, an average of one per 3.3 kb. This was sufficiently detailed to direct the sequencing program without the need for extensive physical mapping. Similarly, the *Saccharomyces cerevisiae* project was supported by a fine-scale genetic map (approximately 1150 genetic markers, on average one per 10 kb). The problem with humans and most other eukaryotes is that it is simply not possible to obtain large numbers of progeny, so relatively few meioses can be studied and the resolving power of linkage analysis is restricted. This means that genes that are several tens of kb apart may appear at the same position on the genetic map.

- *Genetic maps have limited accuracy.* We touched on this point in Section 5.2.3 when we assessed Sturtevant's assumption that crossovers occur at random along chromosomes. This assumption is only partly correct because the presence of recombination hotspots means that crossovers are more likely to occur at some points rather than at others. The effect that this can have on the accuracy of a genetic map was illustrated in 1992 when the complete sequence for *S. cerevisiae* chromosome III was published (Oliver *et al.*, 1992), enabling the first direct comparison to be made between a genetic map and the actual positions of markers as shown by DNA sequencing (*Figure 5.22*). There were considerable discrepancies, even to the extent that one pair of genes had been ordered incorrectly by genetic analysis. Bear in mind that *S. cerevisiae* is one of the two eukaryotes (fruit fly is the second) whose genomes have been subjected to intensive genetic mapping. If the yeast genetic map is inaccurate then how precise are the genetic maps of organisms subjected to less detailed analysis?

These two limitations of genetic mapping mean that for most eukaryotes a genetic map must be checked and supplemented by alternative mapping procedures before large-scale DNA sequencing begins. A plethora of physical mapping techniques has been developed to address this problem, the most important being:

- **Restriction mapping**, which locates the relative positions on a DNA molecule of the recognition sequences for restriction endonucleases;
- **Fluorescent *in situ* hybridization (FISH)**, in which marker locations are mapped by hybridizing a probe containing the marker to intact chromosomes;
- **Sequence tagged site (STS) mapping**, in which the positions of short sequences are mapped by PCR and/or hybridization analysis of genome fragments.

5.3.1 Restriction mapping

Genetic mapping using RFLPs as DNA markers can locate the positions of polymorphic restriction sites within a genome (Section 5.2.2), but very few of the

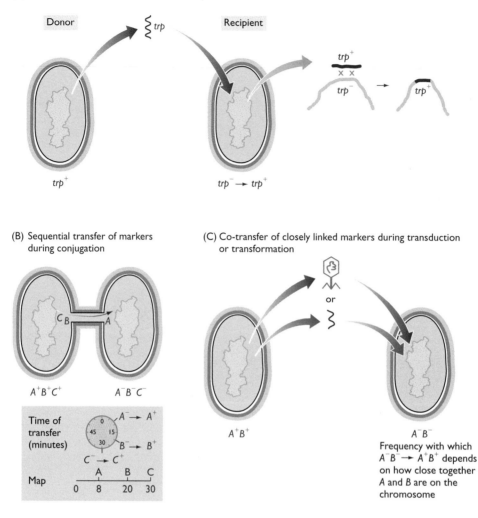

(A) Transfer of DNA between donor and recipient bacteria

Donor   trp   Recipient

trp⁺

trp⁺
x x
trp⁻ → trp⁺

trp⁺   trp⁻ → trp⁺

(B) Sequential transfer of markers during conjugation

(C) Co-transfer of closely linked markers during transduction or transformation

A⁺B⁺C⁺   A⁻B⁻C⁻

or

Time of transfer (minutes)

A⁻ → A⁺
B⁻ → B⁺
C⁻ → C⁺

Map   A   B   C
       0   8   20   30

A⁺B⁺

A⁻B⁻
Frequency with which A⁻B⁻ → A⁺B⁺ depends on how close together A and B are on the chromosome

Figure 5.21   The basis of gene mapping in bacteria.

(A) Transfer of a functional gene for tryptophan biosynthesis from a wild-type bacterium (genotype described as *trp⁺*) to a recipient that lacks a functional copy of this gene (*trp⁻*). The recipient is called a tryptophan **auxotroph** (the word used to describe a mutant bacterium that can survive only if provided with a nutrient – in this case, tryptophan – not required by the wild type; see Section 14.1.2). After transfer, two crossovers (shown as green crosses) are needed to integrate the transferred gene into the recipient cell's chromosome, converting the recipient from *trp⁻* to *trp⁺*. (B) During conjugation, DNA is transferred from donor to recipient in the same way that a string is pulled through a tube. The relative positions of markers on the DNA molecule can therefore be mapped by determining the times at which the markers appear in the recipient cell. In the example shown, markers *A*, *B* and *C* are transferred 8, 20 and 30 minutes after the beginning of conjugation, respectively. The entire *Escherichia coli* chromosome takes approximately 100 minutes to transfer. (C) To be co-transferred during transduction and transformation, two or more markers must be closely linked, because these processes usually result in less than 50 kb of DNA being passed from donor to recipient. Transduction and transformation mapping are used to determine the relative positions of markers that are too close together to be mapped precisely by conjugation analysis. For more details on bacterial gene mapping see Freifelder (1987).

restriction sites in a genome are polymorphic, so many sites are not mapped by this technique (*Figure 5.23*). Could we increase the marker density on a genome map by using an alternative method to locate the positions of some of the non-polymorphic restriction sites? This is

what restriction mapping achieves, although in practice the technique has limitations which mean that it is applicable only to relatively small DNA molecules. We will look first at the technique and then consider its relevance to genome mapping.

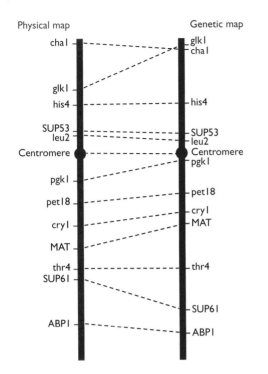

Figure 5.22 Comparison between the genetic and physical maps of *Saccharomyces cerevisiae* chromosome III.

The comparison shows the discrepancies between the genetic and physical maps, the latter determined by DNA sequencing. Note that the order of the upper two markers (glk1 and cha1) is incorrect on the genetic map, and that there are also differences in the relative positioning of other pairs of markers. Reprinted with permission from Oliver SG *et al.*, *Nature*, **357**, 38–46. Copyright 1992 Macmillan Magazines Limited.

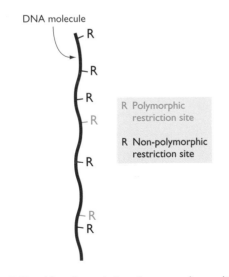

Figure 5.23 Not all restriction sites are polymorphic.

The basic methodology for restriction mapping

The simplest way to construct a restriction map is to compare the fragment sizes produced when a DNA molecule is digested with two different restriction enzymes that recognize different target sequences. An example using the restriction enzymes *Eco*RI and *Bam*HI is shown in *Figure 5.24*. First, the DNA molecule is digested with just one of the enzymes and the sizes of the resulting fragments are measured by agarose gel electrophoresis. Next, the molecule is digested with the second enzyme and the resulting fragments again sized in an agarose gel. The results so far enable the number of restriction sites for each enzyme to be worked out, but do not allow their relative positions to be determined. Additional information is therefore obtained by cutting the DNA molecule with both enzymes together. In the example shown in *Figure 5.24*, this **double restriction** enables three of the sites to be mapped. However, a problem arises with the larger *Eco*RI fragment because this contains two *Bam*HI sites and there are two alternative possibilities for the map location of the outer one of these. The problem is solved by going back to the original DNA molecule and treating it again with *Bam*HI on its own, but this time preventing the digestion from going to completion by, for example, incubating the reaction for only a short time or using a suboptimal incubation temperature. This is called a **partial restriction** and leads to a more complex set of products, the complete restriction products now being supplemented with partially restricted fragments that still contain one or more uncut *Bam*HI sites. In the example shown in *Figure 5.24*, the size of one of the partial restriction fragments is diagnostic and the correct map can be identified.

A partial restriction usually gives the information needed to complete a map, but if there are many restriction sites then this type of analysis becomes unwieldy, simply because there are so many different fragments to consider. An alternative strategy is simpler because it enables the majority of the fragments to be ignored. This is achieved by attaching a radioactive or other type of marker to each end of the starting DNA molecule before carrying out the partial digestion. The result is that many of the partial restriction products become 'invisible' because they do not contain an end-fragment and so do not show up when the agarose gel is screened for labeled products. The sizes of the partial restriction products that are visible enable unmapped sites to be positioned relative to the ends of the starting molecule.

The scale of restriction mapping is limited by the sizes of the restriction fragments

Restriction maps are easy to generate if there are relatively few cut sites for the enzymes being used. However, as the number of cut sites increases, so also do the numbers of single, double and partial restriction products whose sizes must be determined and compared in order for the map to be constructed. Computer analysis can be brought into play but problems still eventually arise. A

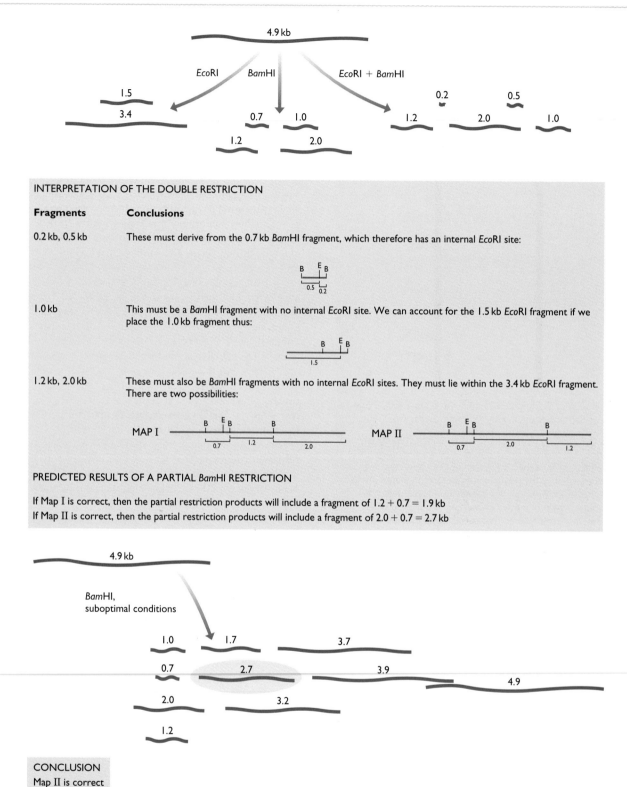

Figure 5.24 Restriction mapping.

The objective is to map the *Eco*RI (E) and *Bam*HI (B) sites in a linear DNA molecule of 4.9 kb. The results of single and double restrictions are shown at the top. The sizes of the fragments given after double restriction enable two alternative maps to be constructed, as explained in the central panel, the unresolved issue being the position of one of the three *Bam*HI sites. The two maps are tested by a partial *Bam*HI restriction (bottom), which shows that Map II is the correct one.

stage will be reached when a digest contains so many fragments that individual bands merge on the agarose gel, increasing the chances of one or more fragments being measured incorrectly or missed out entirely. If several fragments have similar sizes then even if they can all be identified, it may not be possible to assemble them into an unambiguous map.

Restriction mapping is therefore more applicable to small rather than large molecules, with the upper limit for the technique depending on the frequency of the restriction sites in the molecule being mapped. In practice, if a DNA molecule is less than 50 kb in length it is usually possible to construct an unambiguous restriction map for a selection of enzymes with six-nucleotide recognition sequences. Fifty kb is of course way below the minimum size for bacterial or eukaryotic chromosomes, although it does cover a few viral and organelle genomes, and whole-genome restriction maps have indeed been important in directing sequencing projects with these small molecules. Restriction maps are equally useful after bacterial or eukaryotic genomic DNA has been cloned, if the cloned fragments are less than 50 kb, because a detailed restriction map can then be built up as a preliminary to sequencing the cloned region. This is an important application of restriction mapping in sequencing projects with large genomes, but is there any possibility of using restriction analysis for the more general mapping of entire genomes larger than 50 kb?

The answer is a qualified 'yes', because the limitations of restriction mapping can be eased slightly by choosing enzymes expected to have infrequent cut sites in the target DNA molecule. These 'rare cutters' fall into two categories:

- *Enzymes with seven- or eight-nucleotide recognition sequences.* A few restriction enzymes cut at seven- or eight-nucleotide recognition sequences. Examples are *Sap*I (5'–GCTCTTC–3') and *Sgf*I (5'–GCGATCGC–3'). The seven-nucleotide enzymes would be expected, on average, to cut a DNA molecule with a GC content of 50% once every $4^7 = 16\,384$ bp. The eight-nucleotide enzymes should cut once every $4^8 = 65\,536$ bp. These figures compare with $4^6 = 4096$ bp for six-nucleotide enzymes such as *Bam*HI and *Eco*RI. Seven- and eight-nucleotide cutters are often used in restriction mapping of large molecules but the approach is not as useful as it might be simply because not many of these enzymes are known.
- *Enzymes whose recognition sequences contain motifs that are rare in the target DNA.* Genomic DNA molecules do not have random sequences and some are significantly deficient in certain motifs. For example, the sequence 5'–CG–3' is rare in human DNA because human cells possess an enzyme that adds a methyl group to carbon 5 of the C nucleotide in this sequence. The resulting 5-methylcytosine is unstable and tends to undergo deamination to give thymine (*Figure 5.25*). The consequence is that during human evolution

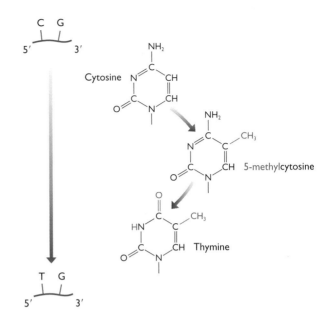

Figure 5.25 The sequence 5'–CG–3' is rare in human DNA because of methylation of the C, followed by deamination to give T.

many of the 5'–CG–3' sequences that were originally in our genome have become converted to 5'–TG–3'. Restriction enzymes that recognize a site containing 5'–CG–3' therefore cut human DNA relatively infrequently. Examples are *Sma*I (5'–CCCGGG–3'), which cuts human DNA on average once every 78 kb, and *Bss*HII (5'–GCGCGC–3') which cuts once every 390 kb. Note that *Not*I, an eight-nucleotide cutter, also targets 5'–CG–3' sequences (recognition sequence 5'–GCGGCCGC–3') and cuts human DNA very rarely – approximately once every 10 Mb.

The potential of restriction mapping is therefore increased by using rare cutters. It is still not possible to construct restriction maps of the genomes of animals and plants, but it is feasible to use the technique with large cloned fragments, and the smaller DNA molecules of prokaryotes and lower eukaryotes such as yeast and fungi.

If a rare cutter is used then it may be necessary to employ a special type of agarose gel electrophoresis to study the resulting restriction fragments. This is because the relationship between the length of a DNA molecule and its migration rate in an electrophoresis gel is not linear, the resolution decreasing as the molecules get longer (*Figure 5.26A*). This means that it is not possible to separate molecules more than about 50 kb in length because all of these longer molecules run as a single slowly migrating band in a standard agarose gel. To separate them it is necessary to replace the linear electric field used in conventional gel electrophoresis with a more complex field. An example is provided by **orthogonal field alternation gel electrophoresis (OFAGE)**, in which the electric field alternates between two pairs of electrodes, each

(A) Standard agarose gel electrophoresis

Poor separation of DNA molecules >50 kb

(B) Orthogonal field alternation gel electrophoresis (OFAGE)

Figure 5.26   Conventional and non-conventional agarose gel electrophoresis.

(A) In standard agarose gel electrophoresis the electrodes are placed at either end of the gel and the DNA molecules migrate directly towards the positive electrode. Molecules longer than about 50 kb cannot be separated from one another in this way. (B) In OFAGE, the electrodes are placed at the corners of the gel, with the field pulsing between the A pair and the B pair. OFAGE enables molecules up to 2 Mb to be separated.

positioned at an angle of 45° to the length of the gel (*Figure 5.26B*). The DNA molecules still move down through the gel, but each change in the field forces the molecules to realign. Shorter molecules realign more

quickly than longer ones and so migrate more rapidly through the gel. The overall result is that molecules much longer than those separated by conventional gel electrophoresis can be resolved. Related techniques include **CHEF** (**contour clamped homogeneous electric fields**) and **FIGE** (**field inversion gel electrophoresis**).

Direct examination of DNA molecules for restriction sites

It is also possible to use methods other than electrophoresis to map restriction sites in DNA molecules. With the technique called **optical mapping** (Schwartz *et al.*, 1993), restriction sites are directly located by looking at the cut DNA molecules with a microscope (*Figure 5.27*). The DNA must first be attached to a glass slide in such a way that the individual molecules become stretched out, rather than clumped together in a mass. There are two ways of doing this: **gel stretching** and **molecular combing**. To prepare gel-stretched DNA fibers (Schwartz *et al.*, 1993), chromosomal DNA is suspended in molten agarose and placed on a microscope slide. As the gel cools and solidifies, the DNA molecules become extended (*Figure 5.28A*). To utilize gel stretching in optical mapping, the microscope slide onto which the molten agarose is placed is first coated with a restriction enzyme. The enzyme is inactive at this stage because there are no magnesium ions, which the enzyme needs in order to function. Once the gel has solidified it is washed with a solution containing magnesium chloride, which activates the restriction enzyme. A fluorescent dye is added, such as DAPI (4,6-diamino-2-phenylindole

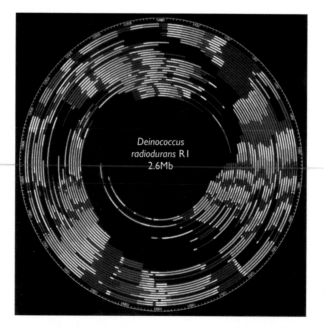

Figure 5.27   Optical mapping.

The image shows a 2.4-Mb segment of the *Deinococcus radiodurans* genome after treatment with the restriction endonuclease *Nhe*I. The positions of the cut sites are visible as gaps in the white strand of DNA. Reprinted with permission from Lin *et al.*, *Science*, **285**, 1558–1562. Copyright 1999 American Association for the Advancement of Science.

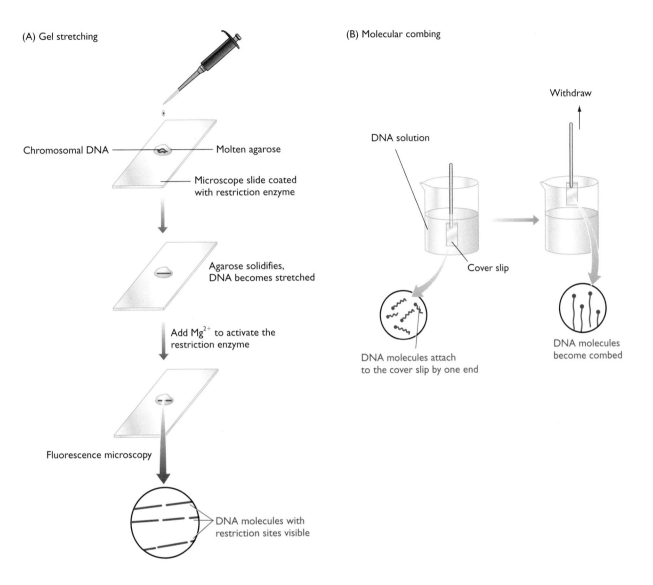

Figure 5.28   Gel stretching and molecular combing.

(A) To carry out gel stretching, molten agarose containing chromosomal DNA molecules is pipetted onto a microscope slide coated with a restriction enzyme. As the gel solidifies, the DNA molecules become stretched. It is not understood why this happens but it is thought that fluid movement on the glass surface during gelation might be responsible. Addition of magnesium chloride activates the restriction enzyme, which cuts the DNA molecules. As the molecules gradually coil up, the gaps representing the cut sites become visible. (B) In molecular combing, a cover slip is dipped into a solution of DNA. The DNA molecules attach to the cover slip by their ends, and the slip is withdrawn from the solution at a rate of 0.3 mm s^{-1}, which produces a 'comb' of parallel molecules.

dihydrochloride), which stains the DNA so that the fibers can be seen when the slide is examined with a high-power fluorescence microscope. The restriction sites in the extended molecules gradually become gaps as the degree of fiber extension is reduced by the natural springiness of the DNA, enabling the relative positions of the cuts to be recorded.

In molecular combing (Michalet *et al.*, 1997), the DNA fibers are prepared by dipping a silicone-coated cover slip into a solution of DNA, leaving it for 5 minutes (during which time the DNA molecules attach to the cover slip by

their ends), and then removing the slip at a constant speed of 0.3 mm s^{-1} (*Figure 5.28B*). The force required to pull the DNA molecules through the meniscus causes them to line up. Once in the air, the surface of the cover slip dries, retaining the DNA molecules as an array of parallel fibers.

Optical mapping was first applied to large DNA fragments cloned in YAC and BAC vectors (Section 4.2.1). More recently, the feasibility of using this technique with genomic DNA has been proven with studies of a 1-Mb chromosome of the malaria parasite *Plasmodium falciparum* (Jing *et al.*, 1999), and the two chromosomes and single megaplasmid of the bacterium *Deinococcus radiodurans* (Lin *et al.*, 1999; see *Table 2.9*, page 53).

5.3.2 Fluorescent *in situ* hybridization (FISH)

The optical mapping method described above provides a link to the second type of physical mapping procedure that we will consider – FISH (Heiskanen *et al.*, 1996). As in optical mapping, FISH enables the position of a marker on a chromosome or extended DNA molecule to be directly visualized. In optical mapping the marker is a restriction site and it is visualized as a gap in an extended DNA fiber. In FISH, the marker is a DNA sequence that is visualized by hybridization with a fluorescent probe.

In situ *hybridization with radioactive or fluorescent probes*

In situ hybridization is a version of hybridization analysis (Section 4.1.2) in which an intact chromosome is examined by probing it with a labeled DNA molecule. The position on the chromosome at which hybridization occurs provides information about the map location of the DNA sequence used as the probe (*Figure 5.29*). For the method to work, the DNA in the chromosome must be made single stranded ('denatured') by breaking the base pairs that hold the double helix together. Only then will the chromosomal DNA be able to hybridize with the probe. The standard method for denaturing chromosomal DNA without destroying the morphology of the chromosome is to dry the preparation onto a glass microscope slide and then treat with formamide.

In the early versions of *in situ* hybridization the probe was radioactively labeled but this procedure was unsatisfactory because it is difficult to achieve both sensitivity and resolution with a radioactive label, two critical requirements for successful *in situ* hybridization. Sensitivity requires that the radioactive label has a high emission energy (an example of such a radiolabel is ^{32}P), but if the radiolabel has a high emission energy then it scatters its signal and so gives poor resolution. High resolution is possible if a radiolabel with low emission energy, such as ^{3}H, is used, but these have such low sensitivity that lengthy exposures are needed, leading to a high background and difficulties in discerning the genuine signal.

These problems were solved in the late 1980s by the development of non-radioactive fluorescent DNA labels. These labels combine high sensitivity with high resolution and are ideal for *in situ* hybridization. Fluorolabels with different colored emissions have been designed, making it possible to hybridize a number of different probes to a single chromosome and distinguish their individual hybridization signals, thus enabling the relative positions of the probe sequences to be mapped. To maximize sensitivity, the probes must be labeled as heavily as possible, which in the past has meant that they must be quite lengthy DNA molecules – usually cloned DNA fragments of at least 40 kb. This requirement is less important

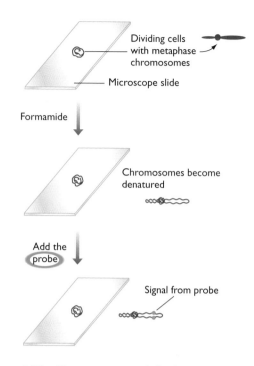

Figure 5.29   Fluorescent *in situ* hybridization.

A sample of dividing cells is dried onto a microscope slide and treated with formamide so that the chromosomes become denatured but do not lose their characteristic metaphase morphologies (see Section 2.2.1). The position at which the probe hybridizes to the chromosomal DNA is visualized by detecting the fluorescent signal emitted by the labeled DNA.

now that techniques for achieving heavy labeling with shorter molecules have been developed. As far as the construction of a physical map is concerned, a cloned DNA fragment can be looked upon as simply another type of marker, although in practice the use of clones as markers adds a second dimension because the cloned DNA is the material from which the DNA sequence is determined. Mapping the positions of clones therefore provides a direct link between a genome map and its DNA sequence.

If the probe is a long fragment of DNA then one potential problem, at least with higher eukaryotes, is that it is likely to contain examples of repetitive DNA sequences (Section 2.4) and so may hybridize to many chromosomal positions, not just the specific point to which it is perfectly matched. To reduce this non-specific hybridization, the probe, before use, is mixed with unlabeled DNA from the organism being studied. This DNA can simply be total nuclear DNA (i.e. representing the entire genome) but it is better if a fraction enriched for repeat sequences is used. The idea is that the unlabeled DNA hybridizes to the repetitive DNA sequences in the probe, blocking these so that the subsequent *in situ* hybridization is driven wholly by the unique sequences (Lichter *et al.*, 1990). Non-specific hybridization is therefore reduced or eliminated entirely (*Figure 5.30*).

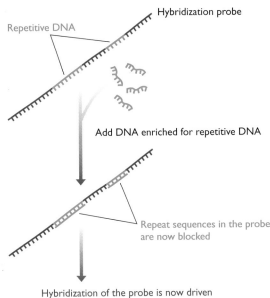

Hybridization probe

Repetitive DNA

Add DNA enriched for repetitive DNA

Repeat sequences in the probe
are now blocked

Hybridization of the probe is now driven
by the unique sequences

Figure 5.30 A method for blocking repetitive DNA sequences in a hybridization probe.

In this example the probe molecule contains two genome-wide repeat sequences (shown in green). If these sequences are not blocked then the probe will hybridize non-specifically to any copies of these genome-wide repeats in the target DNA. To block the repeat sequences, the probe is prehybridized with a DNA fraction enriched for repetitive DNA.

FISH in action

FISH was originally used with metaphase chromosomes (Section 2.2.1). These chromosomes, prepared from nuclei that are undergoing division, are highly condensed and each chromosome in a set takes up a recognizable appearance, characterized by the position of its centromere and the banding pattern that emerges after the chromosome preparation is stained (see *Figure 2.8*, page 40). With metaphase chromosomes, a fluorescent signal obtained by FISH is mapped by measuring its position relative to the end of the short arm of the chromosome (the **FLpter value**). A disadvantage is that the highly condensed nature of metaphase chromosomes means that only low-resolution mapping is possible, two markers having to be at least 1 Mb apart to be resolved as separate hybridization signals (Trask *et al.*, 1991). This degree of resolution is insufficient for the construction of useful chromosome maps, and the main application of metaphase FISH has been in determining the chromosome on which a new marker is located, and providing a rough idea of its map position, as a preliminary to finer scale mapping by other methods.

For several years these 'other methods' did not involve any form of FISH, but since 1995 a range of higher resolution FISH techniques has been developed. With these techniques, higher resolution is achieved by changing the

nature of the chromosomal preparation being studied. If metaphase chromosomes are too condensed for fine-scale mapping then we must use chromosomes that are more extended. There are two ways of doing this (Heiskanen *et al.*, 1996):

- **Mechanically stretched chromosomes** can be obtained by modifying the preparative method used to isolate chromosomes from metaphase nuclei. The inclusion of a centrifugation step generates shear forces which can result in the chromosomes becoming stretched to up to 20 times their normal length. Individual chromosomes are still recognizable and FISH signals can be mapped in the same way as with normal metaphase chromosomes. The resolution is significantly improved and markers that are 200–300 kb apart can be distinguished.

- **Non-metaphase chromosomes** can be used because it is only during metaphase that chromosomes are highly condensed: at other stages of the cell cycle the chromosomes are naturally unpacked. Attempts have been made to use prophase nuclei (see *Figure 5.14*, page 137) because in these the chromosomes are still sufficiently condensed for individual ones to be identified. In practice, however, these preparations provide no advantage over mechanically stretched chromosomes. **Interphase** chromosomes are more useful because this stage of the cell cycle (between nuclear divisions) is when the chromosomes are most unpacked. Resolution down to 25 kb is possible, but chromosome morphology is lost so there are no external reference points against which to map the position of the probe. This technique is therefore used after preliminary map information has been obtained, usually as a means of determining the order of a series of markers in a small region of a chromosome.

Interphase chromosomes contain the most unpacked of all cellular DNA molecules. To improve the resolution of FISH to better than 25 kb it is therefore necessary to abandon intact chromosomes and instead use purified DNA. This approach, called **fiber-FISH**, makes use of DNA prepared by gel stretching or molecular combing (see *Figure 5.28*) and can distinguish markers that are less than 10 kb apart.

5.3.3 Sequence tagged site (STS) mapping

To generate a detailed physical map of a large genome we need, ideally, a high-resolution mapping procedure that is rapid and not technically demanding. Neither of the two techniques that we have considered so far – restriction mapping and FISH – meets these requirements. Restriction mapping is rapid, easy, and provides detailed information, but it cannot be applied to large genomes. FISH can be applied to large genomes, and modified versions such as fiber-FISH can give high-resolution data, but FISH is difficult to carry out and data accumulation is slow, map positions for no more than three or four markers being obtained in a single experiment. If detailed physical maps are to become a reality then we need a more powerful technique.

At present the most powerful physical mapping technique, and the one that has been responsible for generation of the most detailed maps of large genomes, is STS mapping. A **sequence tagged site** or **STS** is simply a short DNA sequence, generally between 100 and 500 bp in length, that is easily recognizable and occurs only once in the chromosome or genome being studied. To map a set of STSs, a collection of overlapping DNA fragments from a single chromosome or from the entire genome is needed. In the example shown in *Figure 5.31* a fragment collection has been prepared from a single chromosome, with each point along the chromosome represented on average five times in the collection. The data from which the map will be derived are obtained by determining which fragments contain which STSs. This can be done by hybridization analysis but PCR is generally used because it is quicker and has proven to be more amenable to automation. The chances of two STSs being present on the same fragment will, of course, depend on how close together they are in the genome. If they are very close then there is a good chance that they will always be on the same fragment; if they are further apart then sometimes they will be on the same fragment and sometimes they will not (*Figure 5.31*). The data can therefore be used to calculate the distance between two markers, in a manner analogous to the way in which map distances are determined by linkage analysis (Section 5.2.3). Remember that in linkage analysis a map distance is calculated from the frequency at which crossovers occur between two markers. STS mapping is essentially the same, except that each map distance is based on the frequency at which *breaks* occur between two markers.

The description of STS mapping given above leaves out some critical questions: What exactly is an STS? How is the DNA fragment collection obtained?

Any unique DNA sequence can be used as an STS

To qualify as an STS, a DNA sequence must satisfy two criteria. The first is that its sequence must be known, so that a PCR assay can be set up to test for the presence or absence of the STS on different DNA fragments. The second requirement is that the STS must have a unique location in the chromosome being studied, or in the genome as a whole if the DNA fragment set covers the entire genome. If the STS sequence occurs at more than one position then the mapping data will be ambiguous. Care must therefore be taken to ensure that STSs do not include sequences found in repetitive DNA.

These are easy criteria to satisfy and STSs can be obtained in many ways, the most common sources being **expressed sequence tags (ESTs)**, SSLPs, and **random genomic sequences**.

■ *Expressed sequence tags (ESTs).* These are short sequences obtained by analysis of cDNA clones (Marra *et al.*, 1998). Complementary DNA is prepared by converting an mRNA preparation into double-stranded DNA (*Figure 5.32*). Because the mRNA in a cell is derived from protein-coding genes, cDNAs and the ESTs obtained from them represent the genes that were being expressed in the cell from which the mRNA was prepared. ESTs are looked upon as a rapid means of gaining access to the sequences of important genes, and they are valuable even if their sequences are incomplete. An EST can

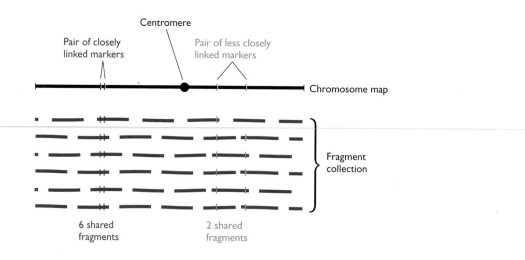

Figure 5.31 A fragment collection suitable for STS mapping.

The fragments span the entire length of a chromosome, with each point on the chromosome present in an average of five fragments. The two blue markers are close together on the chromosome map and there is a high probability that they will be found on the same fragment. The two green markers are more distant from one another and so are less likely to be found on the same fragment.

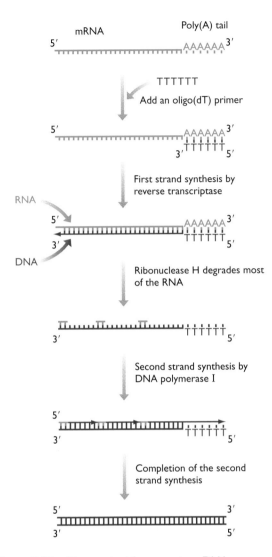

mRNA   Poly(A) tail

5′ ——————————————————— AAAAAA 3′

↓ TTTTTT
Add an oligo(dT) primer

5′ ——————————————————— AAAAAA 3′
                      ↑↑↑↑↑↑
                    3′ TTTTTT 5′

First strand synthesis by
reverse transcriptase

RNA

5′ ————————————————— AAAAAA 3′
3′ ————————————————— ↑↑↑↑↑↑ TTTTTT 5′

DNA

Ribonuclease H degrades most
of the RNA

3′ ————————————————— ↑↑↑↑↑↑ 5′

Second strand synthesis by
DNA polymerase I

5′ ————————————————→
3′ ————————————————— ↑↑↑↑↑↑ 5′

Completion of the second
strand synthesis

5′ ———————————————————————— 3′
3′ ———————————————————————— 5′

Figure 5.32   One method for preparing cDNA.

Most eukaryotic mRNAs have a poly(A) tail at their 3′ end (Section 10.1.2). This series of A nucleotides is used as the priming site for the first stage of cDNA synthesis, carried out by reverse transcriptase – a DNA polymerase that copies an RNA template (Section 4.1.1). The primer is a short synthetic DNA oligonucleotide, typically 20 nucleotides in length, made up entirely of Ts (an 'oligo(dT)' primer). When the first strand synthesis has been completed, the preparation is treated with ribonuclease H, which specifically degrades the RNA component of an RNA–DNA hybrid. Under the conditions used, the enzyme does not degrade all of the RNA, instead leaving short segments that prime the second DNA strand synthesis reaction, this one catalyzed by DNA polymerase I.

also be used as an STS, assuming that it comes from a unique gene and not from a member of a gene family in which all the genes have the same or very similar sequences.

■ *SSLPs.* In Section 5.2.2 we examined the use of microsatellites and other SSLPs in genetic mapping.

SSLPs can also be used as STSs in physical mapping. SSLPs that are polymorphic and have already been mapped by linkage analysis are particularly valuable as they provide a direct connection between the genetic and physical maps.

■ *Random genomic sequences.* These are obtained by sequencing random pieces of cloned genomic DNA, or simply by downloading sequences that have been deposited in the databases.

Fragments of DNA for STS mapping

The second component of an STS mapping procedure is the collection of DNA fragments spanning the chromosome or genome being studied. This collection is sometimes called the **mapping reagent** and at present there are two ways in which it can be assembled: as a clone library and as a panel of **radiation hybrids**. We will consider radiation hybrids first.

A radiation hybrid is a rodent cell that contains fragments of chromosomes from a second organism (McCarthy, 1996). The technology was first developed in the 1970s when it was discovered that exposure of human cells to X-ray doses of 3000–8000 rads causes the chromosomes to break up randomly into fragments, larger X-ray doses producing smaller fragments (*Figure 5.33A*). This treatment is of course lethal for the human cells, but the chromosome fragments can be propagated if the irradiated cells are subsequently fused with non-irradiated hamster or other rodent cells. Fusion is stimulated either chemically with polyethylene glycol or by exposure to Sendai virus (*Figure 5.33B*). Not all of the hamster cells take up chromosome fragments so a means of identifying the hybrids is needed. The routine selection process is to use a hamster cell line that is unable to make either thymidine kinase (TK) or hypoxanthine phosphoribosyl transferase (HPRT), deficiencies in either of these two enzymes being lethal when the cells are grown in a medium containing a mixture of hypoxanthine, aminopterin and thymidine (HAT medium). After fusion, the cells are placed in HAT medium. Those that grow are hybrid hamster cells that have acquired human DNA fragments that include genes for the human TK and HPRT enzymes, which are synthesized inside the hybrids, enabling these cells to grow in the selective medium. The treatment results in hybrid cells that contain a random selection of human DNA fragments inserted into the hamster chromosomes. Typically the fragments are 5–10 Mb in size, with each cell containing fragments equivalent to 15–35% of the human genome. The collection of cells is called a radiation hybrid panel and can be used as a mapping reagent in STS mapping, provided that the PCR assay used to identify the STS does not amplify the equivalent region of DNA from the hamster genome.

A second type of radiation hybrid panel, containing DNA from just one human chromosome, can be constructed if the cell line that is irradiated is not a human one but a second type of rodent hybrid. Cytogeneticists have developed a number of rodent cell lines in which a

RESEARCH

5.1

BRIEFING

The radiation hybrid map of the mouse genome

Physical mapping is a prelude to sequencing of the mouse genome and enables comparisons to be made between mouse and human chromosomes.

Completion of the human genome sequence is not the only objective of the Human Genome Project. The possibility that comparisons with other genomes might be a valuable means of deciphering the human sequence was recognized when the Project was planned in the late 1980s, and the Project has actively stimulated the development of genome projects for other organisms (Section 7.4). The most important of these model organisms is the mouse, whose gene set is very similar to that of humans but which has the major advantage, compared with humans, of being amenable to experimental genetic analysis. Crosses between mice with different phenotypes are a staple of high-school biology; similar crosses, combined with molecular biology techniques such as gene knockouts (Section 7.2.2), will be one of staples of the post-sequencing phase of the Human Genome Project, the aim being to identify the functions of mouse genes and hence to infer the functions of the equivalent human genes.

Radiation hybrid mapping of the mouse genome

As with the human projects, construction of a genetic and physical map has been a preliminary to sequencing of the mouse genome. The physical map has been obtained largely by radiation hybrid mapping, using a panel of 93 mouse–hamster cell lines. To map as many markers as possible, a high throughput screening method based on PCR (Section 4.3) was developed. A pair of primers was designed for every DNA marker that was to be tested, with care taken to ensure that each primer pair was specific for just that one marker, and would not give a PCR product with any other part of the mouse genome. PCRs with different primer pairs would therefore show which markers were present in which cell lines, providing the data from which the physical map would be built up. But one problem remained. Usually, the success or failure of a PCR is determined by agarose gel electrophoresis, the presence in the gel of a band of the expected size indicating that the PCR has worked (see *Figure 4.30*, page 122). Agarose gel electrophoresis is a time-consuming procedure and, importantly for a high-throughput strategy, is difficult to automate. Some other way was therefore needed to determine which PCRs gave products and which did not. The answer was to design a third specific primer for each marker, this primer annealing within the internal region of the resulting PCR product. At the end of a PCR the third primer is added to the reaction mixture, along with a fluorescently labeled dideoxynucleotide (Section 6.1.1). If the PCR has been successful, then

the third primer anneals to the product and is extended by attachment of the dideoxynucleotide to its 3′ end. Attachment to the end of a DNA molecule changes the polarization of the fluorescent signal emitted by the dideoxynucleotide, in a way that can be detected with a suitable polarimeter. The outcome of each PCR is therefore determined by monitoring the fluorescent signal during this final stage of the experiment, a change in polarization indicating that the PCR has been successful and that the marker is present in the cell line being tested.

By automating the analysis, it was possible to perform 10 000 PCRs a day, a sufficient throughput to generate a whole-genome radiation hybrid map in a matter of weeks. The resulting map spanned the entire mouse genome and showed the positions of 11 109 genes, 232 SSLPs and 548 other DNA sequences. The accuracy of the map was assessed by comparing the positions of 685 genes that had

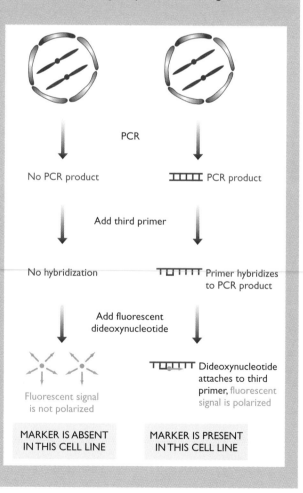

PCR

No PCR product   ⬛⬛⬛⬛ PCR product

Add third primer

No hybridization   ⬛⬛⬛⬛⬛ Primer hybridizes
                                         to PCR product

Add fluorescent
dideoxynucleotide

Fluorescent signal   ⬛⬛⬛⬛⬛ Dideoxynucleotide
is not polarized   attaches to third
                                      primer, fluorescent
                                      signal is polarized

MARKER IS ABSENT   MARKER IS PRESENT
IN THIS CELL LINE   IN THIS CELL LINE

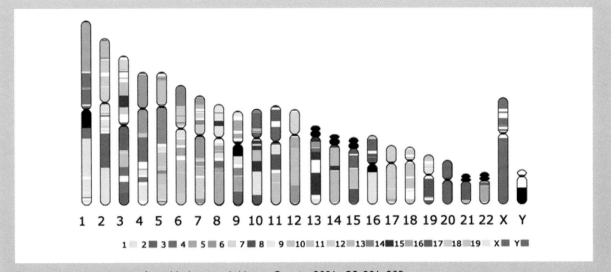

Reprinted with permission from Hudson *et al.*, *Nature Genetics*, 2001, **29**, 201–205.

also been mapped by conventional genetic means. Of these 685 genes, 669 were located at equivalent positions on both the physical and genetic maps, and 15 of the 16 that were in different positions were subsequently shown to be correctly positioned on the physical map, the genetic map being in error. The radiation hybrid map therefore had a high degree of accuracy and could be directly linked to the genetic map.

Insights from the mouse radiation hybrid map

As well as providing a framework for the sequencing phase of the mouse genome project, the radiation hybrid map has revealed, for the first time, the overall distribution of genes among the 20 mouse chromosomes. This distribution is not even, some chromosomes, notably numbers 11 and 19, having a relatively high gene density whereas others, such as chromosomes 14 and 18, are significantly underpopulated. A similar unevenness in gene density is seen with the human chromosomes. As with humans, the most sparsely populated mouse chromosome is the X chromosome.

Completion of the mouse radiation hybrid map has also enabled a comparison to be made between the relative positions of individual genes in the mouse and human genomes. This analysis was possible because 3658 of the mouse genes whose positions were mapped are known to be homologs of human genes, meaning that the mouse and human genes have an evolutionary relationship: they are, in effect the same gene in different organisms (Section 7.1.1). Would homologous genes be on the same chromosomes in both mice and humans? The above figure shows that such a simple relationship holds only for the X chromosomes. In this figure, the 24 human chromosomes are depicted with their centromeres and regions of constitutive heterochromatin (see *Figure 2.8*, page 40) marked in black. Each chromosome is painted with bands of color, each color representing a different mouse chromosome, as indicated by the key. Almost all of the human X chromosome is colored pink which, as the key shows, means that the genes present in this chromosome are homologs of genes in the mouse X chromosome. It therefore appears that the X chromosomes have not undergone any significant evolutionary change since the divergence of the human and mouse lineages several tens of millions of years ago. No other pair of mouse and human chromosomes shows such a close similarity, but clear relationships can still be seen. For example, human chromosome 17 contains many of the genes found in mouse chromosome 11, and human chromosome 20 appears to be a fragment of mouse chromosome 2. But for most chromosomes the relationship is much more complex, extensive rearrangements having occurred in the mouse and human lineages, giving a mosaic appearance to the human chromosomes in the above figure. Human chromosomes 3, 9 and 10, for example, each comprise genes from six or more mouse chromosomes. The comparison therefore suggests that chromosomes evolve by the rearrangement of segments, resulting in blocks of genes being moved from one position to another in the genome. Similar conclusions have been reached following comparisons of genetic maps, particularly those of related plants such as members of the grass family.

Reference

Hudson TJ, Church DM, Greenaway S, et al. (2001) A radiation hybrid map of the mouse genome. *Nature Genet.*, **29**, 201-205.

Web sites

For the latest news relating to the mouse genome project take a look at the following two sites:
http://www.ncbi.nlm.nih.gov/Homology
http://www.informatics.jax.org/

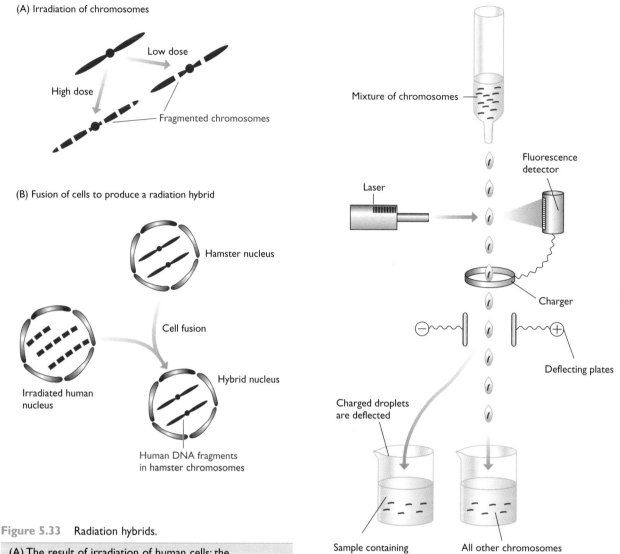

(A) Irradiation of chromosomes

Low dose

High dose

Fragmented chromosomes

(B) Fusion of cells to produce a radiation hybrid

Hamster nucleus

Cell fusion

Irradiated human nucleus

Hybrid nucleus

Human DNA fragments in hamster chromosomes

Mixture of chromosomes

Fluorescence detector

Laser

Charger

Deflecting plates

Charged droplets are deflected

Sample containing just one type of chromosome

All other chromosomes

Figure 5.33   Radiation hybrids.

(A) The result of irradiation of human cells: the chromosomes break into fragments, smaller fragments generated by higher X-ray doses. In (B), a radiation hybrid is produced by fusing an irradiated human cell with an untreated hamster cell. For clarity, only the nuclei are shown.

Figure 5.34   Separating chromosomes by flow cytometry.

A mixture of fluorescently stained chromosomes is passed through a small aperture so that each drop that emerges contains just one chromosome. The fluorescence detector identifies the signal from drops containing the correct chromosome and applies an electric charge to these drops. When the drops reach the electric plates, the charged ones are deflected into a separate beaker. All other drops fall straight through the deflecting plates and are collected in the waste beaker.

single human chromosome is stably propagated in the rodent nucleus. If a cell line of this type is irradiated and fused with hamster cells, then the hybrid hamster cells obtained after selection will contain either human or mouse chromosome fragments, or a mixture of both. The ones containing human DNA can be identified by probing with a human-specific genome-wide repeat sequence, such as the short interspersed nuclear element (SINE) called Alu (Section 2.4.2), which has a copy number of just over 1 million (see *Table 1.2*, page 24) and so occurs on average once every 4 kb in the human genome. Only cells containing human DNA will hybridize to Alu probes, enabling the uninteresting mouse hybrids to be discarded and STS mapping to be directed at the cells containing human chromosome fragments.

Radiation hybrid mapping of the human genome was initially carried out with chromosome-specific rather than whole-genome panels because it was thought that fewer hybrids would be needed to map a single chromosome than would be needed to map the entire genome. It turns out that a high-resolution map of a single human chromosome requires a panel of 100–200 hybrids, which is about the most that can be handled conveniently in a PCR screening program. But whole-genome and single-chromosome panels are constructed differently, the former involving irradiation of just human DNA, and the latter requiring irradiation of a mouse cell containing much mouse DNA and relatively little human DNA. This means that the human DNA content per hybrid is much lower in a single-chromosome panel than in a whole-genome panel. It transpires that detailed mapping of the entire human genome is possible with fewer than 100 whole-genome radiation hybrids, so whole-genome mapping is no more difficult than single-chromosome mapping. Once this was realized, whole-genome radiation hybrids became a central component of the mapping phase of the Human Genome Project (Section 6.3.1). Whole-genome libraries are also being used for STS mapping of other mammalian genomes and for those of the zebra fish and the chicken (McCarthy, 1996).

A clone library can also be used as the mapping reagent for STS analysis

A preliminary to the sequencing phase of a genome project is to break the genome or isolated chromosomes into fragments and to clone each one in a high-capacity vector, one able to handle large fragments of DNA (Section 4.2.1). This results in a clone library, a collection of DNA fragments, which, in this case, have an average size of several hundred kb. As well as supporting the sequencing work, this type of clone library can also be used as a mapping reagent in STS analysis.

As with radiation hybrid panels, a clone library can be prepared from genomic DNA, and so represents the entire genome, or a chromosome-specific library can be

made if the starting DNA comes from just one type of chromosome. The latter is possible because individual chromosomes can be separated by **flow cytometry**. To carry out this technique, dividing cells (ones with condensed chromosomes) are carefully broken open so that a mixture of intact chromosomes is obtained. The chromosomes are then stained with a fluorescent dye. The amount of dye that a chromosome binds depends on its size, so larger chromosomes bind more dye and fluoresce more brightly than smaller ones. The chromosome preparation is diluted and passed through a fine aperture, producing a stream of droplets, each one containing a single chromosome. The droplets pass through a detector that measures the amount of fluorescence, and hence identifies which droplets contain the particular chromosome being sought. An electric charge is applied to these drops, and no others (*Figure 5.34*), enabling the droplets containing the desired chromosome to be deflected and separated from the rest. What if two different chromosomes have similar sizes, as is the case with human chromosomes 21 and 22? These can usually be separated if the dye that is used is not one that binds non-specifically to DNA, but instead has a preference for AT- or GC-rich regions. Examples of such dyes are Hoechst 33258 and chromomycin A_3, respectively. Two chromosomes that are the same size rarely have identical GC contents, and so can be distinguished by the amounts of AT- or GC-specific dye that they bind.

Compared with radiation hybrid panels, clone libraries have one important advantage for STS mapping. This is the fact that the individual clones can subsequently provide the DNA that is actually sequenced. The data resulting from STS analysis, from which the physical map is generated, can equally well be used to determine which

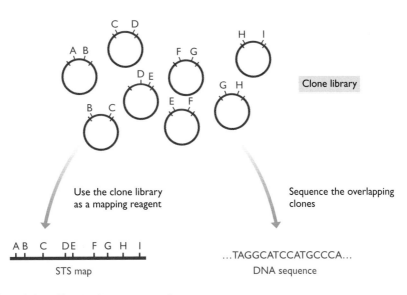

Figure 5.35 The value of clone libraries in genome projects.

The small clone library shown in this example contains sufficient information for an STS map to be constructed, and can also be used as the source of the DNA that will be sequenced.

clones contain overlapping DNA fragments, enabling a **clone contig** to be built up (*Figure 5.35*; for other methods for assembling clone contigs see Section 6.2.2). This assembly of overlapping clones can be used as the base material for a lengthy, continuous DNA sequence, and the STS data can later be used to anchor this sequence precisely onto the physical map. If the STSs also include SSLPs that have been mapped by genetic linkage analysis then the DNA sequence, physical map and genetic map can all be integrated.

References

Dausset J, Cann H, Cohen D, et al. (1990) Program description: Centre d'Étude du Polymorphisme Humaine – collaborative genetic mapping of the human genome. *Genomics*, **6**, 575–577.

Freifelder D (1987) *Microbial Genetics.* Jones & Bartlett, Boston.

Gerhold D, Rushmore T and Caskey CT (1999) DNA chips: promising toys have become powerful tools. *Trends Biochem. Sci.*, **24**, 168–173.

Heiskanen M, Peltonen L and Palotie A (1996) Visual mapping by high resolution FISH. *Trends Genet.*, **12**, 379–382.

Jing JP, Lai ZW, Aston C, et al. (1999) Optical mapping of *Plasmodium falciparum* chromosome 2. *Genome Res.*, **9**, 175–181.

Lichter P, Tang CJ, Call K, et al. (1990) High resolution mapping of human chromosome 11 by *in situ* hybridization with cosmid clones. *Science*, **247**, 64–69.

Lin J, Qi R, Aston C, et al. (1999) Whole-genome shotgun optical mapping of *Deinococcus radiodurans*. *Science*, **285**, 1558–1562.

Marra MA, Hillier L and Waterston RH (1998) Expressed sequence tags – ESTablishing bridges between genomes. *Trends Genet.*, **14**, 4–7.

McCarthy L (1996) Whole genome radiation hybrid mapping. *Trends Genet.*, **12**, 491–493.

Michalet X, Ekong R, Fougerousse F, et al. (1997) Dynamic molecular combing: stretching the whole human genome for high-resolution studies. *Science*, **277**, 1518–1523.

Mir KU and Southern EM (2000) Sequence variation in genes and genomic DNA: methods for large-scale analysis. *Ann. Rev. Genomics Hum. Genet.*, **1**, 329–360.

Morton NE (1955) Sequential tests for the detection of linkage. *Am. J. Hum. Genet.*, **7**, 277–318.

Oliver SG, van der Aart QJM, Agostoni-Carbone ML, et al. (1992) The complete DNA sequence of yeast chromosome III. *Nature*, **357**, 38–46.

Orel V (1995) *Gregor Mendel: The First Geneticist.* Oxford University Press, Oxford.

Schwartz DC, Li X, Hernandez LI, Ramnarain SP, Huff EJ and Wang Y-K (1993) Ordered restriction maps of *Saccharomyces cerevisiae* chromosomes constructed by optical mapping. *Science*, **262**, 110–114.

SNP Group (The International SNP Map Working Group) (2001) A map of human genome sequence variation containing 1.42 million single nucleotide polymorphisms. *Nature*, **409**, 928–933.

Sturtevant AH (1913) The linear arrangement of six sex-linked factors in *Drosophila* as shown by mode of association. *J. Exp. Zool.*, **14**, 39–45.

Trask BJ, Massa H, Kenwrick S and Gitschier J (1991) Mapping of human chromosome Xq28 by 2-color fluorescence *in situ* hybridization of DNA sequences to interphase cell nuclei. *Am. J. Hum. Genet.*, **48**, 1–15.

Tyagi S, Bratu DP and Kramer FR (1998) Multicolor molecular beacons for allele discrimination. *Nature Biotechnol.*, **16**, 49–53.

Wang DG, Fan J-B, Siao C-J, et al. (1998) Large-scale identification, mapping, and genotyping of single-nucleotide polymorphisms in the human genome. *Science*, **280**, 1077–1082.

Yamamoto F, Clausen H, White T, Marken J and Hakamori S (1990) Molecular genetic basis of the histo-blood group ABO system. *Nature*, **345**, 229–233.

Further Reading

Fincham JRS, Day PR and Radford A (1979) *Fungal Genetics*, 4th edition. Blackwell, London. — *The bible for gene mapping in microbial eukaryotes.*

Griffiths AJF, Miller JH, Suzuki DT, Lewontin RC and Gelbart WM (2000) *An Introduction to Genetic Analysis*, 7th edition. W. H. Freeman, New York. — *Particularly good for gene mapping by experimental crosses and for mapping in bacteria.*

Lichter P (1997) Multicolor FISHing: what's the catch? *Trends Genet.*, **13**, 475–479. — *An interesting review of some of the applications of FISH.*

Lodish H, Berk A, Zipursky AL, Matsudaira P, Baltimore D and Darnell J (2000) *Molecular Cell Biology*, 4th edn. W. H. Freeman, New York. — *Contains full details of mitosis and meiosis.*

Primrose SB (1995) *Principles of Genome Analysis.* Blackwell Science, Oxford. — *A little out of date but a good description of mapping strategies.*

Strachan T and Read AP (1999) *Human Molecular Genetics*, 2nd edition. BIOS Scientific Publishers, Oxford. — *Chapters 10 and 11 cover human physical and genetic mapping.*

Sturtevant AH (1965) *A History of Genetics.* Harper and Row, New York. — *Describes the early gene mapping work carried out by Morgan and his colleagues.*

Walter MA, Spillett DJ, Thomas P, Weissenbach J and Goodfellow PN (1994) A method for constructing radiation hybrid maps of whole genomes. *Nature Genet.*, **7**, 22–28. — *An excellent review of this approach to mapping.*

STUDY AIDS FOR CHAPTER 5

Key terms

Give short definitions of the following terms:

Allele
Bivalent
Chromatid
Clone contig
Clone contig method
Codominance
Cohesin
Conjugation
Contour clamped
  homogeneous electric fields
  (CHEF)
Crossing-over
DNA chip
DNA marker
Dominant
Double heterozygote
Double homozygote
Double restriction
Episome transfer
Expressed sequence tag (EST)
Fiber-FISH
Field inversion gel
  electrophoresis (FIGE)
Flow cytometry
FLpter value
Fluorescent *in situ*
  hybridization (FISH)
Gamete
Gel stretching
Genetic linkage
Genetic mapping
Genotype
Heterozygous
Homologous chromosomes
Homozygous
Incomplete dominance
Interphase chromosome
Linkage analysis
lod score
Map
Mapping reagent
Marker

Meiosis
Microarray
Mitosis
Molecular combing
Multiple alleles
Oligonucleotide hybridization
  analysis
Optical mapping
Orthogonal field alternation
  gel electrophoresis (OFAGE)
Parental genotype
Partial linkage
Partial restriction
Pedigree analysis
Phenotype
Phosphorimaging
Photolithography
Physical mapping
Radiation hybrid
Recessive
Recombinant genotype
Recombination frequency
Recombination hotspot
Restriction fragment length
  polymorphism (RFLP)
Restriction mapping
Sequence tagged site (STS)
Shotgun method
Simple sequence length
  polymorphism (SSLP)
Simple tandem repeat (STR)
Single nucleotide
  polymorphism (SNP)
STS mapping
Test cross
Transduction
Transformation
Variable number of tandem
  repeats (VNTR)
Whole-genome shotgun
  method
Wild type

Self study questions

1. Explain why a map is a useful aid to genome sequencing.
2. Distinguish between 'genetic mapping' and 'physical mapping'. What are the strengths and weaknesses of the two techniques?
3. Why are genes not ideal markers for construction of a genetic map?
4. Describe the various types of DNA marker that are used in genetic mapping. How is each type of marker scored?
5. Refer to *Figure 5.5A* (page 131). Draw the appearance of the autoradiograph if the probe hybridized to a region of DNA entirely between restriction sites R_1 and R_2. Would detection of the RFLP still be unambiguous?
6. Explain how Mendel's work led eventually to a method for genetic mapping.
7. Draw diagrams of the key events occurring during (a) mitosis, and (b) meiosis. Annotate your diagrams to highlight the important differences between the two processes.
8. Define the term 'partial linkage' and show how partial linkage is the basis of genetic mapping.
9. Describe how linkage analysis is carried out with (a) mice, (b) humans, and (c) bacteria.
10. What factors are responsible for the inaccuracies that sometimes occur in a genetic map?
11. Explain how a restriction map is obtained. What special procedures can be used to increase the size of DNA molecule for which a restriction map can be obtained?
12. What is FISH and how is it used to construct a physical map?
13. Describe the various types of DNA sequence that can be used in STS mapping.
14. Draw a diagram showing how a double-stranded cDNA is synthesized.
15. Define the term 'mapping reagent' and explain how a panel of radiation hybrids is used as a mapping reagent.
16. Explain how a clone library is used as a mapping reagent.
17. Draw a diagram to show how a sample of a single human chromosome can be obtained by flow cytometry.

Problem-based learning

1. What are the ideal features of a DNA marker that will be used to construct a genetic map. To what extent can RFLPs, SSLPs or SNPs be considered 'ideal' DNA markers?
2. Explore and assess the applications of DNA chip technology in biological research.
3. Evaluate the relative importance of genetic and physical mapping in the Human Genome Project.

Sequencing Genomes

Chapter Contents

6.1 The Methodology for DNA Sequencing 164

6.1.1 Chain termination DNA
sequencing 165

6.1.2 Departures from conventional
DNA sequencing 171

**6.2 Assembly of a Contiguous DNA
Sequence 172**

6.2.1 Sequence assembly by the shotgun
approach 173

6.2.2 Sequence assembly by the clone
contig approach 176

6.2.3 Whole-genome shotgun
sequencing 179

6.3 The Human Genome Projects 181

6.3.1 The mapping phase of the Human
Genome Project 181

6.3.2 Sequencing the human genome 182

6.3.3 The future of the human genome
projects 182

Learning outcomes

When you have read Chapter 6, you should be able to:

■ Distinguish between the two methods used to sequence DNA

■ Give a detailed description of chain termination sequencing and an outline description of the chemical degradation method

■ Describe the key features of automated DNA sequencing and evaluate the importance of automated sequencing in genomics research

■ State the strengths and limitations of the shotgun, whole-genome shotgun and clone contig methods of genome sequencing

■ Describe how a small bacterial genome can be sequenced by the shotgun method, using the *Haemophilus influenzae* project as an example

■ Outline the various ways in which a clone contig can be built up

■ Explain the basis to the whole-genome shotgun approach to genome sequencing, with emphasis on the steps taken to ensure that the resulting sequence is accurate

■ Give an account of the development of the Human Genome Project up to the publication of the draft sequence in February 2001

■ Debate the ethical, legal and social issues raised by the human genome projects

THE ULTIMATE OBJECTIVE of a genome project is the complete DNA sequence for the organism being studied, ideally integrated with the genetic and/or physical maps of the genome so that genes and other interesting features can be located within the DNA sequence. This chapter describes the techniques and research strategies that are used during the sequencing phase of a genome project, when this ultimate objective is being directly addressed. Techniques for sequencing DNA are clearly of central importance in this context and we will begin the chapter with a detailed examination of sequencing methodology. This methodology is of little value however, unless the short sequences that result from individual sequencing experiments can be linked together in the correct order to give the master sequences of the chromosomes that make up the genome. The middle part of this chapter describes the strategies used to ensure that the master sequences are assembled correctly. Finally, we will review the way in which mapping and sequencing were used to produce the two draft human genome sequences that were published in February 2001.

6.1 The Methodology for DNA Sequencing

Rapid and efficient methods for DNA sequencing were first devised in the mid-1970s. Two different procedures were published at almost the same time:

■ The **chain termination method** (Sanger *et al.*, 1977), in which the sequence of a single-stranded DNA molecule is determined by enzymatic synthesis of complementary polynucleotide chains, these chains terminating at specific nucleotide positions;

■ The **chemical degradation method** (Maxam and Gilbert, 1977), in which the sequence of a double-stranded DNA molecule is determined by treatment with chemicals that cut the molecule at specific nucleotide positions.

Both methods were equally popular to begin with but the chain termination procedure has gained ascendancy in recent years, particularly for genome sequencing. This is partly because the chemicals used in the chemical degradation method are toxic and therefore hazardous to the health of the researchers doing the sequencing experiments, but mainly because it has been easier to automate chain termination sequencing. As we will see later in this chapter, a genome project involves a huge number of individual sequencing experiments and it would take many years to perform all these by hand. Automated sequencing techniques are therefore essential if the project is to be completed in a reasonable time-span.

6.1.1 Chain termination DNA sequencing

Chain termination DNA sequencing is based on the principle that single-stranded DNA molecules that dif-fer in length by just a single nucleotide can be sepa-rated from one another by **polyacrylamide gel elec-trophoresis** (Technical Note 6.1). This means that it is possible to resolve a family of molecules, representing all lengths from 10 to 1500 nucleotides, into a series of bands (*Figure 6.1*).

Chain termination sequencing in outline

The starting material for a chain termination sequencing experiment is a preparation of identical single-stranded DNA molecules. The first step is to anneal a short

---

TECHNICAL
6.1
NOTE

Polyacrylamide gel electrophoresis

Separation of DNA molecules differing in length by just one nucleotide.

Polyacrylamide gel electrophoresis is used to examine the families of chain-terminated DNA molecules resulting from a sequencing experiment. Agarose gel electrophoresis (Technical Note 2.1, page 37) cannot be used for this purpose because it does not have the resolving power needed to separate single-stranded DNA molecules that differ in length by just one nucleotide. Polyacrylamide gels have smaller pore sizes than agarose gels and allow precise separations of molecules from 10–1500 bp. As well as DNA sequencing, polyacrylamide gels are also used for other applications where fine-scale DNA separations are required, for instance in the examination of amplification products from PCRs directed at microsatellite loci, where the products of different alleles might differ in size by just two or three base pairs (see *Figure 5.6*, page 132).

A polyacrylamide gel consists of chains of acrylamide monomers ($CH_2=CH–CO–NH_2$) cross-linked with N, N'-methylenebisacrylamide units ($CH_2=CH–CO–NH–CH_2–NH–CO–CH=CH_2$), the latter commonly called 'bis'. The pore size of the gel is determined by both the total con-centration of monomers (acrylamide + bis) and the ratio of acrylamide to bis. For DNA sequencing, a 6% gel with an acrylamide : bis ratio of 19 : 1 is normally used because this allows resolution of single-stranded DNA molecules between 100 and 750 nucleotides in length. About 650 nucleotides of sequence can therefore be read from a single gel. The gel concentration can be increased to 8% in order to read the sequence closer to the primer (resolving molecules 50–400 nucleotides in length) or decreased to 4% to read a more distant sequence (500–1500 nucleotides from the primer). Polymerization of the acrylamide : bis solution is initiated by ammonium persulfate and cata-lyzed by TEMED (N, N, N', N'-tetramethylethylenediamine). Sequencing gels also contain urea, which is a denaturant

that prevents intra-strand base pairs from forming in the chain-terminated molecules. This is important because the change in conformation resulting from base-pairing alters the migration rate of a single-stranded molecule, so the strict equivalence between the length of a molecule and its band position, critical for reading the DNA sequence (see *Figure 6.2D*), is lost.

Polyacrylamide gels are prepared between two glass plates held apart by spacers. This arrangement serves two purposes. First, it enables a very thin (< 1 mm) gel to be made, which facilitates sequence reading by improving the sharpness of the bands. Second, it ensures that polymeriza-tion, which is inhibited by oxygen, occurs evenly throughout the gel. With such a thin gel the amount of DNA per band is small and the banding pattern is only barely visible after ethidium bromide staining. For this reason, a radioactively labeled nucleotide is usually included in the sequencing reactions so that the banding pattern can be visualized by autoradiography.

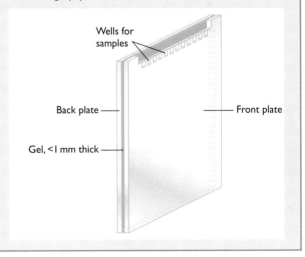

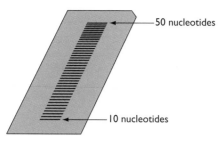

50 nucleotides

10 nucleotides

Figure 6.1 Polyacrylamide gel electrophoresis can resolve single-stranded DNA molecules that differ in length by just one nucleotide.

The banding pattern is produced after separation of single-stranded DNA molecules by denaturing polyacrylamide gel electrophoresis. The molecules are labeled with a radioactive marker and the bands visualized by autoradiography. The bands gradually get closer together towards the top of the ladder. In practice, molecules up to about 1500 nucleotides in length can be separated if the electrophoresis is continued for long enough.

oligonucleotide to the same position on each molecule, this oligonucleotide subsequently acting as the primer for synthesis of a new DNA strand that is complementary to the template (*Figure 6.2A*). The strand synthesis reaction,

which is catalyzed by a DNA polymerase enzyme (Section 4.1.1 and Box 6.1) and requires the four deoxyribonucleotide triphosphates (dNTPs – dATP, dCTP, dGTP and dTTP) as substrates, would normally continue until several thousand nucleotides had been polymerized. This does not occur in a chain termination sequencing experiment because, as well as the four dNTPs, a small amount of a **dideoxynucleotide** (e.g. ddATP) is added to the reaction. The polymerase enzyme does not discriminate between dNTPs and ddNTPs, so the dideoxynucleotide can be incorporated into the growing chain, but it then blocks further elongation because it lacks the 3'-hydroxyl group needed to form a connection with the next nucleotide (*Figure 6.2B*).

If ddATP is present, chain termination occurs at positions opposite thymidines in the template DNA (*Figure 6.2C*). Because dATP is also present the strand synthesis does not always terminate at the first T in the template; in fact it may continue until several hundred nucleotides have been polymerized before a ddATP is eventually incorporated. The result is therefore a set of new chains, all of different lengths, but each ending in ddATP. Now the polyacrylamide gel comes into play. The family of molecules generated in the presence of ddATP is loaded into one lane of the gel, and the families generated with ddCTP, ddGTP and ddTTP loaded into the three adjacent lanes. After electrophoresis, the DNA

Box 6.1: DNA polymerases for chain termination sequencing

Any template-dependent DNA polymerase is capable of extending a primer that has been annealed to a single-stranded DNA molecule, but not all polymerases do this in a way that is useful for DNA sequencing. Three criteria in particular must be fulfilled by a sequencing enzyme:

- **High processivity.** This refers to the length of polynucleotide that is synthesized before the polymerase terminates through natural causes. A sequencing polymerase must have high processivity so that it does not dissociate from the template before incorporating a chain-terminating nucleotide.

- **Negligible or zero 5'→3' exonuclease activity.** Most DNA polymerases also have exonuclease activities, meaning that they can degrade DNA polynucleotides as well as synthesize them (Section 4.1.1; see *Figure 4.7*, page 101). A 5'→3' exonuclease activity enables the polymerase to remove a DNA strand that is already attached to the template. This is a disadvantage in DNA sequencing because removal of nucleotides from the 5' ends of the newly synthesized strands alters the lengths of these strands, making it

impossible to read the sequence from the banding pattern in the polyacrylamide gel.

- **Negligible or zero 3'→5' exonuclease activity** is also desirable so that the polymerase does not remove the chain termination nucleotide once it has been incorporated.

These are stringent requirements and are not entirely met by any naturally occurring DNA polymerase. Instead, artificially modified enzymes are generally used. The first of these to be developed was the Klenow polymerase, which is a version of *Escherichia coli* DNA polymerase I from which the 5'→3' exonuclease activity of the standard enzyme has been removed, either by cleaving away the relevant part of the protein or by genetic engineering (Section 4.1.1). The Klenow polymerase has relatively low processivity, limiting the length of sequence that can be obtained from a single experiment to about 250 bp, and giving non-specific bands on the sequencing gel, these 'shadow' bands representing strands that have terminated naturally rather than by incorporation of a ddNTP. The Klenow enzyme was therefore superseded by a modified version of the DNA polymerase encoded by bacteriophage T7, this enzyme going under the tradename 'Sequenase'. Sequenase has high processivity and no exonuclease activity, and also possesses other desirable features such as rapid reaction rate and the ability to use many modified nucleotides as substrates.

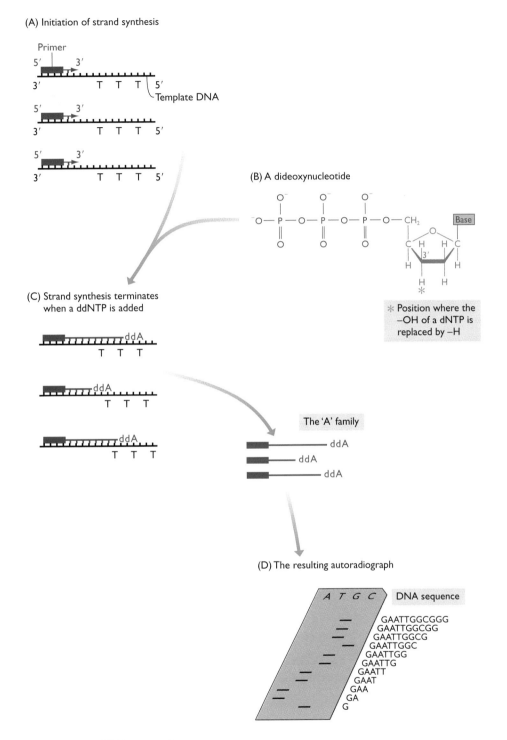

Figure 6.2   Chain termination DNA sequencing.

(A) Chain termination sequencing involves the synthesis of new strands of DNA that are complementary to a single-stranded template. (B) Strand synthesis does not proceed indefinitely because the reaction mixture contains small amounts of a dideoxynucleotide, which blocks further elongation because it has a hydrogen atom rather than a hydroxyl group attached to its 3′-carbon. (C) Strand synthesis in the presence of ddATP results in chains that are terminated opposite Ts in the template. This 'A' family of terminated chains is loaded into one lane of a polyacrylamide gel, alongside the families of terminated chains from the T, G and C reactions. (D) In the methodology shown here, the banding pattern is visualized by autoradiography, the terminated chains having become radioactively labeled by inclusion of a labeled dNTP in the strand synthesis reactions. The sequence, shown on the right, is read by noting which lane each band lies in, starting at the bottom of the autoradiograph and moving band by band towards the top.

sequence can be read directly from the positions of the bands in the gel (*Figure 6.2D*). The band that has moved the furthest represents the smallest piece of DNA, this being the strand that terminated by incorporation of a ddNTP at the first position in the template. In the example shown in *Figure 6.2* this band lies in the 'G' lane (i.e. the lane containing the molecules terminated with ddGTP), so the first nucleotide in the sequence is 'G'. The next band, corresponding to the molecule that is one nucleotide longer than the first, is in the 'A' lane, so the second nucleotide is 'A' and the sequence so far is 'GA'. Continuing up through the gel we see that the next band also lies in the 'A' lane (sequence GAA), then we move to the 'T' lane (GAAT), and so on. The sequence reading can be continued up to the region of the gel where individual bands are not separated.

Chain termination sequencing requires a single-stranded DNA template

The template for a chain termination experiment is a single-stranded version of the DNA molecule to be sequenced. There are several ways in which this can be obtained:

■ **The DNA can be cloned in a plasmid vector** (Section 4.2.1). The resulting DNA will be double stranded so cannot be used directly in sequencing. Instead, it must be converted into single-stranded DNA by denaturation with alkali or by boiling. This is a common method for obtaining template DNA for

DNA sequencing, largely because cloning in a plasmid vector is such a routine technique. A shortcoming is that it can be difficult to prepare plasmid DNA that is not contaminated with small quantities of bacterial DNA and RNA, which can act as spurious templates or primers in the DNA sequencing experiment.

■ **The DNA can be cloned in a bacteriophage M13 vector.** Vectors based on M13 bacteriophage are designed specifically for the production of single-stranded templates for DNA sequencing. M13 bacteriophage has a single-stranded DNA genome which, after infection of *Escherichia coli* bacteria, is converted into a double-stranded **replicative form**. The replicative form is copied until over 100 molecules are present in the cell, and when the cell divides the copy number in the new cells is maintained by further replication. At the same time, the infected cells continually secrete new M13 phage particles, approximately 1000 per generation, these phages containing the single-stranded version of the genome (*Figure 6.3*). Cloning vectors based on M13 vectors are double-stranded DNA molecules equivalent to the replicative form of the M13 genome. They can be manipulated in exactly the same way as a plasmid cloning vector. The difference is that cells that have been transfected with a recombinant M13 vector secrete phage particles containing single-stranded DNA, this DNA comprising the vector molecule plus any additional DNA that has been ligated into it. The phages therefore provide the template DNA for chain termination sequencing.

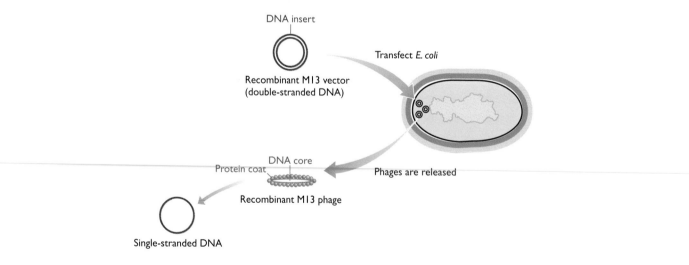

Figure 6.3 Obtaining single-stranded DNA by cloning in a bacteriophage M13 vector.

M13 vectors can be obtained in two forms: the double-stranded replicative molecule and the single-stranded version found in bacteriophage particles. The replicative form can be manipulated in the same way as a plasmid cloning vector (Section 4.2.1) with new DNA inserted by restriction followed by ligation. The recombinant vector is introduced into *Escherichia coli* cells by transfection. Once inside an *E. coli* cell, the double-stranded vector replicates and directs synthesis of single-stranded copies, which are packaged into phage particles and secreted from the cell. The phage particles can be collected from the culture medium after centrifuging to pellet the bacteria. The protein coats of the phages are removed by treating with phenol, and the single-stranded version of the recombinant vector is purified for use in DNA sequencing.

The one disadvantage is that DNA fragments longer than about 3 kb suffer deletions and rearrangements when cloned in an M13 vector, so the system can only be used with short pieces of DNA.

■ ***The DNA can be cloned in a phagemid.*** This is a plasmid cloning vector that contains, in addition to its plasmid origin of replication, the origin from M13 or another phage with a single-stranded DNA genome. If an *E. coli* cell contains both a phagemid and the replicative form of a **helper phage**, the latter carrying genes for the phage replication enzymes and coat proteins, then the phage origin of the phagemid becomes activated, resulting in synthesis of phage particles containing the single-stranded version of the phagemid. The double-stranded plasmid DNA is therefore converted into single-stranded template DNA for DNA sequencing. This system avoids the instabilities of M13 cloning and can be used with fragments up to 10 kb or more.

■ ***PCR can be used to generate single-stranded DNA.*** There are various ways of generating single-stranded DNA by PCR, the most effective being to modify one of the two primers so that DNA strands synthesized from this primer are easily purified. One possibility

is to attach small metallic beads to the primer and then use a magnetic device to purify the resulting strands (*Figure 6.4*).

The primer determines the region of the template DNA that will be sequenced

To begin a chain termination sequencing experiment, an oligonucleotide primer is annealed onto the template DNA. The primer is needed because template-dependent DNA polymerases cannot initiate DNA synthesis on a molecule that is entirely single-stranded: there must be a short double-stranded region to provide a 3′ end onto which the enzyme can add new nucleotides (Section 4.1.1).

The primer also plays the critical role of determining the region of the template molecule that will be sequenced. For most sequencing experiments a 'universal' primer is used, this being one that is complementary to the part of the vector DNA immediately adjacent to the point into which new DNA is ligated (*Figure 6.5A*). The same universal primer can therefore give the sequence of any piece of DNA that has been ligated into the vector. Of course if this inserted DNA is longer than 750 bp or so then only a part of its sequence will be obtained, but usually this is not a problem because the project as a whole simply requires that a large number of short sequences are generated and subsequently assembled into the contiguous master sequence. It is immaterial whether or not the short sequences are the complete or only partial sequences of the DNA

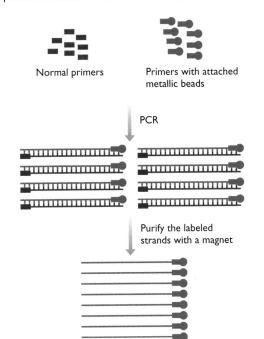

Figure 6.4 One way of using PCR to prepare template DNA for chain termination sequencing.

The PCR is carried out with one normal primer (shown in red), and one primer that is labeled with a metallic bead (shown in brown). After PCR, the labeled strands are purified with a magnetic device. For more details about PCR, see Section 4.3.

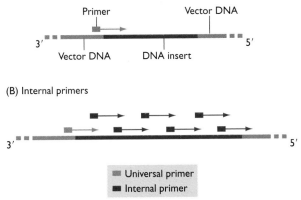

Figure 6.5 Different types of primer for chain termination sequencing.

(A) A universal primer anneals to the vector DNA, adjacent to the position at which new DNA is inserted. A single universal primer can therefore be used to sequence any DNA insert, but only provides the sequence of one end of the insert. (B) One way of obtaining a longer sequence is to carry out a series of chain termination experiments, each with a different internal primer that anneals within the DNA insert.

fragments used as templates. If double-stranded plasmid DNA is being used to provide the template then, if desired, more sequence can be obtained from the other end of the insert. Alternatively, it is possible to extend the sequence in one direction by synthesizing a non-universal primer, designed to anneal at a position within the insert DNA (*Figure 6.5B*). An experiment with this primer will provide a second short sequence that overlaps the previous one.

Thermal cycle sequencing offers an alternative to the traditional methodology

The discovery of thermostable DNA polymerases, which led to the development of PCR (Sections 4.1.1 and 4.3), has also resulted in new methodologies for chain termination sequencing. In particular, the innovation called **thermal cycle sequencing** (Sears *et al.*, 1992) has two advantages over traditional chain termination sequencing. First, it uses double-stranded rather than single-stranded DNA as the starting material. Second, very little

template DNA is needed, so the DNA does not have to be cloned before being sequenced.

Thermal cycle sequencing is carried out in a similar way to PCR but just one primer is used and each reaction mixture includes one of the ddNTPs (*Figure 6.6*). Because there is only one primer, only one of the strands of the starting molecule is copied, and the product accumulates in a linear fashion, not exponentially as is the case in a real PCR. The presence of the ddNTP in the reaction mixture causes chain termination, as in the standard methodology, and the family of resulting strands can be analyzed and the sequence read in the normal manner by polyacrylamide gel electrophoresis.

Fluorescent primers are the basis of automated sequence reading

The standard chain termination sequencing methodology employs radioactive labels, and the banding pattern in the polyacrylamide gel is visualized by autoradiography. Usually one of the nucleotides in the sequencing reaction

---

Box 6.2: The chemical degradation sequencing method

The difference between the two sequencing techniques lies in the way in which the A, C, G and T families of molecules are generated. In the chemical degradation procedure these families are produced by treatment with chemicals that cut specifically at a particular nucleotide, not by enzymatic synthesis. The starting material can be double-stranded DNA but before beginning the sequencing procedure, the double-stranded molecules must be denatured into single-stranded DNA, with each strand labeled at one end.

To illustrate the procedure, we will follow the 'G' reaction. First, the molecules are treated with dimethyl sulfate, which attaches a methyl group to the purine ring of G nucleotides. Only a limited amount of dimethyl sulfate is added, the objective being to modify, on average, just one G per polynucleotide. At this stage the DNA strands are still intact, cleavage not occurring until a second chemical – piperidine – is added. Piperidine removes the modified purine ring and cuts the DNA molecule at the phosphodiester bond immediately upstream of the baseless site that is created. The result is a set of cleaved DNA molecules, some of which are labeled and some of which are not. The labeled molecules all have one end in common and one end determined by the cut sites, the latter indicating the positions of the G nucleotides in the DNA molecules that were cleaved. In other words, the G family of molecules produced by chemical treatment is equivalent to the G family produced by the chain termination method.

The families of cleaved molecules are electrophoresed in a polyacrylamide gel and the sequence read in a similar way to that described for chain termination sequencing. The

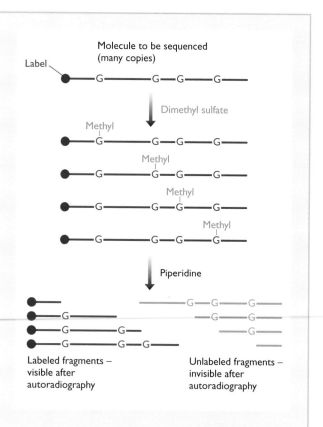

only significant difference is that problems have been encountered in developing chemical treatments to cut specifically at A or T, and so the four reactions that are carried out are usually 'G', 'A + G', 'C' and 'C + T'. This does not affect the accuracy of the sequence that is read from the gel.

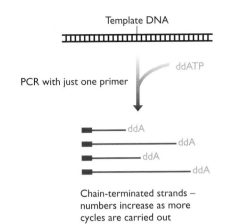

Template DNA

PCR with just one primer

ddATP

ddA

ddA

ddA

ddA

Chain-terminated strands –
numbers increase as more
cycles are carried out

Figure 6.6   Thermal cycle sequencing.

PCR is carried out with just one primer and with a
dideoxynucleotide present in the reaction mixture. The
result is a family of chain-terminated strands – the 'A'
family in the reaction shown. These strands, along with the
products of the C, G and T reactions, are electrophoresed
as in the standard methodology (see *Figure 6.2D*).

is labeled so that the newly synthesized strands contain
radiolabels along their lengths, giving high detection
sensitivity. To ensure good band resolution, ^{33}P or ^{35}S is
generally used, as the emission energies of these isotopes
are relatively low, in contrast to ^{32}P, which has a higher
emission energy and gives poorer resolution because of
signal scattering.

In Section 5.3.2 we saw how the replacement of radioactive labels by fluorescent ones has given a new dimension to *in situ* hybridization techniques. Fluorolabeling has
been equally important in the development of sequencing
methodology, in particular because the detection system
for fluorolabels has opened the way to automated
sequence reading (Prober *et al.*, 1987). The label is
attached to the ddNTPs, with a different fluorolabel used
for each one (*Figure 6.7A*). Chains terminated with A are
therefore labeled with one fluorophore, chains terminated
with C are labeled with a second fluorophore, and so on.
Now it is possible to carry out the four sequencing reactions – for A, C, G and T – in a single tube and to load all
four families of molecules into just one lane of the polyacrylamide gel, because the fluorescent detector can
discriminate between the different labels and hence
determine if each band represents an A, C, G or T. The
sequence can be read directly as the bands pass in front of
the detector and either printed out in a form readable by
eye (*Figure 6.7B*) or sent straight to a computer for storage. When combined with robotic devices that prepare
the sequencing reactions and load the gel, the fluorescent
detection system provides a major increase in throughput
and avoids errors that might arise when a sequence is
read by eye and then entered manually into a computer.
It is only by use of these automated techniques that we
can hope to generate sequence data rapidly enough to
complete a genome project in a reasonable length of time.

(A)

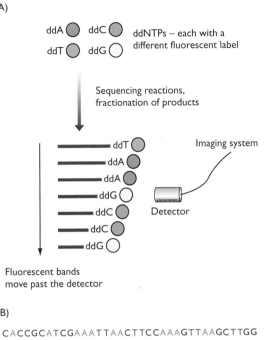

ddA   ddC   ddNTPs – each with a
ddT   ddG   different fluorescent label

Sequencing reactions,
fractionation of products

ddT
ddA
ddA
ddG
ddC
ddC
ddG

Imaging system

Detector

Fluorescent bands
move past the detector

(B)

CACCGCATCGAAATTAACTTCCAAAGTTAAGCTTGG

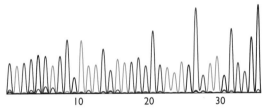

10   20   30

Figure 6.7   Automated DNA sequencing with fluorescently
labeled dideoxynucleotides.

(A) The chain termination reactions are carried out in a
single tube, with each dideoxynucleotide labeled with a
different fluorophore. In the automated sequencer, the
bands in the electrophoresis gel move past a fluorescence
detector, which identifies which dideoxynucleotide is
present in each band. The information is passed to the
imaging system. (B) The printout from an automated
sequencer. The sequence is represented by a series of
peaks, one for each nucleotide position. In this example, a
green peak is an 'A', blue is 'C', black is 'G', and red is 'T'.

6.1.2 Departures from conventional DNA sequencing

In spite of the development of automated techniques, conventional DNA sequencing suffers from the limitation that
only a few hundred bp of sequence can be determined in a
single experiment. In the context of the Human Genome
Project, this means that each experiment provides only one
five-millionth of the total genome sequence. Attempts are
continually being made to modify the technology so that
sequence acquisition is more rapid, a recent example being
the introduction of new automated sequencers that use
capillary separation rather than a polyacrylamide gel.
These have 96 channels so 96 sequences can be determined

in parallel, and each run takes less than 2 hours to complete, enabling up to 1000 sequences to be obtained in a single day (Mullikan and McMurray, 1999). Other systems that are being developed will increase data generation even further by enabling 384 or 1024 sequences to be run at the same time (Rogers, 1999).

There have also been attempts to make sequence acquisition more rapid by devising new sequencing methodologies. One possibility is **pyrosequencing**, which does not require electrophoresis or any other fragment separation procedure and so is more rapid than chain termination sequencing (Ronaghi *et al.*, 1998). In pyrosequencing, the template is copied in a straightforward manner without added ddNTPs. As the new strand is being made, the order in which the dNTPs are incorporated is detected, so the sequence can be 'read' as the reaction proceeds. The addition of a nucleotide to the end of the growing strand is detectable because it is accompanied by release of a molecule of pyrophosphate, which can be converted by the enzyme sulfurylase into a flash of chemiluminescence. Of course, if all four dNTPs were added at once then flashes of light would be seen all the time and no useful sequence information would be obtained. Each dNTP is therefore added separately, one after the other, with a nucleotidase enzyme also present in the reaction mixture so that if a dNTP is not incorporated into the polynucleotide then it is rapidly degraded before the next dNTP is added (*Figure 6.8*). This procedure makes it possible to follow the order in which the dNTPs are incorporated into the growing strand. The technique sounds complicated, but it simply requires that a repetitive series of additions be made to the reaction mixture, precisely the type of procedure that is easily automated, with the possibility of many experiments being carried out in parallel.

A very different approach to DNA sequencing through the use of DNA chips (see Technical Note 5.1, page 133) might one day be possible. A chip carrying an array of different oligonucleotides could be used in DNA sequencing by applying the test molecule – the one whose sequence is to be determined – to the array and detecting the positions at which it hybridizes. Hybridization to an individual oligonucleotide would indicate the presence of that particular oligonucleotide sequence in the test molecule, and comparison of all the oligonucleotides to which hybridization occurs would enable the sequence of the test molecule to be deduced (*Figure 6.9*). The problem with this approach is that the maximum length of the molecule that can be sequenced is given by the square root of the number of oligonucleotides in the array, so if every possible 8-mer oligonucleotide (ones containing eight nucleotides) were attached to the chip – all 65 536 of them – then the maximum length of readable sequence would be only 256 bp (Southern, 1996). Even if the chip carried all the 1 048 576 different 10-mer sequences, it could still only be used to sequence a 1 kb molecule. To sequence a 1 Mb molecule (this being the sort of advance in sequence capability that is really needed) the chip would have to carry all of the 1×10^{12} possible 20-mers.

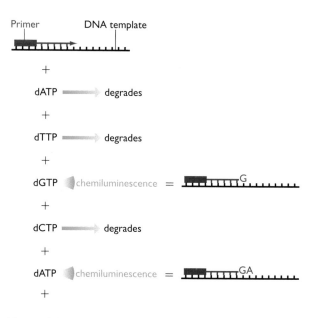

Figure 6.8 Pyrosequencing.

The strand synthesis reaction is carried out in the absence of dideoxynucleotides. Each dNTP is added individually, along with a nucleotidase enzyme that degrades the dNTP if it is not incorporated into the strand being synthesized. Incorporation of a nucleotide is detected by a flash of chemiluminescence induced by the pyrophosphate released from the dNTP. The order in which nucleotides are added to the growing strand can therefore be followed.

This may sound an outlandish proposition but advances in miniaturization, together with the possibility of electronic rather than visual detection of hybridization, could bring such an array within reach in the future.

6.2 Assembly of a Contiguous DNA Sequence

The next question to address is how the master sequence of a chromosome, possibly several tens of Mb in length, can be assembled from the multitude of short sequences generated by chain termination sequencing. We addressed this issue at the start of Chapter 5 and established that the relatively short genomes of prokaryotes can be assembled by shotgun sequencing, but that this approach might lead to errors if applied to larger eukaryotic genomes. The whole-genome shotgun method, which uses a map to aid assembly of the master sequence, has been used with the fruit-fly and human genomes, but it is generally accepted that a greater degree of accuracy is achieved with the clone contig approach, in which the genome is broken down into segments, each with a known position on the genome map, before sequencing is carried out (see *Figure 5.3*, page 128). We will start by examining how shotgun sequencing has been applied to prokaryotic genomes.

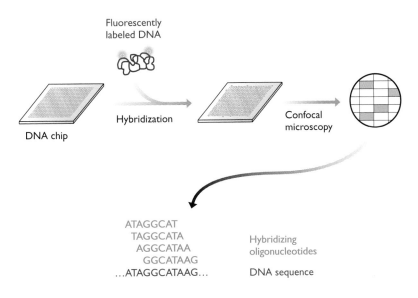

Fluorescently
labeled DNA

DNA chip

Hybridization

Confocal
microscopy

ATAGGCAT
TAGGCATA
AGGCATAA
GGCATAAG
...ATAGGCATAAG...

Hybridizing
oligonucleotides

DNA sequence

Figure 6.9   A possible way of using chip technology in DNA sequencing.

The chip carries an array of every possible 8-mer oligonucleotide. The DNA to be sequenced is labeled with a fluorescent marker and applied to the chip, and the positions of hybridizing oligonucleotides determined by confocal microscopy. Each hybridizing oligonucleotide represents an 8-nucleotide sequence motif that is present in the probe DNA. The sequence of the probe DNA can therefore be deduced from the overlaps between the sequences of these hybridizing oligonucleotides. See Technical Note 5.1, page 133 for more information on DNA chips.

6.2.1 Sequence assembly by the shotgun approach

The straightforward approach to sequence assembly is to build up the master sequence directly from the short sequences obtained from individual sequencing experiments, simply by examining the sequences for overlaps (see *Figure 5.1*, page 127). This is called the shotgun approach. It does not require any prior knowledge of the genome and so can be carried out in the absence of a genetic or physical map.

The potential of the shotgun approach was proven by the Haemophilus influenzae sequence

During the early 1990s there was extensive debate about whether the shotgun approach would work in practice, many molecular biologists being of the opinion that the amount of data handling needed to compare all the mini-sequences and identify overlaps, even with the smallest genomes, would be beyond the capabilities of existing computer systems. These doubts were laid to rest in 1995 when the sequence of the 1830 kb genome of the bacterium *Haemophilus influenzae* was published (Fleischmann *et al.*, 1995).

The *H. influenzae* genome was sequenced entirely by the shotgun approach and without recourse to any genetic or physical map information. The strategy used to obtain the sequence is shown in *Figure 6.10*. The first step was to break the genomic DNA into fragments by **sonication**, a technique which uses high-frequency sound waves to make random cuts in DNA molecules. The frag-

ments were then electrophoresed and those in the range 1.6–2.0 kb purified from the agarose gel and ligated into a plasmid vector. From the resulting library, 19 687 clones were taken at random and 28 643 sequencing experiments carried out, the number of sequencing experiments being greater than the number of plasmids because both ends of some inserts were sequenced. Of these sequencing experiments, 16% were considered to be failures because they resulted in less than 400 bp of sequence. The remaining 24 304 sequences gave a total of 11 631 485 bp, corresponding to six times the length of the *H. influenzae* genome, this amount of redundancy being deemed necessary to ensure complete coverage. Sequence assembly required 30 hours on a computer with 512 Mb of RAM, and resulted in 140 lengthy contiguous sequences, each of these **sequence contigs** representing a different, nonoverlapping portion of the genome.

The next step was to join up pairs of contigs by obtaining sequences from the gaps between them (*Figure 6.11*). First, the library was checked to see if there were any clones whose two end sequences were located in different contigs. If such a clone could be identified, then additional sequencing of its insert would close the 'sequence gap' between the two contigs (*Figure 6.11A*). In fact, there were 99 clones in this category, so 99 of the gaps could be closed without too much difficulty.

This left 42 gaps, which probably consisted of DNA sequences that were unstable in the cloning vector and therefore not present in the library. To close these 'physical gaps' a second clone library was prepared, this one with a different type of vector. Rather than using another plasmid, in which the uncloned sequences would probably

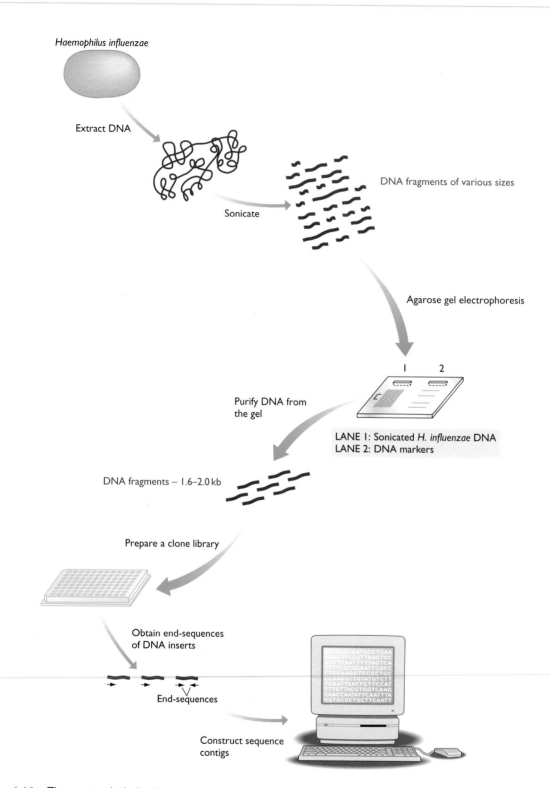

Haemophilus influenzae

Extract DNA

Sonicate

DNA fragments of various sizes

Agarose gel electrophoresis

1   2

Purify DNA from the gel

LANE 1: Sonicated *H. influenzae* DNA
LANE 2: DNA markers

DNA fragments – 1.6–2.0 kb

Prepare a clone library

Obtain end-sequences of DNA inserts

End-sequences

Construct sequence contigs

Figure 6.10 The way in which the shotgun approach was used to obtain the DNA sequence of the *Haemophilus influenzae* genome.

H. influenzae DNA was sonicated and fragments with sizes between 1.6 and 2.0 kb purified from an agarose gel and ligated into a plasmid vector to produce a clone library. End sequences were obtained from clones taken from this library, and a computer used to identify overlaps between sequences. This resulted in 140 sequence contigs, which were assembled into the complete genome sequence, as shown in *Figure 6.11*. For further details, see Fleischmann *et al.* (1995).

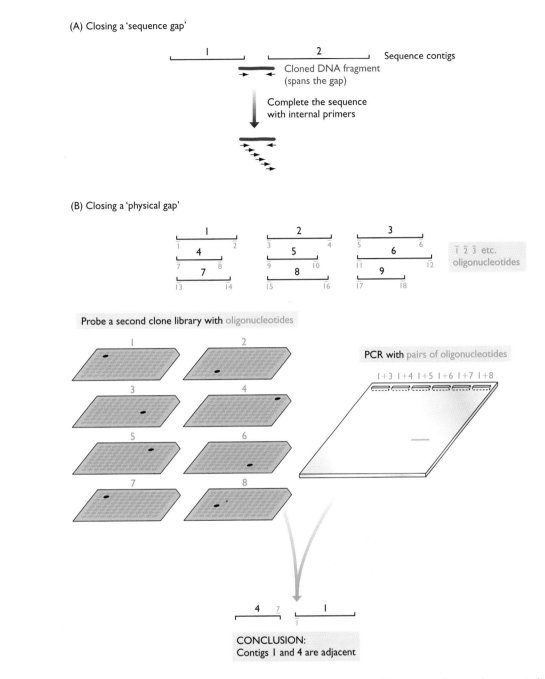

Figure 6.11 Assembly of the complete *Haemophilus influenzae* genome sequence by spanning the gaps between individual sequence contigs.

(A) 'Sequence gaps' are ones which can be closed by further sequencing of clones already present in the library. In this example, the end-sequences of contigs 1 and 2 lie within the same plasmid clone, so further sequencing of this DNA insert with internal primers (see *Figure 6.5B*) will provide the sequence to close the gap. (B) 'Physical gaps' are stretches of sequence that are not present in the clone library, probably because these regions are unstable in the cloning vector that was used. Two strategies for closing these gaps are shown. On the left, a second clone library, prepared with a bacteriophage λ vector rather than a plasmid vector, is probed with oligonucleotides corresponding to the ends of the contigs. Oligonucleotides 1 and 7 both hybridize to the same clone, whose insert must therefore contain DNA spanning the gap between contigs 1 and 4. On the right, PCRs are carried out with pairs of oligonucleotides. Only numbers 1 and 7 give a PCR product, confirming that the contig ends represented by these two oligonucleotides are close together in the genome. The PCR product or the insert from the λ clone could be sequenced to close the gap between contigs 1 and 4.

still be unstable, the second library was prepared in a bacteriophage λ vector (Section 4.2.1). This new library was probed with 84 oligonucleotides, one at a time, these 84 oligonucleotides having sequences identical to the sequences at the ends of the unlinked contigs (*Figure 6.11B*). The rationale was that if two oligonucleotides hybridized to the same λ clone then the ends of the contigs from which they were derived must lie within that clone, and sequencing the DNA in the λ clone would therefore close the gap. Twenty-three of the 42 physical gaps were dealt with in this way.

A second strategy for gap closure was to use pairs of oligonucleotides, from the set of 84 described above, as primers for PCRs of *H. influenzae* genomic DNA. Some oligonucleotide pairs were selected at random and those spanning a gap identified simply from whether or not they gave a PCR product (see *Figure 6.11B*). Sequencing the resulting PCR products closed the relevant gaps. Other primer pairs were chosen on a more rational basis. For example, oligonucleotides were tested as probes with a Southern blot of *H. influenzae* DNA cut with a variety of restriction endonucleases, and pairs that hybridized to similar sets of restriction fragments identified. The two members of an oligonucleotide pair identified in this way must be contained within the same restriction fragments and so are likely to lie close together on the genome. This means that the pair of contigs that the oligonucleotides are derived from are adjacent, and the gap between them can be spanned by a PCR of genomic DNA using the two oligonucleotides as primers, which will provide the template DNA for gap closure.

The demonstration that a small genome can be sequenced relatively rapidly by the shotgun approach led to a sudden plethora of completed microbial genomes. These projects demonstrated that shotgun sequencing can be set up on a production-line basis, with each team member having his or her individual task in DNA preparation, carrying out the sequencing reactions, or analyzing the data. This strategy resulted in the 580 kb genome of *Mycoplasma genitalium* being sequenced by five people in just eight weeks (Fraser *et al.*, 1995), and it is now accepted that a few months should be ample time to generate the complete sequence of any genome less than about 5 Mb, even if nothing is known about the genome before the project begins. The strengths of the shotgun approach are therefore its speed and its ability to work in the absence of a genetic or physical map.

6.2.2 Sequence assembly by the clone contig approach

The clone contig approach is the conventional method for obtaining the sequence of a eukaryotic genome and has also been used with those microbial genomes that have previously been mapped by genetic and/or physical means. In the clone contig approach, the genome is broken into fragments of up to 1.5 Mb, usually by partial restriction (Section 5.3.1), and these cloned in a high-capacity vector such as a BAC or a YAC (Section 4.2.1). A clone contig is built up by identifying clones containing overlapping fragments, which are then individually sequenced by the shotgun method. Ideally the cloned fragments are anchored onto a genetic and/or physical map of the genome, so that the sequence data from the contig can be checked and interpreted by looking for features (e.g. STSs, SSLPs, genes) known to be present in a particular region.

Clone contigs can be built up by chromosome walking, but the method is laborious

The simplest way to build up an overlapping series of cloned DNA fragments is to begin with one clone from a library, identify a second clone whose insert overlaps with the insert in the first clone, then identify a third clone whose insert overlaps with the second clone, and so on. This is the basis of **chromosome walking**, which was the first method devised for assembly of clone contigs.

Chromosome walking was originally used to move relatively short distances along DNA molecules, using clone libraries prepared with λ or cosmid vectors. The most straightforward approach is to use the insert DNA from the starting clone as a hybridization probe to screen all the other clones in the library. Clones whose inserts overlap with the probe give positive hybridization signals, and their inserts can be used as new probes to continue the walk (*Figure 6.12*).

The main problem that arises is that if the probe contains a genome-wide repeat sequence then it will hybridize not only to overlapping clones but also to non-overlapping clones whose inserts also contain copies of the repeat. The extent of this non-specific hybridization can be reduced by blocking the repeat sequences by pre-hybridization with unlabeled genomic DNA (see *Figure 5.30*, page 153). But this does not completely solve the problem, especially if the walk is being carried out with long inserts from high-capacity vectors such as BACs or YACs. For this reason, intact inserts are rarely used for chromosome walks with human DNA and similar DNAs which have a high frequency of genome-wide repeats. Instead, a fragment from the end of an insert is used as the probe, there being less chance of a genome-wide repeat occurring in a short end-fragment compared with the insert as a whole. If complete confidence is required then the end-fragment can be sequenced before use to ensure that no repetitive DNA is present.

If the end-fragment has been sequenced then the walk can be speeded up by using PCR rather than hybridization to identify clones with overlapping inserts. Primers are designed from the sequence of the end-fragment and used in attempted PCRs with all the other clones in the library. A clone that gives a PCR product of the correct size must contain an overlapping insert (*Figure 6.13*). To speed the process up even more, rather than performing a PCR with each individual clone, groups of clones are mixed together in such a way that unambiguous identifi-

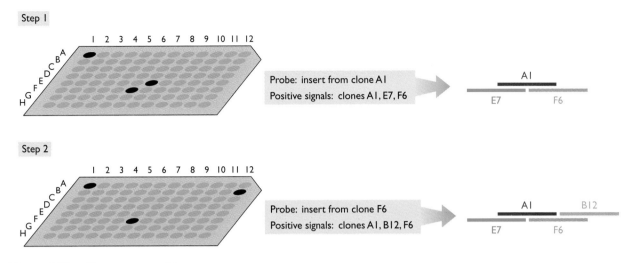

Figure 6.12   Chromosome walking.

The library comprises 96 clones, each containing a different insert. To begin the walk, the insert from one of the clones is used as a hybridization probe against all the other clones in the library. In the example shown, clone A1 is the probe; it hybridizes to itself and to clones E7 and F6. The inserts from the last two clones must therefore overlap with the insert from clone A1. To continue the walk, the probing is repeated but this time with the insert from clone F6. The hybridizing clones are A1, F6 and B12, showing that the insert from B12 overlaps with the insert from F6.

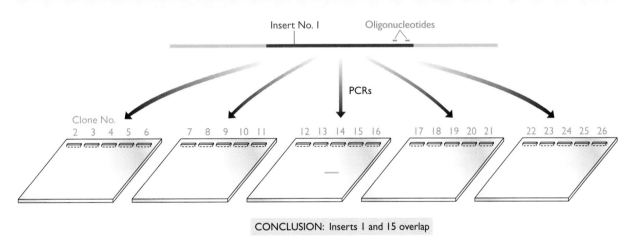

Figure 6.13   Chromosome walking by PCR.

The two oligonucleotides anneal within the end region of insert number 1. They are used in PCRs with all the other clones in the library. Only clone 15 gives a PCR product, showing that the inserts in clones 1 and 15 overlap. The walk would be continued by sequencing the fragment from the other end of clone 15, designing a second pair of oligonucleotides, and using these in a new set of PCRs with all the other clones.

cation of overlapping ones can still be made. The method is illustrated in *Figure 6.14*, in which a library of 960 clones has been prepared in ten microtiter trays, each tray comprising 96 wells in an 8 × 12 array, with one clone per well. PCRs are carried out as follows:

1. Samples of each clone in row A of the first microtiter tray are mixed together and a single PCR carried out. This is repeated for every row of every tray – 80 PCRs in all.

2. Samples of each clone in column 1 of the first microtiter tray are mixed together and a single PCR carried out. This is repeated for every column of every tray – 120 PCRs in all.

3. Clones from well A1 of each of the ten microtiter trays are mixed together and a single PCR carried out. This is repeated for every well – 96 PCRs in all.

As explained in the legend to *Figure 6.14*, these 296 PCRs provide enough information to identify which of the 960 clones give products and which do not. Ambiguities arise only if a substantial number of clones turn out to be positive.

Newer more rapid methods for clone contig assembly

Even when the screening step is carried out by the combinatorial PCR approach shown in *Figure 6.14*, chromosome walking is a slow process and it is rarely possible to assemble contigs of more than 15–20 clones by this method. The procedure has been extremely valuable in **positional cloning**, where the objective is to walk from a mapped site to an interesting gene that is known to be no more than a few Mb distant. It has been less valuable for assembling clone contigs across entire genomes, especially with the complex genomes of higher eukaryotes. So what alternative methods are there?

The main alternative is to use a **clone fingerprinting** technique. Clone fingerprinting provides information on the physical structure of a cloned DNA fragment, this physical information or 'fingerprint' being compared with equivalent data from other clones, enabling those with similarities – possibly indicating overlaps – to be identified. One or a combination of the following techniques is used (*Figure 6.15*):

- **Restriction patterns** can be generated by digesting clones with a variety of restriction enzymes and separating the products in an agarose gel. If two clones contain overlapping inserts then their restriction fingerprints will have bands in common, as both will contain fragments derived from the overlap region.
- **Repetitive DNA fingerprints** can be prepared by blotting a set of restriction fragments and carrying out Southern hybridization (Section 4.1.2) with probes specific for one or more types of genome-wide repeat. As for the restriction fingerprints, overlaps are identified by looking for two clones that have some hybridizing bands in common.

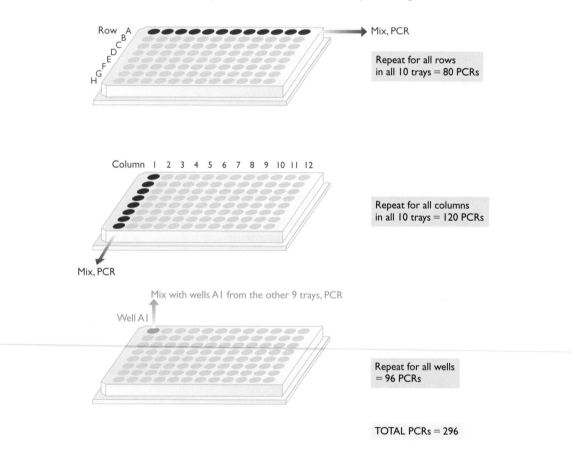

Figure 6.14 Combinatorial screening of clones in microtiter trays.

In this example, a library of 960 clones has to be screened by PCR. Rather than carrying out 960 individual PCRs, the clones are grouped as shown and just 296 PCRs are performed. In most cases, the results enable positive clones to be identified unambiguously. In fact, if there are few positive clones, then sometimes they can be identified by just the 'row' and 'column' PCRs. For example, if positive PCRs are obtained with tray 2 row A, tray 6 row D, tray 2 column 7, and tray 6 column 9, then it can be concluded that there are two positive clones, one in tray 2 well A7 and one in tray 6 well D9. The 'well' PCRs are needed if there are two or more positive clones in the same tray.

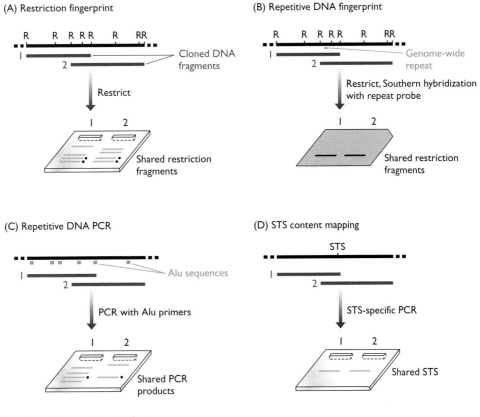

Figure 6.15 Four clone fingerprinting techniques.

- *Repetitive DNA PCR*, or *interspersed repeat element PCR (IRE-PCR)*, uses primers that anneal within genome-wide repeats and so amplify the single-copy DNA between two neighboring repeats. Because genome-wide repeat sequences are not evenly spaced in a genome, the sizes of the products obtained after repetitive DNA PCR can be used as a fingerprint in comparisons with other clones, in order to identify potential overlaps. With human DNA, the genome-wide repeats called Alu elements (Section 2.4.2) are often used because these occur on average once every 4 kb. An **Alu-PCR** of a human BAC insert of 150 kb would therefore be expected to give approximately 38 PCR products of various sizes, resulting in a detailed fingerprint.

- *STS content mapping* is particularly useful because it can result in a clone contig that is anchored onto a physical map of STS locations. PCRs directed at individuals STSs (Section 5.3.3) are carried out with each member of a clone library. Presuming the STS is single copy in the genome, then all clones that give PCR products must contain overlapping inserts.

As with chromosome walking, efficient application of these fingerprinting techniques requires combinatorial screening of gridded clones, ideally with computerized methodology for analyzing the resulting data.

6.2.3 Whole-genome shotgun sequencing

The whole-genome shotgun approach was first proposed by Craig Venter and colleagues as a means of speeding up the acquisition of contiguous sequence data for large genomes such as the human genome and those of other eukaryotes (Venter *et al.*, 1998; Marshall, 1999). Experience with conventional shotgun sequencing (Section 6.2.1) had shown that if the total length of sequence that is generated is between 6.5 and 8 times the length of the genome being studied, then the resulting sequence contigs will span over 99.8% of the genome sequence (Fraser, 1997), with a few gaps that can be closed by methods such as those developed during the *H. influenzae* project (see *Figure 6.11*). This implies that 70 million individual sequences, each 500 bp or so in length, corresponding to a total of 35 000 Mb, would be sufficient if the random approach were taken with the human genome. Seventy million sequences is not an impossibility: in fact, with 75 automatic sequencers, each performing 1000 sequences per day, the task could be achieved in 3 years.

The big question was whether the 70 million sequences could be assembled correctly. If the conventional shotgun approach is used with such a large number of fragments, and no reference is made to a genome map, then the answer is certainly no. The huge amount of computer time needed to identify overlaps between the sequences, and the errors, or at best uncertainties, caused by the extensive repetitive DNA content of most eukaryotic genomes (see *Figure 5.2*, page 127), would make the task impossible. But with reference to a map, Venter argued, it should be possible to assemble the mini-sequences in the correct way.

Key features of whole-genome shotgun sequencing

The most time-consuming part of a shotgun sequencing project is the 'finishing' phase when individual sequence contigs are joined by closure of sequence gaps and physical gaps (see *Figure 6.11*). To minimize the amount of finishing that is needed, the whole-genome shotgun approach makes use of at least two clone libraries, prepared with different types of vector. Two libraries are used because with any cloning vector it is anticipated that some fragments will not be cloned because of incompatibility problems that prevent vectors containing these fragments from being propagated. Different types of vector suffer from different problems, so fragments that cannot be cloned in one vector can often be cloned if a second vector is used. Generating sequence from fragments cloned in two different vectors should therefore improve the overall coverage of the genome.

What about the problems that repeat elements pose for sequence assembly? We highlighted this issue in Chapter 5 as the main argument against the use of shotgun sequencing with eukaryotic genomes, because of the possibility that jumps between repeat units will lead to parts of a repetitive region being left out, or an incorrect connection being made between two separate pieces of the same or different chromosomes (see *Figure 5.2*, page 127). Several possible solutions to this problem have been proposed (Weber and Myers, 1997), but the most successful strategy is to ensure that one of the clone libraries contains fragments that are longer than the longest repeat sequences in the genome being studied. For example, one of the plasmid libraries used when the shotgun approach was applied to the *Drosophila* genome contained inserts with an average size of 10 kb, because most *Drosophila* repeat sequences are 8 kb or fewer. Sequence jumps, from one repeat sequence to another, are avoided by ensuring that the two end-sequences of each 10-kb insert are at their appropriate positions in the master sequence (*Figure 6.16*).

The initial result of sequence assembly is a series of **scaffolds**, each one comprising a set of sequence contigs separated by sequence gaps – ones which lie between the mini-sequences from the two ends of a single cloned fragment and so can be closed by further sequencing of that fragment (*Figure 6.17*). The scaffolds themselves are separated by physical gaps, which are more difficult to close because they represent sequences that are not in the clone libraries. The marker content of each scaffold is used to determine its position on the genome map. For example, if the locations of STSs in the genome map are known then a scaffold can be positioned by determining which STSs it contains. If a scaffold contains STSs from two non-

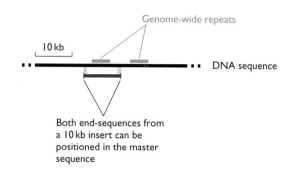

(A) Correct sequence assembly

Genome-wide repeats

10 kb

DNA sequence

Both end-sequences from a 10 kb insert can be positioned in the master sequence

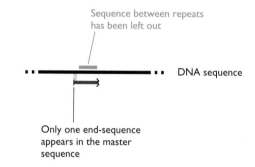

(B) Incorrect sequence assembly

Sequence between repeats has been left out

DNA sequence

Only one end-sequence appears in the master sequence

Figure 6.16 Avoiding errors when the whole-genome shotgun approach is used.

In *Figure 5.2B*, page 127, we saw how easy it would be to 'jump' between repeat sequences when assembling the master sequence by the standard shotgun approach. The result of such an error would be to lose all the sequence between the two repeats that had mistakenly been linked together. This type of error is avoided in the whole-genome shotgun approach by ensuring that the two end-sequences of a cloned DNA fragment (of 10 kb or so) both appear on the master sequence at their expected positions in the unique DNA sequences to either side of a genome-wide repeat. If one of the end-sequences is missing, then an error has been made when assembling the master sequence.

contiguous parts of the genome then an error has occurred during sequence assembly. The accuracy of sequence assembly can be further checked by obtaining end-sequences from fragments of 100 kb or more that have been cloned in a high-capacity vector. If a pair of end-sequences do not fall within a single scaffold at their anticipated positions relative to each other, then again an error in assembly has occurred.

The feasibility of the whole-genome shotgun approach has been demonstrated by its application to the fruit-fly and human genomes (Adams *et al.*, 2000; Venter *et al.*, 2001). The question that remains, and which has been

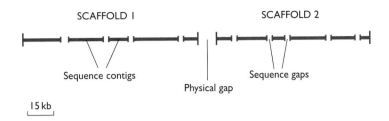

Figure 6.17 Scaffolds are intermediates in sequence assembly by the whole-genome shotgun approach.

Two scaffolds are shown. Each comprises a series of sequence contigs separated by sequence gaps, with the scaffolds themselves separated by physical gaps.

hotly debated (Patterson, 1998), is whether the sequences obtained by the whole-genome shotgun approach have the desired degree of accuracy. Part of the problem is that the random nature of sequence generation means that some parts of the genome are covered by several of the mini-sequences that are obtained, whereas other parts are represented just once or twice (*Figure 6.18*). It is generally accepted that every part of a genome should be sequenced at least four times to ensure an acceptable level of accuracy, and that this coverage should be increased to 8–10 times before the sequence can be looked upon as being complete. A sequence obtained by the whole-genome shotgun approach is likely to exceed this requirement in many regions, but may fall short in other areas. If those areas include genes, then the lack of accuracy could cause major problems when attempts are made to locate the genes and understand their functions (see Chapter 7). These problems have been highlighted by studies of the *Drosophila* genome sequence, which have suggested that as many as 6500 of the 13 600 genes might contain significant sequence errors (Karlin *et al.*, 2001).

6.3 The Human Genome Projects

To conclude our examination of mapping and sequencing we will look at how these techniques were applied to the human genome. Although every genome project is different, with its own challenges and its own solutions to those challenges, the human projects illustrate the general issues that have had to be addressed in order to sequence a large eukaryotic genome, and in many ways illustrate the procedures that are currently regarded as state of the art in this area of molecular biology.

6.3.1 The mapping phase of the Human Genome Project

Until the beginning of the 1980s a detailed map of the human genome was considered to be an unattainable objective. Although comprehensive genetic maps had been constructed for fruit flies and a few other organisms, the problems inherent in analysis of human pedigrees (Section 5.2.4) and the relative paucity of polymorphic genetic markers meant that most geneticists doubted that a human genetic map could ever be achieved. The initial impetus for human genetic mapping came from the discovery of RFLPs, which were the first highly polymorphic DNA markers to be recognized in animal genomes. In 1987 the first human RFLP map was published, comprising 393 RFLPs and ten additional polymorphic markers (Donis-Keller *et al.*, 1987). This map, developed from analysis of 21 families, had an average marker density of one per 10 Mb.

In the late 1980s the Human Genome Project became established as a loose but organized collaboration between geneticists in all parts of the world. One of the goals that the Project set itself was a genetic map with a density of one marker per 1 Mb, although it was thought that a density of one per 2–5 Mb might be the realistic limit. In fact by 1994 an international consortium had met and indeed exceeded the objective, thanks to their use of SSLPs and the large CEPH collection of reference families (Section 5.2.4). The 1994 map contained 5800 markers, of which over 4000 were SSLPs, and had a density of one marker per 0.7 Mb (Murray *et al.*, 1994). A subsequent version of the genetic map (Dib *et al.*, 1996) took the 1994 map slightly further by inclusion of an additional 1250 SSLPs.

Physical mapping did not lag far behind. In the early 1990s considerable effort was put into the generation of

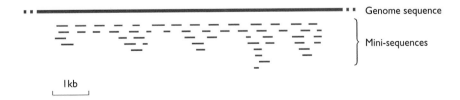

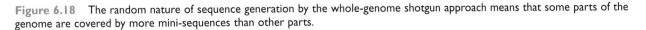

Figure 6.18 The random nature of sequence generation by the whole-genome shotgun approach means that some parts of the genome are covered by more mini-sequences than other parts.

clone contig maps, using STS screening (Section 5.3.3) as well as other clone fingerprinting methods (Section 6.2.2). The major achievement of this phase of the physical mapping project was publication of a clone contig map of the entire genome, consisting of 33 000 YACs containing fragments with an average size of 0.9 Mb (Cohen *et al.*, 1993). However, doubts were raised about the value of YAC contig maps when it was realized that YAC clones can contain two or more pieces of non-contiguous DNA (*Figure 6.19*; Anderson, 1993). The use of these chimeric clones in the construction of contig maps could result in DNA segments that are widely separated in the genome being mistakenly mapped to adjacent positions. These problems led to the adoption of radiation hybrid mapping of STS markers (Section 5.3.3), largely by the Whitehead Institute/MIT Genome Center in Massachusetts, culminating in 1995 with publication of a human STS map containing 15 086 markers, with an average density of one per 199 kb (Hudson *et al.*, 1995). This map was later supplemented with an additional 20 104 STSs, most of these being ESTs and hence positioning protein-coding genes on the physical map (Schuler *et al.*, 1996). The resulting map density approached the target of one marker per 100 kb set as the objective for physical mapping at the outset of the Human Genome Project.

The combined STS maps included positions for almost 7000 polymorphic SSLPs that had also been mapped onto the genome by genetic means. As a result, the physical and genetic maps could be directly compared, and clone contig maps that included STS data could be anchored onto both maps. The net result was a comprehensive, integrated map (Bentley *et al.*, 1998; Deloukas *et al.*, 1998) that could be used as the framework for the DNA sequencing phase of the Human Genome Project.

6.3.2 Sequencing the human genome

The original plan was that the sequencing phase of the Human Genome Project would be based on YAC libraries, because this type of vector can be used with DNA fragments longer than can be handled by any other type of cloning system. This strategy had to be abandoned when it was discovered that some YAC clones contain non-contiguous fragments of DNA. The Project therefore turned its attention to BACs (Section 4.2.1). A library of 300 000 BAC clones was generated and these clones mapped onto the genome, forming a 'sequence-ready' map which could be used as the primary foundation for the sequencing phase of the project, during which the insert from each BAC would be completely sequenced by the shotgun method.

At about the time when the Human Genome Project was gearing itself up to move into the sequence-acquisition phase, the whole-genome shotgun approach was first proposed as an alternative to the more laborious clone contig method that had so far been adopted (Venter *et al.*, 1998). The possibility that the Human Genome Project would not in fact provide the first human genome sequence stimulated the organizers of the Project to bring forward their planned dates for completion of a working draft (Collins *et al.*, 1998). The first sequence of an entire human chromosome (number 22) was published in December 1999 (Dunham *et al.*, 1999) and the sequence of chromosome 21 appeared a few months later (Hattori *et al.*, 2000). Finally, on 26 June 2000, accompanied by the President of the United States, Francis Collins and Craig Venter, the leaders of the two projects, jointly announced completion of their working drafts (Marshall, 2000), which appeared in print eight months later (IHGSC, 2001; Venter *et al.*, 2001).

It is important to understand that the two genome sequences published in 2001 are drafts, not complete final sequences. For example, the version obtained by the clone contig approach covers just 90% of the genome, the missing 320 million bp lying predominantly in **constitutive heterochromatin** (Section 8.1.2) – regions of chromosomes in which the DNA is very tightly packaged and which are thought to contain few, if any genes. Within the 90% of the genome that is covered, each part has been sequenced at least four times, providing an 'acceptable' level of accuracy, but only 25% has been sequenced the 8–10 times that is necessary before the work is considered to be 'finished' (Bork and Copley, 2001). This draft sequence comprises approximately 50 000 scaffolds (see *Figure 6.17*) with an average size of 54.2 kb. Similar statistics apply to the whole-genome shotgun sequence. A substantial number of gaps therefore have to be closed and much additional sequencing must be done in order to bring the two draft sequences to the stage where either is considered to be complete.

6.3.3 The future of the human genome projects

Completion of a finished sequence is not the only goal of the consortia working on the human genome. Understanding the genome sequence is a massive task that will engage many groups around the world, making use of various techniques and approaches which will be described in the next chapter. Important among these are the use of **comparative genomics**, in which two complete genome sequences are compared in order to identify com-

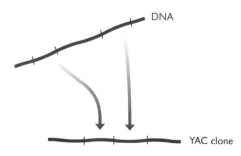

Figure 6.19 Some YAC clones contain segments of DNA from different parts of the human genome.

mon features that, being conserved, are likely to be important (Section 7.4). With the human genome, comparative genomics has the added value that it may allow the animal versions of human disease genes to be located, paving the way for studies of the genetic basis of these diseases using the animal genes as models for the human condition. Genome projects for the mouse and rat are both underway, with draft sequences expected in 2003 (Pennisi, 1999; Denny and Justice, 2000; Marshall, 2001), and plans are being made for a chimpanzee genome project (Normile, 2001). There will also be additional human genome projects aimed at building up a catalog of sequence variability in different populations, the results possibly enabling the ancient origins of these populations to be inferred (Section 16.3.2).

These human diversity projects lead us to the controversial aspects of genome sequencing. Most scientists anticipate that sequence data from different populations will emphasize the unity of the human race by showing that patterns of genetic variability do not reflect the geographic and political groupings that humans have adopted during the last few centuries. But the outcomes of these projects are still certain to stimulate debate in non-scientific circles. Additional controversies center on the question of who, if anyone, will own human DNA sequences. To many, the idea of ownership of a DNA sequence is a peculiar concept, but large sums of money can be made from the information contained in the human genome, for example by using gene sequences to direct development of new drugs and therapies against cancer and other diseases. Pharmaceutical companies involved in genome sequencing naturally want to protect their investments, as they would for any other research enterprise, and currently the only way of doing this is by patenting the DNA sequences that they discover. Unfortunately, in the past, errors have been made in dealing with the financial issues relating to research with human biological material, the individual from whom the material is obtained not always being a party in the profit sharing. These issues have still to be resolved.

The problems relating to the public usage of human genome sequences are even more contentious. A major concern is the possibility that, once the sequence is understood, individuals whose sequences are considered 'sub-standard', for whatever reason, might be discriminated against. The dangers range from increased insurance premiums for individuals whose sequence includes mutations predisposing them to a genetic disease, to the possibility that racists might attempt to define 'good' and 'bad' sequence features, with depressingly predictable implications for the individuals unlucky enough to fall into the 'bad' category.

The two human genome projects, especially in the USA, continue to support research and debate into the ethical, legal and social issues raised by genome sequencing. In particular, great care is being taken to ensure that the genome sequences that result from the projects cannot be identified with any single individual. The DNA that is being cloned and sequenced is taken only from individuals who have given consent for their material to be used in this way and for whom anonymity can be guaranteed. When this policy was first adopted it required a certain amount of realignment of the research effort because older clone libraries had to be destroyed and the existing physical maps checked with the new material. It is accepted, however, that the extra work was necessary and that care must be taken to maintain and enhance public confidence in the projects.

References

Adams MA, Celniker SE, Holt RA, et al. (2000) The genome sequence of *Drosophila melanogaster. Science*, **287**, 2185–2195.

Anderson C (1993) Genome shortcut leads to problems. *Science*, **259**, 1684–1687.

Bentley DR, Pruitt KD, Deloukas P, Schuler GD and Ostell J (1998) Coordination of human genome sequencing via a consensus framework map. *Trends Genet.*, **14**, 381–384.

Bork P and Copley R (2001) Filling in the gaps. *Nature*, **409**, 818–820.

Cohen D, Chumakov I and Weissenbach J (1993) A first-generation map of the human genome. *Nature*, **366**, 698–701.

Collins FS, Patrinos A, Jordan E, et al. (1998) New goals for the US Human Genome Project: 1998–2003. *Science*, **282**, 682–689.

Deloukas P, Schuler GD, Gyapay G, et al. (1998) A physical map of 30,000 genes. *Science*, **282**, 744–746.

Denny P and Justice MJ (2000) Mouse as the measure of man? *Trends Genet.*, **16**, 283–287.

Dib C, Fauré S, Fizames C, et al. (1996) A comprehensive genetic map of the human genome based on 5,264 microsatellites. *Nature*, **380**, 152–154.

Donis-Keller H, Green P, Helms C, et al. (1987) A genetic map of the human genome. *Cell*, **51**, 319–337.

Dunham I, Shimizu N, Roe BA, et al. (1999) The DNA sequence of human chromosome 22. *Nature*, **402**, 489–495.

Fleischmann RD, Adams MD, White O, et al. (1995) Whole-genome random sequencing and assembly of *Haemophilus influenzae* Rd. *Science*, **269**, 496–512.

Fraser CM (1997) How to sequence a small genome. *Trends Genet.*, **13**, poster insert.

Fraser CM, Gocayne JD, White O, et al. (1995) The minimal gene complement of *Mycoplasma genitalium. Science*, **270**, 397–403.

Hattori M, Fujiyama A, Taylor TD, et al. (2000) The DNA sequence of human chromosome 21. *Nature*, **405**, 311–319.

Hudson TJ, Stein LD, Gerety SS, et al. (1995) An STS-based map of the human genome. *Science*, **270**, 1945–1954.

IHGSC (International Human Genome Sequencing Consortium) (2001) Initial sequencing and analysis of the human genome. *Nature*, **409**, 860–921.

Karlin S, Bergman A and Gentles AJ (2001) Annotation of the *Drosophila* genome. *Nature*, **411**, 259–260.

Marshall E (1999) A high-stakes gamble on genome sequencing. *Science*, **284**, 1906–1909.

Marshall E (2000) Rival genome sequencers celebrate a milestone together. *Science*, **288**, 2294–2295.

Marshall E (2001) Rat genome spurs an unusual partnership. *Science*, **291**, 1872.

Maxam AM and Gilbert W (1977) A new method for sequencing DNA. *Proc. Natl Acad. Sci. USA*, **74**, 560–564.

Mullikan JC and McMurray AA (1999) Sequencing the genome, fast. *Science*, **283**, 1867–1868.

Murray JC, Buetow KH, Weber JL, et al. (1994) A comprehensive human linkage map with centimorgan density. *Science*, **265**, 2049–2054.

Normile D (2001) Chimp sequencing crawls forward. *Science*, **291**, 2297.

Patterson M (1998) Politicogenomics takes centre stage. *Trends Genet.*, **14**, 259–260.

Pennisi E (1999) Mouse genome added to sequencing effort. *Science*, **286**, 210.

Prober JM, Trainor GL, Dam RJ, et al. (1987) A system for rapid DNA sequencing with fluorescent chain-terminating dideoxynucleotides. *Science*, **238**, 336–341.

Rogers J (1999) Gels and genomes. *Science*, **286**, 429.

Ronaghi M, Ehleen M and Nyrn P (1998) A sequencing method based on real-time pyrophosphate. *Science*, **281**, 363–365.

Sanger F, Nicklen S and Coulson AR (1977) DNA sequencing with chain terminating inhibitors. *Proc. Natl Acad. Sci. USA*, **74**, 5463–5467.

Schuler GD, Boguski MS, Stewart EA, et al. (1996) A gene map of the human genome. *Science*, **274**, 540–546.

Sears LE, Moran LS, Kisinger C, et al. (1992) CircumVent thermal cycle sequencing and alternative manual and automated DNA sequencing protocols using the highly thermostable Vent (exo⁻) DNA polymerase. *Biotechniques*, **13**, 626–633.

Southern EM (1996) DNA chips: analysing sequence by hybridization to oligonucleotides on a large scale. *Trends Genet.*, **12**, 110–115.

Venter JC, Adams MD, Sutton GG, et al. (1998) Shotgun sequencing of the human genome. *Science*, **280**, 1540–1542.

Venter JC, Adams MD, Myers EW, et al. (2001) The sequence of the human genome. *Science*, **291**, 1304–1351.

Weber JL and Myers EW (1997) Human whole-genome shotgun sequencing. *Genome Res.*, **7**, 401–409.

Further Reading

Brown TA (1994) *DNA Sequencing: The Basics.* Oxford University Press, Oxford. — *Details of DNA sequencing methodology.*

Davies K (2001) *Cracking the Genome: Inside the Race to Unlock Human DNA.* Free Press, New York. (Published in the UK as *The Sequence: Inside the Race for the Human Genome.* Weidenfeld and Nicholson, London.) — *A history of the human genome projects.*

Strachan T and Read AP (1999) *Human Molecular Genetics*, 2nd edition. BIOS Scientific Publishers, Oxford. — *Chapter 13 describes the Human Genome Project.*

Wilkie T (1993) *Perilous Knowledge: The Human Genome Project and its Implications.* Faber and Faber, New York. — *A view of the social impact of the Human Genome Project.*

STUDY AIDS FOR CHAPTER 6

Key terms

Give short definitions of the following terms:

Alu-PCR	Positional cloning
Chain termination method	Processivity
Chemical degradation method	Pyrosequencing
	Repetitive DNA fingerprint
Chromosome walking	Repetitive DNA PCR
Clone contig approach	Replicative form
Clone fingerprinting	Scaffold
Comparative genomics	Sequence contig
Dideoxynucleotide	Sonication
Helper phage	STS content mapping
Interspersed repeat element PCR (IRE-PCR)	Thermal cycle sequencing
	Whole-genome shotgun approach
Phagemid	
Polyacrylamide gel electrophoresis	

Self study questions

1. Draw a diagram illustrating the steps involved in a chain termination sequencing experiment. Your diagram should make clear the process by which the sequence is read from the polyacrylamide gel.

2. Outline how DNA is sequenced by the chemical degradation procedure.

3. What are the desirable features of a DNA polymerase that is to be used for chain termination sequencing?

4. Describe the advantages and disadvantages of M13 vectors as a source of DNA for chain termination sequencing.

5. Describe the methods, other than the use of an M13 vectors, by which single-stranded DNA can be obtained for chain termination sequencing.

6. Describe how automated DNA sequencing differs from the standard chain termination method, and evaluate the importance of automated sequencing in genomics research.

7. Outline the non-conventional methods for DNA sequencing that are currently being explored.

8. Using *Haemophilus influenzae* as an example, describe how the shotgun method is used to sequence a small bacterial genome.

9. Distinguish between the terms 'sequence gap' and 'physical gap'. Describe the various methods that can be used to close physical gaps in a DNA sequence obtained by the shotgun method.

10. Describe the methods used to obtain an overlapping set of clones when the clone contig approach is used to sequence a genome.

11. Explain why the clone contig approach is generally considered to be more accurate than whole-genome shotgun sequencing.

12. Evaluate the strengths and weaknesses of the whole-genome shotgun approach to DNA sequencing. What steps are taken to ensure that a sequence resulting from this approach is accurate?

13. Write a short history of the Human Genome Project.
14. Describe the future directions of research into the human genome sequence.

Problem-based learning

1. In the late 1970s, the chain termination and chemical degradation methods for DNA sequencing appeared to be equally efficacious. But today virtually all sequencing is done by the chain termination method. Write a report that explains why chain termination sequencing has become predominant.
2. You have isolated a new species of bacterium whose genome is a single DNA molecule of approximately 2.6 Mb. Write a detailed project plan to show how you would obtain the genome sequence for this bacterium.
3. Critically evaluate the clone contig approach as a means of sequencing a large eukaryotic genome.
4. From a comparison of the research papers describing the two draft human genome sequences (IHGSC, 2001; Venter et al., 2001) evaluate the success of the whole-genome shotgun approach as applied to the human genome.
5. With the benefit of hindsight, evaluate the decision making that occurred during the course of the Human Genome Project, and assess if any alternative strategies would have resulted in the draft sequence being obtained more quickly.
6. Human genome sequence: friend or foe?

Understanding a Genome Sequence

Chapter Contents

7.1 Locating the Genes in a Genome Sequence 188

7.1.1 Gene location by sequence inspection 188

7.1.2 Experimental techniques for gene location 191

7.2 Determining the Functions of Individual Genes 195

7.2.1 Computer analysis of gene function 196

7.2.2 Assigning gene function by experimental analysis 198

7.2.3 More detailed studies of the activity of a protein coded by an unknown gene 204

7.3 Global Studies of Genome Activity 207

7.3.1 Studying the transcriptome 207

7.3.2 Studying the proteome 208

7.4 Comparative Genomics 213

7.4.1 Comparative genomics as an aid to gene mapping 214

7.4.2 Comparative genomics in the study of human disease genes 215

Learning outcomes

When you have read Chapter 7, you should be able to:

■ Describe the strengths and weaknesses of the computational and experimental methods used to analyze genome sequences

■ Describe the basis of open reading frame (ORF) scanning, and explain why this approach is not always successful in locating genes in eukaryotic genomes

■ Outline the various experimental methods used to identify parts of a genome sequence that specify RNA molecules

■ Define the term 'homology' and explain why homology is important in computer-based studies of gene function

■ Evaluate the limitations of homology analysis, using the yeast genome project as an example

■ Describe the methods used to inactivate individual genes in yeast and mammals, and explain how inactivation can lead to identification of the function of a gene

■ Give outline descriptions of techniques that can be used to obtain more detailed information on the activity of a protein coded by an unknown gene

■ Describe how the transcriptome and proteome are studied

■ Explain how protein interaction maps are constructed and indicate the key features of the yeast map

■ Evaluate the potential and achievements of comparative genomics as a means of understanding a genome sequence

A GENOME SEQUENCE is not an end in itself. A major challenge still has to be met in understanding what the genome contains and how the genome functions. The former is addressed by a combination of computer analysis and experimentation, with the primary aim of locating the genes and their control regions. The first part of this chapter is devoted to these methods. The second question – understanding how the genome functions – is, to a certain extent, merely a different way of stating the objectives of molecular biology over the last 30 years. The difference is that in the past attention has been directed at the expression pathways for individual genes, with groups of genes being considered only when the expression of one gene is linked to that of another. Now the question has become more general and relates to the expression of the genome as a whole. The techniques used to address this topic will be covered in the latter parts of this chapter.

7.1 Locating the Genes in a Genome Sequence

Once a DNA sequence has been obtained, whether it is the sequence of a single cloned fragment or of an entire chromosome, then various methods can be employed to locate the genes that are present. These methods can be divided into those that involve simply inspecting the sequence, by eye or more frequently by computer, to look for the special sequence features associated with genes, and those methods that locate genes by experimental analysis of the DNA sequence. The computer methods form part of the methodology called **bioinformatics**, and it is with these that we begin.

7.1.1 Gene location by sequence inspection

Sequence inspection can be used to locate genes because genes are not random series of nucleotides but instead have distinctive features. These features determine whether a sequence is a gene or not, and so by definition are not possessed by non-coding DNA. At present we do not fully understand the nature of these specific features, and sequence inspection is not a foolproof way of locating genes, but it is still a powerful tool and is usually the first method that is applied to analysis of a new genome sequence.

The coding regions of genes are open reading frames

Genes that code for proteins comprise open reading frames (ORFs) consisting of a series of codons that specify the amino acid sequence of the protein that the gene codes for (see *Figure 1.17*, page 21). The ORF begins with an initiation codon – usually (but not always) ATG – and ends with a termination codon: TAA, TAG or TGA (Section 3.3.2). Searching a DNA sequence for ORFs that begin with an ATG and end with a termination triplet is therefore one way of looking for genes. The analysis is complicated by the fact that each DNA sequence has six **reading frames**, three in one direction and three in the reverse direction on the complementary strand (*Figure 7.1*), but computers are quite capable of scanning all six reading frames for ORFs. How effective is this as a means of gene location?

The key to the success of **ORF scanning** is the frequency with which termination codons appear in the DNA sequence. If the DNA has a random sequence and a GC content of 50% then each of the three termination codons – TAA, TAG and TGA – will appear, on average, once every $4^3 = 64$ bp. If the GC content is > 50% then the termination codons, being AT-rich, will occur less frequently but one will still be expected every 100–200 bp. This means that random DNA should not show many ORFs longer than 50 codons in length, especially if the presence of a starting ATG is used as part of the definition of an 'ORF'. Most genes, on the other hand, are longer than 50 codons: the average lengths are 317 codons for *Escherichia coli*, 483 codons for *Saccharomyces cerevisiae*, and approximately 450 codons for humans.

ORF scanning, in its simplest form, therefore takes a figure of, say, 100 codons as the shortest length of a putative gene and records positive hits for all ORFs longer than this.

How well does this strategy work in practice? With bacterial genomes, simple ORF scanning is an effective way of locating most of the genes in a DNA sequence. This is illustrated by *Figure 7.2*, which shows a segment of the *E. coli* genome with all ORFs longer than 50 codons highlighted. The real genes in the sequence cannot be mistaken because they are much longer than 50 codons in length. With bacteria the analysis is further simplified by the fact that there is relatively little non-coding DNA in the genome (only 11% for *E. coli*, see Section 2.3.2). If we assume that the real genes do not overlap, and that there are no genes-within-genes (see Box 2.2, page 44), which are valid assumptions for most bacterial genomes, then it is only in the non-coding regions that there is a possibility of mistaking a short spurious ORF for a real gene. So if the non-coding component of a genome is small then there is a reduced chance of making mistakes in interpreting the results of a simple ORF scan.

Simple ORF scans are less effective with higher eukaryotic DNA

Although ORF scans work well for bacterial genomes, they are less effective for locating genes in DNA sequences from higher eukaryotes. This is partly because there is substantially more space between the real genes in a eukaryotic genome (62% of the human genome is intergenic – Box 1.4, page 23), increasing the chances of finding spurious ORFs. But the main problem with the human genome and the genomes of higher eukaryotes in general is that their genes are often split by introns (Sections 1.2.1), and so do not appear as continuous ORFs in the DNA sequence. Many exons are shorter than 100 codons, some fewer than 50 codons, and continuing the reading frame into an intron usually leads to a termination sequence that appears to close the ORF (*Figure 7.3*). In other words, the genes of a higher eukaryote do not appear in the genome sequence as long ORFs, and simple ORF scanning cannot locate them.

Solving the problem posed by introns is the main challenge for bioinformaticists writing new software programs for ORF location. Three modifications to the basic procedure for ORF scanning have been adopted (Fickett, 1996):

■ *Codon bias* is taken into account. 'Codon bias' refers to the fact that not all codons are used equally frequently in the genes of a particular organism. For example, leucine is specified by six codons in the genetic code (TTA, TTG, CTT, CTC, CTA and CTG; see *Figure 3.20*, page 86), but in human genes leucine is most frequently coded by CTG and is only rarely specified by TTA or CTA. Similarly, of the four valine codons, human genes use GTG four times

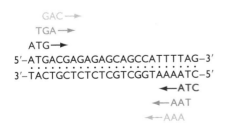

```
           GAC→
          TGA→
      ATG→
5'–ATGACGAGAGAGCAGCCATTTTAG–3'
  ················································
3'–TACTGCTCTCTCGTCGGTAAAATC–5'
                    ←ATC
               ←AAT
            ←AAA
```

Figure 7.1 A double-stranded DNA molecule has six reading frames.

Both strands are read in the 5'→3' direction. Each strand has three reading frames, depending on which nucleotide is chosen as the starting position.

GCGCAACGCAATTAATGTGCGTTAGCTCACTCATTAGGCACCCCAGGCTTTACACTTTATGCTTCCGGCTCGTATGTTGTGGAATTGTGAGCGGATAACAATTTCACACAGGAAACAGCTATGACCATGATTACGGATTCACTGGCCGTCGTTTTACAACGTCGTGACTGGGAAAACCCTGGCGTTACCCAACTTAATCGCCTTGCAGCACATCCCCCTTTCGCCAGCTGGCGTAA

TAGCGAAGAGGCCCGCACCGATCGCCCTTCCCAACAGTTGCGCAGCCTGAATGGCGAATGGCGCTTTGCCTGGTTTCCGGCACCAGAAGCGGTGCCGGAAAGCTGGCTGGAGTGCGATCTTCCTGAGGCCGATACTGTCGTCGTCCCCAAACTGGCAGATGCACGGTTACGATGCGCCCATCTACACCAACGTAACCTATCCCATTACGGTCAATCCGCCGTTTGTTCCCACGGAG

AATCCGACGGGTTGTTACTCGCTCACATTTAATGTTGATGAAAGCTGGCTACAGGAAGGCCAGACGCGAATTATTTTTGATGGCGTTAACTCGGCGTTTCATCTGTGGTGCAACGGGCGCTGGGTCGGTTACGGCCAGGACAGTCGTTTGCCGTCTGAATTTGACCTGAGCGCATTTTTACGCGCCGGAGAAAACCGCCTCGCGGTGATGGTGCTGCGTTGGAGTGACGGCAGTTATC

TGGAAGATCAGGATATGTGGCGGATGAGCGGCATTTTCCGTGACGTCTCGTTGCTGCATAAACCGACTACACAAATCAGCGATTTCCATGTTGCCACTCGCTTTAATGATGATTTCAGCCGGCTGTACTGGAGGCTGAAGTTCAGATGTGCGGCGAGTTGCGTGACTACCTACGGGTAACAGTTTCTTTATGGCAGGGTGAAACGCAGGTCGCCAGCGGCACCGCGCCTTTCGGCGG

TGAAATTATCGATGAGCGTGGTGGTTATGCCGATCGCGTCACACTACGTCTGAACGTCGAAAACCCGAAACTGTGGAGCGCCGAAATCCCGAATCTCTATCGTGCGGTGGTTGAACTGCACACCGCCGACGGCACGCTGATTGAAGCAGAAGCCTGCGATGTCGGTTTCCGCGAGGTGCGGATTGAAAATGGTCTGCTGCTGCTGAACGGCAAGCCGTTGCTGATTCGAGGCGTTAAC

CGTCACGAGCATCATCCTCTGCATGGTCAGGTCATGGATGAGCAGACGATGGTGCAGGATATCCTGCTGATGAAGCAGAACAACTTTAACGCCGTGCGCTGTTCGCATTATCCGAACCATCCGCTGTGGTACACGCTGTGCGACCGGCTACGGGTCGTATGTGGTGGATGAAGCCAATATTGAAACCCACGGCATGGTGCCAATGAACGTCTGACCGATGATCCGCGCTGGCTACCGG

CGATGAGCGAACGCGTAACGCGAATGGTGCAGCGCGATCGTAATCACCCGAGTGTGATCATCTGGTCGCTGGGGAATGAATCAGGCCACGGCGCTAATCACGACGCGCTGTATCGCTGGATCAAATCTGTCGATCCTTCCCGCCCGGTGCAGTATGAAGGCGGCGGAGCCGACACCACGGCCACCGATATTATTTGCCCGATGTACGCGCGCGTGGATGAAGACCAGCCCTTCCCGGC

TGTGCCGAAATGGTCCATCAAAAAATGGCTTTCGCTACCTGGAGAGACGCGCCCGCTGATCCTTTGCGAATACGCCCACGCGATGGGTAACAGTCTTGGCGGTTTCGCTAAATACTGGCAGGCGTTTCGTCAGTATCCCCGTTTACAGGGCGGCTTCGTCTGGGACTGGGTGGATCAGTCGCTGATTAAATATGATGAAAACGGCAACCCGTGGTCGGCTTACGGCGGTGATTTTGGC

GATACGCCGAACGATCGCCAGTTCTGTATGAACGGTCTGGTCTTTGCCGACCGCACGCCGCATCCAGCGCTGACGGAAGCAAAACACCAGCAGCAGTTTTCCAGTTCCGTTTATCCGGGCAAACCATCGAAGTGACCAGCGAATACCTGTTCCGTCATAGCGATAACGAGCTCCTGCACTGGATGGTGGCGCTGGATGGTAAGCCGCTGGCAAGCGGTGAAGTGCCTCTGGATGTCG

CTCCACAAGGTAAACAGTTGATTGAACTGCCTGAACTACCGCAGCCGGAGAGCGCCGGGCAACTCTGGCTCACAGTACGCGTAGTGCAACCGAACGCGACCGCATGGTCAGAAGCCGGGCACATCAGCGCCTGGCAGCAGTGGCGTCTGGCGGAAAACCTCAGTGTGACGCTCCCCGCCGCGTCCCACGCCATCCCGCATCTGACCACCAGCGAAATGGATTTTTGCATCGAGCTGGG

TAATAAGCGTTGGCAATTTAACCGCCAGTCAGGCTTTCTTTCACAGATGTGGATTGGCGATAAAAAACAACTGCTGACGCCGCTGCGCGATCAGTTCACCCGTGCACCGCTGGATAACGACATTGGCGTAAGTGAAGCGACCCGCATTGACCCTAACGCCTGGGTCGAACGCTGGAAGGCGGCGGGCCATTACCAGGCCGAAGCAGCGTTGTTGCAGTGCACGGCAGATACACTTGCT

GATGCGGTGCTGATTACGACCGCTCACGCGTGGCAGCATCAGGGGAAAACCTTATTTATCAGCCGGAAAACCTACCGGATTGATGGTAGTGGTCAAATGGCGATTACCGTTGATGTTGAAGTGGCGAGCGATACACCGCATCCGGCGCGGATTGGCCTGAACTGCCAGCTGGCGCAGGTAGCAGAGCGGGTAAACTGGCTCGGATTAGGGCCGCAAGAAAACTATCCCGACCGCCTTA

CTGCCGCCTGTTTTGACCGCTGGGATCTGCCATTGTCAGACATGTATACCCCGTACGTCTTCCCGAGCGAAAACGGTCTGCGCTGCGGGACGCGCGAATTGAATTATGGCCCACACCAGTGGCGCGGCGACTTCCAGTTCAACATCAGCCGCTACAGTCAACAGCAACTGATGGAAACCAGCCATCGCCATCTGCTGCACGCGGAAGAAGGCACATGGCTGAATATCGACGGTTTCCA

TATGGGGATTGGTGGCGACGACTCCTGGAGCCCGTCAGTATCGGCGGAATTCCAGCTGAGCGCCGGTCGCTACCATTACCAGTTGGTCTGGTGTCAAAAATAATAATAACCGGGCAGGCCATGTCGCCCGTATTTCGCGTAAGGAAATCCATTATGTACTATTTAAAAAACACAAACTTTTGGATGTTCGGTTTATTCTTTTTCTTTTACTTTTTTATCATGGGAGCCTACTTCCCG

TTTTTCCCGATTTGGCTACATGACATCAACCATATCAGCAAAAGTGATACGGGTATTATTTTTGCCGCTATTTCTCTGTTCTCGCTATTATTCCAACCGCTGTTTGGTCTGCTTTCTGACAAACTCGGGCTGCGCAAATACCTGCTGTGGATTATTACCGGCATGTTAGTGATGTTTGCGCCGTTCTTTATTTTTATCTTCGGGCCACGTTACAATAACAACATTTTAGTAGGATCGA

TTGTTGGTGGTATTTATCTAGGCTTTTGTTTTAACGCCGGTGCGCCAGCAGTAGAGGCATTTATTGAGAAAGTCAGCCGTCGCAGTAATTTCGAATTTGGTCGCGCGCGGATGTTTGGCTGTGTTGGGCGCTGTGTGCCTCGATTGTCGGCATCATGTTCACCATCAATAATCAGTTTGTTTTCTGGCTGGGCTCTGGCTGCACTCATCCTCGCCGTTTTACTCTTTTTCGC

CAAACGGATGCGCCCTCTTCTGCCACGGTTGCCAATGCGGTAGGTGCCAACCATTCGGCATTTAGCCTTAAGCTGGCACTGGAACTGTTCAGACAGCCAAAACTGTGGTTTTTGTCACTGTATGTTATTGGCGTTTCCTGCACCTACGATGTTTTTGACCAACAGTTTGCTAATTTCTTTACTTCGTTCTTTGCTACCGGTGAACAGGGTACGCGGGTATTTGGCTACGTAACGACA

ATGGGCGAATTACTTAACGCCTCGATTATGTTCTTTGCGCCACTGATCATTAATCGCATCGGTGGGAAAAACGCCCTGCTGCTGGCTGGCACTATTATGTCTGTACGTATTATTGGCTCATCGTTCGCCACCTCAGCGCTGGAAGTGGTTATTCTGAAAACGCTGCATATGTTTGAAGTACCGTTCCTGCTGGTGGGCTGCTTTAAATATATTACCAGCCAGTTTGAAGTGCGTTTT

CAGCGCGACGATTTATCTGGTCTGTTTCTGCTCTTTAAGCAACTGGCGATGATTTTTATGTCTGTACTGGCGGGCAATATGTATGAAAGCATCGGTTTCCAGGGCGCTTATCTGGTGCTGGGTCTGGTGGCGCTGGGCTTCACCTTAATTTCCGTGTTCACGCTTAGCGCCCCGGCCCGCTTTCCCTGCGTCGTCAGGTGAATGAAGTCGCTTAAGCAATCAATGTCGGATGCGG

Figure 7.2 ORF scanning is an effective way of locating genes in a bacterial genome.

The diagram shows 4522 bp of the lactose operon of *Escherichia coli* with all ORFs longer than 50 codons marked. The sequence contains two real genes – *lacZ* and *lacY* – indicated by the red lines. These real genes cannot be mistaken because they are much longer than the spurious ORFs, shown in blue. See *Figure 2.20A*, page 55 for the detailed structure of the lactose operon.

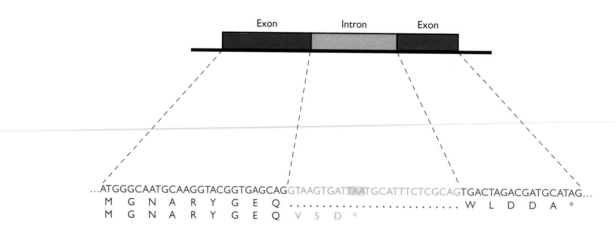

Figure 7.3 ORF scans are complicated by introns.

The nucleotide sequence of a short gene containing a single intron is shown. The correct amino acid sequence of the protein translated from the gene is given immediately below the nucleotide sequence: in this sequence the intron has been left out because it is removed from the transcript before the mRNA is translated into protein. In the lower line, the sequence has been translated without realizing that an intron is present. As a result of this error, the amino acid sequence appears to terminate within the intron. The amino acid sequences have been written using the one-letter abbreviations (see *Table 3.1*, page 84). The genetic code was described in Section 3.3.2; introns are covered in detail in Section 10.1.3.

more frequently than GTA. The biological reason for codon bias is not understood, but all organisms have a bias, which is different in different species. Real exons are expected to display the codon bias whereas chance series of triplets do not. The codon bias of the organism being studied is therefore written into the ORF scanning software.

- *Exon–intron boundaries* can be searched for as these have distinctive sequence features, although unfortunately the distinctiveness of these sequences is not so great as to make their location a trivial task. The sequence of the upstream, exon–intron boundary is usually described as:

5′–AG↓GTAAGT–3′

the arrow indicating the precise boundary point. However, only the 'GT' immediately after the arrow is invariable; elsewhere in the sequence nucleotides other than the ones shown are quite often found. In other words, the sequence shown is a **consensus** – the average of a range of variabilities. The downstream intron–exon boundary is even less well defined:

5′–PyPyPyPyPyPyNCAG↓–3′

where 'Py' means one of the pyrimidine nucleotides (T or C) and 'N' is any nucleotide. Simply searching for the consensus sequences will not locate more than a few exon–intron boundaries because most have sequences other than the ones shown. Writing software that takes account of the known variabilities has proven difficult (Frech *et al.*, 1997), and at present locating exon–intron boundaries by sequence analysis is a hit-and-miss affair.

- *Upstream regulatory sequences* can be used to locate the regions where genes begin. This is because these regulatory sequences, like exon–intron boundaries, have distinctive sequence features that they possess in order to carry out their role as recognition signals for the DNA-binding proteins involved in gene expression (Chapter 9). Unfortunately, as with exon–intron boundaries, the regulatory sequences are variable, more so in eukaryotes than in prokaryotes, and in eukaryotes not all genes have the same collection of regulatory sequences. Using these to locate genes is therefore problematic (Ohler and Niemann, 2001).

These three extensions of simple ORF scanning are generally applicable to all higher eukaryotic genomes. Additional strategies are also possible with individual organisms, based on the special features of their genomes. For example, vertebrate genomes contain **CpG islands** upstream of many genes (Bird, 1986), these being sequences of approximately 1 kb in which the GC content is greater than the average for the genome as a whole. Some 40–50% of human genes are associated with an upstream CpG island. These sequences are distinctive and when one is located in vertebrate DNA, a strong assumption can be made that a gene begins in the region immediately downstream.

Homology searches give an extra dimension to sequence inspection

Most of the various software programs available for gene location can identify up to 95% of the coding regions in a eukaryotic genome, but even the best ones tend to make frequent mistakes in their positioning of the exon–intron boundaries (Reese *et al.*, 2000). Identification of spurious ORFs as real genes is still a major problem. These limitations can be offset to a certain extent by the use of a **homology search** to test whether a series of triplets is a real exon or a chance sequence. In this analysis the DNA databases are searched to determine if the test sequence is identical or similar to any genes that have already been sequenced. Obviously, if the test sequence is part of a gene that has already been sequenced by someone else then an identical match will be found, but this is not the point of a homology search. Instead the intention is to determine if an entirely new sequence is *similar* to any known genes, because if it is then there is a chance that the test and match sequences are **homologous**, meaning that they represent genes that are evolutionarily related. The main use of homology searching is to assign functions to newly discovered genes, and we will therefore return to it when we deal with this aspect of genome analysis later in the chapter (Section 7.2.1). At this point, we will note simply that the technique is also central to *gene location* because it enables tentative exon sequences located by ORF scanning to be tested for functionality. If the tentative exon sequence gives one or more positive matches after a homology search then it is probably a real exon, but if it gives no match then its authenticity must remain in doubt until it is assessed by one or other of the experiment-based gene location techniques.

7.1.2 Experimental techniques for gene location

Most experimental methods for gene location are not based on direct examination of DNA molecules but instead rely on detection of the RNA molecules that are transcribed from genes. All genes are transcribed into RNA, and if the gene is discontinuous then the primary transcript is subsequently processed to remove the introns and link up the exons (Sections 1.2.1 and 10.1.3). Techniques that map the positions of transcribed sequences in a DNA fragment can therefore be used to locate exons and entire genes. The only problem to be kept in mind is that the transcript is usually longer than the coding part of the gene because it begins several tens of nucleotides upstream of the initiation codon and continues several tens or hundreds of nucleotides downstream of the termination codon (see *Figure 1.17*, page 21). Transcript analysis does not therefore give a precise definition of the start and end of the coding region of a gene, but it does tell you that a gene is present in a particular region and it can locate the exon–intron boundaries. Often this is sufficient information to enable the coding region to be delineated.

Hybridization tests can determine if a fragment contains transcribed sequences

The simplest procedures for studying transcribed sequences are based on hybridization analysis. RNA molecules can be separated by specialized forms of agarose gel electrophoresis and transferred to a nitrocellulose or nylon membrane by the process called **northern blotting** (see Technical Note 4.4, page 120). This differs from Southern blotting (Section 4.1.2) only in the precise conditions under which the transfer is carried out, and the fact that it was not invented by a Dr Northern and so does not have a capital 'N'. If a northern blot of cellular RNA is probed with a labeled fragment of the genome, then RNAs transcribed from genes within that fragment will be detected (*Figure 7.4*). Northern hybridization is therefore, theoretically, a means of determining the number of genes present in a DNA fragment and the size of each coding region. There are two weaknesses with this approach:

■ Some individual genes give rise to two or more transcripts of different lengths because some of their exons are optional and may or may not be retained in the mature RNA (Section 10.1.3). If this is the case, then a fragment that contains just one gene could detect two or more hybridizing bands in the northern blot. A similar problem can occur if the gene is a member of a multigene family (Section 2.2.1).

■ With many species, it is not practical to make an mRNA preparation from an entire organism so the extract is obtained from a single organ or tissue. Consequently any genes not expressed in that organ or tissue will not be represented in the RNA population, and so will not be detected when the RNA is probed with the DNA fragment being studied. Even if the whole organism is used, not all genes will give hybridization signals because many are expressed only at a particular developmental stage, and others are weakly expressed, meaning that their RNA products are present in amounts too low to be detected by hybridization analysis.

A second type of hybridization analysis avoids the problems with poorly expressed and tissue-specific genes by searching not for RNAs but for related sequences in the DNAs of other organisms. This approach, like homology searching, is based on the fact that homologous genes in related organisms have similar sequences, whereas the non-coding DNA is usually quite different. If a DNA fragment from one species is used to probe a Southern blot of DNAs from related species, and one or more hybridization signals are obtained, then it is likely that the probe contains one or more genes (*Figure 7.5*). This is called **zoo-blotting**.

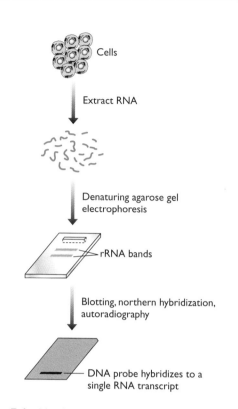

Figure 7.4 Northern hybridization.

An RNA extract is electrophoresed under denaturing conditions in an agarose gel (see Technical Note 4.4, page 120). After ethidium bromide staining, two bands are seen. These are the two largest rRNA molecules (Section 3.2.1) which are abundant in most cells. The smaller rRNAs, which are also abundant, are not seen because they are so short that they run out of the bottom of the gel and, in most cells, none of the mRNAs (the transcripts of protein-coding genes) are abundant enough to form a band visible after ethidium bromide staining. The gel is blotted onto a nylon membrane and, in this example, probed with a radioactively labeled DNA fragment. A single band is visible on the autoradiograph, showing that the DNA fragment used as the probe contains part or all of one transcribed sequence.

cDNA sequencing enables genes to be mapped within DNA fragments

Northern hybridization and zoo-blotting enable the presence or absence of genes in a DNA fragment to be determined, but give no positional information relating to the location of those genes in the DNA sequence. The easiest way to obtain this information is to sequence the relevant cDNAs. A cDNA is a copy of an mRNA (see *Figure 5.32*, page 155) and so corresponds to the coding region of a gene, plus any leader or trailer sequences that are also transcribed. Comparing a cDNA sequence with a genomic DNA sequence therefore delineates the position of the relevant gene and reveals the exon–intron boundaries.

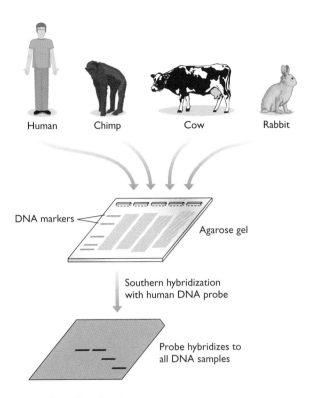

Figure 7.5 Zoo-blotting.

The objective is to determine if a fragment of human DNA hybridizes to DNAs from related species. Samples of human, chimp, cow and rabbit DNAs are therefore prepared, restricted, and electrophoresed in an agarose gel. Southern hybridization is then carried out with a human DNA fragment as the probe. A positive hybridization signal is seen with each of the animal DNAs, suggesting that the human DNA fragment contains an expressed gene. Note that the hybridizing restriction fragments from the cow and rabbit DNAs are smaller than the hybridizing fragments in the human and chimp samples. This indicates that the restriction map around the transcribed sequence is different in cows and rabbits, but does not affect the conclusion that a homologous gene is present in all four species.

In order to obtain an individual cDNA, a cDNA library must first be prepared from all of the mRNA in the tissue being studied. Once the library has been prepared, the success of cDNA sequencing as a means of gene location depends on two factors. The first concerns the frequency of the desired cDNAs in the library. As with northern hybridization, the problem relates to the different expression levels of different genes. If the DNA fragment being studied contains one or more poorly expressed genes, then the relevant cDNAs will be rare in the library and it might be necessary to screen many clones before the desired one is identified. To get around this problem, various methods of **cDNA capture** or **cDNA selection** have been devised, in which the DNA fragment being studied is repeatedly hybridized to the pool of cDNAs in order to enrich the pool for the desired clones (Lovett, 1994).

Because the cDNA pool contains so many different sequences, it is generally not possible to discard all the irrelevant clones by these repeated hybridizations, but it is possible to increase significantly the frequency of those clones that specifically hybridize to the DNA fragment. This reduces the size of the library that must subsequently be screened under stringent conditions to identify the desired clones.

A second factor that determines success or failure is the completeness of the individual cDNA molecules. Usually, cDNAs are made by copying RNA molecules into single-stranded DNA with **reverse transcriptase** and then converting the single-stranded DNA into double-stranded DNA with a DNA polymerase (see *Figure 5.32*, page 155). There is always a chance that one or other of the strand synthesis reactions will not proceed to completion, resulting in a truncated cDNA. The presence of intramolecular base pairs in the RNA can also lead to incomplete copying. Truncated cDNAs may lack some of the information needed to locate the start and end points of a gene and all its exon–intron boundaries.

Methods are available for precise mapping of the ends of transcripts

The problems with incomplete cDNAs mean that more robust methods are needed for locating the precise start and end points of gene transcripts. One possibility is a special type of PCR which uses RNA rather than DNA as the starting material. The first step in this type of PCR is to convert the RNA into cDNA with reverse transcriptase, after which the cDNA is amplified with *Taq* polymerase in the same way as in a normal PCR. These methods go under the collective name of **reverse transcriptase PCR (RT-PCR)** but the particular version that interests us at present is **rapid amplification of cDNA ends (RACE**; Frohman *et al.*, 1988). In the simplest form of this method one of the primers is specific for an internal region close to the beginning of the gene being studied. This primer attaches to the mRNA for the gene and directs the first reverse-transcriptase-catalyzed stage of the process, during which a cDNA corresponding to the start of the mRNA is made (*Figure 7.6*). Because only a small segment of the mRNA is being copied, the expectation is that the cDNA synthesis will not terminate prematurely, so one end of the cDNA will correspond exactly with the start of the mRNA. Once the cDNA has been made, a short poly(A) tail is attached to its 3' end. The second primer anneals to this poly(A) sequence and, during the first round of the normal PCR, converts the single-stranded cDNA into a double-stranded molecule, which is subsequently amplified as the PCR proceeds. The sequence of this amplified molecule will reveal the precise position of the start of the transcript.

Other methods for precise transcript mapping involve **heteroduplex analysis**. If the DNA region being studied is cloned as a restriction fragment in an M13 vector (Section 6.1.1) then it can be obtained as single-stranded DNA. When mixed with an appropriate RNA preparation, the transcribed sequence in the cloned DNA

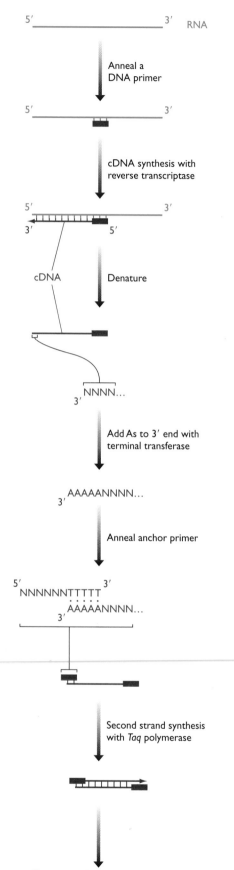

hybridizes with the equivalent mRNA, forming a double-stranded heteroduplex. In the example shown in *Figure 7.7* the start of this mRNA lies within the cloned restriction fragment, so some of the cloned fragment participates in the heteroduplex, but the rest does not. The single-stranded regions can be digested by treatment with a single-strand-specific nuclease such as S1. The size of the heteroduplex is determined by degrading the RNA component with alkali and electrophoresing the resulting single-stranded DNA in an agarose gel. This size measurement is then used to position the start of the transcript relative to the restriction site at the end of the cloned fragment.

Exon–intron boundaries can also be located with precision

Heteroduplex analysis can also be used to locate exon–intron boundaries. The method is almost the same as that shown in *Figure 7.7* with the exception that the cloned restriction fragment spans the exon–intron boundary being mapped rather than the start of the transcript.

A second method for finding exons in a genome sequence is called **exon trapping** (Church *et al.*, 1994). This requires a special type of vector that contains a **minigene** consisting of two exons flanking an intron sequence, the first exon being preceded by the sequence signals needed to initiate transcription in a eukaryotic cell (*Figure 7.8*). To use the vector the piece of DNA to be studied is inserted into a restriction site located within the vector's intron region. The vector is then introduced into a suitable eukaryotic cell line, where it is transcribed and the RNA produced from it is spliced. The result is that any exon contained in the genomic fragment becomes attached between the upstream and downstream exons from the minigene. RT-PCR with primers annealing within the two minigene exons is now used to amplify a DNA fragment, which is sequenced. As the minigene sequence is already known, the nucleotide positions at which the inserted exon starts and ends can be determined, precisely delineating this exon.

Figure 7.6   RACE – rapid amplification of cDNA ends.

The RNA being studied is converted into a partial cDNA by extension of a DNA primer that anneals at an internal position not too distant from the 5′ end of the molecule. The 3′ end of the cDNA is further extended by treatment with terminal deoxynucleotidyl transferase (Section 4.1.4) in the presence of dATP, which results in a series of As being added to the cDNA. This series of As acts as the annealing site for the anchor primer. Extension of the anchor primer leads to a double-stranded DNA molecule which can now be amplified by a standard PCR. This is 5′-RACE, so-called because it results in amplification of the 5′ end of the starting RNA. A similar method – 3′-RACE – can be used if the 3′ end sequence is desired.

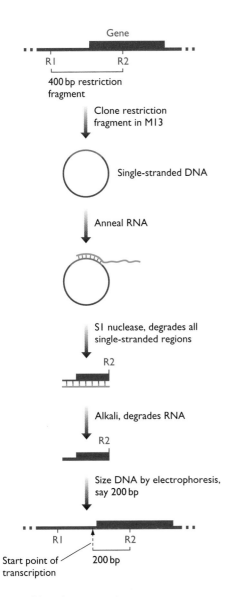

Figure 7.7 S1 nuclease mapping.

This method of transcript mapping makes use of S1 nuclease, an enzyme that degrades single-stranded DNA or RNA polynucleotides, including single-stranded regions in predominantly double-stranded molecules, but has no effect on double-stranded DNA or on DNA–RNA hybrids. In the example shown, a restriction fragment that spans the start of a transcription unit is ligated into an M13 vector and the resulting single-stranded DNA hybridized with an RNA preparation. After S1 treatment, the resulting heteroduplex has one end marked by the start of the transcript and the other by the downstream restriction site (R2). The size of the undigested DNA fragment is therefore measured by gel electrophoresis in order to determine the position of the start of the transcription unit relative to the downstream restriction site.

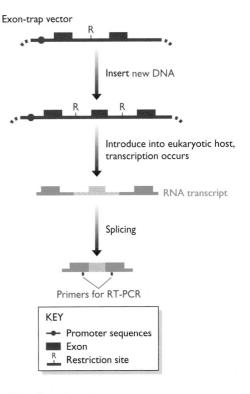

Figure 7.8 Exon trapping.

The exon-trap vector consists of two exon sequences preceded by promoter sequences – the signals required for gene expression in a eukaryotic host (Section 9.2.2). New DNA containing an unmapped exon is ligated into the vector and the recombinant molecule introduced into the host cell. The resulting RNA transcript is then examined by RT-PCR to identify the boundaries of the unmapped exon.

7.2 Determining the Functions of Individual Genes

Once a new gene has been located in a genome sequence, the question of its function has to be addressed. This is turning out to be an important area of genomics research, because completed sequencing projects have revealed that we know rather less than we thought about the content of individual genomes. *E. coli* and *S. cerevisiae*, for example, were studied intensively by conventional genetic analysis before the advent of sequencing projects, and geneticists were at one time fairly confident that most of their genes had been identified. The genome sequences revealed that in fact there are large gaps in our knowledge. Of the 4288 protein-coding genes in the *E. coli* genome sequence, only 1853 (43% of the total) had been previously identified (Blattner *et al.*, 1997). For *S. cerevisiae* the figure was only 30% (Dujon, 1996).

As with gene location, attempts to determine the functions of unknown genes are made by computer analysis and by experimental studies.

7.2.1 Computer analysis of gene function

We have already seen that computer analysis plays an important role in locating genes in DNA sequences, and that one of the most powerful tools available for this purpose is homology searching, which locates genes by comparing the DNA sequence under study with all the other DNA sequences in the databases. The basis of homology searching is that related genes have similar sequences and so a new gene can be discovered by virtue of its similarity to an equivalent, already sequenced, gene from a different organism. Now we will look more closely at homology analysis and see how it can be used to assign a function to a new gene.

Homology reflects evolutionary relationships

Homologous genes are ones that share a common evolutionary ancestor, revealed by sequence similarities between the genes. These similarities form the data on which molecular phylogenies are based, as we will see in Chapter 16. Homologous genes fall into two categories:

- **Orthologous** genes are those homologs that are present in different organisms and whose common ancestor predates the split between the species.
- **Paralogous** genes are present in the same organism, often members of a recognized multigene family (Section 2.2.1), their common ancestor possibly or possibly not predating the species in which the genes are now found.

A pair of homologous genes do not usually have identical nucleotide sequences, because the two genes undergo different random changes by mutation, but they have similar sequences because these random changes have operated on the same starting sequence, the common ancestral gene. Homology searching makes use of these sequence similarities. The basis of the analysis is that if a newly sequenced gene turns out to be similar to a previously sequenced gene, then an evolutionary relationship can be inferred and the function of the new gene is likely to be the same, or at least similar, to the function of the known gene.

It is important not to confuse the words *homology* and *similarity*. It is incorrect to describe a pair of related genes as '80% homologous' if their sequences have 80% nucleotide identity (*Figure 7.9*). A pair of genes are either evolutionarily related or they are not; there are no in-between situations and it is therefore meaningless to ascribe a percentage value to homology.

Homology analysis can provide information on the function of an entire gene or of segments within it

A homology search can be conducted with a DNA sequence but usually a tentative gene sequence is converted into an amino acid sequence before the search is carried out. One reason for this is that there are 20 different amino acids in proteins but only four nucleotides in DNA, so genes that are unrelated usually appear to be more different from one another when their amino acid sequences are compared (*Figure 7.10*). A homology search is therefore less likely to give spurious results if the amino sequence is used. The practicalities of homology searching are not at all daunting. Several software programs exist for this type of analysis, the most popular being BLAST (Basic Local Alignment Search Tool; Altschul *et al.*, 1990). The analysis can be carried out simply by logging on to the web site for one of the DNA databases and entering the sequence into the online search tool.

A positive match to a gene already in the database may give a clear indication of the function of the new gene, or the implications of the match might be more subtle. In particular, genes that have no obvious evolutionary relatedness might have short segments that are similar to one another. The explanation of this is often that, although the genes are unrelated, their proteins have similar functions and the shared sequence encodes a domain within each protein that is central to that shared function. Although the genes themselves have no common ancestor, the domains do, but with their common ancestor occurring at a very ancient time, the homologous domains having subsequently evolved not only by single nucleotide changes, but also by more complex rearrangements that have created new genes within which the domains are found (Section 15.2.1). An interesting example is provided by the tudor domain, an approximately 120-amino-acid motif which was first identified in the sequence of the *Drosophila melanogaster* gene called *tudor* (Ponting, 1997). The protein coded by the *tudor* gene, whose function is unknown, is made up of ten copies of the tudor domain, one after the other (*Figure 7.11*). A homology search using the tudor domain as the test revealed that several known proteins contain this domain. The sequences of these proteins are not highly similar to one another and there is no indication that they are true homologs, but they all possess the tudor domain. These proteins include one involved in RNA transport during *Drosophila* oogenesis, a human protein with a role in RNA metabolism, and others whose activities appear to involve RNA in one way or another. The homology analysis therefore suggests that the tudor sequence plays some part in the interaction between the protein and its RNA substrate. The information from the computer analysis is incomplete by itself,

```
Sequence I    GGTGAGGGTATCATCCCATCTGACTACACCTCATCGGGAGACGGAGCAGT
Sequence 2    GGTCAGGATATGATTCCATCACACTACACCTTATCCCGAGTCGGAGCAGT
Identities    *** *** *** ** ***** ********* *** *** ********
```

Figure 7.9 Two DNA sequences with 80% sequence identity.

```
                  G   A  P   G   M   W   L   R  L   A   A   G   S   F   E  H   A   G
Sequence I   GGTGCACCCGGTATGTGACTGCGATTAGCAGCGGGATCATTTCAGCATGCAGGG
             *  *  *****  ****  ****  **  ***  ****  *****  ***  **  ****  **  *
Sequence 2   GATACACCCCGTATTTGACAGCAATTTGCAGGGGGATGATTGCACCATGGAGCG
                  D   T  P   R   I   W   E   E  F   A   G   G   W   L   H  H   G   A
```

Figure 7.10 Lack of homology between two sequences is often more apparent when comparisons are made at the amino acid level.

Two nucleotide sequences are shown, with nucleotides that are identical in the two sequences given in red and non-identities given in blue. The two nucleotide sequences are 76% identical, as indicated by the asterisks. This might be taken as evidence that the sequences are homologous. However, when the sequences are translated into amino acids the identity decreases to 28%. Identical amino acids are shown in brown, and non-identities in green. The comparison between the amino acid sequences suggests that the genes are not homologous, and that the similarity at the nucleotide level was fortuitous. The amino acid sequences have been written using the one-letter abbreviations (see *Table 3.1*, page 84).

but it points the way to the types of experiment that should be done to obtain more clear-cut data on the function of the tudor domain.

Homology analysis in the yeast genome project

The *S. cerevisiae* genome project has illustrated both the potential and limitations of homology analysis as a means of assigning functions to new genes. The yeast genome contains approximately 6000 genes, 30% of which had been identified by conventional genetic analysis before the sequencing project got underway. The remaining 70% were studied by homology analysis, giving the following results (*Figure 7.12*; Dujon, 1996):

■ Almost another 30% of the genes in the genome could be assigned functions after homology searching of the sequence databases. About half of these were clear homologs of genes whose functions had been established previously, and about half had less striking similarities, including many where the similarities were restricted to discrete domains. For all these genes the homology analysis could be

described as successful, but with various degrees of usefulness (Oliver, 1996a). For some genes the identification of a homolog enabled the function of the yeast gene to be comprehensively determined; examples included identification of yeast genes for DNA polymerase subunits. For other genes the functional assignment could only be to a broad category, such as 'gene for a protein kinase'; in other words, the biochemical properties of the gene product could be inferred, but not the exact role of the protein in the cell. Some identifications were initially puzzling, the best example being the discovery of a yeast homolog of a bacterial gene involved in nitrogen fixation. Yeasts do not fix nitrogen so this could not be the function of the yeast gene. In this case, the discovery of the yeast homolog refocused attention on the previously characterized bacterial gene, with the subsequent realization that, although being involved in nitrogen fixation, the primary role of the bacterial gene product was in the synthesis of metal-containing proteins, which have broad roles in all organisms, not just nitrogen-fixing ones.

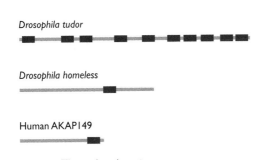

Figure 7.11 The tudor domain.

The top drawing shows the structure of the *Drosophila* tudor protein, which contains ten copies of the tudor domain. The domain is also found in a second *Drosophila* protein, *homeless*, and in the human A-kinase anchor protein (AKAP149), which plays a role in RNA metabolism. The proteins have dissimilar structures other than the presence of the tudor domains. The activity of each protein involves RNA in one way or another.

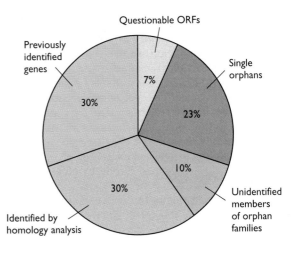

Figure 7.12 Categories of gene in the yeast genome.

■ About 10% of all the yeast genes had homologs in the databases, but the functions of these homologs were unknown. The homology analysis was therefore unable to help in assigning functions to these yeast genes. These yeast genes and their homologs are called **orphan families**.

■ The remaining yeast genes, about 30% of the total, had no homologs in the databases. A proportion of these (about 7% of the total) were questionable ORFs which might not be real genes, being rather short or having an unusual codon bias. The remainder looked like genes but were unique. These are called **single orphans**.

7.2.2 Assigning gene function by experimental analysis

It is clear that homology analysis is not a panacea that can identify the functions of all new genes. Experimental methods are therefore needed to complement and extend the results of homology studies. This is proving to be one of the biggest challenges in genomics research, and most molecular biologists agree that the methodologies and strategies currently in use are not entirely adequate for assigning functions to the vast numbers of unknown genes being discovered by sequencing projects. The problem is that the objective – to plot a course from gene to function – is the reverse of the route normally taken by genetic analysis, in which the starting point is a phenotype and the objective is to identify the underlying gene or genes. The problem we are currently addressing takes us in the opposite direction: starting with a new gene and hopefully leading to identification of the associated phenotype.

Functional analysis by gene inactivation

In conventional genetic analysis, the genetic basis of a phenotype is usually studied by searching for mutant organisms in which the phenotype has become altered. The mutants might be obtained experimentally, for example by treating a population of organisms (e.g. a culture of bacteria) with ultraviolet radiation or a mutagenic chemical (see Section 14.1.1), or the mutants might be present in a natural population. The gene or genes that have been altered in the mutant organism are then studied by genetic crosses (Section 5.2.4), which can locate the position of a gene in a genome and also determine if the gene is the same as one that has already been characterized. The gene can then be studied further by molecular biology techniques such as cloning and sequencing.

The general principle of this conventional analysis is that the genes responsible for a phenotype can be identified by determining which genes are inactivated in organisms that display a mutant version of the phenotype. If the starting point is the gene, rather than the phenotype, then the equivalent strategy would be to mutate the gene and identify the phenotypic change that results. This is

the basis of most of the techniques used to assign functions to unknown genes.

Individual genes can be inactivated by homologous recombination

The easiest way to inactivate a specific gene is to disrupt it with an unrelated segment of DNA (*Figure 7.13*). This can be achieved by **homologous recombination** between the chromosomal copy of the gene and a second piece of DNA that shares some sequence identity with the target gene. Homologous (and other types of) recombination are complex events, which we will deal with in detail in Section 14.3.1. For present purposes it is enough to know that if two DNA molecules have similar sequences, then recombination can result in segments of the molecules being exchanged.

How is gene inactivation carried out in practice? We will consider two examples, the first with *S. cerevisiae*. Since completing the genome sequence in 1996, yeast molecular biologists have embarked on a coordinated, international effort to determine the functions of as many orphan genes as possible (Oliver, 1996b). One technique that is being used is shown in *Figure 7.14* (Wach *et al.*, 1994). The central component is the 'deletion cassette', which carries a gene for antibiotic resistance. This gene is not a normal component of the yeast genome but it will work if transferred into a yeast chromosome, giving rise to a transformed yeast cell that is resistant to the antibiotic geneticin. Before using the deletion cassette, new segments of DNA are attached as tails to either end. These segments have sequences identical to parts of the yeast gene that is going to be inactivated. After the modified cassette is introduced into a yeast cell, homologous recombination occurs between the DNA tails and the chromosomal copy of the yeast gene, replacing the latter with the antibiotic-resistance gene. Cells which have undergone the replacement are therefore selected by plating the culture onto agar medium containing geneticin.

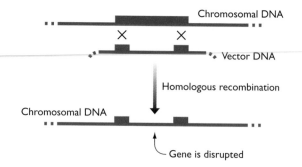

Figure 7.13 Gene inactivation by homologous recombination.

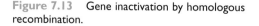

The chromosomal copy of the target gene recombines with a disrupted version of the gene carried by a cloning vector. As a result, the target gene becomes inactivated. For more information on recombination see Section 14.3.

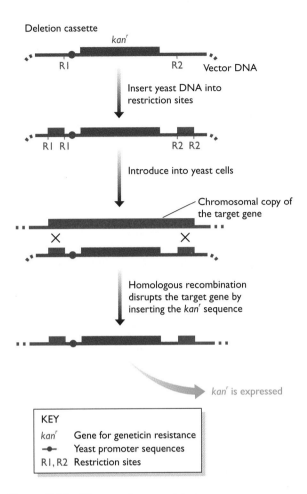

Deletion cassette

KEY

kan'	Gene for geneticin resistance
●	Yeast promoter sequences
R1, R2	Restriction sites

Figure 7.14   The use of a yeast deletion cassette.

The deletion cassette consists of an antibiotic-resistance gene preceded by the promoter sequences needed for expression in yeast, and flanked by two restriction sites. The start and end segments of the target gene are inserted into the restriction sites and the vector introduced into yeast cells. Recombination between the gene segments in the vector and the chromosomal copy of the target gene results in disruption of the latter. Cells in which the disruption has occurred are identified because they now express the antibiotic-resistance gene and so will grow on an agar medium containing geneticin. The gene designation '*kan'*' is an abbreviation for 'kanamycin resistance', kanamycin being the family name of the group of antibiotics that include geneticin.

The resulting colonies lack the target gene activity and their phenotypes can be examined to gain some insight into the function of the gene.

The second example of gene inactivation uses an analogous process but with mice rather than yeast. The mouse is frequently used as a **model organism** for humans because the mouse genome is similar to the human genome, containing many of the same genes. Identifying the functions of unknown human genes is therefore being carried out largely by inactivating the equivalent genes in the mouse, these experiments being ethically unthinkable with humans. The homologous recombination part of the procedure is identical to that described for yeast and once again results in a cell in which the target gene has been inactivated. The problem is that we do not want just one mutated cell, we want a whole mutant mouse, as only with the complete organism can we make a full assessment of the effect of the gene inactivation on the phenotype. To achieve this it is necessary to use a special type of mouse cell, an **embryonic stem** or **ES cell** (Evans *et al.*, 1997). Unlike most mouse cells, ES cells are **totipotent**, meaning that they are not committed to a single developmental pathway and can therefore give rise to all types of differentiated cell. The engineered ES cell is therefore injected into a mouse embryo, which continues to develop and eventually gives rise to a **chimera**, a mouse whose cells are a mixture of mutant ones, derived from the engineered ES cells, and non-mutant ones, derived from all the other cells in the embryo. This is still not quite what we want, so the chimeric mice are allowed to mate with one another. Some of the offspring result from fusion of two mutant gametes, and will therefore be non-chimeric, as every one of their cells will carry the inactivated gene. These are **knockout mice**, and with luck their phenotypes will provide the desired information on the function of the gene being studied. This works well for many gene inactivations but some are lethal and so cannot be studied in a homozygous knockout mouse. Instead, a heterozygous mouse is obtained, the product of fusion between one normal and one mutant gamete, in the hope that the phenotypic effect of the gene inactivation will be apparent even though the mouse still has one correct copy of the gene being studied.

Gene inactivation without homologous recombination

Homologous recombination is not the only way to disrupt a gene in order to study its function. One alternative is use **transposon tagging**, in which inactivation is achieved by the insertion of a transposable element into the gene. Most genomes contain transposable elements (Section 2.4.2) and although the bulk of these are inactive, there are usually a few that retain their ability to transpose. Under normal circumstances, transposition is a relatively rare event, but it is sometimes possible to use recombinant DNA techniques to make modified transposons that change their position in response to an external stimulus. One way of doing this, involving the yeast retrotransposon *Ty1*, is shown in *Figure 7.15*.

Transposon tagging is central to the technique called **genetic footprinting** (Smith *et al.*, 1995), which has been used to inactivate many of the yeast orphans as a first step to assessing their function. Transposon tagging is also important in analysis of the fruit-fly genome, using the endogenous *Drosophila* transposon called the **P element** (Engels, 2000). The weakness with transposon tagging is that it is difficult to target individual genes, because transposition is more or less a random event and it is impossible to predict where a transposon will end up

---

Box 7.1: The phenotypic effect of gene inactivation is sometimes difficult to discern

Once a gene-inactivated yeast strain, knockout mouse, or equivalent with any other organism has been obtained, the next stage is to examine the phenotype of the mutant in order to assign a function to the unknown gene. This can be much more difficult than it sounds. With any organism the range of phenotypes that must be examined is immense. Even with yeast, the list is quite lengthy (see table) and with higher organisms some phenotypes (e.g. behavioral ones) are difficult, if not impossible, to assess in a comprehensive fashion. Furthermore, the effect of gene inactivation can be very subtle and may not be recognized when the phenotype is examined. Almost 5000 of the 6000 yeast genes can be individually inactivated without causing the cells to die, and inactivation of many of these 5000 genes has no detectable effect on the metabolic properties of the cell under normal growth conditions (Cornish-Bowden and Cárdenas, 2001). A good example of the problems that occur was provided by the longest gene on yeast chromosome III which, at 2167 codons and with typical yeast codon bias, simply had to be a functional gene rather than a spurious ORF. Inactivation of this gene had no apparent effect, the mutant yeast cells appearing to have an identical phenotype to normal yeast. For some time it was thought that perhaps this gene is dispensable, its protein product either involved in some com-

pletely non-essential function, or having a function that is duplicated by a second gene. Eventually it was shown that the mutants die when they are grown at low pH in the presence of glucose and acetic acid, which normal yeasts can tolerate, and it was concluded that the gene codes for a protein that pumps acetate out of the cell (Oliver, 1996a). This is definitely an essential function as the gene plays a vital role in protecting yeast from acetic-acid-induced damage, but this essentiality was difficult to track down from the phenotype tests.

Possible biological functions of an unassigned yeast gene

DNA synthesis and the cell cycle	RNA synthesis and processing
Protein synthesis	Stress responses
Cell wall synthesis and morphogenesis	Transport of biochemicals within the cell
Energy and carbohydrate metabolism	Lipid metabolism
Development	DNA repair and recombination
Meiosis	Chromosome structure
Cell architecture	Secretion and protein trafficking

Based on the categories defined by the European Functional Analysis Network (Oliver, 1996b).

---

after it has jumped. If the intention is to inactivate a particular gene then it is necessary to induce a substantial number of transpositions and then to screen the resulting organisms to find one with the correct insertion. Transposon tagging is therefore more applicable to global studies of genome function, in which genes are inactivated at random and groups of genes with similar functions identified by examining the progeny for interesting phenotype changes.

A completely different approach to gene inactivation is provided by **RNA interference**. In this technique, rather than disrupting the gene itself, its mRNA is destroyed. This is accomplished by introducing into the cell short double-stranded RNA molecules whose sequences match that of the mRNA being targeted. The double-stranded RNAs are broken down into shorter molecules which induce degradation of the mRNA (*Figure 7.16*). The process has been shown to work effectively in the worm *Caenorhabditis elegans* (Fire *et al.*, 1998), whose genome has been completely sequenced (see *Table 2.1*, page 31) and which is looked on as an important model organism for higher eukaryotes (Section 12.3.2). Almost 2500 of the 2769 predicted genes on chromosome I of *C. elegans* have been individually inactivated by RNA interference, simply by placing the worms in a solution containing the double-stranded RNA and allowing normal uptake

processes to transport the molecules into the cells (Fraser *et al.*, 2000). Similar projects are being directed at the other *C. elegans* chromosomes.

RNA interference is known to occur naturally in a range of eukaryotes, but applying it to mammalian cells was expected to be difficult because these organisms display a parallel response to double-stranded RNA, in which protein synthesis is generally inhibited, resulting in cell death (Bass, 2001). These worries were unfounded, however, because it has now been shown that introduction of double-stranded RNAs into cultured human cells by fusion with liposomes (*Figure 7.17*) results in inactivation of the target mRNA, with no measurable decrease in overall protein synthesis (Elbashir *et al.*, 2001). The drawback to using this technique with mammals is that it is only possible to work with single cells, rather than whole organisms, because the double-stranded RNAs have a limited lifetime within the cell and cannot be used to engineer permanent changes such as those necessary in the construction of knockout mice.

Gene overexpression can also be used to assess function

So far we have concentrated on techniques that result in inactivation of the gene being studied ('loss of function').

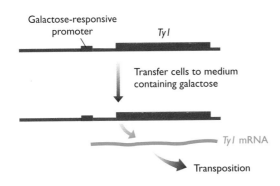

Figure 7.15 Artificial induction of transposition

Recombinant DNA techniques have been used to place a promoter sequence (Section 3.2.2) that is responsive to galactose upstream of a *Ty1* element in the yeast genome. When galactose is absent, the *Ty1* element is not transcribed and so remains quiescent. When the cells are transferred to a culture medium containing galactose, the promoter is activated and the *Ty1* element is transcribed, initiating the transposition process (Smith *et al.*, 1995). For more information on activation of eukaryotic promoters, see Box 9.6, page 267 and for details of the retrotransposition process see Section 14.3.3.

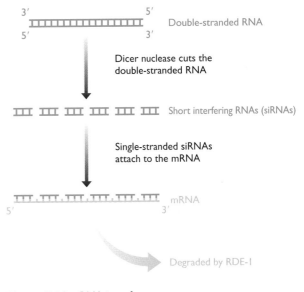

Figure 7.16 RNA interference.

The double-stranded RNA molecule is broken down by the Dicer ribonuclease into 'short interfering RNAs' (siRNAs) of 21–25 bp in length. One strand of each siRNA base pairs to the target mRNA, which is then degraded by the RDE-1 nuclease. For more details on RNA interference, see Section 10.4.2.

The complementary approach is to engineer an organism in which the test gene is much more active than normal ('gain of function') and to determine what changes, if any, this has on the phenotype. The results of these experiments must be treated with caution because of the need to distinguish between a phenotype change that is due to the specific function of an overexpressed gene, and a less specific phenotype change that reflects the abnormality of the situation where a single gene product is being synthesized in excessive amounts, possibly in tissues in which the gene is normally inactive. Despite this qualification, overexpression has provided some important information on gene function.

To overexpress a gene a special type of cloning vector must be used, one designed to ensure that the cloned gene directs the synthesis of as much protein as possible. The vector is therefore **multicopy**, meaning that it multiplies inside the host organism to 40–200 copies per cell, so there are many copies of the test gene. The vector must also contain a highly active promoter (Section 9.2.2) so that each copy of the test gene is converted into large quantities of mRNA, again ensuring that as much protein as possible is made. An example of the technique used with mice genes is shown in *Figure 7.18* (Simonet *et al.*, 1997). In this project the genes to be studied were selected because their sequences suggested that they code for proteins that are secreted into the bloodstream. The cloning vector that was used contained a highly active promoter that is expressed only in the liver, so each **transgenic mouse** overexpressed the test gene in its liver and then secreted the resulting protein into the blood. The phenotype of each transgenic mouse was examined in the search for clues regarding the functions of the cloned genes. An interesting discovery was made when it was realized that one transgenic mouse had bones that were significantly more dense than those of normal mice. This was important for two reasons: first, it enabled the relevant gene to be identified as one involved in bone synthesis; second, the discovery of a protein that increases bone density has implications for the development of treatments for human osteoporosis, a fragile-bone disease.

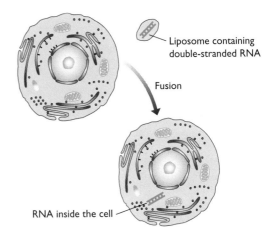

Figure 7.17 Fusion with liposomes can be used to deliver double-stranded RNA into a human cell.

RESEARCH 7.1 BRIEFING

Analysis of chromosome I of *Caenorhabditis elegans* by RNA interference

Functions have been assigned to 339 genes on *C. elegans* chromosome I after individual inactivation by the RNA interference technique.

C. elegans is a tiny nematode worm (see *Figure 12.22*, page 369) which has been extensively used as a model for the development of multicellular eukaryotes (Section 12.3.2). It has many advantages for this type of research, being easy to grow in the laboratory and having a generation time of just a few days. The outer surface of the worm is transparent, which means that its internal development can be followed by microscopy. Every cell division in the developmental pathway from fertilized egg to adult worm has been charted, and every point at which a cell adopts a specialized role has been identified. In addition, the entire connectivity of the 302 cells that comprise the worm's nervous system has been mapped.

The draft sequence of the 97-Mb *C. elegans* genome was completed in 1998 (see *Table 2.1*, page 31). This was an important step forward because many molecular biologists believe that analysis of the *C. elegans* genome will eventually enable a link to be established between the worm's developmental pathway and the activities of its genes. For this goal to become a reality it will first be necessary to identify the functions of all, or at least the majority, of the genes in the *C. elegans* genome. Genetic analysis had identified a few important genes even before the genome sequence was available, and a detailed homology analysis enabled functions to be tentatively assigned to a few more. But over 60% of the 19 099 predicted genes in the draft genome sequence were unidentified. Understanding the genome sequence, and establishing the link between the genome and the worm's developmental pathways, will clearly require a substantial amount of experimental work. During the years since the draft sequence was published, various techniques have been used in attempts to determine the functions of the unidentified *C. elegans* genes. One of these techniques has been RNA interference.

RNA interference with *C. elegans*

The basis of RNA interference is shown in *Figure 7.16*, page 201. The key step is the introduction into a worm of a double-stranded RNA molecule that will give rise to single-stranded interfering RNAs specific for a particular gene. The easiest approach is simply to feed the RNA to the worms. *C. elegans* eats bacteria, including *Escherichia coli*, and is often grown on a lawn of bacteria on an agar plate. If the bacteria are synthesizing a double-stranded RNA with the same sequence as a *C. elegans* gene then, after ingestion, the RNA interference pathway begins to operate.

Cloning techniques (Section 4.2) can be used to prepare an *E. coli* strain that makes a double-stranded RNA specific for a *C. elegans* gene. To begin the procedure, the *C. elegans* gene is amplified by PCR (Section 4.3). The PCR product is then ligated into a special cloning vector, called L4440, which possesses two short DNA sequences, called promoters (Section

9.2.2), that initiate RNA synthesis by the highly active RNA polymerase of the T7 bacteriophage. One of these sequences is located to the right of the ligation site and one to the left, in opposite orientations, so that the RNA that is synthesized is a copy of both strands of the inserted PCR product (see Figure below). A strain of *E. coli* that contains the T7 RNA polymerase is transformed with the recombinant L4440 molecule, which is copied into two complementary RNA strands, which link together by base-pairing to produce a double-stranded RNA. The bacteria are grown in the well of a microtiter tray and three *C. elegans* worms are added. The worms eat the bacteria, ingesting the double-stranded RNA. This initiates the RNA interference process, which subsequently inactivates the target gene in the worms' genomes.

Inactivation of genes on chromosome I

How successful has RNA interference been with *C. elegans*? A study of the 2769 predicted genes on chromosome I provides a good illustration of what has been achieved. This study made use of a library of 2445 bacterial clones, targeting 2416 different genes. In 339 cases, gene inactivation by RNA interference led to a detectable change in the phenotype of the treated worms. The commonest change was 'embryonic lethal', meaning that embryos produced by the treated worms died at an early stage in their development. Other worms became sterile after gene inactivation, or

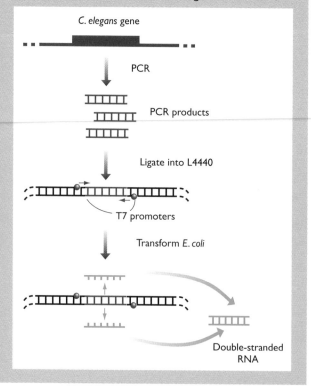

failed to develop beyond the larval stage. Phenotypes that were only detectable in the adult worm were less common, but these included interesting defects such as 'uncoordinated', caused by abnormalities in the neuromuscular system, and 'high incidence of males', which occurs when these normally hermaphroditic organisms develop male characteristics at a higher frequency than usual. The table below lists some of the phenotypes that were observed (note that many worms displayed more than one phenotype).

Phenotype	Number of genes giving the phenotype after inactivation
Embryonic lethal	226
Sterile or sterile progeny	96
Delayed development	145
Larval lethal	38
Uncoordinated	70
Protruding vulva	29
Body morphology defects	27
High incidence of males	13
Molting defects	8
Paralyzed	8
Egg-laying defects	5
Blistering of cuticle	4
Multiple vulvas	2
Hyperactive	1

Many of the genes whose inactivation led to a detectable change in phenotype were ones that had previously been unidentified, but a few had already been assigned biochemical functions by genetic studies or homology analysis. It was therefore possible to begin to examine the relationship between the specific biochemical activities coded by individual genes and the more general phenotype changes observable in the living worm. As expected, inactivation of genes that specified enzymes involved in central metabolic pathways often resulted in one of the more drastic phenotype changes, such as sterility or death of embryos. The less dramatic phenotypes, such as 'uncoordinated', were more frequently associated with genes whose products play a role in cell structure or organization. These comparisons also showed that most of the unidentified genes give rise to a subtle change in phenotype when inactivated, suggesting that they have specialized biochemical functions, rather than coding for important enzymes. To identify these genes it might therefore be necessary to go beyond simple inactivation studies: we really need to find out more about the biochemistry of development before the link can be made with the genome.

Lessons for the functional analysis of other genomes

Only 14% of the genes on *C. elegans* chromosome I could be assigned a function after inactivation by RNA interference. This success rate reflects, to a certain extent, the limited nature of the analysis that was adopted in this particular study. For example, no attempt was made to assess the effect of gene inactivation on the ability of the worms to respond to environmental stress. But the relatively low success rate underlines how difficult it can be to detect a change in phenotype after inactivation of a single gene (see Box 7.1, page 200). The implication is that substantially less than half of the 19 099 genes of *C. elegans* will be identifiable by straightforward phenotype analysis, and that more extensive and time-consuming tests will be needed to assign functions to the remainder. If this view turns out to be correct then the complete annotation of gene functions for the human genome, whether by direct analysis of those genes or by studies of homologous genes in other organisms, will take a great deal of time.

References

Timmons L and Fire A (1998) Specific interference by ingested dsRNA. *Nature*, **395**, 854.
Fraser AG, Kamath RS, Zipperlen P, Martinez-Campos M, Sohrmann M and Ahringer J (2000) Functional genomic analysis of *C. elegans* chromosome I by systematic RNA interference. *Nature*, **408**, 325–330.

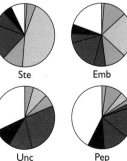

		Ste	Emb	Unc	Pep
DNA synthesis		1.2	1.3	0.0	2.2
RNA metabolism		4.8	11.8	4.7	8.9
Protein metabolism		44.6	22.3	7.8	4.4
Energy/metabolism		10.8	10.9	6.3	2.2
Chromosome dynamics/cell cycle		0.0	6.1	0.0	2.2
Cell structure/organization		18.1	15.7	26.6	15.6
Specific transcription		7.2	4.8	12.5	13.3
Signaling		4.8	4.8	9.4	8.9
Unknown		7.2	18.8	31.3	42.2

KEY   Ste: Sterile   Emb: Embryonic lethal
       Unc: Uncoordinated   Pep: Post-embryonic (adult phenotype)
The table and pie charts show the percentage of genes in each biochemical category

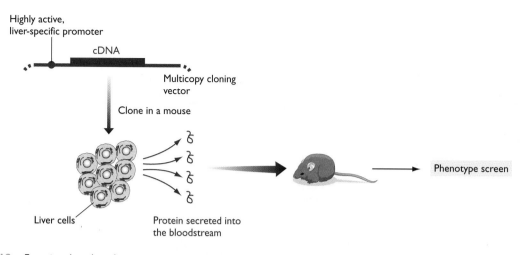

Figure 7.18 Functional analysis by gene overexpression.

The objective is to determine if overexpression of the gene being studied has an effect on the phenotype of a transgenic mouse. A cDNA of the gene is therefore inserted into a cloning vector carrying a highly active promoter sequence that directs expression of the cloned gene in mouse liver cells. The cDNA is used rather than the genomic copy of the gene because the former does not contain introns and so is shorter and easier to manipulate in the test tube.

7.2.3 More detailed studies of the activity of a protein coded by an unknown gene

Gene inactivation and overexpression are the primary techniques used by genome researchers to determine the function of a new gene, but these are not the only procedures that can provide information on gene activity. Other methods can extend and elaborate the results of inactivation and overexpression. These can be used to provide additional information that will aid identification of a gene function, or might form the basis of a more comprehensive examination of the activity of a protein whose gene has already been characterized.

Directed mutagenesis can be used to probe gene function in detail

Inactivation and overexpression can determine the general function of a gene, but they cannot provide detailed information on the activity of a protein coded by a gene. For example, it might be suspected that part of a gene codes for an amino acid sequence that directs its protein product to a particular compartment in the cell, or is responsible for the ability of the protein to respond to a chemical or physical signal. To test these hypotheses it would be necessary to delete or alter the relevant part of the gene sequence, but to leave the bulk unmodified so that the protein is still synthesized and retains the major part of its activity. The various procedures of **site-directed** or *in vitro* **mutagenesis** (Technical Note 7.1) can be used to make these subtle changes. These are important techniques whose applications lie not only with the study of gene activity but also in the area of **protein engineering**, where the objective is to create novel pro-

teins with properties that are better suited for use in industrial or clinical settings.

After mutagenesis the gene sequence must be introduced into the host cell so that homologous recombination can replace the existing copy of the gene with the modified version. This presents a problem because we must have a way of knowing which cells have undergone homologous recombination. Even with yeast this will only be a fraction of the total, and with mice the fraction will be very small. Normally we would solve this problem by placing a marker gene (e.g. one coding for antibiotic resistance) next to the mutated gene and looking for cells that take on the phenotype conferred by this marker. In most cases, cells that insert the marker gene into their genome also insert the closely attached mutated gene and so are the ones we want. The problem is that in a site-directed mutagenesis experiment we must be sure that any change in the activity of the gene being studied is the result of the specific mutation that was introduced into the gene, rather than the indirect result of changing its environment in the genome by inserting a marker gene next to it. The answer is to use a more complex two-step gene replacement (*Figure 7.19*). In this procedure the target gene is first replaced with the marker gene on its own, the cells in which this recombination takes place being identified by selecting for the marker gene phenotype. These cells are then used in the second stage of the gene replacement, when the marker gene is replaced by the mutated gene, success now being monitored by looking for cells that have lost the marker gene phenotype. These cells contain the mutated gene and their phenotypes can be examined to determine the effect of the directed mutation on the activity of the protein product.

Site-directed mutagenesis

Methods for making a precise alteration in a gene sequence in order to change the structure and possibly the activity of a protein.

Changes in protein structure can be engineered by site-directed mutagenesis techniques, which result in defined alterations being made to the nucleotide sequence of the gene coding for a protein of interest. These techniques enable the functions of different parts of a protein to be examined (Section 7.2.3), and also have widespread importance in the development of new enzymes for biotechnological purposes.

Conventional mutagenesis is a random process that introduces changes at unspecified positions in a DNA molecule. Screening of large numbers of mutated organisms is necessary to find a mutation of interest. Even with microbes, which can be screened in huge numbers, the best that can be hoped for is a range of mutations in the correct gene, one of which might affect the part of the protein being studied. Site-directed mutagenesis offers a means of making much more specific mutations.

A variety of DNA manipulations can be used to introduce mutations into cloned genes. The simplest are to delete a restriction fragment, to open the DNA at a restriction site and remove a few nucleotides, or to insert new DNA at a restriction site. These are relatively large-scale changes. To alter a single nucleotide at a specified position the technique called **oligonucleotide-directed mutagenesis** is used, one version of which is described here. First,

a single-stranded version of the gene is obtained by cloning in an M13 vector (Section 6.1.1). A short oligonucleotide is then synthesized, complementary to the relevant region of the gene but containing the desired nucleotide alteration. This oligonucleotide is hybridized to the DNA and used as the primer for a strand synthesis reaction that is allowed to continue all the way around the circular template molecule (see Figure A).

After introduction into *Escherichia coli*, DNA replication produces numerous copies of this recombinant DNA molecule. Half the resulting double-stranded molecules are copies of the original strand of DNA, and half are copies of the strand that contains the mutated sequence. The double-stranded molecules direct synthesis of phage particles so about half the phages released from the infected bacteria carry a single-stranded version of the mutated molecule. The phages are plated onto solid agar so that plaques are produced, and the mutant ones identified by hybridization probing with the original oligonucleotide (see Figure B). The mutated gene can then be placed back in its original host by homologous recombination, as described in Section 7.2.3, or transferred to an *E. coli* vector designed for synthesis of protein from cloned DNA, so that a sample of the mutated protein can be obtained.

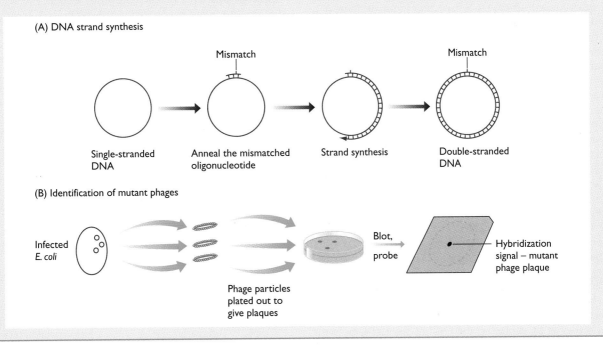

(A) DNA strand synthesis

Single-stranded DNA → Anneal the mismatched oligonucleotide → Strand synthesis → Double-stranded DNA

(B) Identification of mutant phages

Infected *E. coli* → Phage particles plated out to give plaques → Blot, probe → Hybridization signal – mutant phage plaque

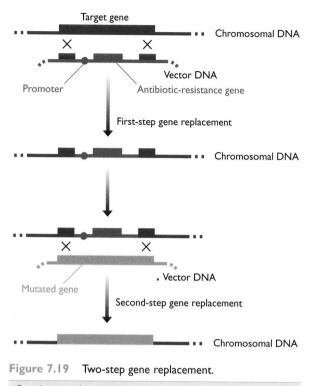

Figure 7.19   Two-step gene replacement.

See the text for details.

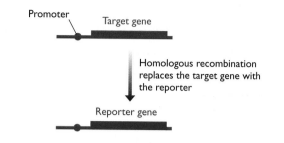

Figure 7.20   A reporter gene.

The open reading frame of the reporter gene replaces the open reading frame of the gene being studied. The result is that the reporter gene is placed under control of the regulatory sequences that usually dictate the expression pattern of the test gene. For more information on these regulatory sequences, see Sections 9.2 and 9.3. Note that the reporter gene strategy assumes that the important regulatory sequences do indeed lie upstream of the gene. This is not always the case for eukaryotic genes.

Reporter genes and immunocytochemistry can be used to locate where and when genes are expressed

Clues to the function of a gene can often be obtained by determining where and when the gene is active. If gene expression is restricted to a particular organ or tissue of a multicellular organism, or to a single set of cells within an organ or tissue, then this positional information can be used to infer the general role of the gene product. The same is true of information relating to the developmental stage at which a gene is expressed. This type of analysis has proved particularly useful in understanding the activities of genes involved in the earliest stages of development in *Drosophila* (Section 12.3.3) and is increasingly being used to unravel the genetics of mammalian development. It is also applicable to those unicellular organisms, such as yeast, which have distinctive developmental stages in their life cycle.

Determining the pattern of gene expression within an organism is possible with a **reporter gene**. This is a gene whose expression can be monitored in a convenient way, ideally by visual examination (*Table 7.1*), cells that express the reporter gene becoming blue, fluorescing or giving off some other visible signal. For the reporter gene to give a reliable indication of where and when a test gene is expressed, the reporter must be subject to the same regulatory signals as the test gene. This is achieved by replacing the ORF of the test gene with the ORF of the reporter gene (*Figure 7.20*). Most of the regulatory signals that control gene expression are contained in the region of DNA upstream of the ORF, so

the reporter gene should now display the same expression pattern as the test gene. The expression pattern can therefore be determined by examining the organism for the reporter signal.

As well as knowing in which cells a gene is expressed, it is often useful to locate the position within the cell where the protein coded by the gene is found. For example, key data regarding gene function can be obtained by showing that the protein product is located in mitochondria, in the nucleus, or on the cell surface. Reporter genes cannot help here because the DNA sequence upstream of the gene – the sequence to which the reporter gene is attached – is not involved in targeting the protein product to its correct intracellular location. Instead it is the amino acid sequence of the protein itself that is important. Therefore the only way to determine where the protein is located is to search for it directly. This is done by **immunocytochemistry**, which makes use of an antibody that is specific for the protein of interest and so binds to this protein and no other. The antibody is labeled so that its position in the cell, and hence the position of the target protein, can be visualized (*Figure 7.21*). Fluorescent labeling and light microscopy are used for low-resolution studies; alternatively, high-resolution immunocytochemistry can be carried out by electron microscopy using an electron-dense label such as colloidal gold.

Table 7.1   Examples of reporter genes

Gene	Gene product	Assay
lacZ	β-galactosidase	Histochemical test
uidA	β-glucuronidase	Histochemical test
lux	Luciferase	Bioluminescence
GFP	Green fluorescent protein	Fluorescence

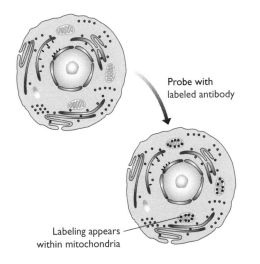

Probe with
labeled antibody

Labeling appears
within mitochondria

Figure 7.21 Immunocytochemistry.

The cell is treated with an antibody that is labeled with a blue fluorescent marker. Examination of the cell shows that the fluorescent signal is associated with the inner mitochondrial membrane. A working hypothesis would therefore be that the target protein is involved in electron transport and oxidative phosphorylation, as these are the main biochemical functions of the inner mitochondrial membrane.

7.3 Global Studies of Genome Activity

Even if every gene in a genome can be identified and assigned a function, a challenge still remains. This is to understand how the genome as a whole operates within the cell, specifying and coordinating the various biochemical activities that take place. These global studies of genome activity must address not the genome itself but the transcriptome and proteome that are synthesized and maintained by the genome (Chapter 3). The objective is to understand the key features of the transcriptomes and proteomes that are present in different tissues and during different developmental stages and, in the case of humans, in different disease states (Section 3.2.3).

7.3.1 Studying the transcriptome

The transcriptome comprises the mRNAs that are present in a cell at a particular time. Transcriptomes can have highly complex compositions, with hundreds or thousands of different mRNAs represented, each making up a different fraction of the overall population (Section 3.2.3). To characterize a transcriptome it is therefore necessary to identify the mRNAs that it contains and, ideally, to determine their relative abundances.

The composition of a transcriptome can be assayed by SAGE

The most direct way to characterize a transcriptome is to convert its mRNA into cDNA (see *Figure 5.32*, page

155), and then to sequence every clone in the resulting cDNA library. Comparisons between the cDNA sequences and the genome sequence will reveal the identities of the genes whose mRNAs are present in the transcriptome. This approach is feasible but it is laborious, with many different cDNA sequences being needed before a near-complete picture of the composition of the transcriptome begins to emerge. If two or more transcriptomes are being compared then the time needed to complete the project increases. Can any shortcuts be used to obtain the vital sequence information more quickly?

Serial analysis of gene expression (SAGE) provides a solution (Velculescu *et al.*, 2000). Rather than studying complete cDNAs, SAGE yields short sequences, as little as 12 bp in length, each of which represents an mRNA present in the transcriptome. The basis of the technique is that these 12-bp sequences, despite their shortness, are sufficient to enable the gene that codes for the mRNA to be identified. The argument is that any particular 12-bp sequence should appear in the genome once every $4^{12} = 16\,777\,216$ bp. The average size of a eukaryotic mRNA is about 1500 bp, so 4^{12} bp is equivalent to the combined length of over 11 000 transcripts. This number is higher than the number of transcripts expected in all but the most complex transcriptomes, so the 12-bp sequence tags should be able to identify unambiguously the genes coding for all the mRNAs that are present.

The procedure used to generate the 12-bp tags is shown in *Figure 7.22*. First, the mRNA is immobilized in a chromatography column by annealing the poly(A) tails present at the 3' ends of these molecules to oligo(dT) strands that have been attached to cellulose beads. The mRNA is converted into double-stranded cDNA and then treated with a restriction enzyme that recognizes a 4-bp target site and so cuts frequently in each cDNA. The terminal restriction fragment of each cDNA remains attached to the cellulose beads, enabling all the other fragments to be eluted and discarded. A short oligonucleotide is now attached to the free end of each cDNA, this oligonucleotide containing a recognition sequence for *Bsm*FI. This is an unusual restriction enzyme in that rather than cutting within its recognition sequence, it cuts 10–14 nucleotides downstream. Treatment with *Bsm*FI therefore removes a fragment with an average length of 12 bp from the end of each cDNA. The fragments are collected, ligated head-to-tail to produce a concatamer, and sequenced. The individual tag sequences are identified within the concatamer and compared with the sequences of the genes in the genome.

Using chip and microarray technology to study a transcriptome

DNA chips and microarrays (see Technical Note 5.1, page 133) can also be used to study transcriptomes. With a small genome such as that of *S. cerevisiae*, chips that carry

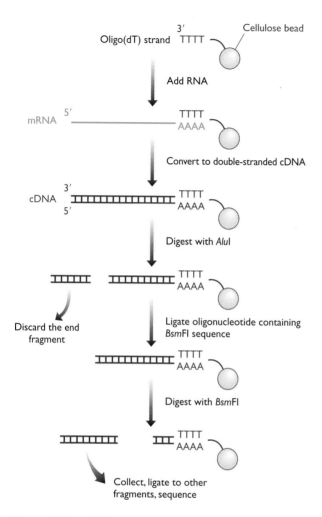

Figure 7.22 SAGE.

See the text for details. In this example, the first restriction enzyme to be used is AluI, which recognizes the 4-bp target site 5′–AGCT–3′ (see *Table 4.3*, page 103). The oligonucleotide that is ligated to the cDNA contains the recognition sequence for BsmFI, which cuts 10–14 nucleotides downstream, and so cleaves off a fragment of the cDNA. Fragments of different cDNAs are ligated to produce the concatamer that is sequenced. Using this method, the concatamer that is formed is made up partly of sequences derived from the BsmFI oligonucleotides. To avoid this, and so obtain a concatamer made up entirely of cDNA fragments, the oligonucleotide can be designed so that the end that ligates to the cDNA contains the recognition sequence for a third restriction enzyme. Treatment with this enzyme cleaves the oligonucleotide from the cDNA fragment.

oligonucleotides representing every gene can be constructed. A transcriptome is then characterized by converting its mRNA into cDNA, labeling the cDNA, and applying it to the chip. The positions at which hybridization occurs indicate the oligonucleotides representing the genes whose transcripts are present in the transcriptome (*Figure 7.23A*). Compared with SAGE, this approach has

the advantage that a rapid evaluation of the differences between two or more transcriptomes can be made by hybridizing the different cDNA preparations to identical chips and comparing the hybridization patterns. A further embellishment can be achieved by probing the chip with cDNA that has been prepared from the mRNA fraction that is bound to ribosomes in the cells being studied, rather than from total mRNA. These mRNAs correspond to the part of the transcriptome that is actively directing protein synthesis, giving a slightly different picture of genome activity (Pradet-Balade *et al.*, 2001).

Microarrays are used in a similar way to chips (Marshall, 1999; Knight 2001), but instead of immobilized oligonucleotides, they carry samples of cloned DNA. Often these are cDNA clones derived from one of the transcriptomes that is being studied. This might appear illogical but the approach enables two related transcriptomes to be compared, differences in their mRNA compositions being visualized as differences in the intensities of the hybridization signals emanating from the immobilized cDNAs when the microarray is probed with each transcriptome in turn (*Figure 7.23B*).

7.3.2 Studying the proteome

Proteome studies are important because of the central role that the proteome plays as the link between the genome and the biochemical capability of the cell (Section 3.3). Characterization of the proteomes of different cells is therefore the key to understanding how the genome operates and how dysfunctional genome activity can lead to diseases such as cancer. Transcriptome studies can only partly address these issues. Examination of the transcriptome gives an accurate indication of which genes are active in a particular cell, but gives a less accurate indication of the proteins that are present. There are several reasons for this lack of equivalence between transcriptome and proteome, the most important being:

- Not all mRNAs are actively translated at any particular time.
- The protein content of the cell is determined by both synthesis of new proteins and degradation of existing ones.

Methods for studying the proteome are therefore needed in order to obtain a complete picture of genome expression.

Proteomics – methodology for characterizing the protein content of a cell

The methodology used to study proteomes is collectively called **proteomics**. It is based on two techniques – **protein electrophoresis** and **mass spectrometry** – both of which have long pedigrees but which were rarely applied together in the pre-genomics era. Today they have been combined into one of the major growth areas of modern research.

(A) Transcriptome analysis with a small genome

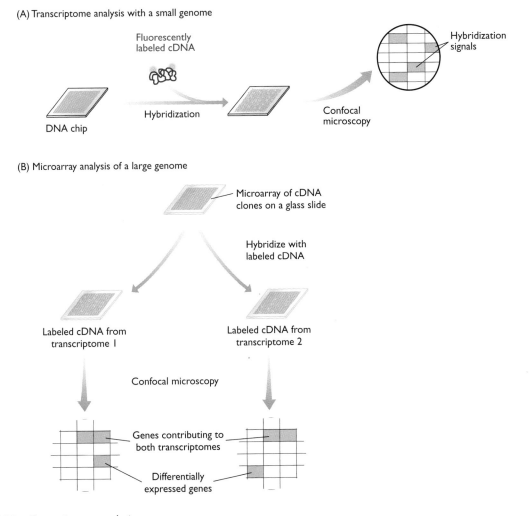

(B) Microarray analysis of a large genome

Figure 7.23 Transcriptome analysis.

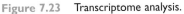

(A) Transcriptome analysis with a DNA chip carrying oligonucleotides representing all the genes in a small genome. After adding labeled cDNA, the positions of the hybridization signals on the chip indicate which genes have contributed to the transcriptome under study. (B) With a larger genome, cDNA clones prepared from the transcriptome of one tissue are immobilized as a microarray and probed with cDNAs representing the same or a different transcriptome. By comparing the hybridization patterns, genes that are expressed differently in the tissues from which the transcriptomes are obtained can be identified.

In order to characterize a proteome it is first necessary to prepare pure samples of its constituent proteins. This is a far from trivial undertaking in view of the complexity of the average proteome: remember that a mammalian cell may contain 10 000–20 000 different proteins (Section 3.3). Polyacrylamide gel electrophoresis (see Technical Note 6.1, page 165) is the standard method for separating proteins, but the usual procedure, in which proteins are separated according to their molecular weights, is unable to resolve the many proteins in an average proteome. To separate individual proteins, the polyacrylamide gel is rotated by 90° and a second electrophoresis carried out at right angles to the first (*Figure 7.24A*). Usually, different conditions are employed in this second run so that the proteins are now separated on the basis of their charges. The result of this **two-dimensional gel electrophoresis** is a series of spots, each one representing a different protein.

Not all the components of the proteome will be visible because the staining methods used to reveal the spots have a detection limit, but a clear picture of the most abundant proteins is obtained. Differences between two proteomes are seen as changes in the position and/or intensity of one or more spots on the gel.

How do we identify which protein is present in which spot? This used to be a difficult procedure but advances in mass spectrometry have provided the rapid and accurate identification procedure dictated by the requirements of genome studies. Mass spectrometry was originally designed as a means of identifying a compound on the basis of the mass-to-charge ratios of the ionized forms that are produced when molecules are exposed to a high-energy field. The standard technique could not be used with proteins because they are too large to be ionized effectively, but a new procedure,

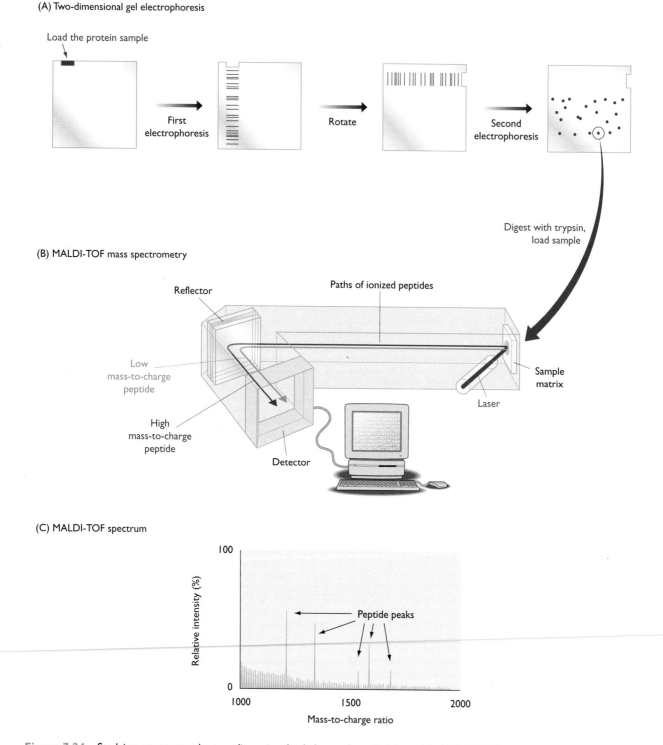

Figure 7.24   Studying a proteome by two-dimensional gel electrophoresis followed by MALDI-TOF.

(A) After two-dimensional gel electrophoresis a protein of interest is excised from the gel and digested with a protease such as trypsin, which cuts immediately after arginine or lysine amino acids. This cleaves the protein into a series of peptides which can be analyzed by MALDI-TOF. (B) In the mass spectrometer the peptides are ionized by a pulse of energy from a laser and then accelerated down the column to the reflector and onto the detector. The time of flight of each peptide depends on its mass-to-charge ratio. The data are visualized as a spectrum (C). The computer contains a database of the predicted molecular weights of every trypsin fragment of every protein encoded by the genome of the organism under study. The computer compares the masses of the detected peptides with the database and identifies the most likely source protein.

called **matrix-assisted laser desorption ionization time-of-flight** (**MALDI-TOF**), gets around this problem, at least with peptides of up to 50 amino acids in length (Yates, 2000). Once ionized, the mass-to-charge ratio of a peptide is determined from its 'time of flight' within the mass spectrometer as its passes from the ionization source to the detector (*Figure 7.24B*). The mass-to-charge ratio enables the molecular weight to be worked out, which in turn allows the amino acid composition of the peptide to be deduced. If a number of peptides from a single protein spot in the two-dimensional gel are analyzed, these peptides obtained by treatment of the protein with a protease such as trypsin, then the resulting compositional information can be related to the genome sequence in order to identify the gene that specifies that protein (*Figure 7.24C*).

Proteomics can also be taken beyond simple characterization of proteome content. For example, the compositions of the peptides derived from a single protein can be used to check a gene sequence (Mann and Pandey, 2001), and in particular to ensure that exon–intron boundaries have been correctly located. This not only helps to delineate the exact position of a gene in a genome (Section 7.1.1), it also allows differential splicing pathways to be identified in cases where two or more proteins are derived from the same gene.

Identifying proteins that interact with one another

Important data pertaining to genome activity can also be obtained by identifying pairs and groups of proteins that interact with one another. At a detailed level, this information is often valuable when attempts are made to assign a function to a newly discovered gene or protein (Section 7.2) because an interaction with a second well-characterized protein can often indicate the role of an unknown protein. For example, an interaction with a protein that is located on the cell surface might indicate that an unknown protein is involved in cell–cell signaling (Section 12.1.2). At a global level, the construction of **protein interaction maps** is looked on as an important step in linking the proteome with the cellular biochemistry.

There are several methods for studying protein–protein interactions, the two most useful being **phage display** and the **yeast two-hybrid system**. In phage display a special type of cloning vector is used, one based on λ bacteriophage or one of the filamentous bacteriophages such as M13 (Clackson and Wells, 1994). The vector is designed so that a new gene that is cloned into it is expressed in such a way that its protein product becomes fused with one of the phage coat proteins (*Figure 7.25A*). The phage protein therefore carries the foreign protein into the phage coat, where it is 'displayed' in a form that enables it to interact with other proteins that the phage encounters. There are several ways in which phage display can be used to study protein interactions. In one method, the test protein is displayed and interactions sought with a series of purified proteins or protein fragments of known function. This approach is limited because it takes time to carry out each test, so is feasible only if some prior information has been obtained about likely interactions. A more powerful strategy is to prepare a **phage display library**, a collection of clones displaying a range of proteins, and identify which members of the library interact with the test protein (*Figure 7.25B*).

The yeast two-hybrid system detects protein interactions in a more complex way (Fields and Sternglanz, 1994). In Section 9.3.2 we will see that proteins called **activators** are responsible for controlling the expression of genes in eukaryotes. To carry out this function an activator must bind to a DNA sequence upstream of a gene and stimulate the RNA polymerase enzyme that copies the gene into RNA. These two abilities – DNA-binding and polymerase activation – are specified by different parts of the activator, and some activators will work even after cleavage into two segments, one segment containing the DNA-binding domain and one the activation domain. In the cell, the two segments interact to form the functional activator.

The two-hybrid system makes use of an *S. cerevisiae* strain that lacks an activator for a reporter gene. This gene is therefore switched off. An artificial gene that codes for the DNA-binding domain of the activator is ligated to the gene for the protein whose interactions we wish to study. This protein can come from any organism, not just yeast: in the example shown in *Figure 7.26A* it is a human protein. After introduction into yeast, this construct specifies synthesis of a fusion protein made up of the DNA-binding domain of the activator attached to the human protein. The recombinant yeast strain is still unable to express the reporter gene because the modified activator only binds to DNA; it cannot influence the RNA polymerase. Activation only occurs after the yeast strain has been cotransformed with a second construct, one comprising the coding sequence for the activation domain fused to a DNA fragment that specifies a protein able to interact with the human protein that is being tested (*Figure 7.26B*). As with phage display, if there is some prior knowledge about possible interactions then individual DNA fragments can be tested one by one in the two-hybrid system. Usually, however, the gene for the activation domain is ligated with a mixture of DNA fragments so that many different constructs are made. After transformation, cells are plated out and those that express the reporter gene identified. These are cells that have taken up a copy of the gene for the activation domain fused to a DNA fragment that encodes a protein able to interact with the test protein.

Protein interaction maps

Protein interaction maps display all of the interactions that occur between the components of a proteome (Legrain *et al.*, 2001). Although a major undertaking, such maps have been constructed for the bacterium *Helicobacter pylori*, comprising over 1200 interactions

(A) Production of a display phage

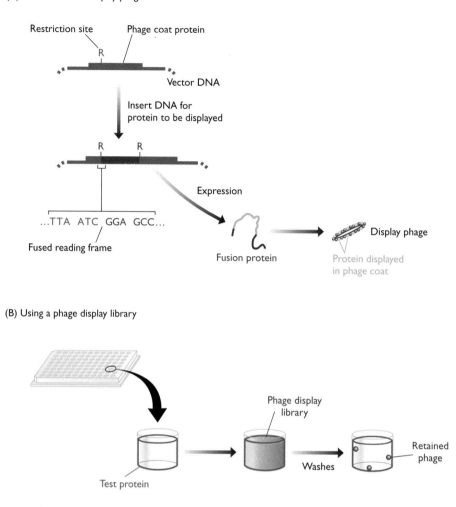

(B) Using a phage display library

Figure 7.25 Phage display.

(A) The cloning vector used for phage display is a bacteriophage genome with a unique restriction site located within a gene for a coat protein. The technique was originally carried out with the gene III coat protein of the filamentous phage called f1, but has now been extended to other phages including λ. To create a display phage, the DNA sequence coding for the test protein is ligated into the restriction site so that a fused reading frame is produced – one in which the series of codons continues unbroken from the coat protein gene into the test gene. After transformation of *Escherichia coli*, this recombinant molecule directs synthesis of a hybrid protein made up of the test protein fused to the coat protein. Phage particles produced by these transformed bacteria therefore display the test protein in their coats. (B) Using a phage display library. The test protein is immobilized within a well of a microtiter tray and the phage display library added. After washing, the phages that are retained in the well are those displaying a protein that interacts with the test protein.

involving almost half of the proteins in the proteome (Rain *et al.*, 2001), and for 2240 interactions between 1870 proteins from the *S. cerevisiae* proteome (Jeong *et al.*, 2001). These two maps were constructed almost entirely from two-hybrid experiments, but various researchers are developing more innovative ways of identifying possible links between proteins. One approach is based on the observation that pairs of proteins that are separate molecules in some organisms are fused into a single polypeptide chain in others. An example is provided by the yeast gene *HIS2*, which codes for an enzyme involved in histidine biosynthesis. In *E. coli*, two genes are homologous to *HIS2*. One of these, itself called *his2*, has sequence similarity with the 5′ region of the yeast gene, and the second, *his10*, is similar to the 3′ region (*Figure 7.27*). The implication is that the proteins coded by *his2* and *his10* interact within the *E. coli* proteome to provide part of the histidine biosynthesis activity. Analysis of the sequence databases reveals many examples of this type, where two proteins in one organism have become fused into a single protein in another organism (Enright *et al.*, 1999; Marcotte *et al.*, 1999), and this infor-

(A) The two-hybrid system

(B) Screening for protein interactions using the two-hybrid system

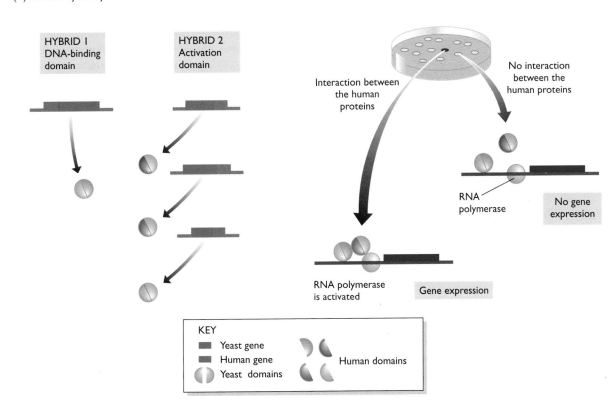

Figure 7.26   The yeast two-hybrid system.

(A) On the left, a gene for a human protein has been ligated to the gene for the DNA-binding domain of a yeast activator. After transformation of yeast, this construct specifies a fusion protein, part human protein and part yeast activator. On the right, various human DNA fragments have been ligated to the gene for the activation domain of the activator: these constructs specify a variety of fusion proteins. (B) The two sets of constructs are mixed and cotransformed into yeast. A colony in which the reporter gene is expressed contains fusion proteins whose human segments interact, thereby bringing the DNA-binding and activation domains into proximity and stimulating the RNA polymerase. See Section 9.3.2 for more information on activators.

mation is proving valuable in extending the results of two-hybrid and other experimental studies.

What are the interesting features of the protein interaction maps that have been generated? The yeast map (*Figure 7.28*) is particularly intriguing because the network is made up of a small number of proteins that have many interactions, and a much larger number of proteins with few individual connections. This architecture, which

is also displayed by the internet, is thought to minimize the effect on the proteome of the disruptive effects of mutations which might inactivate individual proteins. Only if a mutation affects one of the proteins at a highly interconnected node will the network as a whole be damaged. This hypothesis is consistent with the discovery, from gene inactivation studies (Section 7.2.2), that a substantial number of yeast proteins are apparently redundant, meaning that if the protein activity is destroyed the proteome as whole continues to function normally, and there is no discernible impact on the phenotype of the cell (see Box 7.1, page 200).

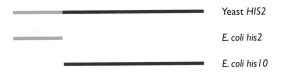

Yeast *HIS2*

E. coli his2

E. coli his10

Figure 7.27   Using homology analysis to deduce protein–protein interactions.

The 5′ region of the yeast *HIS2* gene is homologous to *Escherichia coli his2*, and the 3′ region is homologous to *E. coli his10*.

7.4 Comparative Genomics

The final method for understanding a genome sequence that we will consider is **comparative genomics**. We have already seen how similarities between homologous genes from different organisms provide one way of assigning a function to an unknown gene (Section 7.2.1). This is an

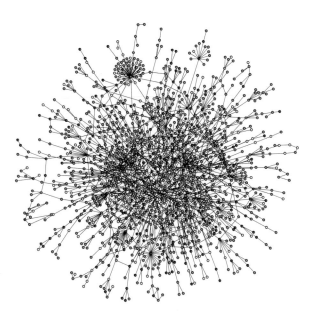

Figure 7.28 The yeast protein interaction map.

Each dot represents a protein, with connecting lines indicating interactions between pairs of proteins. Red dots are essential proteins: an inactivating mutation in the gene for one of these proteins is lethal. Mutations in the genes for proteins indicated by green dots are non-lethal; mutations in genes for proteins shown in orange lead to slow growth. The effects of mutation in genes for proteins shown as yellow dots are not known. From Jeong *et al.*, Nature, **411**, 41–42. Copyright 2001 Macmillan Magazines Limited.

example of how knowledge about the genome of one organism can help in understanding the genome of a second organism. The possibility that a more general comparison with other genomes might be a valuable means of deciphering the human sequence was recognized when the Human Genome Project was planned in the late 1980s, and the Project has actively stimulated the development of genome projects for model organisms such as the mouse and fruit fly. In this section we will explore the extent to which comparisons between different genomes are proving useful.

7.4.1 Comparative genomics as an aid to gene mapping

The basis of comparative genomics is that the genomes of related organisms are similar. The argument is the same one that we considered when looking at homologous genes (Section 7.2.1). Two organisms with a relatively recent common ancestor will have genomes that display species–specific differences built onto the common plan possessed by the ancestral genome. The closer two organisms are on the evolutionary scale, the more related their genomes will be (Nadeau and Sankoff, 1998).

If the two organisms are sufficiently closely related then their genomes might display **synteny**, the partial or complete conservation of gene order. Then it is possible to use map information from one genome to locate genes in the second genome. At one time it was thought that mapping the genomes of the mouse and other mammals, which are at least partially syntenic with the human genome, might provide valuable information that could be used in construction of the human genome map. The problem with this approach is that all the close relatives of humans have equally large genomes that are just as difficult to study, the only advantage being that a genetic map is easier to construct with an animal which, unlike humans, can be subjected to experimental breeding programs (Section 5.2.4). Despite the limitations of human pedigree analysis, progress has been more rapid in mapping the human genome than in mapping those of any of our close relatives, so in this respect comparative genomics is proving more useful in mapping the animal genomes rather than our own. This in itself is a useful corollary to the Human Genome Project because it is revealing animal homologs of human genes involved in diseases, providing animal models for the study of these diseases.

Mapping is significantly easier with a small genome than with a large one. This means that if one member of a pair of syntenic genomes is substantially smaller than the other, then mapping studies with this small genome are likely to provide a real boost to equivalent work with the larger genome. The pufferfish, *Fugu rubripes*, has been proposed in this capacity with respect to the human genome. The pufferfish genome is just 400 Mb, less than one-seventh the size of the human genome but containing approximately the same number of genes. The mapping work carried out to date with the pufferfish indicates that there is some similarity with the human gene order, at least over short distances. This means that it should be possible, to a certain extent, to use the pufferfish map to find human homologs of pufferfish genes, and vice versa. This may be useful in locating undiscovered human genes, but holds greatest promise in identifying essential sequences such as promoters and other regulatory signals upstream of human genes. This is because these signals are likely to be similar in the two genomes, and recognizable because they are surrounded by non-coding DNA that has diverged quite considerably by random mutations (Elgar *et al.*, 1996; Hardison, 2000).

One area where comparative genomics has a definite advantage is in the mapping of plant genomes. Wheat provides a good example. Wheat is the most important food plant in the human diet, being responsible for approximately 20% of the human calorific intake, and is therefore one of the crop plants that we most wish to study and possibly manipulate in the quest for improved crops. Unfortunately, the wheat genome is huge at 16 000 Mb, five times larger than even the human genome. A small model genome with a gene order similar to that of wheat would therefore be useful as a means of mapping desirable genes which might then be obtained from their equivalent positions in the wheat genome. Wheat, and other cereals such as rice, are members of the Gramineae, a large and diverse family of grasses. The rice genome is only 430 Mb, substantially smaller than that of wheat, and there are probably other grasses with even smaller genomes. Comparative mapping of the rice and wheat genomes has revealed many similarities, and the possibility therefore exists that genes from the wheat genome might be isolated by first mapping the positions of the equivalent genes in a smaller Gramineae genome (Gura, 2000).

7.4.2 Comparative genomics in the study of human disease genes

One of the main reasons for sequencing the human genome is to gain access to the sequences of genes involved in human disease. The hope is that the sequence of a disease gene will provide an insight into the biochemical basis of the disease and hence indicate a way of preventing or treating the disease. Comparative genomics has an important role to play in the study of disease genes because the discovery of a homolog of a human disease gene in a second organism is often the key to understanding the biochemical function of the human gene. If the homolog has already been characterized then the information needed to understand the biochemical role of the human gene may already be in place; if it has not been characterized then the necessary research can be directed at the homolog.

To be useful in the study of disease-causing genes, the second genome does not need to be syntenic with the human genome, nor even particularly closely related. *Drosophila* holds great promise in this respect, as the phenotypic effects of many *Drosophila* genes are well known, so the data already exist for inferring the mode of action of human disease genes that have homologs in the *Drosophila* genome (Guffanti *et al.*, 1997). But the greatest success has been with yeast. Several human disease genes have homologs in the *S. cerevisiae* genome (*Table 7.2*). These disease genes include ones involved in cancer, cystic fibrosis, and neurological syndromes, and in several cases the yeast homolog has a known function that provides a clear indication of the biochemical activity of the human gene. In some cases it has even been possible to demonstrate a physiological similarity between the gene activity in humans and yeast. For example, the yeast gene *SGS1* is a homolog of a human gene involved in the diseases called Bloom's and Werner's syndromes, which are characterized by growth disorders. Yeasts with a mutant *SGS1* gene live for shorter periods than normal yeasts and display accelerated onset-of-aging indicators such as sterility (Sinclair *et al.*, 1997). The yeast gene has been shown to code for one of a pair of related DNA helicases that are required for transcription of rRNA genes and for DNA replication (Lee *et al.*, 1999). The link between *SGS1* and the genes for Bloom's and Werner's syndromes, provided by comparative genomics, has therefore indicated the possible biochemical basis of the human diseases.

Table 7.2 Examples of human disease genes that have homologs in *Saccharomyces cerevisiae*

Human disease gene	Yeast homolog	Function of the yeast gene
Amyotrophic lateral sclerosis	*SOD1*	Protein against superoxide (O_2^-)
Ataxia telangiectasia	*TEL1*	Codes for a protein kinase
Colon cancer	*MSH2, MLH1*	DNA repair
Cystic fibrosis	*YCF1*	Metal resistance
Myotonic dystrophy	*YPK1*	Codes for a protein kinase
Type 1 neurofibromatosis	*IRA2*	Codes for a regulatory protein
Bloom's syndrome, Werner's syndrome	*SGS1*	DNA helicase
Wilson's disease	*CCC2*	Copper transport?

Data taken from Bassett *et al.* and Sinclair *et al.* (1997).

References

Altschul SF, Gish W, Miller W, Myers EW and Lipman DJ (1990) Basic local alignment search tool. *J. Mol. Biol.*, **215**, 403–410.

Bass BL (2001) The short answer. *Nature*, **411**, 428–429.

Bassett DE, Boguski MS and Hieter P (1996) Yeast genes and human disease. *Nature*, **379**, 589–590.

Bird A (1986) CpG-rich islands and the function of DNA methylation. *Nature*, **321**, 209–213.

Blattner FR, Plunkett G, Bloch CA, et al. (1997) The complete genome sequence of *Escherichia coli* K-12. *Science*, **277**, 1453–1462.

Church DM, Stotler CJ, Rutter JL, Murrell JR, Trofatter JA and Buckler AJ (1994) Isolation of genes from complex sources of mammalian genomic DNA using exon amplification. *Nature Genet.*, **6**, 98–105.

Clackson T and Wells JA (1994) *In vitro* selection from protein and peptide libraries. *Trends Biotechnol.*, **12**, 173–184.

Cornish-Bowden A and Cárdenas ML (2001) Silent genes given voice. *Nature*, **409**, 571–572.

Dujon B (1996) The yeast genome project: what did we learn? *Trends Genet.*, **12**, 263–270.

Elbashir SM, Harborth J, Lendeckel W, Yalcin A, Weber K and Tuschl T (2001) Duplexes of 21-nucleotide RNAs mediate RNA interference in cultured mammalian cells. *Nature*, **411**, 494–498.

Elgar G, Sandford R, Aparicio S, Macrae A, Vekatesh B and Brenner S (1996) Small is beautiful: comparative genomics with the pufferfish (*Fugu rubripes*). *Trends Genet.*, **12**, 145–150.

Engels WR (2000) Reversal of fortune for *Drosophila* geneticists? *Science*, **288**, 1973–1975.

Enright AJ, Iliopoulos I, Kyrpides NC and Ouzounis CA (1999) Protein interaction maps for complete genomes based on gene fusion events. *Nature*, **402**, 86–90.

Evans MJ, Carlton MBL and Russ AP (1997) Gene trapping and functional genomics. *Trends Genet.*, **13**, 370–374.

Fickett JW (1996) Finding genes by computer: the state of the art. *Trends Genet.*, **12**, 316–320.

Fields S and Sternglanz R (1994) The two-hybrid system: an assay for protein-protein interactions. *Trends Genet.*, **10**, 286–292.

Fire A, Xu S, Montgomery MK, Kostas SA, Driver SE and Mello CC (1998) Potent and specific genetic interference by double-stranded RNA in *Caenorhabditis elegans*. *Nature*, **391**, 806–811.

Fraser AG, Kamath RS, Zipperlen P, Martinez-Campos M, Sohrmann M and Ahringer J (2000) Functional genomic analysis of *C. elegans* chromosome I by systematic RNA interference. *Nature*, **408**, 325–330.

Frech K, Quandt K and Werner T (1997) Finding protein-binding sites in DNA sequences: the next generation. *Trends Biochem. Sci.*, **22**, 103–104.

Frohman MA, Dush MK and Martin GR (1988) Rapid production of full-length cDNAs from rare transcripts: amplification using a single gene-specific oligonucleotide primer. *Proc. Natl Acad. Sci. USA*, **85**, 8998–9002.

Guffanti A, Banfi S, Simon G, Ballabio A and Borsani G (1997) DRES search engine: of flies, men and ESTs. *Trends Genet.*, **13**, 79–80.

Gura T (2000) Reaping the plant gene harvest. *Science*, **287**, 412–414.

Hardison RC (2000) Conserved non-coding sequences are reliable guides to regulatory elements. *Trends Genet.*, **16**, 369–372.

Jeong H, Mason SP, Barabási A-L and Oltvai ZN (2001) Lethality and centrality in protein networks. *Nature*, **411**, 41–42.

Knight J (2001) When the chips are down. *Nature*, **410**, 860–861.

Lee S-K, Johnson RE, Yu S-L, Prakash L and Prakash S (1999) Requirement of yeast SGS1 and SRS2 genes for replication and transcription. *Science*, **286**, 2339–2342.

Legrain P, Wojcik J and Gauthier J-M (2001) Protein-protein interaction maps: a lead towards cellular functions. *Trends Genet.*, **17**, 346–352.

Lovett M (1994) Fishing for complements: finding genes by direct selection. *Trends Genet.*, **10**, 352–357.

Mann M and Pandey A (2001) Use of mass spectrometry-derived data to annotate nucleotide and protein sequence databases. *Trends Biochem. Sci.*, **26**, 54–61.

Marcotte EM, Pellegrini M, Thompson MJ, Yeates TO and Eisenberg D (1999) A combined algorithm for genome-wide prediction of protein function. *Nature*, **402**, 83–86.

Marshall E (1999) Do-it-yourself gene watching. *Science*, **286**, 444–447.

Nadeau JH and Sankoff D (1998) Counting on comparative maps. *Trends Genet.*, **14**, 495–501.

Ohler U and Niemann H (2001) Identification and analysis of eukaryotic promoters: recent computational approaches. *Trends Genet.*, **17**, 56–60.

Oliver SG (1996a) From DNA sequence to biological function. *Nature*, **379**, 597–600.

Oliver SG (1996b) A network approach to the systematic analysis of yeast gene function. *Trends Genet.*, **12**, 241–242.

Ponting CP (1997) Tudor domains in proteins that interact with RNA. *Trends Biochem. Sci.*, **22**, 51–52.

Pradet-Balade B, Boulmé F, Beug H, Müllner EW and Garcia-Sanz JA (2001) Translation control: bridging the gap between genomics and proteomics? *Trends Biochem. Sci.*, **26**, 225–229.

Rain J-C, Selig L, De Reuse H, et al. (2001) The protein-protein interaction map of *Helicobacter pylori*. *Nature*, **409**, 211–215.

Reese MG, Hartzell G, Harris NL, Ohler U, Abril JF and Lewis SE (2000) Genome annotation assessment in *Drosophila melanogaster*. *Genome Res.*, **10**, 483–501.

Simonet WS, Lacey DL, Dunstan CR, et al. (1997) Osteoprotegrin: a novel secreted protein involved in the regulation of bone density. *Cell*, **89**, 309–319.

Sinclair DA, Mills K and Guarente L (1997) Accelerated aging and nucleolar fragmentation in yeast sgs1 mutants. *Science*, **277**, 1313–1316.

Smith V, Botstein D and Brown PO (1995) Genetic footprinting: a genomic strategy for determining a gene's function given its sequence. *Proc. Natl Acad. Sci. USA*, **92**, 6479–6483.

Velculescu VE, Vogelstein B and Kinzler KW (2000) Analysing uncharted transcriptomes with SAGE. *Trends Genet.*, **16**, 423–425.

Wach A, Brachat A, Pohlmann R and Philippsen P (1994) New heterologous modules for classical or PCR-based gene disruptions in *Saccharomyces cerevisiae*. *Yeast*, **10**, 1793–1808.

Yates JR (2000) Mass spectrometry: from genomics to proteomics. *Trends Genet.*, **16**, 5–8.

Further Reading

Ambros V (2001) Dicing up RNAs. *Science*, **293**, 811–813. — *Describes current thinking on the natural role of RNA interference in living organisms.*

Birney E, Bateman A, Clamp ME and Hubbard TJ (2001) Mining the draft human genome. *Nature*, **409**, 827–828. — *A guide to the bioinformatics tools available for analyzing the human genome.*

Fields S (2001) Proteomics in genomeland. *Science*, **291**, 1221–1224. — *Explains the importance of proteomics in understanding the human genome sequence.*

Galas DJ (2001) Making sense of the sequence. *Science*, **291**, 1257–1260. — *Another description of the bioinformatics tools available for analysis of the human genome sequence.*

Mann M, Hendrickson RC and Pandey A (2001) Analysis of proteins and proteomes by mass spectrometry. *Ann. Rev. Biochem.*, **70**, 437–473.

O'Brien SJ, Menotti-Raymond M, Murphy WJ, et al. (1999) The promise of comparative genomics in mammals. *Science*, **286**, 458–481. — *A review that draws out the importance of comparative methods in studies of mammalian genomes.*

Pennisi E (2001) Keeping genome databases clean and up to date. *Science*, **286**, 447–450. — *Highlights some of the problems with the DNA databases.*

Roos DS (2001) Bioinformatics – trying to swim in a sea of data. *Science*, **291**, 1260–1261. — *Outlines the challenges to bioinformatics posed by large sequences such as the human genome.*

Searls DB (2000) Bioinformatics tools for whole genomes. *Ann. Rev. Genomics Hum. Genet.*, **1**, 251–279. — *A review of the applications of computer-based studies in analysis of genome sequences.*

Various authors (2000) Proteomics: A Trends Guide. Elsevier Science, London. — *Reviews and commentaries on various aspects of proteomics.*

STUDY AIDS FOR CHAPTER 7

Key terms

Give short definitions of the following terms:

Activator
Bioinformatics
cDNA capture
cDNA selection
Chimera
Codon bias
Comparative genomics
Consensus sequence
CpG island
Embryonic stem cell
Exon trapping
Genetic footprinting
Heteroduplex analysis
Homologous genes
Homologous recombination
Homology search
Immunocytochemistry
In vitro mutagenesis
Knockout mice
Mass spectrometry
Matrix-assisted laser desorption ionization time-of-flight (MALDI-TOF)
Minigene
Model organism
Multicopy plasmid
Northern blotting
Oligonucleotide-directed mutagenesis

ORF scanning
Orphan family
Orthologous genes
P element
Paralogous genes
Phage display
Phage display library
Protein electrophoresis
Protein engineering
Protein interaction map
Proteomics
RACE
Reading frame
Reporter gene
Reverse transcriptase
RNA interference
RT-PCR
Serial analysis of gene expression (SAGE)
Single orphan
Site-directed mutagenesis
Synteny
Totipotent
Transgenic mouse
Transposon tagging
Two-dimensional gel electrophoresis
Yeast two-hybrid system
Zoo-blotting

Self study questions

1. Explain why ORF scanning is a feasible way of identifying genes in a prokaryotic DNA sequence.
2. What modifications are introduced when ORF scanning is applied to a eukaryotic DNA sequence?
3. Describe how homology searching is used to locate genes in a DNA sequence and to assign possible functions to those genes.
4. Distinguish between northern blotting and zoo-blotting. What are the applications of these two techniques in gene location?
5. Explain how cDNA capture or cDNA selection are used to enrich a clone library for a particular cDNA sequence.
6. Draw a fully annotated diagram illustrating the procedure called 5′-RACE.
7. Describe how S1 nuclease is used to map the positions of the ends of a transcript on to a DNA sequence.
8. What experimental methods can be used to locate exon–intron boundaries in a DNA sequence?

9. Using the yeast genome project as an example, illustrate the strengths and weaknesses of homology analysis as a means of assigning functions to unknown genes.
10. Describe how gene inactivation can be used to determine the function of an unknown gene.
11. Give an example of the use of gene overexpression to determine the function of an unknown gene.
12. Describe how oligonucleotide-directed mutagenesis is carried out and outline the use of this technique in studying the activity of the protein coded by an unknown gene.
13. What is a reporter gene and how is it used?
14. Describe the methods used to study transcriptomes.
15. Explain how two-dimensional gel electrophoresis combined with mass spectrometry is used to study a proteome.
16. Draw diagrams illustrating the techniques called (a) phage display, and (b) the yeast two-hybrid system. What are the similarities and differences between these two techniques?
17. What is a protein interaction map? What has the yeast protein interaction map told us about the construction of the proteome of this organism?
18. Define the term 'synteny' and, using examples, explain how synteny can predict the positions of genes in a genome sequence.
19. Describe the applications of comparative genomics in the study of human disease genes.

Problem-based learning

1. Defend one of the following statements:
   'In future years it will be possible to use bioinformatics to obtain a complete description of the locations and functions of the genes in a genome sequence.'
   'In future years bioinformatics will become obsolete because of the development of rapid and effective experimental methods for locating and assigning functions to the genes in a genome sequence.'
2. Devise a hypothesis to explain the codon biases that occur in the genomes of various organisms. Can your hypothesis be tested?
3. Gene inactivation studies have suggested that at least some genes in a genome are redundant, meaning that they have the same function as a second gene and so can be inactivated without affecting the phenotype of the organism. What evolutionary questions are raised by genetic redundancy? What are the possible answers to these questions?
4. Explore the natural role of RNA interference in living organisms.
5. Gene overexpression has so far provided limited but important information on the function of unknown genes. Assess the overall potential of this approach in functional analysis.
6. 'Comparative genomics has an important role to play in the study of disease genes.' Evaluate this statement.

In order for the cell to utilize the biological information contained within its genome, groups of genes, each gene representing a single unit of information, have to be expressed in a coordinated manner. This coordinated gene expression determines the make-up of the transcriptome, which in turn specifies the nature of the proteome and defines the activities that the cell is able to carry out.

In Part 3 of *Genomes* we examine the events that result in the transfer of biological information from genome to proteome. Our knowledge of these events was initially gained through studies of individual genes, often as 'naked' DNA in test-tube experiments. These experiments provided an interpretation of gene expression that in recent years has been embellished by more sophisticated studies that have taken greater account of the fact that, in reality, it is the genome that is expressed, not individual genes, and that this expression occurs in living cells rather than in a test tube.

We begin our investigation of genome expression in Chapter 8, by examining the substantial and important impact that the nuclear environment has on the utilization of the biological information contained in the genomes of eukaryotes, the accessibility of that information being dependent on the way in which the DNA is packaged into chromatin and being responsive to processes that can silence or inactivate part or all of a chromosome.

Chapter 9 describes the events involved in initiation of transcription, and emphasizes the critical role the DNA-binding proteins play during the early stages of genome expression. The synthesis of transcripts and their subsequent processing into functional RNAs is dealt with in Chapter 10, and Chapter 11 covers the equivalent events that lead to synthesis of the proteome.

As you read Chapters 8–11 you will discover that control over the composition of the transcriptome and of the proteome can be exerted at various stages during the overall chain of events that make up genome expression. These regulatory threads will be drawn together in Chapter 12, where we examine how genome activity changes in response to extracellular signals during differentiation and development.

How Genomes Function

Chapter 8

Accessing the Genome

Chapter 9

Assembly of the Transcription Initiation Complex

Chapter 10

Synthesis and Processing of RNA

Chapter 11

Synthesis and Processing of the Proteome

Chapter 12

Regulation of Genome Activity

8

Accessing the Genome

Chapter Contents

8.1 *Inside the Nucleus* 222

    8.1.1 The internal architecture of
the eukaryotic nucleus 222

    8.1.2 Chromatin domains 224

8.2 *Chromatin Modifications and
Genome Expression* 228

    8.2.1 Activating the genome 228

    8.2.2 Silencing the genome 231

Learning outcomes

When you have read Chapter 8, you should be able to:

- Explain how chromatin structure influences genome expression

- Describe the internal architecture of the eukaryotic nucleus

- Distinguish between the terms 'constitutive heterochromatin', 'facultative heterochromatin' and 'euchromatin'

- Discuss the key features of functional domains, insulators, and locus control regions, and describe the experimental evidence supporting our current knowledge of these structures

- Describe the various types of chemical modification that can be made to histone proteins, and link this information to the concept of the 'histone code'

- State why nucleosome positioning is important in gene expression and give details of a protein complex involved in nucleosome remodeling

- Explain how DNA methylation is carried out and describe the importance of methylation in silencing the genome

- Give details of the involvement of DNA methylation in genomic imprinting and X inactivation

WHEN ONE LOOKS at a genome sequence written out as a series of As, Cs, Gs and Ts, or drawn as a map with the genes indicated by boxes on a string of DNA (as in *Figure 1.14*, page 19, for example), there is a tendency to imagine that all parts of the genome are readily accessible to the DNA-binding proteins that are responsible for its expression. In reality, the situation is very different. The DNA in the nucleus of a eukaryotic cell or the nucleoid of a prokaryote is attached to a variety of proteins that are not directly involved in genome expression and which must be displaced in order for the RNA polymerase and other expression proteins to gain access to the genes. We know very little about these events in prokaryotes, a reflection of our generally poor knowledge about the physical organization of the prokaryotic genome (Section 2.3.1), but we are beginning to understand how the packaging of DNA into chromatin (Section 2.2.1) influences genome expression in eukaryotes. This is an exciting area of molecular biology, with recent research indicating that histones and other packaging proteins are not simply inert structures around which the DNA is wound, but instead are active participants in the processes that determine which parts of the genome are expressed in an individual cell. Many of the discoveries in this area have been driven by new insights into the substructure of the nucleus, and it is with this topic that we begin the chapter.

8.1 Inside the Nucleus

The light microscopy and early electron microscopy studies of eukaryotic cells revealed very few internal features within the nucleus. This apparent lack of structure led to the view that the nucleus has a relatively homogeneous architecture, a typical 'black box' in common parlance. In recent years this interpretation has been overthrown and we now appreciate that the nucleus has a complex internal structure that is related to the variety of biochemical activities that it must carry out. Indeed, the inside of the nucleus is just as complex as the cytoplasm of the cell, the only difference being that, in contrast to the cytoplasm, the functional compartments within the nucleus are not individually enclosed by membranes, and so are not visible when the cell is observed using conventional light or electron microscopy techniques.

8.1.1 The internal architecture of the eukaryotic nucleus

The revised picture of nuclear structure has emerged from two novel types of microscopy analysis. First, conventional electron microscopy has been supplemented by examination of mammalian cells that have been prepared in a special way. After dissolution of membranes by soak-

Box 8.1: Accessing the prokaryotic genome

Chapter 8 is exclusively about eukaryotic genomes. We simply do not know enough about the physical structures of bacterial and archaeal genomes to be able to state, with any degree of confidence, the extent to which expression of a prokaryotic genome is influenced by the packaging systems for these molecules. There are a few intriguing pieces of information. Archaea, for example, possess proteins that are similar to eukaryotic histones (Section 2.3.1), and at least some archaea and bacteria have enzymes similar to the Sir proteins which, in yeast, interact with histones to bring about gene silencing (Section 8.2.2). However, bacteria such as *Escherichia coli* do not possess histones, their packaging proteins (e.g. HU; Section 2.3.1) having a distinctive structure, and neither bacteria nor archaea appear to have enzymes equivalent to the histone acetyltransferases that play a central role in modification of chromatin in eukaryotes (Section 8.2.1).

Our understanding of prokaryotic genome expression therefore begins with assembly of the transcription initiation complex. If you are only interested in prokaryotes, then jump forward to Chapter 9.

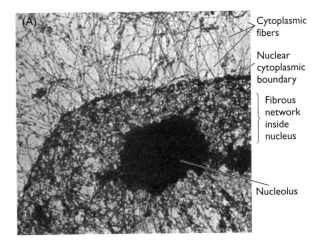

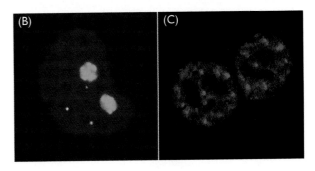

Figure 8.1 The internal architecture of the eukaryotic nucleus.

(A) Transmission electron micrograph showing the nuclear matrix of a cultured human HeLa cell. Cells were treated with a non-ionic detergent to remove membranes, digested with a deoxyribonuclease to degrade most of the DNA, and extracted with ammonium sulfate to remove histones and other chromatin-associated proteins. From *Molecular Cell Biology*, by H Lodish, A Berk, SL Zipursky, P Matsudaira, D Baltimore and J Darnell. ©1986, 1990, 1995, 2000 by WH Freeman and Company. Used with permission. (B) and (C) Images of living nuclei containing fluorescently labeled proteins (see Technical Note 8.1). In (B), the nucleolus is shown in blue and Cajal bodies in yellow. The purple areas in (C) indicate the positions of proteins involved in RNA splicing. B and C from Misteli, *Science*, **291**, 843–847. Copyright 2000 American Association for the Advancement of Science.

ing in a mild non-ionic detergent such as one of the Tween compounds, followed by treatment with a deoxyribonuclease to degrade the nuclear DNA, and salt extraction to remove the chemically basic histone proteins, the nuclear substructure has been revealed as a complex network of protein and RNA fibrils, called the **nuclear matrix** (*Figure 8.1A*). The matrix permeates the entire nucleus and includes regions defined as the **chromosome scaffold**, which changes its structure during cell division, resulting in condensation of the chromosomes into their metaphase forms (see *Figure 2.7*, page 38).

A second novel type of microscopy has involved the use of fluorescent labeling, designed specifically to reveal areas within the nucleus where particular biochemical activities are occurring. The **nucleolus** (*Figure 8.1B*), which is the center for synthesis and processing of rRNA molecules, had been recognized for many years as it is the one structure within the nucleus that can be seen by conventional electron microscopy. Fluorescent labeling directed at the proteins involved in RNA splicing (Sections 1.2.1 and 10.1.3) has shown that this activity is also localized into distinct regions (*Figure 8.1C*), although these are more widely distributed and less well defined than the nucleoli. Other structures, such as Cajal bodies (visible in *Figure 8.1B*), whose functions are not yet understood, are also seen after fluorescent labeling (Lewis and Tollervey, 2000).

The complexity of the nuclear matrix, as shown by *Figure 8.1A*, could be taken as an indication that the nucleus has a static internal environment, with limited movement of molecules from one site to another. Another new microscopy technique, called **fluorescence recovery after photobleaching** (**FRAP**; Technical Note 8.1), which enables the movement of proteins within the nucleus to be visualized, shows that this is not the case. The migration of nuclear proteins does not occur as rapidly as

TECHNICAL

8.1

NOTE

Fluorescence recovery after photobleaching (FRAP)

Visualization of protein mobility in living nuclei.

FRAP is perhaps the most informative of the various innovative microscopy techniques that have opened up our understanding of nuclear substructure. It has enabled, for the first time, the movement of proteins to be visualized inside living nuclei, the resulting data allowing biophysical models of protein dynamism to be tested.

The starting point for a FRAP experiment is a nucleus in which every copy of the protein of interest carries a fluorescent tag. Labeling the protein molecules *in vitro* and then re-introducing them into the nucleus is not possible, so the host organism has to be genetically engineered so that the fluorescent tag is an integral part of the protein that is synthesized *in vivo*. This is achieved by ligating the coding sequence for the **green fluorescent protein** (Tsien, 1998) to the gene for the protein being studied. Standard cloning techniques are then used to insert the modified gene into the host genome (Section 4.2.1), leading to a recombinant cell that synthesizes a fluorescent version of the protein. Observation of the cell using a fluorescence microscope now reveals the distribution of the labeled protein within the nucleus.

To study the mobility of the protein, a small area of the nucleus is **photobleached** by exposure to a tightly focused pulse from a high-energy laser. The laser pulse inactivates the fluorescent signal in the exposed area, leaving a region that appears bleached in the microscopic image. This bleached area gradually retrieves its fluorescent signal, not by a reversal of the bleaching effect, but by migration into the bleached region of fluorescent proteins from the unexposed area of the nucleus. Rapid reappearance of the fluorescent signal in the bleached area therefore indicates that the tagged proteins are highly mobile, whereas a slow recovery indicates that the proteins are relatively static. The kinetics of signal recovery can be used to test theoretical models of protein dynamism derived from biophysical parameters such as binding constants and flux rates (Misteli, 2001).

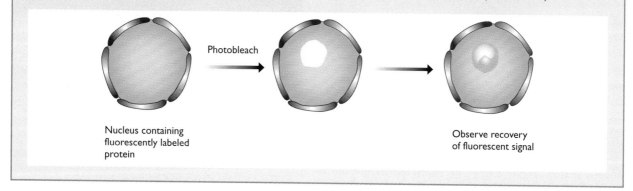

Photobleach

Nucleus containing
fluorescently labeled
protein

Observe recovery
of fluorescent signal

---

would be expected if their movement were totally unhindered, which is entirely expected in view of the large amounts of DNA and RNA in the nucleus, but it is still possible for a protein to traverse the entire diameter of a nucleus in a matter of minutes (Misteli, 2001). Proteins involved in genome expression therefore have the freedom needed to move from one activity site to another, as dictated by the changing requirements of the cell. In particular, the linker histones (Section 2.2.1) continually detach and reattach to their binding sites on the genome (Lever *et al.*, 2000; Misteli *et al.*, 2000). This discovery is important because it emphasizes that the DNA–protein complexes that make up chromatin are dynamic, an observation that has considerable relevance to genome expression, as we will discover in the next section.

8.1.2 Chromatin domains

In Section 2.2.1 we learnt that chromatin is the complex of genomic DNA and chromosomal proteins present in the eukaryotic nucleus. Chromatin structure is hierarchic, ranging from the two lowest levels of DNA packaging – the nucleosome and the 30 nm chromatin fiber (see *Figures 2.5* and *2.6*, page 38) – to the metaphase chromosomes, which represent the most compact form of chromatin in eukaryotes and occur only during nuclear division. After division, the chromosomes become less compact and cannot be distinguished as individual structures. When non-dividing nuclei are examined by light microscopy all that can be seen is a mixture of lightly and darkly staining areas within the nucleus. The dark areas, which tend to be concentrated around the periphery of the nucleus, are called **heterochromatin** and contain DNA that is still in a relatively compact organization, although still less compact than in the metaphase structure. Two types of heterochromatin are recognized:

- **Constitutive heterochromatin** is a permanent feature of all cells and represents DNA that contains no genes and so can always be retained in a compact organization. This fraction includes centromeric and telomeric DNA as well as certain regions of some other chromosomes. For example, most of the human Y chromosome is made of constitutive heterochromatin (see *Figure 2.8*, page 40).
- **Facultative heterochromatin** is not a permanent feature but is seen in some cells some of the time. Facultative heterochromatin is thought to contain genes that are inactive in some cells or at some periods of the cell cycle. When these genes are inactive, their DNA regions are compacted into heterochromatin.

It is assumed that the organization of heterochromatin is so compact that proteins involved in gene expression simply cannot access the DNA. In contrast, the remaining regions of chromosomal DNA, the parts that contain active genes, are less compact and permit entry of the expression proteins. These regions are called **euchromatin** and they are dispersed throughout the nucleus. The exact organization of the DNA within euchromatin is not known, but with the electron microscope it is possible to see loops of DNA within the euchromatin regions, each loop between 40 and 100 kb in length and predominantly in the form of the 30 nm chromatin fiber. The loops are attached to the nuclear matrix via AT-rich DNA segments called **matrix-associated regions** (**MARs**) or **scaffold attachment regions** (**SARs**) (*Figure 8.2*).

The loops of DNA between the nuclear matrix attachment points are called **structural domains**. An intriguing

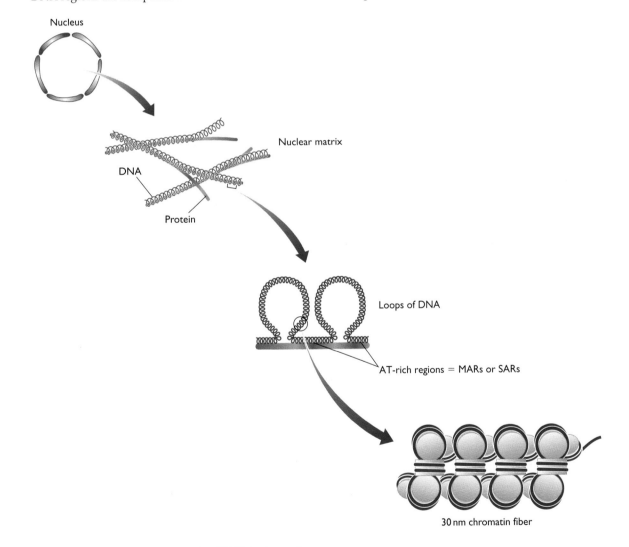

Figure 8.2   A scheme for organization of DNA in the nucleus.

The nuclear matrix is a fibrous protein-based structure whose precise composition and arrangement in the nucleus has not been described. Euchromatin, predominantly in the form of the 30 nm chromatin fiber (see *Figure 2.6*, page 38) is thought to be attached to the matrix by AT-rich sequences called matrix-associated or scaffold attachment regions (MARs or SARs).

question is the precise relationship between these and the **functional domains** that can be discerned when the region of DNA around an expressed gene or set of genes is examined. A functional domain is delineated by treating a region of purified chromatin with deoxyribonuclease I (DNase I) which, being a DNA-binding protein, cannot gain access to the more compacted regions of DNA (*Figure 8.3*). Regions sensitive to DNase I extend to either side of a gene or set of genes that is being expressed, indicating that in this area the chromatin has a more open organization, although it is not clear whether this organization is the 30 nm fiber or the 'beads-on-a-string' structure (see *Figure 2.5A*, page 38). Is there a correspondence between structural and functional domains? Intuition suggests that there should be, and some MARs, which mark the limits of a structural domain, are also located at the boundary of a functional domain. But the correspondence does not seem to be complete because some structural domains contain genes that are not functionally related, and the boundaries of some structural domains lie within genes (Wolffe, 1995).

Functional domains are defined by insulators

The boundaries of functional domains are marked by sequences, 1–2 kb in length, called **insulators** (Bell *et al.*, 2001). Insulator sequences were first discovered in *Drosophila* and have now been identified in a range of eukaryotes. The best studied are the pair of sequences called scs and scs' (scs stands for 'specialized chromatin structure'), which are located either side of the two *hsp70* genes in the fruit-fly genome (*Figure 8.4*).

Insulators display two special properties related to their role as the delimiters of functional domains. The first is their ability to overcome the **positional effect** that occurs during a gene cloning experiment with a eukaryotic host. The positional effect refers to the variability in gene expression that occurs after a new gene has been inserted into a eukaryotic chromosome. It is thought to result from the random nature of the insertion event, which could deliver the gene to a region of highly packaged chromatin, where it will be inactive, or into an area of open chromatin, where it will be expressed (*Figure 8.5A*). The ability of scs and scs' to overcome the positional effect was demonstrated by placing them either side of a fruit-fly gene for eye color (Kellum and Schedl, 1991). When flanked by the insulators, this gene was always highly expressed when

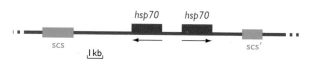

Figure 8.4   Insulator sequences in the fruit-fly genome.

The diagram shows the region of the *Drosophila* genome containing the two *hsp70* genes. The insulator sequences scs and scs' are either side of the gene pair. The arrows below the two genes indicate that they lie on different strands of the double helix and so are transcribed in opposite directions.

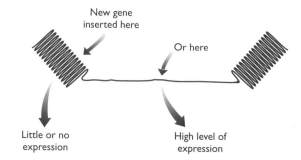

(A) The positional effect

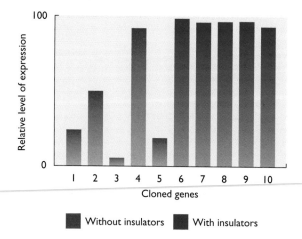

(B) Insulators overcome the positional effect

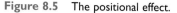

Figure 8.5   The positional effect.

(A) A cloned gene that is inserted into a region of highly packaged chromatin will be inactive, but one inserted into open chromatin will be expressed. (B) The results of cloning experiments without (red) and with (blue) insulator sequences. When insulators are absent, the expression level of the cloned gene is variable, depending on whether it is inserted into packaged or open chromatin. When flanked by insulators, the expression level is consistently high because the insulators establish a functional domain at the insertion site.

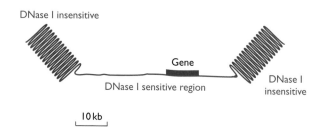

DNase I insensitive

Gene

DNase I sensitive region

DNase I insensitive

10 kb

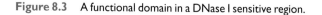

Figure 8.3   A functional domain in a DNase I sensitive region.

inserted back into the *Drosophila* genome, in contrast to the variable expression that was seen when the gene was cloned without the insulators (*Figure 8.5B*). The deduction from this and related experiments is that insulators can bring about modifications to chromatin packaging and hence establish a functional domain when inserted into a new site in the genome.

Insulators also maintain the independence of each functional domain, preventing 'cross-talk' between adjacent domains. If scs or scs' is excised from its normal location and re-inserted between a gene and the upstream regulatory modules that control expression of that gene, then the gene no longer responds to its regulatory modules: it becomes 'insulated' from their effects (*Figure 8.6A*). This observation suggests that, in their normal positions, insulators prevent the genes within a domain from being influenced by the regulatory modules present in an adjacent domain (*Figure 8.6B*).

How insulators carry out their roles is not yet known but it is presumed that the functional component is not the insulating sequence itself but the DNA-binding proteins, such as Su(Hw) in *Drosophila*, that attach specifi-

cally to insulators. As well as binding to insulators, these proteins form associations with the nuclear matrix (Gerasimova *et al.*, 2000), possibly indicating that the functional domains that they define are also structural domains within the chromatin. This is an attractive hypothesis that can be tied in with the ability of insulators to establish open chromatin regions and to prevent cross-talk between functional domains, but it implies that insulators contain MAR sequences, which has not been proven. An equivalence between functional and structural domains therefore remains elusive.

Some functional domains contain locus control regions

The formation and maintenance of an open functional domain, at least for some domains, is the job of a DNA sequence called the **locus control region** or **LCR** (Li *et al.*, 1999). Like insulators, an LCR can overcome the positional effect when linked to a new gene that is inserted into a eukaryotic chromosome. Unlike insulators, an LCR also stimulates the expression of genes contained within its functional domain.

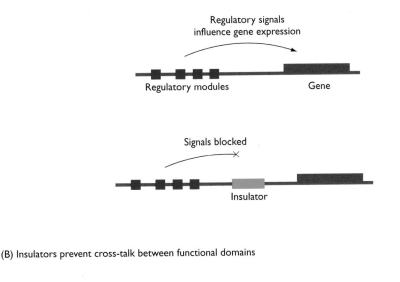

(A) Insulators block the regulatory signals that control gene expression

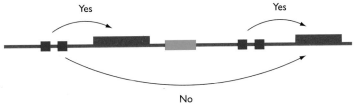

(B) Insulators prevent cross-talk between functional domains

Figure 8.6 Insulators maintain the independence of a functional domain.

(A) When placed between a gene and its upstream regulatory modules, an insulator sequence prevents the regulatory signals from reaching the gene. (B) In their normal positions, insulators prevent cross-talk between functional domains, so the regulatory modules of one gene do not influence expression of a gene in a different domain. For more details about regulatory modules, see Box 9.6, page 267.

LCRs were first discovered during a study of the human β-globin genes (Section 2.2.1) and are now thought to be involved in expression of many genes that are active in only some tissues or during certain developmental stages. The globin LCR is contained in a stretch of DNA some 12 kb in length, positioned upstream of the genes in the 60-kb β-globin functional domain (*Figure 8.7*). The LCR was initially identified during studies of individuals with thalassemia, a blood disease that results from defects in the α- or β-globin proteins. Many thalassemias result from mutations in the coding regions of the globin genes, but a few were shown to map to a 12-kb region upstream of the β-globin gene cluster, the region now called the LCR. The ability of mutations in the LCR to cause thalassemia is a clear indication that disruption of the LCR results in a loss of globin gene expression.

More detailed study of the β-globin LCR has shown that it contains five separate **DNase I hypersensitive sites**, short regions of DNA that are cleaved by DNase I more easily than other parts of the functional domain. These sites are thought to coincide with positions where nucleosomes have been modified or are absent and which are therefore accessible to binding proteins that attach to the DNA. It is these proteins, not the DNA sequence of the LCR, that control the chromatin structure within the functional domain. Exactly how, and in response to what biochemical signals, is not known.

DNase I hypersensitive sites also occur immediately upstream of each of the genes in the β-globin LCR, at the positions where the transcription initiation complex is assembled on the DNA (Section 9.2.3). These assembly positions illustrate an interesting feature of hypersensitive sites: they are not invariant components of a functional domain. Recall that the different β-type globin genes are expressed at different stages of the human developmental cycle, ε being active in the early embryo, G_γ and A_γ in the fetus, and δ and β in the adult (see *Figure 2.14*, page 45). Only when the gene is active is its assembly position for the transcription initiation complex marked by a hypersensitive site. Initially it was thought that this was an *effect* of the differential expression of these genes, in other words

that in the absence of gene activity it was possible for nucleosomes to cover the assembly site, presumably to be pushed to one side when it became time to express the gene. Now it is thought that the presence or absence of nucleosomes is a *cause* of gene expression, the gene being switched off if nucleosomes cover the assembly site, or switched on if access to the site is open.

8.2 Chromatin Modifications and Genome Expression

The previous sections have introduced us to two ways in which chromatin structure can influence genome expression (*Figure 8.8*). First, the degree of chromatin packaging displayed by a segment of a chromosome determines whether or not genes within that segment are expressed. Second, if a gene is accessible, then its transcription is influenced by the precise nature and positioning of the nucleosomes in the region where the transcription initiation complex will be assembled. Significant advances in understanding both types of chromatin modification have been made in recent years, and we now recognize that specific regions of the genome can be either activated or silenced by processes that involve modification of chromatin structure. We know rather more about the activation processes, so we will begin with these.

8.2.1 Activating the genome

Nucleosomes appear to be the primary determinants of genome activity in eukaryotes, not only by virtue of their positioning on a strand of DNA, but also because the precise chemical structure of the histone proteins contained within nucleosomes is the major factor determining the degree of packaging displayed by a segment of chromatin.

Histone modifications determine chromatin structure

Histone proteins can undergo various types of modification, the best studied of these being **histone acetylation** –

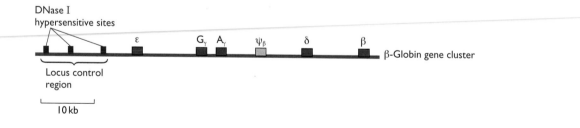

Figure 8.7 DNase I hypersensitive sites indicate the position of the locus control region for the human β-globin gene cluster.

A series of hypersensitive sites are located in the 20 kb of DNA upstream of the start of the β-globin gene cluster. These sites mark the position of the locus control region. Additional hypersensitive sites are seen immediately upstream of each gene, at the position where RNA polymerase attaches to the DNA. These hypersensitive sites are specific to different developmental stages, being seen only during the phase of development when the adjacent gene is active. The 60 kb region shown here represents the entire β-globin functional domain. See *Figure 2.14*, page 45 for more information on the developmental regulation of expression of the β-globin gene cluster.

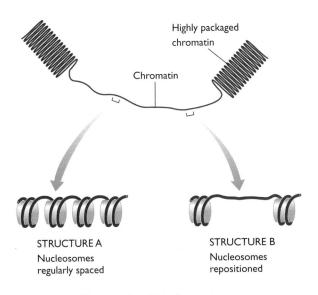

STRUCTURE A
Nucleosomes
regularly spaced

STRUCTURE B
Nucleosomes
repositioned

Figure 8.8 Two ways in which chromatin structure can influence gene expression.

A region of unpackaged chromatin in which the genes are accessible is flanked by two more compact segments. Within the unpackaged region, the positioning of the nucleosomes influences gene expression. On the left, the nucleosomes have regular spacing, as displayed by the typical 'beads-on-a-string' structure. On the right, the nucleosome positioning has changed and a short stretch of DNA, approximately 300 bp, is exposed. See *Figures 2.5* and *2.6*, page 38 for more details on nucleosomes.

the attachment of acetyl groups to lysine amino acids in the N-terminal regions of each of the core molecules. These N termini form tails that protrude from the nucleosome core octamer (*Figure 8.9*) and their acetylation reduces the affinity of the histones for DNA and possibly also reduces the interaction between individual nucleosomes that leads to formation of the 30 nm chromatin fiber. The histones in heterochromatin are generally unacetylated whereas those in functional domains are acetylated, a clear indication that this type of modification is linked to DNA packaging.

The relevance of histone acetylation to genome expression was underlined in 1996 when, after several years of trying, the first examples of **histone acetyltransferases (HATs)** – the enzymes that add acetyl groups to histones – were identified (Pennisi, 1997). It was realized that some proteins that had already been shown to have important influences on genome expression had HAT activity. For example, one of the first HATs to be discovered, the *Tetrahymena* protein called p55, was shown to be a homolog of a yeast protein, GCN5, which was known to activate assembly of the transcription initiation complex (Brownell *et al.*, 1996; Section 9.3.2). Similarly, the mammalian protein called p300/CBP, which had been ascribed a clearly defined role in activation of a variety of genes, was found to be a HAT (Bannister

and Kouzarides, 1996). These observations, plus the demonstration that different types of cell display different patterns of histone acetylation, are clear indications that histone acetylation plays a prominent role in regulating genome expression.

Individual HATs can acetylate histones in the test tube but have negligible activity on intact nucleosomes, indicating that, in the nucleus, HATs almost certainly do not work singly, but instead form multiprotein complexes, such as the ADA and SAGA complexes of yeast and the TFTC complex of humans. Different complexes appear to acetylate different histones and some can also acetylate other proteins involved in genome expression, such as the general transcription factors TFIIE and TFIIF, which we will meet in Section 9.2.3. There are also indications that in addition to local modifications to histone proteins in the regions surrounding expressed genes, HATs can also carry out more general modifications on a global scale throughout the entire genome (Berger, 2000).

Acetylation is not the only type of histone modification. The tails of the core histones also have attachment sites for methyl and phosphate groups and for the common ('ubiquitous') protein called **ubiquitin**. Although information on these modifications is limited, it is clear that they too can influence chromatin structure and have a significant impact on cellular activity. For example, phosphorylation of histone H3 and of the linker histone has been associated with formation of metaphase chromosomes (Bradbury, 1992), and ubiquitination of histone H2B is part of the general role that ubiquitin plays in control of the cell cycle (Robzyk *et al.*, 2000). The effects of methylation of a pair of lysine amino acids at the fourth and ninth positions from the N-terminus of histone H3 are particularly interesting. Methylation of lysine-9 forms a binding site for the HP1 protein which induces chromatin packaging and silences gene expression (Lachner *et al.*, 2001; Bannister *et al.*, 2001), but methylation of lysine-4 has the opposite effect and promotes an open chromatin structure. Within the β-globin functional domain, and probably elsewhere, lysine-4 methylation is closely correlated with acetylation of histone H3 (Litt *et al.*, 2001), and the two types of modification may work hand in hand to activate regions of chromatin. Our growing awareness of the variety of histone modifications that occur, and of the way in which different modifications work together, has led to the suggestion that there is a **histone code**, by which the pattern of chemical modifications specifies which regions of the genome are expressed at a particular time (Strahl and Allis, 2000; Jenuwein and Allis, 2001).

Nucleosome remodeling influences the expression of individual genes

The second type of chromatin modification that can influence genome expression is **nucleosome remodeling**. This term refers to the modification or repositioning of nucleosomes within a short region of the genome, so that DNA-binding proteins can gain access to their attachment sites.

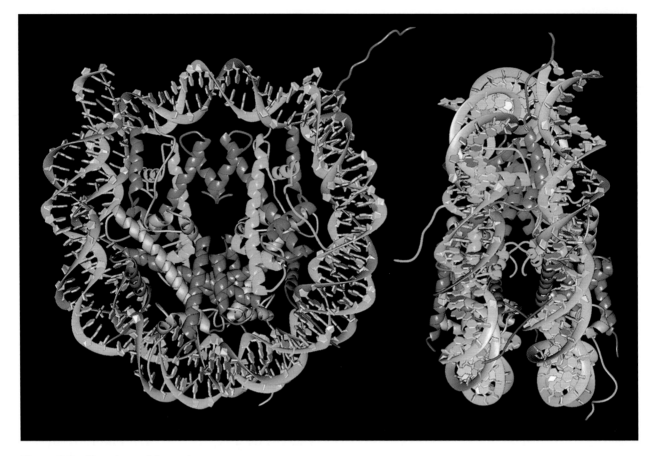

Figure 8.9 Two views of the nucleosome core octamer.

The view on the left is downwards from the top of the barrel-shaped octamer; the view on the right is from the side. The two strands of the DNA double helix wrapped around the octamer are shown in brown and green. The octamer comprises a central tetramer of two histone H3 (blue) and two histone H4 (bright green) subunits plus a pair of H2A (yellow)–H2B (red) dimers, one above and one below the central tetramer. Note the N-terminal tails of the histone proteins protruding from the core octamer. Reprinted with permission from Luger *et al.*, *Nature*, **389**, 251–260. Copyright 1997 Macmillan Magazines Limited.

Unlike acetylation and the other chemical modifications described in the previous section, nucleosome remodeling does not involve covalent alterations to histone molecules. Instead, remodeling is induced by an energy-dependent process that weakens the contact between the nucleosome and the DNA with which it is associated. Three distinct types of change can occur (*Figure 8.10*):

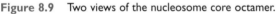

- *Remodeling*, in the strict sense, involves a change in the structure of the nucleosome, but no change in its position. The nature of the structural change is not known, but when induced *in vitro* the outcome is a doubling in size of the nucleosome and an increased DNase sensitivity of the attached DNA.
- *Sliding*, or *cis*-displacement, physically moves the nucleosome along the DNA.
- *Transfer*, or *trans*-displacement, results in the nucleosome being transferred to a second DNA molecule, or to a non-adjacent part of the same molecule.

As with HATs, the proteins responsible for nucleosome remodeling work together in large complexes. One of these is Swi/Snf, made up of at least 11 proteins, which is present in many eukaryotes (Sudarsanam and Winston, 2000). Little is currently known about the way in which Swi/Snf, or any other nucleosome remodeling complex, carries out is role in increasing access to the genome. None

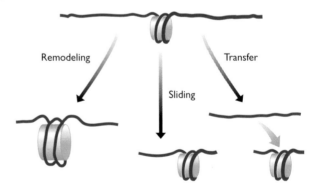

Figure 8.10 Nucleosome remodeling, sliding and transfer.

of the components of Swi/Snf appears to have a DNA-binding capability, so the complex must be recruited to its target site by additional proteins. Interactions have been detected between Swi/Snf and HATs (Syntichaki *et al.*, 2000), suggesting that nucleosome remodeling might occur in conjunction with histone acetylation. This is an attractive hypothesis because it links the two activities that are currently looked on as central to genome activation. But there are problems because Swi/Snf does not appear to have a global effect across an entire genome, but instead influences expression at only a limited number of positions; in the case of yeast, perhaps at no more than 6% of all the genes in the genome. This observation suggests that the more important interactions might be between Swi/Snf and proteins that target a limited set of genes, rather than the HATs which work throughout the genome. The most likely candidates are transcription activators (Section 9.3.2), each of which is specific for a limited set of genes and some of which form associations with Swi/Snf *in vitro*.

8.2.2 Silencing the genome

If mechanisms exist for activating parts of the genome then logic dictates that there must also be complementary processes that silence regions whose expression products are not needed. Silencing of specific genes can be achieved simply by reversing the activating effects of histone acetylation, but acetylation–deacetylation is not the only process that influences genome expression. Direct methylation of DNA also has a silencing effect, and special processes exist for inactivating individual chromosomes. We will look at each of these silencing mechanisms in turn.

Histone deacetylation is one way of repressing gene expression

One way in which silencing can be implemented is by removing acetyl groups from histone tails, and hence reversing the transcription-activating effects of the HATs described above. This is the role of the **histone deacetylases (HDACs)**. The link between HDAC activity and gene silencing was established in 1996, when mammalian HDAC1, the first of these enzymes to be discovered, was shown to be related to the yeast protein called Rpd3, which was known to be a repressor of transcription (Taunton *et al.*, 1996). The link between histone deacetylation and repression was therefore established in the same way as the link between acetylation and activation – by showing that two proteins that were initially thought to have different activities are in fact related. These are good examples of the value of homology analysis in studies of gene and protein function (Section 7.2.1).

In common with HATs and nucleosome remodeling enzymes, HDACs are contained in multiprotein complexes. One of these is the mammalian Sin3 complex, which comprises at least seven proteins, including

---

Box 8.2: Chromatin modification by the HMGN proteins

An additional dimension to chromatin modification is provided by the high mobility group N (HMGN) proteins. These are small proteins, each approximately 10 kDa, present in the nuclei of most vertebrates (Bustin, 2001). They are able to attach to the nucleosome core octamer, probably by interacting with the histone tails, and they also make contacts with the linker histone (Section 2.2.1) on the outer surface of the nucleosome. Attachment of an HMGN protein changes the conformation of the nucleosome, resulting in a decrease in the compactness of the 30 nm chromatin fiber. Attempts to correlate HMGN binding with transcriptional activity have not been entirely successful, most of the supporting evidence being circumstantial. When nuclei are fractionated, HMGN proteins copurify with newly synthesized mRNAs, and injection of antibodies that bind to HMGN proteins results in a decrease in RNA synthesis. But *in vitro* experiments have failed to show a clear connection between HMGN binding and transcription, and as yet there is no evidence linking HMGN proteins with HATs or nucleosome remodeling complexes.

One puzzle concerns the feature of HMG proteins that gives them their name. These proteins are highly mobile and move around the nucleus more rapidly than most other proteins. Their mobility is compatible with a scenario in which HMGN proteins transit rapidly from place to place, modifying chromatin as and when necessary in different parts of the genome. However, their dynamism also suggests that HMGN proteins form only transient associations with nucleosomes, rather than the more stable partnerships anticipated if the proteins are essential components of actively expressed domains.

HDAC1 and HDAC2 along with others that do not have deacetylase activity but which provide ancillary functions essential to the process (Ng and Bird, 2000). Examples of ancillary proteins are RbAp46 and RbAp48, which are members of the Sin3 complex and are thought to contribute the histone-binding capability. RbAp46 and RbAp48 were first recognized through their association with the retinoblastoma protein, which controls cell proliferation by inhibiting expression of various genes until their activities are required and which, when mutated, leads to cancer (Brehm *et al.*, 1998; Magnaghi-Jaulin *et al.*, 1998). This link between Sin3 and a protein implicated in cancer provides a powerful argument for the importance of histone deacetylation in gene silencing. Other deacetylation complexes include NuRD in mammals, which combines HDAC1 and HDAC2 with a different set of ancillary proteins, and yeast Sir2 (Ahringer, 2000), which is different from other HDACs in that it has an energy requirement (Imai *et al.*, 2000). The distinctive features of

RESEARCH 8.1 BRIEFING

Discovery of the mammalian DNA methyltransferases

Innovative experiments have identified the enzymes responsible for *de novo* methylation of mammalian DNA.

Two DNA methylation activities are required in mammalian nuclei (see *Figure 8.11*). The first of these is maintenance methylation, which is responsible for adding methyl groups to the appropriate positions on newly synthesized DNA polynucleotides, ensuring that the methylation pattern of the parent DNA is maintained by the daughter double helices after genome replication. The second type of activity is *de novo* methylation, which changes the methylation pattern of a DNA molecule by adding new methyl groups in a tissue-specific manner, resulting in programmed changes in genome expression patterns. For several years, only one DNA methyltransferase was known in mammalian cells. It was clear that this enzyme, now called DNA methyltransferase 1 (Dnmt1), is responsible for maintenance methylation, but there was controversy regarding its ability to carry out *de novo* methylation as well. Dnmt1 displays *de novo* methylation activity *in vitro*, and a few researchers were tempted to assume that, because no other methyltransferase could be identified, Dnmt1 must also be a *de novo* methylase *in vivo*. This assumption was shown to be incorrect by gene inactivation experiments. The Dnmt1 gene was inactivated in an embryonic stem (ES) cell which was then injected into an embryo so that knockout mice were eventually obtained (Section 7.2.2). These mice were unable to synthesize Dnmt1, but were still capable of carrying out *de novo* methylation, as shown by their ability to add methyl groups to the DNA genome of an infecting retrovirus. This was the clearest possible evidence that Dnmt1 is not a *de novo* methylase *in vivo*, and that there must be one or more additional undiscovered DNA methyltransferases in mammalian cells. But how to find them?

Searches of the sequence databases reveal the undiscovered methyltransferases

One possible way of finding the missing methyltransferases would be to make cell extracts and to try and identify proteins with the relevant biochemical activity. But this type of work is difficult and when used in previous projects had failed to identify any enzyme other than Dnmt1. Instead, a search was made of the DNA databases for genes that could code for DNA methyltransferases, using the sequence of Dnmt1 as a guide to the particular sequence features anticipated in a methyltransferase gene. This search was carried out before the draft human genome sequence had been completed, and therefore made use of the large libraries of expressed sequence tags (ESTs) that had been accumulated from analysis of short cDNAs prepared from mRNA from different tissues (Section 5.3.3). These searches

revealed a number of potential methyltransferases including three – Dnmt2, Dnmt3a and Dnmt3b – that looked particularly promising, their sequences including all the various features that were expected in an enzyme able to add methyl groups to DNA.

Experimental test with new Dnmt genes

To test whether these newly discovered genes did indeed code for active methyltransferases, a series of knockout mice were prepared, each with a different inactivated Dnmt gene. When this experiment was performed with Dnmt2 a disappointing result was obtained: inactivation of the Dnmt2 gene had no observable effect on the phenotype of the mice and affected neither their maintenance nor de novo methylation abilities. It is still thought that Dnmt2 has a DNA methylation role of some description, but exactly what that role is remains unclear.

Inactivation of Dnmt3a and Dnmt3b was much more informative. These knockout mice were unable to complete their full developmental pathways, those with an inactivated Dnmt3b gene dying only a few days after birth, and those lacking Dnmt3a surviving only 1–2 weeks longer. These lethal effects are a clear indication that the proteins coded by the inactivated genes have a critical cellular function. That this function was *de novo* methylation was confirmed by analysis of methylation patterns in the embryos during the period before their death. The graph below compares the extent of DNA methylation in embryos aged between 4.0 and 9.5 days for wild-type mice (with active Dnmt1, Dnmt3a and Dnmt3b) and Dnmt3 knockouts (lacking both Dnmt3a

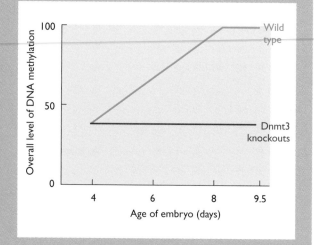

and Dnmt 3b). The overall level of DNA methylation almost doubles in the wild-type mice during this period but stays constant in the Dnmt3 knockouts. The results indicate that maintenance methylation, resulting from Dnmt1, is occurring in the Dnmt3 knockouts, but that de novo methylation is absent, and hence there is no increase in the overall level of DNA methylation in these mice over time.

Further evidence for the roles of Dnmt3a and Dnmt3b

Confirmation that Dnmt3a and Dnmt3b are de novo methylases has been provided by two additional types of study. One of these involved cloning the *Dnmt3a* and *Dnmt3b* genes into fruit flies. The relevance of the fruit fly to this type of work is that cytosine nucleotides, the targets for methylation in mammalian DNA, are not methylated in *Drosophila*. If the DNA of a transgenic fruit fly that contains a gene for a mammalian methyltransferase becomes methylated, then that would be good evidence that the enzyme has de novo methylation activity, as only this activity could add methyl groups to the unmethylated DNA encountered in fruit-fly nuclei. This experiment was tried first with Dnmt1. The *Dnmt1* gene was introduced into fruit flies using a cloning vector based on the *Drosophila* transposon called the P element (Section 7.2.2). Extracts of the insect cells were tested to ensure that the gene was active and that the Dnmt1 enzyme was being synthesized. The enzyme was indeed present, but when the DNA was examined no methylation could be detected. As had become clear from the work with knockout mice, Dnmt1 had no de novo methylation activity. In contrast, transgenic flies containing the *Dnmt3a* gene contained methylated DNA. The de novo methylase activity of Dnmt3a was therefore confirmed.

Why are the phenotypes of Dnmt3a and Dnmt3b knockout mice dissimilar? Both sets of mice die at an early stage of development, but death occurs significantly earlier when Dnmt3a is inactivated. Detailed studies of the patterns of DNA methylation in the two types of knockout mice revealed that the Dnmt3b methylation activity is directed at the centromere regions of chromosomes, whereas Dnmt3a methylates only non-centromeric DNA. As Dnmt3b is responsible for the more lethal knockout, it appears that the deficiency in centromere methylation caused by an absence of this enzyme has a highly deleterious effect on the survival of the embryo.

DNA methylation and human disease

An interesting adjunct to the knockout studies has been provided by studies of a human disease that is associated with abnormal behavior of centromeres. This disease is ICF (immunodeficiency, centromere instability and facial anomalies) which, along with other abnormalities, results in hybrid chromosomes with multiple arms radiating from a single centromere. Similar effects are caused by the drug 5-azacytidine, which is known to demethylate DNA. If the

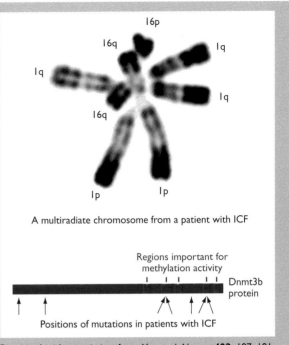

A multiradiate chromosome from a patient with ICF

Positions of mutations in patients with ICF

Reprinted with permission from Xu et al. Nature, **402**, 187–191. Copyright 1999 Macmillan Magazines Limited.

effects caused by the drug are the result of demethylation then might this also be the underlying cause of the disease? Studies of DNA prepared from patients with ICF have shown that their DNA is indeed undermethylated, and examination of their Dnmt3b genes reveals that in each case this gene is mutated, often in a region coding for a part of the enzyme that is thought to be crucial for the DNA methylation activity.

Conclusions

Various lines of evidence have now come together. We have the outcomes of the knockout mice experiments, which show that Dnmt3a and Dnmt3b are needed for de novo methylation in living animals. The experiments with transgenic fruit flies confirm this activity, at least for Dnmt3a, and the studies of ICF indicate that mutation of the Dnmt3b gene in humans is responsible for a disease that is associated with undermethylation of DNA. In combination, these results have solved the long-standing mystery surrounding the de novo methyltransferases of mammalian nuclei.

References

Bird A (1999) DNA methylation de novo. Science, **286**, 2287–2888.
Okano M, Bell DW, Haber DA and Li E (1999) DNA methyltransferases Dnmt3a and Dnmt3b are essential for de novo methylation and mammalian development. Cell, **99**, 247–257.
Xu G-L, Bestor TH, Bourc'his D, et al. (1999) Chromosome instability and immunodeficiency syndrome caused by mutations in a DNA methyltransferase gene. Nature, **402**, 187–191.

Sir2 show that HDACs are more diverse than originally realized, possibly indicating that novel roles for histone deacetylation are waiting to be discovered.

Studies of HDAC complexes are beginning to reveal links between the different mechanisms for genome activation and silencing. Both Sin3 and NuRD contain proteins that bind to methylated DNA (as described in the next section), and NuRD contains proteins that are very similar to components of the nucleosome remodeling complex Swi/Snf. NuRD does in fact act as a classical nucleosome remodeling machine *in vitro*. Further research will almost certainly unveil additional links between what we currently look on as different types of chromatin modification system, but which in reality may simply be different facets of a single grand design.

Genome silencing by DNA methylation

Chromatin modification is not the only process that can bring about genome silencing. DNA methylation can also repress gene activity. Initially this was thought to be quite distinct from histone modification, but now we are beginning to see links between the two activities.

In eukaryotes, cytosine bases in chromosomal DNA molecules are sometimes changed to 5-methylcytosine by the addition of methyl groups by enzymes called **DNA methyltransferases** (see *Figure 5*.25, page 149). Cytosine methylation is relatively rare in lower eukaryotes but in vertebrates up to 10% of the total number of cytosines in a genome are methylated, and in plants the figure can be as high as 30%. The methylation pattern is not random, instead being limited to the cytosine in some copies of the sequences 5′–CG–3′ and, in plants, 5′–CNG–3′. Two types of methylation activity have been distinguished (*Figure 8.11*). The first is **maintenance methylation** which, following genome replication, is responsible for adding methyl groups to the newly synthesized strand of DNA at positions opposite methylated sites on the parent strand (Section 14.2.3). The maintenance activity therefore ensures that the two daughter DNA molecules retain the methylation pattern of the parent molecule. The second activity is *de novo* **methylation**, which adds methyl groups at totally new positions and so changes the pattern of methylation in a localized region of the genome. From the results of test-tube experiments, it was originally thought that the first DNA methyltransferase to be discovered, Dnmt1, was responsible for both types of methylation in mammalian cells. It was subsequently discovered that knockout mice (Section 7.2.2) that have an inactivated gene for Dnmt1 can still carry out *de novo* methylation. This led to the search for new enzymes and the eventual discovery of Dnmt3a and Dnmt3b (see Research Briefing 8.1), which are now considered to be the main *de novo* methylases of mammals, with Dnmt1 primarily responsible for the maintenance activity (Bird, 1999).

Methylation results in repression of gene activity (Jones, 1999). This has been shown by experiments in which methylated or unmethylated genes have been introduced into cells by cloning and their expression

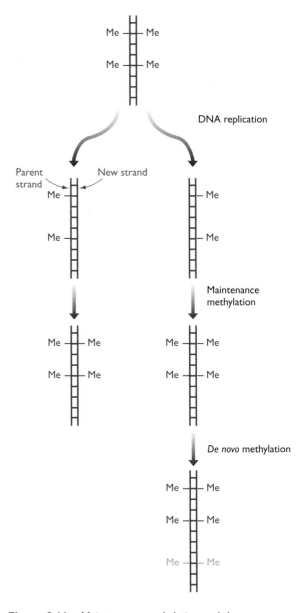

Figure 8.11   Maintenance methylation and *de novo* methylation.

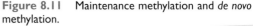

levels measured: expression does not occur if the DNA sequence is methylated. The link with gene expression is also apparent when the methylation patterns in chromosomal DNAs are examined, these showing that active genes are located in unmethylated regions. For example, in humans, 40–50% of all genes are located close to CpG islands (Section 7.1.1), with the methylation status of the CpG island reflecting the expression pattern of the adjacent gene. Housekeeping genes – those that are expressed in all tissues – have unmethylated CpG islands, whereas tissue-specific genes are unmethylated only in those tissues in which the adjacent gene is expressed. Note that because the methylation pattern is maintained after cell division, information specifying which genes should be expressed is inherited by the daughter cells, ensuring that in a differentiated tissue the appropriate pattern of gene expression is retained even though the cells in the tissue are being replaced and/or added to by new cells.

The importance of DNA methylation is underlined by studies of human diseases. The syndrome called ICF (immunodeficiency, centromere instability and facial anomalies) which, as the name suggests, has wide-ranging phenotypic effects, is associated with under-methylation of various genomic regions, and is caused by a mutation in the gene for Dnmt3b (Xu *et al.*, 1999). The opposing situation – hypermethylation – is seen within the CpG islands of genes that exhibit altered expression patterns in certain types of cancer (Baylin and Herman, 2000), although in these cases the abnormal methylation could equally well be a *result* rather than the *cause* of the disease state.

How methylation influences genome expression was a puzzle for many years. Now it is known that **methyl-CpG-binding proteins** (**MeCPs**) are components of both the Sin3 and NuRD histone deacetylase complexes. This discovery has led to a model in which methylated CpG islands are the target sites for attachment of HDAC complexes that modify the surrounding chromatin in order to silence the adjacent genes (*Figure 8.12*).

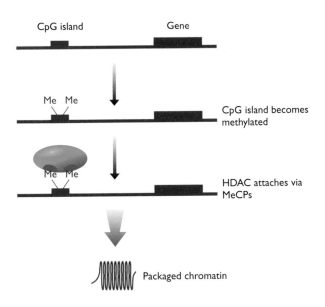

Figure 8.12 A model for the link between DNA methylation and genome expression.

Methylation of the CpG island upstream of a gene provides recognition signals for the methyl-CpG-binding protein (MeCP) components of a histone deacetylase complex (HDAC). The HDAC modifies the chromatin in the region of the CpG island and hence inactivates the gene. Note that the relative positions and sizes of the CpG island and the gene are not drawn to scale.

Methylation is involved in imprinting and X inactivation

Further evidence, if it is needed, of the link between DNA methylation and genome silencing is provided by two intriguing phenomena called **genomic imprinting** and **X inactivation**.

Genomic imprinting is a relatively uncommon but important feature of mammalian genomes in which only one of a pair of genes, present on homologous chromosomes in a diploid nucleus, is expressed, the second being silenced by methylation (Feil and Khosia, 1999). It is always the same member of a pair of genes that is imprinted and hence inactive; for some genes this is the version inherited from the mother, and for other genes it is the paternal version. Thirty genes in humans and mice have been shown to display imprinting. An example is *Igf2*, which codes for a growth factor, a protein involved in signaling between cells (Section 12.1). In mice, only the paternal gene is active (*Figure 8.13*). On the chromosome inherited from the mother various segments of DNA in the region of *Igf2* are methylated, preventing expression of this copy of the gene. Interestingly, a second imprinted gene, *H19*, is located some 90 kb away from *Igf2*, but the imprinting is the other way round: the maternal version of *H19* is active and the paternal version is silent.

The function of imprinting is not known. One possibility is that it has a role in development, because artificially created parthenogenetic mice, which have two copies of the maternal genome, fail to develop properly. More subtle explanations based on the evolutionary conflicts between the males and females of a species have also been proposed (Jaenisch, 1997).

X inactivation is much less enigmatic. This is a special form of imprinting that leads to total inactivation of one of the X chromosomes in a female mammalian cell (Heard *et al.*, 1997). It occurs because females have two X chromosomes whereas males have only one. If both of the female X chromosomes were active then proteins coded by genes on the X chromosome might be synthesized at twice the rate in females compared with males. To avoid this undesirable state of affairs, one of the female X chromosomes is silenced and is seen in the nucleus as a condensed structure called the **Barr body**, which is comprised entirely of heterochromatin. Silencing occurs early in embryo development and is controlled by the X inactivation center, a discrete region present on each X chromosome. In a cell undergoing X inactivation, the inactivation center on one of the X chromosomes initiates the formation of heterochromatin, which spreads out from the nucleation point until the entire chromosome is affected, with the exception of a few short segments containing small clusters of genes that remain active. The process takes several days to complete. The exact mechanism is not understood but it involves, although is not entirely dependent upon, each of the following:

■ A gene called *Xist*, located in the inactivation center, which is transcribed into a 25-kb non-coding RNA, copies of which coat the chromosome as heterochromatin is formed;

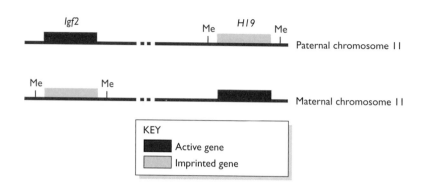

Figure 8.13 A pair of imprinted genes on human chromosome 11.

Igf2 is imprinted on the chromosome inherited from the mother, and *H19* is imprinted on the paternal chromosome. The drawing is not to scale: the two genes are approximately 90 kb apart.

- Replacement of histone H2A, one of the members of the core octamer of the nucleosome (Section 2.2.1), with a special histone, macroH2A1 (Costanzi and Pehrson, 1998);
- Deacetylation of histone H4 (Jeppesen and Turner, 1993), as usually occurs in heterochromatin;

- Hypermethylation of certain DNA sequences, although this appears to occur after the inactive state has been set up.

X inactivation is heritable and is displayed by all cells descended from the initial one within which the inactivation took place.

References

Ahringer J (2000) NuRD and SIN3: histone deacetylase complexes in development. *Trends Genet.*, **16**, 351–356.

Bannister AJ and Kouzarides T (1996) The CBP co-activator is a histone acetyltransferase. *Nature*, **384**, 641–643.

Bannister AJ, Zegerman P, Partridge JF, et al. (2001) Selective recognition of methylated lysine 9 on histone H3 by the HP1 chromo domain. *Nature*, **410**, 120–124.

Baylin SB and Herman JG (2000) DNA hypermethylation in tumorigenesis: epigenetics joins genetics. *Trends Genet.*, **16**, 168–174.

Bell AC, West AG and Felsenfeld G (2001) Insulators and boundaries: versatile regulatory elements in the eukaryotic genome. *Science*, **291**, 447–450.

Berger SL (2000) Local or global? *Nature*, **408**, 412–415.

Bird A (1999) DNA methylation de novo. *Science*, **286**, 2287–2288.

Bradbury EM (1992) Reversible histone modifications and the chromosome cell cycle. *BioEssays*, **14**, 9–16.

Brehm A, Miiska EA, McCance DJ, Reid JL, Bannister AJ and Kouzarides T (1998) Retinoblastoma protein recruits histone deacetylase to repress transcription. *Nature*, **391**, 597–601.

Brownell JE, Zhou J, Ranalli T, et al. (1996) Tetrahymena histone acetyltransferase A: a homolog to yeast Gcn5p linking histone acetylation to gene activation. *Cell*, **84**, 843–851.

Bustin M (2001) Chromatin unfolding and activation by HMGN chromosomal proteins. *Trends Biochem. Sci.*, **26**, 431–437.

Costanzi C and Pehrson JR (1998) Histone macroH2A1 is concentrated in the inactive X chromosome of female mammals. *Nature*, **393**, 599–601.

Feil R and Khosia S (1999) Genomic imprinting in mammals: an interplay between chromatin and DNA methylation? *Trends Genet.*, **15**, 431–434.

Gerasimova TI, Byrd K and Corces VG (2000) A chromatin insulator determines the nuclear localization of DNA. *Mol. Cell*, **6**, 1025–1035.

Heard E, Clerc P and Avner P (1997) X-chromosome inactivation in mammals. *Ann. Rev. Genet.*, **31**, 571–610.

Imai S, Armstrong CM, Kaeberlein M and Guarente L (2000) Transcriptional silencing and longevity protein Sir2 is an NAD-dependent histone deacetylase. *Nature*, **403**, 795–800.

Jaenisch R (1997) DNA methylation and imprinting: why bother? *Trends Genet.*, **13**, 323–329.

Jenuwein T and Allis CD (2001) Translating the histone code. *Science*, **293**, 1074–1080.

Jeppesen P and Turner BM (1993) The inactive X chromosome in female mammals is distinguished by a lack of histone H4 acetylation, a cytogenetic marker for gene expression. *Cell*, **74**, 281–289.

Jones PA (1999) The DNA methylation paradox. *Trends Genet.*, **15**, 34–37.

Kellum R and Schedl P (1991) A position-effect assay for boundaries of higher-order chromosomal domains. *Cell*, **64**, 941–950.

Lachner M, O'Carroll D, Rea S, Mechtler K and Jenuwein T (2001) Methylation of histone H3 lysine 9 creates a binding site for HP1 proteins. *Nature*, **410**, 116–120.

Lever MA, Th'ng JPH, Sun X and Hendzel MJ (2000) Rapid exchange of histone H1.1 on chromatin in living cells. *Nature*, **408**, 873–876.

Lewis JD and Tollervey D (2000) Like attracts like: getting RNA processing together in the nucleus. *Science*, **288**, 1385–1389.

Li Q, Harju S and Peterson KR (1999) Locus control regions: coming of age at a decade plus. *Trends Genet.*, **15**, 403–408.

Litt MD, Simpson M, Gaszner M, Allis CD and Felsenfeld G (2001) Correlation between histone lysine methylation and developmental changes at the chicken β-globin locus. *Science*, **293**, 2453–2455.

Lodish H, Berk A, Zipursky AL, Matsudaira P, Baltimore D and Darnell J (2000) *Molecular Cell Biology*, 4th edition. W. H. Freeman, New York.

Magnaghi-Jaulin L, Groisman R, Naguibneva I, et al. (1998) Retinoblastoma protein represses transcription by recruiting a histone deacetylase. *Nature*, **391**, 601–605.

Misteli T (2001) Protein dynamics: implications for nuclear architecture and gene expression. *Science*, **291**, 843–847.

Misteli T, Gunjan A, Hock R, Bustin M and Brown DT (2000) Dynamic binding of histone H1 to chromatin in living cells. *Nature*, **408**, 877–881.

Ng HH and Bird A (2000) Histone deacetylases: silencers for hire. *Trends Biochem. Sci.*, **25**, 121–126.

Pennisi E (1997) Opening the way to gene activity. *Science*, **275**, 155–157.

Robzyk K, Recht J and Osley MA (2000) Rad6-dependent ubiquitination of histone H2b in yeast. *Science*, **287**, 501–504.

Strahl BD and Allis D (2000) The language of covalent histone modifications. *Nature*, **403**, 41–45.

Sudarsanam P and Winston F (2000) The Swi/Snf family: nucleosome-remodelling complexes and transcriptional control. *Trends Genet.*, **16**, 345–351.

Syntichaki P, Topalidou I and Thireos G (2000) The Gcn5 bromodomain co-ordinates nucleosome remodelling. *Nature*, **404**, 414–417.

Taunton J, Hassig CA and Schreiber SL (1996) A mammalian histone deacetylase related to the yeast transcriptional regulator Rpd3p. *Science*, **272**, 408–411.

Tsien RY (1998) The green fluorescent protein. *Ann. Rev. Biochem.*, **67**, 509–544.

Wolffe AP (1995) Genetic effects of DNA packaging. *Sci. Am.*, **Nov/Dec**, 68–77.

Xu G-L, Bestor TH, Bourc'his D, et al. (1999) Chromosome instability and immunodeficiency syndrome caused by mutations in a DNA methyltransferase gene. *Nature*, **402**, 187–191.

Further Reading

Aalfs JD and Kingston RE (2000) What does 'chromatin remodelling' mean? *Trends Biochem. Sci.*, **25**, 548–555. — *Stimulating discussion of histone modification and nucleosome remodeling.*

Ballabio A and Willard HF (1992) Mammalian X-chromosome inactivation and the XIST gene. *Curr. Opin. Genet. Devel.*, **2**, 439–448.

Brown CE, Lechner T, Howe L and Workman JL (2000) The many HATs of transcription coactivators. *Trends Biochem. Sci.*, **25**, 15–19. — *Details of histone acetyltransferases and their mode of action.*

Jones PA and Takai D (2001) The role of DNA methylation in mammalian epigenetics. *Science*, **293**, 1068–1070. — *A very readable account of our current knowledge of DNA methylation.*

Lee TI and Young RA (2000) Transcription of eukaryotic protein-coding genes. *Ann. Rev. Genet.*, **34**, 77–137. — *Includes extensive details of histone modification and nucleosome remodeling.*

Roth SY, Denu JM and Allis CD (2001) Histone acetyltransferases. *Ann. Rev. Biochem.*, **70**, 81–120.

Wu J and Grunstein M (2000) 25 years after the nucleosome model: chromatin modifications. *Trends Biochem. Sci.*, **25**, 619–623. — *A useful review of this topic.*

STUDY AIDS FOR CHAPTER 8

Key terms

Give short definitions of the following terms:

Barr body
Chromosome scaffold
Cis-displacement
Constitutive heterochromatin
De novo methylation
DNA methyltransferase
DNase I hypersensitive site
Euchromatin
Facultative heterochromatin
Fluorescence recovery after photobleaching (FRAP)
Functional domain
Genomic imprinting
Green fluorescent protein
Heterochromatin
High mobility group N (HMGN) protein
Histone acetylation
Histone acetyltransferase (HAT)

Histone code
Histone deacetylase (HDAC)
Insulator
Locus control region (LCR)
Maintenance methylation
Matrix-associated region (MAR)
Methyl-CpG-binding protein (MeCP)
Nuclear matrix
Nucleolus
Nucleosome remodeling
Photobleaching
Positional effect
Scaffold attachment region (SAR)
Structural domain
Trans-displacement
Ubiquitin
X inactivation

Self study questions

1. How have developments in electron microscopy led to advances in our understanding of the internal structure of the eukaryotic nucleus?
2. Describe the differences between constitutive heterochromatin, facultative heterochromatin and euchromatin.
3. Distinguish between the terms 'structural domain' and 'functional domain'. What experimental techniques are used to delineate a functional domain?
4. What are the special properties of the insulator sequences that mark the boundaries of functional domains?
5. Describe the important features of the β-globin locus control region.
6. Give examples of the various histone modifications that activate or silence regions of the genome. Explain what is meant by 'histone code'.
7. Describe the features and activities of histone acetyltransferases and histone deacetylases.
8. Outline our current knowledge of the interactions between nucleosomes and HMGN proteins.
9. Using the Swi/Snf complex as an example, describe how nucleosome remodeling can influence gene expression
10. Write an essay on DNA methylation. Your essay should distinguish between the roles of maintenance and *de novo* methylation and include an account of the discovery of the mammalian methyltransferases.

11. What is meant by 'genomic imprinting'?
12. Outline the key features of X inactivation.

Problem-based learning

1. To what extent can it be assumed that the picture of nuclear architecture built up by modern electron microscopy is an accurate depiction of the actual structure of the nucleus, as opposed to an artefact of the methods used to prepare cells for examination?

2. In many areas of biology it is difficult to distinguish between *cause* and *effect*. Evaluate this issue with regard to chromatin modifications and genome expression — do chromatin modifications *cause* changes in genome expression or are they the *effect* of these expression changes?

3. Explore and assess the histone code hypothesis.

4. Maintenance methylation ensures that the pattern of DNA methylation displayed by two daughter DNA molecules is the same as the pattern on the parent molecule. In other words, the methylation pattern, and the information on gene expression that it conveys, is inherited. Other aspects of chromatin structure might also be inherited in a similar way. How do these phenomena affect the Mendelian view that inheritance is specified by genes? (*The 10 August 2001 issue of* Science *is a good starting point for your research into this topic.*)

Assembly of the Transcription Initiation Complex

Chapter Contents

9.1 *The Importance of DNA-binding Proteins*   *241*

    9.1.1   Locating the positions of DNA-binding sites in a genome   **242**

    9.1.2   Purifying a DNA-binding protein   **245**

    9.1.3   Studying the structures of proteins and DNA–protein complexes   **246**

    9.1.4   The special features of DNA-binding proteins   **248**

    9.1.5   The interaction between DNA and its binding proteins   **252**

9.2 *DNA–Protein Interactions During Transcription Initiation*   *254*

    9.2.1   RNA polymerases   **254**

    9.2.2   Recognition sequences for transcription initiation   **255**

    9.2.3   Assembly of the transcription initiation complex   **257**

9.3 *Regulation of Transcription Initiation*   *261*

    9.3.1   Strategies for controlling transcription initiation in bacteria   **261**

    9.3.2   Control of transcription initiation in eukaryotes   **265**

Learning outcomes

When you have read Chapter 9, you should be able to:

- Outline the various techniques that are used to locate the position at which a DNA-binding protein attaches to a DNA molecule

- Explain how a DNA-binding protein is purified and how its structure can be determined

- Describe the key structural motifs that enable proteins to make sequence-specific attachments to DNA molecules

- Discuss the features of the double helix that are important in interactions between DNA and its binding proteins, and give details of the chemical events that underlie the interaction

- Identify the key features of the various prokaryotic and eukaryotic RNA polymerases and describe the structures of the promoter sequences that they recognize

- Give a detailed description of how the *Escherichia coli* transcription initiation complex is assembled, and discuss the various ways in which this process can be regulated

- Give a detailed description of the assembly of the RNA polymerase II transcription initiation complex, and explain how assembly of this complex is influenced by proteins that activate or silence gene expression

- Outline the processes of transcription initiation by eukaryotic RNA polymerases I and III

ONCE UPON A TIME the first stage in genome expression was described as 'transcription' or 'DNA makes RNA' (see *Figure 3.2A*, page 71), but we now realize that the process that leads from the genome to the transcriptome is much more complex than simply the synthesis of RNA. This part of genome expression is now divided into two key stages (*Figure 9.1*):

1. *Initiation of transcription*, which results in the complex of proteins, including the RNA polymerase enzyme and its various accessory proteins, that will subsequently copy the gene into an RNA transcript being assembled upstream of the gene. Inherent in this step are the events that determine whether or not the gene is actually expressed.
2. *Synthesis and processing of RNA*, which begins when the RNA polymerase leaves the initiation region and starts to make an RNA copy of the gene, and ends after completion of the processing and modification events that convert the initial transcript into a functional RNA.

This chapter deals with the initiation of transcription, and Chapter 10 covers RNA synthesis and processing.

But before we move on to these topics we must do a little groundwork. The central players in many areas of molecular biology, including transcription, are **DNA-binding proteins** that attach to the genome in order to perform their biochemical functions. Histones are examples of DNA-binding proteins, and we will encounter many others later in this chapter when we look at assembly of the initiation complexes of prokaryotes and eukaryotes. There are also DNA-binding proteins that are involved in DNA replication, repair, and recombination, as well as a large group of related proteins that bind to RNA rather than DNA (*Table 9.1*). Many DNA-binding proteins recognize specific nucleotide sequences and bind predominantly to these target sites, whereas others bind non-specifically at various positions in the genome.

The mode of action of DNA-binding proteins is central to the initiation of transcription, and without a knowledge of how they function we can never hope to understand how the information in the genome is utilized. We will therefore spend some time examining what is known about DNA-binding proteins and how they interact with the genome.

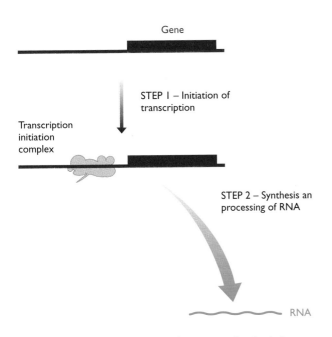

Figure 9.1 The two stages in the process that leads from genome to transcriptome.

9.1 The Importance of DNA-binding Proteins

As in all areas of molecular biology and genetics, the amount we know about a topic depends on the range and effectiveness of the methods available for its study. With regard to DNA-binding proteins we are fortunate in having a number of powerful techniques that can provide information on the interaction between a protein and the DNA sequence or sequences that it binds to. These techniques can be divided into three categories:

- Methods for identifying the region(s) of a DNA molecule to which a protein binds;
- Methods for purifying a DNA-binding protein;
- Methods for studying the tertiary structure of a DNA-binding protein, including the complex formed when the protein is bound to DNA.

Table 9.1 Functions of DNA- and RNA-binding proteins

Function	Examples	Cross-reference
DNA-binding proteins		
Genome expression		
Transcription initiation	Eukaryotic TATA-binding protein	Section 9.2.3
	σ subunit of bacterial RNA polymerase	Section 9.2.3
RNA synthesis	RNA polymerases	Section 9.2.1
Regulation of transcription	Eukaryotic activators and repressors	Section 9.3.2
	Bacterial repressors	Section 9.3.1
DNA packaging	Eukaryotic histones	Section 2.2.1
	Bacterial nucleoid proteins	Section 2.3.1
DNA recombination	RecA	Section 14.3.1
DNA repair	DNA glycosylases, nucleases	Section 14.2.2
DNA replication	Origin recognition proteins	Section 13.2.1
	DNA polymerases and ligases	Sections 4.1.1, 4.1.3 and 13.2.2
	Single-strand binding proteins	Section 13.2.2
	DNA topoisomerases	Section 13.1.2
Others	Prokaryotic restriction endonucleases	Section 4.1.2
RNA-binding proteins		
Genome expression		
Intron splicing	snRNP proteins	Section 10.1.3
mRNA polyadenylation	CPSF, CstF	Section 10.1.2
mRNA editing	Adenosine deaminases	Section 10.3.2
rRNA and tRNA processing	Ribonucleases	Section 10.2.2
Translation	Aminoacyl-tRNA synthetases	Section 11.1.1
	Translation factors	Sections 11.2.2 and 11.2.3
RNA degradation	Ribonucleases	Section 10.4
Ribosome structure	Ribosomal proteins	Section 11.2.1

9.1.1 Locating the positions of DNA-binding sites in a genome

Often the first thing that is discovered about a DNA-binding protein is not the identity of the protein itself but the features of the DNA sequence that the protein recognizes. This is because genetic and molecular biology experiments, which we will deal with later in this chapter, have shown that many of the proteins that are involved in genome expression bind to short DNA sequences immediately upstream of the genes on which they act (*Figure 9.2*). This means that the sequence of a newly discovered gene, assuming that it includes both the coding DNA and the regions upstream of it, provides immediate access to the binding sites of at least some of the proteins responsible for expression of that gene. Because of this, a number of methods have been developed for locating protein binding sites within DNA fragments up to several kb in length, these methods working perfectly well even if the relevant DNA-binding proteins have not been identified.

Gel retardation identifies DNA fragments that bind to proteins

The first of these methods makes use of the substantial difference between the electrophoretic properties of a 'naked' DNA fragment and one that carries a bound protein. Recall that DNA fragments are separated by agarose gel electrophoresis because smaller fragments migrate through the pore-like structure of the gel more quickly than do larger fragments (see Technical Note 2.1, page 37). If a DNA fragment has a protein bound to it then its mobility through the gel will be impeded: the DNA–protein complex therefore forms a band at a position nearer to the starting point (*Figure 9.3*). This is called **gel retardation** (Garner and Revzin, 1981). In practice the technique is carried out with a collection of restriction fragments that span the region thought to contain a protein binding site. The digest is mixed with an extract of nuclear proteins (assuming that a eukary-

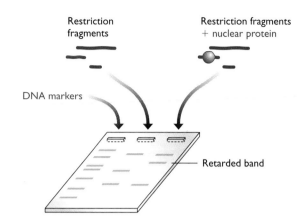

Figure 9.3 Gel retardation analysis.

A nuclear extract has been mixed with a DNA restriction digest and a DNA-binding protein in the extract has attached to one of the restriction fragments. The DNA–protein complex has a larger molecular mass than the 'naked' DNA and so runs more slowly during gel electrophoresis. As a result, the band for this fragment is retarded and can be recognized by comparing with the banding pattern produced by restriction fragments that have not been mixed with the nuclear extract.

ote is being studied) and retarded fragments are identified by comparing the banding pattern obtained after electrophoresis with the pattern for restricted fragments that have not been mixed with proteins. A nuclear extract is used because at this stage of the project the DNA-binding protein has not usually been purified. If, however, the protein is available then the experiment can be carried out just as easily with the pure protein as with a mixed extract.

Protection assays pinpoint binding sites with greater accuracy

Gel retardation gives a general indication of the location of a protein binding site in a DNA sequence, but does not pinpoint the site with great accuracy. Often the retarded fragment is several hundred bp in length, compared with the expected length of the binding site of a few tens of bp at most, and there is no indication of where in the retarded fragment the binding site lies. Also, if the retarded fragment is long then it might contain separate binding sites for several proteins, or if it is quite small then there is the possibility that the binding site also includes nucleotides on adjacent fragments, ones that on their own do not form a stable complex with the protein and so do not lead to gel retardation. Retardation studies are therefore a starting point but other techniques are needed to provide more accurate information.

Modification protection assays can take over where gel retardation leaves off. The basis of these techniques is that if a DNA molecule carries a bound protein then

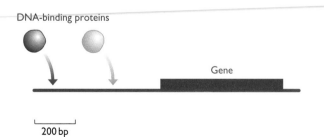

Figure 9.2 Attachment sites for DNA-binding proteins are located immediately upstream of a gene.

See Sections 9.2 and 9.3 for more information on the location and function of these protein attachment sites.

part of its nucleotide sequence will be protected from modification. There are two ways of carrying out the modification:

- By treatment with a nuclease, which cleaves all phosphodiester bonds except those protected by the bound protein;
- By exposure to a methylating agent, such as dimethyl sulfate which adds methyl groups to G nucleotides. Any Gs protected by the bound protein will not be methylated.

The practical details of these two techniques are shown in *Figures 9.4* and *9.5*. Both utilize an experimental approach called **footprinting**. In nuclease footprinting (Galas and Schmitz, 1978), the DNA fragment being examined is labeled at one end, complexed with binding protein (as a nuclear extract or as pure protein), and treated with

deoxyribonuclease I (DNase I). Normally, DNase I cleaves every phosphodiester bond, leaving only the DNA segment protected by the binding protein. This is not very useful because it can be difficult to sequence such a small fragment. It is quicker to use the more subtle approach shown in *Figure 9.4*. The nuclease treatment is carried out under limiting conditions, such as a low temperature and/or very little enzyme, so that on average each copy of the DNA fragment suffers a single 'hit' – meaning that it is cleaved at just one position along its length. Although each fragment is cut just once, in the entire population of fragments all bonds are cleaved except those protected by the bound protein. The protein is now removed, the mixture electrophoresed, and the labeled fragments visualized. Each of these fragments has the label at one end and a cleavage site at the other. The result is a ladder of bands corresponding to fragments that differ in length by one

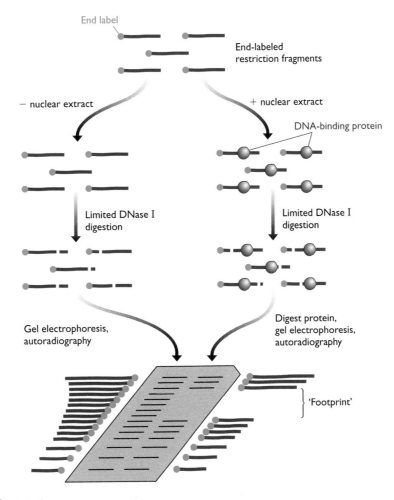

Figure 9.4 DNase I footprinting.

The technique is described in the text. The restriction fragments used at the start of the procedure must be labeled at just one end. This is usually achieved by treating a set of longer restriction fragments with an enzyme that attaches labels at *both* ends, then cutting these labeled molecules with a second restriction enzyme, and purifying one of the sets of end fragments. The DNase I treatment is carried out in the presence of a manganese salt, which induces the enzyme to make random double-stranded cuts in the target molecules, leaving blunt-ended fragments.

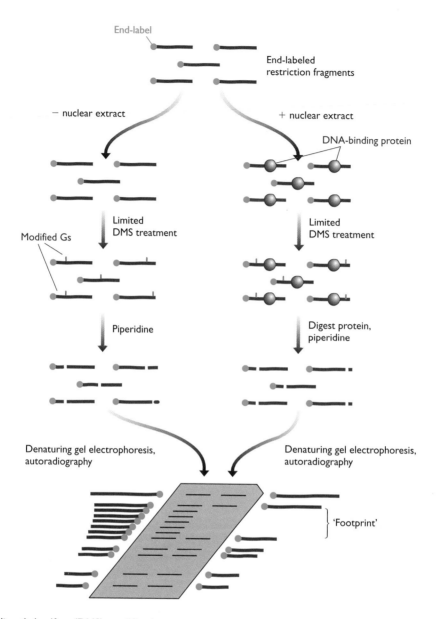

Figure 9.5 The dimethyl sulfate (DMS) modification protection assay.

The technique is similar to DNase I footprinting (see *Figure 9.4*). Instead of DNase I digestion, the fragments are treated with limited amounts of DMS so that a single guanine base is methylated in each fragment. Guanines that are protected by the bound protein cannot be modified. After removal of the protein the DNA is treated with piperidine, which cuts at the modified nucleotide positions. For simplicity, the diagram shows the double-stranded molecules being cleaved at this stage. In fact, they are only nicked, as piperidine only cuts the strand that is modified, rather than making a double-stranded cut across the entire molecule. The samples are therefore examined by *denaturing* gel electrophoresis so that the two strands are separated. The resulting autoradiograph shows the sizes of the strands that have one labeled end and one end created by piperidine nicking. The banding pattern for the control DNA strands – those not incubated with the nuclear extract – indicates the positions of Gs in the restriction fragment, and the footprint seen in the banding pattern for the test sample shows which Gs were protected.

nucleotide, the ladder broken by a blank area in which no labeled bands occur. This blank area, or 'footprint', corresponds to the positions of the protected phosphodiester bonds, and hence of the bound protein, in the starting DNA.

Modification interference identifies nucleotides central to protein binding

Modification protection should not be confused with **modification interference**, a different technique with greater sensitivity in the study of protein binding (Hendrickson

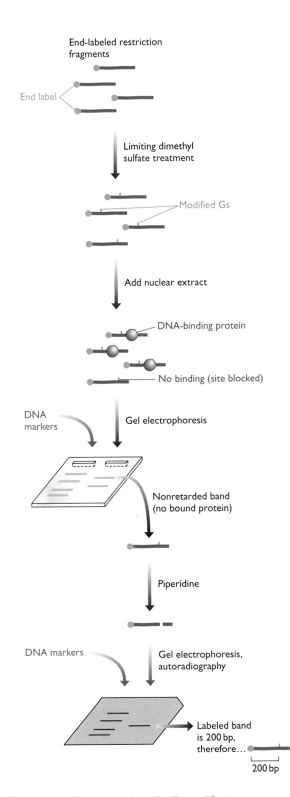

Figure 9.6 Dimethyl sulfate (DMS) modification interference assay.

The method is described in the text. See the legend to *Figure 9.4* for a description of the procedure used to obtain DNA fragments labeled at just one end.

and Schleif, 1985). Modification interference works on the basis that if a nucleotide critical for protein binding is altered, for example by addition of a methyl group, then binding may be prevented. One of this family of techniques is illustrated in *Figure 9.6*. The DNA fragment, labeled at one end, is treated with the modification reagent, in this case dimethyl sulfate, under limiting conditions so that just one guanine per fragment is methylated. Now the binding protein or nuclear extract is added, and the fragments electrophoresed. Two bands are seen, one corresponding to the DNA–protein complex and one containing DNA without bound protein. The latter contains molecules that have been prevented from attaching to the protein because the methylation treatment has modified one or more Gs that are crucial for the binding. To identify which Gs are modified, the fragment is purified from the gel and treated with piperidine, a compound that cleaves DNA at methylguanine nucleotides. The result of this treatment is that each fragment is cut into two segments, one of which carries the label. The length(s) of the labeled fragment(s), determined by a second round of electrophoresis, tells us which nucleotide(s) in the original fragment were methylated and hence identifies the position in the DNA sequence of Gs that participate in the binding reaction. Equivalent techniques can be used to identify the A, C and T nucleotides involved in binding.

9.1.2 Purifying a DNA-binding protein

Once a binding site has been identified in a DNA molecule, this sequence can be used to purify the DNA-binding protein, as a prelude to more detailed structural studies. The purification techniques utilize the ability of the protein to bind to its target site. One possibility is to use a form of **affinity chromatography** (*Figure 9.7A*). A DNA fragment or synthetic oligonucleotide that contains a protein binding site is immobilized in a chromatography column, usually by attaching one end of the DNA to a silica particle (Kadonaga, 1991). The protein extract is then passed through the column in a low-salt buffer, which promotes binding of proteins to their target sites. The binding protein specific for the immobilized sequence is retained in the column while all other proteins pass through. Once these unwanted proteins have been completely washed out, the column is eluted with a high-salt buffer, which destabilizes the DNA–protein complex. The pure binding protein can then be collected.

An alternative is to screen a cloning library (Singh *et al.*, 1988). A library of cDNA clones, each synthesizing a different cloned protein from the organism being studied, is needed. These clones are blotted onto a nylon membrane in such a way that the protein content of each clone is retained (*Figure 9.7B*). The DNA fragment or oligonucleotide containing the protein binding site is labeled, and washed over the membrane. The DNA attaches to a blotted clone only if that clone has been synthesizing the appropriate DNA-binding protein. These clones are identified by detecting where the labeled DNA

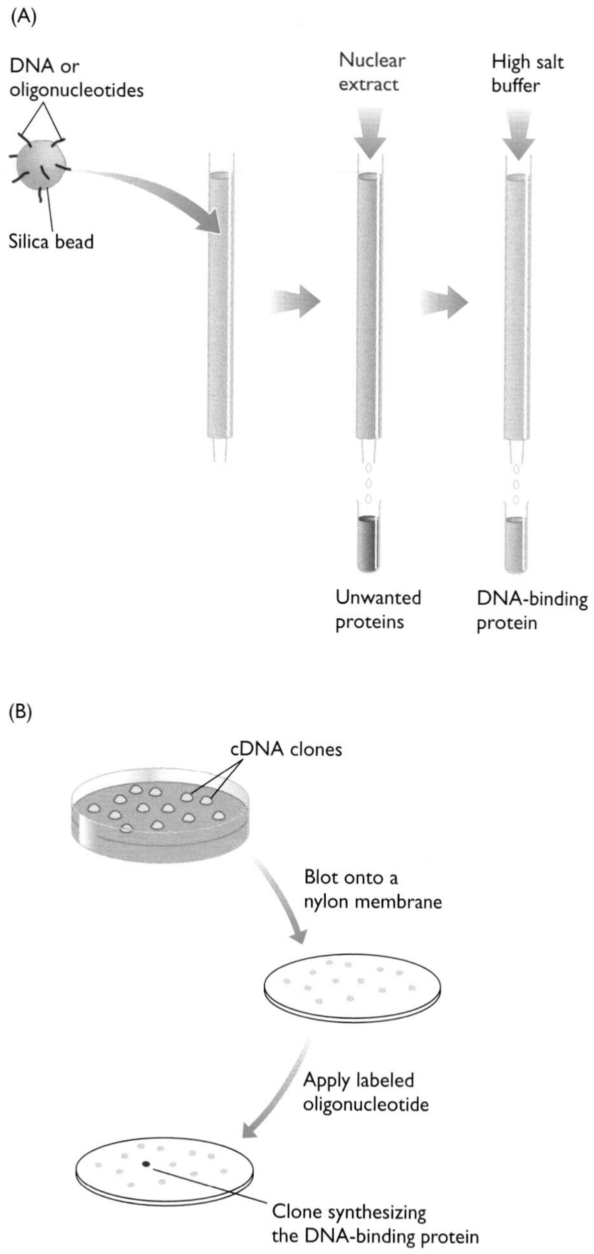

(A)

DNA or
oligonucleotides

Nuclear
extract

High salt
buffer

Silica bead

Unwanted
proteins

DNA-binding
protein

(B)

cDNA clones

Blot onto a
nylon membrane

Apply labeled
oligonucleotide

Clone synthesizing
the DNA-binding protein

Figure 9.7 Two ways of purifying a DNA-binding protein.

(A) Affinity chromatography. DNA fragments or synthetic oligonucleotides containing the attachment site for the binding protein are attached to silica beads and these packed into a chromatography column. The binding protein is retained in the column when a nuclear extract is passed through, and can subsequently be recovered by eluting with a high-salt buffer. **(B) Screening a clone library.** cDNA clones are blotted onto a nylon filter and one that is expressing the binding protein identified through the ability of this colony to bind labeled DNA containing the protein attachment site (see Technical Note 4.3, page 117).

is located on the membrane. Samples of the clones can then be recovered from the master library and used to produce larger quantities of the binding protein.

9.1.3 Studying the structures of proteins and DNA–protein complexes

The availability of a pure sample of a DNA-binding protein makes possible the analysis of its structure, in isolation or attached to its DNA-binding site. This provides the most detailed information on the DNA–protein interaction, enabling the precise structure of the DNA-binding part of the protein to be determined, and allowing the identity and nature of the contacts with the DNA helix to be elucidated. Two techniques – **X-ray crystallography** and **nuclear magnetic resonance (NMR) spectroscopy** – are central to this area of research.

X-ray crystallography has broad applications in structure determination

X-ray crystallography is a long-established technique whose pedigree stretches back to the late 19th century. Indeed, Nobel prizes were awarded as early as 1915 to William and Lawrence Bragg, father and son, for working out the basic methodology and using it to determine the crystal structures of salts such as sodium chloride and zinc sulfide. The technique is based on **X-ray diffraction**. X-rays have very short wavelengths – between 0.01 and 10 nm – which is 4000 times shorter than visible light and comparable with the spacings between atoms in chemical structures. When a beam of X-rays is directed onto a crystal, some of the X-rays pass straight through, but others are diffracted and emerge from the crystal at a different angle from which they entered (*Figure 9.8A*). If the crystal is comprised of many copies of the same molecule, all positioned in a regular array, then different X-rays are diffracted in similar ways, resulting in overlapping circles of diffracted waves which interfere with one another. An X-ray-sensitive photographic film or electronic detector placed across the beam reveals a series of spots (*Figure 9.8B*), an **X-ray diffraction pattern**, from which the structure of the molecule in the crystal can be deduced.

The challenge with X-ray crystallography lies with the complexity of the methodology used to deduce the structure of a molecule from its diffraction pattern. The basic principles are that the relative positioning of the spots indicates the arrangement of the molecules in the crystal, and their relative intensities provide information on the structure of the molecule. The problem is that the more complex the molecule, the greater the number of spots and the larger the number of comparisons that must be made between them. Even with computational help the analysis is difficult and time consuming. If successful, the result is an electron density map (*Figure 9.8C and D*) which, with a protein, provides a chart of the folded polypeptide from which the positioning of structural features such as α-helices and β-sheets can be

(A) Production of a diffraction pattern

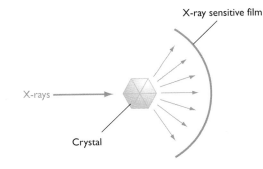

(B) X-ray diffraction pattern for ribonuclease

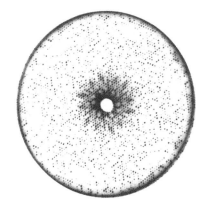

(C) Part of the ribonuclease electron-density map

(D) Interpretation of an electron-density map – a 2 Å resolution electron-density map revealing a tyrosine R group

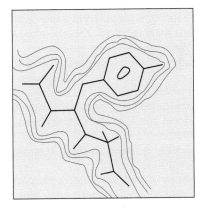

Figure 9.8 X-ray crystallography.

(A) An X-ray diffraction pattern is obtained by passing a beam of X-rays through a crystal of the molecule being studied. (B) The diffraction pattern obtained with crystals of ribonuclease. (C) Part of the electron-density map derived from this diffraction pattern. (D) If the electron-density map has sufficient resolution then it is possible to identify the R groups of individual amino acids, as shown here for tyrosine. (B) and (C) from *Biology*, 3rd edition, by Neil Campbell, copyright 1993 by Benjamin Cummings Publishing Company. Reprinted by permission. (D) reprinted with permission from Zubay G, *Biochemistry*, 4th edition. Copyright 1997 by The McGraw–Hill Companies.

determined. If sufficiently detailed, the R groups of the individual amino acids in the polypeptide can be identified and their orientations relative to one another established, allowing deductions to be made about the hydrogen bonding and other chemical interactions occurring within the protein structure. With luck, these deductions lead to a detailed three-dimensional model of the protein (Rhodes, 1999).

The first protein structures to be determined by X-ray crystallography were for myoglobin and hemoglobin, resulting in further Nobel prizes, for Perutz and Kendrew in 1962. It still takes several months or longer to complete an X-ray crystallography analysis with a new protein, and there are many pitfalls that can prevent a successful conclusion being reached. In particular, it can often be difficult to obtain a suitable crystal of the protein. Despite these problems, the number of completed structures has gradually increased and now includes more than 50 DNA-binding proteins. An important innovation has been to crystallize DNA-binding proteins in the presence of their target sequences, the resulting protein–DNA structures revealing the precise positioning of the proteins relative to the double helix. It is from this type of information that most of our knowledge about the mode of action of DNA-binding proteins has been obtained.

NMR gives detailed structural information for small proteins

Like X-ray crystallography, NMR traces its origins to the early part of the 20th century, first being described in 1936 with the relevant Nobel prizes awarded in 1952. The principle of the technique is that rotation of a charged chemical nucleus generates a magnetic moment. When placed in an applied electromagnetic field, the spinning nucleus orientates in one of two ways, called α and β (*Figure 9.9*), the α-orientation (which is aligned with the magnetic field) having a slightly lower energy. In NMR spectroscopy the magnitude of this energy separation is determined by measuring the frequency of the electromagnetic radiation needed to induce the transition from α to β, the value being described as the resonance frequency of the nucleus being studied. The critical point is that although each type of nucleus (e.g. ^1H, ^{13}C, ^{15}N) has its own specific resonance frequency, the measured frequency is often slightly different from the standard value (typically by less than 10 parts per million) because electrons in the vicinity of the rotating nucleus shield it to a certain extent from the applied magnetic field. This **chemical shift** (the difference between the observed resonance energy and the standard value for the nucleus being studied) enables the chemical environment of the nucleus to be inferred, and hence provides structural information. Particular types of analysis (called COSY and TOCSY) enable atoms linked by chemical bonds to the spinning nucleus to be identified; other analyses (e.g. NOESY) identify atoms that are close to the spinning nucleus in space but not directly connected to it.

Not all chemical nuclei are suitable for NMR. Most protein NMR projects are ^1H studies, the aim being to identify the chemical environments and covalent linkages of every hydrogen atom, and from this information to infer the overall structure of the protein. These studies are frequently supplemented by analyses of substituted proteins in which at least some of the carbon and/or nitrogen atoms have been replaced with the rare isotopes ^{13}C and ^{15}N, these also giving good results with NMR.

When successful, NMR results in the same level of resolution as X-ray crystallography and so provides very detailed information on protein structure (Evans, 1995). The main advantage of NMR is that it works with molecules in solution and so avoids the problems that sometimes occur when attempting to obtain crystals of a protein for X-ray analysis. Solution studies also offer greater flexibility if the aim is to examine changes in protein structure, for example during protein folding or in response to addition of a substrate. The disadvantage of NMR is that it is only suitable for relatively small proteins. There are several reasons for this, one being the need to identify the resonance frequencies for each, or as many as possible, of the ^1H or other nuclei being studied. This depends on the various nuclei having different chemical shifts so that their frequencies do not overlap. The larger the protein, the greater the number of nuclei and the greater the chances that frequencies overlap and structural information is lost. Although this limits the applicability of NMR, the technique is still very valuable. There are many interesting proteins that are small enough to be studied by NMR, and important information can also be obtained by structural analysis of peptides which, although not complete proteins, can act as models for aspects of protein activity such as nucleic acid binding.

9.1.4 The special features of DNA-binding proteins

Now that we have examined the methods used to study DNA-binding proteins, we can turn our attention to the proteins themselves. Our main interest lies with those proteins that are able to target a specific nucleotide sequence and hence bind to a limited number of positions on a DNA molecule, this being the type of interaction that is most important in expression of the genome. To bind in this specific fashion a protein must make contact with the double helix in such a way that the nucleotide sequence can be recognized, which generally requires that part of the protein penetrates into the major and/or minor

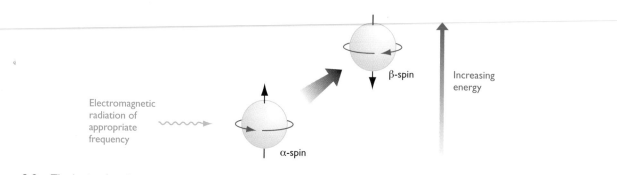

Figure 9.9　The basis of nuclear magnetic resonance (NMR) spectroscopy.

A rotating nucleus can take up either of two orientations in an applied electromagnetic field. The energy separation between the α and β spin states is determined by measuring the frequency of electromagnetic radiation needed to induce an $\alpha\rightarrow\beta$ transition.

grooves of the helix (see *Figures 1.11A* and *1.12*, pages 15 and 17) in order to achieve **direct readout** of the sequence (Section 9.1.5). This is usually accompanied by more general interactions with the surface of the molecule, which may simply stabilize the DNA–protein complex or which may be aimed at accessing indirect information on nucleotide sequence that is provided by the conformation of the helix.

When the structures of sequence-specific DNA-binding proteins are compared, it is immediately evident that the family as a whole can be divided into a limited number of different groups on the basis of the structure of the segment of the protein that interacts with the DNA molecule (*Table 9.2*; Luisi, 1995). Each of these **DNA-binding motifs** is present in a range of proteins, often from very different organisms, and at least some of them probably evolved more than once. We will look at two in detail – the **helix–turn–helix** (HTH) motif and the **zinc finger** – and then briefly survey the others.

The helix–turn–helix motif is present in prokaryotic and eukaryotic proteins

The HTH motif was the first DNA-binding structure to be identified (Harrison and Aggarwal, 1990). As the name suggests, the motif is made up of two α-helices separated by a turn (*Figure 9.10*). The latter is not a random conformation but a specific structure, referred to as a **β-turn**, made up of four amino acids, the second of which is usually glycine. This turn, in conjunction with the first α-helix, positions the second α-helix on the surface of the protein in an orientation that enables it to fit inside the major groove of a DNA molecule. This second α-helix is therefore the **recognition helix** that makes the vital contacts which enable the DNA sequence to be read. The HTH structure is usually 20 or so amino acids in length and so is just a small part of the protein as a whole. Some of the other parts of the protein form attachments with the surface of the DNA molecule, primarily to aid the correct positioning of the recognition helix within the major groove.

Many prokaryotic and eukaryotic DNA-binding proteins utilize an HTH motif. In bacteria, HTH motifs are present in some of the best studied regulatory proteins, which switch on and off the expression of individual genes. An example is the **lactose repressor**, which regulates expression of the lactose operon (Sections 2.3.2 and 9.3.1). The various eukaryotic HTH proteins include many whose DNA-binding properties are important in the developmental regulation of genome expression, such as the **homeodomain** proteins, whose roles we will examine in Section 12.3.3. The homeodomain is an extended

Table 9.2 DNA-binding motifs

Motif	Examples of proteins with this motif
Sequence-specific DNA-binding motifs	
Helix–turn–helix family	
Standard helix–turn–helix	*Escherichia coli* lactose repressor, tryptophan repressor
Homeodomain	*Drosophila* Antennapedia protein
Paired homeodomain	Vertebrate Pax transcription factors
POU domain	Vertebrate regulatory proteins PIT-1, OCT-1 and OCT-2
Winged helix–turn–helix	GABP regulatory protein of higher eukaryotes
High mobility group (HMG) domain	Mammalian sex determination protein SRY
Zinc-finger family	
Cys_2His_2 finger	Transcription factor TFIIIA of eukaryotes
Multi-cysteine zinc finger	Steroid receptor family of higher eukaryotes
Zinc binuclear cluster	Yeast GAL4 transcription factor
Basic domain	Yeast GCN4 transcription factor
Ribbon–helix–helix	Bacterial MetJ, Arc and Mnt repressors
TBP domain	Eukaryotic TATA-binding protein
β-Barrel dimer	Papillomavirus E2 protein
Rel homology domain (RHB)	Mammalian transcription factor NF-κB
Non-specific DNA-binding motifs	
Histone fold	Eukaryotic histones
HU/IHF motif[a]	Bacterial HU and IHF proteins
Polymerase cleft	DNA and RNA polymerases

[a]The HU/IHF motif is a non-specific DNA-binding motif in bacterial HU proteins (the nucleoid packaging proteins; Section 2.3.1) but directs sequence-specific binding of the IHF (integration host factor) protein.

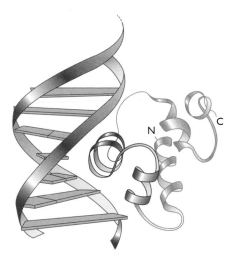

Figure 9.10   The helix–turn–helix motif.

The drawing shows the orientation of the helix–turn–helix motif (in blue) of the *Escherichia coli* bacteriophage 434 repressor in the major groove of the DNA double helix. 'N' and 'C' indicate the N- and C-termini of the motif, respectively. Reprinted from *DNA–Protein Interactions* by Andrew Travers, published by Chapman & Hall, 1993. Reprinted with kind permission of A. Travers.

HTH motif possessed by each of these proteins. It is made up of 60 amino acids which form four α-helices, numbers 2 and 3 separated by a β-turn, with number 3 acting as the recognition helix and number 1 making contacts within the minor groove (*Figure 9.11*). Other versions of the HTH motif found in eukaryotes include:

■   The **POU domain**, which is usually found in proteins that also have a homeodomain, the two motifs probably working together by binding different regions of a double helix. The name 'POU' comes from the initial letters of the names of the first proteins found to contain this motif (Herr *et al.*, 1988).

■   The **winged helix–turn–helix** motif, which is another extended version of the basic HTH structure, this one with a third α-helix on one side of the HTH motif and a β-sheet on the other side.

Many proteins, prokaryotic and eukaryotic, possess an HTH motif, but the details of the interaction of the recognition helix with the major groove are not exactly the same in all cases. The length of the recognition helix varies, generally being longer in eukaryotic proteins, the orientation of the helix in the major groove is not always the same, and the position within the recognition helix of those amino acids that make contacts with nucleotides is different.

Zinc fingers are common in eukaryotic proteins

The second type of DNA-binding motif that we will look at in detail is the zinc finger, which is rare in prokaryotic

---

Box 9.1: RNA-binding motifs

RNA-binding proteins also have specific motifs that form the attachment with the RNA molecule. The most important of these are as follows:

■   The **ribonucleoprotein (RNP) domain** comprises four β-strands and two α-helices in the order β-α-β-β-α-β. The two central β-strands make the critical attachments with the RNA molecule. The RNP domain is the commonest RNA-binding motif and has been found in more than 250 proteins.

■   The **double-stranded RNA binding domain (dsRBD)** is similar to the RNP domain but with the structure α-β-β-β-α. The RNA-binding function lies between the β and α at the end of the structure. As the name implies, the motif is found in proteins that bind double-stranded RNA (Fierro-Monti and Mathews, 2000).

■   The **κ-homology domain** has the structure β-α-α-β-β-α, with the binding function between the pair of α-helices. It is relatively uncommon but present in at least one nuclear RNA-binding protein.

■   The DNA-binding **homeodomain** may also have RNA-binding activity in some proteins. One ribosomal protein uses a structure similar to a homeodomain to attach to rRNA, and some homeodomain proteins such as Bicoid of *Drosophila melanogaster* (Section 12.3.3) can bind both DNA and RNA.

---

proteins but very common in eukaryotes (Mackay and Crossley, 1998). There appear to be more than 500 different zinc-finger proteins in the worm *Caenorhabditis elegans*, out of a total 19 000 proteins (Clarke and Berg, 1998), and it is estimated that 1% of all mammalian genes code for zinc-finger proteins.

There are at least six different versions of the zinc finger. The first to be studied in detail was the **Cys₂His₂ finger**, which comprises a series of 12 or so amino acids, including two cysteines and two histidines, which form a segment of β-sheet followed by an α-helix. These two structures, which form the 'finger' projecting from the surface of the protein, hold between them a bound zinc atom, coordinated to the two cysteines and two histidines (*Figure 9.12*). The α-helix is the part of the motif that makes the critical contacts within the major groove, its positioning within the groove being determined by the β-sheet, which interacts with the sugar–phosphate backbone of the DNA, and the zinc atom, which holds the sheet and helix in the appropriate positions relative to one another. Other versions of the zinc finger differ in the structure of the finger, some lacking the sheet component and consisting simply of one or more α-helices, and the precise way in which the zinc atom is held in place also varies. For example, the **multicysteine zinc fingers** lack histidines, the zinc atom being coordinated between four cysteines.

An interesting feature of the zinc finger is that multiple copies of the finger are sometimes found on a single protein. Several have two, three or four fingers, but there are examples with many more than this – 37 for one toad protein. In most cases, the individual zinc fingers are thought to make independent contacts with the DNA molecule, but in some cases the relationship between different fingers is more complex. In one particular group of proteins – the nuclear or steroid receptor family – two α-helices containing six cysteines combine to coordinate two zinc atoms in a single DNA-binding domain, larger than a standard zinc finger (*Figure 9.13*). Within this motif it appears that one of the α-helices enters the major groove whereas the second makes contacts with other proteins.

Other DNA-binding motifs

The various other DNA-binding motifs that have been discovered in different proteins include:

- The **basic domain**, in which the DNA recognition structure is an α-helix that contains a high number of basic amino acids (e.g. arginine, serine and threonine). A peculiarity of this motif is that the α-helix only forms when the protein interacts with DNA: in the unbound state the helix has a disorganized structure. Basic domains are found in a number of eukaryotic proteins involved in transcription of DNA into RNA.
- The **ribbon–helix–helix** motif, which is one of the few motifs that achieves sequence-specific DNA binding without making use of an α-helix as the

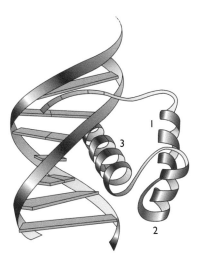

Figure 9.11 The homeodomain motif.

The first three helices of a typical homeodomain are shown with helix 3 orientated in the major groove and helix 1 making contacts in the minor groove. Helices 1–3 run in the N→C direction along the motif. Reprinted from *DNA–Protein Interactions* by Andrew Travers, published by Chapman & Hall, 1993. Reprinted with kind permission of A. Travers.

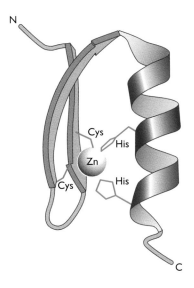

Figure 9.12 The Cys₂His₂ zinc finger.

This particular zinc finger is from the yeast SWI5 protein. The zinc atom is held between two cysteines within the β-sheet of the motif and two histidines in the α-helix. The solid green lines indicate the R groups of these amino acids. 'N' and 'C' indicate the N- and C-termini of the motif, respectively. Reprinted from *DNA–Protein Interactions* by Andrew Travers, published by Chapman & Hall, 1993. Reprinted with kind permission of A. Travers.

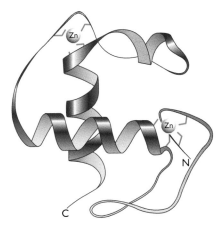

Figure 9.13 The steroid receptor zinc finger.

The R groups of the amino acids involved in the interactions with the zinc atoms are shown as solid green lines. 'N' and 'C' indicate the N- and C-termini of the motif, respectively. Reprinted from *DNA–Protein Interactions* by Andrew Travers, published by Chapman & Hall, 1993. Reprinted with kind permission of A. Travers.

Box 9.2: Can sequence specificity be predicted from the structure of a recognition helix?

An intriguing question is whether the specificity of DNA binding can be understood in sufficient detail for the sequence of a protein's target site to be predicted from examination of the structure of the recognition helix of a DNA-binding motif. To date, this objective has largely eluded us, but it has been possible to deduce some rules for the interaction involving certain types of zinc finger (Choo and Klug, 1997). In these proteins, four amino acids, three in the recognition helix and one immediately adjacent to it, form critical attachments with the nucleotide bases of the target site. Some of these attachments involve a single amino acid with a single base; others involve two amino acids with one base. By comparing the sequences of amino acids in the recognition helices of different zinc fingers with the sequences of nucleotides at the binding sites, it has been possible to identify a set of rules governing the interaction. These enable the nucleotide sequence specificity of a new zinc finger protein to be predicted, admittedly with the possibility of some ambiguity, once the amino acid composition of its recognition helix is known.

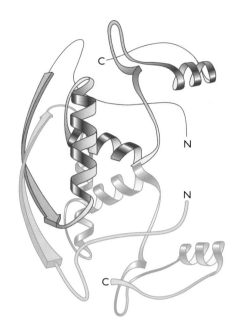

Figure 9.14 The ribbon–helix–helix motif.

The drawing is of the ribbon–helix–helix motif of the *Escherichia coli* MetJ repressor, which consists of a dimer of two identical proteins, one shown in blue and the other in green. The β-strands at the left of the structure make contact with the major groove of the double helix. 'N' and 'C' indicate the N- and C-termini of the motif, respectively. Reprinted from *DNA–Protein Interactions* by Andrew Travers, published by Chapman & Hall, 1993. Reprinted with kind permission of A. Travers.

recognition structure. Instead, the ribbon (i.e. two strands of a β-sheet) makes contact with the major groove (*Figure 9.14*). Ribbon–helix–helix motifs are found in some gene-regulatory proteins in bacteria.

- The **TBP domain** has so far only been discovered in the **TATA-binding protein** (Section 9.2.3), after which it is named (Kim *et al.*, 1993). As with the ribbon–helix–helix motif, the recognition structure is a β-sheet, but in this case the main contacts are with the minor, not major, groove of the DNA molecule.

9.1.5 The interaction between DNA and its binding proteins

In recent years our understanding of the part played by the DNA molecule in the interaction with a binding protein has begun to change. It has always been accepted that proteins that recognize a specific sequence as their binding site can locate this site by forming contacts with chemical groups attached to the nitrogenous bases that are exposed within the major and minor grooves that spiral around the double helix (see *Figure 1.11A*, page 15). It is now recognized that the nucleotide sequence also influences the precise conformation of each region of the helix, and that these conformational features represent a

second, less direct way in which the DNA sequence can influence protein binding.

Direct readout of the nucleotide sequence

It was clear from the double helix structure described by Watson and Crick (Section 1.1.3) that although the nucleotide bases are on the inside of the DNA molecule, they are not entirely buried, and some of the chemical groups attached to the purine and pyrimidine bases are accessible from outside the helix. **Direct readout** of the nucleotide sequence should therefore be possible without breaking the base pairs and opening up the molecule.

In order to form chemical bonds with groups attached to the nucleotide bases, a binding protein must make contacts within one or both of the grooves on the surface of the helix. With the B-form of DNA, the identity and orientation of the exposed parts of the bases within the major groove is such that most sequences can be read unambiguously, whereas within the minor groove it is possible to identify if each base pair is A–T or G–C but difficult to know which nucleotide of the pair is in which strand of the helix (*Figure 9.15*; Kielkopf *et al.*, 1998). Direct readout of the B-form therefore predominantly

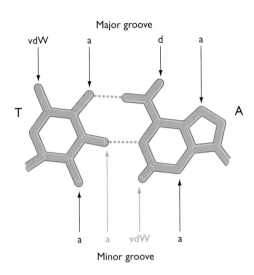

Figure 9.15 Recognition of an A–T base pair in the B-form double helix.

An A–T base pair is shown in outline (see *Figure 1.11B*, page 15), with arrows indicating the chemical features that can be recognized by accessing the base pair via the major groove (above) and minor groove (below). In the major groove the chemical features are asymmetric and the orientation of the A–T pair can be identified by a binding protein. For some time it was believed that this is not possible in the minor groove, because only the two features shown in black were thought to be present, and these are symmetric. Using these two features, the binding protein could recognize the A–T base pair but not know which nucleotide is in which strand of the helix. The asymmetric features shown in green have recently been discovered, suggesting that the orientation of the pair might in fact be discernible via the minor groove. Abbreviations: a, hydrogen bond acceptor; d, hydrogen bond donor; vdW, van der Waals interaction (see Box 3.3, page 82). Reprinted with permission from Kielkopf CL, et al., *Science*, 282, 111–115. Copyright 1998 American Association for the Advancement of Science.

involves contacts in the major groove. With other DNA types there is much less information on the contacts formed with binding proteins, but the picture is likely to be quite different. In the A-form, for example, the major groove is deep and narrow and less easily penetrated by any part of a protein molecule (see *Table 1.1*, page 16). The shallower minor groove is therefore likely to play the main part in direct readout. With Z-DNA, the major groove is virtually non-existent and direct readout is possible to a certain extent without moving beyond the surface of the helix.

The nucleotide sequence has a number of indirect effects on helix structure

The recent change in our view of DNA structure concerns the influence of the nucleotide sequence on the conformation of the helix at different positions along its length.

Originally it was thought that cellular DNA molecules have fairly uniform structures, made up mainly of the B-form of the double helix. Some short segments might be in the A-form, and there might be some Z-DNA tracts, especially near the ends of a molecule, but the majority of the length of a double helix would be unvarying B-DNA. We now recognize that DNA is much more polymorphic, and that it is possible for the A-, B- and Z-DNA configurations, and intermediates between them, to coexist within a single DNA molecule, different parts of the molecule having different structures. These conformational variations are sequence dependent, being largely the result of the base-stacking interactions that occur between adjacent base pairs. As well as being responsible, along with base-pairing, for the stability of helix, the base-stacking also influences the amount of rotation that occurs around the covalent bonds within individual nucleotides and hence determines the conformation of the helix at a particular position. The rotational possibilities in one base pair are influenced, via the base-stacking interactions, by the identities of the neighboring base pairs. This means that the nucleotide sequence indirectly affects the overall conformation of the helix, possibly providing structural information that a binding protein can use to help it locate its appropriate attachment site on a DNA molecule. At present this is just a theoretical possibility as no protein that specifically recognizes a non-B form of the helix has been identified, but many researchers believe that helix conformation is likely to play some role in the interaction between DNA and protein.

A second type of conformational change is **DNA bending** (Travers, 1995). This does not refer to the natural flexibility of DNA which enables it to form circles and supercoils, but instead to localized positions where the nucleotide sequence causes the DNA to bend. Like other conformational variations, DNA bending is sequence dependent. In particular, a DNA molecule in which one polynucleotide contains two or more groups of repeated adenines, each group comprising 3–5 As, with individual groups separated by 10 or 11 nucleotides, will bend at the 3' end of the adenine-rich region (Young and Beveridge, 1998). As with helix conformation, it is not yet known to what extent DNA bending influences protein binding, although protein-induced bending at flexible sites has a clearly demonstrated function in the regulation of some genes (e.g. Falvo *et al.*, 1995; Section 9.3.2).

Contacts between DNA and proteins

The contacts formed between DNA and its binding proteins are non-covalent. Within the major groove, hydrogen bonds form between the nucleotide bases and the R groups of amino acids in the recognition structure of the protein, whereas in the minor groove hydrophobic interactions are more important. On the surface of the helix, the major interactions are electrostatic, between the negative charges on the phosphate component of each nucleotide and the positive charges on the R groups of amino acids such as lysine and arginine, although some hydrogen bonding also occurs. In some cases, hydrogen bonding on the surface of the helix or in the major groove is direct between DNA and protein; in others it is mediated by water molecules. Few generalizations can be made: at this level of DNA–protein interaction each example has its own unique features and the details of the bonding have to be worked out by structural studies rather than by comparisons with other proteins.

Most proteins that recognize specific sequences are also able to bind non-specifically to other parts of a DNA molecule. In fact it has been suggested that the amount of DNA in a cell is so large, and the numbers of each binding protein so small, that the proteins spend most, if not all, of their time attached non-specifically to DNA (Stormo and Fields, 1998). The distinction between the non-specific and specific forms of binding is that the latter is more favorable in thermodynamic terms. As a result, a protein is able to bind to its specific site even though there are literally millions of other sites to which it could attach non-specifically. To achieve this thermodynamic favorability, the specific binding process must involve the greatest possible number of DNA–protein contacts, which explains in part why the recognition structures of many DNA-binding motifs have evolved to fit snugly into the major groove of the helix, where the opportunity for DNA–protein contacts is greatest. It also explains why some DNA–protein interactions result in conformational changes to one or other partner, increasing still further the complementarity of the interacting surfaces, and allowing additional bonding to occur.

The need to maximize contacts in order to ensure specificity is also one of the reasons why many DNA-binding proteins are dimers, consisting of two proteins attached to one another. This is the case for most HTH proteins and many of the zinc-finger type. Dimerization occurs in such a way that the DNA-binding motifs of the two proteins are both able to access the helix, possibly with some degree of cooperativity between them, so that the resulting number of contacts is greater than twice the number achievable by a monomer. As well as their DNA-binding motifs, many proteins contain additional characteristic domains that participate in the protein–protein contacts that result in dimer formation. One of these is the **leucine zipper**, which is an α-helix that coils more tightly than normal and presents a series of leucines on one of its faces. These can form contacts with the leucines of the zipper on a second protein, forming the dimer (*Figure 9.16*). A second dimerization domain is, rather unfortunately, called the **helix–loop–helix** motif, which is distinct from, and should not be confused with, the helix–turn–helix DNA-binding motif.

9.2 DNA–Protein Interactions During Transcription Initiation

Now that we have established that DNA–protein interactions are the key to understanding the initiation of transcription, we can move on to begin our examination of

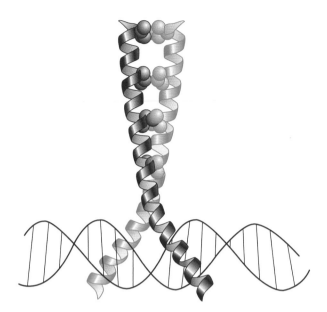

Figure 9.16   A leucine zipper.

This is a bZIP type of leucine zipper. The blue and green structures are parts of different proteins. Each set of spheres represents the R -group of a leucine amino acid. Leucines in the two helices associate with one another via hydrophobic interactions (see Box 3.3, page 82) to hold the two proteins together in a dimer. In this example, the dimerization helices are extended to form a pair of basic-domain DNA-binding motifs, shown making contacts in the major groove. Reprinted from *DNA–Protein Interactions* by Andrew Travers, published by Chapman & Hall, 1993. Reprinted with kind permission of A. Travers.

the events involved in the assembly of the initiation complex. We will do this in two stages. First, we will study the DNA–protein interactions that are involved in transcription initiation. Then, in Section 9.3, we will investigate how assembly of the initiation complex, and its ability to initiate transcription, can be controlled by various additional proteins that respond to stimuli from inside or outside the cell and ensure that the correct genes are transcribed at the appropriate times.

9.2.1 RNA polymerases

In Section 3.2.2 we learnt that the enzymes responsible for transcription of DNA into RNA are called DNA-dependent RNA polymerases. Transcription of eukaryotic nuclear genes requires three different RNA polymerases: **RNA polymerase I**, **RNA polymerase II** and **RNA polymerase III**. Each is a multi-subunit protein (8–12 subunits) with a molecular mass in excess of 500 kDa. Structurally, these polymerases are quite similar to one another, the three largest subunits being closely related and some of the smaller ones being shared by more than one enzyme; functionally, however, they are quite distinct. Each works on a different set of genes, with no

Table 9.3   Functions of the three eukaryotic nuclear RNA polymerases

Polymerase	Genes transcribed*
RNA polymerase I	28S, 5.8S and 18S ribosomal RNA (rRNA) genes
RNA polymerase II	Protein-coding genes; most small nuclear RNA (snRNA) genes
RNA polymerase III	Genes for transfer RNAs (tRNA), 5S rRNA, U6-snRNA, small nucleolar (sno) RNAs, small cytoplasmic (sc) RNA

*See Section 3.2.1 for descriptions of these RNA types.

interchangeability (*Table 9.3*). Most research attention has been directed at RNA polymerase II, as this is the one that transcribes genes that code for proteins. It also works on a set of genes specifying the small nuclear RNAs that are involved in RNA processing. RNA polymerase III transcribes other genes for small RNAs, including those for transfer RNAs (tRNAs). RNA polymerase I transcribes the multicopy repeat units containing the 28S, 5.8S and 18S rRNA genes. The functions of all these RNAs were summarized in Section 3.2.1 and are described in detail in Chapters 10 and 11.

Archaea possess a single RNA polymerase that is very similar to the eukaryotic enzymes (Bult *et al.*, 1996). But this is not typical of the prokaryotes in general because the bacterial RNA polymerase is very different, consisting of just five subunits, described as $\alpha_2\beta\beta'\sigma$ (two α subunits, one each of β and the related β', and one of σ). The α, β and β' subunits are equivalent to the three largest subunits of the eukaryotic RNA polymerases, but the σ subunit has its own special properties, both in terms of its structure and, as we will see in the next section, its function.

9.2.2 Recognition sequences for transcription initiation

It is essential that transcription initiation complexes are constructed at the correct positions on DNA molecules. These positions are marked by target sequences that are recognized either by the RNA polymerase itself or by a DNA-binding protein which, once attached to the DNA, forms a platform to which the RNA polymerase binds (see *Figure 3.6*, page 76).

Bacterial RNA polymerases bind to promoter sequences

In bacteria, the target sequence for RNA polymerase attachment is called the **promoter**. This term was first used by geneticists in 1964 to describe the function of a locus immediately upstream of the three genes in the lactose operon (*Figure 9.17*). When this locus was inactivated by mutation, the genes in the operon were not expressed; the locus therefore appeared to *promote* expression of the genes. We now know that this is because the locus is the binding site for the RNA polymerase that transcribes the operon.

The sequences that make up the *Escherichia coli* promoter were first identified by comparing the regions upstream of over 100 genes. It was assumed that promoter sequences would be very similar for all genes and so should be recognizable when the upstream regions were compared. These analyses showed that the *E. coli* promoter consists of two segments, both of six nucleotides, described as follows (see *Figure 9.17*):

$$\begin{array}{ll} -35\ \text{box} & 5'\text{–TTGACA–}3' \\ -10\ \text{box} & 5'\text{–TATAAT–}3' \end{array}$$

These are consensus sequences and so describe the 'average' of all promoter sequences in *E. coli*; the actual sequences upstream of any particular gene might be slightly different (*Table 9.4*). The names of the boxes indicate their positions relative to the point at which transcription begins. The nucleotide at this point is labeled '+1' and is anything between 20 and 600 nucleotides upstream of the start of the coding region of the gene.

Box 9.3: Mitochondrial and chloroplast RNA polymerases

The RNA polymerases that transcribe organelle genes are unlike their counterparts in the nucleus, reflecting the bacterial origins of mitochondria and chloroplasts (Section 2.2.2).

▪ The mitochondrial RNA polymerase consists of a single subunit with a molecular mass of 140 kDa. Interestingly, this enzyme is more closely related to the RNA polymerases of certain bacteriophages than it is to the standard bacterial version. The polymerase works in conjunction with at least two additional proteins, one of which is a transcription factor – called mtTFI in mammals and ABF2 in *Saccharomyces cerevisiae* – which binds to mitochondrial promoters and enhances initiation of transcription. These promoters are variable in sequence but most contain an AT-rich region.

▪ Chloroplast RNA polymerases appear to be much more similar to the bacterial enzymes and their promoters follow the pattern described for *Escherichia coli*, consisting of two boxes at the −35 and −10 positions relative to the start point for transcription.

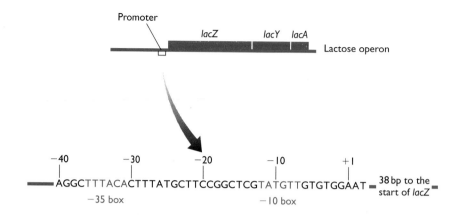

Figure 9.17   The promoter for the lactose operon of *Escherichia coli*.

The promoter is located immediately upstream of *lacZ*, the first gene in the operon. The DNA sequence shows the positions of the −35 and −10 boxes, the two distinct sequence components of the promoter. Compare these sequences with the consensus sequences described in the text. See also *Table 9.4*. For more information on the lactose operon see *Figure 2.20A*, page 55.

The spacing between the two boxes is important because it places the two motifs on the same face of the double helix, facilitating their interaction with the DNA-binding component of the RNA polymerase (Section 9.2.3).

Eukaryotic promoters are more complex

In eukaryotes, the term 'promoter' is used to describe all the sequences that are important in initiation of transcription of a gene. For some genes these sequences can be numerous and diverse in their functions, including not only the **core promoter**, sometimes called the **basal promoter**, which is the site at which the initiation complex is assembled, but also one or more **upstream promoter elements** which, as their name implies, lie upstream of the core promoter. Assembly of the initiation complex on the core promoter can usually occur in the absence of the upstream elements, but only in an inefficient way. This indicates that the proteins that bind to the upstream elements include at least some that are activators of transcription, and which therefore 'promote' gene expression. Inclusion of these sequences in the 'promoter' is therefore justified.

Each of the three types of eukaryotic RNA polymerase recognizes a different type of promoter sequence; indeed, it is the difference between the promoters that defines which genes are transcribed by which polymerases. The details for vertebrates are as follows (*Figure 9.18*):

■ RNA polymerase I promoters consist of a core promoter spanning the transcription start point, between nucleotides −45 and +20, and an **upstream control element** about 100 bp further upstream.

■ RNA polymerase II promoters are variable and can stretch for several kilobases upstream of the transcription start site. The core promoter consists of two segments: the −25 or **TATA box** (consensus 5'–TATAWAW–3', where W is A or T) and the **initiator (Inr) sequence** (consensus 5'–YYCARR–3', where Y is C or T, and R is A or G) located around nucleotide +1. Some genes transcribed by RNA polymerase II have only one of these two components of the core promoter, and some, surprisingly, have neither. The latter are called 'null' genes. They are still transcribed, possibly through interactions between the RNA polymerase and a sequence called MED-1 which lies within the gene (Novina and Roy, 1996), although the start position for transcription is more variable than for a gene with a TATA and/or Inr sequence. As well as the core promoter, genes recognized by RNA polymerase II have various upstream promoter elements, the functions of which are described in Section 9.3.2.

■ RNA polymerase III promoters are unusual in that they are located within the genes whose transcription they promote. These promoters are variable,

Table 9.4   Sequences of *Escherichia coli* promoters

Promoter	Sequence	
	−35 box	**−10 box**
Consensus	5'–TTGACA–3'	5'–TATAAT–3'
Lactose operon	5'–TTTACA–3'	5'–TATGTT–3'
Tryptophan operon	5'–TTGACA–3'	5'–TTAACT–3'

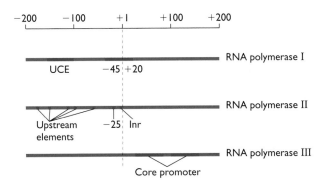

Figure 9.18 Structures of eukaryotic promoters.

Promoter regions are indicated in blue. The RNA polymerase III promoter structure refers to the 5S rRNA genes. Other genes transcribed by RNA polymerase III (see *Table 9.3*) have different promoter structures, including some with a TATA box and upstream elements similar to an RNA polymerase II promoter. See the text for more details. Abbreviations: Inr, initiator sequence of the RNA polymerase II promoter; UCE, upstream control element of the RNA polymerase I promoter.

falling into at least three categories. Usually the core promoter spans 50–100 bp and comprises two sequence boxes. One category of RNA polymerase III promoter is similar to those for RNA polymerase II, having a TATA box and a range of upstream promoter elements. Interestingly, this arrangement is seen with the U6 gene, which is one of a family of genes for small nuclear RNAs, all the other members of which are transcribed by RNA polymerase II.

9.2.3 Assembly of the transcription initiation complex

In a general sense, initiation of transcription operates along the same lines with each of the four types of RNA polymerase that we have been considering (*Figure 9.19*). The bacterial polymerase and the three eukaryotic enzymes all begin by attaching, directly or via accessory proteins, to their promoter or core promoter sequences. Next this **closed promoter complex** is converted into an **open promoter complex** by breakage of a limited number of base pairs around the transcription initiation site. Finally, the RNA polymerase moves away from the promoter. This last step is more complicated than it might appear because some attempts by the polymerase to achieve **promoter clearance** are unsuccessful and lead to truncated transcripts that are degraded soon after they are synthesized. The true completion of the initiation stage of transcription is therefore the establishment of a stable transcription complex that is actively transcribing the gene to which it is attached.

Although the scheme shown in *Figure 9.19* is correct in outline for all four polymerases, the details are different for each one. We will begin with the more straightforward

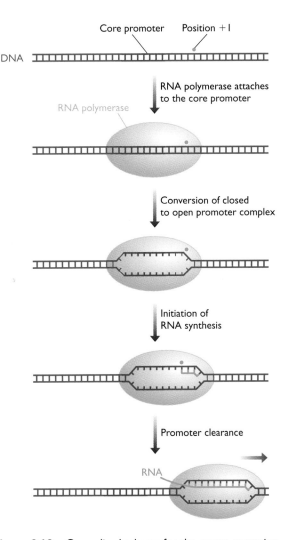

Figure 9.19 Generalized scheme for the events occurring during initiation of transcription.

The core promoter is shown in blue and the transcription initiation site is indicated by a green dot. After RNA polymerase attachment, the closed complex is converted into the open complex by breakage of base pairs within a short region of the DNA double helix. RNA synthesis begins but successful initiation is not achieved until the polymerase moves away from the promoter region.

events occurring in *E. coli* and other bacteria, and then move on to the ramifications of initiation in eukaryotes.

Transcription initiation in E. coli

In *E. coli*, a direct contact is formed between the promoter and RNA polymerase. The sequence specificity of the polymerase resides in its σ subunit: the 'core enzyme', which lacks this component, can only make loose and non-specific attachments to DNA.

Mutational studies of *E. coli* promoters have shown that changes to the sequence of the −35 box affect the ability of RNA polymerase to bind, whereas changes to the −10 box affect the conversion of the closed promoter complex into the open form. These results led to the model for *E. coli* initiation shown in *Figure 9.20*, where recognition of the promoter occurs by an interaction between the σ subunit and the −35 box, forming a closed promoter complex in which the RNA polymerase spans some 60 bp from upstream of the −35 box to downstream of the −10 box. This is followed by breaking of the base pairs within the −10 box to produce the open complex. The model is consistent with the fact that the −10 boxes of different promoters are comprised mainly or entirely of A–T base pairs, which are weaker than G–C pairs, being linked by just two hydrogen bonds as opposed to three (see *Figure 1.11B*, page 15).

Opening up of the helix involves contacts between the polymerase and the non-template strand (i.e. the one that is not copied into RNA), again with the σ subunit playing a central role (Marr and Roberts, 1997). However, the σ

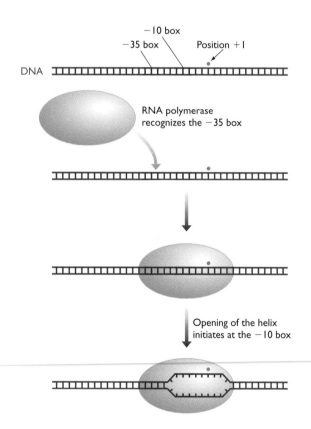

Figure 9.20   Initiation of transcription in *Escherichia coli.*

The *E. coli* RNA polymerase recognizes the −35 box as its binding sequence. After attachment to the DNA, the transition from closed to open complex is initiated by breakage of base pairs in the AT-rich −10 box. Note that although the polymerase is shown as a single structure it is the σ subunit that possesses sequence-specific DNA binding activity and which therefore recognizes the −35 sequence. For subsequent events leading to promoter clearance, refer to *Figure 9.19*.

subunit is not all-important because it dissociates soon after initiation is complete, converting the holoenzyme to the core enzyme which carries out the elongation phase of transcription (Section 10.1.1).

Transcription initiation with RNA polymerase II

How does the easily understandable series of events occurring in *E. coli* compare with the equivalent processes in eukaryotes? RNA polymerase II will show us that eukaryotic initiation involves more proteins and has added complexities.

The first difference between initiation of transcription in *E. coli* and eukaryotes is that eukaryotic polymerases do not directly recognize their core promoter sequences. For genes transcribed by RNA polymerase II, the initial contact is made by the **general transcription factor (GTF)** TFIID, which is a complex made up of the **TATA-binding protein (TBP)** and at least 12 **TBP-associated factors** or **TAFs**. TBP is a sequence-specific protein that binds to DNA via its unusual TBP domain (Section 9.1.4) which makes contact with the minor groove in the region of the TATA box. X-ray crystallography studies of TBP show that it has a saddle-like shape that wraps partially around the double helix (Chasman *et al.*, 1993), forming a platform onto which the remainder of the initiation complex can be assembled. The TAFs are intriguing proteins that appear to play a variety of roles during initiation of transcription and also during other events that involve assembly of multiprotein complexes onto the genome. Five of the yeast TAFs are also present in SAGA, one of the histone acetyltransferase complexes that we met in Section 8.2.1 (Grant *et al.*, 1998), and TAFs have also been implicated in control of the cell cycle in various eukaryotes (Green, 2000) and in regulation of the developmental changes that result in formation of gametes in animals (Verrijzer, 2001). During transcription, TAFs assist in attachment of TFIID to the TATA box and, in conjunction

---

Box 9.4: Initiation of transcription in the archaea

Initiation of transcription is one of the key areas in which the archaea differ from the bacteria, emphasizing the major distinction between these two types of prokaryote. The archaeal RNA polymerase is made of 11 subunits and closely resembles the RNA polymerases of eukaryotes. Promoters are recognized by a TATA-binding protein, and RNA polymerase attachment is mediated by a transcription factor similar to the eukaryotic TFIIB. Archaeal homologs of the other eukaryotic general transcription factors have not been identified, and searches of the completed archaeal genome sequences suggest that they may not be present.

with other proteins called **TAF- and initiator-dependent cofactors (TICs)**, possibly also participate in recognition of the Inr sequence, especially at those promoters that lack a TATA box. A clue as to how TAFs carry out their multifarious roles has been provided by structural studies which have shown that at least three of them contain a histone fold – a DNA-binding domain which, as the name suggests, is also present in histone proteins (*Table 9.2*, page 249; Research Briefing 9.1). It has been proposed that these TAFs might be able to form a DNA-binding structure resembling a nucleosome (Burley and Roeder, 1996), but this idea may not be entirely correct because the TAFs lack certain amino acids that are looked on as essential for stabilizing the contacts between real nucleosomes and DNA (Gangloff *et al.*, 2001).

After TFIID has attached to the core promoter, the **pre-initiation complex (PIC)** is formed by attachment of the remaining GTFs (*Table 9.5*). Test-tube experiments suggest that these GTFs bind to the complex in the order TFIIA, TFIIB, TFIIF/RNA polymerase II, TFIIE and TFIIH (*Figure 9.21*) but it is now thought that *in vivo* assembly involves a more complex set of interactions than indicated by this step-by-step sequence (Lee and Young, 2000). Within the overall process, three events are particularly important:

1. Attachment of TBP induces formation of a bend in the DNA in the region of the TATA box.
2. The bend provides a recognition structure for TFIIB, which ensures correct positioning of RNA polymerase II relative to the transcription start site.
3. The disruption to the base pairing needed to form the open promoter complex is brought about by TFIIH (Kim *et al*, 2000).

The final step in assembly of the initiation complex is the addition of phosphate groups to the **C-terminal domain (CTD)** of the largest subunit of RNA polymerase II. In mammals, this domain consists of 52 repeats of the seven-amino-acid sequence Tyr–Ser–Pro–Thr–Ser–Pro–Ser. Two of the three serines in each repeat unit can be modified by addition of a phosphate group, causing a substantial change in the ionic properties of the polymerase. Once phosphorylated, the polymerase is able to leave the pre-initiation complex and begin synthesizing RNA. Phosphorylation might be carried out by TFIIH, which has the appropriate protein kinase capability, or it might be the function of the **mediator** (Section 9.3.2), which transduces signals from activator proteins that regulate expression of individual genes (Lee and Young, 2000). After departure of the polymerase, at least some of the GTFs detach from the core promoter, but TFIID, TFIIA and TFIIH remain, enabling re-initiation to occur without the need to rebuild the entire assembly from the beginning (Yudkovsky *et al.*, 2000). Re-initiation is therefore a more rapid process than primary initiation, which means that once a gene is switched on, transcripts can be initiated from its promoter with relative ease until such a time as a new set of signals switches the gene off.

Transcription initiation with RNA polymerases I and III

Initiation of transcription at RNA polymerase I and III promoters involves similar events to those seen with RNA polymerase II, but the details are different. One of the most striking similarities is that TBP, first identified as the key sequence-specific DNA-binding component of the RNA polymerase II pre-initiation complex, is also involved in initiation of transcription by the two other eukaryotic RNA polymerases.

The RNA polymerase I initiation complex involves four protein complexes in addition to the polymerase itself. One of these, UBF, is a dimer of identical proteins that interacts with both the core promoter and the upstream control element (see *Figure 9.18*). UBF is another protein which, like some of the RNA polymerase II TAFs, resembles a histone and may form a nucleosome-like structure in the promoter region (Wolffe, 1994). A second protein complex, called SL1 in humans and TIF-IB in mice, contains TBP and, together with UBF, directs RNA polymerase I and the last two complexes, TIF-IA and TIF-IC, to the promoter. Originally it was thought that the

Table 9.5 Functions of human general transcription factors (GTFs)

GTF	Function
TFIID (TBP component)	Recognition of the TATA box and possibly Inr sequence; forms a platform for TFIIB binding
TFIID (TAFs)	Recognition of the core promoter; regulation of TBP binding
TFIIA	Stabilizes TBP and TAF binding
TFIIB	Intermediate in recruitment of RNA polymerase II; influences selection of the start point for transcription
TFIIF	Recruitment of RNA polymerase II
TFIIE	Intermediate in recruitment of TFIIH; modulates the various activities of TFIIH
TFIIH	Helicase activity responsible for the transition from the closed to open promoter complex; possibly influences promoter clearance by phosphorylation of the C-terminal domain of the largest subunit of RNA polymerase II

Based on Lee and Young (2000), which is recommended for further details concerning the functions of the GTFs.

RESEARCH

9.1

BRIEFING

Similarities between TFIID and the histone core octamer

An intriguing insight into the interaction between TFIID and the RNA polymerase II promoter was provided by the discovery that several TAFs have structural similarities with histones.

A key step in initiation of transcription by RNA polymerase II is attachment of the general transcription factor TFIID to the core promoter. The resulting structure forms the platform onto which the pre-initiation complex is constructed. Intuitively, one anticipates that this platform must be a stable structure in order for the bulky pre-initiation complex to be positioned correctly so that productive initiation of transcription can occur. To understand the nature of the attachment, attention has been focused on the TBP-associated factors (TAFs) that, together with the TATA-binding protein, make up the TFIID protein complex.

Homology searching reveals similarities between TAFs and histones

The first link between TAFs and histones was established by homology searching (Section 7.2.1). In 1993–94, a number of TAF genes were cloned, enabling nucleotide sequences to be obtained. The nucleotide sequences were translated into the amino acid sequences of the TAF proteins, and the amino acid sequences used to search the databases for interesting homologies with other proteins. Weak similarities were discovered with the histones of the core octamer of the nucleosome (Section 2.2.1). Part of one of the sequence alignments, between the central regions of TAF$_{II}$62 and histone H4, is shown below. In this central region of 55 amino acids there are 15 positions at which the same amino acid is present in both sequences, and a further 15 positions where the amino acids are 'similar', these being ones which, although different, have equivalent chemical properties. The same kind of pattern is seen when TAF$_{II}$42 is compared with histone H3. The degree of similarity between these two sets of proteins is not so great that a structural and/or functional relationship can immediately be inferred, but it is sufficiently striking to suggest that more detailed comparisons between TAFs and histones should be made.

These more detailed comparisons were carried out by X-ray crystallographic analyses (Section 9.1.3) of TAF$_{II}$42 and TAF$_{II}$62. The results were remarkable, both TAFs being shown to possess a typical histone fold (*Table 9.2*, page 249), a non-sequence-specific DNA-binding motif, comprising a long α-helix flanked on either side by two shorter α-helices, that is a characteristic feature of histones. Even more striking was the discovery that in the complex formed between TAF$_{II}$42

and TAF$_{II}$62, the two histone folds are orientated in almost exactly the same way as in the histone H3/H4 dimer found in the central tetramer of the nucleosome core particle:

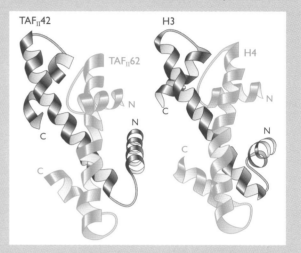

Reprinted with permission from Xie, et al., Nature, 380, 316–322. Copyright 1996 Macmillan Magazines Limited. Figure courtesy of SK Burley.

Can these results be taken as evidence that TFIID forms a nucleosome-like structure at the core promoter, this pseudonucleosome acting as the platform for assembly of the pre-initiation complex? The conclusion is attractive but premature. More research is needed to determine if the similarities between TAFs and histones extend to the details of the attachment between each pre-initiation complex and its DNA target. An alternative is that the similarities merely reflect an equivalence in the protein–protein interactions within the two complexes, and are not directly relevant to the DNA attachment.

References

Kokubo T, Gong D-W, Wootton JC, Horikoshi M, Roeder RG and Nakatani Y (1994) Molecular cloning of *Drosophila* TFIID subunits. *Nature*, **367**, 484–487.

Xie X, Kokubo T, Cohen SL, et al. (1996) Structural similarity between TAFs and the heterotetrameric core of the histone octamer. *Nature*, **380**, 316–322.

TAF$_{II}$62 IAESIGVGSLSDDAAKELAEDVSIKLKRIVQDAAKFMNHAKRQKLSVRDIDMSLK

Histone H4 LARRGGVKRISGLIYEETRGVLKVFLENVIRDAVTYTEHAKRKTVTAMDVVYALK

KEY

| Amino acid identity

: Amino acid similarity

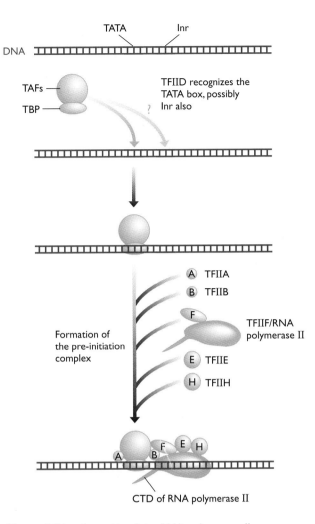

TFIID recognizes the TATA box, possibly Inr also

Formation of the pre-initiation complex

(A) TFIIA
(B) TFIIB
(F) TFIIF/RNA polymerase II
(E) TFIIE
(H) TFIIH

CTD of RNA polymerase II

Figure 9.21 Assembly of the RNA polymerase II pre-initiation complex.

The first step in assembly of the pre-initiation complex is recognition of the TATA box and possibly the Inr sequence by the TATA-binding protein (TBP), in conjunction with the TBP-associated factors (TAFs). Test-tube experiments suggest that the other components of the pre-intiation complex assemble in the order TFIIA, TFIIB, TFIIF/RNA polymerase II, TFIIE and TFIIH. The drawing is purely schematic with regard to the sizes and shapes of the components and their relative positioning within the pre-initiation complex, although account is taken of the protein–protein contacts that are known to occur. The C-terminal domain (CTD) of the largest subunit of RNA polymerase II, which must be phosphorylated before the polymerase can leave the promoter, is depicted as a spike projecting from the bottom of the complex.

initiation complex was built up in a stepwise fashion, but recent results suggest that RNA polymerase I binds the four protein complexes before promoter recognition, the entire assembly attaching to the DNA in a single step (Seither *et al.*, 1998).

RNA polymerase III promoters are variable in structure (see *Figure 9.18*) and this is reflected by a non-uniformity

of the processes through which they are recognized. Initiation at the different categories of RNA polymerase III promoter requires different sets of GTFs, but each type of initiation process involves TFIIIB, one of whose subunits is TBP. With promoters of the type seen with the U6 gene, which contain a TATA sequence, TBP probably binds directly to the DNA. At other RNA polymerase III promoters, which have no TATA sequence, binding is probably via a second protein, the latter making the direct DNA contact.

9.3 Regulation of Transcription Initiation

As we progress through the next few chapters we will encounter a number of strategies that organisms use to regulate expression of individual genes. We will discover that virtually every step in the pathway from genome to proteome is subject to some degree of control. Of all these regulatory systems, it appears that transcription initiation is the stage at which the critical controls over the expression of individual genes (i.e. those controls that have greatest impact on the biochemical properties of the cell) are exerted. This is perfectly understandable. It makes sense that transcription initiation, being the first step in genome expression, should be the stage at which 'primary' regulation occurs, this being the level of regulation that determines which genes are expressed. Later steps in the pathway might be expected to respond to 'secondary' regulation, the function of which is not to switch genes on or off but to modulate expression by making small changes to the rate at which the protein product is synthesized, or possibly by changing the nature of the product in some way (*Figure 9.22*).

In Chapter 8 we looked at how chromatin structure can influence gene expression by controlling the accessibility of promoter sequences to RNA polymerase and its associated proteins. This is just one way in which initiation of transcription can be regulated. To obtain a broader picture we will establish some general principles with bacteria, and then examine the events in eukaryotes.

9.3.1 Strategies for controlling transcription initiation in bacteria

In bacteria such as *E. coli*, we recognize two distinct ways in which transcription initiation is controlled:

- *Constitutive control*, which depends on the structure of the promoter;
- *Regulatory control*, which depends on the influence of regulatory proteins.

Promoter structure determines the basal level of transcription initiation

The consensus sequence for the *E. coli* promoter (Section 9.2.2) is quite variable, with a range of different motifs

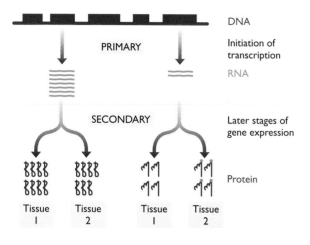

Figure 9.22 Primary and secondary levels of gene regulation.

According to this scheme, 'primary' regulation of genome expression occurs at the level of transcription initiation, this step determining which genes are expressed in a particular cell at a particular time and setting the relative rates of expression of those genes that are switched on. 'Secondary' regulation involves all steps in the gene expression pathway after transcription initiation, and serves to modulate the amount of protein that is synthesized or to change the nature of the protein in some way, for example by chemical modification.

being permissible at both the −35 and −10 boxes (see Table 9.4, page 256). These variations, together with less well-defined sequence features around the transcription start site and in the first 50 or so nucleotides of the transcription unit, affect the efficiency of the promoter. Efficiency is defined as the number of productive initiations that are promoted per second, a productive initiation being one that results in the RNA polymerase clearing the promoter and beginning synthesis of a full-length transcript. The exact way in which the sequence of the promoter affects initiation is not known, but from our discussion of the events involved in transcription initiation (Section 9.2.3) we might, intuitively, expect that the precise sequence of the −35 box would influence recognition by the σ subunit and hence the rate of attachment of RNA polymerase, that the transition from the closed to open promoter complex might be dependent on the sequence of the −10 box, and that the frequency of abortive initiations (ones that terminate before they progress very far into the transcription unit) might be influenced by the sequence at, and immediately downstream of, nucleotide +1. All this is speculation but it is a sound 'working hypothesis'. What is clear is that different promoters vary 1000 fold in their efficiencies, the most efficient promoters (called **strong promoters**) directing 1000 times as many productive initiations as the weakest promoters. We refer to these as differences in the **basal rate** of transcription initiation.

Note that the basal rate of transcription initiation for a gene is preprogrammed by the sequence of its promoter and so, under normal circumstances, cannot be changed. It could be changed by a mutation that alters a critical nucleotide in the promoter, and undoubtedly this happens from time to time, but it is not something that the bacterium has control over. The bacterium can, however, determine which promoter sequences are favored by changing the σ subunit of its RNA polymerase. The σ subunit is the part of the polymerase that has the sequence-specific DNA-binding capability (Section 9.2.3), so replacing one version of this subunit with a different version with a slightly different DNA-binding motif, and hence an altered sequence specificity, would result in a different set of promoters being recognized. In E. coli, the standard σ subunit, which recognizes the consensus promoter sequence shown on page 255 and hence directs transcription of most genes, is called σ^{70} (its molecular mass is approximately 70 kDa). E. coli also has a second σ subunit, σ^{32}, which is made when the bacterium is exposed to a heat shock. During a heat shock, E. coli, in common with other organisms, switches on a set of genes coding for special proteins that help the bacterium withstand the stress (Figure 9.23). These genes have special promoter sequences, ones specifically recognized by the σ^{32} subunit. The bacterium is therefore able to switch on a whole range of different genes by making one simple alteration to the structure of its RNA polymerase. This system is common in bacteria: for example, Klebsiella pneumoniae uses it to control expression of genes involved in nitrogen fixation, this time with the σ^{54} subunit, and Bacillus species use a whole range of different σ subunits to switch on and off groups of genes during the changeover from normal growth to formation of spores (Section 12.3.1).

Regulatory control over bacterial transcription initiation

Promoter structure determines the basal level of transcription initiation for a bacterial gene but, with the exception of alternative σ subunits, does not provide any general means by which the expression of the gene can respond to changes in the environment or to the biochemical requirements of the cell. Other types of regulatory control are needed.

The foundation of our understanding of regulatory control over transcription initiation in bacteria was laid in the early 1960s by François Jacob, Jacques Monod, and other geneticists who studied the lactose operon and other model systems (Burian and Gayon, 1999). We have already seen how this work led to discovery of the promoter for the lactose operon (Section 9.2.2). It also resulted in identification of the **operator**, a region adjacent to the promoter and which regulates initiation of transcription of the operon (Figure 9.24A). The original model envisaged that a DNA-binding protein – the **lactose repressor** – attached to the operator and prevented the RNA polymerase from binding to the promoter, simply by denying it access to the relevant segment of DNA

(A) An *E. coli* heat shock gene

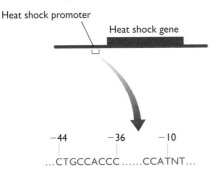

(B) Recognition by the σ³² subunit

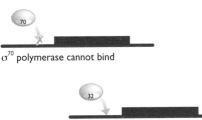

σ⁷⁰ polymerase cannot bind

σ³² polymerase binds to the heat shock promoter

Figure 9.23   Recognition of an *Escherichia coli* heat shock gene by the σ³² subunit.

(A) The sequence of the heat-shock promoter is different from that of the normal *E. coli* promoter (compare with *Table 9.4*, page 256). (B) The heat-shock promoter is not recognized by the normal *E. coli* RNA polymerase containing the σ⁷⁰ subunit, but is recognized by the σ³² RNA polymerase that is active during heat shock. Abbreviation: N, any nucleotide. For more details of the use of novel σ factors by bacteria see Section 12.3.1.

(*Figure 9.24B*). Whether the repressor binds depends on the presence in the cell of allolactose, an isomer of lactose, the latter being the substrate for the biochemical pathway carried out by the enzymes coded by the three genes in the operon. Allolactose is an **inducer** of the lactose operon. When allolactose is present it binds to the lactose repressor, causing a slight structural change which prevents the HTH motifs of the repressor from recognizing the operator as a DNA-binding site. The allolactose–repressor complex therefore cannot bind to the operator, enabling the RNA polymerase to gain access to the promoter. When the supply of lactose is used up and there is no allolactose left to bind to the repressor, the repressor re-attaches to the operator and prevents transcription. The operon is therefore expressed only when the enzymes coded by the operon are needed.

Most of the original scheme for regulation of the lactose operon has been confirmed by DNA sequencing of the control region and by structural studies of the repressor bound to its operator. The one complication has been the discovery that the repressor has three potential binding sites at nucleotide positions −82, +11 and +412, and

attachment at only one of these, +11, would be expected to prevent access of the polymerase to the promoter. The repressor is a tetramer of four identical proteins which work in pairs to attach to a single operator, so it is possible that the repressor has the capacity to bind to two of the three operator sites at once. It is also possible that the repressor can bind to an operator sequence in such a way that it does not block attachment of the polymerase to the promoter, but does prevent a later step in initiation, such as formation of the open promoter complex.

The lactose operon illustrates the basic principle of regulatory control of transcription initiation: attachment of a DNA-binding protein to its specific recognition site can influence the events involved in assembly of the transcription initiation complex and/or initiation of productive RNA synthesis by an RNA polymerase. Several variations on this theme are seen with other bacterial genes:

■ Some repressors respond not to an inducer but to a **co-repressor**. An example is provided by the tryptophan operon of *E. coli*, which codes for a set of genes involved in synthesis of tryptophan (see *Figure 2.20B*, page 55). In contrast to the lactose operon, the regulatory molecule for the tryptophan operon is not a substrate for the relevant biochemical pathway, but the product, tryptophan itself (*Figure 9.25*). Only when tryptophan is attached to the tryptophan repressor can the latter bind to the operator. The

Box 9.5: *Cis* and *trans*

'*Cis*' and '*trans*' are two important terms relevant to the genetic study of gene regulation in bacteria and other organisms.

■ A locus is *cis*-acting on a second locus if it must be on the same DNA molecule in order to have an effect. The operator is a *cis*-acting element because it works only when physically attached to the gene whose expression it regulates.

■ A locus is *trans*-acting if it can affect a second locus even when on a different DNA molecule. The gene for the lactose repressor (*lacI*) is *trans*-acting because it can regulate expression of the lactose operon even when removed from the *Escherichia coli* chromosome and placed on a plasmid.

To a molecular biologist, a *cis*-acting regulatory element is usually a target site for a DNA-binding protein, upstream of the gene whose expression is being regulated. A *trans*-acting element is the regulatory protein itself, which can diffuse through the cell from its site of synthesis to its DNA-binding site.

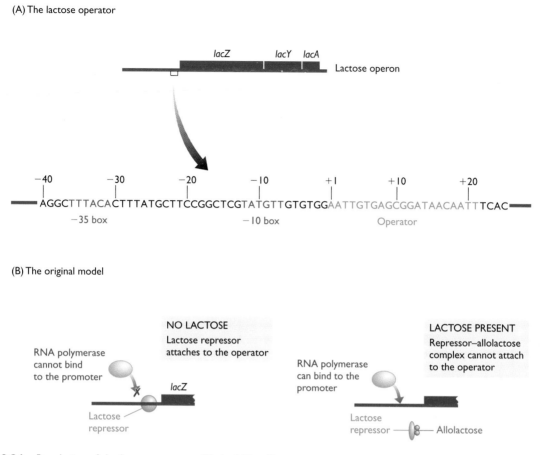

(A) The lactose operator

(B) The original model

Figure 9.24   Regulation of the lactose operon of *Escherichia coli.*

(A) The operator sequence lies immediately downstream of the promoter for the lactose operon. (B) In the original model for lactose regulation, the lactose repressor is looked on as a simple blocking device that binds to the operator and prevents the RNA polymerase gaining access to the promoter. The three genes in the operon are therefore switched off. This is the situation in the absence of lactose, although transcription is not completely blocked because the repressor occasionally detaches, allowing a few transcripts to be made. Because of this basal level of transcription, the bacterium always possesses a few copies of each of the three enzymes coded by the operon (see *Figure 2.20A,* page 256), probably less than five of each. This means that when the bacterium encounters a source of lactose it is able to transport a few molecules into the cell and split these into glucose and galactose. An intermediate in this reaction is allolactose, an isomer of lactose, which induces expression of the lactose operon by binding to the repressor, causing a change in the conformation of the latter so it is no longer able to attach to the operator. This allows the RNA polymerase to bind to the promoter and transcribe the three genes. When fully induced, approximately 5000 copies of each protein product are present in the cell. When the lactose supply is used up and allolactose is no longer present, the repressor re-attaches to the operator and the operon is switched off. Note that the shapes of the repressor and polymerase structures shown here are purely schematic.

tryptophan operon is therefore switched off in the presence of tryptophan, and switched on when tryptophan is needed.

■   Some DNA-binding proteins are **activators** rather than repressors of transcription initiation. The best studied example in *E. coli* is the **catabolite activator protein**, which binds to sites upstream of several operons, including the lactose operon, and increases the efficiency of transcript initiation, probably by forming a direct contact with the RNA polymerase. The biological role of the catabolite activator protein is described in Section 12.1.1.

■   Other DNA-binding proteins work singly or together to increase or repress transcription of genes to which they are not closely linked. These **enhancers** and **silencers** are not common in bacteria but a few examples are known, including an enhancer that acts on the *E. coli* heat-shock genes whose promoters are recognized by the σ^{32} version of the RNA polymerase. Because they are so far from the genes that they control, they can only form a contact with the RNA polymerase if the DNA forms a loop. A characteristic feature is that a single enhancer or silencer can control expression of more than one gene.

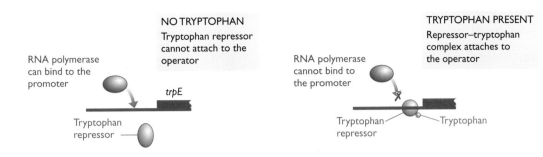

Figure 9.25   Regulation of the tryptophan operon of *Escherichia coli*.

Regulation occurs via a repressor–operator system in a similar way to that described for the lactose operon (*Figure 9.24*) but with the difference that the operon is repressed by the regulatory molecule, tryptophan, which is the product of the biochemical pathway specified by the genes in the operon (see *Figure 2.20B*, page 256). When tryptophan is present, and so does not need to be synthesized, the operon is switched off because the repressor–tryptophan complex binds to the operator. In the absence of tryptophan the repressor cannot attach to the operator, and the operon is expressed.

9.3.2 Control of transcription initiation in eukaryotes

With bacteria, it is possible to make a clear distinction between constitutive and regulatory forms of control over transcription initiation. The former depends on promoter structure and determines the basal rate of transcription initiation; the latter depends on the activity of regulatory proteins and changes the rate of transcription initiation if the basal rate is inappropriate for the prevailing conditions. With eukaryotes, categorization of different types of control system is less easy. This is because of a fundamental difference in transcription initiation between bacteria and eukaryotes. In bacteria, the RNA polymerase has a strong affinity for its promoter and the basal rate of transcription initiation is relatively high for all but the weakest promoters. With most eukaryotic genes, the reverse is true. The RNA polymerase II and III pre-initiation complexes do not assemble efficiently and the basal rate of transcription initiation is therefore very low, regardless of how 'strong' the promoter is. In order to achieve effective initiation, formation of the complex must be activated by additional proteins. Some of these could be defined as 'constitutive' activators, in that they work on many different genes and seem not to respond to any external signals; others could be termed 'regulatory' activators because they target a limited number of genes and do respond to external signals. But there are gray areas between the two types and it is unwise to use this

categorization as anything other than a guide to the types of event that occur.

Activators of eukaryotic transcription initiation

Any protein that stimulates transcription initiation is called an **activator**. Initially it was imagined that all activators were sequence-specific DNA-binding proteins, some recognizing upstream promoter elements and influencing transcription initiation only at the promoter to which these elements are attached, and others targeting sites within enhancers and influencing transcription of several genes at once (*Figure 9.26*). As with bacteria, eukaryotic enhancers can be some distance from their genes; their target specificity is ensured by the presence of insulators at either side of each functional domain, preventing the enhancers within that domain from influencing gene expression in adjacent domains (Section 8.1.2). Whether bound to an upstream promoter element or to a more distant enhancer, the activator, according to the traditional view, stabilizes the pre-initiation complex by making contact with it.

This traditional view still holds for the majority of activators that have been identified but cannot be looked upon as all-encompassing. We have already seen that some proteins that were initially identified as activators are now recognized as components of chromatin modification complexes such as SAGA and Swi/Snf (Section 8.2.1). Other proteins classed as activators influence gene

Figure 9.26   Activators of eukaryotic transcription initiation.

The blue activator is attached to a regulatory module upstream of a gene, and influences transcription initiation only at that single gene. The green activator is attached to a site within an enhancer and is influencing transcription of all three genes.

expression by introducing bends and other distortions into DNA (Thomas and Travers, 2001), possibly as a prelude to chromatin modification, or possibly to bring together proteins attached to non-adjacent sites, enabling the bound factors to work together in a structure that has been called an **enhanceosome**. An example of an activator that works in this way is SRY, which is the primary protein responsible for determining sex in mammals (Wolffe, 1995). Still other activators have no DNA-binding properties and they stimulate transcription simply by forming protein–protein contacts with the pre-initiation complex. As more and more activators are discovered, our appreciation of their diversity will undoubtedly grow (Lee and Young, 2000).

Activators have been looked upon as important in initiation by RNA polymerases II and III, but their role at RNA polymerase I promoters has been less well defined. RNA polymerase I is unusual in that it transcribes just a single set of genes: the multiple copies of the transcription unit containing the 28S, 5.8S and 18S rRNA sequences (Section 3.2.1). These genes are expressed continuously in most cells, but the rate of transcription varies during the cell cycle and is subject to a certain amount of tissue-specific regulation. The regulatory mechanism has not been described in detail but recent research has suggested a role for the RNA polymerase I **termination factor**. This factor, called TTF-1 in mice and Reb1p in *Saccharomyces cerevisiae*, was first identified as an activator of RNA polymerase II transcription. It appears that the termination factor may also activate RNA polymerase I transcription, a binding site for it having been located immediately upstream of the promoter for the rRNA transcription unit (Reeder and Lang, 1997).

Contacts between activators and the pre-initiation complex

A critical feature of the 'traditional' type of activator – those that bind to upstream promoter elements or to enhancers – is the contact that is formed with the pre-initiation complex. The part of the activator that makes this contact is called the **activation domain**. Structural studies have shown that although activation domains are variable, most of them fall into one of three categories:

- **Acidic domains** are ones that are relatively rich in acidic amino acids (aspartic acid and glutamic acid). This is the commonest category of activation domain.
- **Glutamine-rich domains** are often found in activators whose DNA-binding motifs are of the homeodomain or POU type (Section 9.1.4).
- **Proline-rich domains** are less common.

Details of the interaction between activators and the pre-initiation complex were obscure for several years, with apparently conflicting evidence coming from work with different organisms. A number of protein–protein interaction studies had suggested that direct contacts could be made between different activators and various parts of the complex, with TBP, various TAFs, TFIIB, TFIIH and

RNA polymerase II all implicated as partners in different interactions. An alternative possibility was raised when a large protein complex called the **mediator** was identified in yeast. The mediator forms a physical contact between activators and the C-terminal domain of RNA polymerase II (*Figure 9.27*; Thompson *et al.*, 1993; Kim *et al.*, 1994), suggesting that rather than direct interaction between an activator and the pre-initiation complex, the signal is transduced by the mediator. This hypothesis was strengthened when it was shown that the mediator possesses a protein kinase activity that enables it to phosphorylate the CTD of RNA polymerase II, stimulating promoter clearance (Section 9.2.3). The importance of the mediator in yeast transcription initiation was further underlined by the discovery that several of its components were previously looked upon as **coactivators**, proteins that are needed for full activation of the pre-initiation complex but which do not themselves respond directly to any of the external signals that modulate genome expression (as described in the last section in this chapter).

For a few years it appeared that the mediator might not be a common feature of eukaryotic pre-initiation complexes in general, but eventually an equivalent structure was identified in mammalian cells (Kingston, 1999; Malik and Roeder, 2000). Subsequent work has shown that there are several different versions of the mammalian mediator, each one responding to a different, although possibly overlapping, set of activators. Current opinion tends to the view that a mediator is an obligatory component of the RNA polymerase II pre-initiation complex, and that the stimulatory effects of all activators pass through the mediator. The possibility that some activators bypass the mediator and have a direct effect on one or other part of the pre-initiation complex cannot, however, be discounted.

Repressors of eukaryotic transcription initiation

Most of the research on regulation of transcription initiation in eukaryotes has concentrated on activation, partly because the low level of basal initiation occurring at RNA polymerase II and III promoters suggests that the repression of initiation, which is so important in bacteria (Section 9.3.1), is unlikely to play a major part in control of eukaryotic transcription. This view is probably incorrect because a growing number of DNA-binding proteins that repress transcription initiation are being discovered, these proteins binding to upstream promoter

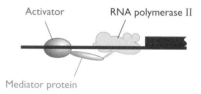

Activator   RNA polymerase II

Mediator protein

Figure 9.27   The role of the mediator.

Box 9.6: The modular structures of RNA polymerase II promoters

The promoter for a gene transcribed by RNA polymerase II can be looked on as a series of modules, each comprising a short sequence of nucleotides and each acting as the binding site for a protein that influences assembly of the transcription initiation complex. Many different genes are transcribed by RNA polymerase II (approximately 35 000 in humans) but there are only a limited number of promoter modules. The expression pattern for a gene is therefore determined not by an individual module but by the combination of modules within its promoter, and possibly by their relative positions. The amount of transcription initiation that occurs is dependent on which modules are occupied by their binding proteins at a particular time.

The modules can be categorized in various ways. One scheme is as follows:

- The **core promoter** modules (Section 9.2.2). The most important of these are the TATA box (consensus 5′–TATAWAW–3′, where W is A or T) and the Inr sequence (consensus 5′–YYCARR–3′, where Y is C or T, and R is A or G).
- **Basal promoter elements** are modules that are present in many promoters and set the basal level of transcription initiation, without responding to any tissue-specific or developmental signals. These include: the **CAAT box** (consensus 5′–GGCCAATCT–3′), recognized by the activators NF-1 and NF-Y; the **GC box** (consensus 5′–GGGCGG–3′) recognized by the Sp1 activator; and the **octamer** module (consensus 5′–ATGCAAAT–3′), recognized by Oct-1.
- **Response modules** are found upstream of various genes and enable transcription initiation to respond to general signals from outside of the cell. Examples are: the cyclic AMP response module CRE (consensus 5′–WCGTCA–3′), recognized by the CREB activator; the heat-shock module (consensus 5′–CTNGAATNTTCTAGA–3′), recognized by HSP70 and other activators; and the serum response module (consensus 5′–CCWWWWWWWGG–3′), recognized by the serum response factor.

- **Cell-specific modules** are located in the promoters of genes that are expressed in just one type of tissue. Examples include: the erythroid module (consensus 5′–WGATAR–3′) which is the binding site for the GATA-1 activator; the pituitary cell module (consensus 5′–ATATTCAT–3′), recognized by Pit-1; the myoblast module (consensus 5′–CAACTGAC–3′), recognized by MyoD; and the lymphoid cell module or κB site (consensus 5′–GGGACTTTCC–3′), recognized by NF-κB. Note that in lymphoid cells the octamer module is recognized by the tissue-specific Oct-2 activator.
- Modules for **developmental regulators** mediate expression of genes that are active at specific developmental stages. Two examples in *Drosophila* are (Section 12.3.3): the Bicoid module (consensus 5′–TCCTAATCCC–3′) and the Antennapedia module (consensus 5′–TAATAATAATAATAA–3′).

The human insulin gene (shown below) illustrates the modular structure of a promoter for RNA polymerase II.

As well as the modules in the region immediately upstream of the gene, the same and other modules can also be contained within enhancers, which are 200–300 bp in length and can be located some distance upstream or downstream of their target gene. A single enhancer can influence transcription initiation of a variety of genes located within a single functional domain (Section 8.1.2). Silencers are similar to enhancers but, as their name suggests, their modules have a negative rather than enhancing influence on transcription initiation.

The modular interpretation of promoter structure is a well-established and useful concept, but its limitations must always be kept in mind. It emphasizes the role of sequence-specific DNA-binding proteins in genome expression and downplays the influence of proteins that either do not bind specifically to the genome, or whose binding sites lie distant from the genes whose expression they influence. The module concept therefore draws attention away from histone-modifying enzymes, nucleosome remodeling complexes, and other proteins that play a critical role in genome expression.

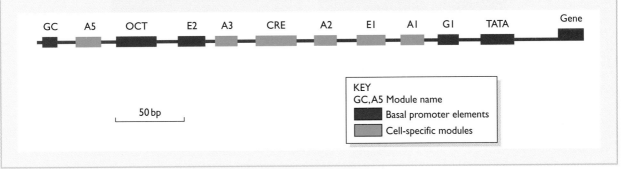

elements or to more distant sites in silencers. Some influence genome expression in a general way through histone deacetylation or DNA methylation (Section 8.2.2), but others have more specific effects at individual promoters. The yeast repressors called Mot1 and NC2, for example, inhibit assembly of the pre-initiation complex by binding directly to TBP and disrupting its activity. Mot1 causes TBP to dissociate from the DNA, and NC2 prevents further assembly of the complex on the bound TBP (Lee and Young, 2000). Both of these repressors have a broad spectrum of activity, inactivating a large set of genes, as does the Ssn6-Tup1 repressor, which is one of the main gene silencers in the yeast *Schizosaccharomyces pombe*, and which has homologs in many other eukaryotes (Smith and Johnson, 2000).

Another indication of the importance of repression in eukaryotic transcription comes from the demonstration that some proteins can exert both activating and repressing effects, depending on the circumstances. NC2, for example, represses initiation of transcription from promoters with a TATA box but has an activating effect on promoters that lack the TATA sequence (Willy *et al.*, 2000). Pit-1, which is the first of the three proteins after which the POU domain is named (Section 9.1.4), activates some genes and represses others, depending on the sequence of its DNA-binding site (Scully *et al.*, 2000). The presence in

this site of two additional nucleotides induces a change in the conformation of Pit-1, enabling it to interact with a second protein called N-CoR and repress transcription of the target gene (*Figures 9.28 and 9.29*).

Relatively little is known about the precise interactions occurring between repressors and the pre-initiation complex. A variety of **inhibition domains** (the converse of an activation domain) have been identified in eukaryotic repressors, several of which are rich in prolines, but no general patterns have emerged (Hanna-Rose and Hansen, 1996). The direct interactions with TBP displayed by Mot1 and NC2 argue against the involvement in repression of a complex equivalent to the mediator that is required for gene activation.

Controlling the activities of activators and repressors

The operation of individual activators and repressors must be controlled in order to ensure that the appropriate set of genes is expressed by a cell. We will return to this topic in Chapter 12, when it will form the central theme of our study of the ways in which genome activity is regulated in response to extracellular signals and during differentiation and development.

There are several ways in which an activator or repressor could be regulated. One possibility is to con-

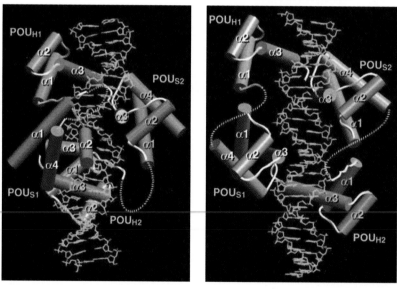

Prolactin   Growth Hormone

Figure 9.28 Conformation of the POU domains of the Pit-1 activator bound to its target sites upstream of the prolactin and growth-hormone genes.

Pit-1 is a dimer, and each monomer has two POU domains (Section 9.1.4). The two domains of one monomer are shown in red and the two domains of the other monomer are shown in blue. The barrels are α-helices, with α3 being the recognition helix of each domain. Note the difference between the conformations of the domains when bound to the two binding sites. The more open structure adopted at the growth hormone site enables the Pit-1 dimer to interact with N-CoR and other proteins to repress transcription of the growth hormone gene. Reprinted with permission from Scully *et al.*, *Science*, **290**, 1127–1131. Copyright 2000 American Association for the Advancement of Science.

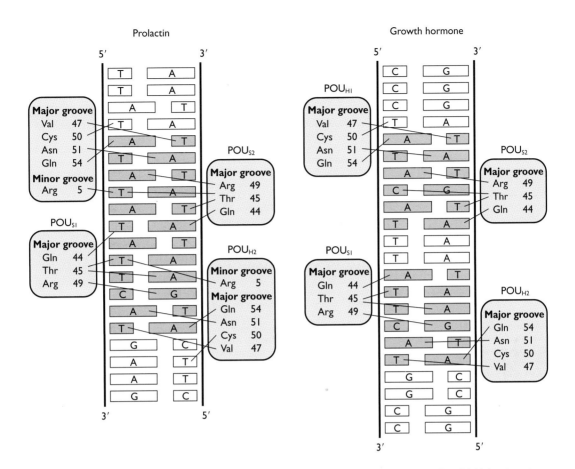

Figure 9.29 Pit-1 can activate or repress transcription initiation depending on the sequence of its DNA-binding site.

Pit-1 activates transcription of the prolactin gene but represses transcription of the growth hormone gene. The drawing shows the contacts made between the amino acids of the POU domain (blue boxes; the numbers indicate the positions of the amino acids in the protein) and the nucleotides at the DNA-binding sites. The lines indicate potential hydrogen bonds and van der Waals contacts (see Box 3.3, page 82). The growth hormone site contains two extra base pairs (shown as open boxes) in the middle of the contact region. The presence of these two base pairs leads to the change in the conformation of the Pit-1 dimer, allowing the interaction with N-CoR. From Scully et al. (2000).

trol its synthesis, but this does not permit rapid changes in genome expression because it takes time to accumulate an activator or repressor in the cell, or to destroy it when it is not needed. This type of control is therefore associated with activators and repressors responsible for maintaining stable patterns of genome expression, for example those underlying cellular differentiation and some aspects of development. An alternative way of controlling an activator or repressor is by chemical modification, for example by phosphorylation, or by inducing a change in its conformation. These changes are much more rapid than *de novo* synthesis, and enable the cell to respond to extracellular signaling compounds that induce transient changes in genome expression. We will examine the details of these various regulatory mechanisms in Chapter 12.

References

Bult CJ, White O, Olsen GJ, et al. (1996) Complete genome sequence of the methanogenic archaeon, *Methanococcus jannaschii. Science*, **273**, 1058–1073.

Burian RM and Gayon J (1999) The French school of genetics: from physiological and population genetics to regulatory molecular genetics. *Ann. Rev. Genet.*, **33**, 313–349.

Burley SK and Roeder RG (1996) Biochemistry and structural biology of transcription factor IID (TFIID). *Ann. Rev. Biochem.*, **65**, 769–799.

Chasman DI, Flaherty KM, Sharp PA and Kornberg RD (1993) Crystal structure of yeast TATA-binding protein and model for interaction with DNA. *Proc. Natl Acad. Sci. USA*, **90**, 8174–8178.

Choo Y and Klug A (1997) Physical basis of a protein-DNA recognition code. *Curr. Opin. Struct. Biol.*, **7**, 117–125.

Clarke ND and Berg JM (1998) Zinc fingers in *Caenorhabditis elegans*: finding families and probing pathways. *Science*, **282**, 2018–2022.

Evans JNS (1995) *Biomolecular NMR Spectroscopy.* Oxford University Press, Oxford.

Falvo JV, Thanos D and Maniatis T (1995) Reversal of intrinsic DNA bends in the IFNb gene enhancer by transcription factors and the architectural protein HMG I(Y). *Cell,* **83,** 1101–1111.

Fierro-Monti I and Mathews MB (2000) Proteins binding to duplexed RNA; one motif, multiple functions. *Trends Biochem. Sci.,* **25,** 241–246.

Galas D and Schmitz A (1978) DNase footprinting: a simple method for the detection of protein-DNA binding specificity. *Nucleic Acids Res.,* **5,** 3157–3170.

Gangloff YG, Romier C, Thuault S, Werten S and Davidson I (2001) The histone fold is a key structural motif of transcription factor TFIID. *Trends Biochem. Sci.,* **26,** 250–257.

Garner MM and Revzin A (1981) A gel electrophoretic method for quantifying the binding of proteins to specific DNA regions: application to components of the *Escherichia coli* lactose operon regulatory system. *Nucleic Acids Res.,* **9,** 3047–3060.

Grant PA, Schieltz D, Pray-Grant MG, et al. (1998) A subset of TAFIIs are integral components of the SAGA complex required for nucleosome acetylation and transcriptional stimulation. *Cell,* **94,** 45–53.

Green MR (2000) TBP-associated factors (TAF$_{II}$s): multiple, selective transcriptional mediators in common complexes. *Trends Biochem. Sci.,* **25,** 59–63.

Hanna-Rose W and Hansen U (1996) Active repression mechanisms of eukaryotic transcription repressors. *Trends Genet.,* **12,** 229–234.

Harrison SC and Aggarwal AK (1990) DNA recognition by proteins with the helix-turn-helix motif. *Ann. Rev. Biochem.,* **59,** 933–969.

Hendrickson W and Schleif R (1985) A dimer of AraC protein contacts three adjacent major groove regions at the Ara I DNA site. *Proc. Natl Acad. Sci. USA,* **82,** 3129–3133.

Herr W, Sturm RA, Clerc RG, et al. (1988) The POU domain: a large conserved region in the mammalian *pit-1, oct-1, oct-2* and *Caenorhabditis elegans unc-86* gene products. *Genes Develop.,* **2,** 1513–1516.

Kadonaga JT (1991) Purification of sequence-specific DNA binding proteins by DNA affinity chromatography. *Methods Enzymol.,* **208,** 10–23.

Kielkopf CL, White S, Szewczyk JW, et al. (1998) A structural basis for recognition of A·T and T·A base pairs in the minor groove of B-DNA. *Science,* **282,** 111–115.

Kim T-K, Ebright RH and Reinberg D (2000) Mechanism of ATP-dependent promoter melting by transcription factor IIH. *Science,* **288,** 1418–1421.

Kim YC, Geiger JH, Hahn S and Sigler PB (1993) Crystal structure of a yeast TBP/TATA-box complex. *Nature,* **365,** 512–520.

Kim YJ, Bjorklund S, Li Y, Sayre MH and Kornberg RD (1994) A multiprotein mediator of transcriptional activation and its interaction with the C-terminal repeat domain of RNA polymerase II. *Cell,* **77,** 599–608.

Kingston RE (1999) A shared but complex bridge. *Nature,* **399,** 199–200.

Lee TI and Young RA (2000) Transcription of eukaryotic protein-coding genes. *Ann. Rev. Genet.,* **34,** 77–137.

Luisi B (1995) DNA–protein interaction at high resolution. In: DMJ Lilley, ed. *DNA-Protein: Structural Interactions,* pp. 1–48. IRL Press, Oxford.

Mackay JP and Crossley M (1998) Zinc fingers are sticking together. *Trends Biochem. Sci.,* **23,** 1–4.

Malik S and Roeder RG (2000) Transcriptional regulation through mediator-like coactivators in yeast and metazoan cells. *Trends Biochem. Sci.,* **25,** 277–283.

Marr MT and Roberts JW (1997) Promoter recognition as measured by binding of polymerase to nontemplate strand oligonucleotide. *Science,* **276,** 1258–1260.

Novina CD and Roy AL (1996) Core promoters and transcriptional control. *Trends Genet.,* **12,** 351–355.

Reeder RH and Lang WH (1997) Terminating transcription in eukaryotes: lessons learned from RNA polymerase I. *Trends Biochem. Sci.,* **22,** 473–477.

Rhodes G (1999) *Crystallography Made Crystal Clear,* 2nd edition. Academic Press, London.

Scully KM, Jacobson EM, Jepsen K, et al. (2000) Allosteric effects of Pit-1 DNA sites on long-term repression in cell type specification. *Science,* **290,** 1127–1131.

Seither P, Iben S and Grummt I (1998) Mammalian RNA polymerase I exists as a holoenzyme with associated basal transcription factors. *J. Mol. Biol,* **275,** 43–53.

Singh H, LeBowitz JH, Baldwin AS and Sharp PA (1988) Molecular cloning of an enhancer binding protein: isolation by screening of an expression library with a recognition site DNA. *Cell,* **52,** 415–423.

Smith RL and Johnson AD (2000) Turning genes off by Ssn6-Tup1: a conserved system of transcriptional repression in eukaryotes. *Trends Biochem. Sci.,* **25,** 325–330.

Stormo GD and Fields DS (1998) Specificity, free energy and information content in protein–DNA interactions. *Trends Biochem. Sci.,* **23,** 109–113.

Thomas JO and Travers AA (2001) HMG1 and 2, and related 'architectural' DNA-binding proteins. *Trends Biochem. Sci.,* **26,** 167–172.

Thompson CM, Koleske AJ, Chao DM and Young RA (1993) A multisubunit complex associated with the RNA polymerase II CTD and TATA-binding protein in yeast. *Cell,* **73,** 1361–1375.

Travers AA (1995) DNA bending by sequence and proteins. In: DMJ Lilley ed. *DNA-Protein: Structural Interactions,* pp. 49–75. IRL Press, Oxford.

Verrijzer CP (2001) Transcription factor IID – not so basal after all. *Science,* **293,** 2010–2011.

Willy PJ, Kobayashi R and Kadonaga JT (2000) A basal transcription factor that activates or represses transcription. *Science,* **290,** 982–984.

Wolffe AP (1994) Architectural transcription factors. *Science,* **264,** 1100–1101.

Wolffe AP (1995) Genetic effects of DNA packaging. *Sci. Am.,* **Nov/Dec,** 68–77.

Young MA and Beveridge DL (1998) Molecular dynamics simulations of an oligonucleotide duplex with adenine tracts phased by a full helix turn. *J. Mol. Biol.,* **281,** 675–687.

Yudkovsky N, Ranish JA and Hahn S (2000) A transcription reinitiation intermediate that is stabilized by activator. *Nature,* **408,** 225–229.

Further Reading

Adhya S (1996) The *lac* and *gal* operons today. In: ECC Lin and AS Lynch, eds. *Regulation of Gene Expression in Escherichia coli,* pp. 181–200. Chapman & Hall, New York. — *A comprehensive description of the lactose operon.*

Geiduschek EP and Kassavetis GA (2001) The RNA polymerase III transcription apparatus. *J. Mol. Biol.,* **310,** 1–26.

Kornberg RD (1999) Eukaryotic transcriptional control. *Trends Genet.,* **15** (millennium issue), M46–M49. — *An excellent overview.*

Latchman DS (1995) *Gene Regulation: A Eukaryotic Perspective.* Stanley Thorne, Cheltenham. — *This is the best general text on this subject.*

Latchman DS (1998) *Eukaryotic Transcription Factors,* 3rd edition. Academic Press, London. — *This is also the best general text on this subject.*

Latchman DS (2001) Transcription factors: bound to activate or repress. *Trends Biochem. Sci.,* **26,** 211–213. — *Short review of proteins that combine activation with repression.*

Lilley DMJ (ed) (1995) *DNA-Protein: Structural Interactions.* IRL Press, Oxford. — *Research-level description of the subject.*

Myers LC and Kornberg RD (2000) Mediator of transcriptional regulation. *Ann. Rev. Biochem.,* **69,** 729–749. — *A detailed review of this topic.*

Nagai K and Mattaj IW (eds) (1994) *RNA-Protein Interactions.* IRL Press, Oxford. — *Complements the material in this chapter by providing details of RNA-binding proteins.*

Neidle S (1994) *DNA Structure and Recognition: In Focus.* IRL Press, Oxford. — *Easy to digest information of DNA-binding proteins.*

Schleif R (2000) Regulation of the L-arabinose operon of *Escherichia coli. Trends Genet.,* 16, 559–565. — *Gives details of one example of bacterial gene regulation.*

Travers A (1993) *DNA-Protein Interactions.* Chapman & Hall, London. — *The most accessible of the various books on this topic.*

STUDY AIDS FOR CHAPTER 9

Key terms

Give short definitions of the following terms:

β-turn
κ-homology domain
Acidic domain
Activation domain
Activator
Affinity chromatography
Basal promoter
Basal promoter element
Basal rate of transcription
 initiation
Basic domain
CAAT box
Catabolite activator protein
Cell-specific module
Chemical shift
Closed promoter complex
Coactivator
Constitutive control
Core promoter
Co-repressor
C-terminal domain (CTD)
Cys$_2$His$_2$ finger
Direct readout
DNA bending
DNA-binding motif
DNA-binding protein
Double-stranded RNA
 binding domain (dsRBD)
Enhanceosome
Enhancer
Footprinting
GC box
Gel retardation
General transcription factor
 (GTF)
Glutamine-rich domain
Helix–loop–helix
Helix–turn–helix
Homeodomain
Inducer
Inhibition domain
Initiator (Inr) sequence

Lactose repressor
Leucine zipper
Mediator
Modification interference
Modification protection
Multicysteine zinc finger
Nuclear magnetic resonance
 (NMR) spectroscopy
Octamer module
Open promoter complex
Operator
POU domain
Pre-initiation complex (PIC)
Proline-rich domain
Promoter
Promoter clearance
Recognition helix
Regulatory control
Response module
Ribbon–helix–helix motif
Ribonucleoprotein (RNP)
 domain
RNA polymerase I
RNA polymerase II
RNA polymerase III
Silencer
Strong promoter
TAF and initiator-dependent
 cofactor (TIC)
TATA box
TATA-binding protein (TBP)
TBP domain
TBP-associated factor (TAF)
Termination factor
Upstream control element
Upstream promoter
 element
Winged helix–turn–helix
X-ray crystallography
X-ray diffraction
X-ray diffraction pattern
Zinc finger

Self study questions

1. Explain why DNA-binding proteins are central to genome expression.
2. Describe how gel retardation can be used to study DNA–protein interactions. What are the limitations of this technique?
3. Draw diagrams to illustrate the modification protection and modification interference techniques. Indicate the key differences and describe how these differences underlie the specific applications of the two techniques.
4. Explain how affinity chromatography is used to purify a DNA-binding protein.
5. Write short essays on (a) X-ray crystallography, and (b) nuclear magnetic resonance spectroscopy, emphasizing the use of these techniques in the study of DNA-binding proteins.
6. Describe, with examples, how proteins that contain the helix–turn–helix motif bind to DNA. List, again with examples, the modified versions of the helix–turn–helix motif that are found in eukaryotic proteins.
7. Using examples, distinguish between two or more types of zinc finger.
8. Compare and contrast the structures used by proteins to bind to DNA and/or RNA molecules.
9. What features of the double helix are important in determining the nature of the interaction between DNA and a binding protein?
10. Describe the types of contact made between DNA and a binding protein. Why are many DNA-binding proteins dimeric?
11. Distinguish between the three nuclear RNA polymerases of eukaryotes. How is the *Escherichia coli* RNA polymerase similar to or different from the eukaryotes enzymes?
12. Define the term 'promoter'. Draw annotated diagrams to illustrate the structures of the promoters for the three eukaryotic RNA polymerases and for the *Escherichia coli* enzyme.
13. Explain the roles of the two components of the *Escherichia coli* promoter during initiation of transcription. Be sure to make clear the difference between the closed and open versions of the promoter–RNA polymerase complex.

14. Write an essay on 'Assembly of the RNA polymerase II initiation complex'. As part of your essay, compile a table giving the names of the main proteins or groups of proteins involved in assembly of this complex, along with a summary of the role of each one.

15. What is the importance of the C-terminal domain of the largest subunit of RNA polymerase II?

16. How does the TATA-binding protein provide a link between the initiation processes of all three eukaryotic RNA polymerases?

17. Describe how promoter structure influences gene expression in *Escherichia coli*.

18. Using examples, outline how the use of alternative σ subunits enables a bacterium to alter its pattern of genome expression.

19. Draw a series of diagrams to show how initiation of transcription is regulated at the lactose and tryptophan operons of *Escherichia coli*. Indicate the key differences between these two control mechanisms.

20. What is an activator? How do activators influence assembly of the RNA polymerase II initiation complex?

21. Explain why the discovery of a mammalian mediator was looked upon as a critical breakthrough in understanding the control of transcription initiation.

22. Describe our current knowledge of proteins that repress eukaryotic transcription initiation.

Problem-based learning

1. The methods for locating the positions of protein binding sites described in Section 9.1.1 assume that these sites are located in the region upstream of a gene. Is this assumption justified?

2. Use your knowledge of DNA chip and microarray technologies (Technical Note 5.1, page 133) to devise a method for identifying the attachment sites for a DNA-binding protein across the entire genome, as opposed to just within the region upstream of a single gene.

3. Write a report that elaborates on, and extends, the discussion presented in Box 9.2 (page 252), concerning the possibility that the amino acid sequence of a recognition helix can be used to deduce the nucleotide sequence of the DNA-binding site for a protein that contains that helix.

4. Construct a hypothesis to explain why eukaryotes have three RNA polymerases. Can your hypothesis be tested?

5. A model for control of transcription of the lactose operon in *Escherichia coli* was first proposed by François Jacob and Jacques Monod in 1961 (Jacob F and Monod J [1961] Genetic regulatory mechanisms in the synthesis of proteins. *J. Mol. Biol.*, **3**, 318–356). Explain the extent to which their work, which was based almost entirely on genetic analysis, provided an accurate description of the molecular events that are now known to occur.

6. To what extent is *E. coli* a good model for the regulation of transcription initiation in eukaryotes? Justify your opinion by providing specific examples of how extrapolations from *E. coli* have been helpful and/or unhelpful in the development of our understanding of equivalent events in eukaryotes.

7. Assess the accuracy and usefulness of the module concept for the structure of an RNA polymerase II promoter.

Synthesis and Processing of RNA

Chapter Contents

10.1 Synthesis and Processing of mRNA — 274

    10.1.1 Synthesis of bacterial mRNAs — 274

    10.1.2 Synthesis of eukaryotic mRNAs by RNA polymerase II — 283

    10.1.3 Intron splicing — 287

10.2 Synthesis and Processing of Non-coding RNAs — 295

    10.2.1 Transcript elongation and termination by RNA polymerases I and III — 295

    10.2.2 Cutting events involved in processing of bacterial and eukaryotic pre-rRNA and pre-tRNA — 295

    10.2.3 Introns in eukaryotic pre-rRNA and pre-tRNA — 296

10.3 Processing of Pre-RNA by Chemical Modification — 298

    10.3.1 Chemical modification of pre-rRNAs — 299

    10.3.2 RNA editing — 303

10.4 Degradation of mRNAs — 304

    10.4.1 Bacterial mRNAs are degraded in the $3' \rightarrow 5'$ direction — 305

    10.4.2 Eukaryotes have more diverse mechanisms for RNA degradation — 306

10.5 Transport of RNA Within the Eukaryotic Cell — 308

Learning outcomes

When you have read Chapter 10, you should be able to:

- Describe the elongation and termination phases of transcription in *Escherichia coli*, and explain how these are regulated by antitermination and attenuation

- Give details of elongation and termination of eukaryotic transcripts, including the processes responsible for capping and polyadenylation of eukaryotic mRNAs

- Distinguish between the splicing pathways of different types of intron, and in particular give a detailed description of splicing of GU–AG introns, including examples of alternative splicing

- Describe the cutting events involved in processing of bacterial and eukaryotic pre-rRNA and pre-tRNA

- Define the term 'ribozyme', and give examples of ribozymes

- Explain how eukaryotic rRNAs are chemically modified at specific nucleotide positions

- Give examples of mRNA editing in mammals and outline the more complex types of RNA editing that occur in various other eukaryotes

- Describe the mRNA degradation processes of bacteria and eukaryotes

- Outline the events involved in transport of eukaryotic RNAs from the nucleus to the cytoplasm

INITIATION OF TRANSCRIPTION, culminating with the RNA polymerase leaving the promoter and beginning synthesis of an RNA molecule, is simply the first step in the genome expression pathway. In this chapter and the next we will follow the process onwards and examine how transcription and translation eventually result in synthesis of the proteome.

10.1 Synthesis and Processing of mRNA

We begin our detailed study of transcription by looking at the synthesis and processing of mRNAs, the molecules that make up the transcriptome and which specify the protein content of the cell. As the central players in genome expression, mRNAs have received the greatest attention from researchers and we now have a detailed picture of how they are produced. Events in bacteria are different in many respects from those in eukaryotes and so we will deal with the two types of organism in different sections. One aspect of eukaryotic mRNA processing – intron splicing – is so important that it requires a section of its own.

10.1.1 Synthesis of bacterial mRNAs

Bacterial mRNAs do not undergo any significant forms of processing: the primary transcript that is synthesized by the RNA polymerase is itself the mature mRNA, and its translation usually begins before transcription is complete (*Figure 10.1*). This coupling of transcription and translation is important in that it allows special types of control to be applied to the regulation of bacterial mRNA synthesis, as will be described on page 277.

The elongation phase of bacterial transcription

Because there is just one bacterial RNA polymerase (Section 9.2.1), the general mechanism of transcription is the same for all bacterial genes. The following descriptions of elongation and termination, given in the context of mRNA synthesis, therefore apply equally well to the synthesis of non-coding RNA.

The chemical basis of the template-dependent synthesis of RNA was shown in *Figure 3.5*, page 75. Ribonucleotides are added one after another to the growing 3' end of the RNA transcript, the identity of each nucleotide specified by the base-pairing rules: A base-pairs with T or

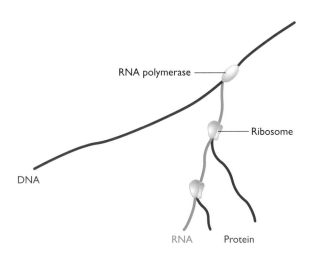

RNA polymerase

Ribosome

DNA

RNA   Protein

Figure 10.1   In bacteria, transcription and translation are often coupled.

U; G base-pairs with C. During each nucleotide addition, the β- and γ-phosphates are removed from the incoming nucleotide (see *Figure 1.6*, page 11), and the hydroxyl group is removed from the 3′-carbon of the nucleotide present at the end of the chain.

At this stage of transcription, the bacterial RNA polymerase is in its core enzyme form, comprising four proteins, two relatively small (approximately 35 kDa) α subunits, and one each of the related subunits β and β′ (both approximately 150 kDa): the σ subunit that has the key role in initiation has now left the complex (Section 9.2.3). The RNA polymerase covers about 30 bp of the template DNA, including the **transcription bubble** of 12–14 bp, within which the growing transcript is held to the template strand of the DNA by approximately eight RNA–DNA base pairs (*Figure 10.2*). The RNA polymerase has to keep a tight grip on both the DNA template and the RNA that it is making in order to prevent

the transcription complex from falling apart before the end of the gene is reached. However, this grip must not be so tight as to prevent the polymerase from moving along the DNA. To understand how these apparently contradictory requirements are met, the interactions between the polymerase, the DNA template and the RNA transcript are being examined by X-ray crystallography studies (Section 9.1.3), combined with crosslinking experiments in which covalent bonds are formed between the DNA or RNA and the polymerase, these bonds enabling the amino acids that are closest to the DNA and RNA to be identified. In the current model the DNA template lies between the β and β′ subunits, within a trough on the enclosed surface of β′. The active site for RNA synthesis also lies between these two subunits, with the non-template strand of DNA held within the β subunit and the RNA transcript extruded from the complex via a channel formed partly by the β and partly by the β′ subunit (Research Briefing 10.1; Korzheva *et al.*, 2000).

Termination of bacterial transcription

Exactly how termination occurs is not known. Current thinking views transcription as a stepwise nucleotide-by-nucleotide process, with the polymerase pausing at each position and making a 'choice' between continuing elongation by adding another ribonucleotide to the transcript, or terminating by dissociating from the template. Which choice is selected depends on which alternative is more favorable in thermodynamic terms (von Hippel, 1998). This model emphasizes that, in order for termination to occur, the polymerase has to reach a position on the template where dissociation is more favorable than continued RNA synthesis.

Bacteria appear to use two distinct strategies for transcription termination. About half the positions in *Escherichia coli* at which transcription terminates correspond to DNA sequences where the template strand con-

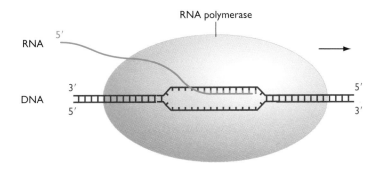

RNA polymerase

RNA   5′

DNA   3′
       5′   5′
                                             3′

Figure 10.2   Schematic representation of the *Escherichia coli* transcription elongation complex.

The RNA polymerase covers approximately 30 bp of DNA, including a transcription bubble of 12–13 bp, with the RNA attached to the template strand of the DNA by eight or so RNA–DNA base pairs. The black arrow shows the direction in which the RNA polymerase moves along the DNA. See Research Briefing 10.1 for a more accurate description of the structure of the elongation complex.

tains an inverted palindrome followed by a run of deoxyadenosine nucleotides (*Figure 10.3*). These **intrinsic terminators** have been thought to promote dissociation of the polymerase by destabilizing the attachment of the growing transcript to the template, in two ways. First, when the inverted palindrome is transcribed, the RNA sequence folds into a stable hairpin, this RNA–RNA base pairing being favored over the DNA–RNA pairing that normally occurs within the transcription bubble. This reduces the number of contacts made between the template and transcript, weakening the overall interaction and favoring dissociation. The interaction is further weakened when the run of As in the template is transcribed, because the resulting A–U base pairs have only two hydrogen bonds each, compared with three for each G–C pair. The net result is that termination is favored over continued elongation (von Hippel, 1998). This model is easy to rationalize with the known properties of DNA–RNA hybrids, but an alternative hypothesis has been prompted by the result of crosslinking experiments, which have shown that the RNA hairpin makes contact with a flap structure on the outer surface of the RNA

polymerase β subunit, adjacent to the exit point of the channel through which the RNA emerges from the complex. Although the flap structure is quite distant (some 6.5 nm) from the active site of the polymerase, a direct connection is made between the two by a segment of β-sheet within the β subunit. Movement of the flap could therefore affect the positioning of amino acids within the active site, possibly leading to breakage of the DNA–RNA base pairs and termination of transcription (Research Briefing 10.1). Additional evidence in support of this model comes from the demonstration that the protein called NusA, which enhances termination at intrinsic promoters, interacts with the hairpin loop and flap structure and may stabilize the contact between the two (Toulokhonov *et al.*, 2001).

The second type of bacterial termination signal is **Rho dependent**. These signals usually retain the hairpin feature of intrinsic terminators, although the hairpin is less stable and there is no run of As in the template. Termination requires the activity of a protein called Rho, which attaches to the transcript and moves along the RNA towards the polymerase. If the polymerase contin-

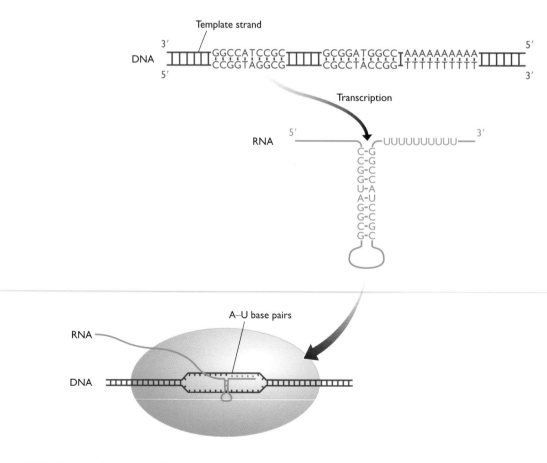

Figure 10.3 Termination at an intrinsic terminator.

The presence of an inverted palindrome in the DNA sequence results in formation of a hairpin loop in the transcript. See Research Briefing 10.1 for a more accurate description of transcription termination.

ues to synthesize RNA then it keeps ahead of the pursuing Rho, but at the termination signal the polymerase stalls (see *Figure 10.4*). Exactly why has not been explained – presumably the hairpin loop that forms in the RNA is responsible in some way – but the result is clear: Rho is able to catch up. Rho is a **helicase**, which means that it actively breaks base pairs, in this case between the template and transcript, resulting in termination of transcription.

Control over the choice between elongation and termination

In bacteria, two mechanisms have evolved for influencing the repeated choice that the polymerase has to make between elongation and termination when copying a template. Both mechanisms are important in regulating the expression of genes contained within operons.

The first process is called **antitermination**. This occurs when the RNA polymerase ignores a termination signal and continues elongating its transcript until a second signal is reached (*Figure 10.5*). It provides a mechanism whereby one or more of the genes at the end of an operon can be switched off or on by the polymerase recognizing or not recognizing a termination signal located upstream of those genes. Antitermination is controlled by an **antiterminator protein**, which attaches to the DNA near the beginning of the operon and then transfers to the RNA polymerase as it moves past en route to the first termination signal. The presence of the antiterminator protein causes the enzyme to ignore the termination signal, presumably by countering the destabilizing properties of an intrinsic terminator or by preventing stalling at a Rho-dependent terminator. Although the mechanics of the process are unclear, the impact that antitermination can have on gene expression has been described in detail, especially during the infection cycle of bacteriophage λ (Box 10.1).

The second type of termination control is called **attenuation**. This system operates primarily with operons that code for enzymes involved in amino acid biosynthesis, but a few other examples are also known. The tryptophan operon of *E. coli* (Section 9.3.1) illustrates how it works. In this operon, two hairpin loops can form in the region between the start of the transcript and the beginning of *trpE*. The smaller of these loops acts as a termination signal, but the larger hairpin loop, which is closer to the start of the transcript, is more stable. The larger loop overlaps with the termination hairpin, so only one of the two hairpins can form at any one time. Which loop forms depends on the relative positioning between the RNA polymerase and a ribosome which attaches to the 5′ end of the transcript as soon as it is synthesized in order to translate the genes into protein (*Figure 10.6*). If the ribosome stalls so that it does not keep up with the polymerase, then the larger hairpin forms and transcription continues. However, if the ribosome keeps pace with the RNA polymerase then it disrupts the larger hairpin by attaching to the RNA that forms part of the stem of this hairpin. When this happens the termination hairpin is able to form, and transcription stops. Ribosome stalling can occur because upstream of the termination signal is a short open reading frame (ORF) coding for a 14-amino-acid peptide that includes two tryptophans. If the amount of free tryptophan is limiting, then the ribosome stalls as it attempts to synthesize this peptide, while the polymerase continues

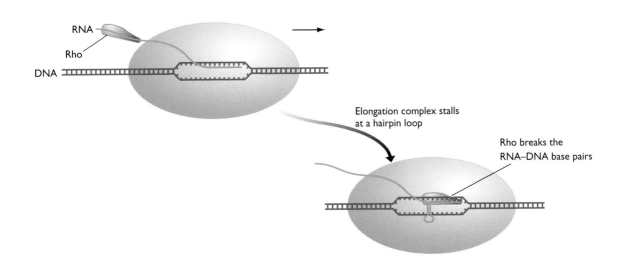

Figure 10.4 Rho-dependent termination.

Rho is a helicase that follows the RNA polymerase along the transcript. When the polymerase stalls at a hairpin, Rho catches up and breaks the RNA–DNA base pairs, releasing the transcript. Note that the diagram is schematic and does not reflect the relative sizes of Rho and the RNA polymerase.

The structure of the bacterial RNA polymerase

Structural studies have provided insights into the mechanism for transcription elongation and termination in bacteria.

One of the most important developments in molecular biology in recent years has been the growing improvement in techniques for structural analysis of the enzymes and other macromolecular assemblies that are involved in genome replication. These techniques have improved our understanding of the structure of the bacterial RNA polymerase and, in particular, have prompted a new model for the events involved in termination of transcription.

The protein–nucleic acid contacts within the bacterial RNA polymerase

A structural model for the transcription elongation complex in bacteria has been built up by combining X-ray crystallography studies with techniques that result in crosslinks being formed between the DNA or RNA components of the complex and the enclosing protein subunits of the RNA polymerase. These crosslinking experiments make use of a variety of photoreactive compounds that are attached to synthetic DNA and RNA molecules. After assembly of the complex, the labels are activated by a pulse of light so that they form crosslinks between the nucleic acid and any

amino acids that are located close to the position of the label within the polymerase. Some labels form crosslinks with any amino acid and others are more discriminatory, forming links only with lysines, for example. After crosslinking, the RNA polymerase is disassembled into subunits and each subunit is treated with cyanogen bromide, which cuts polypeptides specifically at methionine residues, yielding a characteristic set of fragments. Fragments that are crosslinked to the nucleic acid are then identified. These fragments must contain the amino acids that were adjacent to the label in the intact transcription elongation complex. By identifying as many crosslinks as possible, a detailed map can be built up. In the example shown below, the RNA polymerase β and β′ subunits are represented by blue bars (with amino acid positions indicated by the numbers) and the DNA and RNA as colored circles. Crosslinks between nucleotides in the DNA or RNA and amino acids are shown by the thin lines.

The information from the crosslinking map is now combined with the structural data from X-ray crystallography to construct a model of the transcription elongation complex, showing the precise positioning of the different parts of the

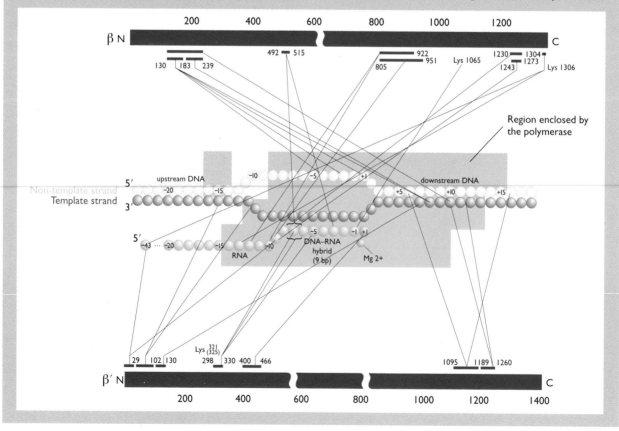

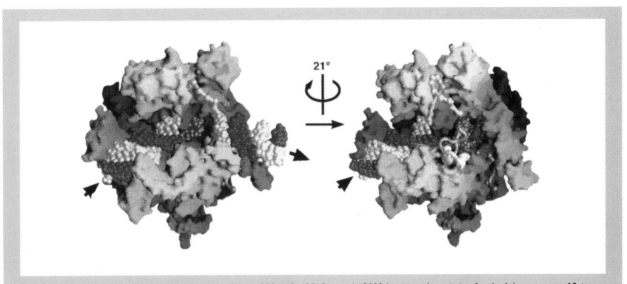

Reproduced with permission from Korzheva *et al.*, *Science*, **289**, 619–625. Copyright 2000 American Association for the Advancement of Science.

DNA double helix and of the RNA transcript within the polymerase. The result of this analysis is shown at the top of the page. The β and β' subunits of the RNA polymerase are depicted in blue/green and pink, respectively, the double helix is colored red (template strand) and yellow (non-template strand), and the RNA transcript is gold. The two views show that the double helix lies between the β and β' subunits, within a trough on the enclosed surface of β'. The active site for RNA synthesis also lies between these two subunits, with the non-template strand of DNA looping away from the active site and held within the β subunit. The RNA transcript extrudes from the complex via a channel formed partly by the β and partly by the β' subunit, at the bottom right of the left-hand structure.

A model for termination

Termination is the stage in bacterial transcription that has caused most problems for molecular biologists. Crosslinking experiments have led to a breakthrough in this area by showing that the hairpin that forms in the RNA transcript when the termination region is reached makes contact with a flap structure on the outer surface of the RNA polymerase β subunit, adjacent to the exit point of the channel through which the RNA emerges from the complex. The part of the β subunit polypeptide that forms the flap is colored dark blue in the structure shown on the right, with the remainder of the β subunit in light blue, the β' subunit in pink, and the α subunit in white.

Note that although the flap structure is located on the surface of the polymerase, the region of the β polypeptide that forms the flap is connected directly to the active site, indicated by the magnesium ion in magenta and the nucleoside 5'-triphosphate substrate in green. An interaction between the RNA hairpin and the flap could therefore

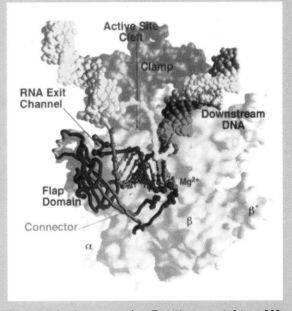

Reproduced with permission from Toulokhonov *et al.*, *Science*, **292**, 730–733. Copyright 2001 American Association for the Advancement of Science.

affect the positioning of amino acids within the active site, possibly causing the DNA–RNA base pairs to break, leading to termination of transcription.

References

Korzheva N, Mustaev A, Kozlov M, et al. (2000) A structural model of transcription elongation. *Science*, **289**, 619–625.
Toulokhonov I, Artsimovitch I and Landick R (2001) Allosteric control of RNA polymerase by a site that contacts nascent RNA hairpins. *Science*, **292**, 730–733.

Box 10.1: Antitermination during the infection cycle of bacteriophage λ

Bacteriophage λ provides the best studied example of the use of antitermination as a means of regulating gene expression (Friedman *et al.*, 1987). Immediately after entering an *Escherichia coli* cell, transcription of the λ genome is initiated by the bacterial RNA polymerase attaching to two promoters, p_L and p_R, and synthesizing two 'immediate-early' mRNAs, these terminating at positions t_{L1} and t_{R1}:

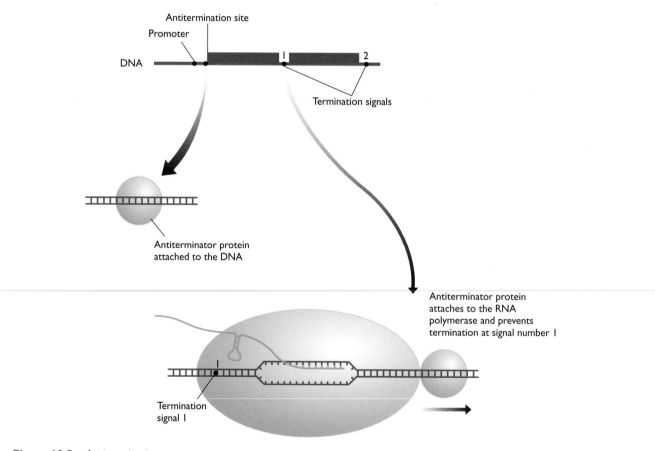

The mRNA transcribed from p_R to t_{R1} codes for a protein called Cro, one of the major regulatory proteins involved in the λ infection cycle. The second mRNA specifies the N protein, which is an antiterminator. The N protein attaches to the λ genome at sites *nutL* and *nutR* and transfers to the RNA polymerase as it passes. Now the RNA polymerase ignores the t_{L1} and t_{R1} terminators and continues transcription downstream of these points. The resulting mRNAs encode the 'delayed-early' proteins:

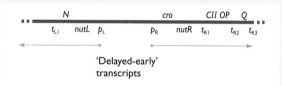

Antitermination controlled by the N protein therefore ensures that the immediate-early and delayed-early proteins are synthesized at the appropriate times during the λ infection cycle. One of the delayed-early proteins, Q, is a second antiterminator that controls the switch to the late stage of the infection cycle.

Figure 10.5 Antitermination.

The antiterminator protein attaches to the DNA and transfers to the RNA polymerase as it moves past, subsequently enabling the polymerase to continue transcription through termination signal number 1, so the second of the pair of genes in this operon is transcribed.

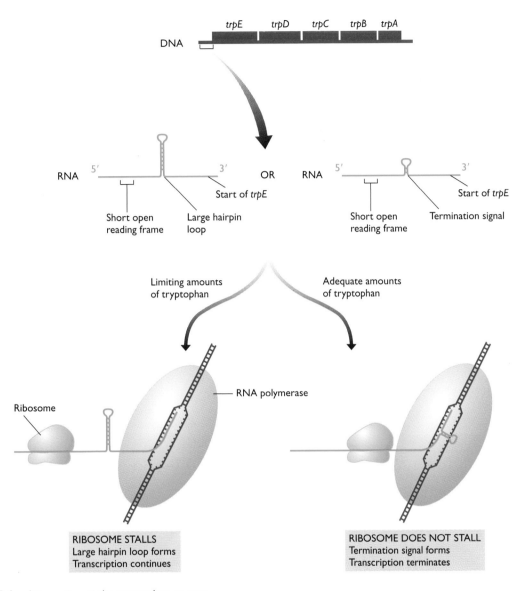

Figure 10.6 Attenuation at the tryptophan operon.

See the text for details.

to make its transcript. Because this transcript contains copies of the genes coding for the biosynthesis of tryptophan, its continued elongation addresses the requirement that the cell has for this amino acid. When the amount of tryptophan in the cell reaches a satisfactory level, the attenuation system prevents further transcription of the tryptophan operon, because now the ribosome does not stall while making the short peptide, and instead keeps pace with the polymerase, allowing the termination signal to form.

The *E. coli* tryptophan operon is controlled not only by attenuation but also by a repressor (Section 9.3.1). Exactly how attenuation and repression work together to regulate expression of the operon is not known, but it is thought that repression provides the basic on–off switch and atten-

uation modulates the precise level of gene expression that occurs. Other *E. coli* operons, such as those for biosynthesis of histidine, leucine and threonine, are controlled solely by attenuation. Interestingly, in some bacteria, including *Bacillus subtilis*, the tryptophan operon is one of those that does not have a repressor system and so is regulated entirely by attenuation. In these bacteria, attenuation is mediated not by the speed at which the ribosome tracks along the mRNA, but by an RNA-binding protein called **trp RNA-binding attenuation protein** (**TRAP**) which, in the presence of tryptophan, attaches to the mRNA in the region equivalent to the short ORF of the *E. coli* transcript (*Figure 10.7*). Attachment of TRAP leads to formation of the termination signal and cessation of transcription (Antson *et al.*, 1999).

(A)

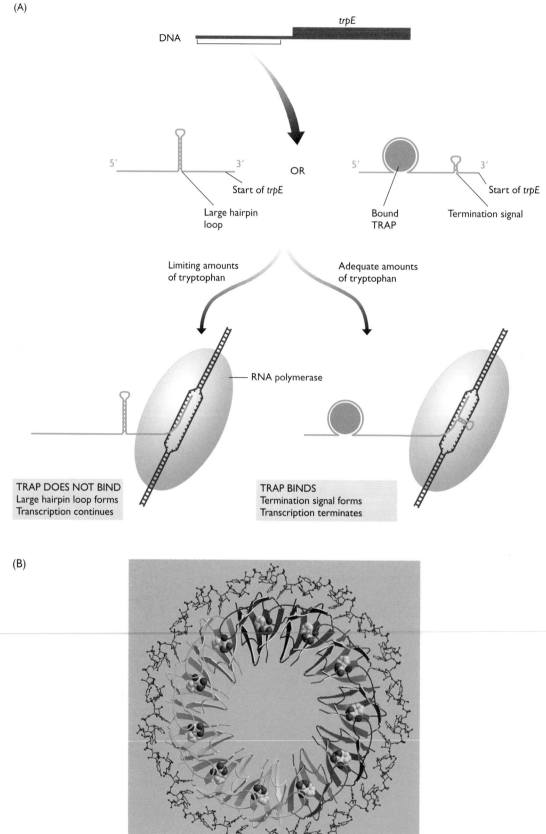

(B)

10.1.2 Synthesis of eukaryotic mRNAs by RNA polymerase II

At the most fundamental level, transcription is similar in bacteria and eukaryotes. The chemistry of RNA polymerization is identical in all types of organism, and the three eukaryotic RNA polymerases are all structurally related to the *E. coli* RNA polymerase, their three largest subunits being equivalent to the α, β and β' subunits of the bacterial enzyme. The contacts between the eukaryotic polymerase II, the template DNA and the RNA transcript, as revealed by X-ray crystallography and crosslinking studies (Klug, 2001), are similar to the interactions described for bacterial transcription (see Research Briefing 10.1), and the basic principle that transcription is a step-by-step competition between elongation and termination (see page 275) also holds.

Despite this equivalence, the overall processes for mRNA synthesis in bacteria and eukaryotes are quite different. The most striking dissimilarity is the extent to which eukaryotic mRNAs are processed during transcription. In bacteria, the transcripts of protein-coding genes are not processed at all: the primary transcripts are mature mRNAs. In contrast, all eukaryotic mRNAs have a cap added to the 5' end, most are also polyadenylated by addition of a series of adenosines to the 3' end, many contain introns and so undergo splicing, and a few are subject to RNA editing. A function has been assigned to capping, but the reason for polyadenylation largely remains a mystery. With splicing and editing we can appreciate why the events occur – the former removes introns that block translation of the mRNA; the latter changes the coding properties of the mRNA – but we do not understand why these mechanisms have evolved. Why do genes have introns in the first place? Why edit an mRNA rather than encoding the desired sequences in the DNA?

Eukaryotic mRNAs are processed while they are being synthesized. The cap is added as soon as transcription has been initiated, splicing and editing begin while the transcript is still being made, and polyadenylation is an inherent part of the termination mechanism for RNA polymerase II. To deal with all of these events together would be confusing, with too many different things being described at once. We will therefore postpone editing until the end of the chapter, which means it can be dealt with in tandem with similar forms of chemical modification occurring during rRNA and tRNA processing, and we will consider splicing after we have studied capping, elongation and polyadenylation.

Capping of RNA polymerase II transcripts occurs immediately after initiation

Although phosphorylation of the C-terminal domain (CTD) of the largest subunit of RNA polymerase II is the final step in initiation of transcription of mRNA-encoding genes in eukaryotes (Section 9.2.3), it is not immediately followed by the onset of elongation. A somewhat gray area exists in our understanding of the events that distinguish **promoter clearance**, which refers to the transition from the pre-initiation complex to a complex that has begun to synthesize RNA, and **promoter escape**, during which the polymerase moves away from the promoter region and becomes committed to making a transcript (*Figure 10.8*). The opposing effects of negative and positive elongation factors influence the ability of the polymerase to begin productive RNA synthesis, and if the negative factors predominate then transcription halts before the polymerase has moved more than 30 nucleotides from the initiation point. Promoter escape could therefore be an important control point, but how regulation is applied at this stage is not yet known (Lee and Young, 2000).

Successful promoter escape could be linked with capping, this processing event being completed before the transcript reaches 30 nucleotides in length. The first step in capping is addition of an extra guanosine to the extreme 5' end of the RNA. Rather than occurring by normal RNA polymerization, capping involves a reaction between the 5' triphosphate of the terminal nucleotide and the triphosphate of a GTP nucleotide. The γ-phosphate of the terminal nucleotide (the outermost phosphate) is removed, as are the β and γ phosphates of the GTP, resulting in a 5'–5' bond (*Figure 10.9*). The reaction is carried out by the enzyme **guanylyl transferase**. The second step of the capping reaction converts the new terminal guanosine into 7-methylguanosine by attachment of a methyl group to nitrogen number 7 of the purine ring, this modification catalyzed by **guanine methyltransferase**. The two capping enzymes make attachments with the CTD and it is possible that they are intrinsic components of the RNA polymerase II complex during promoter clearance (Proudfoot, 2000).

Figure 10.7   Regulation of the tryptophan operon of *Bacillus subtilis*.

(A) Regulation centers on the protein called TRAP, which attaches to the leader region of the transcript when tryptophan levels are adequate. When attached, TRAP *prevents* formation of the large hairpin loop and so *allows* formation of the termination signal. Compare with *Figure 10.6*. (B) Structure of the TRAP–RNA complex. TRAP comprises 11 identical subunits, each one made up mainly of β-sheets, which associate together to form a circular structure with a diameter of 8 nm. The 11 TRAP subunits are shown in different colors, each one with an attached tryptophan molecule indicated by the red–green–blue spherical structures. The RNA molecule is shown as a ball-and-stick structure wound around the TRAP complex. Reprinted with permission from Antson *et al.*, *Nature*, **401**, 235–242. Copyright 1999 Macmillan Magazines Limited. Permission also kindly given by Fred Antson and Paul Gollnick.

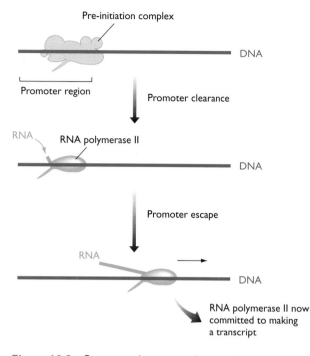

Figure 10.8   Promoter clearance and promoter escape.

Promoter clearance is the transition from the pre-initiation complex to a complex that has begun to synthesize RNA. Promoter escape occurs when the polymerase moves away from the promoter region and becomes committed to making a transcript. Note that the drawing is schematic and is not intended to indicate the shape or subunit composition of the RNA polymerase II complex that synthesizes the transcript.

The 7-methylguanosine structure is called a **type 0 cap** and is the commonest form in yeast. In higher eukaryotes, additional modifications occur (see *Figure 10.9*):

■ A second methylation replaces the hydrogen of the 2′–OH group of what is now the second nucleotide in the transcript. This results in a **type 1 cap**.

■ If this second nucleotide is an adenosine, then a methyl group might be added to nitrogen number 6 of the purine.

■ Another 2′–OH methylation might occur at the third nucleotide position, resulting in a **type 2 cap**.

All RNAs synthesized by RNA polymerase II are capped in one way or another. This means that as well as mRNAs, the snRNAs that are transcribed by this enzyme are also capped (see *Table 9.3*, page 255). The cap may be important for export of mRNAs and snRNAs from the nucleus (Section 10.5), but its best defined role is in translation of mRNAs, which is covered in Section 11.2.2.

Elongation of eukaryotic mRNAs

As mentioned above, the fundamental aspects of transcript elongation are the same in bacteria and eukaryotes. The one major distinction concerns the length of tran-script that must be synthesized. The longest bacterial genes are only a few kb in length and can be transcribed in a matter of minutes by the bacterial RNA polymerase, which has a polymerization rate of several hundred nucleotides per minute. In contrast, RNA polymerase II can take hours to transcribe a single gene, even though it can work at up to 2000 nucleotides per minute. This is because the presence of multiple introns in many eukaryotic genes (Section 10.1.3) means that considerable lengths of DNA must be copied. For example, the pre-mRNA for the human dystrophin gene is 2400 kb in length and takes about 20 hours to synthesize.

The extreme length of eukaryotic genes places demands on the stability of the transcription complex. RNA polymerase II on its own is not able to meet these demands: when the purified enzyme is studied *in vitro* its polymerization rate is less than 300 nucleotides per minute because the enzyme pauses frequently on the template and sometimes stops altogether. In the nucleus, pausing and stopping are reduced because of the action of a series of **elongation factors**, proteins that associate with the polymerase after it has cleared the promoter and left behind the transcription factors involved in initiation (Conaway *et al.*, 2000). Thirteen elongation factors are currently known in mammalian cells, displaying a variety of functions (*Table 10.1*). Their importance is shown by the effects of mutations that disrupt the activity of one or other of the factors (Conaway and Conaway, 1999). Inactivation of CSB, for example, results in Cockayne syndrome, a disease characterized by developmental defects such as mental retardation, and disruption of ELL causes acute myeloid leukemia.

A second difference between bacterial and eukaryotic elongation is that RNA polymerase II, as well as the other eukaryotic nuclear polymerases, has to negotiate the nucleosomes that are attached to the template DNA that is being transcribed. At first glance it is difficult to imagine how the polymerase can elongate its transcript through a region of DNA wound around a nucleosome (see *Figure 2.5*, page 38). The solution to this problem is probably provided by elongation factors that are able to modify the chromatin structure in some way. In mammals, the elongation factor FACT has been shown to interact with histones H2A and H2B, possibly influencing nucleosome positioning, and less well defined interactions have been demonstrated for other factors (Orphanides and Reinberg, 2000). Yeast possesses a factor called **elongator**, which has tentatively been assigned a role in chromatin modification because it contains a subunit that has histone acetyltransferase activity (Section 8.2.1; Wittschieben *et al.*, 1999), but so far a homolog of this complex has not been identified in mammals. An intriguing question is whether the first polymerase to transcribe a particular gene is a 'pioneer' with a special elongation factor complement that opens up the chromatin structure, with subsequent rounds of transcription being performed by standard polymerase complexes that take advantage of the changes induced by the pioneer.

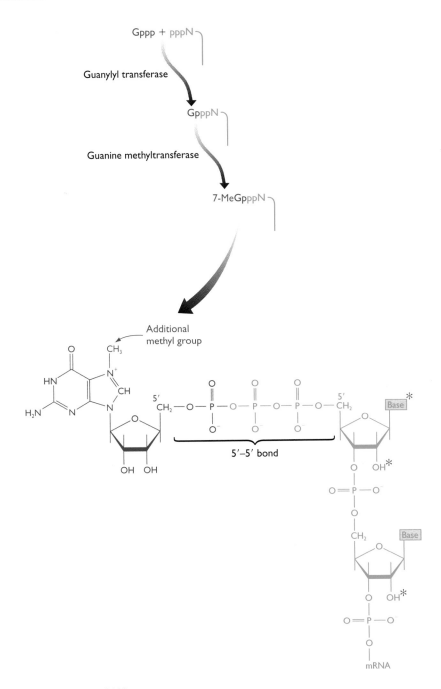

Figure 10.9 Capping of eukaryotic mRNA.

The top part of the diagram shows the capping reaction in outline. A GTP molecule (drawn as Gppp) reacts with the 5′ end of the mRNA to give a triphosphate linkage. In the second step of the process, the terminal G is methylated at nitrogen number 7. The bottom part of the diagram shows the chemical structure of the type 0 cap, with asterisks indicating the positions where additional methylations might occur to produce type 1 and type 2 cap structures.

Table 10.1 Examples of elongation factors for mammalian RNA polymerase II

Elongation factor	Function
CSB, ELL, Elongin	These factors suppress 'pausing' of RNA polymerase II, which can occur when the enzyme transcribes through a region where intra-strand base pairs (e.g. a hairpin loop) can form
SII	Prevents arrest (complete cessation of elongation)
FACT, HMG14	Thought to modify chromatin in order to assist elongation

Termination of mRNA synthesis is combined with polyadenylation

Virtually all eukaryotic mRNAs have a series of up to 250 adenosines at their 3′ ends. These As are not specified by the DNA and are added to the transcript by a template-independent RNA polymerase called **poly(A) polymerase** (Bard *et al.*, 2000). This polymerase does not act at the extreme 3′ end of the transcript, but at an internal site which is cleaved to create a new 3′ end to which the poly(A) tail is added.

The basic features of polyadenylation have been understood for some time. In mammals, polyadenylation is directed by a signal sequence in the mRNA, almost invariably 5′–AAUAAA–3′. This sequence is located between 10 and 30 nucleotides upstream of the polyadenylation site, which is often immediately after the dinucleotide 5′–CA–3′ and is followed 10–20 nucleotides later by a GU-rich region. Both the poly(A) signal sequence and the GU-rich region are binding sites for multi-subunit protein complexes, which are, respectively, the **cleavage and polyadenylation specificity factor (CPSF)** and the **cleavage stimulation factor (CstF)**. Poly(A) polymerase and at least two other protein factors must associate with bound CPSF and CstF in order for polyadenylation to occur (*Figure 10.10*). These additional factors include **polyadenylate-binding protein (PADP)**, which helps the polymerase to add the adenosines, possibly influences the length of the poly(A) tail that is synthesized, and appears to play a role in maintenance of the tail after synthesis. In yeast, the sig-

nal sequences in the transcript are slightly different, but the protein complexes are similar to those in mammals and polyadenylation is thought to occur by more or less the same mechanism (Guo and Sherman, 1996; Manley and Takagaki, 1996).

Polyadenylation was once looked on as a 'post-transcriptional' event but it is now recognized that the process is an inherent part of the mechanism for termination of transcription by RNA polymerase II. CPSF is known to interact with TFIID and is recruited into the polymerase complex during the initiation stage. By riding along the template with RNA polymerase II, CPSF is able to bind to the poly(A) signal sequence as soon as it is transcribed, initiating the polyadenylation reaction (*Figure 10.11*). Both CPSF and CstF form contacts with the CTD of the polymerase. It has been suggested that the nature of these contacts changes when the poly(A) signal sequence is located, and that this change alters the properties of the elongation complex so that termination becomes favored over continued RNA synthesis. As a result, transcription stops soon after the poly(A) signal sequence has been transcribed (Bentley, 1999).

Even though polyadenylation can be identified as an inherent part of the termination process, this does not explain why it is necessary to add a poly(A) tail to the transcript. A role for the poly(A) tail has been sought for several years, but no convincing evidence has been found for any of the various suggestions that have been made. These suggestions include an influence on mRNA stability, which seems unlikely as some stable transcripts have very short poly(A) tails,

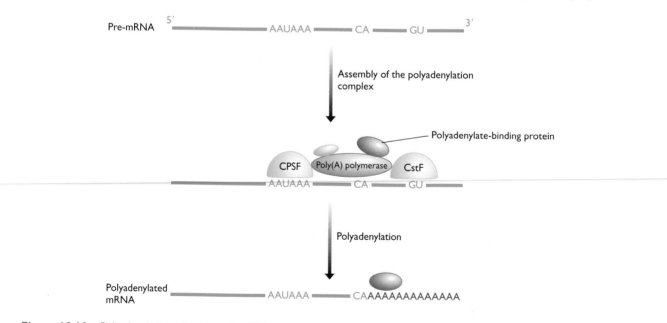

Figure 10.10 Polyadenylation of eukaryotic mRNA.

See the text for details. Note that the diagram is schematic and is not intended to indicate the relative sizes and shapes of the various protein complexes, nor their precise positioning, although CPSF and CstF are thought to bind to the 5′–AAUAAA–3′ and GU-rich sequences, respectively, as shown. Note that 'GU' indicates a GU-rich sequence rather than the dinucleotide 5′–GU–3′.

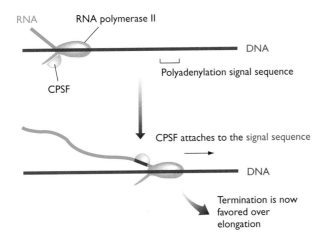

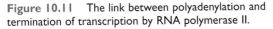

Figure 10.11 The link between polyadenylation and termination of transcription by RNA polymerase II.

CPSF is shown attached to the RNA polymerase II elongation complex that is synthesizing RNA. CPSF binds to the polyadenylation signal sequence as soon as it is transcribed. This changes the interaction between CPSF and the CTD of RNA polymerase II so that termination of transcription is now favored over continued elongation. Note that this is a schematic representation and ignores the possibility that CstF may also be a component of the elongation complex. This representation also shows CPSF leaving the complex in order to bind to the polyadenylation signal, when in reality it may maintain its attachment to RNA polymerase II during the polyadenylation process.

and a role in initiation of translation. The latter proposal is supported by research showing that poly(A) polymerase is repressed during those periods of the cell cycle when relatively little protein synthesis occurs (Colgan *et al.*, 1996).

10.1.3 Intron splicing

The existence of introns was not suspected until 1977 when DNA sequencing was first applied to eukaryotic genes and it was realized that many of these contain 'intervening sequences' that separate different segments

of the coding DNA from one another (*Figure 10.12*). We now recognize seven distinct types of intron in eukaryotes, and additional forms in the archaea (*Table 10.2*). Two of these types – the GU–AG and AU–AC introns – are found in eukaryotic protein-coding genes and are dealt with in this section; the other types will be covered later in the chapter.

Few rules can be established for the distribution of introns in protein-coding genes, beyond the fact that introns are less common in lower eukaryotes: the 6000 genes in the yeast genome contain only 239 introns in total, whereas many individual mammalian genes contain 50 or more introns. When the same gene is compared in related species, we usually find that some of the introns are in identical positions but that each species has one or more unique introns. This implies that some introns remain in place for millions of years, retaining their positions while species diversify, whereas others appear or disappear during this same period. This leads to two competing hypotheses for the evolution of introns:

- **'Introns late'** is the hypothesis that introns evolved relatively recently and are gradually accumulating in eukaryotic genomes.
- **'Introns early'** is the alternative hypothesis, that introns are very ancient and are gradually being lost from eukaryotic genomes.

These are issues that we will return to in Section 15.3.2 when we study molecular evolution. For the time being,

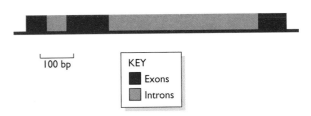

100 bp	KEY
	■ Exons
	■ Introns

Figure 10.12 Introns.

The structure of the human β-globin gene is shown. This gene is 1423 bp in length and contains two introns, one of 131 bp and one of 851 bp, which together make up 69% of the length of the gene.

Table 10.2 Types of intron

Intron type	Where found	Cross-reference
GU–AG introns	Eukaryotic nuclear pre-mRNA	Section 10.1.3
AU–AC introns	Eukaryotic nuclear pre-mRNA	Section 10.1.3
Group I	Eukaryotic nuclear pre-rRNA, organelle RNAs, few bacterial RNAs	Section 10.2.3
Group II	Organelle RNAs, some prokaryotic RNAs	Box 10.2
Group III	Organelle RNAs	Box 10.2
Twintrons	Organelle RNAs	Box 10.2
Pre-tRNA introns	Eukaryotic nuclear pre-tRNA	Section 10.2.3
Archaeal introns	Various RNAs	Box 10.2

Table 10.3   Introns in human genes

Gene	Length (kb)	Number of introns	Amount of the gene taken up by the introns (%)
Insulin	1.4	2	69
β-globin	1.6	2	61
Serum albumin	18	13	79
Type VII collagen	31	117	72
Factor VIII	186	25	95
Dystrophin	2400	78	98

Adapted from Strachan and Read (1999).

what is important is that a eukaryotic pre-mRNA may contain many introns, perhaps over 100, taking up a considerable length of the transcript (*Table 10.3*), and that these introns must be excised and the exons joined together in the correct order before the transcript can function as a mature mRNA.

Conserved sequence motifs indicate the key sites in GU–AG introns

With the vast bulk of pre-mRNA introns, the first two nucleotides of the intron sequence are 5′–GU–3′ and the last two 5′–AG–3′. They are therefore called 'GU–AG' introns and all members of this class are spliced in the same way. These conserved motifs were recognized soon after introns were discovered and it was immediately assumed that they must be important in the splicing process. As intron sequences started to accumulate in the databases it was realized that the GU–AG motifs are merely parts of longer consensus sequences that span the 5′ and 3′ splice sites. These consensus sequences vary in different types of eukaryote; in vertebrates they can be described as:

5′ splice site   5′–AG↓GUAAGU–3′
3′ splice site   5′–PyPyPyPyPyPyNCAG↓–3′

In these designations, 'Py' is one of the two pyrimidine nucleotides (U or C), 'N' is any nucleotide, and the arrow

indicates the exon–intron boundary. The 5′ splice site is also known as the **donor site** and the 3′ splice site as the **acceptor site**.

Other conserved sequences are present in some but not all eukaryotes. Introns in higher eukaryotes usually have a **polypyrimidine tract**, a pyrimidine-rich region located just upstream of the 3′ end of the intron sequence (*Figure 10.13*). This tract is less frequently seen in yeast introns, but these have an invariant 5′–UACUAAC–3′ sequence, located between 18 and 140 bp upstream of the 3′ splice site, which is not present in higher eukaryotes. The polypyrimidine tract and the 5′–UACUAAC–3′ sequence are *not* functionally equivalent, as described in the next two sections.

Outline of the splicing pathway for GU–AG introns

The conserved sequence motifs indicate important regions of GU–AG introns, regions that we would anticipate either acting as recognition sequences for RNA-binding proteins involved in splicing, or playing some other central role in the process. Early attempts to understand splicing were hindered by technical problems (in particular difficulties in developing a cell-free splicing system with which the process could be probed in detail), but during the 1990s there was an explosion of information. This work showed that the splicing pathway can be divided into two steps (*Figure 10.14*):

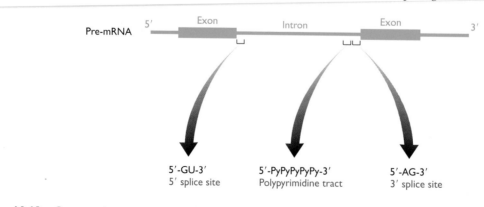

Figure 10.13   Conserved sequences in vertebrate introns.

The longer consensus sequences around the splice sites are given in the text. Abbreviation: Py, pyrimidine nucleotide (U or C).

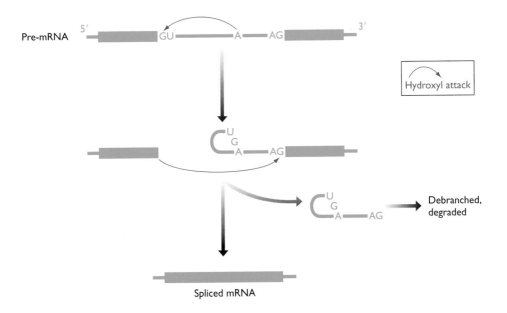

Figure 10.14   Splicing in outline.

Cleavage of the 5′ splice site is promoted by the hydroxyl (OH) attached to the 2′-carbon of an adenosine nucleotide within the intron sequence. This results in the lariat structure and is followed by the 3′-OH group of the upstream exon inducing cleavage of the 3′ splice site. This enables the two exons to be ligated, with the released intron being debranched and degraded.

- *Cleavage of the 5′ splice site* occurs by a transesterification reaction promoted by the hydroxyl group attached to the 2′ carbon of an adenosine nucleotide located within the intron sequence. In yeast, this adenosine is the last one in the conserved 5′–UACUAAC–3′ sequence. The result of the hydroxyl attack is cleavage of the phosphodiester bond at the 5′ splice site, accompanied by formation of a new 5′–2′ phosphodiester bond linking the first nucleotide of the intron (the G of the 5′–GU–3′ motif) with the internal adenosine. This means that the intron has now been looped back on itself to create a **lariat** structure.

- *Cleavage of the 3′ splice site and joining of the exons* result from a second transesterification reaction, this one promoted by the 3′–OH group attached to the end of the upstream exon. This group attacks the phosphodiester bond at the 3′ splice site, cleaving it and so releasing the intron as the lariat structure, which is subsequently converted back to a linear RNA and degraded. At the same time, the 3′ end of the upstream exon joins to the newly formed 5′ end of the downstream exon, completing the splicing process.

In a chemical sense, intron splicing is not a great challenge for the cell. It is simply a double transesterification reaction, no more complicated than many other biochemical reactions that are dealt with by individual enzymes. Why then has such a complex machinery evolved to deal with it? The difficulty lies with the topological problems. The first of these is the substantial distance that might lie between splice sites, possibly a few tens of kb, representing 100 nm or more if the mRNA is in the form of a linear chain. A means is therefore needed of bringing the splice sites into proximity. The second topological problem concerns selection of the correct splice site. All splice sites are similar, so if a pre-mRNA contains two or more introns then there is the possibility that the wrong splice sites could be joined, resulting in **exon skipping** – the loss of an exon from the mature mRNA (*Figure 10.15A*). Equally unfortunate would be selection of a **cryptic splice site**, a site within an intron or exon that has sequence similarity with the consensus motifs of real splice sites (*Figure 10.15B*). Cryptic sites are present in most pre-mRNAs and must be ignored by the splicing apparatus.

snRNAs and their associated proteins are the central components of the splicing apparatus

The central components of the splicing apparatus for GU–AG introns are the snRNAs called U1, U2, U4, U5 and U6. These are short molecules (between 106 nucleotides [U6] and 185 nucleotides [U2] in vertebrates) that associate with proteins to form **small nuclear ribonucleoproteins** (**snRNPs**) (*Figure 10.16*). The snRNPs, together with other accessory proteins, attach to the transcript and form a series of complexes, the last one of which is the **spliceosome**, the structure within which the actual splicing reactions occur (Smith and Valcárcel, 2000). The process operates as follows (*Figure 10.17*):

- The **commitment complex** initiates a splicing activity. This complex comprises U1–snRNP, which binds to the 5′ splice site, partly by RNA–RNA base-pairing, and the protein factors SF1, U2AF^{35} and U2AF^{65}, which make protein–RNA contacts with the branch site, the polypyrimidine tract and the 3′ splice site, respectively.

(A) Exon skipping

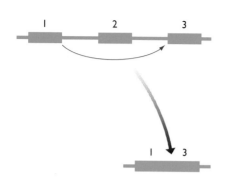

(B) Cryptic splice site selection

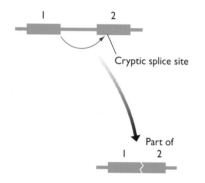

Cryptic splice site

Part of
1   2

Figure 10.15   Two aberrant forms of splicing.

(A) In exon skipping the aberrant splicing results in an exon being lost from the mRNA. (B) When a cryptic splice site is selected, part of an exon might be lost from the mRNA, as shown here, or if the cryptic site lies within an intron then a segment of that intron will be retained in the mRNA.

- The **pre-spliceosome complex** comprises the commitment complex plus U2–snRNP, the latter attached to the branch site. At this stage, an association between U1–snRNP and U2–snRNP brings the 5′ splice site into close proximity with the branch point.
- The **spliceosome** is formed when U4/U6–snRNP (a single snRNP containing two snRNAs) and U5–snRNP attach to the pre-spliceosome complex. This results in additional interactions that bring the 3′ splice site close to the 5′ site and the branch point. All three key positions in the intron are now in proximity and the two transesterifications occur as a linked reaction, possibly catalyzed by U6–snRNP, completing the splicing process.

The series of events shown in *Figure 10.17* provides no clues about how the correct splice sites are selected so that exons are not lost during splicing, and cryptic sites are ignored. This aspect of splicing is still poorly understood but it has become clear that a set of splicing factors called

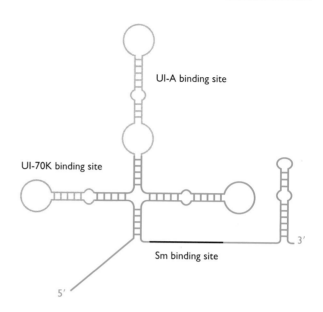

UI-A binding site

UI-70K binding site

Sm binding site

3′

5′

Figure 10.16   Structure of U1–snRNP.

The mammalian U1–snRNP comprises the 165-nucleotide U1–RNA plus ten proteins. Three of these (U1–70K, U1–A and U1–C) are specific to this snRNP, the other seven are Sm proteins that are found in all the snRNPs involved in splicing. The U1–RNA forms a base-paired structure as shown. The U1–70K and U1–A proteins attach to two of the major stem-loops of this base-paired structure, and U1–C attaches via a protein–protein interaction. The Sm proteins attach to the Sm site. Based on Stark *et al.* (2001).

SR proteins are important in splice-site selection. The SR proteins – so-called because their C-terminal domains contain a region rich in serine (abbreviation S) and arginine (R) – were first implicated in splicing when it was discovered that they are components of the spliceosome. They appear to have several functions, including the establishment of a connection between bound U1–snRNP and bound U2AF in the commitment complex (Valcárcel and Green, 1996). This is perhaps the clue to their role in splice-site selection, formation of the commitment complex being the critical stage of the splicing process, as this is the event that identifies which sites will be linked.

SR proteins also interact with **exonic splicing enhancers** (**ESEs**), which are purine-rich sequences located in the exon regions of a transcript (Blencowe, 2000). We are still at an early stage in our understanding of ESEs and their counterparts, the **exonic splicing silencers** (ESSs; Del Gatto-Konczak *et al.*, 1999), but their importance in controlling splicing is clear from the discovery that several human diseases, including one type of muscular dystrophy, are caused by mutations in ESE sequences. The location of ESEs and ESSs indicates that assembly of the spliceosome is driven not simply by contacts within the intron but also by interactions with adjacent exons. In fact, it is possible that an individual commitment complex is not assembled within an intron as shown in *Figure 10.17*, but initially bridges an exon

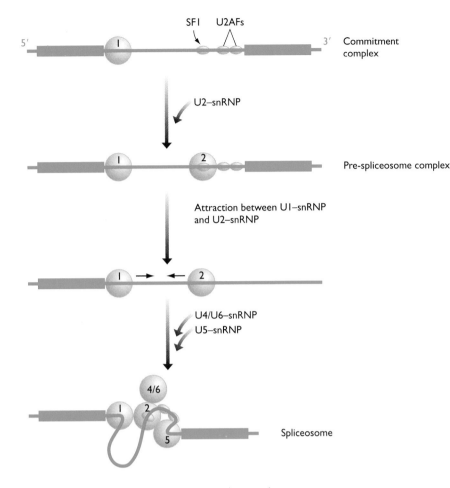

Figure 10.17   The roles of snRNPs and associated proteins during splicing.

See the text for details. There are several unanswered questions about the series of events occurring during splicing and it is unlikely that the scheme shown here is entirely accurate. The key point is that associations between the snRNPs are thought to bring the three critical parts of the intron – the two splice sites and the branch point – into close proximity.

(*Figure 10.18*). This model is attractive not only because it provides a means by which contact between an ESE or ESS and an SR protein could influence splicing, but also because it takes account of the large disparity between the lengths of exons and introns in vertebrate genes. In the human genome, for example, the exons have an average length of 145 bp compared with 3365 bp for introns (IHGSC, 2001). Initial assembly of a commitment complex across an exon might therefore be a less difficult task than assembly across a much longer intron.

There is one final aspect of SR proteins that we should address. This is the possibility that a subset of these SR proteins, called **CASPs** (**CTD-associated SR-like proteins**) or **SCAFs** (**SR-like CTD-associated factors**), form a physical connection between the spliceosome and the CTD of the RNA polymerase II transcription complex, and hence provide a link between transcript elongation and processing. As with some of the polyadenylation proteins (Section 10.1.2), it is probable that these splicing factors ride with the polymerase as it synthesizes the transcript, and are deposited at their appropriate posi-

tions at intron splice sites as soon as these are transcribed. Electron microscopy studies have shown that transcription and splicing occur together, and the discovery of splicing factors that have an affinity for RNA polymerase provides a biochemical basis for this observation (Corden and Patturajan, 1997).

Alternative splicing is common in many eukaryotes

When introns were first discovered it was imagined that each gene always gives rise to the same mRNA: in other words, that there is a single **splicing pathway** for each primary transcript (*Figure 10.19A*). This assumption was found to be incorrect in the 1980s, when it was shown that the primary transcripts of some genes can follow two or more **alternative splicing** pathways, enabling a single transcript to be processed into related but different mRNAs and hence to direct synthesis of a range of proteins (*Figure 10.19B*). In some organisms alternative splicing is uncommon, only three examples being known in *Saccharomyces cerevisiae*, but in higher eukaryotes it is

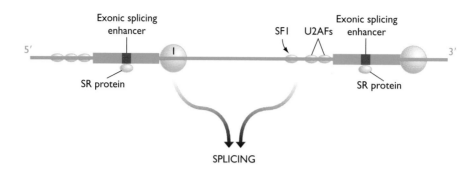

Figure 10.18 An alternative model for assembly of the commitment complex.

In this model, each individual commitment complex (one shown in orange and one in blue) is built up across an exon, bringing the complex into close association with an exonic splicing enhancer or silencer and its attached SR proteins. Once complexes have been built up over adjacent exons, splicing follows the pathway illustrated in *Figure 10.16*, the only difference being that the resulting spliceosome is made up of components of adjacent commitment complexes, rather than being derived from the single commitment complex shown in *Figure 10.16*.

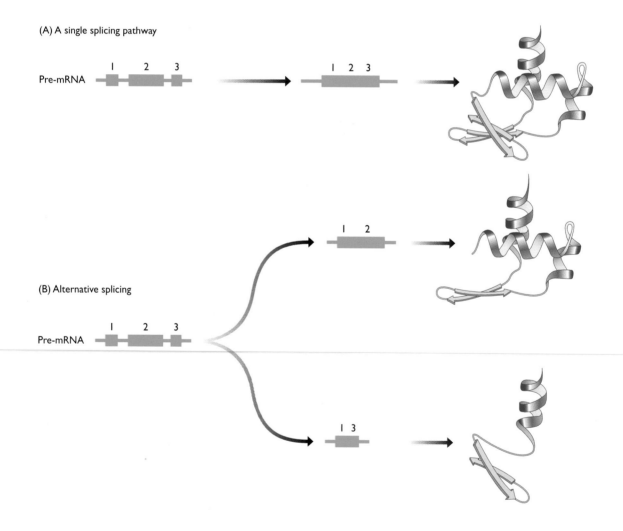

Figure 10.19 The assumption that each pre-mRNA follows a single splicing pathway was shown to be incorrect when alternative splicing was discovered.

(A) Sex-specific alternative splicing of *sxl* pre-mRNA

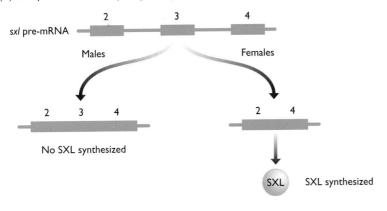

(B) SXL induces cryptic splice site selection in the *tra* pre-mRNA

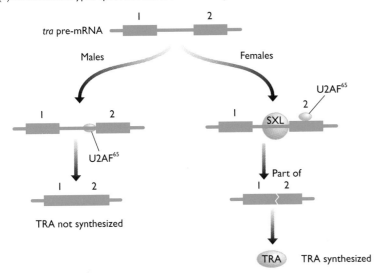

(C) TRA induces alternative splicing of the *dsx* pre-mRNA

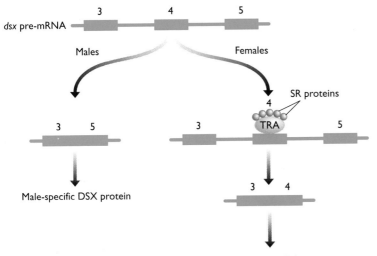

Figure 10.20   Regulation of splicing during expression of genes involved in sex determination in *Drosophila*.

(A) The cascade begins with sex-specific alternative splicing of the *sxl* pre-mRNA. In males all exons are present in the mRNA, but this means that a truncated protein is produced because exon 3 contains a termination codon. In females, exon 3 is skipped, leading to a full-length, functional SXL protein. (B) In females, SXL blocks the 3′ splice site in the first intron of the *tra* pre-mRNA. U2AF^{65} is unable to locate this site and instead directs splicing to a cryptic site in exon 2. This results in an mRNA that codes for a functional TRA protein. In males, there is no SXL so the 3′ splice site is not blocked and a dysfunctional mRNA is produced. (C) In males, exon 4 of the *dsx* pre-mRNA is skipped. The resulting mRNA codes for a male-specific DSX protein. In females, TRA stabilizes the attachment of SR proteins to an exonic splicing enhancer located within exon 4, so this exon is not skipped, resulting in the mRNA that codes for the female-specific DSX protein. The two versions of DSX are the primary determinants of the male and female physiologies. The female *dsx* mRNA ends with exon 4 because the intron between exons 4 and 5 has no 5′ splice site, meaning that exon 5 cannot be ligated to the end of exon 4. Instead a polyadenylation site at the end of exon 4 is recognized in females. See Chabot (1996) for more details. Note that the diagram is schematic and that the introns are not drawn to scale.

much more prevalent. This first became apparent when the draft *Drosophila* sequence was examined (Adams *et al.*, 2000), and it was realized that fruit flies have fewer genes that the microscopic worm *Caenorhabditis elegans* (see *Table 2.1*, page 31), despite the obviously greater physical complexity of *Drosophila*, which should be reflected in a more diverse proteome. The most likely explanation for the lack of congruence between the number of genes in the *Drosophila* genome and the number of proteins in its proteome is that a substantial number of the genes give rise to multiple proteins via alternative splicing. At about the same time, the first human chromosome sequences were obtained and it was recognized that rather than having 80 000–100 000 genes, as suggested by the size of the human proteome, humans have only 35 000 or so genes. It is now believed that at least 35% of the genes in the human genome undergo alternative splicing (Graveley, 2001): the principle 'one gene, one protein', biological dogma since the 1940s, has been completely overthrown.

Alternative splicing is now looked on as a crucial innovation in the genome expression pathway. Two examples will suffice to illustrate its importance. The first of these concerns sex, a fundamental aspect of the biology of any organism, and which in *Drosophila* is determined by an alternative splicing cascade (Chabot, 1996). The first gene in this cascade is *sxl*, whose transcript contains an optional exon which, when spliced to the one preceding it, results in an inactive version of protein SXL. In females the splicing pathway is such that this exon is skipped so that functional SXL is made (*Figure 10.20*). SXL promotes selection of a cryptic splice site in a second transcript, *tra*, by directing U2AF^{65} away from its normal 3' splice site to a second site further downstream. The resulting female-specific TRA protein is again involved in alternative splicing, this time by interacting with SR proteins to form a multifactor complex that attaches to an ESE within an exon of a third pre-mRNA, *dsx*, promoting selection of a secondary, female-specific splice site in this transcript. The male and female versions of the DSX proteins are the primary determinants of *Drosophila* sex.

The second example of alternative splicing illustrates the multiplicity of mRNAs synthesized from some primary transcripts. The human *slo* gene codes for a membrane protein that regulates the entry and exit of potassium ions into and out of cells (Graveley, 2001). The gene has 35 exons, eight of which are involved in alternative splicing events (*Figure 10.21*). The alternative splicing

pathways involve different combinations of the eight optional exons, leading to over 500 distinct mRNAs, each specifying a membrane protein with slightly different functional properties. What are the biological consequences of this example of multiple splicing? The human *slo* genes are active in the inner ear and determine the auditory properties of the hair cells on the basilar membrane of the cochlea. Different hair cells respond to different sound frequencies between 20 and 20 000 Hz, their individual capabilities determined in part by the properties of their Slo proteins. Alternative splicing of *slo* genes in cochlear hair cells therefore determines the auditory range of humans.

At present we do not understand how alternative splicing is regulated and cannot describe the process that determines which of several splicing pathways is followed by a particular transcript. The players are thought to be the SR proteins in conjunction with ESEs and ESSs, but the way in which they control splice site selection is not known.

AU–AC introns are similar to GU–AG introns but require a different splicing apparatus

One of the more surprising events of recent years has been the discovery of a few introns in eukaryotic pre-mRNAs that do not fall into the GU–AG category, having different consensus sequences at their splice sites. These are the **AU–AC introns** which, to date, have been found in approximately 20 genes in organisms as diverse as humans, plants and *Drosophila* (Nilsen, 1996; Tarn and Steitz, 1997).

As well as the sequences at their splice sites, AU–AC introns have a conserved (though not invariant) branch site sequence with the consensus 5'–UCCUUAAC–3', the last adenosine in this motif being the one that participates in the first transesterification reaction. This points us towards the remarkable feature of AU–AC introns: their splicing pathway is very similar to that for GU–AG introns, but involves a different set of splicing factors. Only the U5–snRNP is involved in the splicing mechanisms of both types of intron. The roles of U1–snRNP and U2–snRNP are taken by a previously discovered complex that had never been assigned a function. U11/U12–snRNP, and an entirely new U4atac/U6atac–snRNP have subsequently been isolated, completing the picture.

The splicing pathways for the 'major' and 'minor' types of intron are not identical but many of the interac-

Figure 10.21 The human *slo* gene.

The gene comprises 35 exons, shown as boxes, eight of which (in blue) are optional and appear in different combinations in different *slo* mRNAs. There are 8! = 40 320 possible splicing pathways and hence 40 320 possible mRNAs, but only some 500 of these are thought to be synthesized in the human cochlea. Based on Graveley (2001).

tions between the transcript and the snRNPs and other splicing proteins are remarkably similar. This means that AU–AC introns, rather than simply being a curiosity, are proving useful in testing models for interactions occurring during GU–AG intron splicing. The argument is that a predicted interaction between two components of the GU–AG spliceosome can be checked by seeing if the same interaction is possible with the equivalent AU–AC components. This has already been informative in helping to define a base-paired structure formed between the U2 and U6 snRNAs in the GU–AG spliceosome (Tarn and Steitz, 1996).

10.2 Synthesis and Processing of Non-coding RNAs

In bacteria, the same RNA polymerase synthesizes all types of RNA. The issues that we have already discussed regarding elongation and termination of bacterial mRNA (Section 10.1.1) therefore also hold for rRNA and tRNA synthesis, and the only outstanding areas that we have to cover are the processing of the pre-rRNAs and pre-tRNAs into the mature molecules. This processing involves cutting events and chemical modifications, both types of reaction being similar to equivalent processing events for eukaryotic rRNAs and tRNAs: we will therefore deal with bacterial and eukaryotic processing together, cutting events in Section 10.2.2 and chemical modifications in Section 10.3. The distinctive feature of eukaryotic rRNA and tRNA processing is the presence in some eukaryotic pre-RNAs of introns, different from the pre-mRNA introns described above; these will be covered in Section 10.2.3. First, however, there are issues regarding transcript elongation and termination by RNA polymerases I and III that we must address.

10.2.1 Transcript elongation and termination by RNA polymerases I and III

In general, we know less about transcript elongation and termination by RNA polymerases I and III than we do about equivalent processes for RNA polymerase II. The interaction of the polymerase with the template and transcript during elongation appears to be similar with all three enzymes, a reflection of the structural relatedness of the three largest subunits in each polymerase. One difference is the rate of transcription – RNA polymerase I, for example, being much slower than RNA polymerase II, managing a polymerization rate of only 20 nucleotides per minute, compared with 2000 per minute for mRNA synthesis. A second difference is that neither RNA polymerase I nor RNA polymerase III transcripts are capped. Various proteins that might act as elongation factors for RNA polymerase I or III have been isolated, including SGS1 and SRS2 of yeast, which code for two related DNA helicases. Mutations in the genes for SGS1 and SRS2 cause a reduction in RNA polymerase I transcription as

well as DNA replication (Lee *et al.*, 1999). SGS1 is interesting because it is a homolog of a pair of human proteins that are defective in the growth disorders Bloom's and Werner's syndromes (Section 7.4.2) but the exact involvement of SGS1 and SRS2, and other putative elongation factors, in transcription by RNA polymerases I and III is not known.

The major differences between the three polymerases are seen when the termination processes are compared. The polyadenylation system for RNA polymerase II termination (Section 10.1.2) is unique to that enzyme and no equivalent has been described for the other two polymerases. Termination by RNA polymerase I involves a DNA-binding protein, called Reb1p in *Saccharomyces cerevisiae* and TTF-I in mice, which attaches to the DNA at a recognition sequence located 12–20 bp downstream of the point at which transcription terminates (*Figure 10.22*). Exactly how the bound protein causes termination is not known, but a model in which the polymerase becomes stalled because of the blocking effect of Reb1p/TTF-I has been proposed (Reeder and Lang, 1997). A second protein, PTRF (polymerase I and transcript release factor), is thought to induce dissociation of the polymerase and the transcript from the DNA template (Jansa *et al.*, 1998). Even less is known about RNA polymerase III termination: a run of adenosines in the template is implicated but the process does not involve a hairpin loop and so is not analogous to termination in bacteria.

10.2.2 Cutting events involved in processing of bacterial and eukaryotic pre-rRNA and pre-tRNA

Bacteria synthesize three different rRNAs, called 5S rRNA, 16S rRNA and 23S rRNA, the names indicating the sizes of the molecules as measured by **sedimentation analysis** (see Technical Note 2.2, page 42). The three genes for these rRNAs are linked into a single transcription unit (which is usually present in multiple copies, seven for *E. coli*) and so the pre-rRNA contains copies of all three rRNAs. Cutting events are therefore needed to release the mature rRNAs. These cuts are made by various ribonucleases, at positions specified by double-stranded regions formed by base-pairing between different parts of the pre-rRNA (*Figure 10.23*). The cut ends are subsequently trimmed by exonucleases.

In eukaryotes there are four rRNAs. One of these, for the 5S rRNA, is transcribed by RNA polymerase III and does not undergo processing. The remaining three (the 5.8S, 18S and 28S rRNAs) are transcribed by RNA polymerase I from a single unit, producing a pre-rRNA which, as with the bacterial pre-rRNAs, is processed by cutting and end-trimming. Several nucleases are required, including the multifunctional **ribonuclease MRP** which, as well as 5.8S rRNA processing, is involved in replication of mitochondrial DNA and control of the cell cycle. MRP is particularly interesting because the various subunits of the enzyme include one that is made of RNA rather than

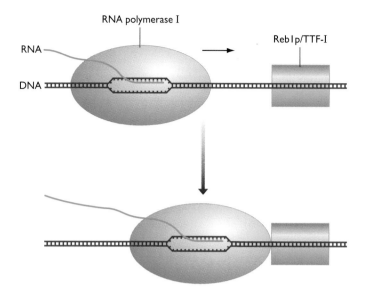

Figure 10.22 A possible scheme for termination of transcription by RNA polymerase I.

DNA (Clayton, 2001). RNA subunits are present in several enzymes involved in RNA processing, including RNase P (see *Figure 10.24*), and RNAs are important components of the snRNPs that participate in splicing of GU–AG and AU–AC introns (Section 10.1.3). The presence of RNA molecules in these processing enzymes and complexes is thought to be a relic of the **RNA world**, the early period of evolution when all biological reactions centered around RNA (Section 15.1.1).

In both bacteria and eukaryotes, tRNA genes occur singly and as multigene transcription units, or, in bacteria, as infiltrators within the rRNA transcription unit. The pre-tRNAs are also processed by a series of ribonucleases, as illustrated in *Figure 10.24*. All mature tRNAs must end with the trinucleotide 5′–CCA–3′. Some tRNAs have this sequence already; those that do not, or from which the 5′–CCA–3′ has been removed by the processing ribonucleases, have the motif added by **tRNA nucleotidyltransferase**.

10.2.3 Introns in eukaryotic pre-rRNA and pre-tRNA

Some eukaryotic pre-rRNAs and pre-tRNAs contain introns which must be spliced during the processing of these transcripts into mature RNAs. Neither type of intron is similar to the GU–AG and AU–AC introns of pre-mRNA.

Eukaryotic pre-rRNA introns are self-splicing

Introns are quite uncommon in eukaryotic pre-rRNAs but a few are known in microbial eukaryotes such as *Tetrahymena*. These introns are members of the Group I family

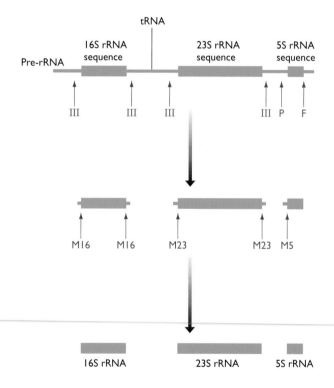

Figure 10.23 Pre-rRNA processing in *Escherichia coli*.

The pre-rRNA, containing copies of the 16S, 23S and 5S rRNAs, is cleaved by ribonucleases III, P and F, and the resulting molecules trimmed by ribonucleases M16, M23 and M5 to give the mature rRNAs. Eukaryotic pre-rRNAs, which contain the rRNA sequences in the order 18S–5.8S–28S, are processed in a similar fashion. Note that a tRNA is located between the 16S and 23S sequences in the *E. coli* pre-rRNA. This tRNA is processed as shown in *Figure 10.24*.

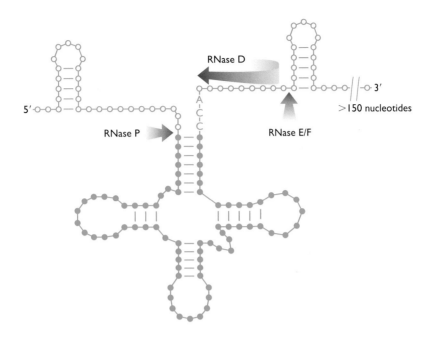

Figure 10.24   Processing of an *Escherichia coli* pre-tRNA.

The example shown results in synthesis of tRNA^{tyr}. The tRNA sequence in the primary transcript adopts its base-paired cloverleaf structure (see *Figure 11.2*, page 315) and two additional hairpin structures form, one on either side of the tRNA. Processing begins with the cut by ribonuclease E or F forming a new 3′ end just upstream of one of the hairpins. Ribonuclease D, which is an exonuclease, trims seven nucleotides from this new 3′ end and then pauses while ribonuclease P makes a cut at the start of the cloverleaf, forming the 5′ end of the mature mRNA. Ribonuclease D then removes two more nucleotides, creating the 3′ end of the mature molecule. With this tRNA the 3′-terminal CCA sequence is present in the RNA and is not removed by ribonuclease D. With some other tRNAs this sequence has to be completely or partly added by tRNA nucleotidyl-transferase. Abbreviation: RNase, ribonuclease. Based on Turner *et al.* (1997).

(see *Table 10.2*, page 287) and are also found in mitochondrial and chloroplast genomes, where they occur in pre-mRNA as well as pre-rRNA. A few isolated examples are known in bacteria, for instance in a tRNA gene of the cyanobacterium *Anabaena* and in the thymidylate synthase gene of the *E. coli* bacteriophage T4.

The splicing pathway for Group I introns is similar to that of pre-mRNA introns in that two transesterifications are involved. The first is induced not by a nucleotide within the intron but by a free nucleoside or nucleotide, any one of guanosine or guanosine mono-, di- or triphosphate (*Figure 10.25*). The 3′–OH of this cofactor attacks the phosphodiester bond at the 5′ splice site, cleaving it, with transfer of the G to the 5′ end of the intron. The second transesterification involves the 3′–OH at the end of the exon, which attacks the phosphodiester bond at the 3′ splice site, causing cleavage, joining of the two exons, and release of the intron. The released intron is linear, rather than the lariat seen with pre-mRNA introns, but may undergo additional transesterifications, leading to circular products, as part of its degradation process.

The remarkable feature of the Group I splicing pathway is that it proceeds in the absence of proteins and hence is autocatalytic, the RNA itself possessing enzymatic activity. This was the first example of an RNA

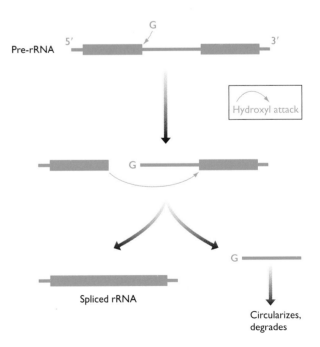

Figure 10.25   The splicing pathway for the *Tetrahymena* rRNA intron.

enzyme or **ribozyme** to be discovered, back in the early 1980s. Initially this caused quite a stir but it is now realized that, although uncommon, there are several examples of ribozymes (*Table 10.4*). The self-splicing activity of Group I introns resides in the base-paired structure taken up by the RNA. This structure was first described in two-dimensional terms by comparing the sequences of different Group I introns and working out a common base-paired arrangement that could be adopted by all versions. This resulted in a model comprising nine major base-paired regions (*Figure 10.26*; Burke *et al.*, 1987). More recently, the three-dimensional structure has been solved by X-ray crystallography (Cate *et al.*, 1996; Golden *et al.*, 1998). The ribozyme consists of a catalytic core made up of two domains, each one comprising two of the base-paired regions, with the splice sites brought into proximity by interactions between two other parts of the secondary structure. Although this RNA structure is sufficient for splicing, it is possible that with some introns the stability of the ribozyme is enhanced by non-catalytic protein factors that bind to it. This has long been suspected with the Group I introns in organelle genes, many of these containing an ORF coding for a protein called a **maturase** that appears to play a role in splicing.

Eukaryotic tRNA introns are variable but all are spliced by the same mechanism

Introns in eukaryotic pre-tRNA are relatively short and are usually found at the same position in the mature tRNA sequence, within the anticodon arm and loop (see *Figure 11.2*, page 315). Their sequences are variable, and no common motifs can be identified. Unlike all other types, splicing of pre-tRNA introns does not involve transesterifications. Instead, the two splice sites are cut by ribonuclease action, leaving a cyclic phosphate structure attached to the 3′ end of the upstream exon, and a

hydroxyl group at the 5′ end of the downstream exon (*Figure 10.27*). These ends are held in proximity by the natural base-pairing adopted by the tRNA sequence and are ligated by an RNA ligase.

10.3 Processing of Pre-RNA by Chemical Modification

The processing events that we have studied so far have been either chemical modifications that affect the ends of transcripts (capping, polyadenylation) or physical changes to the lengths of transcripts (splicing, cutting events). The final type of processing that occurs with pre-RNAs is the chemical modification of nucleotides within the transcript. This occurs with pre-rRNAs and pre-tRNAs of both bacteria and eukaryotes and, to a much lesser extent, with pre-mRNAs of eukaryotes. Equivalent events in the archaea are poorly understood.

A broad spectrum of chemical changes has been identified with different pre-RNAs: over 50 different modifications are known in total (*Table 10.5*). Most of these are carried out directly on an existing nucleotide within the transcript but two modified nucleotides, queosine and wyosine, are put in place by cutting out an entire nucleotide and replacing it with the modified version. Many of these modifications were first identified in tRNAs, within which approximately one in ten nucleotides becomes altered. These modifications are thought to mediate the recognition of individual tRNAs by the enzymes that attach amino acids to these molecules (Section 11.1.1), and to increase the range of the interactions that can occur between tRNAs and codons during translation, enabling a single tRNA to recognize more than one codon (Section 11.1.2).

We know relatively little about how tRNA modifications are carried out, beyond the fact that there are a number of enzymes that catalyze these changes. How the

Table 10.4 Examples of ribozymes

Ribozyme	Description
Self-splicing introns	Some introns of Groups I, II and III splice themselves by an autocatalytic process. There is also growing evidence that the splicing pathway of GU–AG introns includes at least some steps that are catalyzed by snRNAs (Newman, 2001)
Ribonuclease P	The enzyme that creates the 5′ ends of bacterial tRNAs (see Section 10.2.2) consists of an RNA subunit and a protein subunit, with the catalytic activity residing in the RNA
Ribosomal RNA	The peptidyl transferase activity required for peptide bond formation during protein synthesis (Section 11.2.3) is associated with the 23S rRNA of the large subunit of the ribosome
tRNA^{Phe}	Undergoes self-catalyzed cleavage in the presence of divalent lead ions
Virus genomes	Replication of the RNA genomes of some viruses involves self-catalyzed cleavage of chains of newly synthesized genomes linked head to tail. Examples are the plant viroids and virusoids and the animal hepatitis delta virus. These viruses form a diverse group with the self-cleaving activity specified by a variety of different base-paired structures, including a well-studied one that resembles a **hammerhead**.

For more details of ribozymes, see Doherty and Doudna (2000).

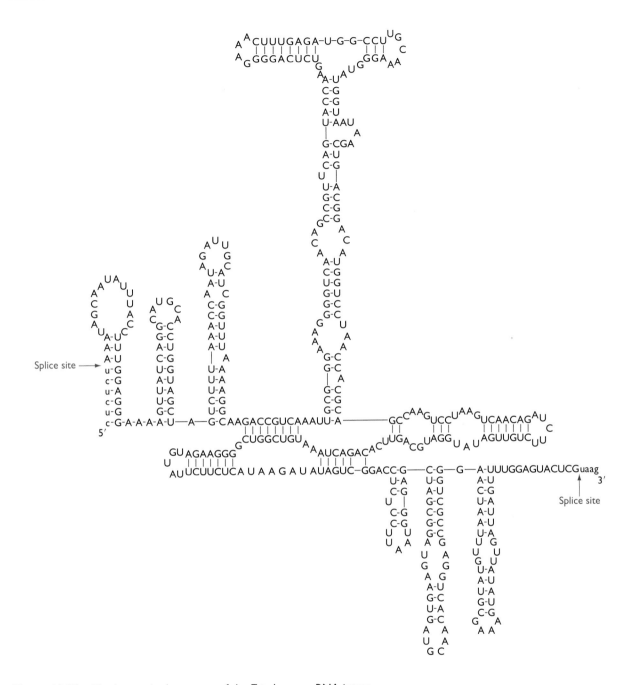

Figure 10.26   The base-paired structure of the *Tetrahymena* rRNA intron.

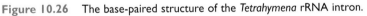

The sequence of the intron is shown in capital letters, with the exons in lower case. Additional interactions fold the intron into a three-dimensional structure that brings the two splice sites close together. Reprinted with permission from Burke *et al.* (1987) *Nucleic Acids Research*, 15, 7217–7221, Oxford University Press.

enzymes are directed to the correct nucleotides on which they must act has not been explained. With the rRNA and mRNA modifications we know rather less about the reasons for the chemical alterations but rather more about how the alterations are carried out. These issues are covered in the next two sections.

10.3.1 Chemical modification of pre-rRNAs

Ribosomal RNAs are modified in two ways: by addition of methyl groups to, mainly, the 2′–OH group on nucleotide sugars, and by conversion of uridine to pseudouridine (see *Table 10.5*). The same modification occurs at the same position on all copies of an rRNA, and

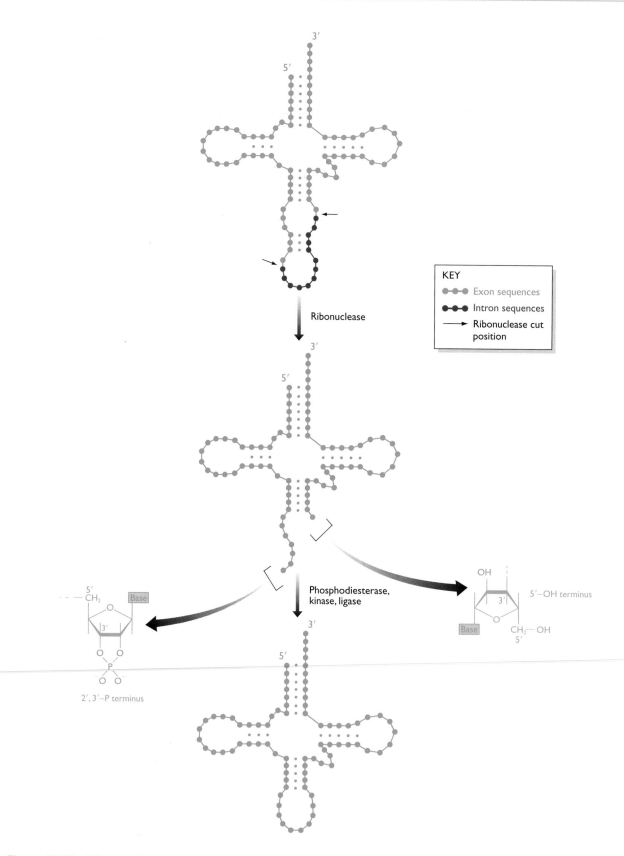

Figure 10.27 Splicing of the *Saccharomyces cerevisiae* pre-tRNA^{Tyr}.

See the text for details. In the second stage of the splicing pathway, the 2′,3′-P terminus is converted to a 3′–OH end by a phosphodiesterase, and the 5′–OH terminus is converted to 5′–P by a kinase. These two ends are then ligated together.

Table 10.5 Examples of chemical modifications occurring with nucleotides in rRNA and tRNA

Modification	Details	Example	
Methylation	Addition of one or more $-CH_3$ groups to the base or sugar		Methylation of guanosine gives 7-methylguanosine
Deamination	Removal of an amino ($-NH_2$) group from the base		Deamination of adenosine gives inosine
Sulfur substitution	Replacement of oxygen with sulfur		4-Thiouridine
Base isomerization	Changing the positions of atoms in the ring component of the base		Isomerization of uridine gives pseudouridine
Double-bond saturation	Converting a double bond to a single bond		Double bond saturation converts uridine to dihydrouridine
Nucleotide replacement	Replacement of an existing nucleotide with a new one		Queosine

---

Box 10.2: Other types of intron

There are eight different types of intron (see *Table 10.2*, page 287). Four types are described in the text: the nuclear pre-mRNA introns of the GU–AG and AU–AC classes, the self-splicing Group I introns, and the introns in eukaryotic pre-tRNA genes. The details of the four other categories are as follows:

■ **Group II introns** are found in the organelle genomes of fungi and plants, and a few are known in prokaryotes. Group II introns take up a characteristic secondary structure and they are able to self-splice *in vitro*, but they are distinct from Group I introns. The secondary structure is different, and the splicing mechanism is more closely allied to that of pre-mRNA introns, the initial transesterification being promoted by the hydroxyl group of an internal adenosine nucleotide and the intron being converted into a lariat structure. These similarities have prompted suggestions that Group II and pre-mRNA introns may have a common evolutionary origin (see Section 15.3.2).

■ **Group III introns** are also found in organelle genomes and they self-splice via a mechanism very similar to that of Group II introns, but Group III introns are smaller and have their own distinctive secondary structure. The resemblance to Group II introns again suggests an evolutionary relationship.

■ **Twintrons** are composite structures made up of two or more Group II and/or Group III introns. The simplest twintrons consist of one intron embedded in another, but more complex ones contain multiple embedded introns. The individual introns that make up a twintron are usually spliced in a defined sequence.

■ **Archaeal introns** are present in tRNA and rRNA genes. They are cleaved by a ribonuclease similar to the one involved in eukaryotic pre-tRNA splicing.

For more details concerning Group II introns see Bonen and Vogel (2001), and for Group III introns and twintrons see Copertino and Hallick (1993). Archaeal introns are described by Lykke-Andersen *et al.* (1997).

---

these modified positions are, to a certain extent, the same in different species. Some similarities in modification patterns are even seen when bacteria and eukaryotes are compared, although bacterial rRNAs are less heavily modified than eukaryotic ones. Functions for the modifications have not been identified, although most occur within those parts of rRNAs thought to be most critical to the activity of these molecules in ribosomes (Section 11.2.1). Modified nucleotides might, for example, be involved in rRNA-catalyzed reactions such as synthesis of peptide bonds.

It is not easy, simply by intuition, to imagine how specificity of rRNA modification can be ensured. Human pre-rRNA, for example, undergoes 106 methylations and 95 pseudouridinylations, each alteration at a specified position, with no obvious sequence similarities that can be inferred as target motifs for the modifying enzymes. Not surprisingly, progress in understanding rRNA modification was slow to begin with. The breakthrough came when it was shown that in eukaryotes the short RNAs called snoRNAs are involved in the modification process. These molecules are 70–100 nucleotides in length and are located in the nucleolus, the region within the nucleus where rRNA processing takes place. The initial discovery was that by base-pairing to the relevant region, snoRNAs pinpoint positions at which the pre-rRNA must be methylated. The base-pairing involves only a few nucleotides, not the entire length of the snoRNA, but these nucleotides are always located immediately upstream of a conserved sequence called the D box (*Figure 10.28A*). The base pair

involving the nucleotide that will be modified is five positions away from the D box. The hypothesis is that the D box is the recognition signal for the methylating enzyme, which is therefore directed towards the appropriate nucleotide (Bachellerie and Cavaillé, 1997). After these initial discoveries with regard to methylation, it was shown that a different family of snoRNAs carries out the same guiding role in conversion of uridines to pseudouridines (Maden, 1997). These snoRNAs do not have D boxes but still have conserved motifs that could be recognized by the modifying enzyme, and each is able to form a specific base-paired interaction with its target site, specifying the nucleotide to be modified.

The implication is that there is a different snoRNA for each modified position in a pre-rRNA, except possibly for a few sites that are close enough together to be dealt with by a single snoRNA. This means that there must be a few hundred snoRNAs per cell. At one time this seemed unlikely because very few snoRNA genes could be located, but now it appears that only a fraction of all the snoRNAs are transcribed from these standard genes, most being specified by sequences within the introns of other genes and released by cutting up the intron after splicing (*Figure 10.28B*).

The snoRNA system provides an elegant solution to site-specific chemical modification but it applies only to eukaryotic rRNAs. In contrast, the modifications made to bacterial rRNAs are carried out by enzymes that directly recognize the sequence and/or structures of the regions of RNA that contain the nucleotides to be modi-

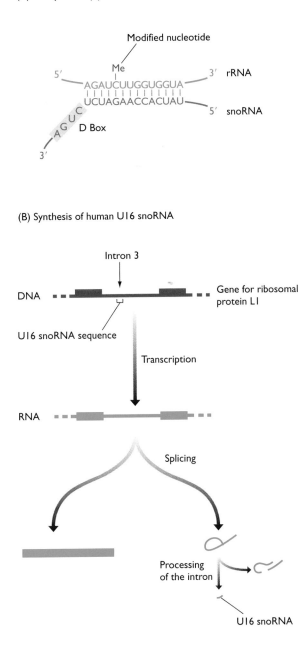

(A) Methylation by yeast U24 snoRNA

(B) Synthesis of human U16 snoRNA

Figure 10.28 Methylation of rRNA by a snoRNA.

(A) This example shows methylation of the C at position 1436 in the *Saccharomyces cerevisiae* 25S rRNA (equivalent to the 28S rRNA of vertebrates), directed by U24 snoRNA. The D box of the snoRNA is highlighted. Modification always occurs at the base pair five positions away from the D box. Note that the interaction between rRNA and snoRNA involves an unusual G–U base pair, which is permissible between RNA polynucleotides (see also *Figure 11.7A*, page 320). Based on Tollervey (1996). (B) Many snoRNAs are synthesized from intron RNA, as shown here for human U16 snoRNA, which is specified by a sequence in intron 3 of the gene for ribosomal protein L1.

fied. Often two or more nucleotides in the same region are modified at once. Bacterial rRNA modification is therefore similar to the systems for modifying tRNAs in both bacteria and eukaryotes.

10.3.2 RNA editing

Because rRNAs and tRNAs are non-coding, chemical modifications to their nucleotides affect only the structural features and, possibly, catalytic activities of the molecules. With mRNAs the situation is very different: chemical modification has the potential to change the coding properties of the transcript, resulting in an equivalent alteration in the amino acid sequence of the protein that is specified. A notable example of **RNA editing** occurs with the human mRNA for apolipoprotein B. The gene for this protein codes for a 456-amino-acid polypeptide, called apolipoprotein B100, which is synthesized in liver cells and secreted into the bloodstream where it transports lipids around the body. A related protein, apolipoprotein B48, is made by intestinal cells. This protein is only 2153 amino acids in length and is synthesized from an edited version of the mRNA for the full-length protein (*Figure 10.29*). In intestinal cells this mRNA is modified by deamination of a cytosine, converting this into a uracil. This changes a CAA codon, specifying glutamine, into a UAA codon, which causes translation to stop, resulting in the truncated protein. The deamination is carried out by an RNA-binding enzyme which, in conjunction with a set of auxiliary protein factors, binds to a sequence immediately downstream of the modification position within the mRNA (Smith and Sowden, 1996).

Although not common, RNA editing occurs in a number of different organisms and includes a variety of different nucleotide changes (*Table 10.6* and Box 10.3). Some editing events have a significant impact on the organism: in humans, editing is partly responsible for the generation of antibody diversity (Neuberger and Scott, 2000; Section 12.2.1) and has also been implicated in control of the HIV-1 infection cycle (Bourara *et al.*, 2000). One particularly interesting type of editing is the deamination of adenosine to inosine, which is carried out by enzymes called **adenosine deaminases acting on RNA (ADARs)** (Reenan, 2001). Some of the target mRNAs for these enzymes are selectively edited at a limited number of

positions. These positions are apparently specified by double-stranded segments of the pre-mRNA, formed by base-pairing between the modification site and sequences from adjacent introns. This type of editing occurs, for example, during processing of the mRNAs for mammalian glutamate receptors (Scott, 1997). Selective editing contrasts with the second type of modification carried out by ADARs, in which the target molecules become extensively deaminated, over 50% of the adenosines in the RNA becoming converted to inosines. Hyperediting has so far been observed mainly, but not exclusively, with viral RNAs and is thought to occur by chance, these RNAs adopting base-paired structures that fortuitously act as substrates for ADAR. It may, however, have physiological importance in the etiology of diseases caused by the edited viruses. This possibility is raised by the discovery that viral RNAs associated with persistent measles infections (as opposed to the more usual transient version of the disease) are hyperedited (Bass, 1997).

10.4 Degradation of mRNAs

So far this chapter has concentrated on synthesis of RNAs. Their degradation is equally important, especially with regard to mRNAs whose presence or absence in the cell determines which proteins will be synthesized. Degradation of specific mRNAs could be a powerful way of regulating genome expression.

The rate of degradation of an mRNA can be estimated by determining its half-life in the cell. The estimates show that there are considerable variations between and within organisms. Bacterial mRNAs are generally turned over very rapidly, their half-lives rarely being longer than a few minutes, a reflection of the rapid changes in protein synthesis patterns that can

Table 10.6 Examples of RNA editing in mammals

Tissue	Target RNA	Change	Comments
Intestine	Apolipoprotein B mRNA	C→U	Converts a glutamine codon to a stop codon
Muscle	α-galactosidase mRNA	U→A	Converts a phenylalanine codon into a tyrosine codon
Testis, tumors	Wilms tumor-1 mRNA	U→C	Converts a leucine codon into a proline codon
Tumors	Neurofibromatosis type-1 mRNA	C→U	Converts an arginine codon into a stop codon
B lymphocytes	Immunoglobulin mRNA	Various	Contributes to the generation of antibody diversity
HIV-infected cells	HIV-1 transcript	G→A, C→U	Involved in regulation of the HIV-1 infection cycle
Brain	Glutamate receptor mRNA	A→inosine	Multiple positions leading to various codon changes

Based on Smith and Sowden (1996), Scott (1997), Bourara *et al.* (2000) and Neuberger and Scott (2000). For more details about the generation of antibody diversity see Section 12.2.1.

Box 10.3: More complex forms of RNA editing

The examples of RNA editing described in the text and in *Table 10.6* are relatively straightforward events that lead to nucleotide changes at a single or limited number of positions in selected mRNAs. More complex types of RNA editing are also known:

■ **Pan-editing** involves the extensive insertion of nucleotides into abbreviated RNAs in order to produce functional molecules. It is particularly common in the mitochondria of trypanosomes, the parasitic protozoa that cause diseases such as sleeping sickness. Many of the RNAs transcribed in trypanosome mitochondria are specified by **cryptogenes** – sequences lacking some of the nucleotides present in the mature RNAs. The pre-RNAs transcribed from these cryptogenes are processed by multiple insertions of U nucleotides, at positions defined by short **guide RNAs**. These are short RNAs that can base-pair to the pre-RNA and which contain As at the positions where Us must be inserted (see figure).

■ Less extensive **insertional editing** occurs with some viral RNAs. For example, the paramyxovirus P gene gives rise to at least two different proteins because of the insertion of Gs at specific positions in the mRNA. These insertions are not

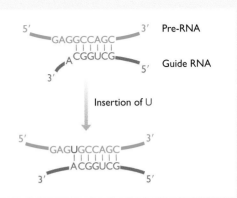

specified by guide RNAs: instead they are added by the RNA polymerase as the mRNA is being synthesized.

■ **Polyadenylation editing** is seen with many animal mitochondrial mRNAs. Five of the mRNAs transcribed from the human mitochondrial genome end with just a U or UA, rather than with one of the three termination codons (UAA, UAG in the human mitochondrial genetic code; see Section 3.3.2). Polyadenylation converts the terminal U or UA into UAAAA . . . , and so creates a termination codon. This is just one of several features that appear to have evolved in order to make vertebrate mitochondrial genomes as small as possible.

occur in an actively growing bacterium with a generation time of 20 minutes or so. Eukaryotic mRNAs are longer lived, with half-lives of, on average, 10–20 minutes for yeast and several hours for mammals. Within individual cells the variations are almost equally striking: some yeast mRNAs have half-lives of only 1 minute whereas for others the figure is more like 35 minutes (Tuite, 1996). These observations raise two questions: what are the processes for mRNA degradation, and how are these processes controlled?

10.4.1 Bacterial mRNAs are degraded in the 3'→5' direction

Studies of mutant bacteria whose mRNAs have extended half-lives have identified a range of ribonucleases and other RNA-degrading enzymes that are thought to be involved in mRNA degradation. These include (Carpousis *et al.*, 1999):

■ RNase E and RNase III, which are endonucleases that make internal cuts in RNA molecules;

■ RNase II, which is an exonuclease that removes nucleotides in the 3'→5' direction;

■ Polynucleotide phosphorylase (PNPase), which also removes nucleotides sequentially from the 3' end of an mRNA but, unlike true nucleases, requires inorganic phosphate as a co-substrate.

No enzyme capable of degrading RNA in the 5'→3' direction has yet been isolated from bacteria. This absence leads to the assumption that the main degradative process for bacterial mRNAs is removal of nucleotides from the 3' end. This is not possible under normal circumstances because most mRNAs have a hairpin structure near the 3' end, the same hairpin that induced termination of transcription (see *Figures 10.3* and *10.4*, pages 276 and 277). This structure blocks the progress of RNase II and PNPase, preventing them from gaining access to the coding part of the transcript (*Figure 10.30*). The model for mRNA degradation therefore begins with removal of the 3' terminal region, including the hairpin, by one of the endonucleases, exposing a new end from which RNase II and PNPase can enter the coding region, destroying the functional activity of the mRNA. Polyadenylation may also have a role. Although looked on primarily as a feature of eukaryotic mRNAs (Section 10.1.2), it has been known since 1975 that many bacterial transcripts have poly(A)

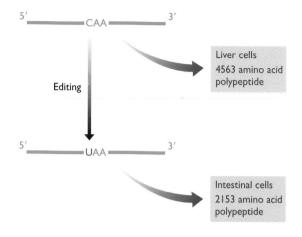

Figure 10.29   Editing of the human apolipoprotein B mRNA.

Conversion of a C to a U creates a termination codon, resulting in a shortened form of apolipoprotein B being synthesized in intestinal cells.

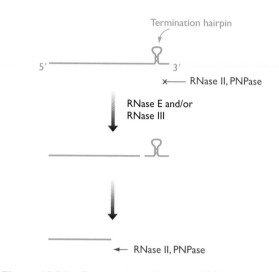

Figure 10.30   Degradation of bacterial RNA.

The termination hairpin blocks the exonuclease activities of RNase II and PNPase, and so must be removed by endonuclease action (RNase E and/or RNase III) before degradation can proceed.

tails at some stage in their existence, but that these tails are rapidly degraded. At present it is not clear whether polyadenylation precedes degradation of an mRNA, or whether it occurs at various intermediate stages after degradation has begun (Carpousis *et al.*, 1999).

In the cell, RNase E and PNPase are located within a multiprotein complex called the **degradosome**. Other components of the degradosome include an RNA helicase, which is thought to aid degradation by unwinding the double-helix structure of the stems of RNA stemloops. Fragments of rRNA occasionally co-purify with the degradosome, suggesting that the complex might be involved in both rRNA and mRNA degradation. But the exact role of the degradosome is still not clear and a few researchers are sceptical about its actual existence, pointing out that proteins not obviously involved in mRNA degradation, such as the glycolysis enzyme enolase, appear to be components of the degradosome, possibly indicating that the complex is an artefact that is produced during extraction of proteins from bacterial cells. A more significant gap in our knowledge concerns the way in which degradation is specifically targeted at individual mRNAs. We know that specific degradation occurs because mRNA degradation has been implicated in the regulation of several sets of bacterial genes, such as the *pap* operon of *E. coli*, which codes for proteins involved in synthesis of the cell surface pili (Baga *et al.*, 1988). Unfortunately, the process by which such control is exerted remains a mystery.

10.4.2 Eukaryotes have more diverse mechanisms for RNA degradation

Among eukaryotes, most progress in understanding mRNA degradation has been made with yeast. At least four pathways have been identified. One of these involves a multiprotein complex called the **exosome**, which degrades transcripts in the 3'→5' direction and contains nucleases related to the enzymes of the bacterial degradosome. Exosomes are probably also present in mammalian cells and are clearly important, but they are not particularly well studied. Their role may not be in mRNA degradation *per se*, but in monitoring polyadenylation and ensuring that transcripts that are about to leave the nucleus have an appropriate poly(A) tail (Hilleren *et al.*, 2001).

Rather more is known about two other eukaryotic mRNA degradation processes. The first of these is **deadenylation-dependent decapping** (*Figure 10.31*), which is triggered by removal of the poly(A) tail, possibly by exonuclease cleavage or possibly by loss of the polyadenylate binding protein which stabilizes the tail (Section 10.1.2). Poly(A) tail removal is followed by cleavage of the 5' cap by the decapping enzyme Dcp1p. Decapping prevents the mRNA from being translated (Section 11.2.2) and so ends its functional life. The mRNA then undergoes rapid exonuclease digestion from its 5' end. Whether or not an individual mRNA is degraded is probably determined by the ability of Dcp1p to gain access to the cap structure, which in turn depends on the association between the cap and the proteins that bind to it in order to initiate translation (Section 11.2.2). Degradation is also influenced, at least with some yeast mRNAs, by sequences called **instability elements**, located within the transcript. The importance of these sequences has been demonstrated by experiments

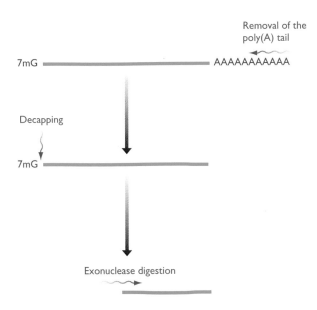

Figure 10.31 The deadenylation-dependent decapping pathway for degradation of an mRNA.

tion or as a result of incorrect splicing. The incorrect codon is thought to be detected by a 'surveillance' mechanism that involves a complex of proteins which scans the mRNA and somehow is able to distinguish between the correct termination codon, located at the end of the coding region of the transcript, and one that is in the wrong place (*Figure 10.32A*; Culbertson, 1999). There are a number of conceptual difficulties with this model because it is not easy to imagine how the surveillance complex could discriminate between correct and incorrect termination codons. Current hypotheses are based on the demonstration that the correct termination codon is recognized as aberrant if the transcript is engineered so that an exon–intron boundary is placed downstream of this termination codon (*Figure 10.32B*). The surveillance enzymes may therefore use exon–intron boundaries as orientation positions in order to distinguish the correct termination codon, which is usually downstream from the last intron (Kim *et al.*, 2001; Lykke-Andersen *et al.*, 2001). Alternative schemes have also been proposed, in which importance is placed not on the position of the termination codon but on the precise nature of the events involved in termination of translation at a premature stop codon compared with one that is at its correct position (Hilleren and Parker, 1999). Whatever the mechanism, identification of an incorrect termination codon induces cap cleavage and 5'→3' exonuclease degradation, without prior removal of the poly(A) tail, by proteins different to those involved in deadenylation-dependent decapping. Although NMD is designed primarily to degrade mRNAs that have become altered by mutation or have been incorrectly spliced, there is evidence that the pathway is also responsible for degradation of normal mRNAs, but probably not in a way that leads to control over expression of any individual gene.

in which an element is artificially deleted, which leads to increased translation and reduced degradation of the mRNA (Tucker and Parker, 2000).

The second well studied system for degradation of eukaryotic mRNAs is called **nonsense-mediated RNA decay (NMD)** or **mRNA surveillance**. The first of these names gives a clue to its function, because in molecular biology jargon a 'nonsense' sequence is a termination codon. NMD results in the specific degradation of mRNAs that have a termination codon at an incorrect position, either because the gene has undergone a muta-

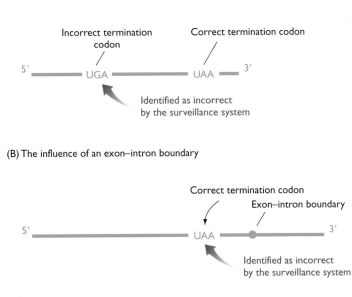

Figure 10.32 mRNA surveillance.

The systems described above represent the eukaryotic processes for controlled degradation of endogenous mRNAs. Eukaryotes also possess other RNA degradation mechanisms that have evolved largely to protect the cell from attack by foreign RNAs such as the genomes of viruses. An example is the pathway called **RNA interference**, a name that will be familiar because RNA interference has been adopted by genome researchers as a means of inactivating selected genes in order to study their function (Section 7.2.2). The target DNA for RNA interference must be double stranded, which excludes cellular mRNAs but encompasses many viral genomes. The double-stranded RNA is cleaved by a ribonuclease called **Dicer** into **short interfering RNAs (siRNAs)** of 21–25 nucleotides in length (Ambros, 2001). This inactivates the virus genome, but what if the virus genes have already been transcribed? If this has occurred then the harmful effects of the virus will already have been initiated and RNA interference would appear to have failed in its attempt to protect the cell from damage. One of the more remarkable discoveries of recent years has revealed a second stage of the interference process that is directed specifically at the viral mRNAs. The siRNAs produced by cleavage of the viral genome are separated into individual strands, one strand of each siRNA subsequently base-pairing to any viral mRNAs that are present in the cell. The double-stranded regions that are formed are target sites for the RDE-1 nuclease, which destroys the mRNAs (see *Figure 7.16*, page 201).

10.5 Transport of RNA Within the Eukaryotic Cell

In a typical mammalian cell, about 14% of the total RNA is present in the nucleus (Alberts *et al.*, 1994). About 80% of this nuclear fraction is RNA that is being processed before leaving for the cytoplasm. The other 20% is snRNAs and snoRNAs, playing an active role in the processing events, at least some of these molecules having already been to the cytoplasm where they were coated with protein molecules before being transported back into the nucleus. In other words, eukaryotic RNAs are continually being moved from nucleus to cytoplasm and possibly back to the nucleus again.

The only way for RNAs to leave or enter the nucleus is via one of the many **nuclear pore complexes** that cover the nuclear membrane (*Figure 10.33*). Initially looked upon as little more than a hole in the membrane, pore complexes are now regarded as complex structures that play an active role in movement of molecules into and out of the nucleus (Wente, 2000). Small molecules can move unimpeded through a pore complex but RNAs and most proteins are too large to diffuse through unaided and so have to be transported across by an energy-dependent process. As in many biochemical systems, the energy is obtained by hydrolysis of one of the high-energy phosphate–phosphate bonds in a ribonu-

cleotide triphosphate, in this case by converting GTP to GDP (other processes use ATP→ADP). Energy generation is carried out by a protein called Ran, and transport requires receptor proteins called **karyopherins**, or **exportins** and **importins** depending on the direction of their transport activity. There are at least 20 different human karyopherins, each responsible for the transport of a different class of molecule – mRNA, rRNA, etc. Examples are exportin-t, which has been identified as the karyopherin for export of tRNAs in yeasts and mammals. Transfer RNAs are directly recognized by exportin-t, but other types of RNA are probably exported by protein-specific karyopherins which recognize the proteins

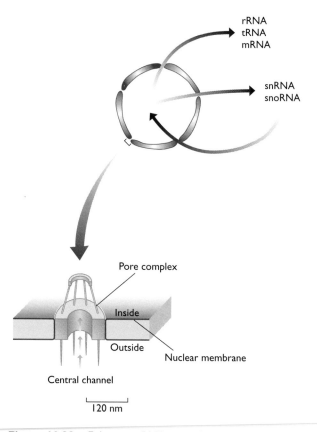

Figure 10.33 Eukaryotic RNAs must be transported through the nuclear pore complexes.

In eukaryotes, rRNAs, tRNAs and mRNAs are transported from the nucleus to the cytoplasm, where these molecules carry out their cellular functions. At least some of the snRNAs and snoRNAs are also transported to the cytoplasm, where they are coated with proteins before returning to the nucleus to carry out their roles in RNA processing. The nuclear pore is not simply a hole in the nuclear membrane. It contains a protein assembly comprising a ring embedded in the pore, with structures radiating into both the nucleus and the cytoplasm. Not shown in this diagram is the central channel complex, a 12 kDa protein that is thought to reside in the channel that connects the cytoplasm to the nucleus.

bound to the RNA, rather than the RNA itself. This also appears to be the case for import of snRNA from cytoplasm to nucleus, which makes use of importin β, a component of one of the protein transport pathways (Nigg, 1997; Weis, 1998).

Export of mRNAs is triggered by completion of the splicing pathway, possibly through the action of the protein called Yra1p in yeast and Aly in animals (Zhou *et al.*, 2000; Keys and Green, 2001). Once outside the nucleus, there are mechanisms that ensure that mRNAs are transported to their appropriate places in the cell. It is not known to what extent protein localization within the cell is due to translation of an mRNA at a specific position or to movement of the protein after it has been synthesized, but it is clear that at least some mRNAs are translated at defined places. For example, those mRNAs coding for proteins that are to be transferred into a mitochondrion are translated by ribosomes located on the surface of the organelle. It is assumed that protein 'address tags' are attached to mRNAs in order to direct them to their correct locations after they are transported out of the nucleus, but very little is known about this process.

References

Adams MA, Celniker SE, Holt RA, et al. (2000) The genome sequence of *Drosophila melanogaster*. *Science*, **287**, 2185–2195.

Alberts B, Bray D, Lewis J, Raff M, Roberts K and Watson JD (1994) *Molecular Biology of the Cell*, 3rd edition. Garland Publishing, New York.

Ambros V (2001) Dicing up RNAs. *Science*, **293**, 811–813.

Antson AA, Dodson EJ, Dodson G, et al. (1999) Structure of the *trp* RNA-binding attenuation protein, TRAP, bound to RNA. *Nature*, **401**, 235–242.

Bachellerie J-P and Cavaillé J (1997) Guiding ribose methylation of rRNA. *Trends Biochem. Sci.*, **22**, 257–261.

Baga M, Goransson M, Normark S and Uhlin BE (1988) Processed mRNA with differential stability in the regulation of *E. coli* pilin gene expression. *Cell*, **52**, 197–206.

Bard J, Zhelkovsky AM, Helmling S, Earnest TN, Moore CL and Bohm A (2000) Structure of yeast poly(A) polymerase alone and in complex with 3′-dATP. *Science*, **289**, 1346–1349.

Bass BL (1997) RNA editing and hypermutation by adenosine deamination. *Trends Biochem. Sci.*, **22**, 157–162.

Bentley D (1999) Coupling RNA polymerase II transcription with pre-mRNA processing. *Curr. Opin. Cell. Biol.*, **11**, 347–351.

Blencowe BJ (2000) Exonic splicing enhancers: mechanism of action, diversity and role in human genetic diseases. *Trends Biochem. Sci.*, **25**, 106–110.

Bonen L and Vogel J (2001) The ins and outs of group II introns. *Trends Genet.*, **17**, 322–331.

Bourara K, Litvak S and Araya A (2000) Generation of G-to-A and C-to-U changes in HIV-1 transcripts by RNA editing. *Science*, **289**, 1564–1566.

Burke JM, Belfort M, Cech TR, et al. (1987) Structural conventions for Group I introns. *Nucleic Acids Res.*, **15**, 7217–7221.

Carpousis AJ, Vanzo NF and Raynal LC (1999) mRNA degradation: a tale of poly(A) and multiprotein machines. *Trends Genet.*, **15**, 24–28.

Cate JH, Gooding AR, Podell E, et al. (1996) Crystal structure of a Group I ribozyme domain: principles of RNA packing. *Science*, **273**, 1678–1685.

Chabot B (1996) Directing alternative splicing: cast and scenarios. *Trends Genet.*, **12**, 472–478.

Clayton DA (2001) A big development for a small RNA. *Nature*, **410**, 29–31.

Colgan DF, Murthy KGK, Prives C and Manley JL (1996) Cell cycle regulation of poly(A) polymerase by phosphorylation. *Nature*, **384**, 282–285.

Conaway JW and Conaway RC (1999) Transcription elongation and human disease. *Ann. Rev. Biochem.*, **68**, 301–309.

Conaway JW, Shilatifard A, Dvir A and Conaway RC (2000) Control of elongation by RNA polymerase II. *Trends Biochem. Sci.*, **25**, 375–380.

Copertino DW and Hallick RB (1993) Group II and Group III introns of twintrons: potential relationships with nuclear pre-mRNA introns. *Trends Biochem. Sci.*, **18**, 467–471.

Corden JL and Patturajan M (1997) A CTD function linking transcription to splicing. *Trends Biochem. Sci.*, **22**, 413–416.

Culbertson MR (1999) RNA surveillance: unforeseen consequences for gene expression, inherited genetic disorders and cancer. *Trends Genet.*, **15**, 74–80.

Del Gatto-Konczak F, Olive M, Gesnel MC and Breathnach R (1999) hnRNP A1 recruited to an exon *in vivo* can function as an exon splicing silencer. *Mol. Cell. Biol.*, **19**, 251–260.

Doherty EA and Doudna JA (2000) Ribozyme structures and mechanisms. *Ann. Rev. Biochem.*, **69**, 597–615.

Friedman DI, Imperiale MJ and Adhya SL (1987) RNA 3′ end formation in the control of gene expression. *Ann. Rev. Genet.*, **21**, 453–488.

Golden BL, Gooding AR, Podell ER and Cech TR (1998) A pre-organized active site in the crystal structure of the *Tetrahymena* ribozyme. *Science*, **282**, 259–264.

Graveley BR (2001) Alternative splicing: increasing diversity in the proteomic world. *Trends Genet.*, **17**, 100–107.

Guo Z and Sherman F (1996) 3′-end forming signals of yeast mRNA. *Trends Biochem. Sci.*, **21**, 477–481.

Hilleren P and Parker R (1999) Mechanisms of mRNA surveillance in eukaryotes. *Ann. Rev. Genet.*, **33**, 229–260.

Hilleren P, McCarthy T, Rosbach M, Parker R and Jensen TH (2001) Quality control of mRNA 3′ end processing is linked to the nuclear exosome. *Nature*, **413**, 538–542.

IHGSC (International Human Genome Sequencing Consortium) (2001) Initial sequencing and analysis of the human genome. *Nature*, **409**, 860–921.

Jansa P, Mason SW, Hoffman-Rohrer U and Grummt I (1998) Cloning and functional characterization of PTRF, a novel protein which induces dissociation of paused ternary transcription complexes. *EMBO J.*, **17**, 2855–2864.

Keys RA and Green MR (2001) The odd coupling. *Nature*, **413**, 583–585.

Kim VN, Kataoka N and Dreyfuss G (2001) Role of nonsense-mediated decay factor hUpf3 in the splicing-dependent exon–exon junction complex. *Science*, **293**, 1832–1836.

Klug A (2001) A marvellous machine for making messages. *Science*, **292**, 1844–1846.

Korzheva N, Mustaev A, Kozlov M, et al. (2000) A structural model of transcription elongation. *Science*, **289**, 619–625.

Lee S-K, Johnson RE, Yu S-L, Prakash L and Prakash S (1999) Requirement of yeast *SGS1* and *SRS2* genes for replication and transcription. *Science*, **286**, 2339–2342.

Lee TI and Young RA (2000) Transcription of eukaryotic protein-coding genes. *Ann. Rev. Genet.*, **34**, 77–137.

Lykke-Andersen J, Aagaard C, Semionenkov M and Garrett RA (1997) Archaeal introns: splicing, intercellular mobility and evolution. *Trends Biochem. Sci.*, **22**, 326–331.

Lykke-Andersen J, Shu M-D and Steitz JA (2001) Communication of the position of exon-exon junctions to the mRNA surveillance machinery by the protein RNPS1. *Science*, **293**, 1836–1839.

Maden BEH (1997) Eukaryotic ribosomal RNA: guides to 95 new angles. *Nature*, **389**, 129–131.

Manley JL and Takagaki Y (1996) The end of the message – another link between yeast and mammals. *Science*, **274**, 1481–1482.

Neuberger MS and Scott J (2000) RNA editing AIDs antibody diversification. *Science*, **289**, 1705–1706.

Newman A (2001) RNA enzymes for RNA splicing. *Nature*, **413**, 695–696.

Nigg EA (1997) Nucleocytoplasmic transport: signals, mechanisms and regulation. *Nature*, **386**, 779–787.

Nilsen TW (1996) A parallel spliceosome. *Science*, **273**, 1813.

Orphanides G and Reinberg D (2000) RNA polymerase II elongation through chromatin. *Nature*, **407**, 471–475.

Proudfoot N (2000) Connecting transcription to messenger RNA processing. *Trends Biochem. Sci.*, **25**, 290–293.

Reeder RH and Lang WH (1997) Terminating transcription in eukaryotes: lessons learned from RNA polymerase I. *Trends Biochem. Sci.*, **22**, 473–477.

Reenan RA (2001) The RNA world meets behavior: A→I pre-mRNA editing in animals. *Trends Genet.*, **17**, 53–56.

Scott J (1997) RNA editing: message change for a fat controller. *Nature*, **387**, 242–243.

Smith CWJ and Valcárcel J (2000) Alternative pre-mRNA splicing: the logic of combinatorial control. *Trends Biochem. Sci.*, **25**, 381–388.

Smith HC and Sowden MP (1996) Base-modification mRNA editing through deamination – the good, the bad and the unregulated. *Trends Genet.*, **12**, 418–424.

Stark H, Dube P, Lührmann R and Kastner B (2001) Arrangement of RNA and proteins in the spliceosomal U1 small nuclear ribonucleoprotein particle. *Nature*, **409**, 539–542.

Strachan T and Read AP (1999) *Human Molecular Genetics*, 2nd edition. BIOS Scientific Publishers, Oxford.

Tarn W-Y and Steitz JA (1996) Highly diverged U4 and U6 small nuclear RNAs required for splicing rare AT–AC introns. *Science*, **273**, 1824–1832.

Tarn W-Y and Steitz JA (1997) Pre-mRNA splicing: the discovery of a new spliceosome doubles the challenge. *Trends Biochem. Sci.*, **22**, 132–137.

Tollervey D (1996) Small nucleolar RNAs guide ribosomal RNA methylation. *Science*, **273**, 1056–1057.

Toulokhonov I, Artsimovitch I and Landick R (2001) Allosteric control of RNA polymerase by a site that contacts nascent RNA hairpins. *Science*, **292**, 730–733.

Tucker M and Packer R (2000) Mechanisms and control of mRNA decapping in *Saccharomyces cerevisiae*. *Ann. Rev. Biochem.*, **69**, 571–595.

Tuite MF (1996) RNA processing: death by decapitation for mRNA. *Nature*, **382**, 577–579.

Turner PC, McLennan AG, Bates AD and White MRH (1997) *Instant Notes in Molecular Biology*. BIOS Scientific Publishers, Oxford.

Valcárcel J and Green MR (1996) The SR protein family: pleiotropic functions in pre-mRNA splicing. *Trends Biochem. Sci.*, **21**, 296–301.

von Hippel PH (1998) An integrated model of the transcription complex in elongation, termination and editing. *Science*, **281**, 660–665.

Weis K (1998) Importins and exportins: how to get in and out of the nucleus. *Trends Biochem. Sci.*, **23**, 185–189.

Wente SR (2000) Gatekeepers of the nucleus. *Science*, **288**, 1374–1377.

Wittschieben BO, Otero G, de Bizemont T, et al. (1999) A novel histone acetyltransferase is an integral subunit of elongating RNA polymerase II holoenzyme. *Mol. Cell*, **4**, 123–128.

Zhou Z, Luo M-J, Straesser K, Katahira J, Hurt E and Reed R (2000) The protein Aly links pre-messenger RNA splicing to nuclear export in metazoans. *Nature*, **407**, 401–405.

Further Reading

Cech TR (1990) Self-splicing of group I introns. *Ann. Rev. Biochem.*, **59**, 543–568. — *Written by one of the discoverers of autocatalytic RNA.*

Geiduschek EP and Kassavetis GA (2001) The RNA polymerase III transcription apparatus. *J. Mol. Biol.*, **310**, 1–26.

Gott JM and Emeson RB (2000) Functions and mechanisms of RNA editing. *Ann. Rev. Genet.*, **34**, 499–531.

Grunberg-Manago M (1999) Messenger RNA stability and its role in control of gene expression in bacteria and phages. *Ann. Rev. Genet.*, **33**, 193–227.

Henkin TM (1996) Control of transcription termination in prokaryotes. *Ann. Rev. Genet.*, **30**, 35–57. — *A detailed account of antitermination and attenuation.*

Losick RL and Sonenshein AL (2001) Turning gene regulation on its head. *Science*, **293**, 2018–2019. — *Describes the attenuation systems at the tryptophan operons of* E. coli *and* B. subtilis.

Maxwell ES and Fournier MJ (1995) The small nucleolar RNAs. *Ann. Rev. Biochem.*, **64**, 897–934.

Nagai K and Mattaj IW (eds) (1994) *RNA–Protein Interactions*. IRL Press, Oxford. — *Chapter 7 describes RNA–protein interactions in snRNPs.*

Venema J and Tollervey D (1999) Ribosome synthesis in *Saccharomyces cerevisiae*. *Ann. Rev. Genet.*, **33**, 261–311. — *Extensive details on rRNA processing.*

STUDY AIDS FOR CHAPTER 10

Key terms

Give short definitions of the following terms:

Acceptor site

Adenosine deaminase acting on RNA (ADAR)

Alternative splicing

Antitermination

Antiterminator protein

Attenuation

AU–AC intron

Cleavage and polyadenylation specificity factor (CPSF)

Cleavage stimulation factor (CstF)

Commitment complex

Cryptic splice site

Cryptogene

CTD-associated SR-like protein (CASP)

Deadenylation-dependent decapping

Degradosome

Dicer

Donor site

Elongation factor

Elongator

Exon skipping

Exonic splicing enhancer (ESE)

Exonic splicing silencer (ESS)

Exosome

Exportin

Group I intron

Group II intron

Group III intron

GU–AG intron

Guanine methyltransferase
Guanylyl transferase
Guide RNA
Hammerhead
Helicase
Importin
Insertional editing
Instability element
Intrinsic terminator
Karyopherin
Lariat
Maturase
mRNA surveillance
Nonsense-mediated RNA
  decay (NMD)
Nuclear pore complex
Pan-editing
Poly(A) polymerase
Polyadenylate-binding protein
  (PADP)
Polyadenylation editing
Polypyrimidine tract
Pre-spliceosome complex
Promoter clearance
Promoter escape

Rho dependent terminator
Ribonuclease MRP
Ribozyme
RNA editing
RNA interference
RNA world
Sedimentation analysis
Short interfering RNA
  (siRNA)
Small nuclear
  ribonucleoprotein (snRNP)
Spliceosome
Splicing pathway
SR protein
SR-like CTD-associated
  factor (SCAF)
Transcription bubble
tRNA nucleotidyltransferase
trp RNA-binding attenuation
  protein (TRAP)
Twintron
Type 0 cap
Type 1 cap
Type 2 cap

Self study questions

1. Outline the important features of the elongation phase of transcription in *Escherichia coli*.
2. Describe how transcription is terminated in *Escherichia coli*.
3. Using diagrams and specific examples, indicate how the processes called antitermination and attenuation influence transcription in bacteria.
4. Describe the series of events that result in capping of a eukaryotic mRNA.
5. Name and outline the functions of three different elongation factors for mammalian RNA polymerase II.
6. Draw a series of diagrams to illustrate how a eukaryotic mRNA becomes polyadenylated.
7. What are the key sequence features of a GU–AG intron?
8. Give a detailed description of the series of events involved in splicing a GU–AG intron.
9. What processes are thought to ensure that the correct splice sites are selected during splicing of a GU–AG intron?
10. Give two examples to illustrate the importance of alternative splicing in genome expression.
11. Why are AU–AC introns remarkable?
12. Outline our current knowledge regarding elongation and termination of transcription by RNA polymerases I and III.
13. Describe the cutting events involved in processing of bacterial and eukaryotic pre-rRNA and pre-tRNA.

14. What is meant by 'self-splicing'? Give details of the types of intron that display self-splicing. In your answer, distinguish between those introns that self-splice *in vivo* and those that only display this property *in vitro*.
15. What is a ribozyme? Compile an annotated list of known ribozymes.
16. List six types of chemical modification that occur with nucleotides in rRNA and tRNA. In each case, draw the structure of an example of a nucleotide resulting from the modification.
17. Write an essay on 'The role of snoRNAs in rRNA processing'.
18. Give details of two examples of mRNA editing that occur in mammals.
19. Outline the more complex forms of RNA editing that are known in various eukaryotes.
20. Describe the processes of mRNA degradation in bacteria. How does the bacterial degradosome compare with the eukaryotic exosome?
21. Distinguish between deadenylation-dependent decapping and nonsense-mediated RNA decay.
22. What is Dicer and what does it do?
23. Outline how eukaryotic RNAs are transported from the nucleus to the cytoplasm.

Problem-based learning

1. 'Current thinking views transcription as a stepwise nucleotide-by-nucleotide process, with the polymerase pausing at each position and making a "choice" between continuing elongation by adding another ribonucleotide to the transcript, or terminating by dissociating from the template. Which choice is selected depends on which alternative is more favorable in thermodynamic terms.' Evaluate this view of transcription.
2. Explore the introns-early and introns-late hypotheses. Is it possible to devise an analysis that will distinguish which of these two hypotheses is correct?
3. *Figure 10.18* (page 292) shows a model for splicing of GU–AG introns in which individual commitment complexes are assembled across exons rather than within introns. According to this model, how might alternative splicing be regulated? Devise experiments to test your ideas.
4. To what extent has the study of AU–AC introns provided insights into the details of GU–AG intron splicing?
5. The existence of ribozymes is looked upon as evidence that RNA evolved before proteins and therefore at one time, during the earliest stages of evolution, all enzymes were made of RNA. Assuming that this hypothesis is correct, explain why some ribozymes persist to the present day.
6. Using the current information on RNA degradation, devise a hypothesis to explain how specific mRNAs could be individually degraded. Can your hypothesis be tested?

Synthesis and Processing of the Proteome

Chapter Contents

11.1 *The Role of tRNA in Protein Synthesis*   *314*

    11.1.1   Aminoacylation: the attachment of amino acids to tRNAs   315

    11.1.2   Codon–anticodon interactions: the attachment of tRNAs to mRNA   317

11.2 *The Role of the Ribosome in Protein Synthesis*   *319*

    11.2.1   Ribosome structure   321

    11.2.2   Initiation of translation   323

    11.2.3   Elongation of translation   328

    11.2.4   Termination of translation   333

11.3 *Post-translational Processing of Proteins*   *335*

    11.3.1   Protein folding   335

    11.3.2   Processing by proteolytic cleavage   339

    11.3.3   Processing by chemical modification   340

    11.3.4   Inteins   342

11.4 *Protein Degradation*   *343*

Learning outcomes

When you have read Chapter 11, you should be able to:

- Draw the general structure of a transfer RNA (tRNA) and explain how this structure enables the tRNA to play both a physical and an informational role during protein synthesis

- Describe how an amino acid becomes attached to a tRNA and outline the processes that ensure that combinations are formed between the correct pairs of amino acids and tRNAs

- Explain how codons and anticodons interact, and discuss the influence of wobble on this interaction

- Outline the techniques that have been used to study the structure of the ribosome, and summarize the information that has resulted from these studies

- Give a detailed description of the process of translation in bacteria and eukaryotes, with emphasis on the roles of the various translation factors, this description including an explanation of how translation is regulated and an outline of the unusual events, such as frameshifting, that can occur during the elongation phase

- Explain why post-translational processing of proteins is an important component of the genome expression pathway, and describe the key features of protein folding, protein processing by proteolytic cleavage and chemical modification, and intein splicing

- Describe the major processes responsible for protein degradation in bacteria and eukaryotes

THE END RESULT of genome expression is the proteome, the collection of functioning proteins synthesized by a living cell. The identity and relative abundance of the individual proteins in a proteome represents a balance between the synthesis of new proteins and the degradation of existing ones. The biochemical capabilities of the proteome can also be changed by chemical modification and other processing events. The combination of synthesis, degradation, and modification/processing enables the proteome to meet the changing requirements of the cell and to respond to external stimuli.

In this chapter we will study the synthesis, processing and degradation of the components of the proteome. To understand protein synthesis we will first examine the role of tRNAs in decoding the genetic code and then investigate the events, occurring at the ribosome, that result in polymerization of amino acids into polypeptides. The ribosomal events are sometimes looked upon as the final stage in expression of an individual gene but the polypeptide that is initially synthesized is inactive until it has been folded, and may also have to undergo cutting and chemical modification before it becomes functional. We will study these pro-

cessing events in Section 11.3. At the end of the chapter we will investigate how the cell degrades proteins that it no longer requires.

11.1 The Role of tRNA in Protein Synthesis

Transfer RNAs play the central role in translation. They are the adaptor molecules, whose existence was predicted by Francis Crick in 1956 (Crick, 1990), which form the link between the mRNA and the polypeptide that is being synthesized. This is both a *physical* link, tRNAs binding to both the mRNA and the growing polypeptide, and an *informational* link, tRNAs ensuring that the polypeptide being synthesized has the amino acid sequence that is denoted, via the genetic code, by the sequence of nucleotides in the mRNA (*Figure 11.1*). To understand how tRNAs play this dual role we must examine **aminoacylation**, the process by which the correct amino acid is attached to each tRNA, and **codon–anticodon recognition**, the interaction between tRNA and mRNA.

11.1.1 Aminoacylation: the attachment of amino acids to tRNAs

Bacteria contain 30–45 different tRNAs and eukaryotes have up to 50. As only 20 amino acids are designated by the genetic code, this means that all organisms have at least some **isoaccepting tRNAs**, different tRNAs that are specific for the same amino acid. The terminology used when describing tRNAs is to indicate the amino acid specificity with a superscript suffix, using the numbers 1, 2, etc., to distinguish different isoacceptors: for example, two tRNAs specific for glycine would be written as tRNA[Gly1] and tRNA[Gly2].

All tRNAs have a similar structure

The smallest tRNAs are only 74 nucleotides in length, and the largest are rarely more than 90 nucleotides. Because of their small size, and because it is possible to purify individual tRNAs, they were among the first nucleic acids to be sequenced, way back in 1965 by Robert Holley's group at Cornell University, New York. The sequences revealed one unexpected feature, that as well as the standard RNA nucleotides (A, C, G and U), tRNAs contain a number of modified nucleotides, 5–10 in any particular tRNA, with over 50 different modifications known altogether (Section 10.3).

Examination of the first tRNA sequence, for tRNA[Ala] of *Saccharomyces cerevisiae*, showed that the molecule could adopt various base-paired secondary structures. After more tRNAs had been sequenced, it became clear that one particular structure could be taken up by all of them. This is the **cloverleaf** (*Figure 11.2*), and has the following features:

■ The **acceptor arm** is formed by seven base pairs between the 5′ and 3′ ends of the molecule. The amino acid is attached to the extreme 3′ end of the tRNA, to the adenosine of the invariant CCA terminal sequence (Section 10.2.2).

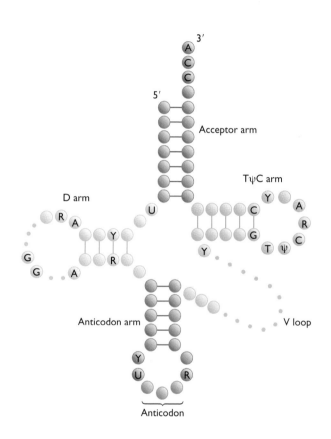

Figure 11.2 The cloverleaf structure of a tRNA.

The tRNA is drawn in the conventional cloverleaf structure, with the different components labeled. Invariant nucleotides (A, C, G, T, U, Ψ, where Ψ = pseudouridine) and semi-invariant nucleotides (abbreviations: R, purine; Y, pyrimidine) are indicated. Optional nucleotides not present in all tRNAs are shown as smaller dots. The standard numbering system places position 1 at the 5′ end and position 76 at the 3′ end; it includes some but not all of the optional nucleotides. The invariant and semi-invariant nucleotides are at positions 8, 11, 14, 15, 18, 19, 21, 24, 32, 33, 37, 48, 53, 54, 55, 56, 57, 58, 60, 61, 74, 75 and 76. The nucleotides of the anticodon are at positions 34, 35 and 36.

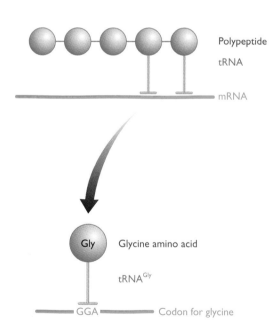

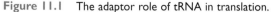

Figure 11.1 The adaptor role of tRNA in translation.

The top drawing shows the physical role of tRNA, forming an attachment between the polypeptide and the mRNA. The lower drawing shows the informational link, the tRNA carrying the amino acid specified by the codon to which it attaches.

- The **D arm**, named after the modified nucleoside dihydrouridine (see *Table 10.5*, page 301), which is always present in this structure.
- The **anticodon arm** contains the triplet of nucleotides called the **anticodon** which base-pair with the mRNA during translation.
- The **V loop** contains 3–5 nucleotides in Class 1 tRNAs or 13–21 nucleotides in Class 2 tRNAs.
- The **TΨC arm**, named after the sequence thymidine–pseudouridine–cytosine, which is always present.

The cloverleaf structure can be formed by virtually all tRNAs, the main exceptions being the tRNAs used in vertebrate mitochondria, which are coded by the mitochondrial genome and which sometimes lack parts of the structure. An example is the human mitochondrial tRNA^{Ser}, which has no D arm. As well as the conserved secondary structure, the identities of nucleotides at some positions are completely invariant (always the same nucleotide) or semi-invariant (always a purine or always a pyrimidine), and the positions of the modified nucleotides are almost always the same.

Many of the invariant nucleotide positions are important in the tertiary structure of tRNA. X-ray crystallography studies have shown that nucleotides in the D and TΨC loops form base pairs that fold the tRNA into a compact L-shaped structure (*Figure 11.3*; Clark, 2001). Each arm of the L-shape is approximately 7 nm long and 2 nm in diameter, with the amino acid binding site at the end of one arm and the anticodon at the end of the other. The additional base-pairing means that the base-stacking (see page 14) is almost continuous from one end of the tRNA to the other, providing stability to the structure.

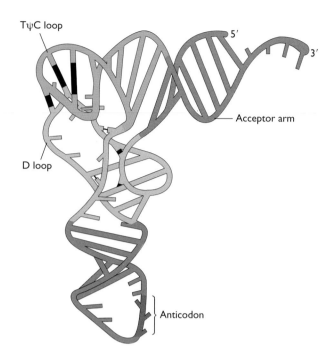

Figure 11.3   The three-dimensional structure of a tRNA.

Additional base pairs, shown in black and mainly between the D and TΨC loops, fold the cloverleaf structure shown in *Figure 11.2* into this L-shaped configuration. Depending on its sequence, the V loop might also form interactions with the D arm, as indicated by thin black lines. The color scheme is the same as in *Figure 11.2*. From Freifelder D, *Molecular Biology*, 2nd edition, 1986, Jones and Bartlett Publishers, Sudbury, MA. Reprinted with permission.

Aminoacyl-tRNA synthetases attach amino acids to tRNAs

The attachment of amino acids to tRNAs – 'charging' in molecular biology jargon – is the function of the group of enzymes called **aminoacyl-tRNA synthetases**. The chemical reaction that results in aminoacylation occurs in two steps. An activated amino acid intermediate is first formed by reaction between the amino acid and ATP, and then the amino acid is transferred to the 3′ end of the tRNA, the link being formed between the –COOH group of the amino acid and the –OH group attached to either the 2′ or 3′ carbon on the sugar of the last nucleotide, which is always an A (*Figure 11.4*).

With a few exceptions, organisms have 20 aminoacyl-tRNA synthetases, one for each amino acid. This means that groups of isoaccepting tRNAs are aminoacylated by a single enzyme. Although the basic chemical reaction is the same for each amino acid, the 20 aminoacyl-tRNA synthetases fall into two distinct groups, Class I and Class II, with several important differences between them (*Table 11.1*). In particular, Class I enzymes attach the amino acid to the 2′–OH group of the terminal nucleotide of the tRNA, whereas Class II enzymes attach the amino acid to the 3′–OH group (Ibba *et al.*, 2000).

Aminoacylation must be carried out accurately: the correct amino acid must be attached to the correct tRNA if the rules of the genetic code are to be followed during protein synthesis. It appears that an aminoacyl-tRNA synthetase has high fidelity for its tRNA, the result of an extensive interaction between the two, covering some 25 nm² of surface area and involving the acceptor arm and anticodon loop of the tRNA, as well as individual nucleotides in the D and TΨC arms. The interaction between enzyme and amino acid is, of necessity, less extensive, amino acids being much smaller than tRNAs, and presents greater problems with regard to specificity because several pairs of amino acids are structurally similar. Errors do therefore occur, at a very low rate for most amino acids but possibly as frequently as one aminoacylation in 80 for difficult pairs such as isoleucine and valine. Most errors are corrected by the aminoacyl-tRNA synthetase itself, by an editing process that is distinct from aminoacylation, involving different contacts with the tRNA (Hale *et al.*, 1997; Silvian *et al.*, 1999).

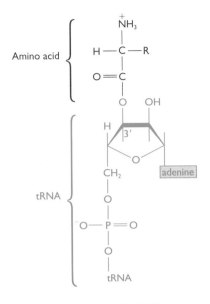

Figure 11.4 Aminoacylation of a tRNA.

The result of aminoacylation by a Class II aminoacyl-tRNA synthetase is shown, the amino acid being attached via its –COOH group to the 3'–OH of the terminal nucleotide of the tRNA. A Class I aminoacyl-tRNA synthetase attaches the amino acid to the 2'–OH group.

In most organisms, aminoacylation is carried out by the process just described, but a few unusual events have been documented. These include a number of instances where the aminoacyl-tRNA synthetase attaches the incorrect amino acid to a tRNA, this amino acid subsequently being transformed into the correct one by a second, separate chemical reaction. This was first discovered in the bacterium *Bacillus megaterium* for synthesis of glutamine-tRNA^{Gln} (i.e. glutamine attached to its tRNA). This aminoacylation is carried out by the enzyme responsible for synthesis of glutamic acid-tRNA^{Glu}, and initially results in attachment of a glutamic acid to the tRNA^{Gln} (*Figure 11.5A*). This glutamic acid is then converted to glutamine by transamidation catalyzed by a second enzyme. The same process is used by various other bacteria (although not *Escherichia coli*) and by the archaea. Some archaea also use transamidation to synthesize asparagine-tRNA^{Asn} from aspartic acid-tRNA^{Asn} (Ibba *et al.*, 2000). In both of these cases, the amino acid that is synthesized by the modification process is one of the 20 that are specified by the genetic code. There are also two examples where the modification results in an unusual amino acid. The first example is the conversion of methionine to *N*-formylmethionine (*Figure 11.5B*), producing the special aminoacyl-tRNA used in initiation of bacterial translation (Section 11.2.2). The second example occurs in both prokaryotes and eukaryotes and results in synthesis of selenocysteine, which is specified in a context-dependent manner by some 5'–UGA–3' codons (Section 3.3.2). These codons are recognized by a special tRNA^{SeCys}, but there is no aminoacyl-tRNA synthetase that is able to attach selenocysteine to this tRNA. Instead, the tRNA is aminoacylated with a serine by the seryl-tRNA synthetase, and then modified by replacement of the –OH group of the serine with an –SeH, to give selenocysteine (*Figure 11.5C*; Low and Berry, 1996).

11.1.2 Codon–anticodon interactions: the attachment of tRNAs to mRNA

Aminoacylation represents the first level of specificity displayed by a tRNA. The second level is the specificity of the interaction between the anticodon of the tRNA and the mRNA being translated. This specificity ensures that protein synthesis follows the rules of the genetic code (see *Figure 3.20*, page 86).

In principle, codon–anticodon recognition is a straightforward process involving base-pairing between the anticodon of the tRNA and a codon in the mRNA (*Figure 11.6*). The specificity of aminoacylation ensures that the tRNA carries the amino acid denoted by the codon that it pairs with, and the ribosome controls the topology of the interaction in such a way that only a single triplet of nucleotides is available for pairing. Because

Table 11.1 Features of aminoacyl-tRNA synthetases

Feature	Class I enzymes	Class II enzymes
Structure of the enzyme active site	Parallel β-sheet	Antiparallel β-sheet
Interaction with the tRNA	Minor groove of the acceptor stem	Major groove of the acceptor stem
Orientation of the bound tRNA	V loop faces away from the enzyme	V loop faces the enzyme
Amino acid attachment	To the 2'–OH of the terminal nucleotide of the tRNA	To the 3'–OH of the terminal nucleotide of the tRNA
Enzymes for *	Arg, Cys, Gln, Glu, Ile, Leu, LysI, Met, Trp, Tyr, Val	Ala, Asn, Asp, Gly, His, LysII, Phe, Pro, Thr, Ser

*The aminoacyl-tRNA synthetase for lysine is a Class I enzyme in some archaea and bacteria and a Class II enzyme in all other organisms. For more details see Arnez and Moras (1997) and Ibba *et al.* (2000).

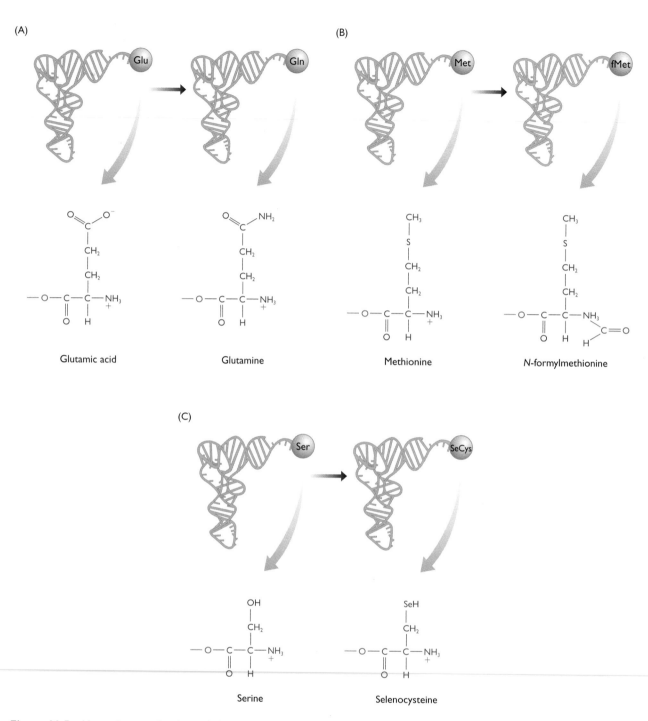

Figure 11.5 Unusual types of aminoacylation.

(A) In some bacteria, tRNA^Gln is aminoacylated with glutamic acid, which is then converted to glutamine by transamidation. (B) The special tRNA used in initiation of translation in bacteria is aminoacylated with methionine, which is then converted to N-formylmethionine. (C) tRNA^SeCys in various organisms is initially aminoacylated with serine.

base-paired polynucleotides are always antiparallel, and because the mRNA is read in the 5'→3' direction, the first nucleotide of the codon pairs with nucleotide 36 of the tRNA, the second with nucleotide 35, and the third with nucleotide 34.

In practice, codon recognition is complicated by the possibility of **wobble**. This is another of the principles of gene expression originally proposed by Crick and subsequently shown to be correct. Because the anticodon is in a loop of RNA, the triplet of nucleotides is slightly curved

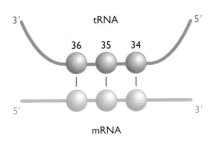

Figure 11.6 The interaction between a codon and an anticodon.

The numbers indicate the nucleotide positions in the tRNA (see *Figure 11.2*, page 315).

(see *Figures 11.2* and *11.3*, pages 315 and 316) and so cannot make an entirely uniform alignment with the codon. As a result, a non-standard base pair can form between the third nucleotide of the codon and the first nucleotide (number 34) of the anticodon. This is called 'wobble'. A variety of pairings is possible, especially if the nucleotide at position 34 is modified. In bacteria, the two main features of wobble are (Ikemura, 1981):

- **G–U base-pairs** are permitted. This means that an anticodon with the sequence 3'–xxG–5' can base-pair with both 5'–xxC–3' and 5'–xxU–3'. Similarly, the anticodon 3'–xxU–5' can base-pair with both 5'–xxA–3' and 5'–xxG–3'. The consequence is that, rather than needing a different tRNA for each codon, the four members of a codon family (e.g. 5'–GCN–3', all coding for alanine) can be decoded by just two tRNAs (*Figure 11.7A*).
- **Inosine**, abbreviated to I, is a modified purine (see *Table 10.5*, page 301) that can base-pair with A, C and U. Inosine can only occur in the tRNA because the mRNA is not modified in this way. The triplet 3'–UAI–5' is sometimes used as the anticodon in a tRNA^{Ile} molecule because it pairs with 5'–AUA–3', 5'–AUC–3' and 5'–AUU–3' (*Figure 11.7B*), which form the three-codon family for this amino acid in the standard genetic code.

Wobble reduces the number of tRNAs needed in a cell by enabling one tRNA to read two or possibly three codons. Hence bacteria can decode their mRNAs with as few as 30 tRNAs. Eukaryotes also make use of wobble but in a restricted way. The human genome, which in this regard is fairly typical of higher eukaryotes, has 48 tRNAs. Of these, 16 are predicted to use wobble to decode two codons each, with the remaining 32 being specific for just a single triplet (*Figure 11.8*; IHGSC, 2001). The distinctive features compared with wobble in bacteria are:

- G–U wobble is used with eight tRNAs but in every case the wobble involves an anticodon with the sequence 3'–xxG–5'. The alternative version of G–U wobble, where the anticodon sequence is 3'–xxU–5', appears not to be used in eukaryotes, possibly

because this could result in a tRNA^{ile} with the anticodon 3'–UAU–5' reading the methionine codon 5'–AUG–3' (*Figure 11.9*). Eukaryotes may therefore have a means of preventing this type of wobble from occurring (Percudani, 2001).

- Eight other human tRNAs have anticodons containing inosine (3'–xxI–5') but these decode only 5'–xxC–3' and 5'–xxU–3'. The base pairing between I and A is weak, which means that 5'–xxA–3' codons are only inefficiently recognized by an 3'–xxI–5' anticodon. To avoid this inefficiency, in every example of wobble involving inosine in the human tRNA set, the 5'–xxA–3' codon is recognized by a separate tRNA. Note, however, that recognition by a separate tRNA does not preclude the 5'–xxA–3' codon from also being decoded by the tRNA containing 3'–xxI–5', albeit inefficiently. This does not compromise the specificity of the genetic code, because wobble involving inosine is limited to those codon families in which all four triplets specify the same amino acid (see *Figure 11.8*).

Other genetic systems use more extreme forms of wobble. Human mitochondria, for example, use only 22 tRNAs. With some of these tRNAs the nucleotide in the wobble position of the anticodon is virtually redundant because it can base-pair with any nucleotide, enabling all four codons of a family to be recognized by the same tRNA. This phenomenon has been called **superwobble**.

11.2 The Role of the Ribosome in Protein Synthesis

An *E. coli* cell contains approximately 20 000 ribosomes, distributed throughout its cytoplasm. The average human cell contains rather more (nobody has ever counted them all), some free in the cytoplasm and some attached to the outer surface of the endoplasmic reticulum, the membranous network of tubes and vesicles that permeates the cell. Originally, ribosomes were looked on as passive partners in protein synthesis, merely the structures on which translation occurs. This view has changed over the years and ribosomes are now considered to play two active roles in protein synthesis:

- Ribosomes *coordinate* protein synthesis by placing the mRNA, aminoacyl-tRNAs and associated protein factors in their correct positions relative to one another.
- Components of ribosomes, including the rRNAs, *catalyze* at least some of the chemical reactions occurring during translation.

To understand how ribosomes play these roles we will first survey the structural features of ribosomes in bacteria and eukaryotes, and then examine the detailed mechanism for protein synthesis in these two types of organism.

(A) G–U base-pairing

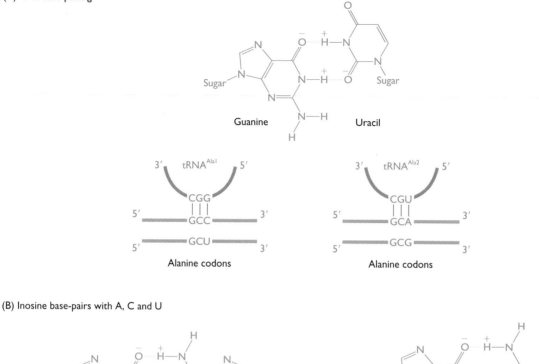

(B) Inosine base-pairs with A, C and U

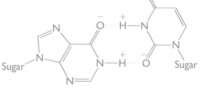

Figure 11.7 Two examples of wobble in bacteria.

(A) Wobble involving a G–U base pair enables the four-codon family for alanine to be decoded by just two tRNAs. Note that wobble involving G–U also enables accurate decoding of a four-codon family that specifies two amino acids. For example, the anticodon 3′–AAG–5′ can decode 5′–UUC–3′ and 5′–UUU–3′, both coding for phenylalanine (see *Figure 3.20*, page 86), and the anticodon 3′–AAU–5′ can decode the other two members of this family, 5′–UUA–3′ and 5′–UUG–3′, which code for leucine. (B) Inosine can base-pair with A, C or U, meaning that a single tRNA can decode all three codons for isoleucine. Dotted lines indicate hydrogen bonds; I, inosine.

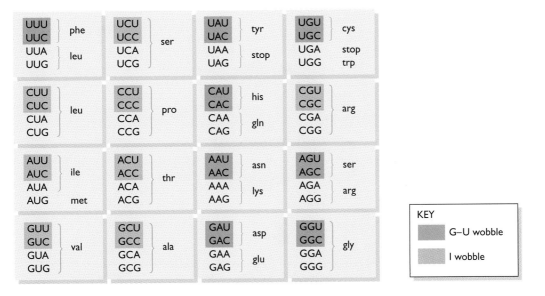

Figure 11.8 The predicted usage of wobble in decoding the human genome.

Pairs of codons that are predicted to be decoded by a single tRNA using G–U wobble are highlighted in red, and those pairs predicted to be decoded by wobble involving inosine are highlighted in green. Codons that are not highlighted have their own individual tRNAs. The predictions are based largely on examination of the anticodon sequences of the tRNAs that have been located in the draft human genome sequence. The analysis shown here implies that there are 45 tRNAs in human cells – the 16 for the wobble pairs and 29 singletons. In fact there are 48 tRNAs. This is because three codons thought to be decoded as part of a wobble pair (5′–AAU–3′, 5′–AUC–3′ and 5′–UAU–3′) also have their own individual tRNAs, although these are present in low abundance.

11.2.1 Ribosome structure

Our understanding of ribosome structure has gradually developed over the last 50 years as more and more powerful techniques have been applied to the problem. Originally called 'microsomes', ribosomes were first observed in the early decades of the 20th century as tiny particles almost beyond the resolving power of light microscopy. In the 1940s and 1950s, the first electron micrographs showed that bacterial ribosomes are oval-shaped, with dimensions of 29 nm × 21 nm, rather smaller than eukaryotic ribosomes, the latter varying a little in size depending on species but averaging about 32 nm × 22 nm. In the mid-1950s, the discovery that ribosomes are the sites of protein synthesis stimulated

attempts to define the structures of these particles in greater detail.

Ultracentrifugation was used to measure the sizes of ribosomes and their components

The initial progress in understanding the detailed structure of the ribosome came not from observing them with the electron microscope but by analyzing their components by ultracentrifugation (Technical Note 2.2, page 42). Intact ribosomes have sedimentation coefficients of 80S for eukaryotes and 70S for bacteria, and each can be broken down into smaller components (*Figure 11.10*):

■ Each ribosome comprises two subunits. In eukaryotes these subunits are 60S and 40S; in bacteria they are 50S and 30S. Note that sedimentation coefficients are not additive because they depend on shape as well as mass; it is perfectly acceptable for the intact ribosome to have an S value less than the sum of its two subunits.

■ The large subunit contains three rRNAs in eukaryotes (the 28S, 5.8S and 5S rRNAs) but only two in bacteria (23S and 5S rRNAs). In bacteria the equivalent of the eukaryotic 5.8S rRNA is contained within the 23S rRNA.

■ The small subunit contains a single rRNA in both types of organism: an 18S rRNA in eukaryotes and a 16S rRNA in bacteria.

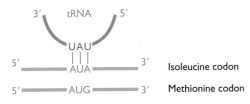

Figure 11.9 A tRNA with the anticodon 3′–UAU–5′ could read the isoleucine codon 5′–AUA–3′ as well as the methionine codon.

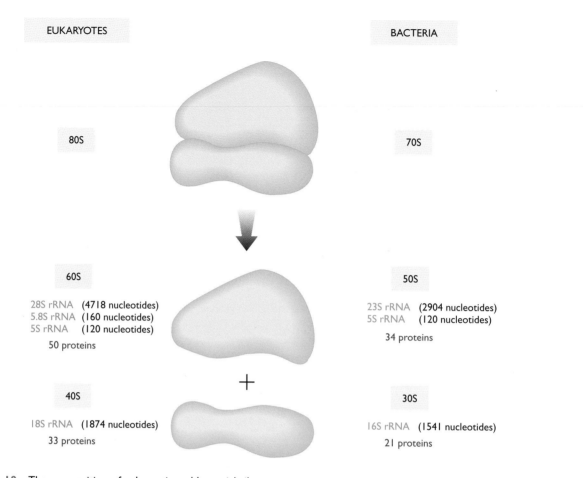

EUKARYOTES

BACTERIA

80S

70S

60S

28S rRNA (4718 nucleotides)
5.8S rRNA (160 nucleotides)
5S rRNA (120 nucleotides)
50 proteins

50S

23S rRNA (2904 nucleotides)
5S rRNA (120 nucleotides)
34 proteins

40S

18S rRNA (1874 nucleotides)
33 proteins

30S

16S rRNA (1541 nucleotides)
21 proteins

Figure 11.10 The composition of eukaryotic and bacterial ribosomes.

The details refer to a 'typical' eukaryotic ribosome and the *Escherichia coli* ribosome. Variations between different species mainly concern the numbers of ribosomal proteins.

- Both subunits contain a variety of **ribosomal proteins**, the numbers detailed in *Figure 11.10*. The ribosomal proteins of the small subunit are called S1, S2, etc.; those of the large subunit are L1, L2, etc. There is just one of each protein per ribosome, except for L7 and L12, which are present as dimers.

Probing the fine structure of the ribosome

Once the basic composition of eukaryotic and bacterial ribosomes had been worked out, attention was focused on the way in which the various rRNAs and proteins fit together. Important information was provided by the first rRNA sequences, comparisons between these identifying conserved regions that can base-pair to form complex two-dimensional structures (*Figure 11.11*). This suggested that the rRNAs provide a scaffolding within the ribosome, to which the proteins are attached, an interpretation that under-emphasizes the active role that rRNAs play in protein synthesis but which nonetheless was a useful foundation on which to base subsequent research.

Much of that subsequent research has concentrated on the bacterial ribosome, which is smaller than the eukaryotic version and available in large amounts from extracts of cells grown to high density in liquid cultures. A number of technical approaches have been used to study the bacterial ribosome:

- *Nuclease protection studies* (Section 2.2.1) enable contacts between rRNAs and proteins to be identified.
- *Protein–protein crosslinking* identifies pairs or groups of proteins that are located close to one another in the ribosome.
- *Electron microscopy* has gradually become more sophisticated, enabling the overall structure of the ribosome to be resolved in greater detail. For example, innovations such as **immunoelectron microscopy**, in which ribosomes are labeled with antibodies specific for individual ribosomal proteins before examination, have been used to locate the positions of these proteins on the surface of the ribosome.

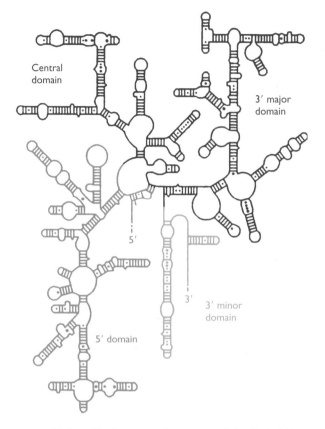

Figure 11.11   The base-paired structure of the *Escherichia coli* 16S rRNA.

> In this representation, standard base pairs (G–C, A–U) are shown as bars; non-standard base pairs (e.g. G–U) are shown as dots.

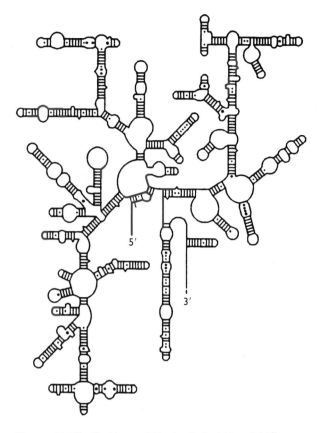

Figure 11.12   Positions within the *Escherichia coli* 16S rRNA that form contacts with ribosomal protein S5.

> The distribution of the contact positions (shown in red) for this single ribosomal protein emphasizes the extent to which the base-paired secondary structure of the rRNA is further folded within the three-dimensional structure of the ribosome. For details of the work that led to these results, see Heilek and Noller (1996).

■ *Site-directed hydroxyl radical probing* makes use of the ability of Fe(II) ions to generate hydroxyl radicals that cleave RNA phosphodiester bonds located within 1 nm of the site of radical production. This technique was used to determine the exact positioning of ribosomal protein S5 in the *E. coli* ribosome. Different amino acids within S5 were labeled with Fe(II) and hydroxyl radicals induced in reconstituted ribosomes. The positions at which the 16S rRNA was cleaved were then used to infer the topology of the rRNA in the vicinity of S5 protein (*Figure 11.12*; Heilek and Noller, 1996).

In recent years these techniques have been increasingly supplemented by X-ray crystallography (Section 9.1.3), which has been responsible for the most exciting insights into ribosome structure. Analyzing the massive amounts of X-ray diffraction data that are produced by crystals of an object as large as a ribosome is a huge task, particularly at the level needed to obtain a structure that is detailed enough to be informative about the way in which the ribosome works (Pennisi, 1999). This challenge has been met, and structures have been deduced

for ribosomal proteins bound to their segments of rRNA (Conn *et al.*, 1999; Agalarov *et al.*, 2000), for the large and small subunits (Ban *et al.*, 2000; Wimberly *et al.*, 2000), and for the entire bacterial ribosome attached to mRNA and tRNAs (Yusupov *et al.*, 2001). As well as revealing the structure of the ribosome (*Figure 11.13*), this recent explosion of information has had an important impact on our understanding of the translation process, as we will see in the next section.

11.2.2 Initiation of translation

Although ribosomal architecture is similar in bacteria and eukaryotes, there are distinctions in the way in which translation is carried out in the two types of organism. The most important of these differences occurs during the first stage of translation, when the ribosome is assembled on the mRNA at a position upstream of the initiation codon.

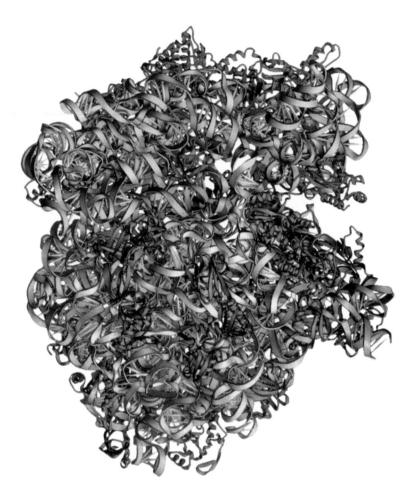

Figure 11.13 The bacterial ribosome.

The picture shows the ribosome of the bacterium *Thermus thermophilus*. The small subunit is at the top, with the 16S rRNA in light blue and the small subunit ribosomal proteins in dark blue. The large subunit rRNAs are in grey and the proteins in purple. The gold area is the A site (Section 11.2.3) – the point at which aminoacylated tRNAs enter the ribosome during protein synthesis. This site, and most of the region within which protein synthesis actually occurs, is located in the cleft between the two subunits. Reprinted with permission from Mathews and Pe'ery (2001) *Trends Biochem. Sci.*, **26**, 585–587.

Initiation in bacteria requires an internal ribosome binding site

The main difference between initiation of translation in bacteria and eukaryotes is that in bacteria the translation initiation complex is built up directly at the initiation codon, the point at which protein synthesis will begin, whereas eukaryotes use a more indirect process for locating the initiation point, as we will see in the next section.

When not actively participating in protein synthesis, ribosomes dissociate into their subunits, which remain in the cytoplasm waiting to be used for a new round of translation. In bacteria, the process initiates when a small subunit, in conjunction with the translation **initiation factor** IF-3 (*Table 11.2*), attaches to the **ribosome binding site** (also called the **Shine–Dalgarno sequence**). This is a short target site, consensus 5'–AGGAGGU–3' in *E. coli* (*Table 11.3*), located about 3–10 nucleotides upstream of the initiation codon, the point at which translation will begin (*Figure 11.14*). The ribosome binding site is complementary to a region at the 3' end of the 16S rRNA, the one present in the small subunit, and it is thought that base-pairing between the two is involved in the attachment of the small subunit to the mRNA.

Attachment to the ribosome binding site positions the small subunit of the ribosome over the initiation codon (*Figure 11.15*). This codon is usually 5'–AUG–3', which codes for methionine, although 5'–GUG–3' and 5'–UUG–3' are sometimes used. All three codons can be recognized by the same initiator tRNA, the last two by wobble. This initiator tRNA is the one that was aminoacylated with methionine and subsequently modified by conversion of the methionine to N-formylmethionine (see *Figure 11.5B*, page 318). The modification attaches a formyl group (–COH) to the amino group, which means

Table 11.2 Functions of the bacterial translation factors

Factor	Function
Initiation factors	
IF-1	Unclear; X-ray crystallography studies show that binding of IF-1 blocks the A site (see page 329), so its function may be to prevent premature entry of tRNAs into the A site. Alternatively IF-1 may cause conformational changes that prepare the small subunit for attachment to the large subunit
IF-2	Directs the initiator tRNA$_i^{Met}$ to its correct position in the initiation complex
IF-3	Prevents premature reassociation of the large and small subunits of the ribosome
Elongation factors	
EF-Tu	Directs the next tRNA to its correct position in the ribosome
EF-Ts	Regenerates EF-Tu after the latter has yielded the energy contained in its attached GTP molecule
EF-G	Mediates translocation
Release factors	
RF-1	Recognizes the termination codons 5′–UAA–3′ and 5′–UAG–3′
RF-2	Recognizes 5′–UAA–3′ and 5′–UGA–3′
RF-3	Stimulates dissociation of RF1 and RF2 from the ribosome after termination
Ribosome recycling factor	
RRF	Responsible for disassociating the ribosome subunits after translation has terminated

Table 11.3 Examples of ribosome binding sequences in *Escherichia coli*

Gene	Codes for	Ribosome binding sequence	Nucleotides to the start codon
E. coli consensus	–	5′–AGGAGGU–3′	10
Lactose operon	Lactose utilization enzymes	5′–A G G A –3′	7
galE	Hexose-1-phosphate uridyltransferase	5′–G G A G –3′	6
rplJ	Ribosomal protein L10	5′–A G G A G –3′	8

that only the carboxyl group of the initiator methionine is free to participate in peptide bond formation. This ensures that polypeptide synthesis can take place only in the N→C direction. The initiator tRNA$_i^{Met}$ is brought to the small subunit of the ribosome by a second initiation factor, IF-2, along with a molecule of GTP, the latter acting as a source of energy for the final step of initiation. Note that the tRNA$_i^{Met}$ is only able to decode the initiation

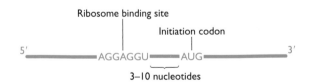

Figure 11.14 The ribosome binding site for bacterial translation.

In *Escherichia coli*, the ribosome binding site has the consensus sequence 5′–AGGAGGU–3′ and is located between 3 and 10 nucleotides upstream of the initiation codon.

codon; it cannot enter the complete ribosome during the elongation phase of translation during which internal 5′–AUG–3′ codons are recognized by a different tRNA^{Met} carrying an unmodified methionine.

Completion of the initiation phase occurs when IF-1 binds to the initiation complex. The precise role of IF-1 is unclear (see *Table 11.2*), but it may induce a conformational change in the initiation complex, enabling the large subunit of the ribosome to attach. Attachment of the large subunit requires energy, which is generated by hydrolysis of the bound GTP, and results in release of the initiation factors.

Initiation in eukaryotes is mediated by the cap structure and poly(A) tail

Only a small number of eukaryotic mRNAs have internal ribosome binding sites (see the next section). Instead, with most mRNAs the small subunit of the ribosome makes its initial attachment at the 5′-end of the molecule and then **scans** along the sequence until it locates the initiation codon. The process requires a plethora of initiation factors and there is still some confusion over the functions

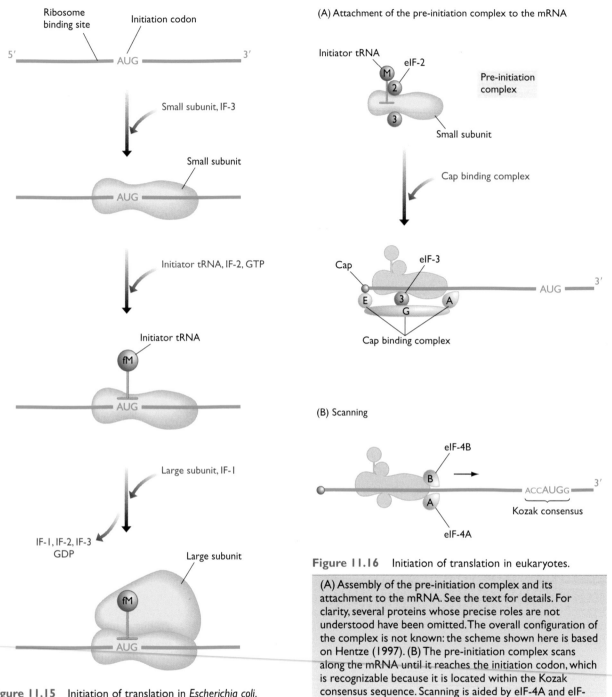

Figure II.15 Initiation of translation in *Escherichia coli*.

See the text for details. Note that the different components of the initiation complex are not drawn to scale. Abbreviation: fM, *N*-formylmethionine.

Figure II.16 Initiation of translation in eukaryotes.

(A) Assembly of the pre-initiation complex and its attachment to the mRNA. See the text for details. For clarity, several proteins whose precise roles are not understood have been omitted. The overall configuration of the complex is not known: the scheme shown here is based on Hentze (1997). (B) The pre-initiation complex scans along the mRNA until it reaches the initiation codon, which is recognizable because it is located within the Kozak consensus sequence. Scanning is aided by eIF-4A and eIF-4B, which are thought to have helicase activity. It is probable that eIF-3 remains attached to the pre-initiation complex during scanning, as shown here. It is not clear whether eIF-4E and eIF-4G also remain attached at this stage. Note that scanning is an energy-dependent process that requires hydrolysis of ATP. Abbreviation: M, methionine.

of all of these (*Table 11.4*). The details are as follows (*Figure 11.16*; Dever, 1999).

The first step involves assembly of the **pre-initiation complex**. This structure comprises the 40S subunit of the ribosome, a 'ternary complex' made up of the initiation factor eIF-2 bound to the initiator tRNA^{Met} and a molecule of GTP, and three additional initiation factors, eIF-1, eIF-1A, and eIF-3. As in bacteria, the initiator tRNA is distinct from the normal tRNA^{Met} that recognizes internal 5'–AUG–3' codons but, unlike bacteria, it

Table 11.4 Eukaryotic translation factors

Factor	Function
Initiation factors	
eIF-1	Component of the pre-initiation complex
eIF-1A	Component of the pre-initiation complex
eIF-2	Binds to the initiator tRNA^{Met} within the ternary complex component of the pre-initiation complex; phosphorylation of eIF-2 results in a global repression of translation
eIF-3	Component of the pre-initiation complex; makes direct contact with eIF-4G and so forms the link with the cap binding complex
eIF-4A	Component of the cap binding complex; a helicase that aids scanning by breaking intramolecular base pairs in the mRNA
eIF-4B	Aids scanning, possibly by acting as a helicase that breaks intramolecular base pairs in the mRNA
eIF-4E	Component of the cap binding complex, possibly the component that makes direct contact with the cap structure at the 5' end of the mRNA
eIF-4F	The cap binding complex, comprising eIF-4A, eIF-4E and eIF-4G, which makes the primary contact with the cap structure at the 5' end of the mRNA
eIF-4G	Component of the cap binding complex; forms a bridge between the cap binding complex and eIF-3 in the pre-initiation complex; in at least some organisms, eIF-4G also forms an association with the poly(A) tail, via the polyadenylate-binding protein
eIF-5	Aids release of the other initiation factors at the completion of initiation
eIF-6	Associated with the large subunit of the ribosome; prevents large subunits from attaching to small subunits in the cytoplasm
Elongation factors	
eEF-1	Complex of four subunits (eEF-1a, eEF-1b, eEF-1d and eEF-1g); directs the next tRNA to its correct position in the ribosome
eEF-2	Mediates translocation
Release factors	
eRF-1	Recognizes the termination codon
eRF-3	Possibly stimulates dissociation of eRF-1 from the ribosome after termination; possibly causes the ribosome subunits to disassociate after termination of translation

is aminoacylated with normal methionine, not the formylated version.

After assembly, the pre-initiation complex associates with the 5' end of the mRNA. This step requires the **cap binding complex** (sometimes called eIF-4F), which comprises the initiation factors eIF-4A, eIF-4E and eIF-4G. The contact with the cap might be made by eIF-4E alone (as shown in *Figure 11.16*) or might involve a more general interaction with the cap binding complex (Pestova and Hellen, 1999). The factor eIF-4G acts as a bridge between eIF-4E, bound to the cap, and eIF-3, attached to the pre-initiation complex (Hentze, 1997). The result is that the pre-initiation complex becomes attached to the 5' region of the mRNA. Attachment of the pre-initiation complex to the mRNA is also influenced by the poly(A) tail, at the distant 3' end of the mRNA. This interaction is thought to be mediated by the polyadenylate-binding protein (PADP), which is attached to the poly(A) tail (Section 10.1.2). In yeast and plants it has been shown that PADP can form an association with eIF-4G, this association requiring that the mRNA bends back on itself. With artificially uncapped mRNAs, the PADP

interaction is sufficient to load the pre-initiation complex onto the 5' end of the mRNA, but under normal circumstances the cap structure and poly(A) tail probably work together (Preiss and Hentze, 1998). The poly(A) tail could have an important regulatory role, as the length of the tail appears to be correlated with the extent of initiation that occurs with a particular mRNA.

After becoming attached to the 5' end of the mRNA, the **initiation complex**, as it is now called, has to scan along the molecule and find the initiation codon. The leader regions of eukaryotic mRNAs can be several tens, or even hundreds, of nucleotides in length and often contain regions that form hairpins and other base-paired structures. These are probably removed by a combination of eIF-4A and eIF-4B. eIF-4A, and possibly also eIF-4B, has a helicase activity and so is able to break intramolecular base pairs in the mRNA, freeing the passage for the initiation complex (*Figure 11.16B*). The initiation codon, which is usually 5'–AUG–3' in eukaryotes, is recognizable because it is contained in a short consensus sequence, 5'–ACCAUGG–3', referred to as the **Kozak consensus**.

Once the initiation complex is positioned over the initiation codon, the large subunit of the ribosome attaches. As in bacteria, this requires hydrolysis of GTP and leads to release of the initiation factors. Two final initiation factors are involved at this stage: eIF-5, which aids release of the other factors, and eIF-6, which is associated with the unbound large subunit and prevents it from attaching to a small subunit in the cytoplasm.

Initiation of eukaryotic translation without scanning

The scanning system for initiation of translation does not apply to every eukaryotic mRNA. This was first recognized with the picornaviruses, a group of viruses with RNA genomes which includes the human poliovirus and rhinovirus, the latter being responsible for the common cold. Transcripts from these viruses are not capped but instead have an **internal ribosome entry site** (**IRES**) which is similar in function to the ribosome binding site of bacteria, although the sequences of IRESs and their positions relative to the initiation codon are more variable than the bacterial versions (Mountford and Smith, 1995). The presence of IRESs on their transcripts means that picornaviruses can block protein synthesis in the host cell by inactivating the cap binding complex, without affecting translation of their own transcripts, although this is not a normal part of the infection strategy of all picornaviruses.

Remarkably, no virus proteins are required for recognition of an IRES by a host ribosome. In other words, the normal eukaryotic cell possesses proteins and/or other factors that enable it to initiate translation by the IRES method (Holcik *et al.*, 2000). Because of their variability, IRESs are difficult to identify by inspection of DNA sequences, but it is becoming clear that a few nuclear gene transcripts possess them and that these are translated, at least under some circumstances, via their IRES rather than by scanning. Examples are the mRNAs for the mammalian immunoglobulin heavy-chain binding protein and the *Drosophila* Antennapedia protein (Section 12.3.3). IRESs are also found on several mRNAs whose protein products are translated when the cell is put under stress, for example by exposure to heat, irradiation, or low oxygen conditions. Under these circumstances, cap-dependent translation is globally suppressed (as described in the next section). The presence of IRESs on the 'survival' mRNAs therefore enables these to undergo preferential translation at the time when their products are needed.

Regulation of translation initiation

The initiation of translation is an important control point in protein synthesis, at which two different types of regulation can be exerted. The first of these is **global regulation**, which involves a general alteration in the amount of protein synthesis occurring, with all mRNAs translated by the cap mechanism being affected to a similar extent. In eukaryotes this is commonly achieved by phosphorylation of eIF-2, which results in repression of translation initiation by preventing eIF-2 from binding the molecule of GTP that it needs before it can transport the initiator tRNA to the small subunit of the ribosome. Phosphorylation of eIF-2 occurs during stresses such as heat shock, when the overall level of protein synthesis is decreased and IRES-mediated translation takes over.

Transcript-specific regulation involves mechanisms that act on a single transcript or a small group of transcripts coding for related proteins. The most frequently cited example of transcript-specific regulation involves the operons for the ribosomal protein genes of *E. coli* (*Figure 11.17A*). The leader region of the mRNA transcribed from each operon contains a sequence that acts as a binding site for one of the proteins coded by the operon. When this protein is synthesized it can either attach to its position on the ribosomal RNA, or bind to the leader region of the mRNA. The rRNA attachment is favored and occurs if there are free rRNAs in the cell. Once all the free rRNAs have been assembled into ribosomes, the ribosomal protein binds to its mRNA, blocking translation initiation and hence switching off further synthesis of the ribosomal proteins coded by that particular mRNA. Similar events involving other mRNAs ensure that synthesis of each ribosomal protein is coordinated with the amount of free rRNA in the cell.

A second example of transcript-specific regulation, one occurring in mammals, involves the mRNA for ferritin, an iron-storage protein (*Figure 11.17B*). In the absence of iron, ferritin synthesis is inhibited by proteins that bind to sequences called **iron-response elements** located in the leader region of the ferritin mRNA. The bound proteins block the ribosome as it attempts to scan along the mRNA in search of the initiation codon. When iron is present, the binding proteins detach and the mRNA is translated. Interestingly, the mRNA for a related protein – the transferrin receptor involved in the uptake of iron – also has iron-response elements, but in this case detachment of the binding proteins in the presence of iron results not in translation of the mRNA but in its degradation. This is logical because when iron is present in the cell, there is less requirement for transferrin receptor activity because there is less need to import iron from outside.

11.2.3 Elongation of translation

The main differences between translation in bacteria and eukaryotes occur during the initiation phase; the events after the large subunit of the ribosome becomes associated with the initiation complex are similar in both types of organism. We can therefore deal with them together, by looking at what happens in bacteria and referring to the distinctive features of eukaryotic translation where appropriate.

Elongation in bacteria and eukaryotes

Attachment of the large subunit results in two sites at which aminoacyl-tRNAs can bind. The first of these, the

(A) Autoregulation of ribosomal protein synthesis

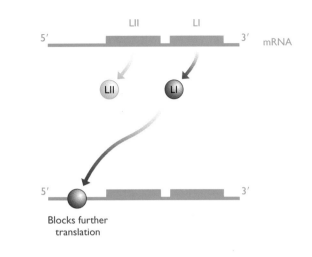

(B) Regulation by iron-response elements

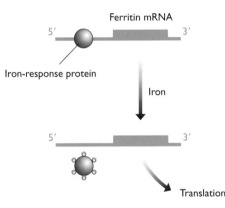

Figure 11.17 Transcript-specific regulation of translation initiation.

(A) Regulation of ribosomal protein synthesis in bacteria. The L11 operon of *Escherichia coli* is transcribed into an mRNA carrying copies of the genes for the L11 and L1 ribosomal proteins. When the L1 binding sites on the available 23S rRNA molecules have been filled, L1 binds to the 5′ untranslated region of the mRNA, blocking further initiation of translation.
(B) Regulation of ferritin protein synthesis in mammals. The iron-response protein binds to the 5′ untranslated region of the ferritin mRNA when iron is absent, preventing the synthesis of ferritin.

P or **peptidyl site**, is already occupied by the initiator tRNA$_i^{Met}$, charged with *N*-formylmethionine or methionine, and base-paired with the initiation codon. The second site, the **A** or **aminoacyl site**, covers the second codon in the open reading frame (*Figure 11.18*). The structures revealed by X-ray crystallography show that these sites are located in the cavity between the large and small subunits of the ribosome, the codon–anticodon interaction being associated with the small subunit and the aminoacyl end of the tRNA with the large subunit (*Figure 11.19*; Yusupov *et al.*, 2001).

The A site becomes filled with the appropriate aminoacyl-tRNA, which in *E. coli* is brought into position by the **elongation factor** EF-Tu, which ensures that only tRNAs that carry the correct amino acid are able to enter the ribosome, mischarged tRNAs being rejected at this point (Ibba, 2001). EF-Tu is an example of a G protein, meaning that it binds a molecule of GTP which it can hydrolyze to release energy. In eukaryotes the equivalent factor is called eEF-1, which is a complex of four subunits: eEF-1a, eEF-1b, eEF-1d and eEF-1g (see *Table 11.4*). The first of these exists in at least two forms, eEF-1a1 and eEF-1a2,

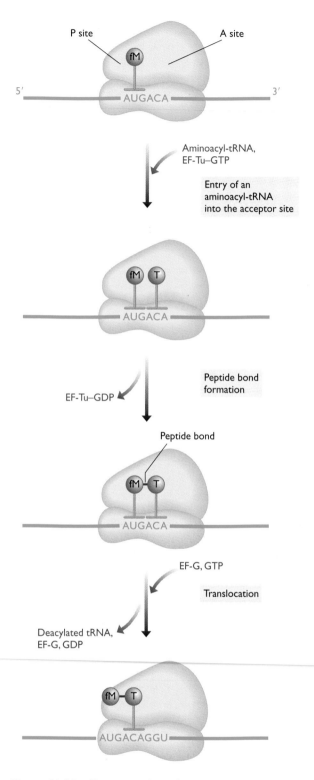

P site　　　　A site

5′　　　　　　　　　　　　　3′

AUGACA

Aminoacyl-tRNA,
EF-Tu–GTP

Entry of an
aminoacyl-tRNA
into the acceptor site

AUGACA

Peptide bond
formation

EF-Tu–GDP

Peptide bond

AUGACA

EF-G, GTP

Translocation

Deacylated tRNA,
EF-G, GDP

AUGACAGGU

Figure 11.18　Elongation of translation.

The diagram shows the events occurring during a single
elongation cycle in *Escherichia coli*. See the text for details
regarding eukaryotic translation. Abbreviations: fM, *N*-
formylmethionine; T, threonine.

which are highly similar proteins that probably have
equivalent functions in different tissues (Hafezparast and
Fisher, 1998). Specific contacts between the tRNA, mRNA
and the 16S rRNA within the A site ensure that only the
correct tRNA is accepted. These contacts are able to discriminate between a codon–anticodon interaction in
which all three base pairs have formed, and one in which
one or more mis-pairs are present, the latter signaling that
the wrong tRNA is present (Yoshizawa *et al.*, 1999). This is
probably just one part of a series of safeguards that
ensure the accuracy of the translation process (Rodnina
and Wintermeyer, 2001a; 2001b).

When the aminoacyl-tRNA has entered the A site, a
peptide bond is formed between the two amino acids.
This involves a **peptidyl transferase** enzyme, which
releases the amino acid from the initiator $tRNA_i^{Met}$ and
then forms a peptide bond between this amino acid and
the one attached to the second tRNA. In bacteria, the peptidyl transferase activity resides in the 23S rRNA of the
large subunit, and so is an example of a ribozyme (Section
10.2.3; see Research Briefing 11.1). The reaction is energy
dependent and requires hydrolysis of the GTP attached to
EF-Tu (eEF-1 in eukaryotes). This inactivates EF-Tu,
which is ejected from the ribosome and regenerated by
EF-Ts. A eukaryotic equivalent of EF-Ts has not been
identified, and it is possible that one of the subunits of
eEF-1 has the regenerative activity.

Now the dipeptide corresponding to the first two
codons in the open reading frame is attached to the tRNA
in the A site. The next step is **translocation**, during which
three things happen at once (see *Figure 11.18*):

- The ribosome moves along three nucleotides, so the
next codon enters the A site.
- The dipeptide-tRNA in the A site moves to the P site.
- The deacylated tRNA in the P site moves to a third
position, the **E** or **exit site**, in bacteria or, in eukaryotes, is simply ejected from the ribosome.

Translocation requires hydrolysis of a molecule of GTP
and is mediated by EF-G in bacteria and by eEF-2 in
eukaryotes. Electron microscopy of ribosomes at different
intermediate stages in translocation suggests that the two
subunits rotate slightly in opposite directions, opening up
the space between them and enabling the ribosome to
slide along the mRNA (Frank and Agrawal, 2000).
Translocation results in the A site becoming vacant, allowing a new aminoacyl-tRNA to enter. The elongation cycle
is now repeated, and continues until the end of the open
reading frame is reached.

Frameshifting and other unusual events during elongation

The straightforward codon-by-codon translation of an
mRNA is looked upon as the standard way in which proteins are synthesized. But an increasing number of
unusual elongation events are being discovered. One of
these is **frameshifting**, which occurs when a ribosome
pauses in the middle of an mRNA, moves back one

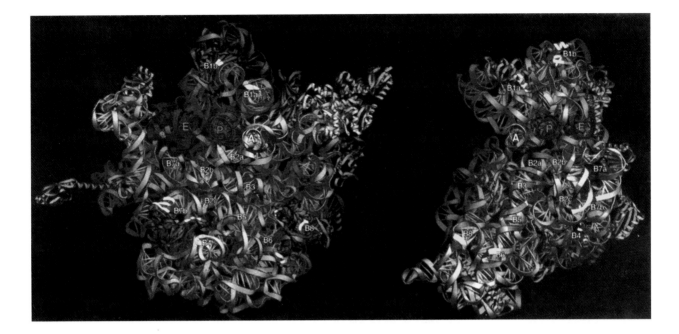

Figure 11.19 The important sites in the ribosome.

The structure on the left is the large subunit of the *Thermus thermophilus* ribosome; that on the right is the small subunit. The views look down onto the two surfaces that contact one another when the subunits are placed together to make the intact ribosome. The A, P and E sites are labeled, and each one is occupied by a tRNA shown in red or orange. The main part of each tRNA is embedded within the large subunit, with just the anticodon arms and loops associated with the small subunit. Those parts of the ribosome that make the important bridging contacts between the subunits are labeled as B1a, etc. Reprinted with permission from Yusupov *et al.*, *Science*, **292**, 883–896. Copyright 2001 American Association for the Advancement of Science.

nucleotide or, less frequently, forward one nucleotide, and then continues translation (Farabaugh, 1996). The result is that the codons that are read after the pause are not contiguous with the preceding set of codons: they lie in a different reading frame (*Figure 11.20A*).

Spontaneous frameshifts occur randomly and are deleterious because the polypeptide synthesized after the frameshift has the incorrect amino acid sequence. But not all frameshifts are spontaneous: a few mRNAs utilize **programmed frameshifting** to induce the ribosome to change frame at a specific point within the transcript. Programmed frameshifting occurs in all types of organism, from bacteria through to humans, as well as during expression of a number of viral genomes. An example occurs during synthesis of DNA polymerase III in *E. coli*, the main enzyme involved in replication of DNA (Section 13.2.2). Two of the DNA polymerase III subunits, γ and τ, are coded by a single gene, *dnaX*. Subunit τ is the full-length translation product of the *dnaX* mRNA, and subunit γ is a shortened version. Synthesis of γ involves a frameshift in the middle of the *dnaX* mRNA, the ribosome encountering a termination codon immediately after the frameshift and so producing the truncated γ version of the translation product. It is thought that the frameshift is induced by three features of the *dnaX* mRNA:

- A hairpin loop, located immediately after the frameshift position, which stalls the ribosome;
- A sequence similar to a ribosome binding site immediately upstream of the frameshift position, which is thought to base-pair with the 16S rRNA (as does an authentic ribosome binding site), again causing the ribosome to stall;
- The codon 5'–AAG–3' at the frameshift position. The presence of a modified nucleotide at the wobble position of the tRNA^{Lys} that decodes 5'–AAG–3' means that the codon–anticodon interaction is relatively weak at this position, enabling the frameshift to occur.

A similar phenomenon – translational **slippage** – enables a single ribosome to translate an mRNA that contains copies of two or more genes (*Figure 11.20B*). This means that, for example, a single ribosome can synthesize each of the five proteins coded by the mRNA transcribed from the tryptophan operon of *E. coli* (see *Figure 2.20B*, page 55). When the ribosome reaches the end of one series of codons it releases the protein it has just made, slips to the next initiation codon, and begins synthesizing the next protein. A more extreme form of slippage is **translational bypassing** (Herr *et al.*, 2000) in which a larger part of the transcript, possibly a few tens of base pairs, is skipped,

RESEARCH 11.1 BRIEFING

Peptidyl transferase is a ribozyme

A ribosome-associated protein that has the peptidyl transferase activity needed to synthesize peptide bonds during translation has never been isolated. The reason for this lack of success is now known: the enzyme activity is specified by part of the 23S rRNA.

When the base-paired structures of rRNAs (see *Figure 11.11*, page 323) were first determined in the early 1980s, the possibility that an RNA molecule could have enzymatic activity was unheard of, the breakthrough discoveries with regard to ribozymes not being made until the period 1982–86. Ribosomal RNAs were therefore initially assigned purely structural roles in the ribosome, their base-paired conformations being looked upon as scaffolds to which the important components of the ribosome – the proteins – were attached. Problems with this interpretation began to arise in the late 1980s when difficulties were encountered in identifying the protein or proteins responsible for the central catalytic activity of the ribosome – the formation of peptide bonds. By now the existence of ribozymes had been established and molecular biologists began to take seriously the possibility that rRNAs might have an enzymatic role in protein synthesis.

Locating the site of peptidyl transferase activity in the ribosome

Over the years, antibiotics and other inhibitors of protein synthesis have played an important role in studies of ribosome function. In 1995, a new inhibitor called CCdA-phosphate-puromycin was synthesized, this compound being an analog of the intermediate structure formed when two amino acids are joined by formation of a peptide bond during protein synthesis. CCdA-phosphate-puromycin binds tightly to the bacterial ribosome and, because of its structure, this binding site must be at precisely the position where peptide bonds are formed in the functioning ribosome. Would it be possible to use the inhibitor to find out where in the ribosome peptide bonds are made?

X-ray crystallography (Section 9.1.3) has revealed exactly where CCdA-phosphate-puromycin binds within the 50S subunit. Its position is deep down within the body of the subunit. The view shown here depicts the critical part of CCdA-phosphate-puromycin as a red dot, marking the position where the chemical reaction that creates a dipeptide must occur. This position is closely associated with the 23S rRNA of the large subunit (the rRNA is not shown in the figure) but is 18.4 Å away from the nearest protein, L3, and slightly more distant from L2, L4 and L10 (10 Å = 1 nm).

In atomic terms, 18–24 Å is a massive distance and it is inconceivable that any biochemical activity occurring at such a position could be catalyzed by one of the four proteins shown in the figure. The positioning of CCdA-phosphate-puromycin, and hence of the active site for peptide bond formation, provides convincing evidence that peptidyl transferase must be a ribozyme.

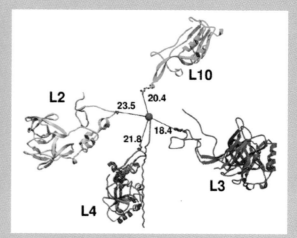

Reprinted with permission from Nissen *et al.*, *Science*, **289**, 920–930. Copyright 2000 American Association for the Advancement of Science.

Now that the evidence has finally been obtained, researchers are moving on to determine exactly how the rRNA backbone acts as a ribozyme in peptide bond formation. Attention was initially concentrated on an adenine nucleotide at position 2451 in the *E. coli* 23S rRNA, because this adenine has unusual charge properties compared with other nucleotides. The hypothesis was that an interaction between this adenine and a nearby guanine, at position 2447, is the key to protein synthesis. But this model has been thrown into disarray by mutational studies, which have shown that both A2451 and G2447 can be replaced by other nucleotides without a detectable effect on the ability of the ribosome to carry out peptide bond synthesis.

These results have prompted a re-evaluation of the roles of A2451 and G2447 in peptide bond formation, and attention is now turning to other nucleotides present in the parts of the 23S rRNA that are located in the vicinity of the active site. Much work still needs to be done, but the ribozymal basis for peptidyl transferase activity is gradually being tracked down.

References

Nissen P, Hansen J, Ban N, Moore PB and Steitz TA (2000) The structural basis of ribosome activity in peptide bond synthesis. *Science*, **289**, 920–930.

Polacek N, Gaynor M, Yassin A and Mankin AS (2001) Ribosomal peptidyl transferase can withstand mutations at the putative catalytic nucleotide. *Nature*, **411**, 498–501.

(A) Programmed frameshifting in the *dnaX* mRNA

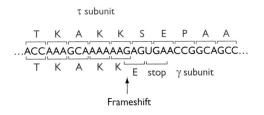

(B) Translational slippage in the lactose operon mRNA

(C) Translational bypassing in the T4 gene *60* mRNA

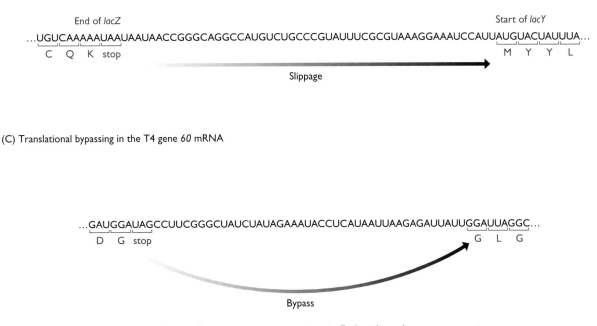

Figure 11.20   Three unusual translation elongation events occurring in *Escherichia coli.*

(A) Programmed frameshifting during translation of the *dnaX* mRNA. During synthesis of the γ subunit the ribosome shifts back one nucleotide, immediately after a series of As coding for two lysine amino acids. The ribosome inserts a glutamic acid into the polypeptide and then encounters a termination codon. (B) Slippage between the *lacZ* and *lacY* genes of the lactose operon mRNA. (C) Bypassing during translation of the T4 gene *60* mRNA involves a jump between two glycine codons. For the one-letter abbreviations of the amino acids see *Table 3.1*, page 84.

and elongation of the original protein continues after the bypassing event (*Figure 11.20C*). The bypass starts and ends either at two identical codons or at two codons that can be translated by the same tRNA by wobble. This suggests that the jump is controlled by the tRNA attached to the growing polypeptide, which scans the mRNA as the ribosome tracks along, and halts the bypass when a new codon to which it can base-pair is reached. Translational bypassing of 44 nucleotides occurs in *E. coli* during translation of the mRNA for gene *60* of T4 bacteriophage, which codes for a DNA topoisomerase subunit. Similar

events have also been identified in a variety of other bacteria. Bypassing could result in two different proteins being synthesized from one mRNA – one protein from normal translation and one from bypassing – but whether this is its general function is not yet known.

11.2.4 Termination of translation

Protein synthesis ends when one of the three termination codons is reached. The A site is now entered not by a tRNA

Box 11.1: Translation in the archaea

In most respects, translation in the archaea more closely resembles the equivalent events in the eukaryotic cytoplasm rather than in bacteria. The one apparent exception is that the archaeal ribosome, at 70S, is comparable in size to the bacterial ribosome and, like bacterial ribosomes, contains 23S, 16S and 5S rRNAs. This apparent similarity is illusory because the archaeal rRNAs form base-paired secondary structures that are significantly different from the equivalent bacterial structures. The archaeal structures are also different from the eukaryotic versions, but the ribosomal proteins that attach to the rRNAs are homologs of the eukaryotic proteins. Archaeal mRNAs are capped and polyadenylated, and translation initiation is thought to involve a scanning process similar to that described for eukaryotic mRNAs. Archaeal tRNAs display a few unique features, including the absence of thymine in the so-called TΨC arm of the cloverleaf, and the presence at various positions of modified nucleotides not seen in either bacteria or eukaryotes. The methionine carried by the initiator tRNA is not N-formylated and the initiation and elongation factors resemble the eukaryotic molecules.

but by a protein **release factor** (*Figure 11.21*). Bacteria have three of these: RF-1 which recognizes the termination codons 5'–UAA–3' and 5'–UAG–3', RF-2 which recognizes 5'–UAA–3' and 5'–UGA–3', and RF-3 which stimulates release of RF1 and RF2 from the ribosome after termination, in a reaction requiring energy from the hydrolysis of GTP. Eukaryotes have just two release factors (see *Table 11.4*): eRF-1, which recognizes the termination codon, and eRF-3, which might play the same role as RF-3 although this has not been proven (Kisselev and Buckingham, 2000). The structure of eRF-1 has been solved by X-ray crystallography, showing that the shape of this protein is very similar to that of a tRNA (*Figure 11.22*). This gives an indication of how the release factor is able to enter the A site when the termination codon is reached.

The release factors terminate translation but they do not appear to be responsible for disassociation of the ribosomal subunits, at least not in bacteria. This is the function of an additional protein called **ribosome recycling factor** (**RRF**) which, like eRF-1, has a tRNA-like structure (Selmer *et al.*, 1999). RRF probably enters the P or A site and 'unlocks' the ribosome (see *Figure 11.21*). Disassociation requires energy, which is released from GTP by EF-G, one of the elongation factors (see *Table 11.2*), and also requires the initiation factor IF-3 to prevent the subunits from attaching together again. A eukaryotic equivalent of RRF has not been identified, and this may be one of the

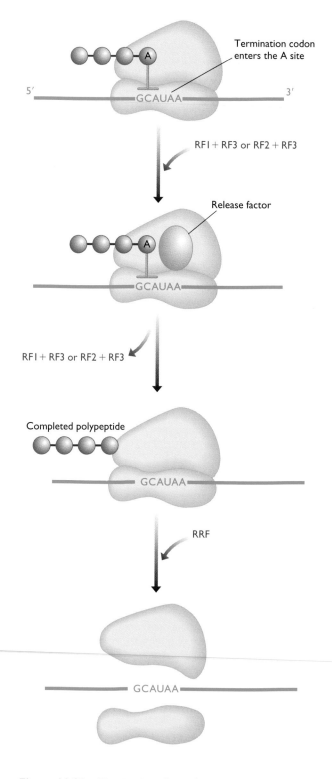

Figure 11.21 Termination of translation.

Termination in *Escherichia coli* is illustrated. For differences in eukaryotes, see the text. The amino acid labeled with an 'A' is an alanine. Abbreviations: RF, release factor; RRF, ribosome recycling factor.

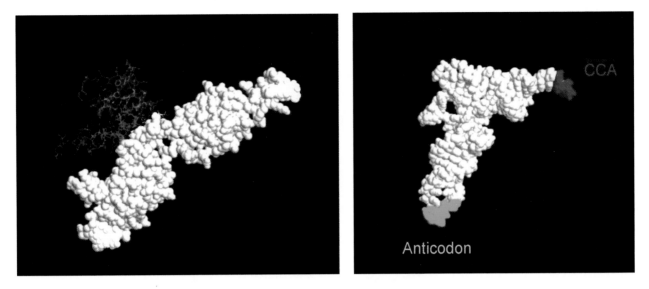

Figure 11.22 The structure of the eukaryotic release factor eRF-1 is similar to that of a tRNA.

The left panel shows eRF-1 and the right panel shows a tRNA. The part of eRF-1 that resembles the tRNA is highlighted in white. The purple segment of eRF-1 interacts with the second eukaryotic release factor, eRF-3. Reproduced with permission from Kisselev and Buckingham (2000) *Trends Biochem. Sci.*, **25**, 561–566.

functions of eRF-3. The disassociated ribosome subunits enter the cytoplasmic pool, where they remain until used again in another round of translation.

11.3 Post-translational Processing of Proteins

Translation is not the end of the genome expression pathway. The polypeptide that emerges from the ribosome is inactive, and before taking on its functional role in the cell must undergo at least the first of the following four types of post-translational processing (*Figure 11.23*):

- *Protein folding.* The polypeptide is inactive until it is folded into its correct tertiary structure.
- *Proteolytic cleavage.* Some proteins are processed by cutting events carried out by enzymes called **proteases**. These cutting events may remove segments from one or both ends of the polypeptide, resulting in a shortened form of the protein, or they may cut the polypeptide into a number of different segments, all or some of which are active.
- *Chemical modification.* Individual amino acids in the polypeptide might be modified by attachment of new chemical groups.
- *Intein splicing.* Inteins are intervening sequences in some proteins, similar in a way to introns in mRNAs. They have to be removed and the **exteins** ligated in order for the protein to become active.

Often these different types of processing occur together, the polypeptide being cut and modified at the same time that it is folded. If this is the case then the cutting, modi-

fication and/or splicing events may be necessary for the polypeptide to take up its correct three-dimensional conformation, because this is dependent in part on the relative positioning of the various chemical groups along the molecule. Alternatively, a cutting event or a chemical modification may occur after the protein has been folded, possibly as part of a regulatory mechanism that converts a folded but inactive protein into an active form.

11.3.1 Protein folding

Protein folding was introduced in Chapter 3 when we examined the four levels of protein structure (primary, secondary, tertiary and quaternary) and learnt that all of the information that a polypeptide needs in order to adopt its correct three-dimensional structure is contained within its amino acid sequence (Section 3.3.3). This is one of the central principles of molecular biology. We must therefore examine its experimental basis and consider how the information contained in the amino acid sequence is utilized during the folding process for a newly translated polypeptide.

Not all proteins fold spontaneously in the test tube

The notion that the amino acid sequence contains all the information needed to fold the polypeptide into its correct tertiary structure derives from experiments carried out with ribonuclease in the 1960s (Anfinsen, 1973). Ribonuclease is a small protein, just 124 amino acids in length, containing four disulfide bridges and with a tertiary structure that is made up predominantly of β-sheet,

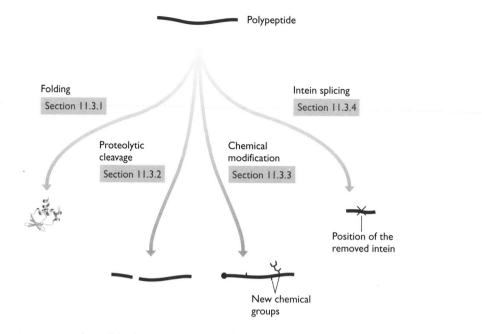

Figure 11.23 Schematic representation of the four types of post-translational processing event.

Not all events occur in all organisms – see the text for details.

with very little α-helix. Studies of its folding were carried out with ribonuclease that had been purified from cow pancreas and resuspended in buffer. Addition of urea, a compound that disrupts hydrogen bonding, resulted in a decrease in the activity of the enzyme (measured by test-ing its ability to cut RNA) and an increase in the viscosity of the solution (*Figure 11.24*), indicating that the protein was being **denatured** by unfolding into an unstructured polypeptide chain. The critical observation was that when the urea was removed by dialysis, the viscosity

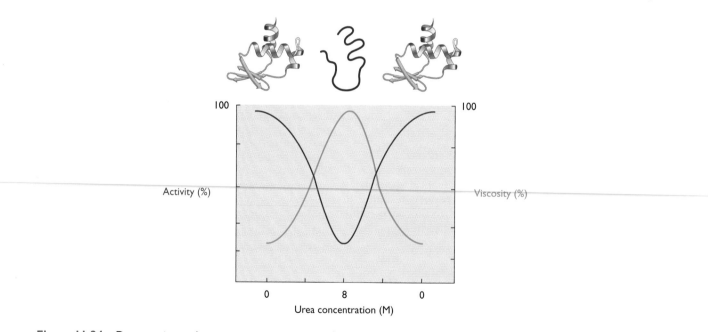

Figure 11.24 Denaturation and spontaneous renaturation of a small protein.

As the urea concentration increases to 8M, the protein becomes denatured by unfolding: its activity decreases and the viscosity of the solution increases. When the urea is removed by dialysis, this small protein re-adopts its folded conformation. The activity of the protein increases back to the original level and the viscosity of the solution decreases.

decreased and the enzyme activity reappeared. The conclusion is that the protein refolds spontaneously when the denaturant (in this case, urea) is removed. In these initial experiments the four disulfide bonds remained intact because they were not disrupted by urea, but the same result occurred when the urea treatment was combined with addition of a reducing agent to break the disulfide bonds: the activity was still regained on renaturation. This shows that the disulfide bonds are not critical to the protein's ability to refold, they merely stabilize the tertiary structure once it has been adopted.

More detailed study of the spontaneous folding pathways for ribonuclease and other small proteins has led to the following general two-step description of the process (Hartl, 1996):

1. The secondary structural motifs along the polypeptide chain form within a few milliseconds of the denaturant being removed. This step is accompanied by the protein collapsing into a compact, but not folded, organization, with its hydrophobic groups on the inside, shielded from water.

2. During the next few seconds or minutes, the secondary structural motifs interact with one another and the tertiary structure gradually takes shape, often via a series of intermediate conformations. In other words, the protein follows a **folding pathway**. There may, however, be more than one possible pathway that a protein can follow to reach its correctly folded structure (Radford, 2000). The pathways may also have side-branches into which the protein can be diverted, leading to an incorrect structure. If an incorrect structure is sufficiently unstable then partial or complete unfolding may occur, allowing the protein a second opportunity to pursue a productive route towards its correct conformation (*Figure 11.25*).

For several years it was more or less assumed that all proteins would fold spontaneously in the test tube, but experiments have shown that only smaller proteins with less complex structures possess this ability. Two factors seem to prevent larger proteins from folding spontaneously. The first of these is their tendency to form insoluble aggregates when the denaturant is removed: the polypeptides may collapse into interlocked networks when they attempt to protect their hydrophobic groups from water in step 1 of the general folding pathway. Experimentally, this can be avoided by using a low dilution of the protein, but this is not an option that the cell can take to prevent its unfolded proteins from aggregating. The second factor that prevents folding is that a large protein tends to get stuck in non-productive side branches of its folding pathway, taking on an intermediate form that is incorrectly folded but which is too stable to unfold to any significant extent. Concerns have also been raised about the relevance of *in vitro* folding, as studied with ribonuclease, to the folding of proteins in the cell, because a cellular protein might begin to fold before it has been fully synthesized. If the initial folding

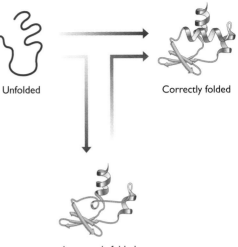

Figure 11.25 An incorrectly folded protein might be able to refold into its correct conformation.

The blue arrow represents the correct folding pathway, leading from the unfolded protein on the left to the active protein on the right. The red arrow leads to an incorrectly folded conformation, but this conformation is unstable and the protein is able to unfold partially, return to its correct folding pathway and, eventually, reach its active conformation.

occurs when only part of the polypeptide is available, then there might be an increased possibility of incorrect branches of the folding pathway being followed. These various considerations prompted research into folding in living cells.

In cells, folding is aided by molecular chaperones

Most of our current understanding of protein folding in the cell is founded on the discovery of proteins that help other proteins to fold. These are called **molecular chaperones** and have been studied in most detail in *E. coli*. It is clear that both eukaryotes and archaea possess equivalent proteins, although some of the details of the way they work are different (Hartl, 1996; Slavotinek and Biesecker, 2001).

The molecular chaperones in *E. coli* can be divided into two groups:

■ The **Hsp70 chaperones**, which include the proteins called Hsp70 (coded by the *dnaK* gene and sometimes called DnaK protein), Hsp40 (coded by *dnaJ*) and GrpE;

■ The **chaperonins**, the main version of which in *E. coli* is the GroEL/GroES complex.

Molecular chaperones do not specify the tertiary structure of a protein, they merely help the protein find that correct structure. The two types of chaperone do this in different ways. The Hsp70 family bind to hydrophobic

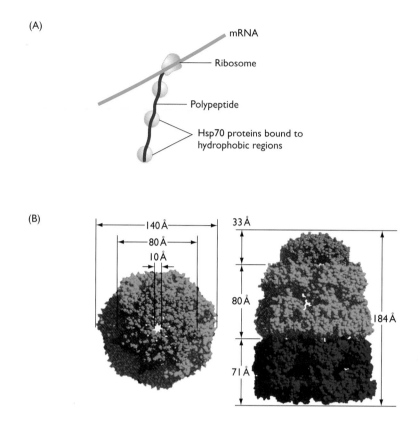

Figure 11.26 Molecular chaperones of *Escherichia coli*.

(A) Hsp70 chaperones bind to hydrophobic regions in unfolded polypeptides, including those that are still being translated, and hold the protein in an open conformation until it is ready to be folded. (B) The structure of the GroEL/GroES chaperonin. On the left is a view from the top and on the right a view from the side. 1Å is equal to 0.1 nm. The GroES part of the structure is made up of seven identical protein subunits and is shown in gold. The GroEL components consist of 14 identical proteins arranged into two rings (shown in red and green), each containing seven subunits. The main entrance into the central cavity is through the bottom of the structure shown on the right. Reprinted with permission from Xu *et al.*, *Nature*, **388**, 741–750. Copyright 1997 Macmillan Magazines Limited. Original image kindly supplied by Dr Zhaohui Xu, Department of Biological Chemistry, The University of Michigan.

regions of proteins, including proteins that are still being translated (*Figure 11.26A*). They prevent protein aggregation by holding the protein in an open conformation until it is completely synthesized and ready to fold. The Hsp70 chaperones are also involved in other processes that require shielding of hydrophobic regions in proteins, such as transport through membranes and disaggregation of proteins that have been damaged by heat stress.

The chaperonins work in a quite different way. GroEL and GroES form a multi-subunit structure that looks like a hollowed-out bullet with a central cavity (*Figure 11.26B*; Xu *et al.*, 1997). A single unfolded protein enters the cavity and emerges folded. The mechanism for this is not known but it is postulated that GroEL/GroES acts as a cage that prevents the unfolded protein from aggregating with other proteins, and that the inside surface of the cav-

ity changes from hydrophobic to hydrophilic in such a way as to promote the burial of hydrophobic amino acids within the protein. This is not the only hypothesis: other researchers hold that the cavity unfolds proteins that have folded incorrectly, passing these unfolded proteins back to the cytoplasm so they can have a second attempt at adopting their correct tertiary structure (Shtilerman *et al.*, 1999).

Eukaryotic proteins equivalent to both the Hsp70 family of chaperones and the GroEL/GroES chaperonins have been found, but it seems that in eukaryotes protein folding depends mainly on the action of the Hsp70 proteins. This is probably true also of bacteria (Ellis, 2000) even though the GroEL/GroES chaperonins play a major role in the folding of metabolic enzymes, and proteins involved in transcription and translation (Houry *et al.*, 1999).

11.3.2 Processing by proteolytic cleavage

Proteolytic cleavage has two functions in post-translational processing of proteins (*Figure 11.27*):

■ It is used to remove short pieces from the N- and/or C-terminal regions of polypeptides, leaving a single shortened molecule that folds into the active protein.

■ It is used to cut **polyproteins** into segments, all or some of which are active proteins.

These events are relatively common in eukaryotes but less frequent in bacteria.

Cleavage of the ends of polypeptides

Processing by cleavage is common with secreted polypeptides whose biochemical activities might be deleterious to the cell producing the protein. An example is provided by melittin, the most abundant protein in bee venom and the one responsible for causing cell lysis after injection of the bee sting into the person or animal being stung. Melittin lyses cells in bees as well as animals and so must initially be synthesized as an inactive precursor. This precursor, promelittin, has 22 additional amino acids at its N terminus. The pre-sequence is removed by an extracellular protease that cuts it at 11 positions, releasing the active venom protein. The protease does not cleave within the active sequence because its mode of action is to release dipeptides with the sequence X–Y, where X is alanine, aspartic acid or glutamic acid, and Y is alanine or proline; these motifs do not occur in the active sequence (*Figure 11.28A*).

A similar type of processing occurs with insulin, the protein made in the islets of Langerhans in the vertebrate pancreas and responsible for controlling blood sugar levels. Insulin is synthesized as preproinsulin, which is 105 amino acids in length (*Figure 11.28B*). The processing pathway involves the removal of the first 24 amino acids to give proinsulin, followed by two additional cuts which excise a central segment, leaving two active parts of the protein, the A and B chains, which link together by formation of two disulfide bonds to form mature insulin. The first segment to be removed, the 24 amino acids from the N terminus, is a **signal peptide**, a highly hydrophobic stretch of amino acids that attaches the precursor protein to a membrane prior to transport across that membrane and out of the cell. Signal peptides are commonly found on proteins that bind to and/or cross membranes, in both eukaryotes and prokaryotes.

Proteolytic processing of polyproteins

In the examples shown in *Figure 11.28*, proteolytic processing results in a single mature protein. This is not always the case. Some proteins are initially synthesized as polyproteins, long polypeptides that contain a series of mature proteins linked together in head-to-tail fashion. Cleavage of the polyprotein releases the individual proteins, which may have very different functions from one another.

Polyproteins are not uncommon in eukaryotes. Several types of virus that infect eukaryotic cells use them as a way of reducing the sizes of their genomes, a single polyprotein gene with one promoter and one terminator taking up less space than a series of individual genes. Polyproteins are

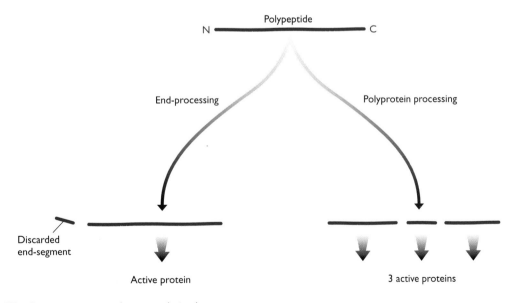

Figure 11.27   Protein processing by proteolytic cleavage.

On the left, the protein is processed by removal of the N-terminal segment. C-terminal processing also occurs with some proteins. On the right, a polyprotein is processed to give three different proteins. Not all proteins undergo proteolytic cleavage.

(A)

Cut sites

APEPEPAPEPEAEADAEADPEAGIGAVLKVLTTGLPALISWIKRKRQQG

(B)

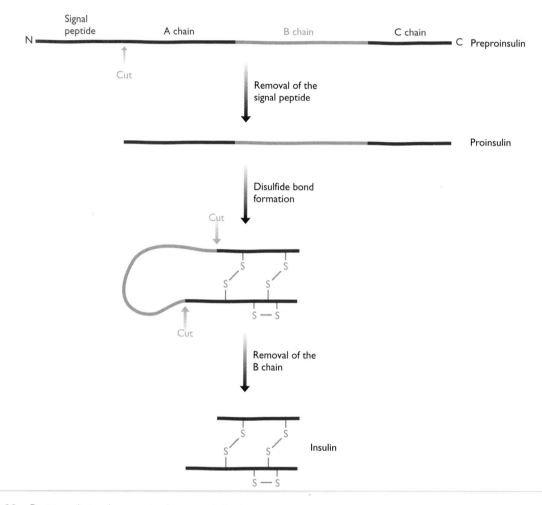

Figure 11.28 Post-translational processing by proteolytic cleavage.

(A) Processing of promelittin, the bee-sting venom. Arrows indicate the cut sites. For the one-letter abbreviations of the amino acids see *Table 3.1*, page 84. (B) Processing of preproinsulin. See the text for details.

also involved in the synthesis of peptide hormones in vertebrates. For example, the polyprotein called pro-opiomelanocortin, made in the pituitary gland, contains at least ten different peptide hormones. These are released by proteolytic cleavage of the polyprotein (*Figure 11.29*), but not all can be produced at once because of overlaps between individual peptide sequences. Instead, the exact cleavage pattern is different in different cells.

11.3.3 Processing by chemical modification

The genome has the capacity to code for 21 different amino acids: the 20 specified by the standard genetic code, and selenocysteine, which is inserted into polypeptides by the context-dependent reading of a 5′–UGA–3′ codon (Section 3.3.2). This repertoire is increased dramatically by post-translational chemical modification of proteins, which

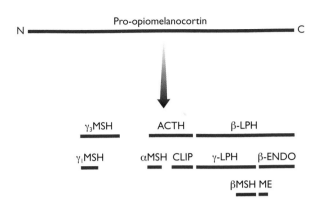

Figure 11.29 Processing of the pro-opiomelanocortin polyprotein.

Abbreviations: ACTH, adrenocorticotropic hormone; CLIP, corticotropin-like intermediate lobe protein; ENDO, endorphin; LPH, lipotropin; ME, met-encephalin; MSH, melanotropin.

results in a vast array of different amino acid types. The simpler types of modification occur in all organisms; the more complex ones, especially glycosylation, are rare in bacteria.

The simplest types of chemical modification involve addition of a small chemical group (e.g. an acetyl, methyl or phosphate group; *Table 11.5*) to an amino acid side chain, or to the amino or carboxyl groups of the terminal amino acids in a polypeptide (for an example see Bradshaw *et al.*, 1998). Over 150 different modified amino acids have been documented in different proteins, with each modification carried out in a highly specific manner, the same amino acids being modified in the same way in every copy of the protein. This is illustrated in *Figure 11.30* for histone H3. The example reminds us that chemical modification often plays an important role in determining the precise biochemical activity of the target protein: we saw in Section 8.2.1 how acetylation and methylation of H3 and other histones have an important influence on chromatin structure and hence on genome expression. Other types of chemical modification have important regulatory roles, an example being phosphorylation, which is used to activate many proteins involved in signal transduction (Section 12.1.2).

```
      Me     Me    Ac        Ac    Me
      |      |     |         |     |
  ARTKQTARKSTGGKAPRKQLATKAARKSAP━━━━━━━━━━━
```

Figure 11.30 Post-translational chemical modification of calf histone H3.

The first 30 amino acids of this 135-amino-acid protein are listed using the one-letter abbreviations (see *Table 3.1*, page 84). Five modifications occur: three methylations and two acetylations. For the role of methylation and acetylation of histones in determining chromatin structure see Section 8.2.1.

Table 11.5 Examples of post-translational chemical modifications

Modification	Amino acids that are modified	Examples of proteins
Addition of small chemical groups		
Acetylation	Lysine	Histones
Methylation	Lysine	Histones
Phosphorylation	Serine, threonine, tyrosine	Some proteins involved in signal transduction
Hydroxylation	Proline, lysine	Collagen
N-formylation	N-terminal glycine	Melittin
Addition of sugar side chains		
O-linked glycosylation	Serine, threonine	Many membrane proteins and secreted proteins
N-linked glycosylation	Asparagine	Many membrane proteins and secreted proteins
Addition of lipid side chains		
Acylation	Serine, threonine, cysteine	Many membrane proteins
N-myristoylation	N-terminal glycine	Some protein kinases involved in signal transduction
Addition of biotin		
Biotinylation	Lysine	Various carboxylase enzymes

See Section 12.1.2 for more information on the role of chemical modification during signal transduction.

A more complex type of modification is **glycosylation**, the attachment of large carbohydrate side chains to polypeptides (Drickamer and Taylor, 1998). There are two general types of glycosylation (*Figure 11.31*):

- *O-linked glycosylation* is the attachment of a sugar side chain via the hydroxyl group of a serine or threonine amino acid.
- *N-linked glycosylation* involves attachment through the amino group on the side chain of asparagine.

Glycosylation can result in attachment to the protein of grand structures comprising branched networks of 10–20 sugar units of various types. These side chains help to target proteins to particular sites in cells and determine the stability of proteins circulating in the bloodstream. Another type of large-scale modification involves attachment of long-chain lipids, often to serine or cysteine amino acids. This process is called **acylation** and occurs with many proteins that become associated with membranes. A less common modification is **biotinylation**, in which a molecule of biotin is attached to a small number of enzymes that catalyze the carboxylation of organic acids such as acetate and propionate (Chapman-Smith and Cronan, 1999).

11.3.4 Inteins

The final type of post-translational processing that we must consider is intein splicing, a protein version of the more extensive intron splicing that occurs with pre-RNAs. Inteins are internal segments of proteins that are removed soon after translation, the two external segments or exteins becoming linked together (*Figure 11.32*). The first intein was discovered in *S. cerevisiae* in 1990, and there have been only 100 confirmed identifications so far. Despite their scarcity, inteins are widespread. Most are known in bacteria and archaea but there are also examples in lower eukaryotes. In a few cases there is more than one intein in a single protein.

Most inteins are approximately 150 amino acids in length and, like pre-mRNA introns (Section 10.1.3), the sequences at the splice junctions of inteins have some similarity in most of the known examples. In particular, the first amino acid of the downstream extein is cysteine, serine or threonine. A few other amino acids within the intein sequence are also conserved. These conserved amino acids are involved in the splicing process, which is self-catalyzed by the intein itself (Paulus, 2000).

Two interesting features of inteins have recently come to light. The first of these was discovered when the structures of two inteins were determined by X-ray crystallography (Duan *et al.*, 1997; Klabunde *et al.*, 1998). These structures are similar in some respects to that of a *Drosophila* protein called Hedgehog, which is involved in development of the segmentation pattern of the fly embryo. Hedgehog is an autoprocessing protein that cuts itself in two. The structural similarity with inteins lies in

the part of the Hedgehog protein that catalyzes its self-cleavage. Possibly the same protein structure has evolved twice, or possibly inteins and Hedgehog shared a common link at some stage in the evolutionary past.

(A) O-linked glycosylation

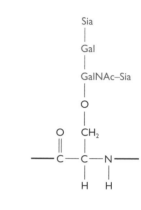

(B) N-linked glycosylation

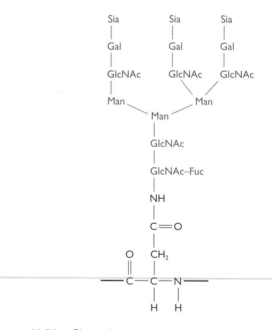

Figure 11.31   Glycosylation.

(A) O-linked glycosylation. The structure shown is found in a number of glycoproteins. It is drawn here attached to a serine amino acid but it can also be linked to a threonine. (B) N-linked glycosylation usually results in larger sugar structures than are seen with O-linked glycosylation. The drawing shows a typical example of a complex glycan attached to an asparagine amino acid. Abbreviations: Fuc, fucose; Gal, galactose; GalNAc, N-acetylgalactosamine; GlcNAc, N-acetylglucosamine; Man, mannose; Sia, sialic acid.

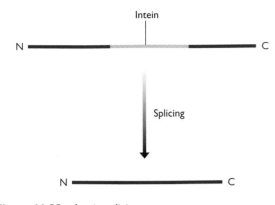

Intein

N ———————————— C

↓ Splicing

N ———————————— C

Figure 11.32   Intein splicing.

The second interesting feature is that with some inteins the excised segment is a sequence-specific endonuclease. The intein cuts DNA at the sequence corresponding to its insertion site in a gene coding for an intein-free version of the protein from which it is derived (*Figure 11.33*). If the cell also contains a gene coding for the intein-containing protein, then the DNA sequence for the intein is able to jump into the cut site, converting the intein-minus gene into an intein-plus version, a process called **intein homing** (Pietrokovski, 2001). The same type of event occurs with some Group I introns (Section 10.2.3), which code for proteins that direct **intron homing**. It is possible that

transfer of inteins and Group I introns might also occur between cells or even between species (Cooper and Stevens, 1995). This is thought to be a mechanism by which **selfish DNA** is able to propagate (see Box 15.3, page 479).

11.4 Protein Degradation

The protein synthesis and processing events that we have studied so far in this chapter result in new active proteins that take up their place in the cell's proteome. These proteins either replace existing ones that have reached the end of their working lives or provide new protein functions in response to the changing requirements of the cell. The concept that the proteome of a cell can change over time requires not only *de novo* protein synthesis but also the removal of proteins whose functions are no longer required. This removal must be highly selective so that only the correct proteins are degraded, and must also be rapid in order to account for the abrupt changes that occur under certain conditions, for example during key transitions in the cell cycle (Hunt, 1997).

For many years, protein degradation was an unfashionable subject and it was not until the 1990s that real progress was made in understanding how specific proteolysis events are linked with processes such as the cell cycle and differentiation. Even now, our knowledge cen-

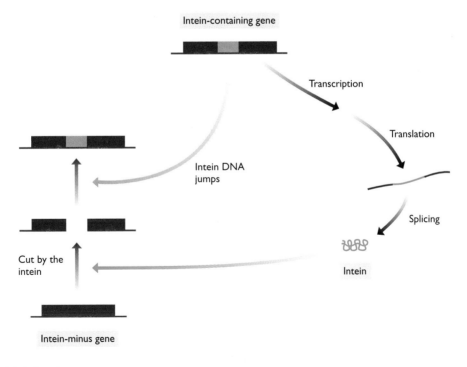

Figure 11.33   Intein homing.

The cell is heterozygous for the intein-containing gene, possessing one allele with the intein and one allele without the intein. After protein splicing, the intein cuts the intein-minus gene at the appropriate place, allowing a copy of the intein DNA sequence to jump into this gene, converting it into the intein-plus version.

ters largely on descriptions of general protein breakdown pathways and less on the regulation of the pathways and the mechanisms used to target specific proteins. There appear to be a number of different types of breakdown pathway whose interconnectivities have not yet been traced. This is particularly true in bacteria, which seem to have a range of proteases that work together in controlled degradation of proteins. In eukaryotes, most breakdown involves a single system, involving **ubiquitin** and the **proteasome**.

A link between ubiquitin and protein degradation was first established in 1975 when it was shown that this abundant 76-amino-acid protein is involved in energy-dependent proteolysis reactions in rabbit cells (Varshavsky, 1997). Subsequent research identified a series of three enzymes that attach ubiquitin molecules, singly or in chains, to lysine amino acids in proteins that are targeted for breakdown. Whether or not a protein becomes ubiquitinated depends on the presence or absence within it of amino acid motifs that act as degradation-susceptibility signals. These signals have not been completely characterized but there are thought to be at least ten different types in S. *cerevisiae*, including:

- **The N-degron**, a sequence element present at the N terminus of a protein;
- **PEST sequences**, internal sequences that are rich in proline (P), glutamic acid (E), serine (S) and threonine (T).

These sequences are permanent features of the proteins that contain them and so cannot be straightforward 'degradation signals': if they were then these proteins would be broken down as soon as they are synthesized. Instead, they must determine susceptibility to degradation and hence the general stability of a protein in the cell. How this might be linked to the controlled breakdown of selected proteins at specific times, for instance during the cell cycle, is not yet clear.

The second component of the ubiquitin-dependent degradation pathway is the proteasome, the structure within which ubiquitinated proteins are broken down. In eukaryotes the proteasome is a large, multi-subunit structure with a sedimentation coefficient of 26S, comprising a hollow cylinder of 20S and two 'caps' of 19S (Groll *et al.*, 1997; Ferrell *et al.*, 2000). Archaea also have proteasomes of about the same size but these are less complex, being composed of multiple copies of just two proteins; eukaryotic proteasomes contain 14 different types of protein subunit. The entrance into the cavity within the proteasome is narrow, and a protein must be unfolded before it can enter. This unfolding probably occurs through an energy-dependent process and may involve structures similar to chaperonins (Section 11.3.1) but with unfolding rather than folding activity (Lupas *et al.*, 1997). After unfolding, the protein can enter the proteasome within which it is cleaved into short peptides 4–10 amino acids in length. These are released back into the cytoplasm where they are broken down into individual amino acids which can be re-utilized in protein synthesis.

References

Agalarov SC, Prasad GS, Funke PM, Stout CD and Williamson JR (2000) Structure of the S15, S6, S18–rRNA complex: assembly of the 30S ribosome central domain. *Science*, **288**, 107–112.

Anfinsen CB (1973) Principles that govern the folding of protein chains. *Science*, **181**, 223–230.

Arnez JG and Moras D (1997) Structural and functional considerations of the aminoacylation reaction. *Trends Biochem. Sci.*, **22**, 211–216.

Ban N, Nissen P, Hansen J, Moore PB and Steitz TA (2000) The complete atomic structure of the large ribosomal subunit at 2.4 angstrom resolution. *Science*, **289**, 905–920.

Bradshaw RA, Brickey WW and Walker KW (1998) N-terminal processing: the methionine aminopeptidase and Nα-acetyl transferase families. *Trends Biochem. Sci.*, **23**, 263–267.

Chapman-Smith A and Cronan JE (1999) The enzymatic biotinylation of proteins: a post-translational modification of exceptional specificity. *Trends. Biochem. Sci.*, **24**, 359–363.

Clark BFC (2001) The crystallization and structural determination of tRNA. *Trends Biochem. Sci.*, **26**, 511–514.

Conn GL, Draper DE, Lattman EE and Gittis AG (1999) Crystal structure of a conserved protein–RNA complex. *Science*, **284**, 1171–1174.

Cooper AA and Stevens TH (1995) Protein splicing: self-splicing of genetically mobile elements at the protein level. *Trends Biochem. Sci.*, **20**, 352–357.

Crick FHC (1990) *What Mad Pursuit: a Personal View of Scientific Discovery.* Penguin Books, London.

Dever TE (1999) Translation initiation: adept at adapting. *Trends Biochem. Sci.*, **24**, 398–403.

Drickamer K and Taylor ME (1998) Evolving views of protein glycosylation. *Trends Biochem. Sci.* **23**, 321–324.

Duan XQ, Gimble FS and Quiocho FA (1997) Crystal structure of PI-SceI, a homing endonuclease with protein splicing activity. *Cell*, **89**, 555–564.

Ellis RJ (2000) Chaperone substrates inside the cell. *Trends Biochem. Sci.*, **25**, 210–212.

Farabaugh PJ (1996) Programmed translational frameshifting. *Ann. Rev. Genet.*, **30**, 507–528.

Ferrell K, Wilkinson CRM, Dubiel W and Gordon C (2000) Regulatory subunit interactions of the 26S proteasome, a complex problem. *Trends Biochem. Sci.*, **25**, 83–88.

Frank J and Agarwal RK (2000) A ratchet-like inter-subunit reorganization of the ribosome during translocation. *Nature*, **406**, 318–322.

Groll M, Ditzel L, Löwe J, et al. (1997) Structure of 20S proteasome from yeast at 2.4 Å resolution. *Nature*, **386**, 463–471.

Hafezparast M and Fisher E (1998) Wasted by an elongation factor. *Trends Genet.*, **14**, 215–217.

Hale SP, Auld DS, Schmidt E and Schimmel P (1997) Discrete determinants in transfer RNA for editing and aminoacylation. *Science*, **276**, 1250–1252.

Hartl FU (1996) Molecular chaperones in cellular protein folding. *Nature*, **381**, 571–580.

Heilek GM and Noller HF (1996) Site-directed hydroxyl radical probing of the rRNA neighbourhood of ribosomal protein S5. *Science*, **272**, 1659–1662.

Hentze MW (1997) eIF4G: a multipurpose ribosome adapter? *Science*, **275**, 500–501; **275**, 1553.

Herr AJ, Atkins JF and Gesteland RF (2000) Coupling of open reading frames by translational bypassing. *Ann. Rev. Biochem.*, **69**, 343–372.

Holcik M, Sonenberg N and Korneluk RG (2000) Internal ribosome initiation of translation and the control of cell death. *Trends Genet.*, **16**, 469–473.

Houry WA, Frishman D, Eckerskorn C, Lottspeich F and Hartl FU (1999) Identification of *in vivo* substrates of chaperonin GroEL. *Nature*, **402**, 147–154.

Hunt T (1997) Extinction is forever. *Trends Biochem. Sci.*, **22**, 371.

Ibba M (2001) Discriminating right from wrong. *Science*, **294**, 70–71.

Ibba M, Becker HD, Stathopoulos C, Tumbula DL and Söll D (2000) The adaptor hypothesis revisited. *Trends Biochem. Sci.*, **25**, 311–316.

IHGSC (International Human Genome Sequencing Consortium) (2001) Initial sequencing and analysis of the human genome. *Nature*, **409**, 860–921.

Ikemura T (1981) Correlation between the abundance of *Escherichia coli* transfer RNAs and the occurrence of respective codons in protein genes: a proposal for synonymous codon choice that is optimal for the *E. coli* translational system. *J. Mol. Biol.*, **151**, 389–409.

Kisselev LL and Buckingham RH (2000) Translational termination comes of age. *Trends Biochem. Sci.*, **25**, 561–566.

Klabunde T, Sharma S, Telenti A, Jacobs WR and Sacchettini JC (1998) Crystal structure of *gyrA* intein from *Mycobacterium xenopi* reveals structural basis of protein splicing. *Nature Struct. Biol.*, **5**, 31–36.

Low SC and Berry MJ (1996) Knowing when not to stop: selenocysteine incorporation in eukaryotes. *Trends Biochem. Sci.*, **21**, 203–208.

Lupas A, Flanaghan JM, Tamaru T and Baumeister W (1997) Self-compartmentalizing proteases. *Trends Biochem. Sci.*, **22**, 399–404.

Mountford PS and Smith AG (1995) Internal ribosome entry sites and dicistronic RNAs in mammalian transgenesis. *Trends Genet.*, **11**, 179–184.

Paulus H (2000) Protein splicing and related forms of protein autoprocessing. *Ann. Rev. Biochem.*, **69**, 447–496.

Pennisi E (1999) The race to the ribosome structure. *Science*, **285**, 2048–2051.

Percudani R (2001) Restricted wobble rules for eukaryotic genomes. *Trends Genet.*, **17**, 133–135.

Pestova TV and Hellen CUT (1999) Ribosome recruitment and scanning: what's new? *Trends Biochem. Sci.*, **24**, 85–87.

Pietrokovski S (2001) Intein spread and extinction in evolution. *Trends Genet.*, **17**, 465–472.

Preiss T and Hentze MW (1998) Dual function of the messenger RNA cap structure in poly(A)-tail-promoted translation in yeast. *Nature*, **392**, 516–520.

Radford S (2000) Protein folding: progress made and promises ahead. *Trends Biochem. Sci.*, **25**, 611–618.

Rodnina MV and Wintermeyer W (2001a) Ribosome fidelity: tRNA discrimination, proofreading and induced fit. *Trends Biochem. Sci.*, **26**, 124–130.

Rodnina MV and Wintermeyer W (2001b) Fidelity of aminoacyl-tRNA selection on the ribosome: kinetic and structural mechanisms. *Ann. Rev. Biochem.*, **70**, 415–435.

Selmer M, Al-Karadaghi S, Hirokawa G, Kaji A and Liljas A (1999) Crystal structure of *Thermatoga maritima* ribosome recycling factor: a tRNA mimic. *Science*, **286**, 2349–2352.

Shtilerman M, Lorimer GH and Englander SW (1999) Chaperonin function: folding by forced unfolding. *Science*, **284**, 822–825.

Silvian LF, Wang J and Steitz TA (1999) Insights into editing from an Ile-tRNA synthetase structure with tRNA^{Ile} and mucoprotein. *Science*, **285**, 1074–1077.

Slavotinek AM and Biesecker LG (2001) Unfolding the role of chaperones and chaperonins in human disease. *Trends Genet.*, **17**, 528–535.

Varshavsky A (1997) The ubiquitin system. *Trends Biochem. Sci.*, **22**, 383–387.

Wimberly BT, Brodersen DE, Clemons WM, et al. (2000) Structure of the 30S ribosomal subunit. *Nature*, **407**, 327–339.

Xu Z, Horwich AL and Sigler PB (1997) The crystal structure of the asymmetric GroEL-GroES-(ADP)$_7$ chaperonin complex. *Nature*, **388**, 741–750.

Yoshizawa S, Fourmy D and Puglisi JD (1999) Recognition of the codon–anticodon helix by ribosomal RNA. *Science*, **285**, 1722–1725.

Yusupov MM, Yusupova GZ, Baucom A, et al. (2001) Crystal structure of the ribosome at 5.5Å resolution. *Science*, **292**, 883–896.

Further Reading

Arnstein HRV and Cox RA (1992) *Protein Biosynthesis: In Focus.* IRL Press, Oxford. — *Excellent source of general information on translation.*

Frydman J (2001) Folding of newly translated proteins *in vivo*: the role of molecular chaperones. *Ann. Rev. Biochem.*, **70**, 603–649.

Gingras A-C, Raught B and Sonenberg N (1999) eIF4 initiation factors: effectors of mRNA recruitment to ribosomes and regulators of translation. *Ann. Rev. Biochem.*, **68**, 913–963. — *Comprehensive details on this subject.*

Ibba M and Söll D (1999) Quality control mechanisms during translation. *Science*, **286**, 1893–1897. — *Describes the various processes that prevent errors from occurring during protein synthesis.*

Ibba M and Söll D (2000) Aminoacyl-tRNA synthetases. *Ann. Rev. Biochem.*, **69**, 617–650.

Liu XQ (2000) Protein-splicing intein: genetic mobility, origin and evolution. *Ann. Rev. Genet.*, **34**, 61–76. — *Includes speculations on the evolution of inteins.*

McCarthy JEG (1998) Post-transcriptional control of gene expression in yeast. *Microbiol. Mol. Biol. Rev.*, **62**, 1492–1553. — *Detailed review of translation and its control in yeast.*

Voges D, Zwickl P and Baumeister W (1999) The 26S proteasome: a molecular machine designed for controlled proteolysis. *Ann. Rev. Biochem.*, **68**, 1015–1068.

Weisblum B (1999) Back to Camelot: defining the specific role of tRNA in protein synthesis. *Trends Biochem. Sci.*, **24**, 247–250. — *A historical account.*

Wickner S, Maurizi MR and Gottesman S (1999) Post-translational quality control: folding, refolding and degrading proteins. *Science*, **286**, 1888–1893. — *General review on protein folding.*

STUDY AIDS FOR CHAPTER 11

Key terms

Give short definitions of the following terms:

Acceptor arm
Acceptor site
Acylation
Aminoacylation
Aminoacyl-tRNA synthetase
Anticodon
Anticodon arm
Biotinylation
Cap binding complex
Chaperonin
Cloverleaf
Codon–anticodon recognition
D arm
Denaturation of proteins
Elongation factor
Exit site
Extein
Folding pathway
Frameshifting
Global regulation
Glycosylation
Hsp70 chaperone
Immunoelectron microscopy
Initiation complex
Initiation factor
Inosine
Intein
Intein homing
Internal ribosome entry site
  (IRES)
Intron homing
Iron-response element
Isoaccepting tRNAs

Kozak consensus
Molecular chaperone
N-degron
N-linked glycosylation
O-linked glycosylation
Peptidyl site
Peptidyl transferase
PEST sequences
Polyprotein
Pre-initiation complex
Protease
Proteasome
Protein–protein crosslinking
Release factor
Ribosomal protein
Ribosome binding site
Ribosome recycling factor
  (RRF)
Selfish DNA
Shine–Dalgarno sequence
Signal peptide
Site-directed hydroxyl radical
  probing
Slippage
Superwobble
TΨC arm
Transcript-specific regulation
Translational bypassing
Translocation
Ubiquitin
V loop
Wobble

Self study questions

1. Outline the terminology used to distinguish between isoaccepting tRNAs.
2. Draw the cloverleaf structure of a tRNA. Indicate the key features of the structure, including those parts of the molecule that form attachments with the amino acid and the codon.
3. Write a short essay on aminoacyl-tRNA synthetases. Make sure that you cover the following points: the two classes of synthetase; fidelity of aminoacylation; modification of attached amino acids after aminoacylation.
4. Draw a series of diagrams to illustrate the codon–anticodon interactions that occur during wobble involving G–U base pairs and inosine.
5. Compare and contrast the usage of wobble during translation of mRNAs in (a) *Escherichia coli*, and (b) humans.
6. Construct a table giving details of the RNA and protein components of bacterial and eukaryotic ribosomes.

7. List the various techniques that have been used to study the structure of the bacterial ribosome. What are the applications and limitations of each technique?
8. Give a detailed description of initiation of translation in (a) *Escherichia coli*, and (b) eukaryotes.
9. How are eukaryotic mRNAs that lack a cap structure translated?
10. How is translation initiation regulated?
11. Give a detailed description of the elongation phase of translation in bacteria and eukaryotes.
12. Using examples, distinguish between the terms 'frameshifting', 'programmed frameshifting', 'slippage' and 'translational bypassing'.
13. Outline the roles of the three release factors and the ribosome recycling factor during termination of translation in *Escherichia coli*. Which proteins play the equivalent roles during termination in eukaryotes?
14. Describe the experiments which showed that a small protein such as ribonuclease can fold spontaneously *in vitro*. Why are larger proteins unable to fold spontaneously?
15. Distinguish between the activities of Hsp70 chaperones and chaperonins in protein folding.
16. Give examples of proteins that are processed by proteolytic cleavage.
17. Construct a table listing, with examples, the various types of post-translational chemical modification that occur with different proteins.
18. What is an intein? How are inteins spliced?
19. Describe the processes thought to be responsible for protein degradation in eukaryotes.

Problem-based learning

1. Why are there two classes of aminoacyl-tRNA synthetases? (*Hint: a good starting point for tackling this difficult problem is de Pouplana LR and Schimmel P [2001] Aminoacyl-tRNA synthetases: potential markers of genetic code development. Trends Biochem. Sci., 26, 591–596.*)
2. To what extent have studies of ribosome structure been of value in understanding the detailed process by which proteins are synthesized?
3. Evaluate the information suggesting that peptidyl transferase is a ribozyme.
4. It is thought that translational bypassing 'is controlled by the tRNA attached to the growing polypeptide, which scans the mRNA as the ribosome tracks along, and halts the bypass when a new codon to which it can base-pair is reached.' Devise an experiment to test this hypothesis.
5. Are protein folding studies that are conducted *in vitro* good models for protein folding *in vivo*?
6. Are the similarities between inteins and introns purely coincidental?
7. Using the current information on protein degradation, devise a hypothesis to explain how specific proteins could be individually degraded. Can your hypothesis be tested?

Regulation of Genome Activity

Chapter Contents

12.1 Transient Changes in Genome
Activity   350

    12.1.1   Signal transmission by import of
the extracellular signaling
compound   351

    12.1.2   Signal transmission mediated by
cell surface receptors   354

12.2 Permanent and Semipermanent
Changes in Genome Activity   360

    12.2.1   Genome rearrangements   360

    12.2.2   Changes in chromatin structure   362

    12.2.3   Genome regulation by feedback
loops   363

12.3 Regulation of Genome Activity
During Development   365

    12.3.1   Sporulation in *Bacillus*   366

    12.3.2   Vulva development in
Caenorhabditis elegans   369

    12.3.3   Development in *Drosophila
melanogaster*   372

Learning outcomes

When you have read Chapter 12, you should be able to:

■ Distinguish between differentiation and development, and outline how regulation of genome expression underlies these two processes

■ Describe, with examples, the various ways in which extracellular signaling compounds can bring about transient changes in genome activity, making clear distinction between those signaling compounds that enter the cell and those that bind to a cell surface receptor

■ Describe, with examples, the various ways in which permanent and semipermanent changes in genome activity can be brought about, making clear distinction between those processes that involve rearrangement of the genome, those that involve changes in chromatin structure, and those that involve feedback loops

■ Discuss how studies of sporulation in *Bacillus subtilis*, vulva development in *Caenorhabditis elegans*, and embryogenesis in *Drosophila melanogaster* have contributed to our understanding of genome regulation during development, and explain why lower organisms can act as models for development in higher eukaryotes such as humans

WE HAVE FOLLOWED the pathway by which expression of the genome specifies the content of the proteome, which in turn determines the biochemical signature of the cell. In no organism is this biochemical signature entirely constant. Even the simplest unicellular organisms are able to alter their proteomes to take account of changes in the environment, so that their biochemical capabilities are continually in tune with the available nutrient supply and the prevailing physical and chemical conditions. Cells in multicellular organisms are equally responsive to changes in the extracellular environment, the only difference being that the major stimuli include hormones and growth factors as well as nutrients. The resulting *transient* changes in genome activity enable the proteome to be remodeled continuously to satisfy the demands that the outside world places on the cell (*Figure 12.1*). Other changes in genome activity are *permanent* or at least *semipermanent*, and result in the cell's biochemical signature becoming altered in a way that is not readily reversible. These changes lead to cellular **differentiation**, the adoption by the cell of a specialized physiological role. Differentiation pathways are known in many unicellular organisms, an example being the production of spore cells by bacteria such as *Bacillus*, but we more frequently associate differentiation with multicellular organisms, in which a variety of specialized cell types (over 250 types in humans) are organized into tissues and organs. Assembly of these complex multicellular structures, and of the organism as a whole, requires coordination of the activities of the genomes in different cells. This coordination involves both transient and permanent changes, and must continue over a long period of time during the **development** of the organism.

There are many steps within the expression pathways for individual genes at which regulation can be exerted (*Table 12.1*). Examples of the biological roles of different mechanisms were given at the appropriate places in Chapters 8–11. The objective of this chapter is not to reiterate these gene-specific control systems, but to explain how the activity of the genome as a whole is regulated. In doing this we should bear in mind that the biosphere is so diverse, and the numbers of genes in individual genomes so large, that it is reasonable to assume that any mechanism that could have evolved to regulate genome expression is likely to have done so. It is therefore no surprise that we can nominate examples of regulation for every point in the genome expression pathway. But are all these control points of equal importance in regulating the activity of the genome as a whole? Our current perception is that they are not. Our understanding may be imperfect, based as it is on investigation of just a limited number of genes in a few organisms, but it appears that the critical controls over genome expression – the decisions about which genes are switched on and which are switched off – are exerted at the level of transcription initiation. For most genes, control that is exerted at later steps serves to modulate expression but does not act as the primary determinant of whether the gene is on or off (see *Figure 9.22*, page 262). Most, but not all, of what we will discuss in this chapter therefore concerns control of genome activity by mechanisms that

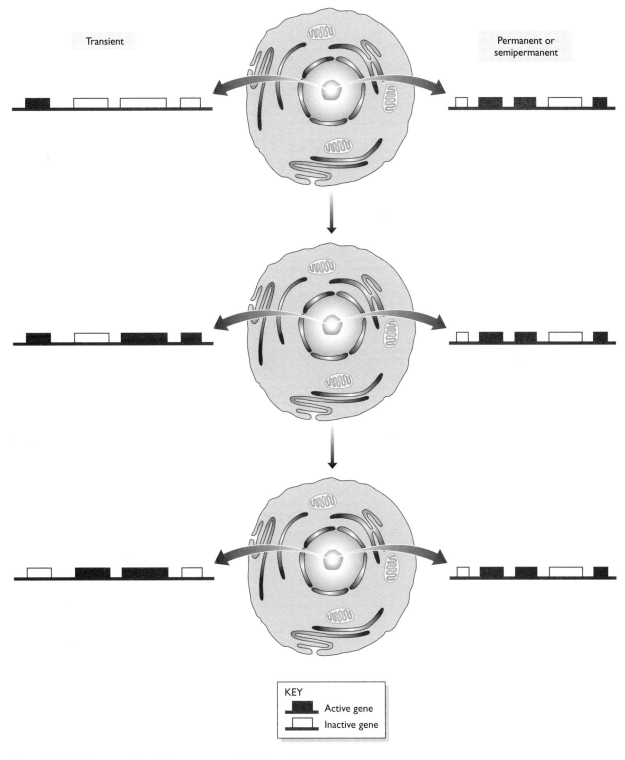

Figure 12.1 Two ways in which genome activity is regulated.

The genes on the left are subject to transient regulation and are switched on and off in response to changes in the extracellular environment. The genes on the right have undergone a permanent or semipermanent change in their expression pattern, resulting in the same three genes being expressed continuously.

Table 12.1 Examples of steps in the genome expression pathway at which regulation can be exerted

Step	Example of regulation	Cross-reference
Transcription		
Gene accessibility	Locus control regions determine chromatin structure in areas that contain genes	Section 8.1.2
	Histone modifications influence chromatin structure and determine which genes are accessible	Section 8.2.1
	Nucleosome positioning controls access of RNA polymerase and transcription factors to the promoter region	Section 8.2.1
	DNA methylation silences regions of the genome	Section 8.2.2
Initiation of transcription	Productive initiation is influenced by activators, repressors and other control systems	Section 9.3
Synthesis of RNA	Prokaryotes use antitermination and attenuation to control the amount and nature of individual transcripts	Section 10.1.1
Eukaryotic mRNA processing		
Capping	Some animals use capping as a means of regulating protein synthesis during egg maturation	–
Polyadenylation	Translation of *bicoid* mRNA in *Drosophila* eggs is activated after fertilization by extension of the poly(A) tail	Section 12.3.3
Splicing	Alternative splice site selection controls sex determination in *Drosophila*	Section 10.1.3
Chemical modification	RNA editing of apolipoprotein-B mRNA results in liver- and intestine-specific versions of this protein	Section 10.3.2
mRNA degradation	Iron controls degradation of transferrin receptor mRNA	Section 11.2.2
Protein synthesis and processing		
Initiation of translation	Phosphorylation of eIF-2 results in a general reduction in translation initiation in eukaryotes	Section 11.2.2
	Ribosomal proteins in bacteria control their own synthesis by modulating ribosome attachment to their mRNAs	Section 11.2.2
	In some eukaryotes, iron controls ribosome scanning on ferritin mRNAs	Section 11.2.2
Protein synthesis	Frameshifting enables two DNA polymerase III subunits to be translated from the *Escherichia coli dnaX* gene	Section 11.2.3
Cutting events	Alternative cleavage pathways for polyproteins result in tissue-specific protein products	Section 11.3.2
Chemical modification	Many proteins involved in signal transduction are activated by phosphorylation	Section 12.1.2

specify which genes are transcribed and which are silent. We will address two issues: the ways in which transient and permanent changes in genome activity are brought about, and the ways in which these changes are linked in time and space in developmental pathways.

12.1 Transient Changes in Genome Activity

Transient changes in genome activity occur predominantly in response to external stimuli. For unicellular organisms, the most important external stimuli relate to nutrient availability, these cells living in variable environ-

ments in which the identities and relative amounts of the nutrients change over time. The genomes of unicellular organisms therefore include genes for uptake and utilization of a range of nutrients, and changes in nutrient availability are shadowed by changes in genome activity, so that at any one time only those genes needed to utilize the available nutrients are expressed. Most cells in multicellular organisms live in less variable environments, but an environment whose maintenance requires coordination between the activities of different cells. For these cells, the major external stimuli are therefore hormones, growth factors, and related compounds that convey signals within the organism and stimulate coordinated changes in genome activity.

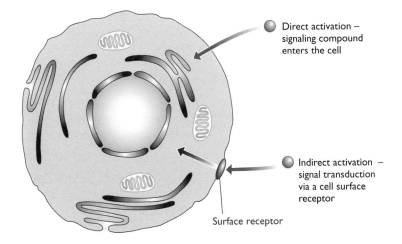

Figure 12.2   Two ways in which an extracellular signaling compound can influence events occurring within a cell.

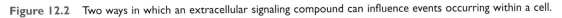

To exert an effect on genome activity, the nutrient, hormone, growth factor, or other extracellular compound that represents the external stimulus must influence events within the cell. There are two ways in which it can do this (*Figure 12.2*):

- *Directly*, by acting as a signaling compound that is transported across the cell membrane and into the cell;
- *Indirectly*, by binding to a cell surface receptor which transmits a signal into the cell.

Signal transmission, by direct or indirect means, is one of the major research areas in cell biology (Lodish *et al.*, 2000), with attention focused in particular on its relevance to the abnormal biochemical activities that underlie cancer. Many examples of signal transmission have been discovered, some of general importance in a variety of organisms and others restricted to just a few species. In the first part of this chapter we will survey the field.

12.1.1 Signal transmission by import of the extracellular signaling compound

In the direct method of signal transmission, the extracellular compound that represents the external stimulus crosses the cell membrane and enters the cell. After import into the cell, the signaling compound could influence genome activity by any one of three routes (*Figure 12.3*):

- If the signaling compound is a protein, then it could act in the same way as one of the various protein factors that we met in Chapters 8–11, for example by activating or repressing assembly of the transcription initiation complex (Section 9.3), or by interacting with a splicing enhancer or silencer (Section 10.1.3).
- The signaling compound could influence the activity of an existing protein factor. Such a signaling compound need not be a protein: it could, theoretically, be any type of compound.

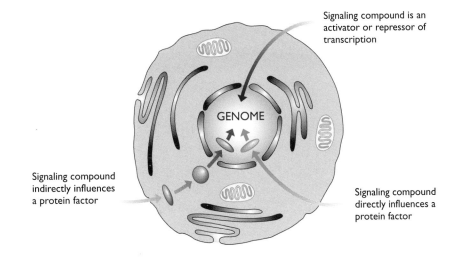

Figure 12.3   Three ways in which an extracellular signaling compound could influence genome activity.

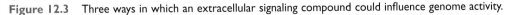

■ The signaling compound could influence the activity of an existing protein factor via one or more intermediates, rather than by interacting with it directly.

Examples of each of these three modes of action are described below.

Lactoferrin is an extracellular signaling protein which acts as a transcription activator

If the extracellular signaling compound that is imported into the cell is a protein with suitable properties then it could directly affect the activity of its target genes by acting as an activator or repressor of some stage in the genome expression pathway. This might appear to be an attractively straightforward way of regulating genome activity, but it is not a common mechanism. The reason for this is not clear but probably relates, at least partly, to the difficulty in designing a protein that combines the hydrophobic properties needed for effective transport across a membrane with the hydrophilic properties needed for migration through the aqueous cytoplasm to the protein's site of action in the nucleus or on a ribosome.

The one clear example of a signaling compound that can function in this way is provided by lactoferrin, a mammalian protein found mainly in milk and, to a lesser extent, in the bloodstream. Lactoferrin is a transcription activator (Section 9.3.2). Its specific function has been difficult to pin down, but it seems to play a role in the body's defenses against microbial attack. As its name suggests, lactoferrin is able to bind iron, and it is thought that at least part of its protective role arises from its ability to reduce free-iron levels in milk, thereby starving invading microbes of this essential cofactor. It might therefore appear unlikely that lactoferrin would have a role in genome expression, but it has been known since the early 1980s that the protein is multi-talented and, among other things, can bind to DNA. This property was linked to a second function of lactoferrin – stimulation of the blood cells involved in the immune response – when in 1992 it was shown that the protein is taken up by immune cells, enters their nuclei, and attaches to the genome (Garre *et al.*, 1992). Subsequently the DNA binding was shown to be sequence specific and to stimulate transcription, confirming that lactoferrin is a true transcription activator (He and Furmanski, 1995).

Some imported signaling compounds directly influence the activity of pre-existing protein factors

Although only a few imported signaling compounds are able themselves to act as activators or repressors of genome expression, many have the ability to influence directly the activity of factors that are already present in the cell. We encountered one example of this type of regulation in Section 9.3.1 when we studied the lactose operon of *Escherichia coli*. This operon responds to extracellular levels of lactose, the latter acting as a signaling molecule which enters the cell and, after conversion to its isomer allolactose, influences the DNA-binding properties of the

lactose repressor and hence determines whether or not the lactose operon is transcribed (see *Figure 9.24*, page 264). Many other bacterial operons coding for genes involved in sugar utilization are controlled in this way.

Direct interaction between a signaling compound and a transcription activator or repressor is also a common means of regulating genome activity in eukaryotes. A good example is provided by the control system that maintains the intracellular metal-ion content at an appropriate level. Cells need metal ions such as copper and zinc as cofactors in biochemical reactions, but these metals are toxic if they accumulate in the cell above a certain level. Their uptake therefore has to be carefully controlled so that the cell contains sufficient metal ions when the environment is lacking in metal compounds, but does not over-accumulate metal ions when the environmental concentrations are high. The strategies used

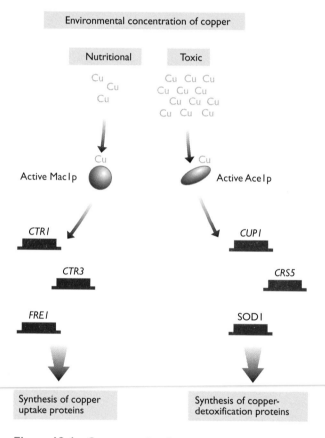

Figure 12.4 Copper-regulated gene expression in *Saccharomyces cerevisiae*.

Yeast requires low amounts of copper because a few of its enzymes (e.g. cytochrome c oxidase and tyrosinase) are copper-containing metalloproteins, but too much copper is toxic for the cell. When copper levels are low, the Mac1p protein factor is activated by copper binding and switches on expression of genes for copper uptake. When the copper levels are too high, a second factor, Ace1p, is activated, switching on expression of a different set of genes, these coding for proteins involved in copper detoxification.

are illustrated by the copper-control system of *Saccharomyces cerevisiae*. This yeast has two copper-dependent transcription activators, Mac1p and Ace1p. Both of these activators bind copper ions, the binding inducing a conformational change that enables the factor to stimulate expression of its target genes (*Figure 12.4*). For Mac1p these target genes code for copper-uptake proteins, whereas for Ace1p they are genes coding for proteins such as superoxide dismutase that are involved in copper detoxification. The metal-controlled balance between the activities of Mac1p and Ace1p ensures that the copper content of the cell remains within acceptable levels (Pena *et al.*, 1998; Winge *et al.*, 1998).

Transcription activators are also the targets for **steroid hormones** such as progesterone, estrogen and glucocorticoid hormones. Steroid hormones are signaling compounds that coordinate a range of physiological activities in the cells of higher eukaryotes (Tsai and O'Malley, 1994). Steroids are hydrophobic and so easily penetrate the cell membrane. Once inside the cell, each hormone binds to a specific **steroid receptor** protein, which is usually located in the cytoplasm (*Figure 12.5*). After binding, the activated receptor migrates into the nucleus, where it attaches to a response element (see Box 9.6, page 267) upstream of a target gene. The typical **hormone response element** is a 15 bp sequence comprising a 6 bp inverted palindrome separated by a 3 bp spacer to which the steroid receptor binds via a special version of the zinc finger (see *Figure 9.13*, page 251). Response elements for each receptor are located upstream of 50–100 genes and, once bound, the receptor acts as a transcription activator. A steroid hormone can therefore induce a large-scale change in the biochemical properties of the cell.

All steroid receptors are structurally similar, not just with regard to their DNA-binding domains but also in other parts of their protein structures (*Figure 12.6*). Recognition of these similarities has led to the identification of a number of putative or orphan steroid receptors whose hormonal partners and cellular functions are not yet known. The structural similarities have also shown that a second set of receptor proteins, the **nuclear receptor superfamily**, belongs to the same general class as steroid receptors, although the hormones that they work with are not steroids. As their name suggests, these receptors are located in the nucleus rather than the cytoplasm. They include the receptors for vitamin D3, whose roles include control of bone development, and thyroxine, which stimulates the tadpole-to-frog metamorphosis.

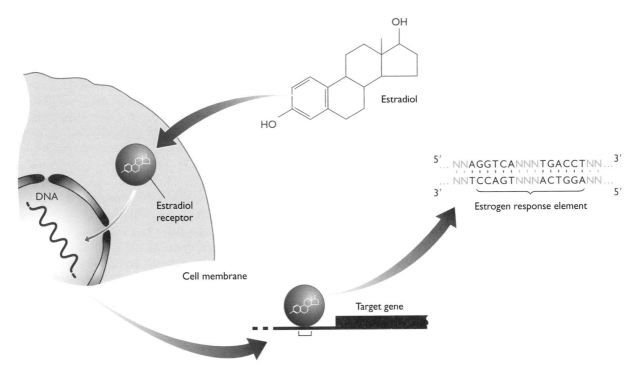

Figure 12.5 Gene activation by a steroid hormone.

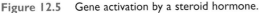

Estradiol is one of the estrogen steroid hormones. After entering the cell, estradiol attaches to its receptor protein and the complex enters the nucleus where it binds to the 15-bp estrogen response element (abbreviation: N, any nucleotide), which is located upstream of those genes activated by estradiol and other estrogens. Other steroid hormone receptors recognize other response elements. For example, glucocorticoid hormones target the sequence 5′-AGAACANNNTGTTCT-3′. Note that this sequence, and that of the estrogen response element, is an inverted palindrome. The response element for vitamin D3, which is a steroid derivative that activates transcription via a nuclear receptor (see the text), has the sequence 5′-AGGTCANNNAGGTCA-3′, which is a direct repeat rather than an inverted palindrome.

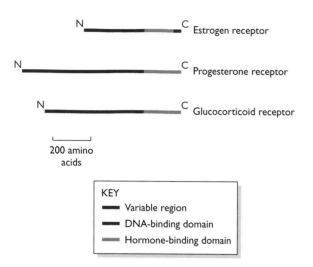

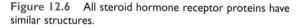

Figure 12.6 All steroid hormone receptor proteins have similar structures.

Three receptor proteins are compared. Each one is shown as an unfolded polypeptide with the two conserved functional domains aligned. The DNA-binding domain is very similar in all steroid receptors, displaying 50–90% amino acid sequence identity. The hormone-binding domain is less well conserved, with 20–60% sequence identity. The activation domain (Section 9.3.2) lies between the N terminus and the DNA-binding domain, but this region displays little sequence similarity in different receptors.

Some imported signaling compounds influence genome activity indirectly

The link between a signaling molecule and the protein factors involved in genome expression does not have to be as direct as in the examples described in the previous section. Signaling molecules can also influence genome activity in an indirect manner via one or more intermediates. An example is provided by the **catabolite repression** system of bacteria. This is the means by which extracellular and intracellular glucose levels dictate whether or not operons for utilization of other sugars are switched on when those alternative sugars are present in the medium.

This phenomenon was discovered by Jacques Monod in 1941, who showed that if *E. coli* or *Bacillus subtilis* are provided with a mixture of sugars, then one sugar will be metabolized first, the bacteria turning to the second sugar only when the first has been used up. Monod used a French word to describe this: **diauxie** (Brock, 1990). One combination of sugars that elicits a diauxic response is glucose plus lactose, glucose being used before the lactose (*Figure 12.7A*). When the details of the lactose operon were worked out some 20 years later (Section 9.3.1) it became clear that the diauxie between glucose and lactose must involve a mechanism whereby the presence of glucose can override the normal inductive effect that lactose has on its operon. In the presence of

lactose plus glucose, the lactose operon is switched off, even though some of the lactose in the mixture is converted into allolactose which binds to the lactose repressor, so that under normal circumstances the operon would be transcribed (*Figure 12.7B*).

The explanation for the diauxic response is that glucose acts as a signaling compound that represses expression of the lactose operon, as well as other sugar utilization operons, through an indirect influence on the **catabolite activator protein**. This protein binds to various sites in the bacterial genome and activates transcription initiation at downstream promoters. Productive initiation of transcription at these promoters is dependent on the presence of the bound protein: if the protein is absent then the genes it controls are not transcribed.

Glucose does not itself interact with the catabolite activator protein. Instead, glucose controls the level in the cell of the modified nucleotide **cyclic AMP** (**cAMP**; *Figure 12.7C*). It does this by inhibiting the activity of **adenylate cyclase**, the enzyme that synthesizes cAMP from ATP. This means that if glucose levels are high, the cAMP content of the cell is low. The catabolite activator protein can bind to its target sites only in the presence of cAMP, so when glucose is present the protein remains detached and the operons it controls are switched off. In the specific case of diauxie involving glucose plus lactose, the indirect effect of glucose on the catabolite activator protein means that the lactose operon remains inactivated, even though the lactose repressor is not bound, and so the glucose in the medium is used up first. When the glucose is gone, the cAMP level rises and the catabolite activator protein binds to its target sites, including the site upstream of the lactose operon, and transcription of the lactose genes is activated.

12.1.2 Signal transmission mediated by cell surface receptors

Many extracellular signaling compounds are unable to enter the cell because they are too hydrophilic to penetrate the lipid membrane and the cell lacks a specific transport mechanism for their uptake. In order to influence genome activity these signaling compounds must bind to cell surface receptors that carry their signals across the cell membrane. These receptors are proteins that span the membrane, with a site for binding the signaling compound on the outer surface. Binding of the signaling compound results in a conformational change in the receptor, inducing a biochemical event within the cell, often phosphorylation of an intracellular protein. This event forms the first step in the intracellular stage of the **signal transduction** pathway (*Figure 12.8*). Several types of cell surface receptor have been discovered (*Table 12.2*) and the intracellular events that they initiate are diverse, with many variations on each theme, not all of these specifically involved in regulating genome activity. Three examples will help us to appreciate the complexity of the system.

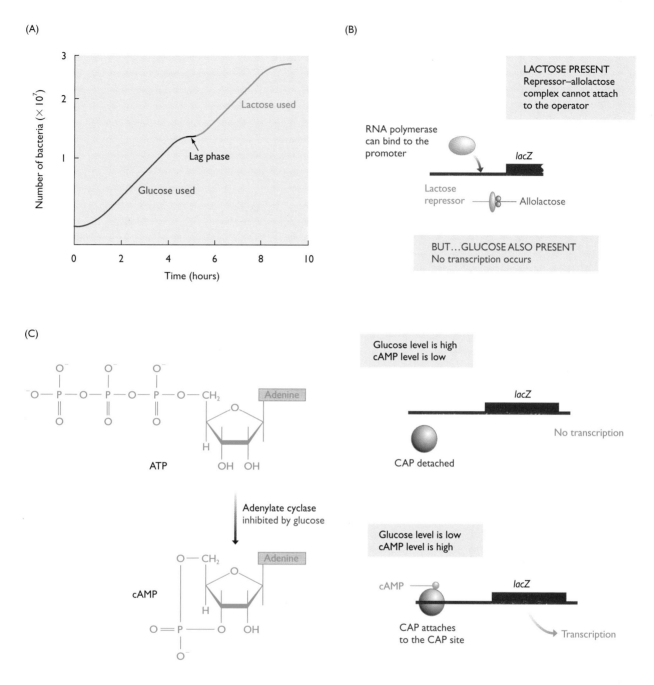

Figure 12.7 Catabolite repression.

(A) A typical diauxic growth curve, as seen when *Escherichia coli* is grown in a medium containing a mixture of glucose and lactose. During the first few hours the bacteria divide exponentially, using the glucose as the carbon and energy source. When the glucose is used up there is a brief lag period while the *lac* genes are switched on before the bacteria return to exponential growth, now using up the lactose. (B) Glucose overrides the lactose repressor. If lactose is present then the repressor detaches from the operator and the lactose operon should be transcribed, but it remains silent if glucose is also present. Refer to *Figure 9.24B*, page 264 for details of how the lactose repressor controls expression of the lactose operon. (C) Glucose exerts its effect on the lactose operon and other target genes by controlling the activity of adenylate cyclase and hence regulating the amount of cAMP in the cell. The catabolite activator protein (CAP) can attach to its DNA-binding site only in the presence of cAMP. If glucose is present, the cAMP level is low, so CAP does not bind to the DNA and does not activate the RNA polymerase. Once the glucose has been used up, the cAMP level rises, allowing CAP to bind to the DNA and activate transcription of the lactose operon and its other target genes.

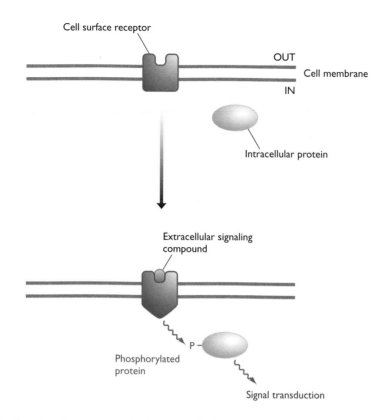

Figure 12.8   The role of a cell surface receptor in signal transduction.

Binding of the extracellular signaling compound to the outer surface of the receptor protein causes a conformational change that results in activation of an intracellular protein, for example by phosphorylation. The events occurring 'downstream' of this initial protein activation are diverse, as described in the text. 'P' indicates a phosphate group, PO_3^{2-}.

Signal transduction with one step between receptor and genome

With some signal transduction systems, stimulation of the cell surface receptor by attachment of the extracellular signaling compound results in the direct activation of a protein that influences genome activity. This is the simplest system by which an extracellular signal can be transduced into a genomic response.

The direct system is used by many cytokines such as interleukins and interferons, which are extracellular signaling polypeptides that control cell growth and division. Binding of these polypeptides to their cell surface receptors results in activation of a type of transcription factor called a **STAT** (signal transducer and activator of transcription). Activation is by phosphorylation of a single tyrosine amino acid at a position near to the C terminus

Table 12.2   Cell surface receptor proteins involved in signal transmission into eukaryotic cells

Receptor type	Description	Signals
G-protein-coupled receptors	Activate intracellular G-proteins, which bind GTP and control biochemical activities by conversion of this GTP to GDP with the release of energy	Diverse: epinephrine, peptides (e.g. glucagon), protein hormones, odorants, light
Tyrosine kinases	Activate intracellular proteins by tyrosine phosphorylation	Hormones (e.g. insulin), various growth factors
Tyrosine-kinase-associated	Similar to tyrosine kinase receptors but activate intracellular proteins indirectly (e.g. see description of STATs in the text)	Hormones, growth factors
Serine–threonine kinases	Activate intracellular proteins by serine and/or threonine phosphorylation	Hormones, growth factors
Ion channels	Control intracellular activities by regulating the movement of ions and other small molecules into and out of cells	Chemical stimuli (e.g. glutamate), electrical charges

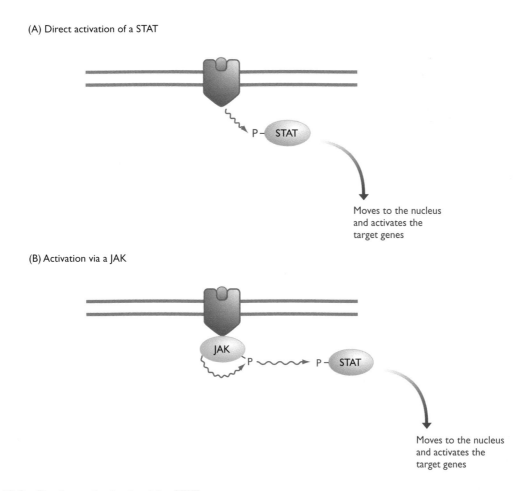

(A) Direct activation of a STAT

Moves to the nucleus
and activates the
target genes

(B) Activation via a JAK

Moves to the nucleus
and activates the
target genes

Figure 12.9 Signal transduction involving STATs.

(A) If the receptor is a member of the tyrosine kinase family then it can activate the STAT directly. (B) If the receptor is a tyrosine-kinase-associated type then it acts via a Janus kinase (JAK), which autophosphorylates when the extracellular signal binds and then activates the STAT. Note that activation of the JAK usually involves dimerization, the extracellular signal inducing two subunits to associate, resulting in the version of the JAK with phosphorylation activity. Dimerization is also central to activation of a STAT, phosphorylation causing two STATs, not necessarily of the same type, to form a dimer. This dimer is able to act as a transcription activator. 'P' indicates a phosphate group, PO_3^{2-}.

of the STAT polypeptide. If the cell surface receptor is a member of the tyrosine kinase family (see *Table 12.2*) then it is able to activate the STAT directly (*Figure 12.9A*). If it is a tyrosine-kinase-associated receptor then it does not itself have the ability to phosphorylate a STAT, or any other intracellular protein, but acts through intermediaries called **Janus kinases** (**JAKs**). Binding of the signaling molecule to a tyrosine-kinase-associated receptor causes a change in the conformation of the receptor, often by inducing dimerization. This causes a JAK that is associated with the receptor to phosphorylate itself, this autoactivation being followed by phosphorylation of the STAT by the JAK (*Figure 12.9B*).

Seven STATs have so far been identified in mammals (Horvath, 2000). Three of these – STATs 2, 4 and 6 – are specific for just one or two extracellular cytokines, but the others are broad spectrum and can be activated by several different interleukins and interferons. Discrimination is provided by the cell surface receptors: a particular receptor binds just one type of cytokine, and most cells have only one or a few types of receptor. Different cells therefore respond in different ways to the presence of particular cytokines, even though the internal signaling process involves only a limited number of STATs.

The consensus sequence of the DNA-binding sites for STATs has been defined as $5'–TTN_{5-6}AA–3'$, largely by studies in which purified STATs have been tested against oligonucleotides of known sequence. The DNA-binding domain of the STAT protein is made up of three loops emerging from a barrel-shaped β-sheet structure (Becker *et al.*, 1998). This is an unusual type of DNA-binding domain and has not been identified in precisely the same form in any other type of protein, although it has similarities with the DNA-binding domains of the NK-κB and Rel transcription activators. These similarities refer only to the tertiary structures of the DNA-binding domains because STATs, NK-κB and Rel, as a whole, have very little amino acid sequence identity. Many target genes are activated by STATs but the overall genomic response is modulated by other proteins which interact with STATs and influence which genes are switched on under a particular set of circumstances. Complexity is entirely expected because the cellular processes that STATs mediate – growth and division – are themselves complex and we anticipate that changes in these processes will require extensive remodeling of the proteome and hence large-scale alterations in genome activity.

Signal transduction with many steps between receptor and genome

The simplicity of the system whereby the cell surface receptor activates a STAT, directly or through a JAK associated with the receptor, contrasts with the more prevalent forms of signal transduction, in which the receptor represents just the first in a series of steps that lead eventually to one or more transcription activators or repressors being switched on or off. A number of these **cascade** pathways have been delineated in different organisms. The following are the important ones in mammals:

- The **MAP** (mitogen activated protein) **kinase system** (*Figure 12.10*) responds to many extracellular signals, including mitogens – compounds with similar effects to cytokines but that specifically stimulate cell division (Robinson and Cobb, 1997). Binding of the signaling compound causes the internal parts of the mitogen receptor to become phosphorylated. Phosphorylation stimulates attachment to the receptor, on the internal side of the membrane, of various cytoplasmic proteins, one of which is Raf, a protein kinase that is activated when it becomes membrane bound. Raf initiates a cascade of phosphorylation reactions. It phosphorylates Mek, activating this protein so that it, in turn, phosphorylates the MAP kinase. The activated MAP kinase now moves into the nucleus where it switches on, again by phosphorylation, a series of transcription activators. The MAP kinase also phosphorylates another protein kinase, this one called Rsk, which phosphorylates and activates a second set of factors. Additional flexibility is provided by the possibility of replacing one or more of the proteins in the MAP kinase pathway

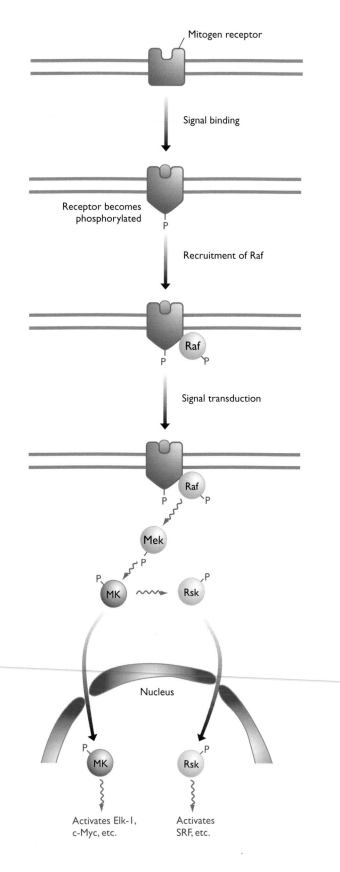

Figure 12.10 Signal transduction by the MAP kinase pathway.

See the text for details. 'MK' is the MAP kinase and 'P' indicates a phosphate group, PO_3^{2-}. Elk-1, c-Myc and SRF (serum response factor) are examples of transcription factors activated at the end of the pathway.

with related proteins, ones with slightly different specificities and so activating another suite of factors. The MAP kinase pathway is used by vertebrate cells; equivalent pathways, using intermediates similar to those identified in mammals, are known in other organisms (see Section 12.3.2 for an example).

■ The **Ras system** is centered around the Ras proteins, three of which are known in mammalian cells (H-, K- and N-Ras), and similar proteins such as Rac and Rho. These proteins are involved in regulation of cell growth and differentiation and, as with many proteins in this category, when dysfunctional they can give rise to cancer. The Ras family proteins are not limited to mammals, examples being known in other eukaryotes such as the fruit fly. Ras proteins are intermediates in signal transduction pathways that initiate with autophosphorylation of a tyrosine kinase receptor in response to an extracellular signal. The phosphorylated version of the receptor forms protein–protein complexes with **GNRPs (guanine nucleotide releasing proteins)** and **GAPs (GTPase activating proteins)** which activate and inactivate Ras, respectively (*Figure 12.11*; Schlessinger, 1993). The extracellular signals can therefore switch Ras-mediated signal transduction on or off, the choice between the two depending on the nature of the signal and the relative amounts of active GNRPs and GAPs in the cell. When activated, Ras stimulates Raf

activity, so in effect Ras provides a second entry point into the MAP kinase pathway, although this is unlikely to be the only function of Ras and it probably also activates proteins involved in signal transduction by second messengers (as described in the next section).

■ The **SAP (stress activated protein) kinase system** is induced by stress-related signals such as ultraviolet radiation, and growth factors associated with inflammation. The pathway has not been described in detail but is similar to the MAP kinase system although targeting a different set of transcription activators.

Signal transduction via second messengers

Some signal transduction cascades do not involve the direct transfer of the external signal to the genome but instead utilize an indirect means of influencing transcription. The indirectness is provided by **second messengers**, which are less specific internal signaling compounds that transduce the signal from a cell surface receptor in several directions so that a variety of cellular activities, not just transcription, respond to the one signal.

In Section 12.1.1 we saw how glucose modulates the catabolite activator protein by influencing cAMP levels in bacteria (see *Figure 12.7*). Cyclic nucleotides are also important second messengers in eukaryotic cells. Some cell surface receptors have guanidylate cyclase activity, and so convert GTP to cGMP, but most receptors in this family work indirectly by influencing the activity of cytoplasmic cyclases and decyclases. These cyclases and decyclases determine the cellular levels of cGMP and cAMP, which in turn control the activities of various target enzymes. The latter include protein kinase A, which is stimulated by cAMP. One of the activities of protein kinase A is to phosphorylate, and hence activate, a transcription activator called **CREB**. This is one of several proteins that influence the activity of a variety of genes by interacting with a second activator, p300/CBP, which is able to modify histone proteins and so affect chromatin structure and nucleosome positioning (Sections 8.2.1).

As well as being activated indirectly by cAMP, p300/CBP responds to another second messenger, calcium (Chawla *et al.*, 1998). The calcium-ion concentration is substantially lower inside than outside the cell, so proteins that open calcium channels in the cell membrane allow calcium ions to enter (Berridge *et al.*, 2000). This can be induced by extracellular signals that activate tyrosine kinase receptors which in turn activate phospholipases that cleave phosphatidylinositol-4,5-bisphosphate, a lipid component of the inner cell membrane, into inositol-1,4,5-trisphosphate (Ins(1,4,5)P$_3$) and 1,2-diacylglycerol (DAG). Ins(1,4,5)P$_3$ opens calcium channels (*Figure 12.12*). Ins(1,4,5)P$_3$ and DAG are themselves second messengers that can initiate other signal transduction cascades (Spiegel *et al.*, 1996; Toker and Cantley, 1997). Both the

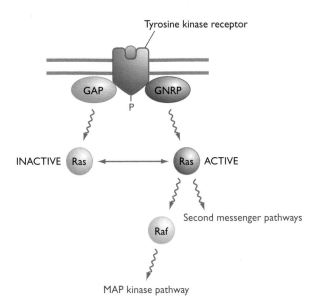

Figure 12.11 The Ras signal transduction system.

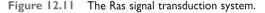

See the text for details. Abbreviations: GAP, GTPase activating protein; GNRP, guanine nucleotide releasing protein. 'P' indicates a phosphate group, PO$_3^{2-}$.

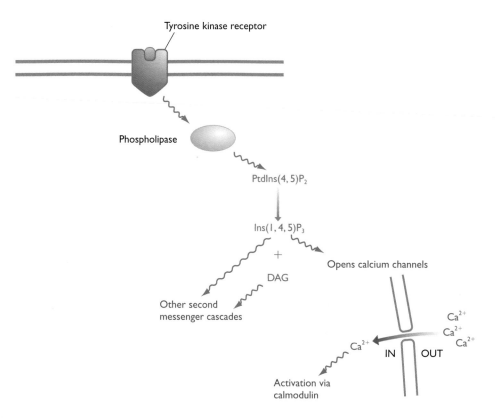

Tyrosine kinase receptor

Phospholipase

PtdIns(4, 5)P₂

Ins(1, 4, 5)P₃

+

DAG

Opens calcium channels

Other second messenger cascades

Ca²⁺

Ca²⁺

Ca²⁺

Ca²⁺

IN OUT

Activation via calmodulin

Figure 12.12 Induction of the calcium second messenger system.

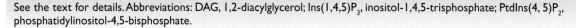

See the text for details. Abbreviations: DAG, 1,2-diacylglycerol; Ins(1,4,5)P₃, inositol-1,4,5-trisphosphate; PtdIns(4, 5)P₂, phosphatidylinositol-4,5-bisphosphate.

calcium- and the lipid-induced cascades target transcription activators, but only indirectly: the primary targets are other proteins. Calcium, for example, binds to and activates the protein called calmodulin, which regulates a variety of enzyme types, including protein kinases, ATPases, phosphatases and nucleotide cyclases.

12.2 Permanent and Semipermanent Changes in Genome Activity

Transient changes in genome activity are, by definition, readily reversible, the gene expression pattern reverting to its original state when the external stimulus is removed or replaced by a contradictory stimulus. In contrast, the permanent and semipermanent changes in genome activity that underlie cellular differentiation must persist for long periods, and ideally should be maintained even when the stimulus that originally induced them has disappeared. We therefore anticipate that the regulatory mechanisms bringing about these longer term changes will involve systems additional to the modulation of transcription activators and repressors. This expectation is correct. We will look at three mechanisms:

■ Changes resulting from physical rearrangement of the genome;

■ Changes due to chromatin structure;

■ Changes maintained by feedback loops.

12.2.1 Genome rearrangements

Changing the physical structure of the genome is an obvious, although drastic, way to bring about a permanent change in genome expression. It is not a common regulatory mechanism, but several important examples are known. These are described below.

Yeast mating types are determined by gene conversion events

Mating type is the equivalent of sex in yeasts and other eukaryotic microorganisms. Because these organisms reproduce mainly by vegetative cell division, there is the possibility that a population, being derived from just one or a few ancestral cells, will be largely or completely composed of a single mating type and so will not be able to reproduce sexually. To avoid this problem, cells are able to change sex by the process called **mating-type switching**. In *Saccharomyces cerevisiae* and some other species this switching involves a genome rearrangement called **gene conversion**.

The two *S. cerevisiae* mating types are called a and α. The mating type is specified by the *MAT* gene, located on chromosome III. This gene has two alleles, *MATa* and *MATα*, a haploid yeast cell displaying the mating type corresponding to whichever allele it possesses. Elsewhere on chromosome III are two additional *MAT*-like genes, called *HMLα* and *HMRa* (*Figure 12.13*). These have the same sequences as *MATα* and *MATa* respectively, but neither gene is expressed because upstream of each one is a silencer that represses transcription initiation. These two genes are called 'silent mating-type cassettes'.

Mating-type switching is initiated by the HO endonuclease, which makes a double-stranded cut at a 24-bp sequence located within the *MAT* gene. This enables a gene conversion event to take place. We examine the details of gene conversion in Section 14.3.1; all that concerns us at the moment is that one of free 3' ends produced by the endonuclease can be extended by DNA synthesis, using one of the two silent cassettes as the template. The newly synthesized DNA subsequently replaces the DNA currently at the *MAT* locus. The silent cassette chosen as the template is usually the one that is different to the allele originally at *MAT* (Haber, 1998), so replacement with the newly synthesized strand converts the *MAT* gene from *MATa* to *MATα* or vice versa (see *Figure 12.13*). This results in mating-type switching.

The *MAT* gene codes for a regulatory protein that interacts with a transcription activator, MCM1, thus determining which set of genes are switched on by this factor. The *MATa* and *MATα* gene products have different effects on MCM1, and so specify different allele-specific gene expression patterns. These expression patterns are maintained in a semipermanent fashion until another *MAT* gene conversion occurs.

Genome rearrangements are responsible for immunoglobulin and T-cell receptor diversities

In vertebrates there are two striking examples of the use of DNA rearrangements to achieve permanent changes in genome activity. These two examples, which are very similar, are responsible for the generation of immunoglobulin and T-cell receptor diversities.

Immunoglobulins and T-cell receptors are proteins that are synthesized by B and T lymphocytes, respectively. Both types of protein become attached to the outer surfaces of their cells, and immunoglobulins are also released into the bloodstream. The proteins help to protect the body against invasion by bacteria, viruses and other unwanted substances by binding to these **antigens**, as they are called. During its lifetime, an organism could be exposed to any number of a vast range of antigens, which means that the immune system must be able to synthesize an equally vast range of immunoglobulin and T-cell receptor proteins. In fact, humans can make approximately 10^8 different immunoglobulin and T-cell receptor proteins. But there are only 3.5×10^4 genes in the human genome, so where do all these proteins come from?

To understand the answer we will look at the structure of a typical immunoglobulin protein. Each immunoglobulin is a tetramer of four polypeptides linked by disulfide bonds (*Figure 12.14*). There are two long 'heavy' chains and two short 'light' chains. When the sequences of different heavy chains are compared it becomes clear that

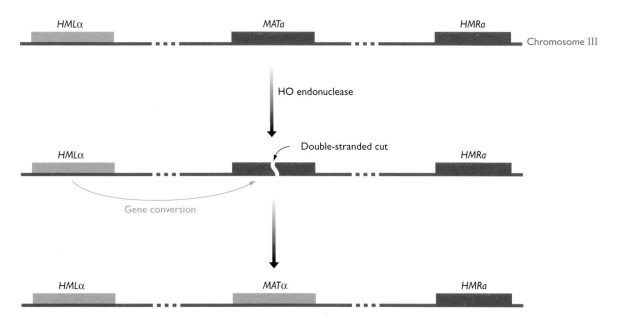

Figure 12.13   Mating-type switching in yeast.

In this example, the cell begins as mating-type a. The HO endonuclease cuts the *MATa* locus, initiating gene conversion by the *HMLα* locus. The result is that the mating type switches to type α. For details of the molecular basis of gene conversion, see Section 14.3.1.

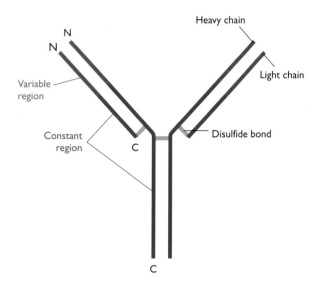

Figure 12.14 Immunoglobulin structure.

Each immunoglobulin protein is made up of two heavy and two light chains, linked by disulfide bonds. Each heavy chain is 446 amino acids in length and consists of a variable region (shown in red) spanning amino acids 1–108 followed by a constant region. Each light chain is 214 amino acids, again with an N-terminal variable region of 108 amino acids. Additional disulfide bonds form between different parts of individual chains: these and other interactions fold the protein into a more complex three-dimensional structure.

the variability between them lies mainly in the N-terminal regions of these polypeptides, the C-terminal parts being very similar, or 'constant', in all heavy chains. The same is true for the light chains, except that two families, κ and λ, can be distinguished, differing in the sequences of their constant regions.

In the human genome there are no complete genes for the immunoglobulin heavy and light polypeptides. Instead, these proteins are specified by gene segments. The heavy-chain segments are on chromosome 14 and comprise 11 C_H gene segments, preceded by 86 V_H gene segments, 30 D_H gene segments and 9 J_H gene segments, these last three coding for different versions of the V (variable), D (diverse) and J (joining) components of the variable part of the heavy chain (*Table 12.3*; *Figure 12.15A*).

The entire heavy-chain locus stretches over several megabase pairs. A similar arrangement is seen with the light-chain loci on chromosomes 2 (κ locus) and 22 (λ locus), the only difference being that the light chains do not have D segments (*Table 12.3*).

As a B lymphocyte develops, the immunoglobulin loci in its genome undergo rearrangements. Within the heavy-chain loci, these rearrangements link one of the V_H gene segments with one of D_H gene segments, and then links this V–D combination with a J_H gene segment, and finally attaches the resulting sequence to a C_H gene segment (*Figure 12.15B*). The end result is a complete heavy-chain gene, but one that is specific for just that one lymphocyte. A similar series of DNA rearrangements results in the lymphocyte's light-chain gene, and transcription of the two genes produces one of the 10^8 immunoglobulins that the human body needs.

Diversity of T-cell receptors is based on similar rearrangements which link V, D, J and C gene segments in different combinations to produce cell-specific genes. We met two small components of this system – the Tβ gene segments V28 and V29-1 – in Chapter 1 in the 50-kb segment of the human genome (see *Figure 1.14*, page 19) with which we began our exploration of genomes.

12.2.2 Changes in chromatin structure

Some of the effects that chromatin structure can have on gene expression were described in Section 8.2. These range from the modulation of transcription initiation at an individual promoter by nucleosome positioning, through to the silencing of large segments of DNA locked up in higher order chromatin structure. The latter is an important means of bringing about long-term changes in genome activity and is implicated in a number of regulatory events. One of these concerns the yeast mating-type loci that we looked at earlier in this section, the silencing of the *HML*α and *HMR*α cassettes resulting mainly from these loci being buried in inaccessible chromatin in response to the influence of their upstream silencer sequences (Haber, 1998). X inactivation (Section 8.2.2) also involves the formation of inaccessible chromatin, in this case along virtually the entire length of one of the two X chromosomes in a female nucleus.

One other example of chromatin silencing merits attention. This is a system that we will meet again later in the

Table 12.3 Immunoglobulin gene segments in the human genome

Component	Locus	Chromosome	Number of gene segments			
			V	**D**	**J**	**C**
Heavy chain	*IGH*	14	86	30	9	11
Light κ chain	*IGK*	2	76	0	5	1
Light λ chain	*IGL*	22	52	0	7	7

It is not known if all the gene segments are functional. For further details see Strachan and Read (1999).

(A) Organization of the *IGH* locus

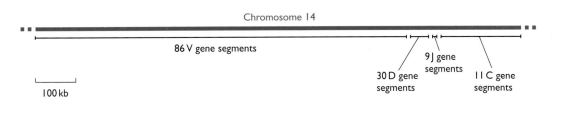

(B) Construction of an immunoglobulin gene by genome rearrangement

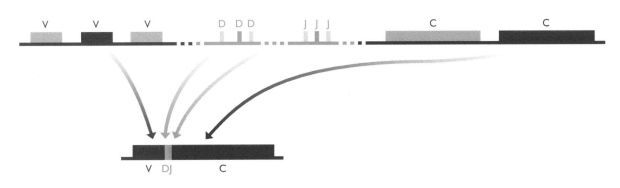

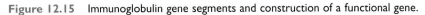

Figure 12.15   Immunoglobulin gene segments and construction of a functional gene.

(A) Organization of the human *IGH* locus on chromosome 14, containing gene segments for the immunoglobulin heavy chain. For a more detailed map of this region see Strachan and Read (1999). (B) A functional heavy-chain gene is constructed by genome rearrangements which link V, D, J and C gene segments.

chapter when we look at development processes in the fruit fly. It concerns the *Polycomb* gene family. The proteins coded by these genes bind to DNA sequences called Polycomb response elements and induce formation of heterochromatin, the condensed form of chromatin that prevents transcription of the genes that it contains (*Figure 12.16*). The heterochromatin nucleates around the attached Polycomb protein and then propagates along the DNA for tens of kilobases (Pirrotta, 1997). The regions that become silenced contain homeotic genes which, as we will see in Section 12.3.3, specify the development of the individual body parts of the fly. As only one body part must be specified at a particular position in the fruit fly, it is important that a cell expresses only the correct homeotic gene. This is ensured by the action of Polycomb, which permanently silences the homeotic genes that must be switched off. An important point is that the heterochromatin induced by Polycomb is heritable: after division, the two new cells retain the heterochromatin established in the parent cell. This type of regulation of genome activity is therefore permanent not only in a single cell, but also in a cell lineage.

12.2.3 Genome regulation by feedback loops

The final mechanism that we will consider for bringing about long-term changes in genome activity involves the use of a feedback loop. In this system a regulatory protein activates its own transcription so that once its gene has been switched on, it is expressed continuously (*Figure 12.17*). A number of examples of this type of feedback regulation are known:

■ **The MyoD transcription activator**, which is involved in muscle development, is one of the best understood examples of cellular differentiation in vertebrates. A cell becomes committed to becoming a muscle cell when it begins to express the *myoD* gene. The product of this gene is a transcription activator that targets a number of other genes coding for muscle-specific proteins, such as actin and myosin, and is also indirectly responsible for one of the key features of muscle cells – the absence of a normal cell cycle, these cells being stopped in the G1 phase (Section 13.3.1). The MyoD protein also binds upstream of *myoD*, ensuring that its own gene is continuously expressed. The result of this positive-feedback loop is that the cell continues to synthesize the muscle-specific proteins and remains a muscle cell. The differentiated state is heritable because cell division is accompanied by transmission of MyoD to the daughter cells, ensuring that these are also muscle cells.

■ **Deformed of Drosophila** is one of several proteins coded by homeotic selector genes and is responsible

RESEARCH 12.1 BRIEFING

Unraveling a signal transduction pathway

A typical set of experiments for studying the functions of proteins involved in a signal transduction pathway.

One of the most important extracellular signaling compounds is transforming growth factor-β (TGF-β), a family of some 30 related polypeptides that control processes such as cell division and differentiation in vertebrates. The cell surface receptors for TGF-β are serine–threonine kinases (see *Table 12.2*, page 356), which activate a variety of target proteins within the cell. Part of the signal transduction process initiated by TGF-β binding involves a set of proteins called the SMAD family, the name being an abbreviation of 'SMA/MAD related', referring to the proteins in *Drosophila melanogaster* and *Caenorhabditis elegans*, respectively, which were the original members of the family to be isolated.

SMAD signaling in vertebrates

Initially, five SMADs were discovered in vertebrate cells. Four of these – Smad1, Smad2, Smad3 and Smad5 – are called receptor-regulated SMADs because they associate directly with the cell surface receptor. Each of these SMADs is specific for a different type of serine–threonine receptor and hence responds to different members of the TGF-β family of signaling compounds. Binding of the extracellular signal induces a receptor to phosphorylate its SMAD, which then associates with Smad4, moves to the nucleus and, via interactions with DNA-binding proteins, activates a set of target genes. Smad4 is therefore a co-mediator that participates in the signaling pathway of each of the other four SMADs.

This interpretation of the SMAD pathway was complicated by the discovery of two additional SMADs – numbers 6 and 7 – that do not fit into the scheme. These SMADs lack the amino acid sequence motif serine–serine–X–serine (where X is valine or methionine), present in the C-terminal region of Smad1, 2, 3 and 5, that is phosphorylated by the receptor. Clearly, therefore, Smad6 and Smad7 do not respond directly to attachment of the extracellular signals to the receptor protein. Are they co-mediators similar to Smad4, or do they have some other role in TGF-β signal transduction?

The first step in understanding the functions of Smad6 and Smad7 was to determine the effect of overexpression of these proteins on TGF-β signal transduction. Overexpression was achieved by attaching the Smad6 or Smad7 gene to a strong promoter and then using cloning techniques to introduce the gene into cultured cells. The outcome was that nuclear genes normally switched on by TGF-β became non-responsive to the extracellular signal in cells overexpressing Smad6 or Smad7. This result gave the first indication that Smad6 and Smad7 have inhibitory effects on the TGF-β pathway.

Two models have been proposed to explain how the inhibitory SMADs, as Smad 6 and Smad7 are now called, repress the TGF-β pathway. The first model is based on the observation that in cell extracts the Smad6 and Smad7 proteins are associated with the intracellular parts of the cell surface receptors. The hypothesis is that Smad6 and Smad7 inhibit signal transduction by preventing the activated receptors from phosphorylating the other SMADs.

This model probably explains the inhibitory effects of overexpression of Smad6 and Smad7, but in normal cells there may not be enough copies of these proteins to block the cell surface receptors entirely. An alternative model has therefore been proposed, in which the inhibitory SMADs bind to one or more of the other SMADs, removing these from the pathway and hence stopping signal transduction. There is good evidence that this type of interaction is the explanation for the inhibitory effect that Smad6 has on Smad1. Yeast two-hybrid studies (Section 7.3.2) have shown that these two SMADs interact, and after binding to Smad6, Smad1 is unable to influence the transcriptional activators that it normally stimulates, even after it has been phosphorylated by the cell surface receptors.

Whatever the mechanism for Smad6 and Smad7 activity, the discovery of these inhibitory SMADs shows that TGF-β signaling via the SMAD pathway is more flexible than was originally envisaged. Rather than being an all-or-nothing response, the activity of the receptor-regulated SMADs can be modified by the inhibitory effects of Smad6 and Smad7, these proteins presumably responding to as-yet-unidentified intracellular signals in order to modulate the effects of TGF-β binding in an appropriate way.

References

Imamura T, Takase M, Nishihara A, et al. (1998) Smad6 inhibits signalling by the TGF-β superfamily. *Nature*, **389**, 622–626.

Whitman M (1998) Feedback from inhibitory SMADs. *Nature*, **389**, 549–551.

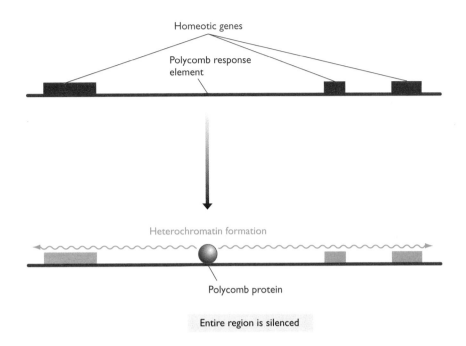

Figure 12.16   Polycomb silences regions of the *Drosophila* genome by initiating heterochromatin formation.

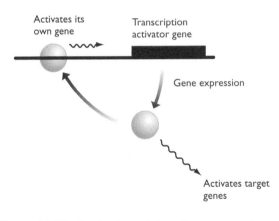

Figure 12.17   Feedback regulation of gene expression.

for specifying segment identity in the fruit fly (Section 12.3.3). The Deformed (Dfd) protein is responsible for the identity of the head segments. To perform this function, Dfd must be continuously expressed in the relevant cells. This is achieved by a feedback system, Dfd binding to an enhancer located upstream of the *Dfd* gene (Regulski *et al.*, 1991). Feedback autoregulation also controls the expression of at least some homeotic selector genes of vertebrates (Popperl *et al.*, 1995).

12.3 Regulation of Genome Activity During Development

The developmental pathway of a multicellular eukaryote begins with a fertilized egg cell and ends with an adult form of the organism. In between lies a complex series of genetic, cellular and physiological events that must occur in the correct order, in the correct cells, and at the appropriate times if the pathway is to reach a successful culmination. With humans, this developmental pathway results in an adult containing 10^{13} cells differentiated into approximately 250 specialized types, the activity of each individual cell coordinated with that of every other cell. Developmental processes of such complexity might appear intractable, even to the powerful investigative tools of modern molecular biology, but remarkably good progress towards understanding them has been made in recent years. The research that has underpinned this progress has been designed around three guiding principles:

1. It should be possible to describe and comprehend the genetic and biochemical events that underlie differentiation of individual cell types. This in turn means that an understanding of how specialized tissues, and even complex body parts, are constructed should be within reach.
2. The signaling processes that coordinate events in different cells should be amenable to study. We saw in Section 12.1 that a start is being made to describing these systems at the molecular level.
3. There should be similarities and parallels between developmental processes in different organisms, reflecting common evolutionary origins. This means that information relevant to human development can be obtained from studies of model organisms chosen for the relative simplicity of their developmental pathways.

Developmental biology encompasses areas of genetics, molecular biology, cell biology, physiology and biochemistry. We are concerned only with the role of the genome in development and so will not attempt a wide-ranging overview of developmental research in all its guises. Instead we will concentrate on three model systems of increasing complexity in order to investigate the types of change in genome activity that occur during development.

12.3.1 Sporulation in *Bacillus*

The first developmental pathway that we will examine is formation of spores by the bacterium *Bacillus subtilis* (Grossman, 1995; Errington, 1996; Stragier and Losick, 1996). Strictly speaking, this is not a developmental pathway, merely a type of cellular differentiation, but the process illustrates two of the fundamental issues that have to be addressed when genuine development in multicellular organisms is studied. These issues are how a series of changes in genome activity over time is controlled, and how signaling establishes coordination between events occurring in different cells. The advantages of *Bacillus* as a model system are that it is easy to grow in the laboratory and is amenable to study by genetic and molecular biology techniques such as analysis of mutants and sequencing of genes.

Sporulation involves coordinated activities in two distinct cell types

Bacillus is one of several genera of bacteria that produce endospores in response to unfavorable environmental conditions. These spores are highly resistant to physical and chemical abuse and can survive for decades or even centuries – the possibility of infection with anthrax spores produced by *B. anthracis* is taken very seriously by archaeologists excavating sites containing human and animal remains. Resistance is due to the specialized nature of the spore coat, which is impermeable to many chemicals and to biochemical changes that retard the decay of DNA and other polymers and enable the spore to survive a prolonged period of dormancy.

In the laboratory, sporulation is usually induced by nutrient starvation. This causes the bacteria to abandon their normal vegetative mode of cell division, which involves synthesis of a septum (or cross-wall) in the center of the cell. Instead the cells construct an unusual septum, one that is thinner than normal, at one end of the cell (*Figure 12.18*). This produces two cellular compartments, the smaller of which is called the prespore and the larger the mother cell. As sporulation proceeds, the prespore becomes entirely engulfed by the mother cell. By now the two cells are committed to different but coordinated differentiation pathways, the prespore undergoing the biochemical changes that enable it to become dormant, and the mother cell constructing the resistant coat around the spore and eventually dying.

Special σ subunits control genome activity during sporulation

Changes in genome activity during sporulation are controlled largely by the synthesis of special σ subunits that change the promoter specificity of the *Bacillus* RNA polymerase. Recall that the σ subunit is the part of the RNA polymerase that recognizes the bacterial promoter sequence, and that replacement of one σ subunit with another with a different DNA-binding specificity can result in a different set of genes being transcribed (Section 9.3.1). We have seen how this simple control system is used by *E. coli* in response to heat stress (see *Figure 9.23*, page 263). It is also the key to the changes in genome activity that occur during sporulation.

The standard *B. subtilis* σ subunits are called σ^A and σ^H. These subunits are synthesized in vegetative cells and enable the RNA polymerase to recognize promoters for all the genes it needs to transcribe in order to maintain normal growth and cell division. In the prespore and mother cell these subunits are replaced by σ^F and σ^E, respectively, which recognize different promoter sequences and so result in large-scale changes in gene expression patterns. The master switch from vegetative growth to spore formation is provided by a protein called SpoOA, which is present in vegetative cells but in an inactive form. This protein is activated by phosphorylation, the protein kinases that phosphorylate it responding to various extracellular signals that indicate the presence of an environmental stress such as lack of nutrients (Sonenshein, 2000). Activated SpoOA is a transcription factor that modulates the expression of various genes transcribed by the vegetative RNA polymerase and hence recognized by the regular σ^A and σ^H subunits. The genes that are switched on include those for σ^F and σ^E, resulting in the switch to prespore and mother cell differentiation (*Figure 12.19*).

Initially, both σ^F and σ^E are present in each of the two differentiating cells. This is not exactly what is wanted because σ^F is the prespore-specific subunit and so should be active only in this cell, and σ^E is mother-cell specific. A means is therefore needed of activating or inactivating the appropriate subunit in the correct cell. This is thought to be achieved as follows (*Figure 12.20*; Errington, 1996):

■ **σ^F is activated by release from a complex with a second protein, SpoIIAB.** This is controlled by a third protein, SpoIIAA, which, when unphosphorylated, can also attach to SpoIIAB and prevent the latter from binding to σ^F. If SpoIIAA is unphosphorylated then σ^F is released and is active; when SpoIIAA is phosphorylated then σ^F remains bound to SpoIIAB and so is inactive. In the mother cell, SpoIIAB phosphorylates SpoIIAA and so keeps σ^F in its bound inactive state. But in the prespore, SpoIIAB's attempts to phosphorylate SpoIIAA are antagonized by yet another protein, SpoIIE, and so σ^F is released and becomes active. SpoIIE's ability to antagonize SpoIIAB in the prespore but not the mother cell derives from the fact that SpoIIE mole-

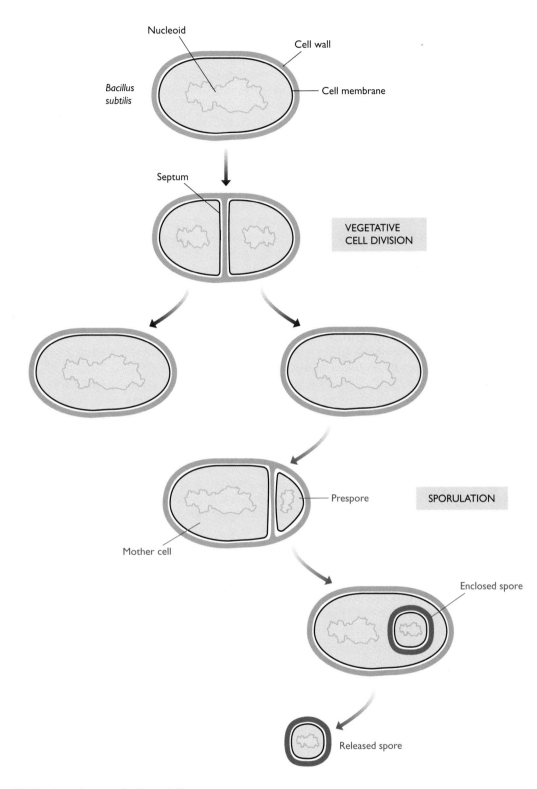

Figure 12.18 Sporulation in *Bacillus subtilis*.

The top part of the diagram shows the normal vegetative mode of cell division, involving formation of a septum across the center of the bacterium and resulting in two identical daughter cells. The lower part of the diagram shows sporulation, in which the septum forms near one end of the cell, leading to a mother cell and prespore of different sizes. Eventually the mother cell completely engulfs the prespore. At the end of the process, the mature resistant spore is released.

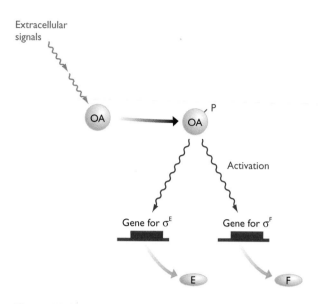

Figure 12.19 Role of SpoOA in *Bacillus* sporulation.

SpoOA is phosphorylated in response to extracellular signals derived from environmental stresses. It is a transcription activator with roles that include activation of the genes for the σ^E and σ^F RNA polymerase subunits. Abbreviations: E, σ^E; F, σ^F; OA, SpoOA. 'P' indicates a phosphate group, PO_3^{2-}.

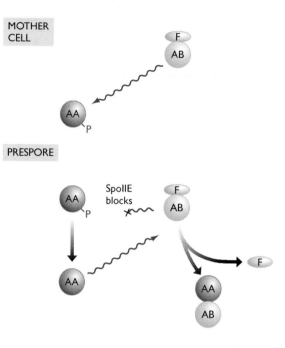

(A) Activation of σ^F in the prespore

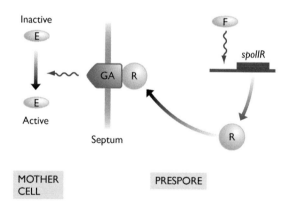

(B) Activation of σ^E in the mother cell

cules are bound to the membrane on the surface of the septum. Because the prespore is much smaller than the mother cell, but the septum surface area is similar in both, the concentration of SpoIIE is greater in the prespore, and this enables it to antagonize SpoIIAB.

■ *σ^E is activated by proteolytic cleavage of a precursor protein.* The protease that carries out this cleavage is the SpoIIGA protein, which spans the septum between the prespore and mother cell. The protease domain, which is on the mother-cell side of the septum, is activated by binding of SpoIIR to a receptor domain on the prespore side. It is a typical receptor-mediated signal transduction system (Section 12.1.2). SpoIIR is one of the genes whose promoter is recognized specifically by σ^F, so activation of the protease, and conversion of pre-σ^E to active σ^E, occurs once σ^F-directed transcription is underway in the prespore.

Activation of σ^F and σ^E is just the beginning of the story. In the prespore, about 1 hour after its activation, σ^F responds to an unknown signal (possibly from the mother cell) which results in a slight change in genome activity in the spore. This includes transcription of a gene for another σ subunit, σ^G, which recognizes promoters upstream of genes whose products are required during the later stages of spore differentiation. One of these proteins is SpoIVB, which activates another septum-bound protease, SpoIVF (*Figure 12.21*). This protease then acti-

Figure 12.20 Activation of the prespore- and mother-cell-specific σ subunits during *Bacillus* sporulation.

(A) In the mother cell, σ^F is inactive because it is bound to SpoAB, which phosphorylates SpoAA and prevents the latter releasing σ^F. Activation of σ^F in the prespore occurs by release from its complex with SpoAB, which is indirectly influenced by the concentration of membrane-bound SpoIIE. (B) In the mother cell, σ^E is activated by proteolytic cleavage by SpoIIGA, which responds to the presence in the prespore of the σ^F-dependent protein SpoIIR. See the text for more details. Abbreviations: AA, SpoIIAA; AB, SpoIIAB; E, σ^E; F, σ^F; GA, SpoIIGA; R, SpoIIR.

vates a second mother cell σ subunit, σ^K, which is coded by a σ^E-transcribed gene but retained in the mother cell in an inactive form until the signal for its activation is received from the prespore. σ^K directs transcription of the

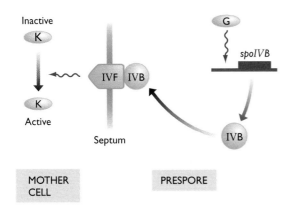

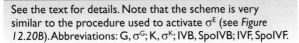

Figure 12.21 Activation of σ^K during *Bacillus* sporulation.

See the text for details. Note that the scheme is very similar to the procedure used to activate σ^E (see *Figure 12.20B*). Abbreviations: G, σ^G; K, σ^K; IVB, SpoIVB; IVF, SpoIVF.

genes whose products are needed during the later stages of the mother-cell differentiation pathway.

To summarize, the key features of *Bacillus* sporulation are as follows:

■ The master protein, SpoOA, responds to external stimuli to determine if and when the switch to sporulation should occur.

■ A cascade of σ subunits in prespore and mother cell brings about time-dependent changes in genome activity in the two cells.

■ Cell–cell signaling ensures that the events occurring in prespore and mother cell are coordinated.

12.3.2 Vulva development in *Caenorhabditis elegans*

B. subtilis is a unicellular organism and, although sporulation involves the coordinated differentiation of two cell types, it can hardly be looked upon as comparable to the developmental processes that occur in multicellular organisms. Sporulation provides pointers to the general ways in which genome activity might be regulated during the development of a multicellular organism, but it does not indicate the specific events to expect. We therefore need to examine development in a simple multicellular eukaryote.

C. elegans *is a model for multicellular eukaryotic development*

Research with the microscopic nematode worm *C. elegans* (*Figure 12.22*) was initiated by Sydney Brenner in the 1960s with the aim of utilizing it as a simple model for multicellular eukaryotic development. *C. elegans* is easy to grow in the laboratory and has a short generation time, measured in days but still convenient for genetic analysis. The worm is transparent at all stages of its life cycle, so internal examination is possible without killing the animal. This is an important point because it has enabled researchers to follow the entire developmental process of the worm at the cellular level. Every cell division in the pathway from fertilized egg to adult worm has been charted, and every point at which a cell adopts a specialized role has been identified. In addition, the complete connectivity of the 302 cells that comprise the nervous system of the worm has been mapped.

The genome of *C. elegans* is relatively small, just 97 Mb (see *Table 2.1*, page 31), and the entire sequence is known (CESC, 1998). Analysis of the sequence, using many of the techniques described in Chapter 7, is beginning to assign functions to the unknown genes and to establish links between genome activity and the developmental pathways. The objective is a complete genetic description of development in *C. elegans*, a goal that is attainable in the not-too-distant future.

Determination of cell fate during development of the C. elegans *vulva*

A critical feature that underpins the usefulness of *C. elegans* as a tool for research is the fact that its development is more or less invariant: the pattern of cell division and differentiation is virtually the same in every individual. This appears to be due in large part to cell–cell signaling, which induces each cell to follow its appropriate differentiation pathway. To illustrate this we will look at development of the *C. elegans* vulva (Sharma-Kishore *et al.*, 1999).

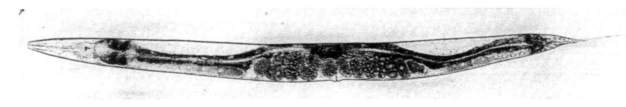

Figure 12.22 The nematode worm *Caenorhabditis elegans*.

The micrograph shows an adult hermaphrodite worm, approximately 1 mm in length. The vulva is the small projection located on the underside of the animal, about halfway along. Egg cells can be seen inside the worm's body in the region either side of the vulva. Reprinted with permission from Kendrew J (1994) *The Encyclopaedia of Molecular Biology*, p. 127, Blackwell Science, Oxford.

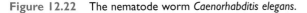

The link between genome replication and sporulation in *Bacillus*

A combination of genetic and biochemical studies have shown how sporulation in *Bacillus subtilis* is coordinated with genome replication.

A prerequisite for sporulation in *B. subtilis* is that the bacterium replicates its genome. It must do this in order to have two complete copies, one to be retained by the mother cell and one to packaged into the spore. How are the two processes – genome replication and sporulation – coordinated?

Genetic studies indicate a central role for sda

Many advances in understanding sporulation in *Bacillus* have depended on the study of mutants that either cannot sporulate or follow an aberrant sporulation pathway. One such mutant is *dnaA1*, which is unable to initiate the sporulation process. The function of the DnaA protein, specified by *dnaA1*, is well known. It is a DNA-binding protein that attaches to the origin of replication of the *Bacillus* genome and initiates the series of events that lead to genome replication (Section 13.2.1). The *dnaA1* mutation has no effect on replication, showing that the resulting change in DnaA does not disrupt this protein's ability to bind to the replication origin, but it does prevent sporulation. Further study of *dnaA1* might therefore reveal the link between replication and sporulation.

One of the standard ways of finding out more about a bacterial mutant is to search for additional mutations that reverse the effects of the one being studied. These are called **suppressor mutations** and they can be generated simply by exposing the mutant bacteria to ultraviolet radiation or to a chemical mutagen (Section 14.1.1), and then examining the cells for ones that have regained the wild-type characteristics. When this experiment was tried with *dnaA1* mutants, five such revertants were obtained. Genetic mapping (Section 5.2.4) showed that all five of these revertants carried mutations in a short previously unrecognized gene that the researchers named *sda*, standing for *suppressor of dnaA1*. The Sda protein must be the intermediary between genome replication and sporulation.

Biochemical studies of Sda

The next step was to identify the biochemical activity of the Sda protein. Attention was focused on a possible interaction between Sda and one of the proteins involved in the cascade of events that begins with receipt by the bacterium of an appropriate extracellular signal and culminates in phosphorylation of SpoOA, the master switch that sets the sporulation pathway in motion (see *Figure 12.19*, page 368). The initial response to the extracellular signals that trigger

sporulation is provided by three kinases – KinA, KinB and KinC – which phosphorylate themselves and then pass the phosphate via SpoOF and SpoOB to SpoOA.

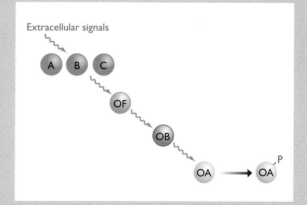

It was immediately discovered that Sda inhibits the activity of KinA, and possibly also has an effect on KinB. By inhibiting one or both of these kinases, Sda could indirectly prevent SpoOA from becoming phosphorylated, and hence could block initiation of the sporulation pathway.

The link between replication and sporulation

Identification of the *sda* gene as key player in sporulation, and elucidation of the effects of the Sda protein on the KinA and KinB kinases, has enabled models for the link between genome replication and sporulation to be constructed. These models envisage that a signal passes between DnaA and Sda, indicating that genome replication has been successfully completed and that sporulation can commence. In response to this signal, Sda lifts its inhibition of KinA and KinB, enabling these kinases to respond, in turn, to the extracellular signals that initiate sporulation. The model is tentative and there are a number of unresolved issues, but it is a good working hypothesis on which to base future research.

References

Burkholder WF, Kurtser I and Grossman AD (2001) Replication initiation proteins regulate a developmental checkpoint in *Bacillus subtilis*. *Cell*, **104**, 269–279.

Michael WM (2001) Cell cycle: Connecting DNA replication to sporulation in *Bacillus*. *Curr. Biol.*, **11**, R443–R445.

Most *C. elegans* worms are hermaphrodites, meaning that they have both male and female sex organs. The vulva is part of the female sex apparatus, being the tube through which sperm enter and fertilized eggs are laid. The adult vulva comprises 22 cells which are the progeny of three ancestral cells originally located in a row on the undersurface of the developing worm (*Figure 12.23*). Each of these ancestral cells becomes committed to the differentiation pathway that leads to production of vulva cells. The central cell, called P6.p, adopts the 'primary vulva cell fate' and divides to produce eight new cells. The other two cells – P5.p and P7.p – take on the 'secondary vulva cell fate' and divide into seven cells each. These 22 cells then reorganize their positions to construct the vulva (Sulston and Horwitz, 1977).

A critical aspect of vulva development is that it must occur in the correct position relative to the gonad, the structure containing the egg cells. If the vulva develops in the wrong place then the gonad will not receive sperm and the egg cells will never be fertilized. The positional information needed by the vulva progenitor cells is provided by a cell within the gonad called the anchor cell (*Figure 12.24*). The importance of the anchor cell has been demonstrated by experiments in which it is artificially destroyed in the embryonic worm: in the absence of the anchor cell, a vulva does not develop. The implication is that the anchor cell secretes an extracellular signaling compound that induces P5.p, P6.p and P7.p to differentiate. This signaling compound is the protein called LIN-3, coded by the *lin-3* gene (Hill and Sternberg, 1992).

Why does P6.p adopt the primary cell fate whereas P5.p and P7.p take on secondary cell fates? There are two possibilities. The first is that LIN-3 forms a concentration gradient and therefore has different effects on P6.p, the cell which is closest to it, and the more distant P5.p and P7.p, as shown in *Figure 12.24*. Evidence in favor of this idea comes from studies showing that isolated cells adopt the secondary fate when exposed to low levels of LIN-3 (Katz *et al.*, 1995). Alternatively, the signal that commits P5.p and P7.p to their secondary fates might not come

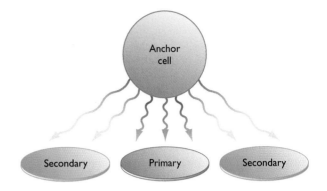

Figure 12.24 The postulated role of the anchor cell in determining cell fate during vulva development in *Caenorhabditis elegans*.

It is thought that release of the signaling compound LIN-3 by the anchor cell commits P6.p (shown in blue), the cell closest to the anchor cell, to the primary vulva cell fate. P5.p and P7.p (shown in green) are further away from the anchor cell and so are exposed to a lower concentration of LIN-3 and become secondary vulva cells. As described in the text, there is evidence that commitment of the secondary cells to their fates is also influenced by signals from the primary vulva cell.

directly from the anchor cell but via P6.p in the form of a different extracellular signaling compound whose synthesis by P6.p is switched on by LIN-3 activation (Kornfeld, 1997). This hypothesis is supported by the abnormal features displayed by certain mutants in which more than three cells become committed to vulva development. With these mutants there is more than one primary cell, but each one is invariably surrounded by two secondary cells, suggesting that in the living worm adoption of the secondary cell fate is dependent on the presence of an adjacent primary cell.

There are other instructive features of vulva development in *C. elegans*. The first is that the signaling process

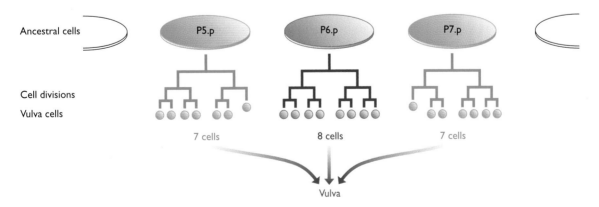

Figure 12.23 Cell divisions resulting in production of the vulva cells of *Caenorhabditis elegans*.

Three ancestral cells divide in a programmed manner to produce 22 progeny cells, which re-organize their positions relative to one another to construct the vulva.

that commits P6.p to its primary cell fate has many similarities with the MAP kinase signal transduction system of vertebrates (see *Figure 12.10*, page 358). The cell surface receptor for LIN-3 is a protein kinase called LET-23 (Aroian *et al.*, 1990) which, when activated by binding LIN-3, initiates a series of intracellular reactions that leads to activation of a MAP-kinase-like protein, which in turn switches on a variety of transcription activators (Sternberg and Han, 1998). Unfortunately the target genes have not yet been delineated in either the primary or secondary vulva progenitor cells, but the system is open to study.

A second noteworthy feature is that as well as the activation signal provided by the anchor cell in the form of LIN-3, the vulva progenitor cells are also subject to the deactivating effects of a second signaling compound secreted by the hypodermal cell, a multinuclear sheath that surrounds most of the worm's body. This repressive signal is overcome by the positive signals that induce P5.p, P6.p and P7.p to differentiate, but prevents the unwanted differentiation of three adjacent cells, P3.p, P4.p and P8.p, each of which can become committed to vulva development if the repressive signal malfunctions, for example in a mutant worm.

In summary, the general concepts to emerge from vulva development in *C. elegans* are as follows:

■ In a multicellular organism, positional information is important: the correct structure must develop at the appropriate place.

■ The commitment to differentiation of a small number of progenitor cells can lead to construction of a multicellular structure.

■ Cell–cell signaling can utilize a concentration gradient to induce different responses in cells at different positions relative to the signaling cell.

■ A cell might be subject to competitive signaling, where one signal tells it to do one thing and a second signal tells it to do the opposite.

12.3.3 Development in *Drosophila melanogaster*

The last organism whose development we will study is *Drosophila melanogaster*. The experimental history of the fruit fly dates back to 1910 when Thomas Hunt Morgan first used this organism as a model system in genetic research. For Morgan the advantages of *Drosophila* were its small size, enabling large numbers to be studied in a single experiment, its minimal nutritional requirements (the flies like bananas), and the presence in natural populations of occasional variants with easily recognized genetic characteristics such as unusual eye colors. Morgan was not aware that other advantages are a small genome (180 Mb; see *Table 2.1*, page 31) and the fact that gene isolation is aided by the presence in the salivary glands of 'giant' chromosomes. These are made up of multiple copies of the same DNA molecule laid side by side, and display banding patterns that can be correlated with the physical map of each chromosome to pinpoint

the positions of desired genes. But Morgan did foresee that *Drosophila* might become an important organism for developmental research, a topic that he was as interested in as we are today.

The major contribution that *Drosophila* has made to our understanding of development has been through the insights it has provided into how an undifferentiated embryo acquires positional information that eventually results in the construction of complex body parts at the correct places in the adult organism. Although in some respects *Drosophila* is quite unusual in its embryonic organization (as we will see in the next section), the genetic mechanisms that specify the fly's body plan are similar to those in other organisms, including humans. Knowledge gained from *Drosophila* has therefore directed research into areas of human development that for a long time were thought to be inaccessible. To explore this story we must start with the events that occur in the developing *Drosophila* embryo.

Maternal genes establish protein gradients in the Drosophila embryo

The unusual feature of the early *Drosophila* embryo is that it is not made up of lots of cells, as in most organisms, but instead is a single **syncytium** comprising a mass of cytoplasm and multiple nuclei (*Figure 12.25*). This structure persists until successive rounds of nuclear division have produced some 1500 nuclei: only then do individual uninucleate cells start to appear around the outside of the syncytium, producing the structure called the blastoderm. Before the blastoderm stage has been reached, the positional information has begun to be established.

Initially the positional information that the embryo needs is a definition of which end is the front (anterior) and which the back (posterior), as well as similar information relating to up (dorsal) and down (ventral). This information is provided by concentration gradients of proteins that become established in the syncytium. The bulk of these proteins are not synthesized from genes in the embryo, but are translated from mRNAs injected into the embryo by the mother. To see how these **maternal-effect genes** work we will examine the synthesis of Bicoid, one of the four proteins involved in determining the anterior–posterior axis.

The *bicoid* gene is transcribed in the maternal nurse cells, which are in contact with the egg cells, and the mRNA is injected into the anterior end of the unfertilized egg. This position is defined by the orientation of the egg cell in the egg chamber. The *bicoid* mRNA remains in the anterior region of the egg cell, attached by its 3' untranslated region to the cell's cytoskeleton. It is not translated immediately, probably because its poly(A) tail is too short. This is inferred because translation, which occurs after fertilization of the egg, is preceded by extension of the poly(A) tail through the combined efforts of the Cortex, Grauzone and Staufen proteins, all of which are synthesized from genes in the egg. Bicoid protein then diffuses through the syncytium, setting up a concentra-

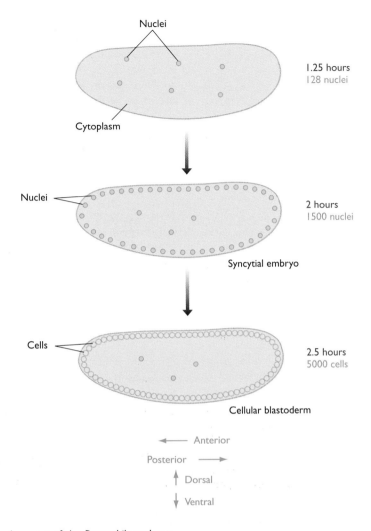

Figure 12.25 Early development of the *Drosophila* embryo.

To begin with, the embryo is a single syncytium containing a gradually increasing number of nuclei. These nuclei migrate to the periphery of the embryo after about 2 hours, and within another 30 minutes cells begin to be constructed. The embryo is approximately 500 μm in length and 170 μm in diameter.

tion gradient, highest at the anterior end and lowest at the posterior end (*Figure 12.26*).

Three other maternal-effect gene products are also involved in setting up the anterior–posterior gradient. These are the Hunchback, Nanos and Caudal proteins. All are injected as mRNAs into the anterior region of the unfertilized egg. The *nanos* mRNA is transported to the posterior part of the egg and attached to the cytoskeleton while it awaits translation. The *hunchback* and *caudal* mRNAs become distributed evenly through the cytoplasm, but their proteins subsequently form gradients through the action of Bicoid and Nanos:

■ Bicoid activates the *hunchback* gene in the embryonic nuclei and represses translation of the maternal *caudal* mRNA, increasing the concentration of the Hunchback protein in the anterior region and decreasing that of Caudal.

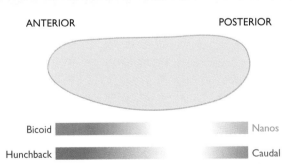

Figure 12.26 Establishment of the anterior–posterior axis in a *Drosophila* embryo.

The anterior–posterior axis is established by gradients of Bicoid, Nanos, Caudal and Hunchback proteins, as described in the text. In this diagram, the concentration gradients are indicated by the colored bars under the outline of the embryo.

■ Nanos represses translation of *hunchback* mRNA, contributing further to the anterior–posterior gradient of the Hunchback protein.

The net result is a gradient of Bicoid and Hunchback, greater at the anterior end, and of Nanos and Caudal, greater at the posterior end (see *Figure 12.26*). The gradient is supplemented with Torso protein, another maternal-effect gene product, which accumulates at the extreme anterior and posterior ends. Similar events result in a dorsal-to-ventral gradient, predominantly of the protein called Dorsal.

A cascade of gene expression converts positional information into a segmentation pattern

The body plan of the adult fly, as well as that of the larva, is built from a series of segments, each with a different structural role. This is clearest in the thorax, which has three segments, each carrying one pair of legs, and the abdomen, which is made up of eight segments, but is also true for the head, even though in the head the segmented structure is less visible (*Figure 12.27*). The objective of embryo development is therefore production of a young larva with the correct segmentation pattern.

The gradients established in the embryo by the maternal-effect gene products are the first stage in formation of the segmentation pattern. These gradients provide the interior of the embryo with a basic amount of positional information, each point in the syncytium now having its own unique chemical signature defined by the relative amounts of the various maternal-effect gene

products. This positional information is made more precise by expression of the **gap genes**.

Three of the anterior–posterior gradient proteins – Bicoid, Hunchback and Caudal – are transcription activators that target the gap genes in the nuclei that now line the inside of the embryo (see *Figure 12.25*). The identities of the gap genes expressed in a particular nucleus depend on the relative concentrations of the gradient proteins and hence on the position of the nucleus along the anterior–posterior axis. Some gaps genes are activated directly by Bicoid, Hunchback and Caudal, examples being *buttonhead*, *empty spiracles* and *orthodenticle* which are activated by Bicoid. Other gap genes are switched on indirectly, as is the case with *hucklebein* and *tailless*, which respond to transcription activators that are switched on by Torso. There are also repressive effects (e.g. Bicoid represses expression of *knirps*) and the gap gene products regulate their own expression in various ways. This complex interplay results in the positional information in the embryo, now carried by the relative concentrations of the gap gene products, becoming more detailed (*Figure 12.28*).

The next set of genes to be activated, the **pair-rule genes**, establish the basic segmentation pattern. Transcription of these genes responds to the relative concentrations of the gap gene products and occurs in nuclei that have become enclosed in cells. The pair-rule gene products therefore do not diffuse through the syncytium but remain localized within the cells that express them. The result is that the embryo can now be looked upon as comprising a series of stripes, each stripe consisting of a set of cells expressing a particular pair-rule gene. In a further round of gene activation, the **segment polarity genes**

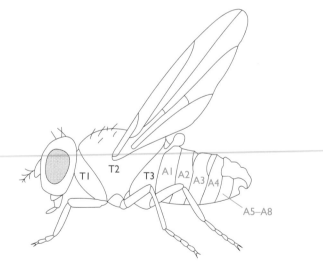

Figure 12.27 The segmentation pattern of the adult *Drosophila melanogaster*.

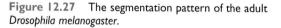

Note that the head is also segmented, but the pattern is not easily discernible from the morphology of the adult fly. Reprinted with permission from Lewis EB, *Nature*, **276**, 565. Copyright 1978 Macmillan Magazines Limited.

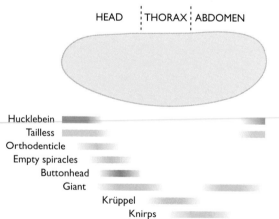

Figure 12.28 The role of the gap gene products in conferring positional information during embryo development in *Drosophila melanogaster*.

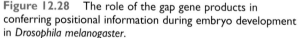

As in *Figure 12.26*, the concentration gradient of each gap gene product is denoted by the colored bars. The parts of the embryo that give rise to the head, thorax and abdomen regions of the adult fly are indicated.

become switched on, providing greater definition to the stripes by setting the sizes and precise locations of what will eventually be the segments of the larval fly. Gradually we have converted the imprecise positional information of the maternal-effect gradients into a sharply defined segmentation pattern.

Segment identity is determined by the homeotic selector genes

The pair-rule and segment polarity genes establish the segmentation pattern of the embryo but do not determine the identities of the individual segments. This is the job of the **homeotic selector genes**, which were first discovered by virtue of the extravagant effects that mutations in these genes have on the appearance of the adult fly. The *antennapedia* mutation, for example, transforms the head segment that usually produces an antenna into one that makes a leg, so the mutant fly has a pair of legs where its antennae should be. The early geneticists were fascinated by these monstrous **homeotic mutants** and many were collected during the first few decades of the 20th century.

Genetic mapping of homeotic mutations has revealed that the selector genes are clustered in two groups on chromosome 3. These clusters are called the Antennapedia complex (ANT-C), which contains genes involved in determination of the head and thorax segments, and the Bithorax complex (BX-C), which contains genes for the abdomen segments (*Figure 12.29*). Some additional non-selector development genes, such as *bicoid*, are also located in ANT-C. One interesting feature of the ANT-C and BX-C clusters, which is still not understood, is that the order of genes corresponds to the order of the segments in the fly, the first gene in ANT-C being *labial palps*,

---

Box 12.1: The genetic basis of flower development

Developmental processes in plants are, in most respects, very different from those of fruit flies and other animals, but at the genetic level there are certain similarities, sufficient for the knowledge gained about *Drosophila* development to be of value in interpreting similar research carried out with plants. In particular, the recognition that a limited number of homeotic selector genes control the *Drosophila* body plan has led to a model for plant development which postulates that the structure of the flower is determined by a small number of homeotic genes.

All flowers are constructed along similar lines, made up of four concentric whorls, each comprising a different floral organ. The outer whorl, number 1, contains sepals, which are modified leaves that envelop and protect the bud during its early development. The next whorl, number 2, contains the distinctive petals, and within these are whorls 3 (stamens, the male reproductive organs) and 4 (carpels, the female reproductive organs).

Most of the research on plant development has been carried out with *Antirrhinum* (the snapdragon) and *Arabidopsis*

thaliana, a small vetch that has been adopted as a model species, partly because it has a genome of only 125 Mb (see *Table 2.1*, page 31), one of the smallest known among flowering plants. Although these plants do not appear to contain homeodomain proteins, they do have genes which, when mutated, lead to homeotic changes in the floral architecture, such as replacement of sepals by carpels. Analysis of these mutants has led to the 'ABC model', which states that there are three types of homeotic genes – A, B and C – controlling flower development:

- Whorl 1 is specified by A-type genes: examples in *Arabidopsis* are *apetala1* and *apetala2*.
- Whorl 2 is specified by A genes acting in concert with B genes, examples of the latter including *apetala3* and *pistillata*.
- Whorl 3 is specified by the B genes plus the C gene, *agamous*.
- Whorl 4 is specified by the C gene acting on its own.

As anticipated from the work with *Drosophila*, the A, B and C homeotic gene products are transcription activators. All except the APETALA2 protein contain the same DNA-binding domain, the **MADS box**, which is also found in other proteins involved in plant development, including SEPALLATA1, 2 and 3, which work with the A, B and C proteins in defining the detailed structure of the flower (Theissen and Saedler, 2001). Other components of the flower development system include at least one master gene, called *floricaula* in *Antirrhinum* and *leafy* in *Arabidopsis*, which controls the switch from vegetative to reproductive growth, initiating flower development, and also has a role in establishing the pattern of homeotic gene expression (Ma, 1998; Parcy *et al.*, 1998). In *Arabidopsis* there is also a gene, called *curly leaf*, whose product acts like Polycomb of *Drosophila* (Section 12.2.2), maintaining the differentiated state of each cell by repressing those homeotic genes that are inactive in a particular whorl (Goodrich *et al.*, 1997).

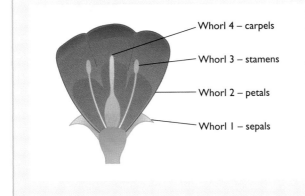

Whorl 4 – carpels

Whorl 3 – stamens

Whorl 2 – petals

Whorl 1 – sepals

Antennapedia complex (ANT-C)

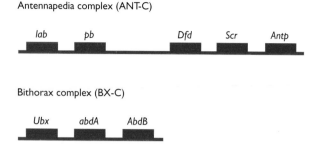

Bithorax complex (BX-C)

Figure 12.29 The Antennapedia and Bithorax gene complexes of *Drosophila melanogaster*.

Both complexes are located on the fruit-fly chromosome 3, ANT-C upstream of BX-C. The genes are usually drawn in the order shown, although this means that they are transcribed from right to left. The diagram does not reflect the actual lengths of the genes. The full gene names are as follows: *lab*, labial palps; *pb*, proboscipedia; *Dfd*, Deformed; *Scr*, Sex combs reduced; *Antp*, Antennapedia; *Ubx*, Ultrabithorax; *abdA*, abdominal A; *AbdB*, Abdominal B. In ANT-C, the nonselector genes *zerknüllt* and *bicoid* occur between *pb* and *Dfd*, and *fushi tarazu* lies between *Scr* and *Antp*.

a coactivator such as Extradenticle, switches on the set of genes needed to initiate development of the specified segment. Maintenance of the differentiated state is ensured partly by the repressive effect that each selector gene product has on expression of the other selector genes, and partly by the work of Polycomb which, as we saw in Section 12.2.2, constructs inactive chromatin over the selector genes that are not expressed in a particular cell (Pirrotta, 1997).

Homeotic selector genes are universal features of higher eukaryotic development

The homeodomains of the various *Drosophila* selector genes are strikingly similar. This observation led researchers in the 1980s to search for other homeotic genes by using the homeodomain as a probe in hybridization experiments. First, the *Drosophila* genome was searched, resulting in isolation of several previously unknown homeodomain-containing genes. These have turned out not to be selector genes but other types of gene coding for transcription activators involved in development. Examples include the pair-rule genes *even-skipped* and *fushi tarazu*, and the segment polarity gene *engrailed*.

The real excitement came when the genomes of other organisms were probed and it was realized that homeodomains are present in genes in a wide variety of animals, including humans. Even more unexpected was the discovery that some of the homeodomain genes in these other organisms are homeotic selector genes organized into clusters similar to ANT-C and BX-C, and that these genes have equivalent functions to the *Drosophila* versions, specifying construction of the body plan.

We now look on the ANT-C and BX-C clusters of selector genes in *Drosophila* as two parts of a single complex,

which controls the most anterior segment of the fly, and the last gene in BX-C being *Abdominal B*, which specifies the most posterior segment.

The correct selector gene is expressed in each segment because the activation of each one is responsive to the positional information represented by the distributions of the gap and pair-rule gene products. The selector gene products are themselves transcription activators, each containing a homeodomain version of the helix–turn–helix DNA-binding structure (Section 9.1.4). Each selector gene product, possibly in conjunction with

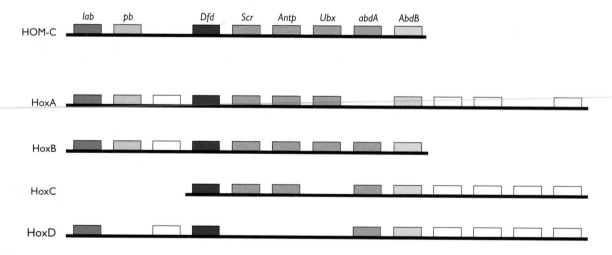

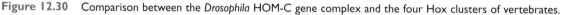

Figure 12.30 Comparison between the *Drosophila* HOM-C gene complex and the four Hox clusters of vertebrates.

Genes that code for proteins with related structures and functions are indicated by the colors. For more details on the evolution of the Hox clusters, see Section 15.2.1. The diagram does not reflect the actual lengths of the genes. Gene names are given in the legend to *Figure 12.29*.

usually referred to as the homeotic gene complex or HOM-C. In vertebrates there are four homeotic gene clusters, called HoxA to HoxD. When these four clusters are aligned with one another and with HOM-C (*Figure 12.30*) similarities are seen between the genes at equivalent positions, such that the evolutionary history of the homeotic selector gene clusters can be traced from insects through to humans (see Section 15.2.1). The genes in the vertebrate clusters specify the development of body structures and, as in *Drosophila*, the order of genes reflects the order of

these structures in the adult body plan. This is clearly seen with the mouse HoxB cluster, which controls development of the nervous system (*Figure 12.31*).

The remarkable conclusion is that, at this fundamental level, developmental processes in fruit flies and other 'simple' eukaryotes are similar to the processes occurring in human and other 'complex' organisms. The discovery that studies of fruit flies are directly relevant to human development opens up vast vistas of future research possibilities.

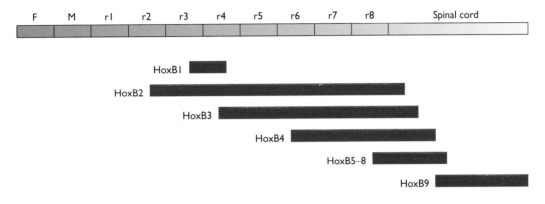

Figure 12.31   Specification of the mouse nervous system by selector genes of the HoxB cluster.

The nervous system is shown schematically and the positions specified by the individual HoxB genes (HoxB1 to HoxB9) indicated by the red bars. The components of the nervous system are: F, forebrain; M, midbrain; r1–r8, rhombomeres 1–8; followed by the spinal cord. Rhombomeres are segments of the hindbrain seen during development.

References

Aroian RV, Koga M, Mendel JE, Ohshima Y and Sternberg PW (1990) The *let-23* gene necessary for *Caenorhabditis elegans* vulval induction encodes a tyrosine kinase of the EGF receptor subfamily. *Nature*, **348**, 693–699.

Becker S, Groner B and Müller CW (1998) Three-dimensional structure of the Stat3b homodimer bound to DNA. *Nature*, **394**, 145–151.

Berridge MJ, Lipp P and Bootman MD (2000) The calcium entry pas de deux. *Science*, **287**, 1604–1605.

Brock TD (1990) *The Emergence of Bacterial Genetics*. Cold Spring Harbor Laboratory Press, Cold Spring Harbor, New York.

CESC (The C. elegans Sequencing Consortium) (1998) Genome sequence of the nematode *C. elegans*: a platform for investigating biology. *Science*, **282**, 2012–2018.

Chawla S, Hardingham GE, Quinn DR and Bading H (1998) CBP: a signal-regulated transcriptional coactivator controlled by nuclear calcium and CaM kinase IV. *Science*, **281**, 1505–1509.

Errington J (1996) Determination of cell fate in *Bacillus subtilis*. *Trends Genet.*, **12**, 31–34.

Garre C, Bianchiscarra G, Sirito M, Musso M and Ravazzolo R (1992) Lactoferrin binding sites and nuclear localization in K562(S) cells. *J. Cell Physiol.*, **153**, 477–482.

Goodrich J, Puangsomlee P, Martin M, Long D, Meyerowitz EM and Coupland G (1997) A Polycomb-group gene regulates homeotic gene expression in *Arabidopsis*. *Nature*, **386**, 44–51.

Grossman AD (1995) Genetic networks controlling the initiation of sporulation and the development of genetic competence in *Bacillus subtilis*. *Ann. Rev. Genet.*, **29**, 477–508.

Haber JE (1998) A locus control region regulates yeast recombination. *Trends Genet.*, **14**, 317–321.

He J and Furmanski P (1995) Sequence specificity and transcriptional activation in the binding of lactoferrin to DNA. *Nature*, **373**, 721–724.

Hill RJ and Sternberg PW (1992) The *lin-3* gene encodes an inductive signal for vulval development in *C. elegans*. *Nature*, **358**, 470–476.

Horvath CM (2000) STAT proteins and transcriptional responses to extracellular signals. *Trends Biochem. Sci.*, **25**, 496–502.

Katz WS, Hill RJ, Clandinin TR and Sternberg PW (1995) Different levels of the *C. elegans* growth factor LIN-3 promote distinct vulval precursor fates. *Cell*, **82**, 297–307.

Kornfeld K (1997) Vulval development in *Caenorhabditis elegans*. *Trends Genet.*, **13**, 55–61.

Lodish H, Berk A, Zipursky AL, Matsudaira P, Baltimore D and Darnell J (2000) *Molecular Cell Biology*, 4th edition. W. H. Freeman, New York.

Ma H (1998) To be, or not to be, a flower – control of floral meristem identity. *Trends Genet.*, **14**, 26–32.

Parcy F, Nilsson O, Busch MA, Lee I and Weigel D (1998) A genetic framework for floral patterning. *Nature*, **395**, 561–566.

Pena MMO, Koch KA and Thiele DJ (1998) Dynamic regulation of copper uptake and detoxification genes in *Saccharomyces cerevisiae*. *Mol. Cell. Biochem.*, **18**, 2514–2523.

Pirrotta V (1997) Chromatin-silencing mechanisms in *Drosophila* maintain patterns of gene expression. *Trends Genet.*, **13**, 314–318.

Popperl H, Bienz M, Studer M, et al. (1995) Segmental expression of HoxB-1 is controlled by a highly conserved autoregulatory loop dependent upon exd/pbx. *Cell*, **81**, 1031–1042.

Regulski M, Dessain S, McGinnis N and McGinnis W (1991) High affinity binding sites for the Deformed protein are required for the function of an autoregulatory enhancer of the *deformed* gene. *Genes Devel.*, **5**, 278–286.

Robinson MJ and Cobb MH (1997) Mitogen-activated kinase pathways. *Curr. Opin. Cell Biol.*, **9**, 180–186.

Schlessinger J (1993) How receptor tyrosine kinases activate Ras. *Trends Biochem. Sci.*, **18**, 273–275.

Sharma-Kishore R, White JG, Southgate E and Podbilewicz B (1999) Formation of the vulva in *Caenorhabditis elegans*: a paradigm for organogenesis. *Development*, **126**, 691–699.

Sonenshein AL (2000) Control of sporulation initiation in *Bacillus subtilis*. *Curr. Opin. Microbiol.*, **3**, 561–566.

Spiegel S, Foster D and Kolesnick R (1996) Signal transduction through lipid second messengers. *Curr. Opin. Cell Biol.*, **8**, 159–167.

Sternberg PW and Han M (1998) Genetics of RAS signaling in *C. elegans*. *Trends Genet.*, **14**, 466–472.

Strachan T and Read AP (1999) *Human Molecular Genetics*, 2nd edition. BIOS Scientific Publishers, Oxford.

Stragier P and Losick R (1996) Molecular genetics of sporulation in *Bacillus subtilis*. *Ann. Rev. Genet.*, **30**, 297–341.

Sulston J and Horwitz HR (1977) Postembryonic cell lineages of the nematode *Caenorhabditis elegans*. *Dev. Biol.*, **56**, 110–156.

Theissen G and Saedler H (2001) Floral quartets. *Nature*, **409**, 469–471.

Toker A and Cantley LC (1997) Signaling through the lipid products of phosphoinositide-3-OH kinase. *Nature*, **387**, 673–676.

Tsai M-J and O'Malley BW (1994) Molecular mechanisms of action of steroid/thyroid receptor superfamily members. *Ann. Rev. Biochem.*, **63**, 451–486.

Winge DR, Jensen LT and Srinivasan C (1998) Metal ion regulation of gene expression in yeast. *Curr. Opin. Chem. Biol.*, **2**, 216–221.

Further Reading

Gehring WJ, Affolter M and Bürglin T (1994) Homeodomain proteins. *Ann. Rev. Biochem.*, **63**, 487–526. — *Details of the* Drosophila *homeodomain proteins, with emphasis on the DNA–protein interactions.*

Karin M and Hunter T (1995) Transcriptional control by protein phosphorylation: signal transmission from the cell surface to the nucleus. *Curr. Biol.*, **5**, 747–757.

Labouesse M and Mango SE (1999) Patterning the *C. elegans* embryo: moving beyond the cell lineage. *Trends Genet.*, **15**, 307–313. — *Reviews the developmental pathways of* C. elegans.

Maconochie M, Nonchev S, Morrison A and Krumlauf R (1996) Paralogous Hox genes: function and regulation. *Ann. Rev. Genet.*, **30**, 529–556. — *Describes homeotic selector genes in vertebrates.*

Maruta H and Burgess AW (1994) Regulation of the Ras signaling network. *Bioessays*, **16**, 489–496.

Tan PBO and Kim SK (1999) Signaling specificity: the RTK/RAS/MAP kinase pathway in metazoans. *Trends Genet.*, **15**, 145–149. — *Describes how a single signal transduction pathway can activate different genes in different cells.*

STUDY AIDS FOR CHAPTER 12

Key terms

Give short definitions of the following terms:

Adenylate cyclase
Antigen
Catabolite activator protein
Catabolite repression
CREB
Cyclic AMP (cAMP)
Development
Diauxie
Differentiation
Gap gene
Gene conversion
GTPase activating protein (GAP)
Guanine nucleotide releasing protein (GNRP)
Homeotic mutant
Homeotic selector gene
Hormone response element

Janus kinase (JAK)
MADS box
Map kinase system
Maternal-effect gene
Mating-type switching
Nuclear receptor superfamily
Pair-rule gene
Ras system
SAP kinase system
Second messenger
Segment polarity gene
Signal transducer and activator of transcription (STAT)
Signal transduction
Steroid hormone
Steroid receptor
Syncytium

Self study questions

1. Distinguish between the terms 'differentiation' and 'development'.
2. Outline the various ways by which transient changes in genome activity can be brought about.
3. Give details of one example of an extracellular signaling protein that can act as a transcription activator.
4. Describe how copper influences gene expression in *Saccharomyces cerevisiae*.
5. Outline the process by which steroid hormones regulate genome activity in higher eukaryotes.
6. Draw a series of diagrams to illustrate the catabolite repression system of bacteria.
7. Distinguish between a STAT and a JAK and explain how the two work together to regulate genome activity.
8. Outline the MAP kinase, Ras and SAP kinase signal transduction systems.
9. What is a second messenger? Give two examples of signal transduction via second messengers.
10. Describe how genome rearrangements underlie the processes involved in (a) mating-type switching in *Saccharomyces cerevisiae*, and (b) the generation of immunoglobulin diversity in vertebrates.

11. How do the Polycomb proteins influence genome activity?

12. Give two examples of feedback autoregulation of gene expression.

13. Describe how *Bacillus subtilis* uses sigma factor replacement as a means of regulating genome activity during sporulation.

14. How is cell–cell signaling used to control genome activity during sporulation in *Bacillus subtilis*?

15. Write a short essay on 'The importance of protein phosphorylation during sporulation in *Bacillus subtilis*'.

16. Why is *Caenorhabditis elegans* a good model organism for development in higher eukaryotes?

17. Discuss the key features of genome regulation during vulva development in *C. elegans*. Your answer should emphasize those features of vulva development that have parallels in developmental processes in higher eukaryotes.

18. Why is *Drosophila melanogaster* a good model organism for development in higher eukaryotes?

19. Describe how the undifferentiated fruit-fly embryo acquires positional information.

20. Discuss the importance of homeotic selector genes in development in fruit flies and vertebrates.

Problem-based learning

1. 'We should bear in mind that the biosphere is so diverse, and the numbers of genes in individual genomes so large, that it is reasonable to assume that any mechanism that could have evolved to regulate genome expression is likely to have done so.' To what extent is this statement supported by our current knowledge of genome regulation?

2. Describe how studies of signal transduction have improved our understanding of the abnormal biochemical activities that underlie cancer.

3. Explore the influence of signal transduction by second messengers on the regulation of genome activity.

4. Evaluate the three 'guiding principles' that have underpinned research into developmental processes (as listed on page 365).

5. Are *Caenorhabditis elegans* and *Drosophila melanogaster* good model organisms for development in higher eukaryotes?

6. What would be the key features of an ideal model organism for development in higher eukaryotes?

7. Is there any need for a model organism for development in higher eukaryotes?

How Genomes Replicate and Evolve

The primary function of a genome is to specify the biochemical signature of the cell in which it resides. In Part 2 we saw that the genome achieves this objective by the coordinated expression of genes and groups of genes, resulting in maintenance of a proteome whose individual protein components carry out and regulate the cell's biochemical activities. In order to continue carrying out this function, the genome must replicate every time that the cell divides. This means that the entire DNA content of the cell must be copied at the appropriate period in the cell cycle, and the resulting DNA molecules must be distributed to the daughter cells so that each one receives a complete copy of the genome. This elaborate process, which spans the interface between molecular biology, biochemistry and cell biology, is described in Chapter 13.

Normally we think of replication as producing two identical copies of the genome, this being a requirement if the daughter cells are to have the same biochemical capabilities as the parent cell. In a general sense, genome replication does result in identical copies, but over time the genome undergoes change, nucleotide sequence alterations accumulating as a result of mutations, occasional errors in replication, and sequence rearrangements caused by recombination and related events.

In Chapter 14, you will learn how these sequence alterations occur and how some of them are repaired. You will realize that the accumulation of unrepaired sequence alterations is the basis for evolution of the genome, which in turn underlies the evolution of organisms, although in a complex manner that is not yet fully understood. These evolutionary themes are explored in the last two chapters of *Genomes*. Chapter 15 examines how molecular evolution has resulted in the vast variety of genomes present in the different organisms alive today, and Chapter 16 explains how the techniques of molecular phylogenetics can be used to make comparisons between the sequences of genes and of entire genomes, enabling evolutionary relationships to be inferred.

Chapter 13
Genome Replication

Chapter 14
Mutation, Repair and Recombination

Chapter 15
How Genomes Evolve

Chapter 16
Molecular Phylogenetics

13

Genome Replication

Chapter Contents

13.1 The Topological Problem   384

13.1.1   Experimental proof for the Watson–Crick scheme for DNA replication   385

13.1.2   DNA topoisomerases provide a solution to the topological problem   388

13.1.3   Variations on the semiconservative theme   389

13.2 The Replication Process   391

13.2.1   Initiation of genome replication   391

13.2.2   The elongation phase of replication   393

13.2.3   Termination of replication   401

13.2.4   Maintaining the ends of a linear DNA molecule   404

13.3 Regulation of Eukaryotic Genome Replication   409

13.3.1   Coordination of genome replication and cell division   409

13.3.2   Control within S phase   412

Learning outcomes

When you have read Chapter 13, you should be able to:

- State what is meant by the topological problem and explain how DNA topoisomerases solve this problem

- Describe the key experiment that proved that DNA replication occurs by the semiconservative process, and outline the exceptions to semiconservative replication that are known in nature

- Discuss how replication is initiated in bacteria, yeast and mammals

- Give a detailed description of the events occurring at the bacterial replication fork, and indicate how these events differ from those occurring in eukaryotes

- Describe what is currently known about termination of replication in bacteria and eukaryotes

- Explain how telomerase maintains the ends of a chromosomal DNA molecule in eukaryotes, and appraise the possible links between telomere length, cell senescence and cancer

- Describe how genome replication is coordinated with the cell cycle

GENOME REPLICATION has been studied since Watson and Crick first discovered the double helix structure of DNA back in 1953. In the years since then research has been driven by three related but distinct issues:

- *The topological problem* was the primary concern in the years from 1953 to 1958. This problem arises from the need to unwind the double helix in order to make copies of its two polynucleotides (see *Figure 13.1*). The issue assumed center stage in the mid-1950s because it was the main stumbling block to acceptance of the double helix as the correct structure for DNA, but moved into the background in 1958 when Matthew Meselson and Franklin Stahl demonstrated that, despite the perceived difficulties, DNA replication in *Escherichia coli* occurs by the method predicted by the double helix structure. The **Meselson–Stahl experiment** enabled research into genome replication to move forward, even though the topological problem itself was not solved until the early 1980s when the mode of action of **DNA topoisomerases** was first understood (Section 13.1.2).
- *The replication process* has been studied intensively since 1958. During the 1960s, the enzymes and proteins involved in replication in *E. coli* were identified and their functions delineated, and in the following years similar progress was made in understanding the details of eukaryotic genome replication. This work is ongoing, with research today centered on topics such as the initiation of replication and the

precise modes of action of the proteins active at the replication fork.
- *The regulation of genome replication*, particularly in the context of the cell cycle, has become the predominant area of research in recent years. This work has shown that initiation is the key control point in genome replication and has begun to explain how replication is synchronized with the cell cycle so that daughter genomes are available when the cell divides.

Our study of genome replication will deal with each of these three topics in the order listed above.

13.1 The Topological Problem

In their paper in *Nature* announcing the discovery of the double helix structure of DNA, Watson and Crick (1953a) made one of the most famous statements in molecular biology:

> It has not escaped our notice that the specific pairing we have postulated immediately suggests a possible copying mechanism for the genetic material.

The pairing process that they refer to is one in which each strand of the double helix acts as a template for synthesis of a second complementary strand, the end result being that both of the daughter double helices are identical to the parent molecule (*Figure 13.1*). The scheme is almost implicit in the double helix structure, but it presents problems, as admitted by Watson and Crick in a second paper published in *Nature* just a month after the report of the structure. This paper (Watson and Crick, 1953b) describes the postulated replication process in more detail, but points out the difficulties that arise from the need to unwind the double helix. The most trivial of these difficulties is the possibility of the daughter molecules getting tangled up. More critical is the rotation that would accompany the unwinding: with one turn occurring for every 10 bp of the double helix, complete replication of the DNA molecule in human chromosome 1, which is 250 Mb in length, would require 25 million rotations of

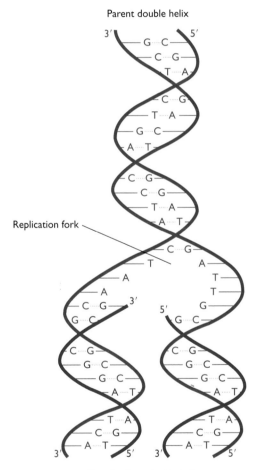

Parent double helix

Replication fork

Two daughter double helices

Figure 13.1 DNA replication, as predicted by Watson and Crick.

The polynucleotides of the parent double helix are shown in red. Both act as templates for synthesis of new strands of DNA, shown in blue. The sequences of these new strands are determined by base-pairing with the template molecules. The topological problem arises because the two polynucleotides of the parent helix cannot simply be pulled apart: the helix has to be unwound in some way.

the chromosomal DNA. It is difficult to imagine how this could occur within the constrained volume of the nucleus, but the unwinding of a linear chromosomal DNA molecule is not physically impossible. In contrast, a circular double-stranded molecule, for example a bacterial or bacteriophage genome, having no free ends, would not be able to rotate in the required manner and so, apparently, could not be replicated by the Watson–Crick scheme. Finding an answer to this dilemma was a major preoccupation of molecular biology during the 1950s.

13.1.1 Experimental proof for the Watson–Crick scheme for DNA replication

The topological problem was considered so serious by some molecular biologists, notably Max Delbrück, that there was initially some resistance to accepting the double helix as the correct structure of DNA (Holmes, 1998). The difficulty relates to the **plectonemic** nature of the double helix, this being the topological arrangement that prevents the two strands of a coil being separated without unwinding. The problem would therefore be resolved if the double helix was in fact **paranemic**, because this would mean that the two strands could be separated simply by moving each one sideways without unwinding the molecule. It was suggested that the double helix could be converted into a paranemic structure by supercoiling (see *Figure 2.17*, page 50) in the direction opposite to the turn of the helix itself, or that within a DNA molecule the right-handed helix proposed by Watson and Crick might be 'balanced' by equal lengths of a left-handed helical structure. The possibility that double-stranded DNA was not a helix at all, but a side-by-side ribbon structure, was also briefly considered, this idea surprisingly being revived in the late 1970s (e.g. Rodley *et al.*, 1976) and receiving a rather acerbic response from Crick and his colleagues (Crick *et al.*, 1979). Each of these proposed solutions to the topological problem were individually rejected for one reason or another, most of them because they required alterations to the double helix structure, alterations that were not compatible with the X-ray diffraction results and other experimental data pertaining to DNA structure.

The first real progress towards a solution of the topological problem came in 1954 when Delbrück proposed a 'breakage-and-reunion' model for separating the strands of the double helix (Holmes, 1998). In this model, the strands are separated not by unwinding the helix with accompanying rotation of the molecule, but by breaking one of the strands, passing the second strand though the gap, and rejoining the first strand. This scheme is in fact very close to the correct solution to the topological problem, being one of the ways in which DNA topoisomerases work (see *Figure 13.4A*), but unfortunately Delbrück overcomplicated the issue by attempting to combine breakage and reunion with the DNA synthesis that occurs during the actual replication process. This led him to a model for

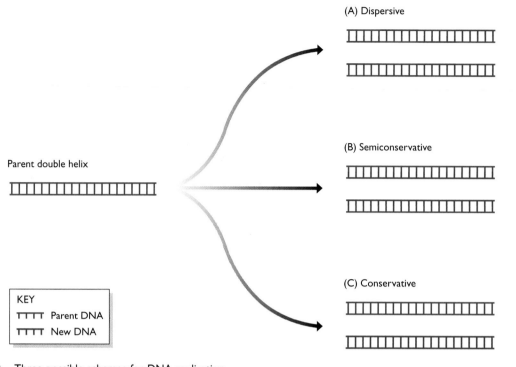

Figure 13.2 Three possible schemes for DNA replication.

For the sake of clarity, the DNA molecules are drawn as ladders rather than helices.

DNA replication which results in each polynucleotide in the daughter molecule being made up partly of parental DNA and partly of newly synthesized DNA (*Figure 13.2A*). This **dispersive** mode of replication contrasts with the **semiconservative** system proposed by Watson and Crick (*Figure 13.2B*). A third possibility is that replication is fully **conservative**, one of the daughter double helices being made entirely of newly synthesized DNA and the other comprising the two parental strands (*Figure 13.2C*). Models for conservative replication are difficult to devise, but one can imagine that this type of replication might be accomplished without unwinding the parent helix.

The Meselson–Stahl experiment

Delbrück's breakage-and-reunion model was important because it stimulated experiments designed to test between the three modes of DNA replication illustrated in *Figure 13.2*. Radioactive isotopes had recently been introduced into molecular biology so attempts were made to use DNA labeling (Technical Note 4.1, page 98) to distinguish newly synthesized DNA from the parental polynucleotides. Each mode of replication predicts a different distribution of newly synthesized DNA, and hence of radioactive label, in the double helices resulting after ʾtwo or more rounds of replication.

Figure 13.3 The Meselson–Stahl experiment.

(A) The experiment carried out by Meselson and Stahl involved growing a culture of *Escherichia coli* in a medium containing $^{15}NH_4Cl$ (ammonium chloride labeled with the heavy isotope of nitrogen). Cells were then transferred to normal medium (containing $^{14}NH_4Cl$) and samples taken after 20 minutes (one cell division) and 40 minutes (two cell divisions). DNA was extracted from each sample and the molecules analyzed by density gradient centrifugation. After 20 minutes all the DNA contained similar amounts of ^{14}N and ^{15}N, but after 40 minutes two bands were seen, one corresponding to hybrid ^{14}N-^{15}N-DNA, and the other to DNA molecules made entirely of ^{14}N. (B) The predicted outcome of the experiment is shown for each of the three possible modes of DNA replication. The banding pattern seen after 20 minutes enables conservative replication to be discounted because this scheme predicts that after one round of replication there will be two different types of double helix, one containing just ^{15}N and the other containing just ^{14}N. The single ^{14}N-^{15}N-DNA band that was actually seen after 20 minutes is compatible with both dispersive and semiconservative replication, but the two bands seen after 40 minutes are consistent only with semiconservative replication. Dispersive replication continues to give hybrid ^{14}N-^{15}N molecules after two rounds of replication, whereas the granddaughter molecules produced at this stage by semiconservative replication include two that are made entirely of ^{14}N-DNA.

(A) The experiment

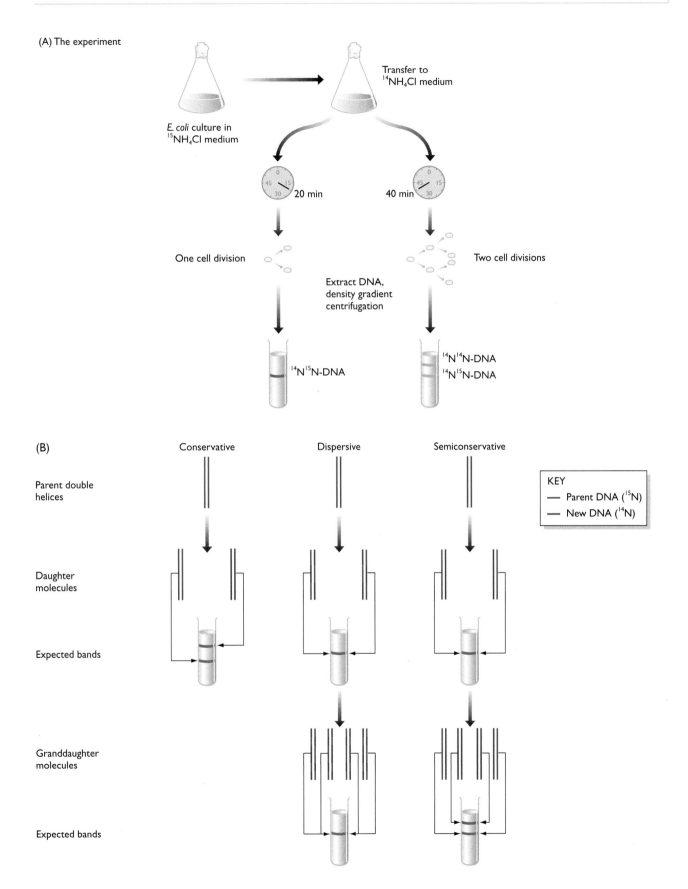

E. coli culture in
$^{15}NH_4Cl$ medium

Transfer to
$^{14}NH_4Cl$ medium

20 min

40 min

One cell division

Two cell divisions

Extract DNA,
density gradient
centrifugation

$^{14}N^{15}N$-DNA

$^{14}N^{14}N$-DNA
$^{14}N^{15}N$-DNA

(B)

Conservative

Dispersive

Semiconservative

KEY
— Parent DNA (^{15}N)
— New DNA (^{14}N)

Parent double
helices

Daughter
molecules

Expected bands

Granddaughter
molecules

Expected bands

Analysis of the radioactive contents of these molecules should therefore determine which replication scheme operates in living cells. Unfortunately, it proved impossible to obtain a clearcut result, largely because of the difficulty in measuring the precise amount of radioactivity in the DNA molecules, the analysis being complicated by the rapid decay of the ^{32}P isotope that was used as the label.

The breakthrough was eventually made by Matthew Meselson and Franklin Stahl who, in 1958, carried out the required experiment not with a radioactive label but with ^{15}N, the non-radioactive 'heavy' isotope of nitrogen. Now it was possible to analyze the replicated double helices by density gradient centrifugation (Technical Note 2.2, page 42), because a DNA molecule labeled with ^{15}N has a higher buoyant density than an unlabeled molecule. Meselson and Stahl (1958) started with a culture of *E. coli* cells that had been grown with $^{15}NH_4Cl$ and whose DNA molecules therefore contained heavy nitrogen. The cells were transferred to normal medium, and samples taken after 20 minutes and 40 minutes, corresponding to one and two cell divisions, respectively. DNA was extracted from each sample and the molecules examined by density gradient centrifugation (*Figure 13.3A*). After one round of DNA replication, the daughter molecules synthesized in the presence of normal nitrogen formed a single band in the density gradient, indicating that each double helix was made up of equal amounts of newly synthesized and parental DNA. This result immediately enabled the conservative mode of replication to be discounted, as this predicts that there would be two bands after one round of replication (*Figure 13.3B*), but did not provide a distinction between Delbrück's dispersive model and the semiconservative process favored by Watson and Crick. The distinction was, however, possible when the DNA molecules resulting from two rounds of replication were examined. Now the density gradient revealed two bands of DNA, the first corresponding to a hybrid composed of equal parts of newly synthesized and old DNA, and the second corresponding to molecules made up entirely of new DNA. This result agrees with the semiconservative scheme but is incompatible with dispersive replication, the latter predicting that after two rounds of replication all molecules would be hybrids.

13.1.2 DNA topoisomerases provide a solution to the topological problem

The Meselson–Stahl experiment proved that DNA replication in living cells follows the semiconservative scheme proposed by Watson and Crick, and hence indicated that the cell must have a solution to the topological problem. This solution was not understood by molecular biologists until some 25 years later, when the activities of the groups of enzymes called DNA topoisomerases were characterized.

DNA topoisomerases are enzymes that carry out breakage-and-reunion reactions similar but not identical to that envisaged by Delbrück. Three types of DNA topoisomerase are recognized (*Table 13.1*; Champoux, 2001):

- *Type IA topoisomerases* introduce a break in one polynucleotide and pass the second polynucleotide through the gap that is formed (*Figure 13.4A*). The two ends of the broken strand are then re-ligated (Lima *et al.*, 1994). This mode of action results in the linking number (the number of times one strand crosses the other in a circular molecule) being changed by one.
- *Type IB topoisomerases* act in a similar way to the Type IA enzymes, although the detailed mechanism is different (Redinbo *et al.*, 1998; Stewart *et al.*, 1998). Type IA and IB topoisomerases probably evolved separately.
- *Type II topoisomerases* break both strands of the double helix, creating a 'gate' through which a second segment of the helix is passed (*Figure 13.4B*; Berger *et al.*, 1996; Cabral *et al.*, 1997). This changes the linking number by two.

DNA topoisomerases do not themselves *unwind* the double helix. Instead they solve the topological problem by counteracting the overwinding that otherwise would be introduced into the molecule by the progression of the replication fork. The result is that the helix can be 'unzipped', with the two strands pulled apart sideways without the molecule having to rotate (*Figure 13.5*).

Replication is not the only activity that is complicated by the topology of the double helix, and it is becoming increasingly clear that DNA topoisomerases have equally important roles during transcription, recombination and

Table 13.1 DNA topoisomerases

Type	Substrate	Examples
Type IA	Single-stranded DNA	*Escherichia coli* topoisomerases I and III; yeast and human topoisomerase III; archaeal reverse gyrase
Type IB	Single-stranded DNA	Eukaryotic topoisomerase I
Type II	Double-stranded DNA	*E. coli* topoisomerases II (DNA gyrase) and IV; eukaryotic topoisomerases II and IV

(A) Type I   (B) Type II

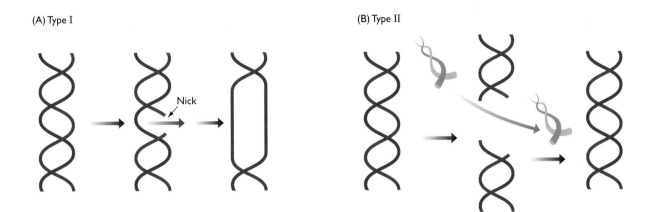

Figure 13.4 The mode of action of Type I and Type II DNA topoisomerases.

(A) A Type I topoisomerase makes a nick in one strand of a DNA molecule, passes the intact strand through the nick, and reseals the gap. (B) A Type II topoisomerase makes a double-stranded break in the double helix, creating a gate through which a second segment of the helix is passed.

other processes that can result in over- or underwinding of DNA. In eukaryotes, topoisomerases form a major part of the nuclear matrix, the scaffold-like network that permeates the nucleus (Section 8.1.1), and are responsible for maintaining chromatin structure and unlinking DNA molecules during chromosome division. Most topoisomerases are only able to relax DNA, but prokaryotic Type II enzymes, such as the bacterial DNA gyrase and the archaeal reverse gyrase, can carry out the reverse reaction and introduce supercoils into DNA molecules.

13.1.3 Variations on the semiconservative theme

No exceptions to the semiconservative mode of DNA replication are known but there are several variations on this basic theme. DNA copying via a replication fork, as shown in *Figure 13.1*, is the predominant system, being used by chromosomal DNA molecules in eukaryotes and by the circular genomes of prokaryotes. Some smaller

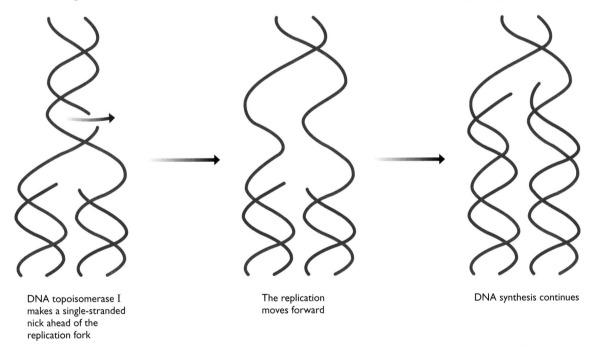

DNA topoisomerase I makes a single-stranded nick ahead of the replication fork

The replication moves forward

DNA synthesis continues

Figure 13.5 Unzipping the double helix.

During replication, the double helix is 'unzipped' as a result of the action of DNA topoisomerases. The replication fork is therefore able to proceed along the molecule without the helix having to rotate.

circular molecules, such as the human mitochondrial genome (Section 2.2.2), use a slightly different process called **displacement replication**, which involves continuous copying of one strand of the helix, the second strand being displaced and subsequently copied after synthesis of the first daughter genome has been completed (*Figure 13.6A*). The advantage of displacement

replication as performed by human mitochondrial DNA is not clear. In contrast, the special type of displacement process called **rolling circle replication** is an efficient mechanism for the rapid synthesis of multiple copies of a circular genome (Novick, 1998). Rolling circle replication, which is used by λ and various other bacteriophages, initiates at a nick which is made in one of the

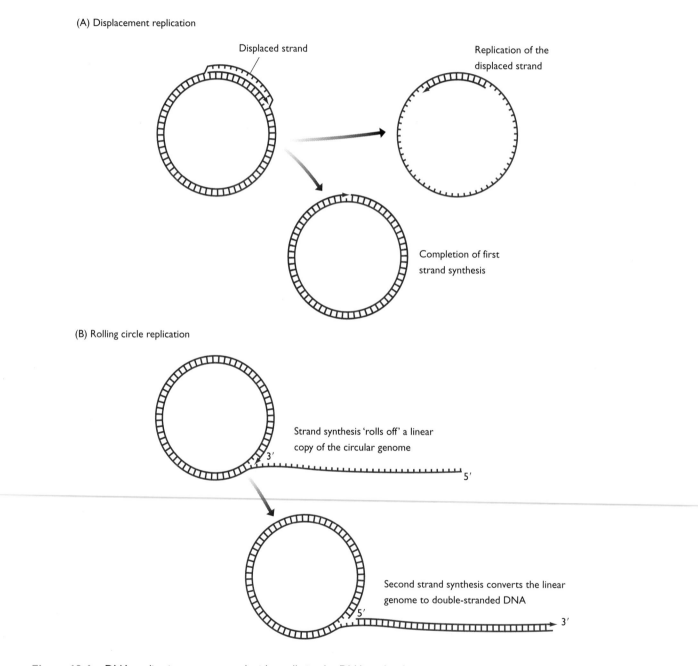

(A) Displacement replication

Displaced strand

Replication of the displaced strand

Completion of first strand synthesis

(B) Rolling circle replication

Strand synthesis 'rolls off' a linear copy of the circular genome

3′

5′

Second strand synthesis converts the linear genome to double-stranded DNA

5′

3′

Figure 13.6 DNA replication systems used with small circular DNA molecules.

(A) Displacement replication, as displayed by the human mitochondrial genome. (B) Rolling circle replication, used by various bacteriophages.

parent polynucleotides. The free 3′ end that results is extended, displacing the 5′ end of the polynucleotide. Continued DNA synthesis 'rolls off' a complete copy of the genome, and further synthesis eventually results in a series of genomes linked head to tail (*Figure 13.6B*). These genomes are single stranded and linear, but can easily be converted to double-stranded circular molecules by complementary strand synthesis followed by cleavage at the junction points between genomes and circularization of the resulting segments.

13.2 The Replication Process

As with many processes in molecular biology, we conventionally look on genome replication as being made up of three phases – initiation, elongation and termination:

- *Initiation* (Section 13.2.1) involves recognition of the position(s) on a DNA molecule where replication will begin.
- *Elongation* (Section 13.2.2) concerns the events occurring at the replication fork, where the parent polynucleotides are copied.
- *Termination* (Section 13.2.3), which in general is only vaguely understood, occurs when the parent molecule has been completely replicated.

As well as these three stages in replication, one additional topic demands attention. This relates to a limitation in the replication process that, if uncorrected, would lead to linear double-stranded DNA molecules getting shorter each time they are replicated (see *Figure 13.24*). The solution to this problem, which concerns the structure and synthesis of the telomeres at the ends of chromosomes (Section 2.2.1), is described in Section 13.2.4.

13.2.1 Initiation of genome replication

Initiation of replication is not a random process and always begins at the same position or positions on a DNA molecule, these points being called the **origins of replication**. Once initiated, two replication forks can emerge from the origin and progress in opposite directions along the DNA: replication is therefore bidirectional with most genomes (*Figure 13.7*). A circular bacterial genome has a single origin of replication, meaning that several thousand kb of DNA are copied by each replication fork. This situation differs from that seen with eukaryotic chromosomes, which have multiple origins and whose replication forks progress for shorter distances. The yeast *Saccharomyces cerevisiae*, for example, has about 300 origins, corresponding to 1 per 40 kb of DNA, and humans have some 20 000 origins, or 1 for every 150 kb.

Initiation at the E. coli *origin of replication*

We know substantially more about initiation of replication in bacteria than in eukaryotes. The *E. coli* origin of

(A) Replication of a circular bacterial chromosome

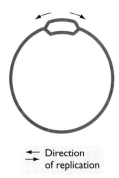

← Direction
→ of replication

(B) Replication of a linear eukaryotic chromosome

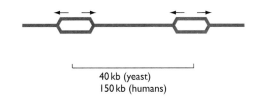

40 kb (yeast)
150 kb (humans)

Figure 13.7 Bidirectional DNA replication of (A) a circular bacterial chromosome and (B) a linear eukaryotic chromosome.

replication is referred to as *oriC*. By transferring segments of DNA from the *oriC* region into plasmids that lack their own origins, it has been estimated that the *E. coli* origin of replication spans approximately 245 bp of DNA. Sequence analysis of this segment shows that it contains two short repeat motifs, one of nine nucleotides and the other of 13 nucleotides (*Figure 13.8A*). The nine-nucleotide repeat, five copies of which are dispersed throughout *oriC*, is the binding site for a protein called DnaA. With five copies of the binding sequence, it might be imagined that five copies of DnaA attach to the origin, but in fact bound DnaA proteins cooperate with unbound molecules until some 30 are associated with the origin. Attachment occurs only when the DNA is negatively supercoiled, as is the normal situation for the *E. coli* chromosome (Section 2.3.1).

The result of DnaA binding is that the double helix opens up ('melts') within the tandem array of three AT-rich, 13-nucleotide repeats located at one end of the *oriC* sequence (*Figure 13.8B*). The exact mechanism is unknown but DnaA does not appear to possess the enzymatic activity needed to break base pairs, and it is therefore assumed that the helix is melted by torsional stresses introduced by attachment of the DnaA proteins. An attractive model imagines that the DnaA proteins form a barrel-like structure around which the helix is wound. Melting the helix is promoted by HU, the most abundant of the DNA packaging proteins of *E. coli* (Section 2.3.1).

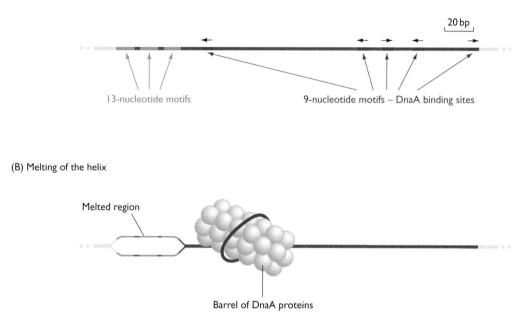

Figure 13.8 The *Escherichia coli* origin of replication.

(A) The *E. coli* origin of replication is called *oriC* and is approximately 245 bp in length. It contains three copies of a 13-nucleotide repeat motif, consensus sequence 5′–GATCTNTTNTTTT–3′ where 'N' is any nucleotide, and five copies of a nine-nucleotide repeat, consensus 5′–TT^A_TT^A_C CA^A_CA–3′. The 13-nucleotide sequences form a tandem array of direct repeats at one end of *oriC*. The nine-nucleotide sequences are distributed through *oriC*, three units forming a series of direct repeats and two units in the inverted configuration, as indicated by the arrows. Three of the nine-nucleotide repeats – numbers 1, 3 and 5 when counted from the left-hand end of *oriC* as drawn here – are regarded as major sites for DnaA attachment; the other two repeats are minor sites. The overall structure of the origin is similar in all bacteria and the sequences of the repeats do not vary greatly. (B) Model for the attachment of DnaA proteins to *oriC*, resulting in melting of the helix within the AT-rich 13-nucleotide sequences.

Melting of the helix initiates a series of events that culminates in the start of the elongation phase of replication. The first step is attachment of a complex of two proteins, DnaBC, forming the **prepriming complex**. DnaC has a transitory role and is released from the complex soon after it is formed, its function probably being simply to aid the attachment of DnaB. The latter is a **helicase**, an enzyme which can break base pairs (see Section 13.2.2). DnaB begins to increase the single-stranded region within the origin, enabling the enzymes involved in the elongation phase of genome replication to attach. This represents the end of the initiation phase of replication in *E. coli* as the replication forks now start to progress away from the origin and DNA copying begins.

Origins of replication in yeast have been clearly defined

The technique used to delineate the *E. coli oriC* sequence, involving transfer of DNA segments into a non-replicating plasmid, has also proved valuable in identifying origins of replication in the yeast *Saccharomyces cerevisiae*. Origins identified in this way are called **autonomously replicat-**

ing sequences or **ARSs**. A typical yeast ARS is shorter than the *E. coli* origin, usually less than 200 bp in length, but, like the *E. coli* origin, contains discrete segments with different functional roles, these 'subdomains' having similar sequences in different ARSs (*Figure 13.9A*). Four subdomains are recognized. Two of these – subdomains A and B1 – make up the **origin recognition sequence**, a stretch of some 40 bp in total that is the binding site for the **origin recognition complex (ORC)**, a set of six proteins that attach to the ARS (*Figure 13.9B*). ORCs have been described as yeast versions of the *E. coli* DnaA proteins (Kelman, 2000), but this interpretation is probably not strictly correct because ORCs appear to remain attached to yeast origins throughout the cell cycle (Bell and Stillman, 1992; Diffley and Cocker, 1992). Rather than being genuine initiator proteins, it is more likely that ORCs are involved in the regulation of genome replication, acting as mediators between replication origins and the regulatory signals that coordinate the initiation of DNA replication with the cell cycle (Section 13.3; Stillman, 1996).

We must therefore look elsewhere in yeast ARSs for sequences with functions strictly equivalent to that of

(A) Structure of a yeast origin of replication

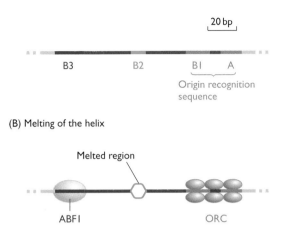

(B) Melting of the helix

Figure 13.9 Structure of a yeast origin of replication.

(A) Structure of ARS1, a typical autonomously replicating sequence (ARS) that acts as an origin of replication in *Saccharomyces cerevisiae*. The relative positions of the functional sequences A, B1, B2 and B3 are shown. For more details see Bielinsky and Gerbi (1998). (B) Melting of the helix occurs within subdomain B2, induced by attachment of the ARS binding protein 1 (ABF1) to subdomain B3. The proteins of the origin replication complex (ORC) are permanently attached to subdomains A and B1.

oriC of *E. coli*. This leads us to the two other conserved sequences in the typical yeast ARS, subdomains B2 and B3 (see *Figure 13.9A*). Our current understanding suggests that these two subdomains function in a manner similar to the *E. coli* origin. Subdomain B2 appears to correspond to the 13-nucleotide repeat array of the *E. coli* origin, being the position at which the two strands of the helix are first separated. This melting is induced by torsional stress introduced by attachment of a DNA-binding protein, ARS binding factor 1 (ABF1), which attaches to subdomain B3 (see *Figure 13.9B*). As in *E. coli*, melting of the helix within a yeast replication origin is followed by attachment of the helicase and other replication enzymes to the DNA, completing the initiation process and enabling the replication forks to begin their progress along the DNA, as described in Section 13.2.2.

Replication origins in higher eukaryotes have been less easy to identify

Attempts to identify replication origins in humans and other higher eukaryotes have, until recently, been less successful (Gilbert, 2001). **Initiation regions** (parts of the chromosomal DNA where replication initiates) were delineated by various biochemical methods, for example by allowing replication to initiate in the presence of labeled nucleotides, then arresting the process, purifying

the newly synthesized DNA, and determining the positions of these nascent strands in the genome. These experiments suggested that there are specific regions in mammalian chromosomes where replication begins, but some researchers were doubtful whether these regions contained replication origins equivalent to yeast ARSs. One alternative hypothesis was that replication is initiated by protein structures that have specific positions in the nucleus, the chromosome initiation regions simply being those DNA segments located close to these protein structures in the three-dimensional organization of the nucleus.

Doubts about mammalian replication origins were increased by the failure of mammalian initiation regions to confer replicative ability on replication-deficient plasmids, although these experiments were not considered conclusive because it was recognized that a mammalian origin might be too long to be cloned in a plasmid or might function only when activated by distant sites in the chromosomal DNA. The breakthrough eventually came when an 8 kb segment of a human initiation region was transferred to the monkey genome, where it still directed replication despite being removed from any hypothetical protein structure in the human nucleus (Aladjem *et al.*, 1998). Analysis of this transferred initiation region showed that there are primary sites within the region where initiation occurs at high frequency, surrounded by secondary sites, spanning the entire 8 kb region, at which replication initiates with lower frequency. The presence of discrete functional domains within the initiation region could also be demonstrated by examining the effects of deletions of parts of the region on the efficiency of replication initiation.

The demonstration that the human genome contains replication origins equivalent to those in yeast raises the question of whether mammals possess an equivalent of the yeast ORC. The answer appears to be yes, as several genes whose protein products have similar sequences to the yeast ORC proteins have been identified in higher eukaryotes and some of these have been shown to be able to replace the equivalent yeast protein in the yeast ORC (Carpenter *et al.*, 1996). These results indicate that initiation of replication in yeast is a good model for events occurring in mammals, a conclusion that is very relevant to studies of the control of replication initiation, as we will see in Section 13.3.

13.2.2 The elongation phase of replication

Once replication has been initiated, the replication forks progress along the DNA and participate in the central activity of genome replication – the synthesis of new strands of DNA that are complementary to the parent polynucleotides. At the chemical level the template-dependent synthesis of DNA (*Figure 13.10*) is very similar to the template-dependent synthesis of RNA that occurs during transcription (compare with *Figure 3.5*, page 75). However, this similarity should not mislead us into making an extensive analogy between transcription and replication. The mechanics of the two processes are quite different, replication being complicated by two factors that do not apply to transcription:

■ During DNA replication both strands of the double helix must be copied. This is an important complication because, as noted in Section 1.1.2, DNA polymerase enzymes are only able to synthesize DNA in

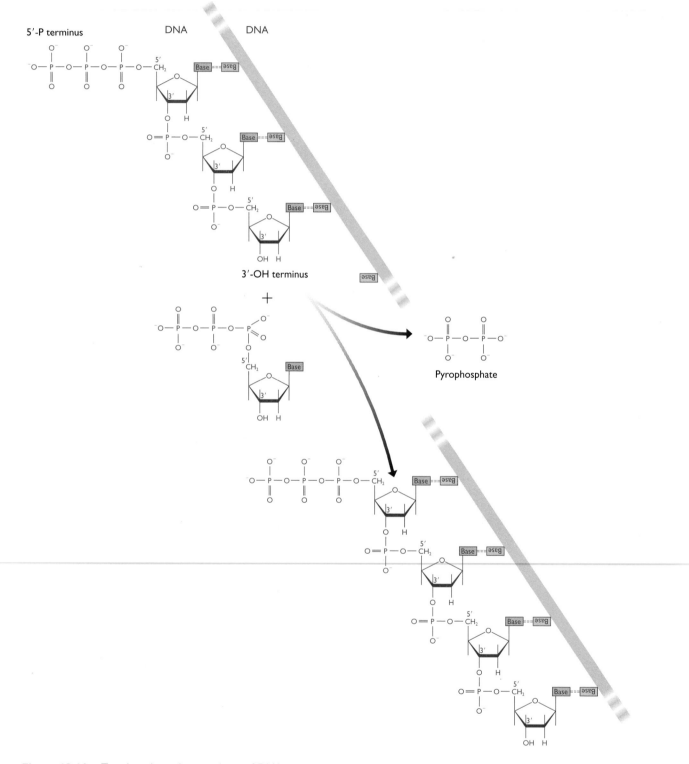

Figure 13.10 Template-dependent synthesis of DNA.

Compare this reaction with template-dependent synthesis of RNA, shown in *Figure 3.5*, page 75.

the 5′→3′ direction. This means that one strand of the parent double helix, called the **leading strand**, can be copied in a continuous manner, but replication of the **lagging strand** has to be carried out in a discontinuous fashion, resulting in a series of short segments that must be ligated together to produce the intact daughter strand (*Figure 13.11*).

- The second complication arises because template-dependent DNA polymerases cannot initiate DNA synthesis on a molecule that is entirely single-stranded: there must be a short double-stranded region to provide a 3′ end onto which the enzyme can add new nucleotides. This means that **primers** are needed, one to initiate complementary strand synthesis on the leading polynucleotide, and one for every segment of discontinuous DNA synthesized on the lagging strand (*Figure 13.11*).

Before dealing with these two complications we will first examine the DNA polymerase enzymes themselves.

The DNA polymerases of bacteria and eukaryotes

The principal chemical reaction catalyzed by a DNA polymerase is the 5′→3′ synthesis of a DNA polynucleotide, as shown in *Figure 13.10*. We learnt in Section

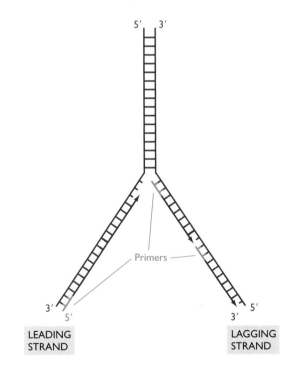

LEADING STRAND

LAGGING STRAND

Figure 13.11 Complications with DNA replication.

Two complications have to be solved when double-stranded DNA is replicated. First, only the leading strand can be continuously replicated by 5′→3′ DNA synthesis; replication of the lagging strand has to be carried out discontinuously. Second, initiation of DNA synthesis requires a primer. This is true both of cellular DNA synthesis, as shown here, and DNA synthesis reactions that are carried out in the test tube (Section 4.1.1).

4.1.1 that some DNA polymerases combine this function with at least one exonuclease activity, which means that these enzymes can degrade polynucleotides as well as synthesize them (see *Figure 4.7*, page 101):

- A 3′→5′ **exonuclease** is possessed by many bacterial and eukaryotic template-dependent DNA polymerases (*Table 13.2*). This activity enables the enzyme to remove nucleotides from the 3′ end of the strand that it has just synthesized. It is looked on as a **proofreading** activity whose function is to correct the occasional base-pairing error that might occur during strand synthesis (see Section 14.1.1).

- A 5′→3′ **exonuclease** activity is less common but is possessed by some polymerases whose function in replication requires that they must be able to remove at least part of a polynucleotide that is already attached to the template strand that the polymerase is copying. This activity is utilized during the process that joins together the discontinuous DNA fragments synthesized on the lagging strand during bacterial DNA replication (see *Figure 13.17*).

The search for DNA polymerases began in the mid-1950s, as soon as it was realized that DNA synthesis was the key to replication of genes. It was thought that bacteria would probably have just a single DNA polymerase, and when the enzyme now called **DNA polymerase I** was isolated by Arthur Kornberg in 1957 there was a widespread assumption that this was the main replicating enzyme. The discovery that inactivation of the *E. coli polA* gene, which codes for DNA polymerase I, was not lethal (cells were still able to replicate their genomes) therefore came as something of a surprise, especially when a similar result was obtained with inactivation of *polB*, coding for a second enzyme, **DNA polymerase II**, which we now know is mainly involved in repair of damaged DNA rather than genome replication (Section 14.2.5). It was not until 1972 that the main replicating polymerase of *E. coli*, **DNA polymerase III**, was eventually isolated. Both DNA polymerases I and III are involved in genome replication, as we will see in the next section.

The properties of the two *E. coli* DNA polymerases involved in genome replication are described in *Table 13.2*. DNA polymerases I and II are single polypeptides but DNA polymerase III, befitting its role as the main replicating enzyme, is multi-subunit, with a molecular mass of approximately 900 kDa. The three main subunits, which form the core enzyme, are called α, ε and θ, with the polymerase activity specified by the α subunit and the 3′→5′ exonuclease by ε. The function of θ is not clear: it may have a purely structural role in bringing together the other two core subunits and in assembling the various accessory subunits. The latter include τ and γ, both coded by the same gene, with synthesis of γ involving translational frameshifting (Section 11.2.3), β, which acts as a 'sliding clamp' and holds the polymerase complex tightly to the template, δ, δ′, χ and ψ.

Eukaryotes have at least nine DNA polymerases (Hübscher *et al.*, 2000), which in mammals are distin-

Table 13.2   DNA polymerases involved in replication of bacterial and eukaryotic genomes

Enzyme	Subunits	Exonuclease activities		Function
		$3' \rightarrow 5'$	$5' \rightarrow 3'$	
Bacterial DNA polymerases				
DNA polymerase I	1	Yes	Yes	DNA repair, replication
DNA polymerase III	At least 10	Yes	No	Main replicating enzyme
Eukaryotic DNA polymerases				
DNA polymerase α	4	No	No	Priming during replication
DNA polymerase γ	2	Yes	No	Mitochondrial DNA replication
DNA polymerase δ	2 or 3	Yes	No	Main replicative enzyme
DNA polymerase ε	At least 1	Yes	No	Required for detection of DNA damage during genome replication (Section 13.3.2)
DNA polymerase κ	1 or 2?	?	?	Required for attachment of cohesin proteins which hold sister chromatids together until the anaphase stage of nuclear division (Section 13.2.3)

Bacteria and eukaryotes possess other DNA polymerases involved primarily in repair of damaged DNA. These enzymes include DNA polymerases II, IV and V of *Escherichia coli* and the eukaryotic DNA polymerases β, ζ, η, θ and ι. Repair processes are described in Section 14.2.

guished by Greek suffices (α, β, γ, δ, etc.), an unfortunate choice of nomenclature as it tempts confusion with the identically named subunits of *E. coli* DNA polymerase III. The main replicating enzyme is **DNA polymerase δ** (*Table 13.2*), which has two subunits (three according to some researchers) and works in conjunction with an accessory protein called the **proliferating cell nuclear antigen** (**PCNA**). PCNA is the functional equivalent of the β subunit of *E. coli* DNA polymerase III and holds the enzyme tightly to the template. **DNA polymerase α** also has an important function in DNA synthesis, being the enzyme that primes eukaryotic replication (see *Figure 13.12B*). **DNA polymerase γ**, although coded by a nuclear gene, is responsible for replicating the mitochondrial genome.

Discontinuous strand synthesis and the priming problem

The limitation that DNA polymerases can synthesize polynucleotides only in the $5' \rightarrow 3'$ direction means that

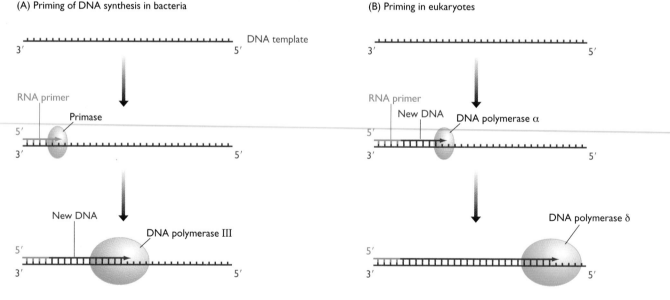

Figure 13.12   Priming of DNA synthesis in (A) bacteria and (B) eukaryotes.

In eukaryotes the primase forms a complex with DNA polymerase α, which is shown synthesizing the RNA primer followed by the first few nucleotides of DNA.

the lagging strand of the parent molecule must be copied in a discontinuous fashion, as shown in *Figure 13.11*. The implication of this model – that the initial products of lagging-strand replication are short segments of polynucleotide – was confirmed in 1969 when **Okazaki fragments**, as these segments are now called, were first isolated from *E. coli* (Okazaki and Okazaki, 1969). In bacteria, Okazaki fragments are 1000–2000 nucleotides in length, but in eukaryotes the equivalent fragments appear to be much shorter, perhaps less than 200 nucleotides. This is an interesting observation that might indicate that each round of discontinuous synthesis replicates the DNA associated with a single nucleosome (140 and 150 bp wound around the core particle plus 50–70 bp of linker DNA: Section 2.2.1).

The second difficulty illustrated in *Figure 13.11* is the need for a primer to initiate synthesis of each new polynucleotide. It is not known for certain why DNA polymerases cannot begin synthesis on an entirely single-stranded template, but it may relate to the proofreading activity of these enzymes, which is essential for the accuracy of replication. As described in Section 14.1.1, a nucleotide that has been inserted incorrectly at the extreme 3′ end of a growing DNA strand, and hence is not base-paired to the template polynucleotide, can be removed by the 3′→5′ exonuclease activity of a DNA polymerase. This means that the 3′→5′ exonuclease activity must be more effective than the 5′→3′ polymerase activity when the 3′ nucleotide is not base-paired to the template. The implication is that the polymerase can extend a polynucleotide efficiently only if its 3′ nucleotide is base-paired, which in turn could be the reason why an entirely single-stranded template, which by definition lacks a base-paired 3′ nucleotide, cannot be used by a DNA polymerase.

Whatever the reason, priming is a necessity in DNA replication but does not present too much of a problem. Although DNA polymerases cannot deal with an entirely single-stranded template, RNA polymerases have no difficulty in this respect, so the primers for DNA replication are made of RNA. In bacteria, primers are synthesized by **primase**, a special RNA polymerase unrelated to the transcribing enzyme, with each primer 4–15 nucleotides in length (Frick and Richardson, 2001). Once the primer has been completed, strand synthesis is continued by DNA polymerase III (*Figure 13.12A*). In eukaryotes the situation is slightly more complex because the primase is tightly bound to DNA polymerase α, and cooperates with this enzyme in synthesis of the first few nucleotides of a new polynucleotide. This primase synthesizes an RNA primer of 8–12 nucleotides, and then hands over to DNA polymerase α, which extends the RNA primer by adding about 20 nucleotides of DNA. This DNA stretch often has a few ribonucleotides mixed in, but it is not clear if these are incorporated by DNA polymerase α or by intermittent activity of the primase. After completion of the RNA–DNA primer, DNA synthesis is continued by the main replicative enzyme, DNA polymerase δ (*Figure 13.12B*).

Priming needs to occur just once on the leading strand, within the replication origin, because once primed, the leading-strand copy is synthesized continuously until replication is completed. On the lagging strand, priming is a repeated process that must occur every time a new Okazaki fragment is initiated. In *E. coli*, which makes Okazaki fragments of 1000–2000 nucleotides in length, approximately 4000 priming events are needed every time the genome is replicated. In eukaryotes the Okazaki fragments are much shorter and priming is a highly repetitive event.

Events at the bacterial replication fork

Now that we have considered the complications introduced by discontinuous strand synthesis and the priming problem, we can move on to study the combination of events occurring at the replication fork during the elongation phase of genome replication.

In Section 13.2.1 we identified attachment of the DnaB helicase, followed by extension of the melted region of the replication origin, as representing the end of the initiation phase of replication in *E. coli*. To a large extent, the division between initiation and elongation is artificial, the two processes running seamlessly one into the other. After the helicase has bound to the origin to form the prepriming complex, the primase is recruited, resulting in the **primosome**, which initiates replication of the leading strand. It does this by synthesizing the RNA primer that DNA polymerase III needs in order to begin copying the template.

DnaB is the main helicase involved in genome replication in *E. coli*, but it is by no means the only helicase that this bacterium possesses: in fact there were eleven at the last count (van Brabant *et al.*, 2000). The size of the collection reflects the fact that DNA unwinding is required not only during replication but also during diverse processes such as transcription, recombination and DNA repair. The mode of action of a typical helicase has not been precisely defined, but it is thought that these enzymes bind to single-stranded rather than double-stranded DNA, and migrate along the polynucleotide in either the 5′→3′ or 3′→5′ direction, depending on the specificity of the helicase. Breakage of base pairs in advance of the helicase requires energy, which is generated by hydrolysis of ATP. According to this model, a single DnaB helicase could migrate along the lagging strand (DnaB is a 5′→3′ helicase), unzipping the helix and generating the replication fork, the torsional stress generated by the unwinding activity being relieved by DNA topoisomerase action (*Figure 13.13*). This model is probably a good approximation of what actually happens, although it does not provide a function for the two other *E. coli* helicases thought to be involved in genome replication. Both of these, PriA and Rep, are 3′→5′ helicases and so could conceivably complement DnaB activity by migrating along the leading strand, but they may have lesser roles. The involvement of Rep in DNA replication might in fact be limited to participation in the rolling circle process used by λ and a few other *E. coli* bacteriophages (Section 13.1.3).

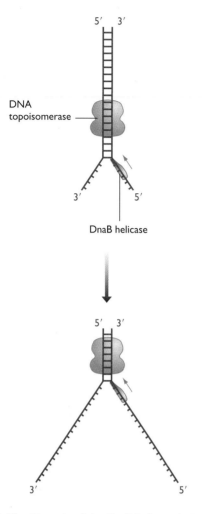

5′ 3′

DNA topoisomerase

DnaB helicase

3′ 5′

5′ 3′

3′ 5′

Figure 13.13 The role of the DnaB helicase during DNA replication in *Escherichia coli*.

DnaB is a 5′→3′ helicase and so migrates along the lagging strand, breaking base pairs as it goes. It works in conjunction with a DNA topoisomerase (see *Figure 13.4*, page 389) to unwind the helix. To avoid confusion, the primase enzyme normally associated with the DnaB helicase is not shown in this drawing.

(A) SSBs attach to the unpaired polynucleotides

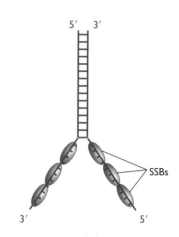

5′ 3′

SSBs

3′ 5′

(B) Structure of RPA, a eukaryotic SSB

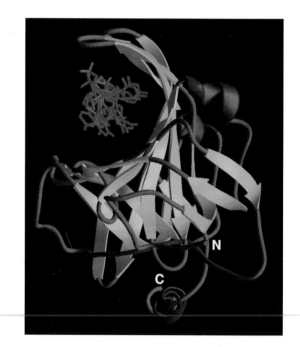

N

C

Figure 13.14 The role of single-strand binding proteins (SSBs) during DNA replication.

(A) SSBs attach to the unpaired polynucleotides produced by helicase action and prevent the strands from base-pairing with one another or being degraded by single-strand-specific nucleases. (B) Structure of the eukaryotic SSB called RPA. The protein contains a β-sheet structure that forms a channel in which the DNA (shown in orange, viewed from the end) is bound. Reproduced with permission from Bochkarev *et al.*, *Nature* **385**, 176–181. Copyright 1997 Macmillan Magazines Limited. Image supplied courtesy of Dr Lori Frappier, Department of Medical Genetics and Microbiology at the University of Toronto, Canada.

Single-stranded DNA is naturally 'sticky' and the two separated polynucleotides produced by helicase action would immediately reform base pairs after the enzyme has passed, if allowed to. The single strands are also highly susceptible to nuclease attack and are likely to be degraded if not protected in some way. To avoid these unwanted outcomes, **single-strand binding proteins (SSBs)** attach to the polynucleotides and prevent them from reassociating or being degraded (*Figure 13.14A*). The *E. coli* SSB is made up of four identical subunits and probably works in a similar way to the major eukaryotic SSB, called **replication protein A (RPA)**, by enclosing the polynucleotide in a channel formed by a series of SSBs attached side by side on the strand (*Figure 13.14B*; Bochkarev *et al.*, 1997). Detachment of the SSBs, which must occur when the replication complex arrives to copy the single strands, is brought about by a second set of proteins called **replication mediator proteins** (**RMPs**; Beernick and Morrical, 1999). As with helicases, SSBs have diverse roles in different processes involving DNA unwinding.

After 1000–2000 nucleotides of the leading strand have been replicated, the first round of discontinuous strand synthesis on the lagging strand can begin. The primase, which is still associated with the DnaB helicase in the primosome, makes an RNA primer which is then extended by DNA polymerase III (*Figure 13.15*). This is the same DNA polymerase III complex that is synthesizing the leading-strand copy, the complex comprising, in effect, two copies of the polymerase. It is not two complete copies because there is only a single γ **complex**, containing subunit γ in association with δ, δ′, χ and ψ. The main

function of the γ complex is to interact with the β subunit (the 'sliding clamp') and hence control the attachment and removal of the enzyme from the template, a function that is required primarily during lagging-strand replication when the enzyme has to attach and detach repeatedly at the start and end of each Okazaki fragment. Some models of the DNA polymerase III complex place the two enzymes in opposite orientations to reflect the different directions in which DNA synthesis occurs, towards the replication fork on the leading strand and away from it on the lagging strand. It is more likely, however, that the pair of enzymes face the same direction and the lagging strand forms a loop, so that DNA synthesis can proceed in parallel as the polymerase complex moves forwards in pace with the progress of the replication fork (*Figure 13.16*).

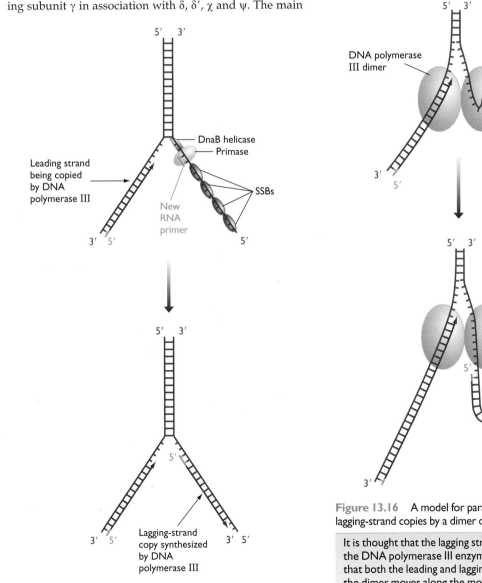

Figure 13.15 Priming and synthesis of the lagging-strand copy during DNA replication in *Escherichia coli*.

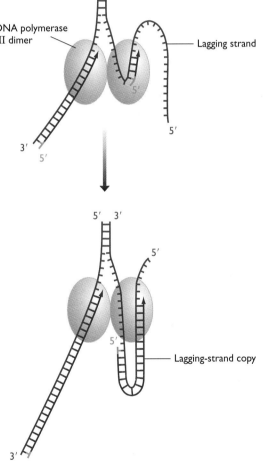

Figure 13.16 A model for parallel synthesis of the leading- and lagging-strand copies by a dimer of DNA polymerase III enzymes.

It is thought that the lagging strand loops through its copy of the DNA polymerase III enzyme, in the manner shown, so that both the leading and lagging strands can be copied as the dimer moves along the molecule being replicated. The two components of the DNA polymerase III dimer are not identical because there is only one copy of the γ complex.

The combination of the DNA polymerase III dimer and the primosome, migrating along the parent DNA and carrying out most of the replicative functions, is called the **replisome**. After its passage, the replication process must be completed by joining up the individual Okazaki fragments. This is not a trivial event because one member of each pair of adjacent Okazaki fragments still has its RNA primer attached at the point where ligation should take place (*Figure 13.17*). *Table 13.2* shows us that this primer cannot be removed by DNA polymerase III, because this enzyme lacks the required 5'→3' exonuclease activity. At this point, DNA polymerase III releases the lagging strand and its place is taken by DNA polymerase I, which does have a 5'→3' exonuclease and so removes the primer, and usually the start of the DNA component of the Okazaki fragment as well, extending the 3' end of the adjacent fragment into the region of the template that is exposed. The two Okazaki fragments now abut, with the terminal regions of both composed entirely of DNA. All that remains is for the missing phosphodiester bond to be put in place by a **DNA ligase**, linking the two fragments and completing replication of this region of the lagging strand.

The eukaryotic replication fork: variations on the bacterial theme

The elongation phase of genome replication is similar in bacteria and eukaryotes, although the details differ. The progress of the replication fork in eukaryotes is maintained by helicase activity, although which of the several eukaryotic helicases that have been identified are primarily responsible for DNA unwinding during replication has not been established. The separated polynucleotides are prevented from reattaching by single-strand binding proteins, the main one of these in eukaryotes being RPA.

We begin to encounter unique features of the eukaryotic replication process when we examine the method used to prime DNA synthesis. As described on page 397, the eukaryotic DNA polymerase α cooperates with the primase enzyme to put in place the RNA–DNA primers at the start of the leading-strand copy and at the beginning of each Okazaki fragment. However, DNA polymerase α is not capable of lengthy DNA synthesis, presumably because it lacks the stabilizing effect of a sliding clamp equivalent to the β subunit of *E. coli* DNA polymerase III or the PCNA accessory protein that aids the eukaryotic DNA polymerase δ. This means that although DNA polymerase α can extend the initial RNA primer with about 20 nucleotides of DNA, it must then be replaced by the main replicative enzyme, DNA polymerase δ (see *Figure 13.12B*, page 396).

The DNA polymerase enzymes that copy the leading and lagging strands in eukaryotes do not associate into a dimeric complex equivalent to the one formed by DNA polymerase III during replication in *E. coli*. Instead, the two copies of the polymerase remain separate. The function performed by the γ complex of the *E. coli* polymerase

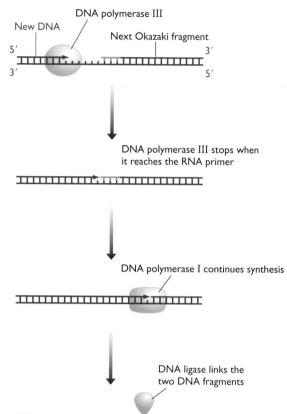

Figure 13.17 The series of events involved in joining up adjacent Okazaki fragments during DNA replication in *Escherichia coli*.

DNA polymerase III lacks a 5'→3' exonuclease activity and so stops making DNA when it reaches the RNA primer of the next Okazaki fragment. At this point DNA synthesis is continued by DNA polymerase I, which does have a 5'→3' exonuclease activity, and which works in conjunction with RNase H to remove the RNA primer and replace it with DNA. DNA polymerase I usually also replaces some of the DNA from the Okazaki fragment before detaching from the template. This leaves a single missing phosphodiester bond, which is synthesized by DNA ligase, completing this step in the replication process.

– controlling attachment and detachment of the enzyme from the lagging strand – appears to be carried out by a multi-subunit accessory protein called **replication factor C (RFC)**.

As in *E. coli*, completion of lagging-strand synthesis requires removal of the RNA primer from each Okazaki fragment. There appears to be no eukaryotic DNA polymerase with the 5'→3' exonuclease needed for this purpose and the process is therefore very different to that described for bacterial cells. The central player is the 'flap endonuclease', **FEN1** (previously called MF1), which associates with the DNA polymerase δ complex at the 3' end of an Okazaki fragment, in order to degrade the primer from the 5′ end of the adjacent fragment. Understanding exactly how this occurs is complicated by the inability of FEN1 to initiate primer degradation because it is unable to remove the ribonucleotide at the extreme 5′ end of the primer, because this ribonucleotide carries a 5′-triphosphate group which blocks FEN1 activity (*Figure 13.18*). Two alternative models to circumvent this problem have been proposed (Waga and Stillman, 1998):

- The first possibility is that a helicase breaks the base pairs holding the primer to the template strand, enabling the primer to be pushed aside by DNA polymerase δ as it extends the adjacent Okazaki fragment into the region thus exposed (*Figure 13.19A*). The flap that results can be cut off by FEN1, whose endonuclease activity can cleave the phosphodiester bond at the branch point where the displaced region attaches to the part of the fragment that is still base-paired.
- Alternatively, most of the RNA component of the primer could be removed by RNase H, which can degrade the RNA part of a base-paired RNA–DNA hybrid, but cannot cleave the phosphodiester bond between the last ribonucleotide and the first deoxyribonucleotide. However, this ribonucleotide will carry a 5′-monophosphate rather than triphosphate and so can be removed by FEN1 (*Figure 13.19B*).

Both schemes are made attractive by the possibility that both the RNA primer and all of the DNA originally synthesized by DNA polymerase α are removed. This is because DNA polymerase α has no 3′→5′ proofreading activity (see *Table 13.2*, page 396) and therefore synthesizes DNA in a relatively error-prone manner. To prevent these errors from becoming permanent features of the daughter double helix, this region of DNA might be degraded and resynthesized by DNA polymerase δ, which does have a proofreading activity and so makes a highly accurate copy of the template. At present this possibility remains speculative.

The final difference between replication in bacteria and eukaryotes is that in eukaryotes there is no equivalent of the bacterial replisome. Instead, the enzymes and proteins involved in replication form sizeable structures within the nucleus, each containing hundreds or thousands of individual replication complexes. These structures are immobile because of attachments with the nuclear matrix, so DNA molecules are threaded through the complexes as they are replicated. The structures are referred to as **replication factories** (*Figure 13.20*) and may in fact also be features of the replication process in at least some bacteria (Lemon and Grossman, 1998; Cook, 1999).

13.2.3 Termination of replication

Replication forks proceed along linear genomes, or around circular ones, generally unimpeded except when a region that is being transcribed is encountered. DNA synthesis occurs at approximately five times the rate of RNA synthesis, so the replication complex can easily

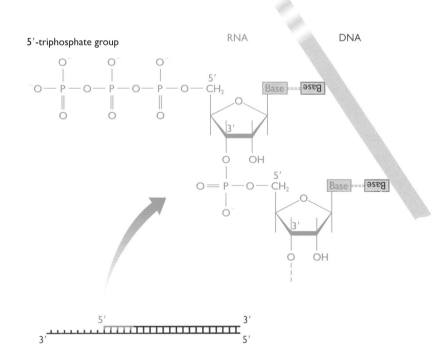

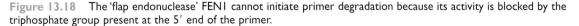

Figure 13.18 The 'flap endonuclease' FEN1 cannot initiate primer degradation because its activity is blocked by the triphosphate group present at the 5′ end of the primer.

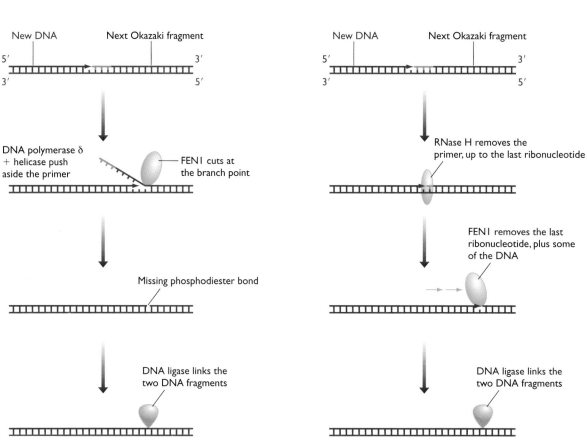

(A) The flap model

New DNA · Next Okazaki fragment

5'
3'

DNA polymerase δ
+ helicase push
aside the primer

FEN1 cuts at
the branch point

Missing phosphodiester bond

DNA ligase links the
two DNA fragments

(B) The RNase H model

New DNA · Next Okazaki fragment

5'
3'

RNase H removes the
primer, up to the last ribonucleotide

FEN1 removes the last
ribonucleotide, plus some
of the DNA

DNA ligase links the
two DNA fragments

Figure 13.19   Two models for completion of lagging strand replication in eukaryotes.

See the text for details. The new DNA (blue strand) is synthesized by DNA polymerase δ but this enzyme is not shown in order to increase the clarity of the diagrams.

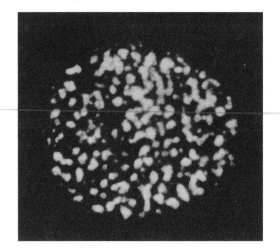

Figure 13.20   Replication factories in a eukaryotic nucleus.

Equivalent **transcription factories** are responsible for RNA synthesis. Reproduced with permission from Nakamura H et al. (1986) *Exp. Cell Res.*, **165**, 291–297, Academic Press, Inc., Orlando, FL.

overtake an RNA polymerase, but this probably does not happen: instead it is thought that the replication fork pauses behind the RNA polymerase, proceeding only when the transcript has been completed (Deshpande and Newlon, 1996).

Eventually the replication fork reaches the end of the molecule or meets a second replication fork moving in the opposite direction. What happens next is one of the least understood aspects of genome replication.

Replication of the E. coli genome terminates within a defined region

Bacterial genomes are replicated bidirectionally from a single point (see *Figure 13.7*, page 391), which means that the two replication forks should meet at a position diametrically opposite the origin of replication on the genome map. However, if one fork is delayed, possibly because it has to replicate extensive regions where transcription is occurring, then it might be possible for the other fork to overshoot the halfway point and continue replication on the 'other side' of the genome (*Figure 13.21*). It is not

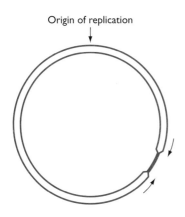

Origin of replication

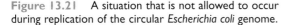

Figure 13.21 A situation that is not allowed to occur during replication of the circular *Escherichia coli* genome.

One of the replication forks has proceeded some distance past the halfway point. This does not happen during DNA replication in *E. coli* because of the action of the Tus proteins (see *Figure 13.22B*).

immediately apparent why this should be undesirable, the daughter molecules presumably being unaffected, but it is not allowed to happen because of the presence of **terminator sequences**. Seven of these have been identified in the *E. coli* genome (*Figure 13.22A*), each one acting as the recognition site for a sequence-specific DNA-binding protein called **Tus**.

The mode of action of Tus is quite unusual. When bound to a terminator sequence, a Tus protein allows a replication fork to pass if the fork is moving in one direction, but blocks progress if the fork is moving in the opposite direction around the genome. The directionality is set by the orientation of the Tus protein on the double helix. When approached from one direction, Tus blocks the passage of the DnaB helicase, which is responsible for progression of the replication fork, because the helicase is

faced with a 'wall' of β-strands which it is unable to penetrate. But when approaching from the other direction, DnaB is able to disrupt the structure of the Tus protein, probably because of the effect that unwinding of the double helix has on Tus, and so is able to pass by (*Figure 13.22B*; Kamada *et al.*, 1996).

The orientation of the termination sequences, and hence of the bound Tus proteins, in the *E. coli* genome is such that both replication forks become trapped within a relatively short region on the opposite side of the genome to the origin (see *Figure 13.22A*). This ensures that termination always occurs at or near the same position. Exactly what happens when the two replication forks meet is unknown, but the event is followed by disassembly of the replisomes, either spontaneously or in a controlled fashion. The result is two interlinked daughter molecules, which are separated by topoisomerase IV.

Little is known about termination of replication in eukaryotes

No sequences equivalent to bacterial terminators are known in eukaryotes, and proteins similar to Tus have not been identified. Quite possibly, replication forks meet at random positions and termination simply involves ligation of the ends of the new polynucleotides. We do know that the replication complexes do not break down, because these factories are permanent features of the nucleus (see *Figure 13.20*).

Rather than concentrating on the molecular events occurring when replication forks meet, attention has been focused on the difficult question of how the daughter DNA molecules produced in a eukaryotic nucleus do not become impossibly tangled up. Although DNA topoisomerases have the ability to untangle DNA molecules, it is generally assumed that tangling is kept to a minimum so that extensive breakage-and-reunion reactions, as catalyzed by topoisomerases (see *Figure 13.4*, page 389), can be avoided. Various models have been proposed to solve this problem (Falaschi, 2000). One of these (Cook, 1998;

Box 13.1: Genome replication in the archaea

We have little direct information about DNA replication in archaea, most of what we know having been deduced by searching archaeal genomes for genes and other sequences similar to the components of the replication apparatus in bacteria and/or eukaryotes. Initial attempts to locate origins of replication in archaeal genomes by searching for sequence motifs found at bacterial or eukaryotic origins were unsuccessful. Subsequently, potential origins for a range of species were identified by statistic analysis of the frequencies of the four nucleotides in different parts of each archaeal genome, the rationale being that these frequencies might be significantly different to either side of an origin, as is the case in bacteria

(Kelman, 2000). For one species, *Pyrococcus abyssi*, a potential origin identified by nucleotide frequency analysis lies in the region of the genome that is replicated first, and so may be the true origin (Myllykallio *et al.*, 2000). The sequences of most of the proteins involved in the elongation phase of archaeal genome replication, as predicted from their genes, are similar to the equivalent eukaryotic versions. In particular, archaea have proteins that appear to be homologs of the eukaryotic RFC and PCNA. The archaeal DNA polymerase is interesting because the subunit that specifies the DNA synthesis activity is similar to the equivalent subunit of the eukaryotic DNA polymerase δ, whereas the proofreading function is conferred by a protein that appears to be a homolog of subunit ε of *Escherichia coli* DNA polymerase III.

(A) Terminator sequences in the *E. coli* genome

(B) The role of Tus

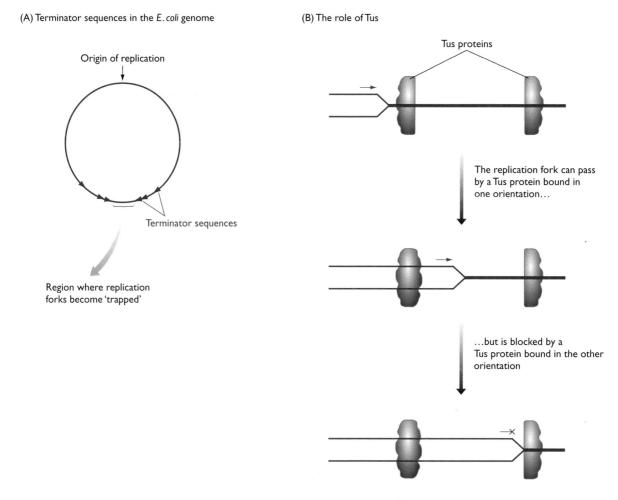

Origin of replication

Terminator sequences

Region where replication forks become 'trapped'

Tus proteins

The replication fork can pass by a Tus protein bound in one orientation…

…but is blocked by a Tus protein bound in the other orientation

Figure 13.22 The role of terminator sequences during DNA replication in *Escherichia coli*.

(A) The positions of the six terminator sequences on the *E. coli* genome are shown, with the arrowheads indicating the direction that each terminator sequence can be passed by a replication fork. (B) Bound Tus proteins allow a replication fork to pass when the fork approaches from one direction but not when it approaches from the other direction. The diagram shows a replication fork passing by the left-hand Tus, because the DnaB helicase that is moving the fork forwards can disrupt the Tus when it approaches it from this direction. The fork is then blocked by the second Tus, because this one has its impenetrable wall of β-strands facing towards the fork.

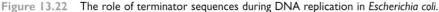

Wei *et al.*, 1998) suggests that a eukaryotic genome is not randomly packed into the nucleus, but is ordered around the replication factories, which appear to be present in only limited numbers. It is envisaged that each factory replicates a single region of the DNA, maintaining the daughter molecules in a specific arrangement that avoids their entanglement. Initially, the two daughter molecules are held together by **cohesin** proteins, which are attached immediately after passage of the replication fork by a process that appears to involve DNA polymerase κ (Takahashi and Yanagida, 2000), an enigmatic enzyme that is essential for replication but whose only known role does not obviously require a DNA polymerase activity. The cohesins maintain the alignment of the sister chromatids until the anaphase stage of nuclear division, when they are cleaved by cutting proteins, enabling the daughter chromosomes to separate (*Figure 13.23*; Murray, 1999).

13.2.4 Maintaining the ends of a linear DNA molecule

There is one final problem that we must consider before leaving the replication process. This concerns the steps that have to be taken to prevent the ends of a linear double-stranded molecule from gradually getting shorter during successive rounds of chromosomal DNA replication. There are two ways in which this shortening might occur:

■ The extreme 3′ end of the lagging strand might not be copied because the final Okazaki fragment cannot be primed, the natural position for the priming site being beyond the end of the template (*Figure 13.24A*). The absence of this Okazaki fragment means that the lagging-strand copy is shorter than it should be. If the copy remains this length then when it acts as a

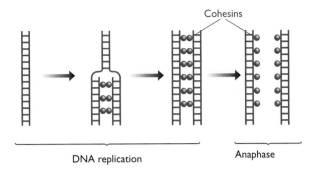

DNA replication   Anaphase

Figure 13.23   Cohesins.

Cohesin proteins attach immediately after passage of the replication fork and hold the daughter molecules together until anaphase. During anaphase, the cohesins are cleaved, enabling the replicated chromosomes to separate prior to their distribution into daughter nuclei (see *Figure 5.14*, page 137).

parental polynucleotide in the next round of replication the resulting daughter molecule will be shorter than its grandparent.

■ If the primer for the last Okazaki fragment is placed at the extreme 3' end of the lagging strand, then shortening will still occur, although to a lesser extent, because this terminal RNA primer cannot be converted into DNA by the standard processes for primer removal (*Figure 13.24B*). This is because the methods for primer removal (as illustrated in *Figure 13.17*, page 400 for bacteria and *Figure 13.19*, page 402 for eukaryotes) require extension of the 3' end of an adjacent Okazaki fragment, which cannot occur at the very end of the molecule.

Once this problem had been recognized, attention was directed at the telomeres, the unusual DNA sequences at the ends of eukaryotic chromosomes. We noted in Section 2.2.1 that telomeric DNA is made up of a type of minisatellite sequence, being comprised of multiple copies of a short repeat motif, 5'–TTAGGG–3' in most higher eukaryotes, a few hundred copies of this sequence occurring in tandem repeats at each end of every chromosome. The solution to the end-shortening problem lies with the way in which this telomeric DNA is synthesized.

Telomeric DNA is synthesized by the telomerase enzyme

Most of the telomeric DNA is copied in the normal fashion during DNA replication but this is not the only way in which it can be synthesized. To compensate for the limitations of the replication process, telomeres can be extended by an independent mechanism catalyzed by the enzyme **telomerase**. This is an unusual enzyme in that it consists of both protein and RNA. In the human enzyme the RNA component is 450 nucleotides in length and contains near its 5' end the sequence 5'–CUAACCCUAAC–3', whose central region is the reverse complement of the human telomere repeat sequence 5'–TTAGGG–3' (Feng *et al.*, 1995). This enables telomerase to extend the telomeric DNA at the 3' end of a polynucleotide by the copying mechanism shown in *Figure 13.25*, in which the telomerase RNA is used as a template for each extension step, the DNA synthesis being carried out by the protein component of the enzyme, which is a reverse transcriptase (Lingner *et al.*, 1997). The correctness of this model is indicated by comparisons between telomere repeat sequences and the telomerase RNAs of other species (*Table 13.3*): in all organisms that have been looked at, the telomerase RNA contains a sequence that enables it to make copies of the repeat motif present at the organism's telomeres. An interesting feature is that in all organisms the strand synthesized by telomerase has a preponderance of G nucleotides; it is therefore referred to as the G-rich strand.

Telomerase can only synthesize this G-rich strand. It is not clear how the other polynucleotide – the C-rich strand – is extended, but it is presumed that when the G-rich strand is long enough, the primase–DNA polymerase α complex attaches at its end and initiates synthesis of complementary DNA in the normal way (*Figure 13.26*). This requires the use of a new RNA primer, so the C-rich strand will still be shorter than the G-rich one, but the important point is that the overall length of the chromosomal DNA has not been reduced.

Telomere length is implicated in senescence and cancer

Perhaps surprisingly, telomerase is not active in all mammalian cells. The enzyme is functional in the early embryo, but after birth is active only in the reproductive and **stem cells**. The latter are progenitor cells that divide continually throughout the lifetime of an organism,

Table 13.3   Sequences of telomere repeats and telomerase RNAs in various organisms

Species	Telomere repeat sequence	Telomerase RNA template sequence
Human	5'–TTAGGG–3'	5'–CUAACCCUAAC–3'
Oxytricha	5'–TTTTGGGG–3'	5'–CAAAACCCCAAAACC–3'
Tetrahymena	5'–TTGGGG–3'	5'–CAACCCCAA–3'

Oxytricha and *Tetrahymena* are protozoans which are particularly useful for telomere studies because at certain developmental stages their chromosomes break into small fragments, all of which have telomeres: they therefore have many telomeres per cell (Greider, 1996).

(A) The final Okazaki fragment cannot be primed

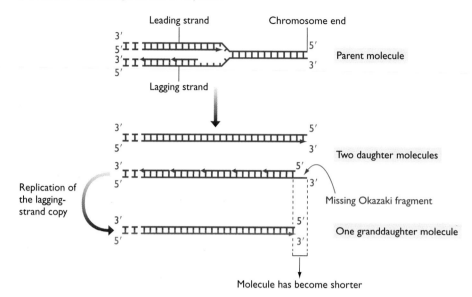

(B) The primer for the last Okazaki fragment is at the extreme 3' end of the lagging strand

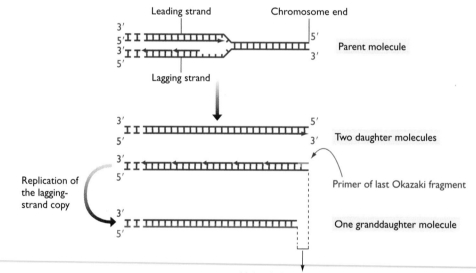

Figure 13.24   Two of the reasons why linear DNA molecules could become shorter after DNA replication.

In both examples, the parent molecule is replicated in the normal way. A complete copy is made of its leading strand, but in (A) the lagging-strand copy is incomplete because the last Okazaki fragment is not made. This is because primers for Okazaki fragments are synthesized at positions approximately 200 bp apart on the lagging strand. If one Okazaki fragment begins at a position less than 200 bp from the 3' end of the lagging strand then there will not be room for another priming site, and the remaining segment of the lagging strand is not copied. The resulting daughter molecule therefore has a 3' overhang and, when replicated, gives rise to a granddaughter molecule that is shorter than the original parent. In (B) the final Okazaki fragment can be positioned at the extreme 3' end of the lagging strand, but its RNA primer cannot be converted into DNA because this would require extension of another Okazaki fragment positioned beyond the end of the lagging strand. It is not clear if a terminal RNA primer can be retained throughout the cell cycle, nor is it clear if a retained RNA primer can be copied into DNA during a subsequent round of DNA replication. If the primer is not retained or is not copied into DNA, then one of the granddaughter molecules will be shorter than the original parent.

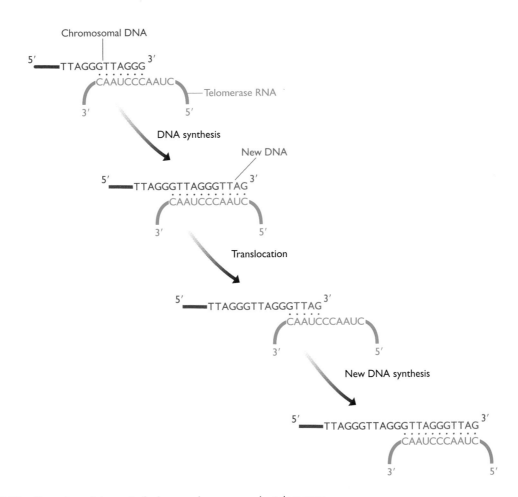

Figure 13.25   Extension of the end of a human chromosome by telomerase.

The 3′ end of a human chromosomal DNA molecule is shown. The sequence comprises repeats of the human telomere motif 5′–TTAGGG–3′. The telomerase RNA base-pairs to the end of the DNA molecule which is extended a short distance, the length of this extension possibly determined by the presence of a stem-loop structure in the telomerase RNA (Tzfati *et al.*, 2000). The telomerase RNA then translocates to a new base-pairing position slightly further along the DNA polynucleotide and the molecule is extended by a few more nucleotides. The process can be repeated until the chromosome end has been extended by a sufficient amount.

producing new cells to maintain organs and tissues in a functioning state. The best-studied examples are the hemopoietic stem cells of the bone marrow, which generate new blood cells.

Cells that lack telomerase activity undergo chromosome shortening every time they divide. Eventually, after many cell divisions, the chromosome ends could become so truncated that essential genes are lost, but this is unlikely to be a major cause of the defects that can occur in cells lacking telomerase activity. Instead, the critical factor is the need to maintain a protein 'cap' on each chromosome end, to protect these ends from the effects of the DNA repair enzymes that join together the uncapped ends that are produced by accidental breakage of a chromosome (Section 2.2.1). The proteins that form this protective cap, such as TRF2 in humans, recognize the telomere repeats as their binding

sequences, and so have no attachment points after the telomeres have been deleted. If these proteins are absent then the repair enzymes can make inappropriate linkages between the ends of intact, although shortened, chromosomes; it is this that is probably the underlying cause of the disruption to the cell cycle that results from telomere shortening.

Telomere shortening will therefore lead to the termination of a cell lineage. For several years biologists have attempted to link this process with **cell senescence**, a phenomenon originally observed in cell cultures. All normal cell cultures have a limited lifetime: after a certain number of divisions the cells enter a senescent state in which they remain alive but cannot divide (*Figure 13.27*). With some mammalian cell lines, notably fibroblast cultures (connective tissue cells), senescence can be delayed by engineering the cells so

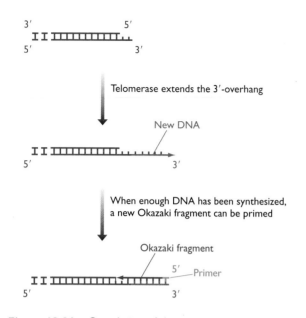

Figure 13.26 Completion of the extension process at the end of a chromosome.

It is believed that after telomerase has extended the 3' end by a sufficient amount, as shown in *Figure 13.25*, a new Okazaki fragment is primed and synthesized, converting the 3' extension into a completely double-stranded end.

that they synthesize active telomerase (Reddel, 1998). These experiments suggest a clear relationship between telomere shortening and senescence, but the exactness of the link has been questioned (Blackburn, 2000), and any extrapolation from cell senescence to aging of the organism is fraught with difficulties (Kipling and Faragher, 1999).

Not all cell lines display senescence. Cancerous cells are able to divide continuously in culture, their immortality being looked upon as analogous to tumor growth in an intact organism. With several types of cancer, this absence of senescence is associated with activation of

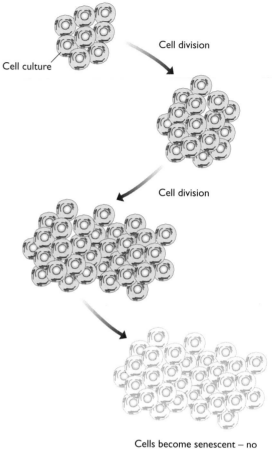

Figure 13.27 Cultured cells become senescent after multiple cell divisions.

telomerase, sometimes to the extent that telomere length is maintained through multiple cell divisions, but often in such a way that the telomeres become longer than normal because the telomerase is overac-

---

Box 13.2: Telomeres in *Drosophila*

The text describes the reverse transcriptase reaction carried out by the protein component of telomerase. When the amino acid sequences of telomerases are compared with those of other reverse transcriptases, similarities are seen with the enzymes coded by the non-LTR (long terminal repeat) retroelements called retroposons (Section 2.4.2; Eickbush, 1997). This is a fascinating observation when taken in conjunction with the unusual structure of the telomeres of *Drosophila*. These telomeres are not made up of the short repeated sequences seen in most other organisms, but instead consist of tandem arrays of much longer repeats, 6 or 10 kb in length. These repeats

are full-length copies of two typical retroposons, related to LINE-1 of humans, called *HeT-A* and *TART* (Pardue et al., 1996). It is not known how these telomeres are maintained, but it is conceivable that the process is analogous to that carried out by telomerase, with a template RNA obtained by transcription of the telomeric retroposons being copied by the reverse transcriptase coded by the *TART* sequences (*HeT-A* does not have a reverse transcriptase gene).

The unusual structure of the *Drosophila* telomere could simply be a quirk of nature, but the attractive possibility that the telomeres of other organisms are degraded retroposons, as suggested by the similarities between telomerase and retroposon reverse transcriptases, cannot be discounted.

tive. It is not clear if telomerase activation is a *cause* or an *effect* of cancer, although the former seems more likely because at least one type of cancer, dyskeratosis congenita, appears to result from a mutation in the gene specifying the RNA component of human telomerase (Marciniak and Guarente, 2001). The question is critical to understanding the etiology of the cancer but is less relevant to the therapeutic issue, which centers on whether telomerase could be a target for drugs designed to combat the cancer. Such a therapy could be successful even if telomerase activation is an effect of the cancer, because inactivation by drugs would induce senescence of the cancer cells and hence prevent their proliferation.

13.3 Regulation of Eukaryotic Genome Replication

Genome replication in eukaryotic cells is regulated at two levels:

1. Replication is coordinated with the cell cycle so that two copies of the genome are available when the cell divides.
2. The replication process itself can be arrested under certain circumstances, for example if the DNA is damaged and must be repaired before copying can be completed.

We will end this chapter by looking at these regulatory mechanisms.

13.3.1 Coordination of genome replication and cell division

The concept of a **cell cycle** emerged from light microscopy studies carried out by the early cell biologists. Their observations showed that dividing cells pass through repeated cycles of mitosis (see *Figure 5.14*, page 137) – the period when nuclear and cell division occurs – and interphase, a less dramatic period when few dynamic changes can be detected with the light microscope. It was understood that chromosomes divide during interphase, so when DNA was identified as the genetic material, interphase took on a new importance as the period when genome replication must take place. This led to a re-interpretation of the cell cycle as a four-stage process (*Figure 13.28*), comprising:

- **Mitosis** or **M phase**, the period when the nucleus and cell divide;
- **Gap 1** or **G1 phase**, an interval when transcription, translation and other general cellular activities occur;
- **Synthesis** or **S phase**, when the genome is replicated;
- **Gap 2** or **G2 phase**, a second interval period.

It is clearly important that the S and M phases are coordinated so that the genome is completely replicated, but replicated only once, before mitosis occurs. The periods

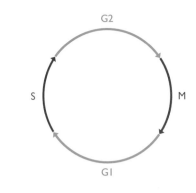

Figure 13.28   The cell cycle.

The lengths of the individual phases vary in different cells. Abbreviations: G1 and G2, gap phases; M, mitosis; S synthesis phase.

immediately before entry into S and M phases are looked upon as key **cell cycle checkpoints**, and it is at one of these two points that the cycle becomes arrested if critical genes involved in cell-cycle control are mutated, or if the cell undergoes trauma such as extensive DNA damage. Attempts to understand how genome replication is coordinated with mitosis have therefore concentrated on these two checkpoints, especially the pre-S checkpoint, the period immediately before replication.

Establishment of the pre-replication complex enables genome replication to commence

Studies primarily with *Saccharomyces cerevisiae* have led to a model for controlling the timing of S phase which postulates that genome replication requires construction of **pre-replication complexes** (**pre-RCs**) at origins of replication, these pre-RCs being converted to **post-RCs** as replication proceeds. A post-RC is unable to initiate replication and so cannot accidentally re-copy a piece of the genome before mitosis has occurred (Stillman, 1996). The ORC, the complex of six proteins that is assembled onto domains A and B1 of a yeast ARS (see *Figure 13.9B*, page 393), was an early contender for the pre-RC but is probably not a central component because ORCs are present at origins of replication at all stages of the cell cycle. Instead, the ORC is looked on as the 'landing pad' on which the pre-RC is constructed.

Various types of protein have been implicated as components of the pre-RC. The first is Cdc6p, which was originally identified in yeast and subsequently shown to have homologs in higher eukaryotes. Yeast Cdc6p is synthesized at the end of G2, as the cell enters mitosis, and becomes associated with chromatin in early G1 before disappearing at the end of G1, when replication begins (*Figure 13.29*). The involvement of Cdc6p in the pre-RC is suggested by experiments in which its gene is repressed, which results in an absence of pre-RCs, and other experiments in which Cdc6p is over-produced, which leads to multiple genome replications in the absence of mitosis.

Replication of the yeast genome

RESEARCH 13.1 BRIEFING

Microarray analysis is being used to examine the dynamics of origin firing and fork migration in replicating yeast chromosomes.

Research into genome replication is entering a new phase. Attention is still focused on the detailed biochemical events occurring at individual replication forks, but these studies are now being supplemented by new experimental strategies that have been made possible by the availability of complete genome sequences. These new strategies, which are being pioneered with the yeast *Saccharomyces cerevisiae*, are providing a global view of the pattern and dynamics of replication across entire genomes. Research during the 1990s had shown that, as in all eukaryotes, replication of individual yeast chromosomes follows a consistent pattern from one cell cycle to the next, the region around the centromere usually being one of the first parts of a chromosome to be replicated, and the telomeres being replicated towards the end of S phase. But these features of genome replication had been extrapolated from studies of just a few origins, and it was generally agreed that further progress would require a new experimental strategy that enabled the activities of many origins to be studied at the same time. The new strategy that has been devised combines an established approach to genome replication – labeling with heavy isotopes – with the latest advances in microarray technology.

Compared with other eukaryotes, yeast has a relatively small genome (see *Table 2.1*, page 31). This is a major advantage in global studies of genome activity because it means that microarrays representing all of the genome can be prepared quite easily, using a set of cDNAs or oligonucleotides specific for every one of the 6000 yeast genes. In Section 3.2.3 we saw how cDNA probing of microarrays has been used to follow the global pattern of gene expression during the yeast sporulation pathway. Can a similar approach be used to follow the pattern of origin activation during S phase?

Microarray analysis is based on hybridization probing, so to use a microarray to follow the pattern of genome replication it will be necessary to devise a way of separating unreplicated DNA from replicated DNA, so that one or the other fraction can be used as a hybridization probe. If, for example, a sample of replicated DNA could be obtained from cells that had just entered S phase, then this DNA could be used to probe the microarray in order to identify those genes that have already been replicated at this early stage. A second probe, prepared from replicated DNA from slightly later in S phase, would identify the next set of genes

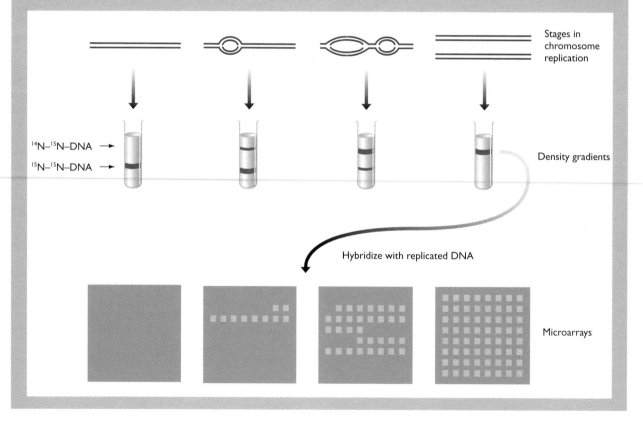

Stages in chromosome replication

$^{14}N-^{15}N-DNA \rightarrow$
$^{15}N-^{15}N-DNA \rightarrow$

Density gradients

Hybridize with replicated DNA

Microarrays

to be replicated, and so on. When data have been collected for all of S phase, the gene identities could be correlated with the genome map to chart the pattern of replication for each chromosome. But how can samples of replicated DNA be prepared? The problem has already been solved, way back in 1958 by Meselson and Stahl, who used heavy-isotope labeling to distinguish between old and new DNA during replication of the *Escherichia coli* genome (Section 13.1.1). By growing bacteria in medium containing $^{15}NH_4Cl$ they prepared cells containing ^{15}N-labeled double-stranded DNA. Transfer to normal medium, followed by one round of genome replication, gave rise to daughter cells whose DNA molecules were hybrids, comprised of one strand of ^{15}N-DNA and one strand of ^{14}N-DNA (see *Figure 13.3A*, page 387). A similar approach can be used with yeast: grow the cells in medium with heavy nitrogen, transfer to normal medium, and allow the cells to enter S phase. Extract the DNA, treat it with a restriction endonuclease, and fractionate by density gradient centrifugation. Two bands are seen. One is made up of fragments of ^{15}N-^{15}N-DNA, derived from the unreplicated component of the genome. The other band is ^{14}N-^{15}N-DNA and hence is derived from the regions that have undergone replication. Purify the ^{14}N-^{15}N-DNA, label it with a radioactive or fluorescent marker, and apply to the microarray (see figure on facing page).

Early and late origins

The analysis is simple but remarkably informative, as illustrated by the two graphs on the right. Graph A charts the dynamics of origin firing along yeast chromosome VI and identifies a region midway along the shorter arm as the area where replication of this chromosome commences. As indicated by previous experiments, the centromere (indicated by the circle on the x-axis) is replicated early in S phase, and the telomeres at the end of the cycle. Note that both telomeres are replicated at approximately the same time. This observation holds true for all yeast chromosomes, but not all chromosomes complete their replication cycle at the same time. Some, such as chromosomes XI and XV, are completely replicated early in S phase, and others, such as VIII and IX, are not fully replicated until much later. These time differences are not related to the length of the chromosome but instead indicate real variations in the kinetics of chromosome replication. Graph B emphasizes this point, showing that 15 minutes after the start of S phase, most of the origins in chromosome XV have been activated whereas replication of chromosome XII has only just begun. Microarray analysis also enables the migration rate of individual forks to be inferred, again showing variability. The mean speed is 2.9 kb per minute, but some forks move much more quickly, up to 11 kb per minute for the most active ones.

More questions

As with many novel techniques, microarray analysis of yeast genome replication answers some questions but raises many new ones. Why do some replication forks migrate much more quickly than others? The variations do not appear to be the result of chromatin structure because adjacent forks that are moving towards one another often migrate at quite different speeds. What sets the time of firing for a particular origin? Previous research has suggested that timing is not an intrinsic property of the origin itself, but instead is determined by other controlling sequences in the genome. Are all origins used in every round of genome replication? Again, previous work has suggested that some origins are used in every S phase whereas others are optional. Microarray analysis will enable all of these questions to be addressed in more detail.

Reference

Raghuraman MK, Winzeler EA, Collingwood D, et al. (2001) Replication dynamics of the yeast genome. *Science*, **294**, 115–121.

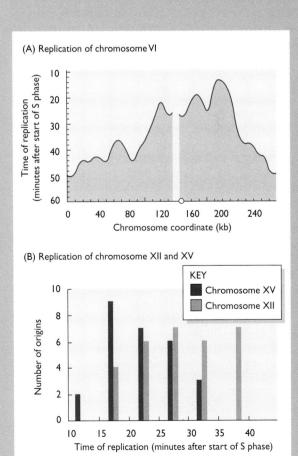

(A) Replication of chromosome VI

(B) Replication of chromosome XII and XV

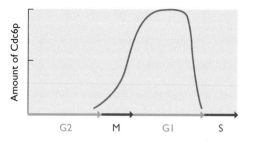

Figure 13.29 Graph showing the amount of Cdc6p in the nucleus at different stages of the cell cycle.

There is also biochemical evidence for a direct interaction between Cdc6p and yeast ORCs.

A second component of the pre-RC is thought to be the group of proteins called **replication licensing factors (RLFs)**. As with Cdc6p, the first examples of these proteins were identified in yeast (the MCM family of proteins; Tye, 1999) with homologs in higher eukaryotes discovered at a later date. RLFs become bound to chromatin towards the end of M phase and remain in place until the start of S phase, after which they are gradually removed from the DNA as it is replicated. Their removal may be the key event that converts a pre-RC into a post-RC and so prevents re-initiation of replication at an origin that has already directed a round of replication (Blow and Tada, 2000).

Regulation of pre-RC assembly

Identification of the components of the pre-RC takes us some distance towards understanding how genome replication is initiated, but still leaves open the question of how replication is coordinated with other events in the cell cycle. Cell cycle control is a complex process, mediated largely by protein kinases which phosphorylate and activate enzymes and other proteins that have specific functions during the cell cycle. The same protein kinases are present in the nucleus throughout the cell cycle, so they must themselves be subject to control. This control is exerted partly by proteins called **cyclins** (whose abundance varies at different stages of the cell cycle), partly by other protein kinases that activate the cyclin-dependent protein kinases, and partly by inhibitory proteins. Even before we start looking for regulators of pre-RC assembly we can anticipate that the control system will be convoluted.

A number of cyclins have been linked with activation of genome replication and prevention of pre-RC reassembly after replication has been completed (Stillman, 1996). These include the mitotic cyclins, whose main function was originally thought to be activation of mitosis but which also repress genome replication. When the effects of these cyclins are blocked by, for example, overproduction of proteins that inhibit their activity, the cell is not only incapable of entering M phase but also undergoes repeated genome replication. There are also

more specific S-phase cyclins, such as Clb5p and Clb6p in *S. cerevisiae*, inactivation of which delays or prevents genome replication, and other mitotic cyclins that are active during G2 phase and prevent the assembly of pre-RCs in the period after genome replication and before cell division (*Figure 13.30*).

In addition to these cyclin-dependent control systems, genome replication is also regulated by a cyclin-independent protein kinase, Cdc7p-Dbf4p, found in organisms as diverse as yeasts and mammals. The proteins activated by this kinase have not been identified, separate lines of evidence suggesting that both RLFs and ORCs are targeted. Whatever the mechanism, Cdc7p-Dbf4p activity is a prerequisite for replication, the cyclin-dependent processes on their own being insufficient to push the cell into S phase.

13.3.2 Control within S phase

Regulation of the G1–S transition can be looked upon as the major control process affecting genome regulation, but it is not the only one. The specific events occurring during S phase are also subject to regulation.

Early and late replication origins

Initiation of replication does not occur at the same time at all replication origins, nor is 'origin firing' an entirely random process. Some parts of the genome are replicated early in S phase and some later, the pattern of replication being consistent from cell division to cell division (Fangman and Brewer, 1992). The general pattern is that actively transcribed genes and the centromere are replicated early in S phase, and non-transcribed regions of the genome later on. Early-firing origins are therefore tissue specific and reflect the pattern of gene expression occurring in a particular cell.

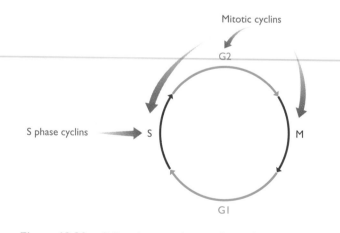

Figure 13.30 Cell-cycle control points for cyclins involved in regulation of genome replication.

See the text for details.

Understanding what determines the firing time of a replication origin is proving quite difficult. It is not simply the sequence of the origin, because transfer of a DNA segment from its normal position to another site in the same or a different chromosome can result in a change in the firing pattern of origins contained in that segment. This positional effect may be linked with chromatin organization and hence influenced by structures such as locus control regions (Section 8.1.2) that control DNA packaging. The position of the origin in the nucleus may also be important as origins that become active at similar periods within S phase appear to be clustered together, at least in mammals.

Checkpoints within S phase

The final aspect of the regulation of genome replication that we will consider is the function of the checkpoints that exist within S phase. These were first identified when it was shown that one of the responses of yeast cells to DNA damage is a slowing down and possibly a complete halting of the genome replication process (Paulovich and Hartwell, 1995). This is linked with the activation of genes whose products are involved in DNA repair (Section 14.2; Zhou and Elledge, 2000).

As with entry into S phase, cyclin-dependent kinases are implicated in the regulation of S-phase checkpoints. These kinases respond to signals from proteins associated with the replication fork. The identity of these damage-detection proteins has not yet been confirmed, although DNA polymerase ε, which has not been assigned a precise function during DNA synthesis (*Table 13.2*, page 396), is a particularly strong candidate because mutant yeast cells that have abnormal DNA polymerase ε enzymes do not respond to DNA damage in the same way as normal cells. Other replication-fork proteins, including components of the PCNA and the accessory protein RFC, have also been assigned roles in damage detection (Waga and Stillman, 1998). The signals from these proteins are mediated by kinases such as ATM, ATR, Chk1 and Chk2, which elicit the appropriate cellular response. The replication process can be arrested by repressing the firing of origins of replication that are usually activated at later stages in S phase (Santocanale and Diffley, 1998) or by slowing the progression of existing replication forks. If the damage is not excessive then DNA repair processes are activated (Section 14.2); alternatively the cell may be shunted into the pathway of programmed cell death called **apoptosis**, the death of a single somatic cell as a result of DNA damage usually being less dangerous than allowing that cell to replicate its mutated DNA and possibly give rise to a tumor or other cancerous growth. In mammals, a central player in induction of cell cycle arrest and apoptosis is the protein called p53. This is classified as a tumor-suppressor protein, because when this protein is defective, cells with damaged genomes can avoid the S-phase checkpoints and possibly proliferate into a cancer. p53 is a sequence-specific DNA-binding protein that activates a number of genes thought to be directly responsible for arrest and apoptosis, and also represses expression of others that must be switched off to facilitate these processes.

References

Aladjem MI, Rodewald LW, Kolman JL and Wahl GM (1998) Genetic dissection of a mammalian replicator in the human β-globin locus. *Science*, **281**, 1005–1009.

Bae S-H, Bae K-H, Kim J-A and Seo Y-S (2001) RPA governs endonuclease switching during processing of Okazaki fragments in eukaryotes. *Nature*, **412**, 456–461.

Beernick HTH and Morrical SW (1999) RMPs: recombination/replication mediator proteins. *Trends Biochem. Sci.*, **24**, 385–389.

Bell SP and Stillman B (1992) ATP-dependent recognition of eukaryotic origins of DNA replication by a multiprotein complex. *Nature*, **357**, 128–134.

Berger JM, Gamblin SJ, Harrison SC and Wang JC (1996) Structure and mechanism of DNA topoisomerase II. *Nature*, **379**, 225–232; **380**, 179.

Bielinsky A-K and Gerbi SA (1998) Discrete start sites for DNA synthesis in the yeast *ARS1* origin. *Science*, **279**, 95–98.

Blackburn EH (2000) Telomere states and cell fates. *Nature*, **408**, 53–56.

Blow JJ and Tada S (2000) A new check on issuing the licence. *Nature*, **404**, 560–561.

Bochkarev A, Pfuetzner RA, Edwards AM and Frappier L (1997) Structure of the single-stranded-DNA-binding domain of replication protein A bound to DNA. *Nature*, **385**, 176–181.

Cabral JHM, Jackson AP, Smith CV, Shikotra N, Maxwell A and Liddington RC (1997) Crystal structure of the breakage-reunion domain of DNA gyrase. *Nature*, **388**, 903–906.

Carpenter PB, Mueller PR and Dunphy WG (1996) Role for a *Xenopus* Orc2-related protein in controlling DNA replication. *Nature*, **379**, 357–360.

Champoux JJ (2001) DNA topoisomerases: structure, function, and mechanism. *Ann. Rev. Biochem.*, **70**, 369–413.

Cook P (1998) Duplicating a tangled genome. *Science*, **281**, 1466–1467.

Cook PR (1999) The organization of replication and transcription. *Science*, **284**, 1790–1795.

Crick FHC, Wang JC and Bauer WR (1979) Is DNA really a double helix? *J. Mol. Biol.*, **129**, 449–461.

Deshpande AM and Newlon CS (1996) DNA replication fork pause sites dependent on transcription. *Science*, **272**, 1030–1033.

Diffley JFX and Cocker JH (1992) Protein–DNA interactions at a yeast replication origin. *Nature*, **357**, 169–172.

Eickbush TH (1997) Telomerase and retrotransposons: which came first? *Science*, **277**, 911–912.

Falaschi A (2000) Eukaryotic DNA replication: a model for a foxed double replisome. *Trends Genet.*, **16**, 88–92.

Fangman WL and Brewer BJ (1992) A question of time – replication origins of eukaryotic chromosomes. *Cell*, **71**, 363–366.

Feng J, Funk WD, Wang S-S, et al. (1995) The RNA component of human telomerase. *Science*, **269**, 1236–1241.

Frick DN and Richardson CC (2001) DNA primases. *Ann. Rev. Biochem.*, **70**, 39–80.

Gilbert DM (2001) Making sense of eukaryotic replication origins. *Science*, **294**, 96–100.

Greider CW (1996) Telomere length regulation. *Ann. Rev. Biochem.*, **65**, 337–365.

Holmes FL (1998) The DNA replication problem, 1953–1958. *Trends Biochem. Sci.*, **23**, 117–120.

Hübscher U, Nasheuer H-P and Syväoja JE (2000) Eukaryotic DNA polymerases: a growing family. *Trends Biochem. Sci.*, **25**, 143–147.

Kamada K, Horiuchi T, Ohsumi K, Shimamoto N and Morikawa K (1996) Structure of a replicator-terminator protein complexed with DNA. *Nature*, **383**, 598–603.

Kelman Z (2000) The replication origin of archaea is finally revealed. *Trends Biochem. Sci.*, **25**, 521–523.

Kipling D and Faragher RGA (1999) Ageing hard or hardly ageing? *Nature*, **398**, 191–193.

Lemon KP and Grossman AD (1998) Localization of bacterial DNA polymerase: evidence for a factory model of replication. *Science*, **282**, 1516–1519.

Lima CD, Wang JC and Mondragón A (1994) Three-dimensional structure of the 67K N-terminal fragment of *E. coli* DNA topoisomerase I. *Nature*, **367**, 138–146.

Lingner J, Hughes TR, Shevchenko A, Mann M, Lundblad V and Cech TR (1997) Reverse transcriptase motifs in the catalytic subunit of telomerase. *Science*, **276**, 561–567.

Marciniak R and Guarente L (2001) Testing telomerase. *Nature*, **413**, 370–373.

Meselson M and Stahl F (1958) The replication of DNA in *Escherichia coli. Proc. Natl Acad. Sci. USA*, **44**, 671–682.

Murray A (1999) A snip separates sisters. *Nature*, **400**, 19–21.

Myllykallio H, Lopez P, López-Garcia P, et al. (2000) Bacterial mode of replication with eukaryotic-like machinery in a hyperthermophilic archaeon. *Science*, **288**, 2212–2215.

Novick RP (1998) Contrasting lifestyles of rolling-circle phages and plasmids. *Trends Biochem. Sci.*, **23**, 434–438.

Okazaki T and Okazaki R (1969) Mechanisms of DNA chain growth. *Proc. Natl Acad. Sci. USA*, **64**, 1242–1248.

Pardue ML, Danilevskaya ON, Lowenhaupt K, Slot F and Traverse KL (1996) *Drosophila* telomeres: new views on chromosome evolution. *Trends Genet.*, **12**, 48–52.

Paulovich AG and Hartwell LH (1995) A checkpoint regulates the rate of progression through S phase in *S. cerevisiae* in response to DNA damage. *Cell*, **82**, 841–847.

Reddel RR (1998) A reassessment of the telomere hypothesis of senescence. *BioEssays*, **20**, 977–984.

Redinbo MR, Stewart L, Kuhn P, Champoux JJ and Hol WGJ (1998) Crystal structures of human topoisomerase I in covalent and noncovalent complexes with DNA. *Science*, **279**, 1504–1513.

Rodley GA, Scobie RS, Bates RHT and Lewitt RM (1976) A possible conformation for double-stranded polynucleotides. *Proc. Natl Acad. Sci. USA*, **73**, 2959–2963.

Santocanale C and Diffley JFX (1998) A Mec1- and Rad53-dependent checkpoint controls late-firing origins of DNA replication. *Nature*, **395**, 615–618.

Stewart L, Redinbo MR, Qiu X, Hol WGJ and Champoux JJ (1998) A model for the mechanism of human topoisomerase I. *Science*, **279**, 1534–1541.

Stillman B (1996) Cell cycle control of DNA replication. *Science*, **274**, 1659–1664.

Takahashi K and Yanagida M (2000) Replication meets cohesion. *Science*, **289**, 735–736.

Tye BK (1999) MCM proteins in DNA replication. *Ann. Rev. Biochem.*, **68**, 649–686.

Tzfati, Y, Fulton TB, Roy J and Blackburn EH (2000) Template boundary in a yeast telomerase specified by RNA structure. *Science*, **288**, 863–867.

van Brabant AJ, Stan R and Ellis NA (2000) DNA helicases, genome instability, and human genetic disease. *Ann. Rev. Genomics Hum. Genet.*, **1**, 409–459.

Waga S and Stillman B (1998) The DNA replication fork in eukaryotic cells. *Ann. Rev. Biochem.*, **67**, 721–751.

Watson JD and Crick FHC (1953a) Molecular structure of nucleic acids: a structure for deoxyribose nucleic acid. *Nature*, **171**, 737–738.

Watson JD and Crick FHC (1953b) Genetical implications of the structure of deoxyribonucleic acid. *Nature*, **171**, 964–967.

Wei X, Samarabandu J, Devdhar RS, Siegel AJ, Acharya R and Berezney R (1998) Segregation of transcription and replication sites into higher order domains. *Science*, **281**, 1502–1505.

Zhou B-BS and Elledge SJ (2000) The DNA damage response: putting checkpoints in perspective. *Nature*, **408**, 433–439.

Further Reading

Adams RLP (1991) *DNA Replication: In Focus.* IRL Press, Oxford. — *A concise introduction to the subject.*

Benkovic SJ, Valentine AM and Salinas F (2001) Replisome-mediated DNA replication. *Ann. Rev. Biochem.*, **70**, 181–208. — *Details of events at the bacterial replication fork.*

Collins K (1999) Ciliate telomerase biochemistry. *Ann. Rev. Biochem.*, **68**, 187–218. — *Details of telomerase action in ciliates, which are used as model organisms in telomere studies.*

DePamphilis ML (ed.) (1996) *DNA Replication in Eukaryotic Cells.* Cold Spring Harbor Laboratory Press, Cold Spring Harbor, NY. — *A more advanced and detailed treatment.*

Judson HF (1979) *The Eighth Day of Creation: Makers of the Revolution in Biology.* Penguin Books, London. — *Includes an account of the topological problems and the Meselson–Stahl experiment.*

Kelly TJ and Brown GW (2000) Regulation of chromosome replication. *Ann. Rev. Biochem.*, **69**, 829–880.

Kornberg A (1989) *For the Love of Enzymes: The Odyssey of a Biochemist.* Harvard University Press, Boston. — *A fascinating autobiography by the discoverer of DNA polymerase.*

McEachern MJ, Krauskopf A and Blackburn EH (2000) Telomeres and their control. *Ann. Rev. Genet.*, **34**, 331–358. — *Describes the processes involved in regulation of telomere length.*

Nasmyth K, Peters J-M and Uhlmann F (2000) Splitting the chromosome: cutting the ties that bind sister chromatids. *Science*, **288**, 1379–1384. — *Reviews our current understanding of the role of cohesins during the final stages of DNA replication.*

Patel SS and Picha KM (2000) Structure and function of hexameric helicases. *Ann. Rev. Biochem.*, **69**, 651–697. — *A detailed review.*

Soultanas P and Wigley DB (2001) Unwinding the 'Gordian knot' of helicase action. *Trends Biochem. Sci.*, **26**, 47–54 — *Recent developments in the study of helicase action.*

Wang JC (1996) DNA topoisomerases. *Ann. Rev. Biochem.*, **65**, 635–692.

STUDY AIDS FOR CHAPTER 13

Key terms

Give short definitions of the following terms:

γ complex
Apoptosis
Autonomously replicating
  sequence (ARS)
Cell cycle
Cell-cycle checkpoint
Cell senescence
Cohesin
Conservative replication
Cyclin
Dispersive replication
Displacement replication
DNA ligase
DNA polymerase α
DNA polymerase δ
DNA polymerase γ
DNA polymerase I
DNA polymerase II
DNA polymerase III
DNA topoisomerase
FEN1
G1 phase
G2 phase
Helicase
Initiation region
Lagging strand
Leading strand
M phase
Meselson–Stahl experiment
Okazaki fragment
Origin of replication
Origin recognition complex
  (ORC)
Origin recognition sequence

Paranemic
Plectonemic
Post-replication complex
  (post-RC)
Prepriming complex
Pre-replication complex
  (pre-RC)
Primase
Primer
Primosome
Proliferating cell nuclear
  antigen (PCNA)
Proofreading
Replication factor C (RFC)
Replication factory
Replication licensing factor
  (RLF)
Replication mediator protein
  (RMP)
Replication protein A (RPA)
Replisome
Rolling circle replication
S phase
Semiconservative replication
Single-strand binding protein
  (SSB)
Stem cell
Telomerase
Terminator sequence
Transcription factory
Transition
Transversion
Tus

Self study questions

1. Distinguish between the terms 'dispersive', 'semiconservative' and 'conservative', as applied to DNA replication.
2. Draw a fully annotated diagram illustrating the Meselson–Stahl experiment. What conclusions can be drawn from the results of this experiment?
3. Explain why the discovery of DNA topoisomerases was an important step in the development of knowledge about DNA replication.
4. With the aid of diagrams, indicate how displacement replication and rolling circle replication differ from the semiconservative process.

5. Give a detailed description of the structure of the *Escherichia coli* origin of replication and outline the role of each component of the origin in the initiation of replication.
6. Compare and contrast the *Escherichia coli* origin of replication with those of yeast and mammals.
7. What impact does the inability of DNA polymerases to synthesize DNA in the $3' \rightarrow 5'$ direction have on DNA replication?
8. Explain why DNA replication must be primed, and describe how the priming problem is solved by *Escherichia coli* and by eukaryotes.
9. Construct a table listing the DNA polymerases involved in DNA replication in *Escherichia coli* and in eukaryotes, and summarize the function and key features of each enzyme.
10. Give a detailed description of the events occurring at the replication fork in *Escherichia coli*.
11. In what ways do events at the eukaryotic replication fork differ from those occurring in *Escherichia coli*?
12. Outline how replication terminates in *Escherichia coli*. What is currently known about the termination of replication in eukaryotes?
13. Explain why the ends of a chromosomal DNA molecule could become shortened after repeated rounds of DNA replication, and show how telomerase prevents this from occurring.
14. Discuss the links between telomeres, cell senescence and cancer.
15. Draw a diagram of the eukaryotic cell cycle and indicate the periods that act as cell-cycle checkpoints.
16. Outline our current knowledge concerning the composition of pre-replication complexes, the factors that influence their assembly, and the events that lead to their conversion into post-replication complexes.
17. Describe how genome replication is regulated during S phase.

Problem-based learning

1. Evaluate the status of current research into mammalian replication origins.
2. Why is inactivation of the *Escherichia coli* polA gene, coding for DNA polymerase I, not lethal?
3. Write an extended report on 'DNA helicases'.
4. Our current knowledge of genome replication in eukaryotes is biased towards the events occurring at the replication fork. The next challenge is to convert this DNA-centered description of replication into a model that describes how replication is organized within the nucleus, addressing issues such as the role of replication factories and the processes used to avoid tangling of the daughter molecules. Devise a research plan to address one or more of these issues.
5. Explore the links between telomeres, cell senescence and cancer.

14

Mutation, Repair and Recombination

Chapter Contents

14.1 Mutations   420

    14.1.1 The causes of mutations   420

    14.1.2 The effects of mutations   428

    14.1.3 Hypermutation and the possibility
    of programmed mutations   433

14.2 DNA Repair   434

    14.2.1 Direct repair systems fill in
    nicks and correct some types of
    nucleotide modification   434

    14.2.2 Excision repair   437

    14.2.3 Mismatch repair: correcting errors
    of replication   440

    14.2.4 Repair of double-stranded DNA
    breaks   441

    14.2.5 Bypassing DNA damage during
    genome replication   442

    14.2.6 Defects in DNA repair underlie
    human diseases, including
    cancers   444

14.3 Recombination   444

    14.3.1 Homologous recombination   444

    14.3.2 Site-specific recombination   448

    14.3.3 Transposition   450

Learning outcomes

When you have read Chapter 14, you should be able to:

■ Distinguish between the terms 'mutation' and 'recombination', and define the various terms that are used to identify different types of mutation

■ Describe, with specific examples, how mutations are caused by spontaneous errors in replication and by chemical and physical mutagens

■ Recount, with specific examples, the effects of mutations on genomes and organisms

■ Discuss the biological significance of hypermutation and programmed mutations

■ Distinguish between the various types of DNA repair mechanism, and give detailed descriptions of the molecular events occurring during each type of repair

■ Outline the link between DNA repair and human disease

■ Draw diagrams, with detailed annotation, illustrating the processes of homologous recombination, gene conversion, site-specific recombination, conservative and replicative transposition, and retrotransposition, and discuss the biological significance of each of these mechanisms

GENOMES ARE DYNAMIC entities that change over time as a result of the cumulative effects of small-scale sequence alterations caused by **mutation** and larger scale rearrangements arising from **recombination**. Mutation and recombination can both be defined as *processes that result in changes to a genome*, but they are unrelated and we must make a clear distinction between them:

■ A **mutation** (Section 14.1) is a change in the nucleotide sequence of a short region of a genome (*Figure 14.1A*). Many mutations are **point mutations** that replace one nucleotide with another; others involve **insertion** or **deletion** of one or a few nucleotides. Mutations result either from errors in DNA replication or from the damaging effects of **mutagens**, such as chemicals and radiation, which react with DNA and change the structures of individual nucleotides. All cells possess **DNA-repair** enzymes that attempt to minimize the number of mutations that occur (Section 14.2). These enzymes work in two ways. Some are pre-replicative and search the DNA for nucleotides with unusual structures, these being replaced before replication occurs; others are post-replicative and check newly synthesized DNA for errors, correcting any errors that they find (*Figure 14.1B*). A possible definition of mutation is therefore *a deficiency in DNA repair*.

■ **Recombination** (Section 14.3) results in a restructuring of part of a genome, for example by exchange of segments of homologous chromosomes during meiosis or by transposition of a mobile element from one position to another within a chromosome or between chromosomes (*Figure 14.1C*). Various other events that we have studied, including mating-type switching in yeast (see *Figure 12.13*, page 361) and construction of immunoglobulin genes (see *Figure 12.15*, page 363), are also the results of recombination. Recombination is a cellular process which, like other cellular processes involving DNA (e.g. transcription and replication), is carried out and regulated by enzymes and other proteins.

Both mutation and recombination can have dramatic effects on the cell in which they occur. A mutation in a key gene may cause the cell to die if the protein coded by the mutant gene is defective (Section 14.1.2), and some recombination events lead to defining changes in the biochemical capabilities of the cell, for example by determining the mating type of a yeast cell or the immunological properties of a mammalian B or T lymphocyte. Other mutation and recombination events have a less significant impact on the phenotype of the cell and many have none at all. As we will see in Chapter 15, all events that are not lethal have the potential to contribute to the evolution of the genome but for this to happen they must be inherited when the organism reproduces. With a single-celled organism such as a bacterium or yeast, all genome alterations that are not lethal or reversible are inherited by

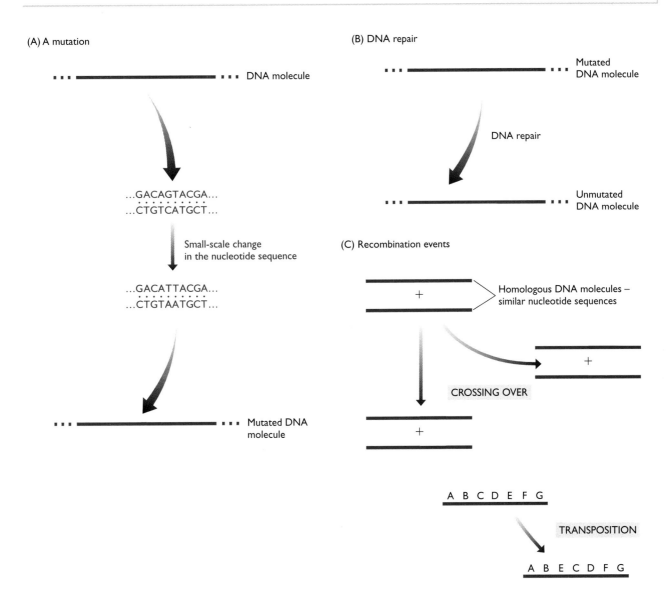

(A) A mutation

DNA molecule

...GACAGTACGA...
...CTGTCATGCT...

Small-scale change
in the nucleotide sequence

...GACATTACGA...
...CTGTAATGCT...

Mutated DNA
molecule

(B) DNA repair

Mutated
DNA molecule

DNA repair

Unmutated
DNA molecule

(C) Recombination events

+

Homologous DNA molecules –
similar nucleotide sequences

+

CROSSING OVER

+

A B C D E F G

TRANSPOSITION

A B E C D F G

Figure 14.1 Mutation, repair and recombination.

(A) A mutation is a small-scale change in the nucleotide sequence of a DNA molecule. A point mutation is shown but there are several other types of mutation, as described in the text. (B) DNA repair corrects mutations that arise as errors in replication and as a result of mutagenic activity. (C) Recombination events include exchange of segments of DNA molecules, as occurs during meiosis (see *Figure 5.15*, page 138), and the movement of a segment from one position in a DNA molecule to another, for example by transposition (Section 14.3.3).

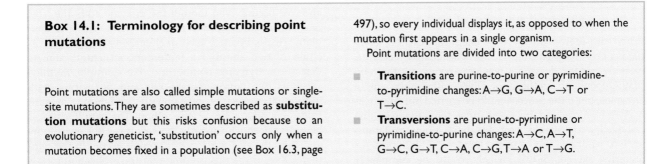

Box 14.1: Terminology for describing point mutations

Point mutations are also called simple mutations or single-site mutations. They are sometimes described as **substitution mutations** but this risks confusion because to an evolutionary geneticist, 'substitution' occurs only when a mutation becomes fixed in a population (see Box 16.3, page

497), so every individual displays it, as opposed to when the mutation first appears in a single organism.

Point mutations are divided into two categories:

- **Transitions** are purine-to-purine or pyrimidine-to-pyrimidine changes: A→G, G→A, C→T or T→C.
- **Transversions** are purine-to-pyrimidine or pyrimidine-to-purine changes: A→C, A→T, G→C, G→T, C→A, C→G, T→A or T→G.

daughter cells and become permanent features of the lineage that descends from the original cell in which the alteration occurred. In a multicellular organism, only those events that occur in germ cells are relevant to genome evolution. Changes to the genomes of somatic cells are unimportant in an evolutionary sense, but they will have biological relevance if they result in a deleterious phenotype that affects the health of the organism.

14.1 Mutations

With mutations, the issues that we have to consider are: how they arise; the effects they have on the genome and on the organism in which the genome resides; whether it is possible for a cell to increase its mutation rate and induce programmed mutations under certain circumstances; and how mutations are repaired.

14.1.1 The causes of mutations

Mutations arise in two ways:

■ Some mutations are **spontaneous** errors in replication that evade the proofreading function of the DNA polymerases that synthesize new polynucleotides at the replication fork (Section 13.2.2). These mutations are called **mismatches** because they are positions where the nucleotide that is inserted into the daughter polynucleotide does not match, by base-pairing, the nucleotide at the corresponding position in the template DNA (*Figure 14.2A*). If the mismatch is retained in the daughter double helix then *one* of the granddaughter molecules produced during the next round of DNA replication will carry a permanent double-stranded version of the mutation.

■ Other mutations arise because a mutagen has reacted with the parent DNA, causing a structural change that affects the base-pairing capability of the altered nucleotide. Usually this alteration affects only one strand of the parent double helix, so only one of the daughter molecules carries the mutation, but two of the granddaughter molecules produced during the next round of replication will have it (*Figure 14.2B*).

Errors in replication are a source of point mutations

When considered purely as a chemical reaction, complementary base-pairing is not particularly accurate. Nobody has yet devised a way of carrying out the template-dependent synthesis of DNA without the aid of enzymes, but if the process could be carried out simply as a chemical reaction in a test tube then the resulting polynucleotide would probably have point mutations at 5–10 positions out of every hundred. This represents an error rate of 5–10%, which would be completely unacceptable during genome replication. The template-

dependent DNA polymerases that carry out DNA replication must therefore increase the accuracy of the process by several orders of magnitude. This improvement is brought about in two ways:

■ The DNA polymerase operates a nucleotide selection process that dramatically increases the accuracy of template-dependent DNA synthesis (*Figure 14.3A*). This selection process probably acts at three different stages during the polymerization reaction, discrimination against an incorrect nucleotide occurring when the nucleotide is first bound to the DNA polymerase, when it is shifted to the active site of the enzyme, and when it is attached to the 3' end of the polynucleotide that is being synthesized.

■ The accuracy of DNA synthesis is increased still further if the DNA polymerase possesses a 3'→5' exonuclease activity and so is able to remove an incorrect nucleotide that evades the base selection process and becomes attached to the 3' end of the new polynucleotide (see *Figure 4.7B*, page 101). This is called **proofreading** (Section 13.2.2), but the name is a misnomer because the process is not an active checking mechanism. Instead, each step in the synthesis of a polynucleotide should be viewed as a competition between the polymerase and exonuclease functions of the enzyme, the polymerase usually winning because it is more active than the exonuclease, at least when the 3'-terminal nucleotide is base-paired to the template. But the polymerase activity is less efficient if the terminal nucleotide is not base-paired, the resulting pause in polymerization allowing the exonuclease activity to predominate so that the incorrect nucleotide is removed (see *Figure 14.3B*).

Escherichia coli is able to synthesize DNA with an error rate of only 1 per 10^7 nucleotide additions. Interestingly, these errors are not evenly distributed between the two daughter molecules, the product of lagging-strand replication being prone to about 20 times as many errors as the leading-strand replicate. This asymmetry might indicate that DNA polymerase I, which is involved only in lagging-strand replication (Section 13.2.2), has a less effective base selection and proofreading capability compared with DNA polymerase III, the main replicating enzyme (Francino and Ochman, 1997).

Not all of the errors that occur during DNA synthesis can be blamed on the polymerase enzymes: sometimes an error occurs even though the enzyme adds the 'correct' nucleotide, the one that base-pairs with the template. This is because each nucleotide base can occur as either of two alternative **tautomers**, structural isomers that are in dynamic equilibrium. For example, thymine exists as two tautomers, the *keto* and *enol* forms, with individual molecules occasionally undergoing a shift from one tautomer to the other. The equilibrium is biased very much towards the *keto* form but every now and then the *enol* version of thymine occurs in the template DNA at the precise time that the replication fork is moving past. This will lead to an 'error', because *enol-*

(A) An error in replication

(B) One possible effect of a mutagen

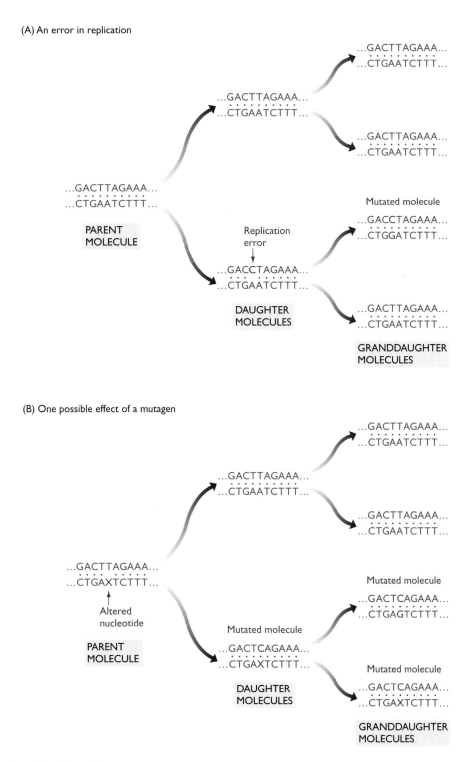

Figure 14.2 Examples of mutations.

(A) An error in replication leads to a mismatch in one of the daughter double helices, in this case a T-to-C change because one of the As in the template DNA was miscopied. When the mismatched molecule is itself replicated it gives one double helix with the correct sequence and one with a mutated sequence. (B) A mutagen has altered the structure of an A in the lower strand of the parent molecule, giving nucleotide X, which does not base-pair with the T in the other strand so, in effect, a mismatch has been created. When the parent molecule is replicated, X base-pairs with C, giving a mutated daughter molecule. When this daughter molecule is replicated, both granddaughters inherit the mutation.

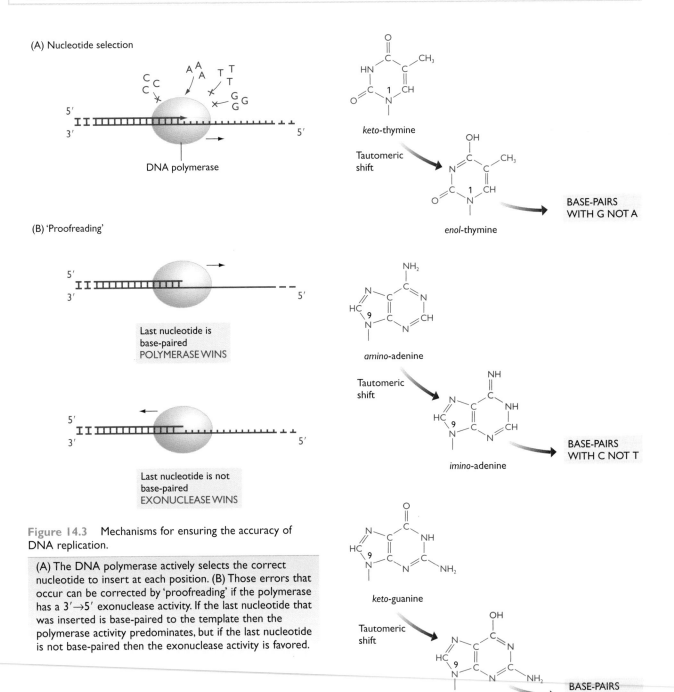

Figure 14.3 Mechanisms for ensuring the accuracy of DNA replication.

(A) The DNA polymerase actively selects the correct nucleotide to insert at each position. (B) Those errors that occur can be corrected by 'proofreading' if the polymerase has a 3'→5' exonuclease activity. If the last nucleotide that was inserted is base-paired to the template then the polymerase activity predominates, but if the last nucleotide is not base-paired then the exonuclease activity is favored.

Figure 14.4 The effects of tautomerism on base-pairing.

In each of these three examples, the two tautomeric forms of the base have different pairing properties. Cytosine also has *amino* and *imino* tautomers but both pair with guanine.

thymine base-pairs with G rather than A (*Figure 14.4*). The same problem can occur with adenine, the rare *imino* tautomer of this base preferentially forming a pair with C, and with guanine, *enol*-guanine pairing with thymine. After replication, the rare tautomer will inevitably revert to its more common form, leading to a mismatch in the daughter double helix.

As stated above, the error rate for DNA synthesis in *E. coli* is 1 in 10⁷. The overall error rate for replication of the *E. coli* genome is only 1 in 10^{10} to 1 in 10^{11}, the improvement compared with the polymerase error rate being the result of the mismatch repair system (Section 14.2.3) that scans newly replicated DNA for positions where the bases are unpaired and hence corrects the few mistakes that the replication enzymes make. The implication is that only one uncorrected replication error occurs every 1000 times that the *E. coli* genome is copied.

Replication errors can also lead to insertion and deletion mutations

Not all errors in replication are point mutations. Aberrant replication can also result in small numbers of extra nucleotides being inserted into the polynucleotide being synthesized, or some nucleotides in the template not being copied. Insertions and deletions are often called **frameshift** mutations because when one occurs within a coding region it can result in a shift in the reading frame used for translation of the protein specified by the gene (see *Figure 14.12*, page 430). However, it is inaccurate to use 'frameshift' to describe all insertions and deletions because they can occur anywhere, not just in genes, and not all insertions or deletions in coding regions result in frameshifts: an insertion or deletion of three nucleotides, or multiples of three, simply adds or removes codons or parts of adjacent codons without affecting the reading frame.

Insertion and deletion mutations can affect all parts of the genome but are particularly prevalent when the template DNA contains short repeated sequences, such as those found in microsatellites (Section 2.4.1). This is because repeated sequences can induce **replication slippage**, in which the template strand and its copy shift their relative positions so that part of the template is either copied twice or missed out. The result is that the new

polynucleotide has a larger or smaller number, respectively, of the repeat units (*Figure 14.5*). This is the main reason why microsatellite sequences are so variable, replication slippage occasionally generating a new length variant, adding to the collection of alleles already present in the population.

Replication slippage is probably also responsible for the **trinucleotide repeat expansion diseases** that have been discovered in humans in recent years (Ashley and Warren, 1995). Each of these neurodegenerative diseases is caused by a relatively short series of trinucleotide repeats becoming elongated to two or more times its normal length. For example, the human *HD* gene contains the sequence 5'–CAG–3' repeated between 6 and 35 times in tandem, coding for a series of glutamines in the protein product. In Huntington's disease this repeat expands to a copy number of 36–121, increasing the length of the polyglutamine tract and resulting in a dysfunctional protein (Perutz, 1999). Several other human diseases are also caused by expansions of polyglutamine codons (*Table 14.1*). Some diseases associated with mental retardation result from trinucleotide expansions in the leader region of a gene, giving a **fragile site**, a position where the chromosome is likely to break (Sutherland *et al.*, 1998). Expansions involving intron and trailer regions are also known.

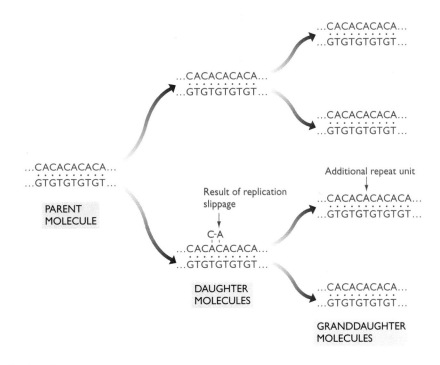

Figure 14.5   Replication slippage.

The diagram shows replication of a five-unit CA repeat microsatellite. Slippage has occurred during replication of the parent molecule, inserting an additional repeat unit into the newly synthesized polynucleotide of one of the daughter molecules. When this daughter molecule replicates it gives a granddaughter molecule whose microsatellite array is one unit longer than that of the original parent.

Table 14.1   Examples of human trinucleotide repeat expansions

Locus	Repeat sequence		Associated disease
	Normal	**Mutated**	
Polyglutamine expansions (all in coding regions of genes)			
HD	$(CAG)_{6-35}$	$(CAG)_{36-121}$	Huntington's disease
AR	$(CAG)_{9-36}$	$(CAG)_{38-62}$	Spinal and bulbar muscular atrophy
DRPLA	$(CAG)_{6-35}$	$(CAG)_{49-88}$	Dentatoribral-pallidoluysian atrophy
SCA1	$(CAG)_{6-44}$	$(CAG)_{39-82}$	Spinocerebellar ataxia type 1
SCA3	$(CAG)_{12-40}$	$(CAG)_{55-84}$	Machado–Joseph disease
Fragile site expansions (both in the untranslated leader regions of genes)			
FRM1	$(CGG)_{6-53}$	$(CGG)_{60-over\ 230}$	Fragile X syndrome
FRM2	$(GCC)_{6-35}$	$(GCC)_{61-over\ 200}$	Fragile XE mental retardation
Other expansions (positions described below)			
DMPK	$(CTG)_{5-37}$	$(CTG)_{50-3000}$	Myotonic dystrophy
X25	$(GAA)_{7-34}$	$(GAA)_{34-over\ 200}$	Friedreich's ataxia

For more details see Cummings and Zoghbi (2000). The DMPK and X25 expansions are in the trailer and intron regions of their genes, respectively, and are thought to affect RNA processing (Campuzano et al., 1996; Cummings and Zoghbi, 2000). There are also a few disease-causing mutations that involve expansions of longer sequences, such as progressive myoclonus epilepsy caused by a $(CCCCGCCCCGCG)_{2-3}$ to $(CCCCGCCCCGCG)_{over\ 12}$ expansion in the promoter region of the EPM1 locus (Mandel, 1997).

How triplet expansions are generated is not precisely understood. The size of the insertion is much greater than occurs with normal replication slippage, such as that seen with microsatellite sequences, and once the expansion reaches a certain length it appears to become susceptible to further expansion in subsequent rounds of replication, so that the disease becomes increasingly severe in succeeding generations. The possibility that expansion involves formation of hairpin loops in the DNA has been raised, based on the observation that only a limited number of trinucleotide sequences are known to undergo expansion, and all of these sequences are GC-rich and so might form stable secondary structures. There is also evidence that at least one triplet expansion region – for Friedreich's ataxia – can form a triple helix structure (Gacy et al., 1998). Studies of similar triplet expansions in yeast have shown that these are more prevalent when the RAD27 gene is inactivated (Freudenreich et al., 1998), an interesting observation as RAD27 is the yeast version of the mammalian gene for FEN1, the protein involved in processing of Okazaki fragments (Section 13.2.2). This might indicate that a trinucleotide repeat expansion is caused by an aberration in lagging-strand synthesis.

Mutations are also caused by chemical and physical mutagens

Many chemicals that occur naturally in the environment have mutagenic properties and these have been supplemented in recent years with other chemical mutagens that result from human industrial activity. Physical agents such as radiation are also mutagenic. Most organisms are exposed to greater or lesser amounts of these various mutagens, their genomes suffering damage as a result.

The definition of the term 'mutagen' is *a chemical or physical agent that causes mutations*. This definition is important because it distinguishes mutagens from other types of environmental agent that cause damage to cells in ways other than by causing mutations (*Table 14.2*). There are overlaps between these categories (for example, some mutagens are also carcinogens) but each type of agent has a distinct biological effect. The definition of 'mutagen' also makes a distinction between true mutagens and other agents that damage DNA without causing mutations, for example by causing breaks in DNA molecules. This type of damage may block replication and cause the cell to die, but it is not a mutation in the strict sense of the term and the causative agents are therefore not mutagens.

Mutagens cause mutations in three different ways:

■ Some act as **base analogs** and are mistakenly used as substrates when new DNA is synthesized at the replication fork.

■ Some react directly with DNA, causing structural changes that lead to miscopying of the template strand when the DNA is replicated. These structural

Table 14.2   Categories of environmental agent that cause damage to living cells

Agent	Effect on living cells
Carcinogen	Causes cancer – the neoplastic transformation of eukaryotic cells
Clastogen	Causes fragmentation of chromosomes
Mutagen	Causes mutations
Oncogen	Induces tumor formation
Teratogen	Results in developmental abnormalities

changes are diverse, as we will see when we look at individual mutagens.

■ Some mutagens act indirectly on DNA. They do not themselves affect DNA structure, but instead cause the cell to synthesize chemicals such as peroxides that have a direct mutagenic effect.

The range of mutagens is so vast that it is difficult to devise an all-embracing classification. We will therefore restrict our study to the most common types. For chemical mutagens these are as follows:

■ *Base analogs* are purine and pyrimidine bases that are similar enough to the standard bases to be incor-

porated into nucleotides when these are synthesized by the cell. The resulting unusual nucleotides can then be used as substrates for DNA synthesis during genome replication. For example, **5-bromouracil** (**5-bU**; *Figure 14.6A*) has the same base-pairing properties as thymine, and nucleotides containing this base can be added to the daughter polynucleotide at positions opposite As in the template. The mutagenic effect arises because the equilibrium between the two tautomers of 5-bU is shifted more towards the rarer *enol* form than is the case with thymine. This means that during the next round of replication there is a relatively high chance of the polymerase encountering

(A) 5-Bromouracil

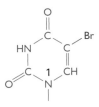

(B) Base-pairing with 5-bromouracil

(C) The mutagenic effect of 5-bromouracil

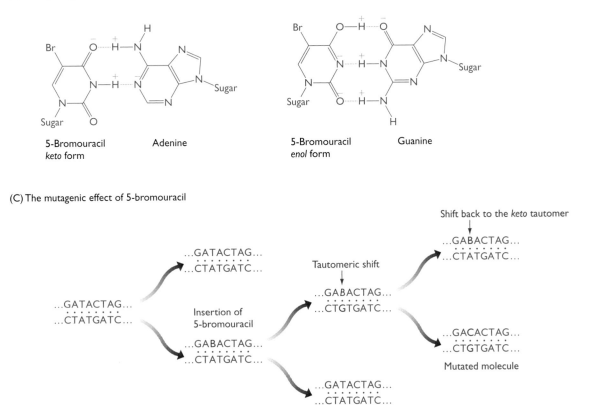

Figure 14.6   5-Bromouracil and its mutagenic effect.

See the text for details.

enol-5bU, which (like *enol*-thymine) pairs with G rather than A (*Figure 14.6B*). This results in a point mutation (*Figure 14.6C*). **2-Aminopurine** acts in a similar way: it is an analog of adenine with an *amino*-tautomer that pairs with thymine and an *imino*-tautomer that pairs with cytosine, the *imino* form being more common than *imino*-adenine and hence inducing T-to-C transitions during DNA replication.

- *Deaminating agents* also cause point mutations. A certain amount of base deamination (removal of an amino group) occurs spontaneously in genomic DNA molecules, with the rate being increased by chemicals such as nitrous acid, which deaminates adenine, cytosine and guanine (thymine has no amino group and so cannot be deaminated), and sodium bisulfite, which acts only on cytosine. Deamination of guanine is not mutagenic because the resulting base, xanthine, blocks replication when it appears in the template polynucleotide. Deamination of adenine gives hypoxanthine (*Figure 14.7*), which pairs with C rather than T, and deamination of cytosine gives uracil, which pairs with A rather than G. Deamination of these two bases therefore results in point mutations when the template strand is copied.

- *Alkylating agents* are a third type of mutagen that can give rise to point mutations. Chemicals such as **ethylmethane sulfonate (EMS)** and dimethylnitrosamine add alkyl groups to nucleotides in DNA molecules, as do methylating agents such as methyl halides which are present in the atmosphere, and the products of nitrite metabolism. The effect of alkylation depends on the position at which the nucleotide is modified and the type of alkyl group that is added. Methylations, for example, often result in modified nucleotides with altered base-pairing properties and so lead to point mutations. Other alkylations block replication by forming crosslinks between the two strands of a DNA molecule, or by adding large alkyl groups that prevent progress of the replication complex.

- *Intercalating agents* are usually associated with insertion mutations. The best known mutagen of this type is **ethidium bromide**, which fluoresces when exposed to UV radiation and so is used to reveal the positions of DNA bands after agarose gel electrophoresis (see Technical Note 2.1, page 37). Ethidium bromide and other intercalating agents are flat molecules that can slip between base pairs in the double helix, slightly unwinding the helix and hence increasing the distance between adjacent base pairs (*Figure 14.8*).

The most important types of physical mutagen are as follows:

- **UV radiation** of 260 nm induces dimerization of adjacent pyrimidine bases, especially if these are both thymines (*Figure 14.9A*), resulting in a **cyclobutyl dimer**. Other pyrimidine combinations also form dimers, the order of frequency being 5′–CT–3′ > 5′–TC–3′ > 5′–CC–3′. Purine dimers are much less common. UV-induced dimerization usually results in

(A) Ethidium bromide

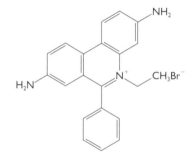

(B) The mutagenic effect

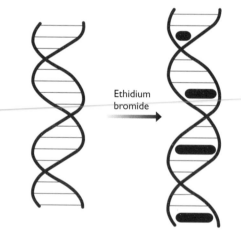

Figure 14.8 The mutagenic effect of ethidium bromide.

(A) Ethidium bromide is a flat plate-like molecule that is able to slot in between the base pairs of the double helix. (B) Ethidium bromide molecules are shown intercalated into the helix: the molecules are viewed sideways on. Note that intercalation increases the distance between adjacent base pairs.

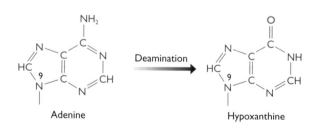

Figure 14.7 Hypoxanthine is a deaminated version of adenine.

The nucleoside that contains hypoxanthine is called inosine (see Table 10.5, see page 301).

a deletion mutation when the modified strand is copied. Another type of UV-induced **photoproduct** is the **(6–4) lesion** in which carbons number 4 and 6 of adjacent pyrimidines become covalently linked (*Figure 14.9B*).

- *Ionizing radiation* has various effects on DNA depending on the type of radiation and its intensity. Point, insertion and/or deletion mutations might arise, as well as more severe forms of DNA damage that prevent subsequent replication of the genome. Some types of ionizing radiation act directly on DNA, others act indirectly by stimulating the formation of reactive molecules such as peroxides in the cell.

- *Heat* stimulates the water-induced cleavage of the β-*N*-glycosidic bond that attaches the base to the sugar component of the nucleotide (*Figure 14.10A*). This occurs more frequently with purines than with pyrimidines and results in an **AP** (apurinic/apyrimidinic) or **baseless site**. The sugar–phosphate that is left is unstable and rapidly degrades, leaving a gap if the DNA molecule is double stranded (*Figure 14.10B*). This reaction is not normally mutagenic because cells have effective systems for repairing nicks (Section 14.2.1), which is reassuring when one considers that 10 000 AP sites are generated in each human cell per day. Gaps do, however, lead to mutations under certain circumstances, for example in *E. coli* when the SOS response is activated, when gaps are filled with As regardless of the identity of the nucleotide in the other strand (Section 14.1.3).

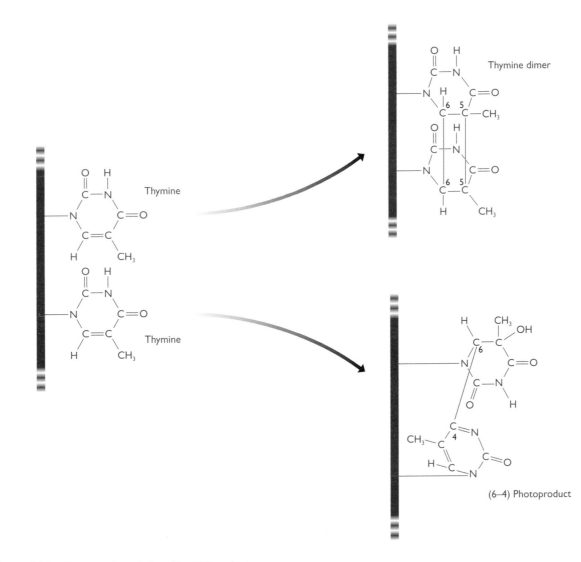

Figure 14.9 Photoproducts induced by UV irradiation.

A segment of a polynucleotide containing two adjacent thymine bases is shown. (A) A thymine dimer contains two UV-induced covalent bonds, one linking the carbons at position 6 and the other linking the carbons at position 5. (B) The (6–4) lesion involves formation of a covalent bond between carbons 4 and 6 of the adjacent nucleotides.

(A) Heat-induced hydrolysis of a β-*N*-glycosidic bond

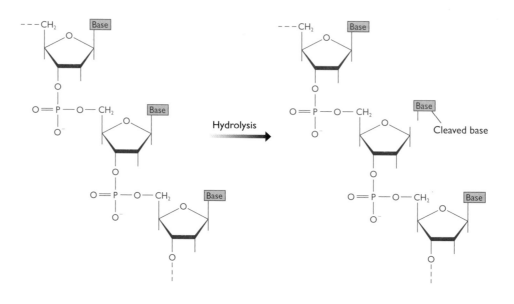

(B) The effect of hydrolysis on double-stranded DNA

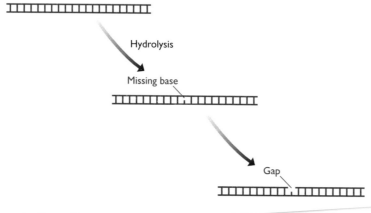

Figure 14.10   The mutagenic effect of heat.

(A) Heat induces hydrolysis of β-*N*-glycosidic bonds, resulting in a baseless site in a polynucleotide. (B) Schematic representation of the effect of heat-induced hydrolysis on a double-stranded DNA molecule. The baseless site is unstable and degrades, leaving a gap in one strand.

14.1.2 The effects of mutations

When considering the effects of mutations we must make a distinction between the *direct* effect that a mutation has on the functioning of a genome and its *indirect* effect on the phenotype of the organism in which it occurs. The direct effect is relatively easy to assess because we can use our understanding of gene structure and expression to predict the impact that a mutation will have on genome function. The indirect effects are more complex because these relate to the phenotype of the mutated organism which, as described in Section 7.2.2, is often difficult to correlate with the activities of individual genes.

The effects of mutations on genomes

Many mutations result in nucleotide sequence changes that have no effect on the functioning of the genome. These **silent mutations** include virtually all of those that occur in intergenic DNA and in the non-coding components of genes and gene-related sequences. In other words, some 98.5% of the human genome (see Box 1.4, page 23) can be mutated without significant effect.

Mutations in the coding regions of genes are much more important. First, we will look at point mutations that change the sequence of a triplet codon. A mutation of this type will have one of four effects (*Figure 14.11*):

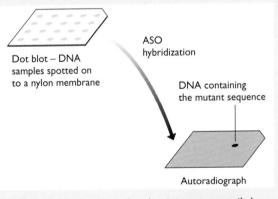

TECHNICAL 14.1 NOTE

Mutation detection

Rapid procedures for detecting mutations in DNA molecules.

Many genetic diseases are caused by point mutations that result in modification or inactivation of a gene product. Methods for detecting these mutations are important in two contexts. First, when a new gene responsible for a genetic disease is identified it is usually necessary to examine many versions of that gene from different individuals in order to identify the mutation or mutations responsible for the disease state. Second, when a disease-causing mutation has been characterized, high-throughput methods are needed so that clinicians can screen many DNA samples in order to identify individuals who have the mutation and are at risk of developing the disease or passing it on to their children.

Any mutation can be identified by DNA sequencing but sequencing is relatively slow and would be inappropriate for screening a large number of samples. DNA chip technology (Technical Note 5.1, page 133) could also be employed, but this is not yet a widely available option. For these reasons, a number of 'low technology' methods have been devised. These can be divided into two categories: **mutation scanning** techniques, which require no prior information about the position of a mutation, and **mutation screening** techniques, which determine whether a specific mutation is present.

Most scanning techniques involve analysis of the heteroduplex formed between a single strand of the DNA being examined and the complementary strand of a control DNA that has the unmutated sequence:

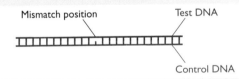

Mismatch position   Test DNA

Control DNA

If the test DNA contains a mutation then there will be a single mismatched position in the heteroduplex, where a base pair has not formed. Various techniques can be used to detect whether this mismatch is present or not (Cotton, 1997):

■ **Electrophoresis** or **high-performance liquid chromatography (HPLC)** can detect the mismatch by identifying the difference in the mobility of the mismatched hybrid compared with the fully base-paired one in a polyacrylamide gel or HPLC column. This approach determines if a mismatch is present but does not provide information on where in the test DNA the mutation is located.

■ **Cleavage** of the heteroduplex at the mismatch position followed by gel electrophoresis will locate the position of a mismatch. If the heteroduplex stays intact then no mismatch is present; if it is cleaved then it contains a mismatch, the position of the mutation in the test DNA being indicated by the sizes of the cleavage products. Cleavage is carried out by treatment with enzymes or chemicals that cut at single-stranded regions of mainly double-stranded DNA, or with a single-strand-specific ribonuclease such as S1 (see *Figure 7.7*, page 195) if the hybrid has been formed between the control DNA and an RNA version of the test DNA.

Most *screening* methods for detection of specific mutations make use of the ability of oligonucleotide hybridization to distinguish between target DNAs whose sequences differ at just one nucleotide position (see *Figure 5.8*, page 132). In **allele-specific oligonucleotide (ASO) hybridization** the DNA samples are screened by probing with an oligonucleotide that hybridizes only to the mutant sequence:

ASO hybridization

Dot blot – DNA samples spotted on to a nylon membrane

DNA containing the mutant sequence

Autoradiograph

This is an efficient procedure but it is unnecessarily long-winded. The DNA samples are usually obtained by PCR of clinical isolates so a more rapid alternative is to use the diagnostic oligonucleotide as one of the PCR primers, so that the presence or absence of the mutation in the test DNA is indicated by the synthesis or otherwise of a PCR product.

■ It may result in a **synonymous** change, the new codon specifying the same amino acid as the unmutated codon. A synonymous change is therefore a silent mutation because it has no effect on the coding function of the genome: the mutated gene codes for exactly the same protein as the unmutated gene.

■ It may result in a **non-synonymous** change, the mutation altering the codon so that it specifies a different amino acid. The protein coded by the mutated gene therefore has a single amino acid change. This often has no significant effect on the biological activity of the protein because most proteins can tolerate

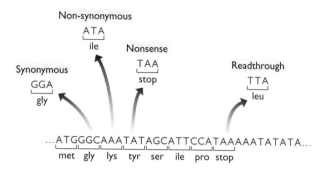

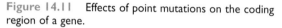

Figure 14.11 Effects of point mutations on the coding region of a gene.

Four different effects of point mutations are shown, as described in the text. The readthrough mutation results in the gene being extended beyond the end of the sequence shown here, the leucine codon created by the mutation being followed by AAA = lys, TAT = tyr and ATA = ile. See *Figure 3.20*, page 86 for the genetic code.

at least a few amino acid changes without noticeable effect on their ability to function in the cell, but changes to some amino acids, such as those at the active site of an enzyme, have a greater impact. A non-synonymous change is also called a **missense** mutation.

■ The mutation may convert a codon that specifies an amino acid into a termination codon. This is a **nonsense** mutation and it results in a shortened protein because translation of the mRNA stops at this new termination codon rather than proceeding to the correct termination codon further downstream. The effect of this on protein activity depends on how much of the polypeptide is lost: usually the effect is drastic and the protein is non-functional.

■ The mutation could convert a termination codon into one specifying an amino acid, resulting in **readthrough** of the stop signal so the protein is extended by an additional series of amino acids at its C terminus. Most proteins can tolerate short extensions without an effect on function, but longer extensions might interfere with folding of the protein and so result in reduced activity.

Deletion and insertion mutations also have distinct effects on the coding capabilities of genes (*Figure 14.12*). If the number of deleted or inserted nucleotides is three or a multiple of three then one or more codons are removed or added, the resulting loss or gain of amino acids having varying effects on the function of the encoded protein. Deletions or insertions of this type are often inconsequential but will have an impact if, for example, amino acids involved in an enzyme's active site are lost, or if an insertion disrupts an important secondary structure in the protein. On the other hand, if the number of deleted or inserted nucleotides is not three or a multiple of three

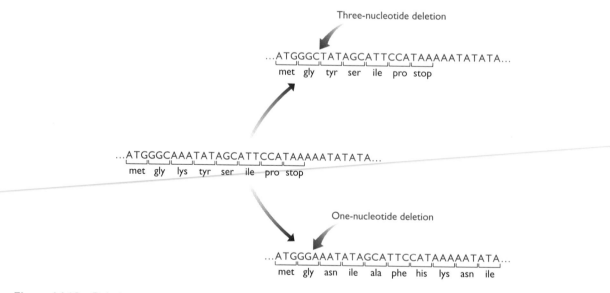

Figure 14.12 Deletion mutations.

In the top sequence three nucleotides comprising a single codon are deleted. This shortens the resulting protein product by one amino acid but does not affect the rest of its sequence. In the lower section, a single nucleotide is deleted. This results in a frameshift so that all the codons downstream of the deletion are changed, including the termination codon which is now read through. See *Figure 3.20*, page 86, for the genetic code. Note that if a three-nucleotide deletion removes parts of adjacent codons then the result is more complicated than shown here. Consider, for example, deletion of the trinucleotide GCA from the sequence . . . ATGGGCAAATAT . . . coding for met–gly–lys–tyr. The new sequence is . . . ATGGAATAT . . . , coding for met–glu–tyr. Two amino acids have been replaced by a single, different one.

then a frameshift results, all of the codons downstream of the mutation being taken from a different reading frame from that used in the unmutated gene. This usually has a significant effect on the protein function, because a greater or lesser part of the mutated polypeptide has a completely different sequence to the normal polypeptide.

It is less easy to make generalizations about the effects of mutations that occur outside of the coding regions of the genome. Any protein binding site is susceptible to point, insertion or deletion mutations that change the identity or relative positioning of nucleotides involved in the DNA–protein interaction. These mutations therefore have the potential to inactivate promoters or regulatory sequences, with predictable consequences for gene expression (*Figure 14.13*; Sections 9.2 and 9.3). Origins of replication could conceivably be made non-functional by mutations that change, delete or disrupt sequences recognized by the relevant binding proteins (Section 13.2.1) but these possibilities are not well documented. There is also little information about the potential impact on gene expression of mutations that affect nucleosome positioning (Section 8.2.1).

One area that has been better researched concerns mutations that occur in introns or at intron–exon boundaries. In these regions, single point mutations will be important if they change nucleotides involved in the RNA–protein and RNA–RNA interactions that occur during splicing of different types of intron (Sections 10.1.3 and 10.2.3). For example, mutation of either the G or T in the DNA copy of the 5' splice site of a GU–AG intron, or of the A or G at the 3' splice site, will disrupt splicing because the correct intron–exon boundary will no longer be recognized. This may mean that the intron is not removed from the pre-mRNA, but it is more likely that a cryptic splice site (see page 289) will be used as an alternative. It is also possible for a mutation within an intron or an exon to create a new cryptic site that is preferred over a genuine splice site that is not itself mutated. Both types of event have the same result: relocation of the active splice site, leading to aberrant splicing. This might delete part of the resulting protein, add a new stretch of amino acids, or lead to a frameshift. Several versions of the blood disease β-thalassemia are caused by mutations that lead to cryptic splice site selection during processing of β-globin transcripts.

The effects of mutations on multicellular organisms

Now we turn to the indirect effects that mutations have on organisms, beginning with multicellular diploid eukaryotes such as humans. The first issue to consider is the relative importance of the same mutation in a somatic cell compared with a germ cell. Because somatic cells do not pass copies of their genomes to the next generation, a somatic cell mutation is important only for the organism in which it occurs: it has no potential evolutionary impact. In fact, most somatic cell mutations have no significant effect, even if they result in cell death, because there are many other identical cells in the same tissue and the loss of one cell is immaterial. An exception is when a mutation causes a somatic cell to malfunction in a way that is harmful to the organism, for instance by inducing tumor formation or other cancerous activity.

Mutations in germ cells are more important because they can be transmitted to members of the next generation and will then be present in all the cells of any individual who inherits the mutation. Most mutations, including all silent ones and many in coding regions, will still not change the phenotype of the organism in any significant way. Those that do have an effect can be divided into two categories:

■ **Loss-of-function** is the normal result of a mutation that reduces or abolishes a protein activity. Most loss-of-function mutations are recessive (Section 5.2.3), because in a heterozygote the second chromosome copy carries an unmutated version of the gene coding for a fully functional protein whose presence

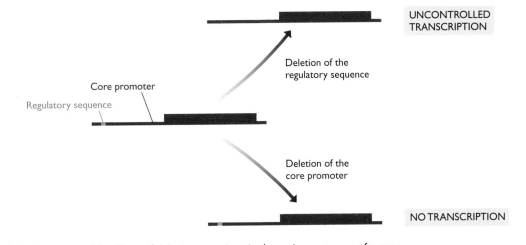

Figure 14.13   Two possible effects of deletion mutations in the region upstream of a gene.

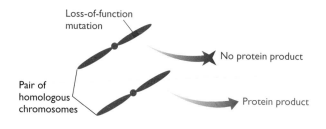

Figure 14.14 A loss-of-function mutation is usually recessive because a functional version of the gene is present on the second chromosome copy.

compensates for the effect of the mutation (*Figure 14.14*). There are some exceptions where a loss-of-function mutation is dominant, one example being **haploinsufficiency**, where the organism is unable to tolerate the approximately 50% reduction in protein activity suffered by the heterozygote. This is the explanation for a few genetic diseases in humans, including Marfan syndrome which results from a mutation in the gene for the connective tissue protein called fibrillin.

■ **Gain-of-function** mutations are much less common. The mutation must be one that confers an abnormal activity on a protein. Many gain-of-function mutations are in regulatory sequences rather than in coding regions, and can therefore have a number of consequences. For example, a mutation might lead to one or more genes being expressed in the wrong tissues, these tissues gaining functions that they normally lack. Alternatively the mutation could lead to overexpression of one or more genes involved in control of the cell cycle, thus leading to uncontrolled cell division and hence to cancer. Because of their nature, gain-of-function mutations are usually dominant.

Assessing the effects of mutations on the phenotypes of multicellular organisms can be difficult. Not all mutations have an immediate impact: some are **delayed onset** and only confer an altered phenotype later in the individual's life. Others display **non-penetrance** in some individuals, never being expressed even though the individual has a dominant mutation or is a homozygous recessive. With humans, these factors complicate attempts to map disease-causing mutations by pedigree analysis (Section 5.2.4) because they introduce uncertainty about which members of a pedigree carry a mutant allele.

The effects of mutations on microorganisms

Mutations in microbes such as bacteria and yeast can also be described as loss-of-function or gain-of-function, but with microorganisms this is neither the normal nor the most useful classification scheme. Instead, a more detailed description of the phenotype is usually attempted on the basis of the growth properties of mutated cells in various culture media. This enables most mutations to be assigned to one of four categories:

■ **Auxotrophs** are cells that will only grow when provided with a nutrient not required by the unmutated organism. For example, *E. coli* normally makes its own tryptophan, courtesy of the enzymes coded by the five genes in the tryptophan operon (*Figure 2.20B*, page 55). If one of these genes is mutated in such a way that its protein product is inactivated, then the cell is no longer able to make tryptophan and so becomes a tryptophan auxotroph. It cannot survive on a medium that lacks tryptophan, and can grow only when this amino acid is provided as a nutrient (*Figure 14.15*). Unmutated bacteria, which do not require extra supplements in their growth media, are called **prototrophs**.

■ **Conditional-lethal** mutants are unable to withstand certain growth conditions: under **permissive conditions** they appear to be entirely normal but when transferred to **restrictive conditions** the mutant phenotype is seen. **Temperature-sensitive** mutants are typical examples of conditional-lethal mutants. Temperature-sensitive mutants behave like wild-type cells at low temperatures but exhibit their mutant phenotype when the temperature is raised above a certain threshold, which is different for each mutant. Usually this is because the mutation reduces the stability of a

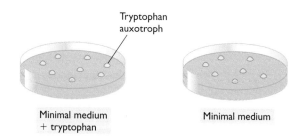

Figure 14.15 A tryptophan auxotrophic mutant.

Two Petri-dish cultures are shown. Both contain **minimal medium**, which provides just the basic nutritional requirements for bacterial growth (nitrogen, carbon and energy sources, plus some salts). The medium on the left is supplemented with tryptophan but the medium on the right is not. Unmutated bacteria, plus tryptophan auxotrophs, can grow on the plate on the left, the auxotrophs growing because the medium supplies the tryptophan that they cannot make themselves. Tryptophan auxotrophs cannot grow on the plate on the right, because this does not contain tryptophan. To identify a tryptophan auxotroph, colonies are first grown on the minimal medium + tryptophan plate and then transferred to the minimal medium plate by replica plating (see *Figure 4.18*, page 111). After incubation, colonies appear on the minimal medium plate in the same relative positions as on the plate containing tryptophan, except for the tryptophan auxotrophs which do not grow. These colonies can therefore be identified and samples of the tryptophan auxotrophic bacteria recovered from the minimal medium + tryptophan plate.

protein, so the protein becomes unfolded and hence inactive when the temperature is raised.

- **Inhibitor-resistant** mutants are able to resist the toxic effects of an antibiotic or another type of inhibitor. There are various molecular explanations for this type of mutant. In some cases the mutation changes the structure of the protein that is targeted by the inhibitor, so the latter can no longer bind to the protein and interfere with its function. This is the basis of streptomycin resistance in *E. coli*, which results from a change in the structure of ribosomal protein S12. Another possibility is that the mutation changes the properties of a protein responsible for transporting the inhibitor into the cell, this often being the way in which resistance to toxic metals is acquired.

- **Regulatory** mutants have defects in promoters and other regulatory sequences. This category includes **constitutive** mutants, which continuously express genes that are normally switched on and off under different conditions. For example, a mutation in the operator sequence of the lactose operon (Section 9.3.1) can prevent the repressor from binding and so results in the lactose operon being expressed all the time, even when lactose is absent and the genes should be switched off (*Figure 14.16*).

In addition to these four categories, many mutations are lethal and so result in death of the mutant cell, whereas others have no effect. The latter are less common in microorganisms than in higher eukaryotes, because most microbial genomes are relatively compact, with little non-coding DNA. Mutations can also be **leaky**, meaning that

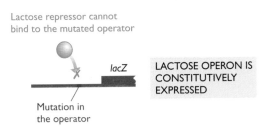

Figure 14.16 The effect of a constitutive mutation in the lactose operator.

The operator sequence has been altered by a mutation and the lactose repressor can no longer bind to it. The result is that the lactose operon is transcribed all the time, even when lactose is absent from the medium. This is not the only way in which a constitutive *lac* mutant can arise. For example, the mutation could be in the gene coding for the lactose repressor, changing the tertiary structure of the repressor protein so that its DNA-binding motif is disrupted and it can no longer recognize the operator sequence, even when the latter is unmutated. See *Figure 9.24*, page 264, for more details about the lactose repressor and its regulatory effect on expression of the lactose operon.

a less extreme form of the mutant phenotype is expressed. For example, a leaky version of the tryptophan auxotroph illustrated in *Figure 14.15* would grow slowly on minimal medium, rather than not growing at all.

14.1.3 Hypermutation and the possibility of programmed mutations

Is it possible for cells to utilize mutations in a positive fashion, either by increasing the rate at which mutations appear in their genomes, or by directing mutations towards specific genes? Both types of event might appear, at first glance, to go against the accepted wisdom that mutations occur randomly but, as we shall see, **hypermutation** and **programmed mutations** are possible without contravening this dogma.

Hypermutation occurs when a cell allows the rate at which mutations occur in its genome to increase. Several examples of hypermutation are known, one of these forming part of the mechanism used by vertebrates, including humans, to generate a diverse array of immunoglobulin proteins. We have already touched on this phenomenon in Section 12.2.1 when we examined the genome rearrangements that result in joining of the V, D, J and H segments of the immunoglobulin heavy- and light-chain genes (see *Figure 12.15*, page 363). Additional diversity is produced by hypermutation of the V-gene segments after assembly of the intact immunoglobulin gene (*Figure 14.17*), the mutation rate for these segments being 6–7 orders of magnitude greater than the background mutation rate experienced by the rest of the genome (Shannon and Weigert, 1998). This enhanced mutation rate appears to result from the unusual behavior of the mismatch repair system which normally corrects replication errors. At all other positions within the genome, the mismatch repair system corrects errors of replication by searching for mismatches and replacing the nucleotide in the daughter strand, this being the strand that has just been synthesized and so contains the error (see Section 14.2.3). At V-gene segments, the repair system changes the nucleotide in the parent strand, and so stabilizes the mutation rather than correcting it (Cascalho *et al.*, 1998). The mechanism by which this is achieved has not yet been described.

An apparent increase in mutation rate arising from modifications to the normal DNA repair process does not contradict the dogma regarding the randomness of mutations. However, problems have arisen with reports, dating back to 1988 (Cairns *et al.*, 1988), which suggested that *E. coli* is able to direct mutations towards genes whose mutation would be advantageous under the environmental conditions that the bacterium is encountering. The original experiments involved a strain of *E. coli* that has a nonsense mutation in the lactose operon, inactivating the proteins needed for utilization of this sugar (Research Briefing 14.1). The bacteria were spread on an agar medium in which the only carbon source was lactose. This meant that a cell could grow and divide only if a

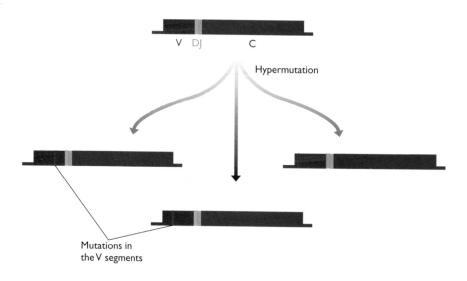

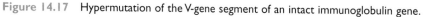

Figure 14.17 Hypermutation of the V-gene segment of an intact immunoglobulin gene.

See Figure 12.15, page 363, for a description of the events leading to assembly of an immunoglobulin gene.

second mutation occurred in the lactose operon, reversing the effects of the nonsense mutation and therefore allowing the lactose enzymes to be synthesized. Mutations with this effect appeared to occur significantly more frequently than expected, and at a rate that was greater than mutations in other parts of the genomes of these *E. coli* cells.

These experiments suggested that bacteria can program mutations according to the selective pressures that they are placed under. In other words, the environment can directly affect the phenotype of the organism, as suggested by Lamarck, rather than operating through the random processes postulated by Darwin. With such radical implications, it is not surprising that the experiments have been debated at length, with numerous attempts to discover flaws in their design or alternative explanations for the results. Variations of the original experimental system have suggested that the results are authentic, and similar events in other bacteria have been described. Models based on gene amplification rather than selective mutation are being tested (Andersson *et al.*, 1998), and attention has also been directed at the possible roles of recombination events such as transposition of insertion elements in the generation of programmed mutations (Foster, 1999).

14.2 DNA Repair

In view of the thousands of damage events that genomes suffer every day, coupled with the errors that occur when the genome replicates, it is essential that cells possess efficient repair systems. Without these repair systems a genome would not be able to maintain its essential cellular functions for more than a few hours before key genes became inactivated by DNA damage. Similarly, cell line-

ages would accumulate replication errors at such a rate that their genomes would become dysfunctional after a few cell divisions.

Most cells possess four different categories of DNA repair system (*Figure 14.18*; Lindahl and Wood, 1999):

- **Direct repair systems** (Section 14.2.1), as the name suggests, act directly on damaged nucleotides, converting each one back to its original structure.
- **Excision repair** (Section 14.2.2) involves excision of a segment of the polynucleotide containing a damaged site, followed by resynthesis of the correct nucleotide sequence by a DNA polymerase.
- **Mismatch repair** (Section 14.2.3) corrects errors of replication, again by excising a stretch of single-stranded DNA containing the offending nucleotide and then repairing the resulting gap.
- **Recombination repair** (Section 14.2.4) is used to mend double-strand breaks.

Most if not all organisms also possess systems that enable them to replicate damaged regions of their genome without prior repair. We will examine these systems in Section 14.2.5, and in Section 14.2.6 we will survey the human diseases that result from defects in DNA repair processes.

14.2.1 Direct repair systems fill in nicks and correct some types of nucleotide modification

Most of the types of DNA damage that are caused by chemical or physical mutagens (Section 14.1.1) can only be repaired by excision of the damaged nucleotide followed by resynthesis of a new stretch of DNA, as shown in *Figure 14.18B*. Only a few types of damaged nucleotide can be repaired directly:

RESEARCH 14.1 BRIEFING

Programmed mutations?

In 1988 startling results were published suggesting that under some circumstances *Escherichia coli* bacteria are able to mutate in a directed way that enables cells to adapt to an environmental stress.

The randomness of mutations is an important concept in biology because it is a requirement of the Darwinian view of evolution, which holds that changes in the characteristics of an organism occur by chance and are not influenced by the environment in which the organism is placed. In contrast, the Lamarckian theory of evolution, which biologists rejected well over a century ago, states that organisms acquire changes that enable them to adapt to their environment. The Darwinian view requires that mutations occur at random, whereas Lamarckian evolution demands that programmed mutations occur in response to the environment.

Random mutations in E. coli

The randomness of mutations in bacteria was first demonstrated by Luria and Delbrück in 1943. They grew a series of *E. coli* cultures in different flasks and then added T1 bacteriophages to each one. Most of the bacteria were killed by the phages, but a few T1-resistant mutants were able to survive. These were identified by plating samples from each culture, soon after T1 infection, onto an agar medium. If mutations leading to T1 resistance occurred randomly in the cultures before the bacteriophages were added, each culture would contain a different number of resistant mutants, the number depending on how early during the growth period the first mutant cells arose. Those that arose early would divide many times to give rise to a large number of resistant progeny in the culture at the end of the growth period, whereas those that arose later would give rise to just a few progeny. Some cultures would therefore contain many T1-resistant cells and others would contain just a few. Alternatively, if resistant bacteria arose by programmed mutation only when the T1 phage was added, then all cultures would have similar numbers of mutants. Luria and Delbrück found that each of their cultures contained a different number of T1-resistant bacteria; thus, they concluded that mutations occur randomly and not in response to T1 phage.

Programmed mutations in E. coli

The possibility that Luria and Delbrück's conclusion might not be universally true for *E. coli* mutations was first suggested by studies of an *E. coli* strain that carries a nonsense mutation in its *lacZ* gene. The presence of the termination codon in *lacZ* means that these cells are unable to synthesize functional β-galactosidase enzymes and so cannot use lactose as a carbon and energy source – they are therefore lactose auxotrophs. This is not necessarily a permanent situation because a cell could undergo a mutation that converts the termination codon back into one specifying an amino acid. These new mutants would be able to make β-galactosidase and use any lactose that is available. According to Luria and Delbrück's results, such mutations should occur at random and should not be influenced by the pres-

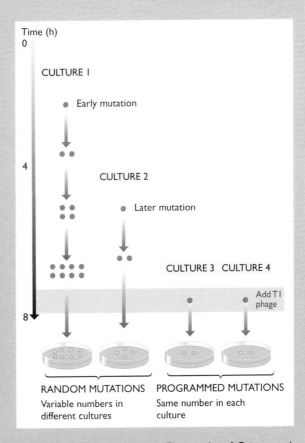

ence of lactose in the medium. The results of Cairns *et al.* (1988) showed that when the lactose auxotrophs were plated onto a minimal medium containing lactose as the only sugar – circumstances that require that the bacteria must mutate into lactose prototrophs in order to survive – then the number of lactose prototrophs that arose was significantly higher than that expected if mutations occurred randomly. In other words, some cells underwent programmed mutation and acquired the specific change in DNA sequence needed to withstand the selective pressure.

Since 1988, a number of examples of what appear to be programmed mutations have been published, but the notion that bacteria, and possibly other organisms, can program mutations in response to environmental stress is by no means accepted by the scientific community. It is quite possible that these mutations will eventually be disproved or be shown to have an orthodox basis. However, until this happens we are left with the tantalizing possibility that even at this fundamental level our knowledge about genomes might be far from complete.

Reference

Cairns J, Overbaugh J and Miller S (1988) The origin of mutants. *Nature*, **335**, 142–145.

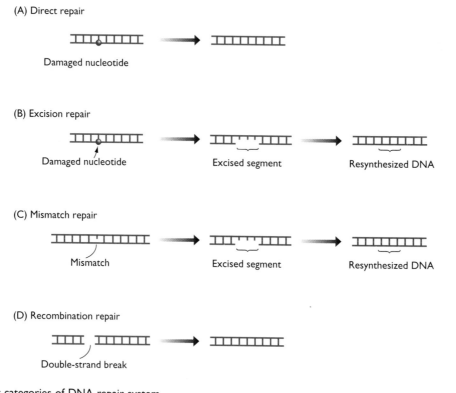

(A) Direct repair

Damaged nucleotide

(B) Excision repair

Damaged nucleotide   Excised segment   Resynthesized DNA

(C) Mismatch repair

Mismatch   Excised segment   Resynthesized DNA

(D) Recombination repair

Double-strand break

Figure 14.18 Four categories of DNA repair system.
See the text for details.

■ **Nicks** can be repaired by a DNA ligase if all that has happened is that a phosphodiester bond has been broken, without damage to the 5′-phosphate and 3′-hydroxyl groups of the nucleotides either side of the nick (*Figure 14.19*). This is often the case with nicks resulting from the effects of ionizing radiation.

■ Some forms of **alkylation** damage are directly reversible by enzymes that transfer the alkyl group from the nucleotide to their own polypeptide chains.

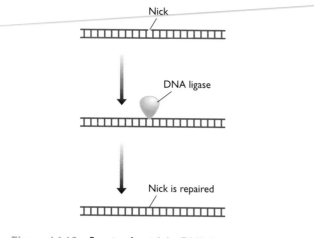

Nick

DNA ligase

Nick is repaired

Figure 14.19 Repair of a nick by DNA ligase.

Enzymes capable of doing this are known in many different organisms and include the **Ada enzyme** of *E. coli*, which is involved in an <u>ada</u>ptive process that this bacterium is able to activate in response to DNA damage. Ada removes alkyl groups attached to the oxygen groups at positions 4 and 6 of thymine and guanine, respectively, and can also repair phosphodiester bonds that have become methylated. Other alkylation repair enzymes have more restricted specificities, an example being human **MGMT** (O^6-methylguanine–DNA methyltransferase) which, as its name suggests, only removes alkyl groups from position 6 of guanine.

■ **Cyclobutyl dimers** are repaired by a light-dependent direct system called **photoreactivation**. In *E. coli*, the process involves the enzyme called **DNA photolyase** (more correctly named deoxyribodipyrimidine photolyase). When stimulated by light with a wavelength between 300 and 500 nm the enzyme binds to cyclobutyl dimers and converts them back to the original monomeric nucleotides. Photoreactivation is a widespread but not universal type of repair: it is known in many but not all bacteria and also in quite a few eukaryotes, including some vertebrates, but is absent in humans and other placental mammals. A similar type of photoreactivation involves the **(6–4) photoproduct photolyase** and results in repair of (6–4) lesions. Neither *E. coli* nor humans have this enzyme but it is possessed by a variety of other organisms.

14.2.2 Excision repair

The direct types of damage reversal described above are important, but they form a very minor component of the DNA repair mechanisms of most organisms. This point is illustrated by the draft human genome sequences, which appear to contain just a single gene coding for a protein involved in direct repair (the *MGMT* gene), but which have at least 40 genes for components of the excision repair pathways (Wood *et al.*, 2001). These pathways fall into two categories:

- *Base excision repair* involves removal of a damaged nucleotide base, excision of a short piece of the polynucleotide around the AP site thus created, and resynthesis with a DNA polymerase.
- *Nucleotide excision repair* is similar to base excision repair but is not preceded by removal of a damaged base and can act on more substantially damaged areas of DNA.

We will examine each of these pathways in turn.

Base excision repairs many types of damaged nucleotide

Base excision is the least complex of the various repair systems that involve removal of one or more damaged nucleotides followed by resynthesis of DNA to span the resulting gap. It is used to repair many modified nucleotides whose bases have suffered relatively minor damage resulting from, for example, exposure to alkylating agents or ionizing radiation (Section 14.1.1). The process is initiated by a **DNA glycosylase** which cleaves the β-*N*-glycosidic bond between a damaged base and the sugar component of the nucleotide (*Figure 14.20A*). Each DNA glycosylase has a limited specificity (*Table 14.3*), the specificities of the glycosylases possessed by a cell determining the range of damaged nucleotides that can be repaired by the base excision pathway. Most organisms are able to deal with deaminated bases such as uracil (deaminated cytosine) and hypoxanthine (deaminated adenine), oxidation products such as 5-hydroxycytosine and thymine glycol, and methylated bases such as 3-methyladenine, 7-methylguanine and 2-methylcytosine (Seeberg *et al.*, 1995). Other DNA glycosylases remove normal bases as part of the mismatch repair system (Section 14.2.3). Most of the DNA glycosylases involved in base excision repair are thought to diffuse along the minor groove of the DNA double helix in search of damaged nucleotides, but some may be associated with the replication enzymes.

A DNA glycosylase removes a damaged base by 'flipping' the structure to a position outside of the helix and then detaching it from the polynucleotide (Kunkel and Wilson, 1996; Roberts and Cheng, 1998). This creates an AP or baseless site (see *Figure 14.10*, page 428) which is converted into a single nucleotide gap in the second step of the repair pathway (*Figure 14.20B*). This step can be carried out in a variety of ways. The standard method makes use of an **AP endonuclease**, such as exonuclease III or endonuclease IV of *E. coli* or human APE1, which cuts the phosphodiester bond on the 5′ side of the AP site. Some AP endonucleases can also remove the sugar from the AP site, this being all that remains of the damaged nucleotide, but others lack this ability and so work in conjunction with a separate **phosphodiesterase**. An alternative pathway for converting the AP site into a gap utilizes the endonuclease activity possessed by some DNA glycosylases, which can make a cut at the 3′ side of the AP site, probably at the same time that the damaged base is removed, followed again by removal of the sugar by a phosphodiesterase.

The single nucleotide gap is filled by a DNA polymerase, using base-paring with the undamaged base in the other strand of the DNA molecule to ensure that the correct nucleotide is inserted. In *E. coli* the gap is filled by DNA polymerase I and in mammals by DNA polymerase β (see *Table 13.2*, page 396; Sobol *et al.*, 1996). Yeast seems to be unusual in that it uses its main DNA replicating enzyme, DNA polymerase δ, for this purpose (Seeberg *et al.*, 1995). After gap filling, the final phosphodiester bond is put in place by a DNA ligase.

Nucleotide excision repair is used to correct more extensive types of damage

Nucleotide excision repair has a much broader specificity than the base excision system and is able to deal with more extreme forms of damage such as intrastrand crosslinks and bases that have become modified by attachment of large chemical groups. It is also able to correct cyclobutyl dimers by a **dark repair** process, providing those organisms that do not have the photoreactivation system (such as humans) with a means of repairing this type of damage.

In nucleotide excision repair, a segment of single-stranded DNA containing the damaged nucleotide(s) is excised and replaced with new DNA. The process is therefore similar to base excision repair except that it is not preceded by selective base removal, and a longer stretch of polynucleotide is excised. The best studied example of nucleotide excision repair is the **short patch** process of *E. coli*, so called because the region of polynucleotide that is excised and subsequently 'patched' is relatively short, usually 12 nucleotides in length.

Short patch repair is initiated by a multienzyme complex called the **UvrABC endonuclease**, sometimes also referred to as the 'excinuclease'. In the first stage of the process a trimer comprising two UvrA proteins and one copy of UvrB attaches to the DNA at the damaged site. How the site is recognized is not known but the broad specificity of the process indicates that individual types of damage are not directly detected and that the complex must search for a more general attribute of DNA damage

(A) Removal of a damaged base by DNA glycosylase

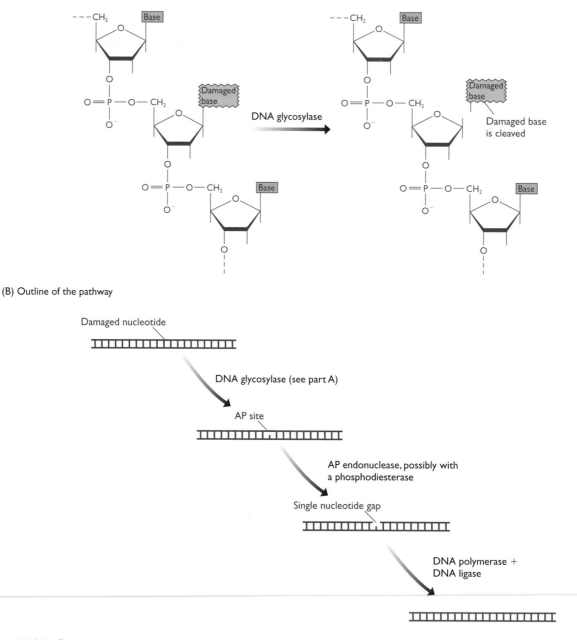

(B) Outline of the pathway

Figure 14.20 Base excision repair.

(A) Excision of a damaged nucleotide by a DNA glycosylase. (B) Schematic representation of the base excision repair pathway. Alternative versions of the pathway are described in the text.

such as distortion of the double helix. UvrA may be the part of the complex most involved in damage location because it dissociates once the site has been found and plays no further part in the repair process. Departure of UvrA allows UvrC to bind (*Figure 14.21*), forming a UvrBC dimer that cuts the polynucleotide either side of the damaged site. The first cut is made by UvrB at the fifth phosphodiester bond downstream of the damaged

nucleotide, and the second cut is made by UvrC at the eighth phosphodiester bond upstream, resulting in the 12 nucleotide excision, although there is some variability, especially in the position of the UvrB cut site. The excised segment is then removed, usually as an intact oligonucleotide, by DNA helicase II, which presumably detaches the segment by breaking the base pairs holding it to the second strand. UvrC also detaches at this stage, but UvrB

Table 14.3   Examples of human DNA glycosylases

DNA glycosylase	Specific for
MBD4	Uracil
MPG	Ethenoadenine, hypoxanthine, 3-methyladenine
NTH1	Cytosine glycol, dihydrouracil, formamidopyrimidine, thymine glycol
OGG1	Formamidopyrimidine, 8-oxoguanine
SMUG1	Uracil
TDG	Ethenocytosine, uracil
UNG	Uracil, 5-hydroxyuracil

Based on Lindahl and Wood (1999) and Wood et al. (2001).

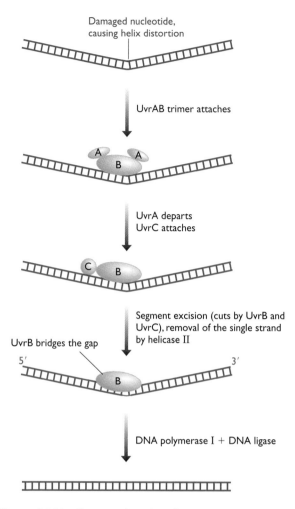

Figure 14.21   Short patch nucleotide excision repair in *Escherichia coli*.

The damaged nucleotide is shown distorting the helix because this is thought to be one of the recognition signals for the UvrAB trimer that initiates the short patch process. See the text for details of the events occurring during the repair pathway.

remains in place and bridges the gap produced by the excision. The bound UvrB is thought to prevent the single-stranded region that has been exposed from base-pairing with itself, but alternative roles could be to prevent this strand from becoming damaged, or possibly to direct the DNA polymerase to the site that needs to be repaired. As in base excision repair, the gap is filled by DNA polymerase I and the last phosphodiester bond is synthesized by DNA ligase.

E. coli also has a **long patch** nucleotide excision repair system that involves Uvr proteins but differs in that the piece of DNA that is excised can be anything up to 2 kb in length. Long patch repair has been less well studied and the process is not understood in detail, but it is presumed to work on more extensive forms of damage, possibly regions where groups of nucleotides, rather than just individual ones, have become modified. The eukaryotic nucleotide excision repair process is also called 'long patch' but results in replacement of only 24–29 nucleotides of DNA. In fact, there is no 'short patch' system in eukaryotes and the name is used to distinguish the process from base excision repair. The system is more complex than in *E. coli* and the relevant enzymes do not seem to be homologs of the Uvr proteins. In humans at least 16 proteins are involved, with the downstream cut being made at the same position as in *E. coli* – the fifth phosphodiester bond – but with a more distant upstream cut, resulting in the longer excision. Both cuts are made by endonucleases that attack single-stranded DNA specifically at its junction with a double-stranded region, indicating that before the cuts are made the DNA around the damage site has been melted, presumably by a helicase (*Figure 14.22*). This activity is provided at least in part by TFIIH, one of the components of the RNA polymerase II initiation complex (see *Table 9.5*, page 259). At first it was assumed that TFIIH simply had a dual role in the cell, functioning separately in both transcription and repair, but now it is thought that there is a more direct link between the two processes (Lehmann, 1995; Svejstrup *et al.*, 1996). This view is supported by the discovery of

transcription-coupled repair, which repairs some forms of damage in the template strands of genes that are being actively transcribed. The first type of transcription-coupled repair to be discovered was a modified version of nucleotide excision, but it now known that base-excision repair is also coupled with transcription (Cooper *et al.*, 1997). These discoveries do not imply that non-transcribed regions of the genome are not repaired. The excision repair processes protect the entire genome from damage, but it is entirely logical that special mechanisms should exist for directing the processes at genes that are being transcribed. The template strands of these genes contain the genome's biological information and maintaining their integrity should be the highest priority for the repair systems.

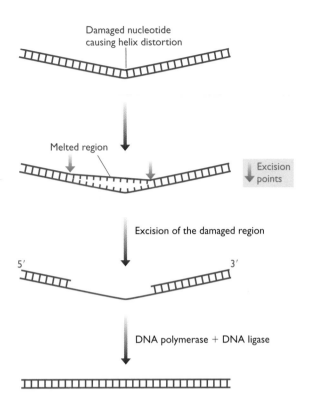

Figure 14.22 Outline of the events involved during nucleotide excision repair in eukaryotes.

The endonucleases that remove the damaged region make cuts specifically at the junction between single-stranded and double-stranded regions of a DNA molecule. The DNA is therefore thought to melt either side of the damaged nucleotide, as shown in the diagram, possibly as a result of the helicase activity of TFIIH.

14.2.3 Mismatch repair: correcting errors of replication

Each of the repair systems that we have looked at so far – direct, base excision and nucleotide excision repair – recognize and act upon DNA damage caused by mutagens. This means that they search for abnormal chemical structures such as modified nucleotides, cyclobutyl dimers and intrastrand crosslinks. They cannot, however, correct mismatches resulting from errors in replication because the mismatched nucleotide is not abnormal in any way, it is simply an A, C, G or T that has been inserted at the wrong position. As these nucleotides look exactly like any other nucleotide, the mismatch repair system that corrects replication errors has to detect not the mismatched nucleotide itself but the absence of base-pairing between the parent and daughter strands. Once it has found a mismatch, the repair system excises part of the daughter polynucleotide and fills in the gap, in a manner similar to base and nucleotide excision repair.

The scheme described above leaves one important question unanswered. The repair must be made in the daughter polynucleotide because it is in this newly syn-thesized strand that the error has occurred; the parent polynucleotide has the correct sequence. How does the repair process know which strand is which? In *E. coli* the answer is that the daughter strand is, at this stage, under-methylated and can therefore be distinguished from the parent polynucleotide, which has a full complement of methyl groups. *E. coli* DNA is methylated because of the activities of the **DNA adenine methylase (Dam)**, which converts adenines to 6-methyladenines in the sequence 5'–GATC–3', and the **DNA cytosine methylase (Dcm)**, which converts cytosines to 5-methylcytosines in 5'–CCAGG–3' and 5'–CCTGG–3'. These methylations are not mutagenic, the modified nucleotides having the same base-pairing properties as the unmodified versions. There is a delay between DNA replication and methylation of the daughter strand, and it is during this window of opportunity that the repair system scans the DNA for mismatches and makes the required corrections in the undermethylated, daughter strand (*Figure 14.23*).

E. coli has at least three mismatch repair systems, called 'long patch', 'short patch and 'very short patch', the names indicating the relative lengths of the excised and resynthesized segments. The long patch system replaces up to a kb or more of DNA and requires the MutH, MutL and MutS proteins, as well as the DNA helicase II that we met during nucleotide excision repair. MutS recognizes the mismatch and MutH distinguishes the two strands by binding to unmethylated 5'–GATC–3' sequences (*Figure 14.24*). The role of MutL is unclear but it might coordinate the activities of the other two proteins so that MutH binds to 5'–GATC–3' sequences only in the vicinity of mismatch sites recognized by MutS. After binding, MutH cuts the phosphodiester bond immediately upstream of the G in the methylation sequence and DNA helicase II detaches the single strand. There does not appear to be an enzyme that cuts the strand downstream of the mismatch; instead the detached single-stranded region is degraded by an exonuclease that follows the helicase and continues beyond the mismatch site. The gap is then filled in by DNA polymerase I and DNA ligase. Similar events are thought to occur during short and very short mismatch repair, the difference being the specificities of the proteins that recognize the mismatch. The short patch system, which results in excision of a segment less than 10 nucleotides in length, begins when MutY recognizes an A–G or A–C mismatch, and the very short repair system corrects G–T mismatches which are recognized by the Vsr endonuclease.

Eukaryotes have homologs of the *E. coli* Mut proteins and their mismatch repair processes probably work in a similar way (Kolodner, 2000). The one difference is that methylation might not be the method used to distinguish between the parent and daughter polynucleotides. Methylation has been implicated in mismatch repair in mammalian cells, but the DNA of some eukaryotes, including fruit flies and yeast, is not extensively methy-lated; it is thought that these organisms must therefore use a different method. Possibilities include an association between the repair enzymes and the replication

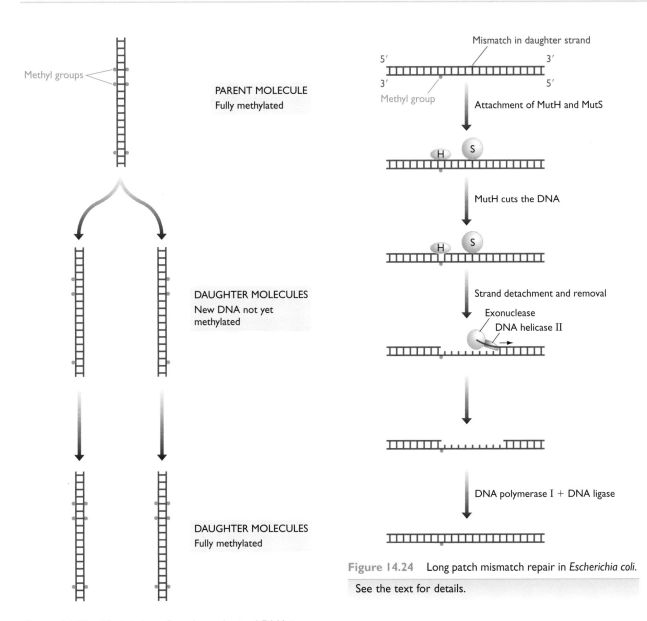

Figure 14.23 Methylation of newly synthesized DNA in *Escherichia coli* does not occur immediately after replication, providing a window of opportunity for the mismatch repair proteins to recognize the daughter strands and correct replication errors.

Figure 14.24 Long patch mismatch repair in *Escherichia coli*. See the text for details.

complex, so that repair is coupled with DNA synthesis, or use of single-strand binding proteins that mark the parent strand.

14.2.4 Repair of double-stranded DNA breaks

A single-stranded break in a double-stranded DNA molecule, such as is produced during the base and nucleotide excision repair processes and by some types of oxidative damage, does not present the cell with a critical problem. The double helix retains its overall intactness and the

break can be repaired by template-dependent DNA synthesis (*Figure 14.25A*). A double-stranded break is more serious because this converts the original double helix into two separate fragments which have to be brought back together again in order for the break to be repaired (*Figure 14.25B*). The two broken ends must be protected from further degradation, which could result in a deletion mutation appearing at the repaired break point. The repair processes must also ensure that the correct ends are joined: if there are two broken chromosomes in the nucleus, then the correct pairs must be brought together so that the original structures are restored. Experimental studies of mouse cells indicate that achieving this outcome is difficult and if two chromosomes are broken then misrepair resulting in hybrid structures occurs relatively frequently (Richardson and Jasin, 2000). Even if only one chromosome is broken, there is still a possibility that a natural chromosome end could be confused as a break

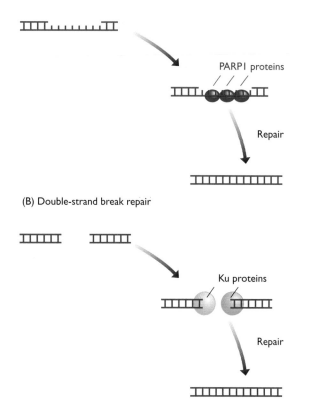

(A) Single-strand break repair

PARP1 proteins

Repair

(B) Double-strand break repair

Ku proteins

Repair

Figure 14.25   Single- and double-strand-break repair.

(A) A single-strand break does not disrupt the integrity of the double helix. The exposed single strand is coated with PARP1 proteins, which protect this intact strand from breaking and prevent it from participating in unwanted recombination events. The break is then filled in by the enzymes involved in the excision repair pathways. (B) A double-stranded break is more serious because the double helix is cleaved into two segments. Various proteins bind to the broken ends, notably Ku (see the text), to protect the ends and initiate double-strand break repair.

and an incorrect repair made. This type of error is not unknown, despite the presence of special telomere-binding proteins that mark the natural ends of chromosomes (Section 2.2.1).

Double-strand breaks are generated by exposure to ionizing radiation and some chemical mutagens, and are also made by the cell, in a controlled fashion, during recombination events such as the genome rearrangements that join together immunoglobulin gene segments and T-cell receptor gene segments in B and T lymphocytes (Section 12.2.1). Progress in understanding the break repair system has been stimulated by studies of mutant human cell lines, which have resulted in the identification of various sets of genes involved in the process (Critchlow and Jackson, 1998). These genes specify a multi-component protein complex that directs a DNA ligase to the break (*Figure 14.26*). The complex includes a

protein called Ku, made up of two non-identical subunits, which binds the DNA ends either side of the break (Walker *et al.*, 2001). Ku binds to the DNA in association with the DNA-PK$_{CS}$ protein kinase, which activates a third protein, XRCC4, which interacts with the mammalian DNA ligase IV, directing this repair protein to the double-strand break.

The repair process is called **non-homologous end-joining** (**NHEJ**), the name indicating that there is no need for homology between the two molecules whose ends are being joined, unlike other end-joining mechanisms that we will encounter when we study recombination in Section 14.3. NHEJ is looked on as a type of recombination because, as well as repairing breaks, it can be used to join molecules or fragments that were not previously joined, producing new combinations. A version of the NHEJ system is probably used during construction of immunoglobulin and T-cell receptor genes, but the details are likely to be different because these programmed rearrangements of the genome involve intermediate structures, such as DNA hairpin loops, that are not seen during the repair of DNA breaks resulting from damage.

14.2.5 Bypassing DNA damage during genome replication

If a region of the genome has suffered extensive damage then it is conceivable that the repair processes will be overwhelmed. The cell then faces a stark choice between dying or attempting to replicate the damaged region even though this replication may be error-prone and result in mutated daughter molecules. When faced with this choice *E. coli* cells invariably take the second option, by inducing one of several emergency procedures for bypassing sites of major damage. The best studied of these bypass processes is the **SOS response**, which enables the cell to replicate its DNA even though the template polynucleotides contain AP sites and/or cyclobutyl dimers and other photoproducts resulting from exposure to chemical mutagens or UV radiation that would normally block, or at least delay, the replication complex. Bypass of these sites requires construction of a **mutasome**, comprising the UmuD'$_2$C complex (also called DNA polymerase V, a trimer made up of two UmuD' proteins and one copy of UmuC) and several copies of the **RecA protein** (Goodman, 2000). The latter is a single-stranded DNA-binding protein that coats the damaged strands, enabling the UmuD'$_2$C complex to displace DNA polymerase III and carry out error-prone DNA synthesis until the damaged region has been passed and DNA polymerase III can take over once again (*Figure 14.27*).

The SOS response is primarily looked on as the last best chance that the bacterium has to replicate its DNA and hence survive under adverse conditions. However, the price of survival is an increased mutation rate because the mutasome does not repair damage, it simply allows a damaged region of a polynucleotide to be replicated. When it encounters a damaged position in the template

(A) The non-homologous end-joining repair process

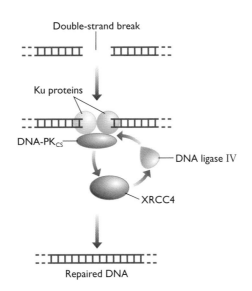

(B) The structure of the Ku–DNA complex

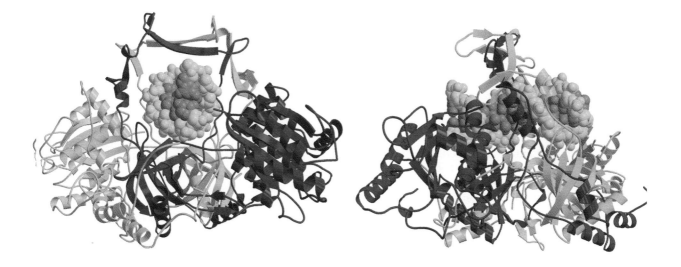

Figure 14.26 Non-homologous end-joining (NHEJ) in humans.

(A) The repair process. Additional proteins not shown in the diagram are also involved in NHEJ. These include the protein kinases ATM and ATR (Section 13.3.2), whose main role may be to signal to the cell the fact that a double-strand break has occurred and the cell cycle should be arrested until the break is repaired. If the cell enters mitosis with a broken chromosome then part of that chromosome will be lost, because only one of the fragments will contain a centromere, and centromeres are essential for distribution of chromosomes to the daughter nuclei during anaphase (see *Figure 5.14*, page 137). (B) Structure of the Ku–DNA complex. On the left is the view looking down onto the broken end of the DNA double helix, and on the right is the side view, with the broken end of the DNA molecule on the left. Ku is a heterodimer made up of the Ku70 and Ku80 subunits, the numbers indicating the molecular masses in kDa. Ku70 is colored red and Ku80 orange. The DNA molecule is shown in gray. Reprinted with permission from Walker *et al.*, *Nature*, **412**, 607–614. Copyright Macmilllan Magazines Limited.

DNA, the polymerase selects a nucleotide more or less at random, although with some preference for placing an A opposite an AP site: in effect the error rate of the replication process increases. It has been suggested that this increased mutation rate is the purpose of the SOS response,

mutation being in some way an advantageous response to DNA damage, but this idea remains controversial (Chicurel, 2001).

For some time, the SOS response was thought to be the only damage-bypass process in bacteria, but we now

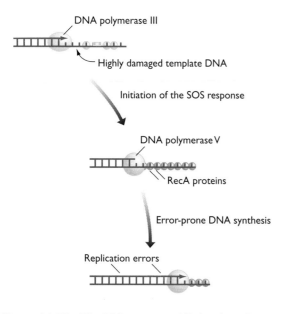

Figure 14.27 The SOS response of *Escherichia coli*. See the text for details.

appreciate that at least two other *E. coli* polymerases act in a similar way, although with different types of damage. These are DNA polymerase II, which can bypass nucleotides bound to mutagenic chemicals such as *N*-2-acetylaminofluorene, and DNA polymerase IV (also called DinB), which can replicate through a region of template DNA in which the two parent polynucleotides have become misaligned (Lindahl and Wood, 1999; Hanaoka, 2001). Bypass polymerases have also been discovered in eukaryotic cells. These include DNA polymerase η, which can bypass cyclobutyl dimers (Johnson *et al.*, 1999), and DNA polymerases ι and ζ, which work together to replicate through photoproducts and AP sites (Johnson *et al.*, 2000).

14.2.6 Defects in DNA repair underlie human diseases, including cancers

The importance of DNA repair is emphasized by the number and severity of inherited human diseases that have been linked with defects in one of the repair processes. One of the best characterized of these is xeroderma pigmentosum, which results from a mutation in any one of several genes for proteins involved in nucleotide excision repair. Nucleotide excision is the only way in which human cells can repair cyclobutyl dimers and other photoproducts, so it is no surprise that the symptoms of xeroderma pigmentosum include hypersensitivity to UV radiation, patients suffering more mutations than normal on exposure to sunlight, which often leads to skin cancer (Lehmann, 1995). Trichothiodystrophy is also caused by defects in nucleotide excision repair, but this is

a more complex disorder which, although not involving cancer, usually includes problems with both the skin and nervous system.

A few diseases have been linked with defects in the transcription-coupled component of nucleotide excision repair. These include breast and ovarian cancers, the *BRCA1* gene that confers susceptibility to these cancers coding for a protein that has been implicated, at least indirectly, with transcription-coupled repair (Gowen *et al.*, 1998), and Cockayne syndrome, a complex disease manifested by growth and neurologic disorders (Hanawalt, 2000). A deficiency in transcription-coupled repair has also been identified in humans suffering from the cancer-susceptibility syndrome called HNPCC (hereditary non-polyposis colorectal cancer; Mellon *et al.*, 1996), although this disease was originally identified as a defect in mismatch repair (Kolodner, 1995). Ataxia telangiectasia, the symptoms of which include sensitivity to ionizing radiation, results from defects in the *ATX* gene, which is involved in the damage-detection process (Section 13.3.2). Other diseases that are associated with a breakdown in DNA repair are Bloom's and Werner's syndromes, which are caused by inactivation of a DNA helicase that may have a role in NHEJ (Shen and Loeb, 2000; Wu and Hickson, 2001), and Fanconi's anemia, which confers sensitivity to chemicals that cause crosslinks in DNA but whose biochemical basis is not yet known.

14.3 Recombination

Without recombination, genomes would be relatively static structures, undergoing very little change. The gradual accumulation of mutations over a long period of time would result in small-scale alterations in the nucleotide sequence of the genome, but more extensive restructuring, which is the role of recombination, would not occur, and the evolutionary potential of the genome would be severely restricted.

Recombination was first recognized as the process responsible for crossing-over and exchange of DNA segments between homologous chromosomes during meiosis of eukaryotic cells (see *Figure 5.15*, page 138), and was subsequently implicated in the integration of transferred DNA into bacterial genomes after conjugation, transduction or transformation (Section 5.2.4). The biological importance of these processes stimulated the first attempts to describe the molecular events involved in recombination and led to the Holliday model (Holliday, 1964), with which we will begin our study of recombination.

14.3.1 Homologous recombination

The Holliday model refers to a type of recombination called **general** or **homologous recombination**. This is the most important version of recombination in nature, being responsible for meiotic crossing-over and the integration of transferred DNA into bacterial genomes.

The Holliday model for homologous recombination

The Holliday model describes recombination between two homologous double-stranded molecules, ones with identical or nearly identical sequences, but is equally applicable to two different molecules that share a limited region of homology, or a single molecule that recombines with itself because it contains two separate regions that are homologous with one another.

The central feature of the model is formation of a **heteroduplex** resulting from the exchange of polynucleotide segments between the two homologous molecules (*Figure 14.28*). The heteroduplex is initially stabilized by base-pairing between each transferred strand and the intact polynucleotide of the recipient molecule, this base-pairing being possible because of the sequence similarity between the two molecules. Subsequently the gaps are sealed by DNA ligase, giving a **Holliday structure**. This structure is dynamic, **branch migration** resulting in exchange of longer segments of DNA if the two helices rotate in the same direction.

Separation, or **resolution**, of the Holliday structure back into individual double-stranded molecules occurs by cleavage across the branch point. This is the key to the entire process because the cut can be made in either of two orientations, as becomes apparent when the three-dimensional configuration or **chi form** of the Holliday

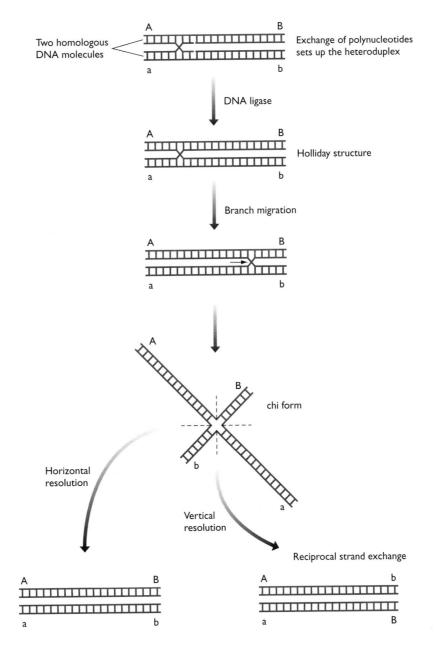

Figure 14.28   The Holliday model for homologous recombination.

structure is examined (see *Figure 14.28*). These two cuts have very different results. If the cut is made left–right across the chi form as drawn in *Figure 14.28*, then all that happens is that a short segment of polynucleotide, corresponding to the distance migrated by the branch of the Holliday structure, is transferred between the two molecules. On the other hand, an up–down cut results in **reciprocal strand exchange**, double-stranded DNA being transferred between the two molecules so that the end of one molecule is exchanged for the end of the other molecule. This is the DNA transfer seen in crossing-over.

So far we have ignored one aspect of the Holliday model. This is the way in which the two double-stranded molecules interact at the beginning of the process to produce the heteroduplex. In the original scheme, the two molecules lined up with one another and single-stranded nicks appeared at equivalent positions in each helix. This produced free single-stranded ends that could be exchanged, resulting in the heteroduplex (*Figure 14.29A*).

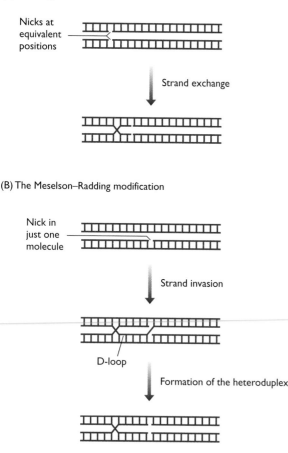

(A) The original model

Nicks at equivalent positions

Strand exchange

(B) The Meselson–Radding modification

Nick in just one molecule

Strand invasion

D-loop

Formation of the heteroduplex

Figure 14.29 Two schemes for initiation of homologous recombination.

(A) Initiation as described by the original model for homologous recombination. (B) The Meselson–Radding modification, which proposes a more plausible series of events for formation of the heteroduplex.

This feature of the model was criticized because no mechanism could be proposed for ensuring that the nicks occurred at precisely the same position on each molecule. The Meselson–Radding modification (Meselson and Radding, 1975) proposes a more satisfactory scheme whereby a single-stranded nick occurs in just one of the double helices, the free end that is produced 'invading' the unbroken double helix at the homologous position and displacing one of its strands, forming a **D-loop** (*Figure 14.29B*). Subsequent cleavage of the displaced strand at the junction between its single-stranded and base-paired regions produces the heteroduplex.

Proteins involved in homologous recombination in E. coli

The Holliday model and Meselson–Radding modification refer to homologous recombination in all organisms but, as with many areas of molecular biology, the initial progress in understanding how the process is carried out in the cell was made with *E. coli*. The specific recombination system that has been studied has the circular *E. coli* genome as one partner and a linear chromosome fragment as the second partner, this being the situation that occurs during conjugation, transduction or transformation of bacterial cells (Section 5.2.4).

Mutation studies have identified a number of *E. coli* genes that, when inactivated, give rise to defects in homologous recombination, indicating that their protein products are involved in the process in some way. Three distinct recombination systems have been described, these being the RecBCD, RecE and RecF pathways, with RecBCD apparently being the most important in the bacterium (Camerini-Otero and Hsieh, 1995). In this pathway, recombination is initiated by the **RecBCD enzyme**, which has both nuclease and helicase activities. Its precise mode of action is uncertain: in the simplest model the enzyme binds to one end of the linear molecule and unwinds it until it reaches the first copy of the eight-nucleotide consensus sequence 5'–GCTGGTGG–3' (rather confusingly called the **chi site**), which occurs once every 6 kb in *E. coli* DNA (Blattner *et al.*, 1997). The nuclease activity of the enzyme then makes the single-stranded nick at a position approximately 56 nucleotides to the 3' side of the chi site (*Figure 14.30*). Alternative proposals have the RecBCD enzyme making nicks as it progresses along the linear DNA, this activity being inhibited when the chi site is reached, the last of these progressive nicks being equivalent to the single nick envisaged in the first model (Eggleston and West, 1996).

Whatever the precise mechanism, the RecBCD enzyme produces the free single-stranded end which, according to the Meselson–Radding modification, invades the intact partner, in this case the circular *E. coli* genome. This stage is mediated by the RecA protein, which forms a protein-coated DNA filament that is able to invade the intact double helix and set up the D-loop (see *Figure 14.30*). An intermediate in formation of the D-loop is probably a **triplex** structure, a three-stranded DNA helix in

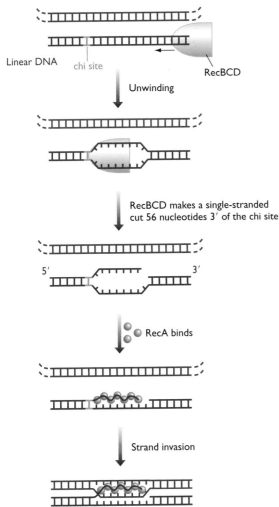

Circular DNA

Linear DNA   chi site

RecBCD

Unwinding

RecBCD makes a single-stranded
cut 56 nucleotides 3′ of the chi site

5′   3′

RecA binds

Strand invasion

Figure 14.30 The RecBCD pathway for homologous recombination in *Escherichia coli*.

The events leading to formation of the heteroduplex are shown. In the bottom structure the RecA-coated DNA filament has formed a triplex structure, which is thought to be an intermediate in the process. See the text for details.

which the invading polynucleotide lies within the major groove of the intact helix and forms hydrogen bonds with the base pairs it encounters (Camerini-Otero and Hsieh, 1995).

Branch migration is catalyzed by the RuvA and RuvB proteins, both of which attach to the branch point of the Holliday structure. X-ray crystallography studies suggest that four copies of RuvA bind directly to the branch, forming a core to which two RuvB rings, each consisting of eight proteins, attach, one to either side (*Figure 14.31*; Rafferty *et al.*, 1996). The resulting structure might act as a 'molecular motor', rotating the helices in the required manner so that the branch point moves. The RecG protein also has a role in branch migration but it is not clear if this

is in conjunction with RuvAB, or as part of an alternative mechanism (Eggleston and West, 1996).

Branch migration does not appear to be a random process, but instead stops preferentially at the sequence $5'-\frac{A}{T}$ TT $\frac{G}{C}$ $-3'$. This sequence occurs frequently in the *E. coli* genome, so presumably migration does not halt at the first instance of the motif that is reached. When branch migration has ended, the RuvAB complex detaches and is replaced by two RuvC proteins (see *Figure 14.31*) which carry out the cleavage that resolves the Holliday structure. The cuts are made between the second T and the $\frac{G}{C}$ components of the recognition sequence.

The double-strand break model for recombination in yeast

Although the Holliday model for homologous recombination, either in its original form or as modified by Meselson and Radding, explains most of the results of recombination in all organisms, it has a few inadequacies, which prompted the development of alternative schemes. In particular, it was thought that the Holliday model could not explain **gene conversion**, a phenomenon first described in yeast and fungi but now known to occur with many eukaryotes. In yeast, fusion of a pair of gametes results in a zygote that gives rise to an ascus containing four haploid spores whose genotypes can be individually determined. If the gametes have different alleles at a particular locus then under normal circumstances two of the spores will display one genotype and two will display the other genotype, but sometimes this expected 2 : 2 segregation pattern is replaced by an

---

Box 14.2: The RecE and RecF recombination pathways of *Escherichia coli*

The RecBCD, RecE and RecF pathways involve similar mechanisms and share several of the same proteins. RecA is involved in each pathway, RecF is a component of both the RecE and RecF pathways, and other proteins such as RecJ, RecO and RecQ are common to two or more of the systems. The precise functions of the different pathways are not entirely clear. In normal *E. coli* cells most recombination takes place via RecBCD, but if this pathway is inactivated by mutation the RecE system is able to take over, to be replaced by RecF when RecE is inactivated. These observations suggest that the functions of the three pathways overlap, but that there are dissimilarities between them. For example, only the RecBCD pathway initiates recombination at the chi sites scattered around the *E. coli* genome, and only RecF is able to induce recombination between a pair of plasmids.

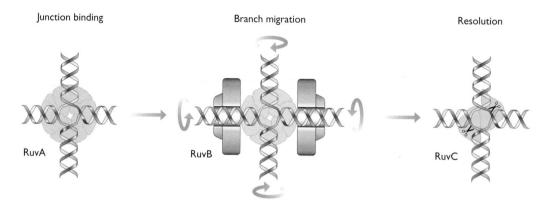

Junction binding   Branch migration   Resolution

RuvA   RuvB   RuvC

Figure 14.31 The role of the Ruv proteins in homologous recombination in *Escherichia coli*.

Branch migration is induced by a structure comprising four copies of RuvA bound to the Holliday junction with an RuvB ring on either side. After RuvAB has detached, two RuvC proteins bind to the junction, the orientation of their attachment determining the direction of the cuts that resolve the structure. Reprinted with permission from Rafferty JB *et al.* (1996) *Science*, **274**, 415–421. Copyright 1996 American Association for the Advancement of Science.

unexpected 3 : 1 ratio (*Figure 14.32*). This is called gene conversion because the ratio can only be explained by one of the alleles 'converting' from one type to the other, presumably by recombination during the meiosis that occurs after the gametes have fused.

The double-strand break model provides an opportunity for gene conversion to take place during the recombination process. It initiates not with a single-strand nick, as in the Holliday scheme, but with a double-strand cut that breaks one of the partners in the recombination into two pieces (*Figure 14.33*). This might appear to be a drastic move to make but it has been shown that the protein responsible for the cut is a Type II DNA topoisomerase (Section 13.1.2) which forms covalent linkages with the two pieces of DNA and hence prevents them drifting completely apart. After the double-stranded cut, one strand in each half of the molecule is trimmed back by a 5′→3′ exonuclease, so each end now has a 3′ overhang of approximately 500 nucleotides. One of these invades the homologous DNA molecule in a manner similar to that envisaged by the Meselson–Radding scheme, setting up a Holliday junction that can migrate along the heteroduplex if the invading strand is extended by a DNA polymerase. To complete the heteroduplex, the other broken strand (the one not involved in the Holliday junction) is also extended. Note that both DNA syntheses involve extension of strands from the partner that suffered the double-stranded cut, using as templates the equivalent regions of the uncut partner. This is the basis of the gene conversion because it means that the polynucleotide segments removed from the cut partner by the exonuclease have been replaced with copies of the DNA from the uncut partner.

The resulting heteroduplex has a pair of Holliday structures that can be resolved in a number of ways, some resulting in gene conversion and others giving a standard reciprocal strand exchange. An example leading to gene conversion is shown in *Figure 14.33*.

The double-strand break model has been sufficiently well characterized in yeast for there to be little doubt that it occurs, at least in a form approximating to that shown in *Figure 14.33*. Some of the proteins involved in recombination in yeast are very similar to their counterparts in *E. coli* – eukaryotic RAD51, for example, has sequence similarity with RecA and is believed to work in the same way (Baumann and West, 1998) – prompting the suggestion that recombination in all organisms follows the double-strand break system. As yet there is little evidence to support this idea, particularly for the larger chromosomes of higher eukaryotes, and many geneticists resist the suggestion that vertebrate DNA undergoes frequent double-strand breaks during meiosis.

14.3.2 Site-specific recombination

A region of extensive homology is not a prerequisite for recombination: the process can also be initiated between two DNA molecules that have only very short sequences in common. This is called **site-specific recombination** and it has been extensively studied because of the part that it plays during the infection cycle of bacteriophage λ.

Integration of λ DNA into the E. coli *genome involves site-specific recombination*

After injecting its DNA into an *E. coli* cell, bacteriophage λ can follow either of two infection pathways (Section 4.2.1). One of these, the lytic pathway, results in the rapid synthesis of λ coat proteins, combined with replication of the λ genome, leading to death of the bacterium and release of new phages within about 20 minutes of the initial infection. In contrast, if the phage follows the lysogenic pathway, new phages do not immediately appear. The bacterium divides as normal, possibly for many cell divisions, with the phage in a quiescent form called the

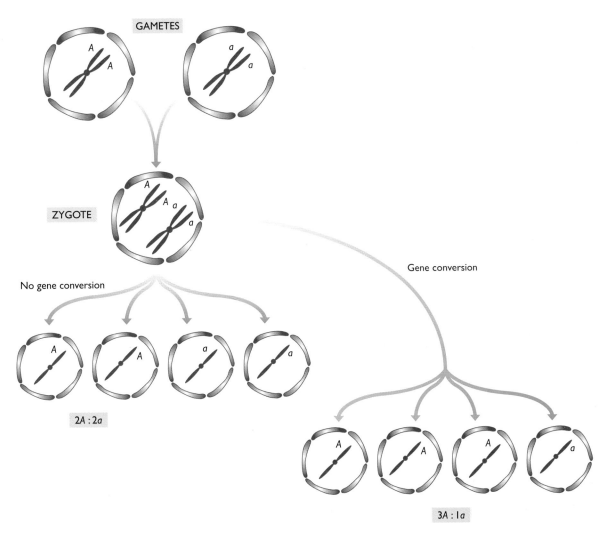

GAMETES

ZYGOTE

No gene conversion

Gene conversion

2A : 2a

3A : 1a

Figure 14.32   Gene conversion.

One gamete contains allele A and the other contains allele a. These fuse to produce a zygote that gives rise to four haploid spores, all contained in a single ascus. Normally, two of the spores will have allele A and two will have allele a, but if gene conversion occurs the ratio will be changed, possibly to 3A : 1a as shown here.

prophage. Eventually, possibly as the result of DNA damage or some other stimulus, the phage becomes active again, replicating its genome, directing synthesis of coat proteins, and bursting from the cell.

During the lysogenic phase the λ genome becomes integrated into the *E. coli* chromosome. It is therefore replicated whenever the *E. coli* DNA is copied, and so is passed on to daughter cells as if a standard part of the bacterium's genome. Integration occurs by site-specific recombination between the *att* sites, one on the λ genome and one on the *E. coli* chromosome, which have at their center an identical 15-bp sequence (*Figure 14.34*). Because this is recombination between two circular molecules, the result is that one bigger circle is formed; in other words the λ DNA becomes integrated into the bacterial genome. A second site-specific recombination between the two *att*

sites, now both contained in the same molecule, reverses the original process and releases the λ DNA, which can now return to the lytic mode of infection and direct synthesis of new phages.

The recombination event is catalyzed by a specialized Type I topoisomerase (Section 13.1.2) called **integrase** (Kwon *et al.*, 1997), a member of a diverse family of **recombinases** present in bacteria, archaea and yeast. The enzyme makes a staggered double-stranded cut at equivalent positions in the λ and bacterial *att* sites. The two short single-stranded overhangs are then exchanged between the DNA molecules, producing a Holliday junction which migrates a few base pairs along the heteroduplex before being cleaved. This cleavage, providing that it is made in the appropriate orientation, resolves the Holliday structure in such a way that the λ DNA becomes

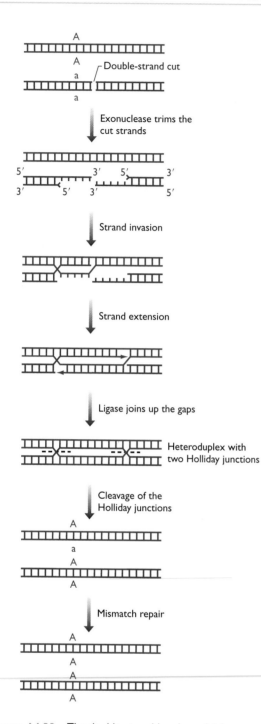

Figure 14.33 The double-strand break model for recombination in yeast.

This model explains how gene conversion can occur. See the text for details.

inserted into the *E. coli* genome. A similar process underlies excision, which is also carried out by integrase, but in conjunction with a second protein, 'excisionase', coded by the λ *xis* gene. If integrase could carry out excision on its own then it would probably excise the λ DNA as soon as it had integrated it.

14.3.3 Transposition

Transposition is not a type of recombination but a process that utilizes recombination, the end result being the transfer of a segment of DNA from one position in the genome to another. A characteristic feature of transposition is that the transferred segment is flanked by a pair of short direct repeats (*Figure 14.35*) which, as we will see, are formed during the transposition process.

In Section 2.4.2 we examined the various types of transposable element known in eukaryotes and prokaryotes and discovered that these could be broadly divided into three categories on the basis of their transposition mechanism (*Figure 14.36*):

- DNA transposons that transpose replicatively, the original transposon remaining in place and a new copy appearing elsewhere in the genome;
- DNA transposons that transpose conservatively, the original transposon moving to a new site by a cut-and-paste process;
- Retroelements, all of which transpose via an RNA intermediate.

Box 14.3: DNA methylation and transposition

Transposition can have deleterious effects on a genome. These effects go beyond the obvious disruption of gene activity that will occur if a transposable element takes up a new position that lies within the coding region of a gene. Some elements, notably retrotransposons, contain promoter and enhancer sequences that can modify the expression patterns of adjacent genes, and transposition often involves the creation of double-stranded breaks which, as we saw in Section 14.2.4, can have seriously harmful effects on the integrity of a genome. It might therefore be expected that genomes have evolved mechanisms for limiting the movement of their transposable elements. One way in which this could be achieved is by methylation of DNA sequences within transposable elements, methylation being a common means of silencing genomic regions (Section 8.2.2). Many transposable element sequences are indeed hypermethylated, but experimental evidence linking this methylation with suppression of transposition has been difficult to obtain. Recently it has been shown that mutants of the plant *Arabidopsis thaliana* that have deficient methylation systems suffer from a higher than normal amount of transposition (Symer and Bender, 2001), but this heightened transposition might not be displayed by all types of transposon in the plant genome. Further work is underway to explore the role of methylation in *Arabidopsis* and to extend the studies to other types of organism.

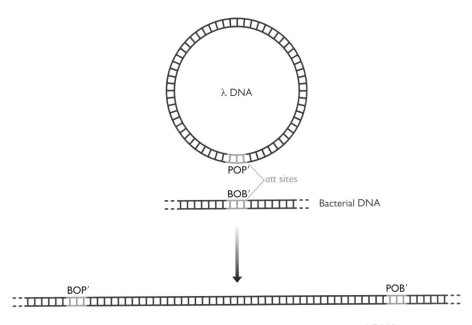

Figure 14.34 Integration of the bacteriophage λ genome into *Escherichia coli* chromosomal DNA.

Both λ and *E. coli* DNA have a copy of the *att* site, each one comprising an identical central sequence called 'O' and flanking sequences P and P′ (for the phage *att* site) or B and B′ (bacterial *att* site). Recombination between the O regions integrates the λ genome into the bacterial DNA.

We will now examine the recombination events that are responsible for each of these three types of transposition.

Replicative and conservative transposition of DNA transposons

A number of models for replicative and conservative transposition have been proposed over the years but most are modifications of a scheme originally outlined by Shapiro (1979). According to this model, the replicative transposition of a bacterial element such as a Tn3-type transposon or a transposable phage (Section 2.4.2) is initiated by one or more endonucleases that make single-stranded cuts either side of the transposon and in the target site where the new copy of the element will be inserted (*Figure 14.37*). At the target site the two cuts are separated by a few base pairs, so that the cleaved double-stranded molecule has short 5′ overhangs.

Ligation of these 5′ overhangs to the free 3′ ends either side of the transposon produces a hybrid molecule in which the original two DNAs – the one containing the transposon and the one containing the target site – are linked together by the transposable element flanked by a pair of structures resembling replication forks. DNA synthesis at these replication forks copies the transposable element and converts the initial hybrid into a **co-integrate**,

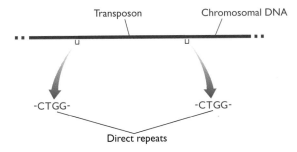
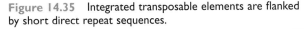

Figure 14.35 Integrated transposable elements are flanked by short direct repeat sequences.

This particular transposon is flanked by the tetranucleotide repeat 5′–CTGG–3′. Other transposons have different direct repeat sequences.

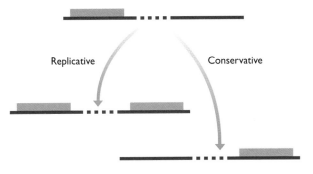

Figure 14.36 Replicative and conservative transposition.

DNA transposons use either the replicative or conservative pathway (some can use both). Retroelements transpose replicatively via an RNA intermediate.

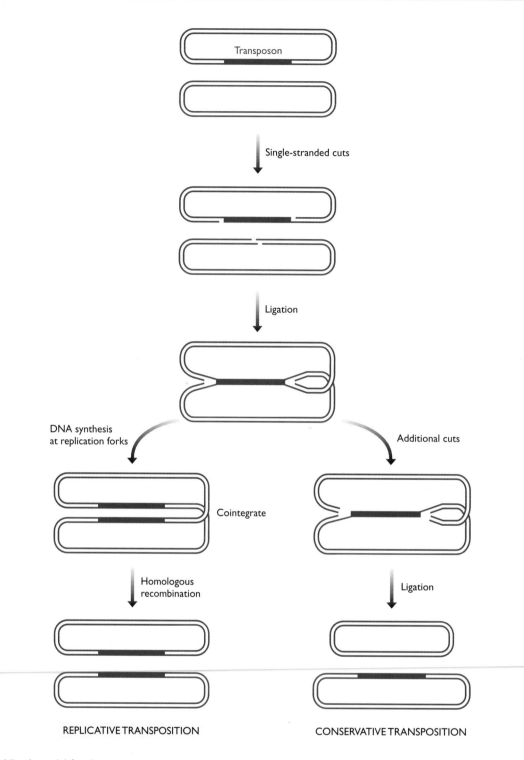

Figure 14.37 A model for the process resulting in replicative and conservative transposition.

See the text for details.

in which the two original DNAs are still linked. Homologous recombination between the two copies of the transposon uncouples the co-integrate, separating the original DNA molecule (with its copy of the transposon still in place) from the target molecule, which now contains a copy of the transposon. Replicative transposition has therefore occurred.

A modification of the process just described changes the mode of transposition from replicative to conservative (see *Figure 14.37*). Rather than carrying out DNA

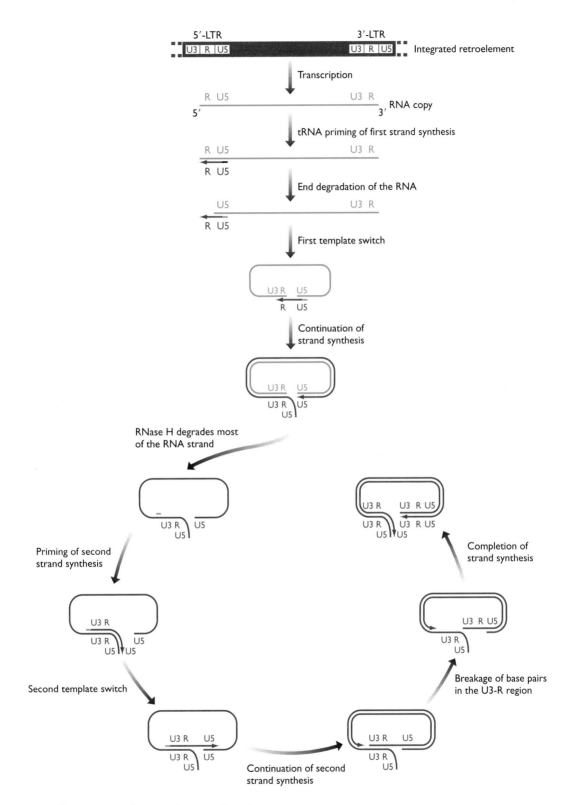

Figure 14.38 Transposition of a retroelement – Part 1.

This diagram shows how an integrated retroelement is copied into a free double-stranded DNA version. The first step is synthesis of an RNA copy, which is then converted to double-stranded DNA by a series of events that involves two template switches, as described in the text.

synthesis, the hybrid structure is converted back into two separate DNA molecules simply by making additional single-stranded nicks either side of the transposon. This cuts the transposon out of its original molecule, leaving it 'pasted' into the target DNA.

Transposition of retroelements

From the human perspective, the most important retroelements are the retroviruses, which include the human immunodeficiency viruses that cause AIDS and various other virulent types. Most of what we know about retrotransposition refers specifically to retroviruses, although it is believed that other retroelements, such as retrotransposons of the *Ty1/copia* and *Ty3/gypsy* families, transpose by similar mechanisms.

The first step in retrotransposition is synthesis of an RNA copy of the inserted retroelement (*Figure 14.38*). The long terminal repeat (LTR) at the 5' end of the element contains a TATA sequence which acts as a promoter for transcription by RNA polymerase II (Section 9.2.2). Some retroelements also have enhancer sequences (Section 9.3) that are thought to regulate the amount of transcription that occurs. Transcription continues through the entire length of the element, up to a polyadenylation sequence (Section 10.1.2) in the 3' LTR. The transcript now acts as the template for RNA-dependent DNA synthesis, catalyzed by a reverse transcriptase enzyme coded by part of the *pol* gene of the retroelement (see *Figure 2.26*, page 61). Because this is synthesis of DNA, a primer is required (Section 13.2.2), and as during genome replication, the primer is made of RNA rather than DNA. During genome replication, the primer is synthesized *de novo* by a polymerase enzyme (see *Figure 13.12*, page 396), but retroelements do not code for RNA polymerases and so cannot make primers in this way. Instead they use one of the cell's tRNA molecules as a primer, which one depending on the retroelement: the *Ty1/copia* family of elements always use tRNA^Met but other retroelements use different tRNAs.

The tRNA primer anneals to a site within the 5' LTR (see *Figure 14.38*) At first glance this appears to be a strange location for the priming site because it means that DNA synthesis is directed away from the central region of the retroelement and so results in only a short copy of part of the 5' LTR. In fact, when the DNA copy has been extended to the end of the LTR, a part of the RNA template is degraded and the DNA overhang that is produced re-anneals to the 3' LTR of the retroelement which, being a long terminal *repeat*, has the same sequence as the 5' LTR and so can base-pair with the DNA copy. DNA synthesis now continues along the RNA template, eventually displacing the tRNA primer. Note that the result is a DNA copy of the entire template, including the priming site: the template switching is, in effect, the strategy that the retroelement uses to solve the 'end-shortening' problem, the same problem that chromosomal DNAs address through telomere synthesis (Section 13.2.4).

Completion of synthesis of the first DNA strand results in a DNA–RNA hybrid. The RNA is partially degraded by an RNase H enzyme, coded by another part of the *pol* gene. The RNA that is not degraded, usually just a single fragment attached to a short polypurine sequence adjacent to the 3' LTR, primes synthesis of the second DNA strand, again by reverse transcriptase, which is able to act as both an RNA- and DNA-dependent DNA polymerase. As with the first round of DNA synthesis, second-strand synthesis initially results in a DNA copy of just the LTR, but a second template switch, to the other end of the molecule, enables the DNA copy to be extended until it is full length. This creates a template for further extension of the first DNA strand, so that the resulting double-stranded DNA is a complete copy of the internal region of the retroelement plus the two LTRs.

All that remains is to insert the new copy of the retroelement into the genome. It was originally thought that insertion occurred randomly, but it now appears that

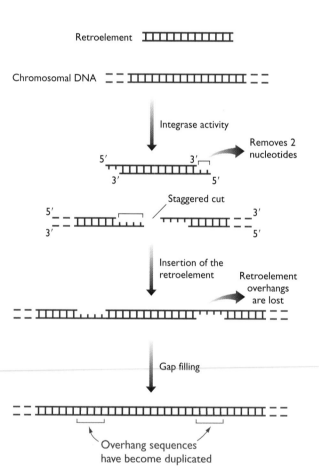

Figure 14.39 Transposition of a retroelement – Part 2.

Integration of the retroelement into the genome results in a four-nucleotide direct repeat either side of the inserted sequence. With retroviruses, this stage in the transposition pathway requires both integrase and the DNA-PK$_{CS}$ protein kinase and Ku proteins that are also involved in NHEJ (Section 14.2.4; Daniel *et al.*, 1999).

although no particular sequence is used as a target site, integration occurs preferentially at certain positions (Devine and Boeke, 1996). Insertion involves removal of two nucleotides from the 3′ ends of the double-stranded retroelement by the integrase enzyme (coded by yet another part of *pol*). The integrase also makes a staggered cut in the genomic DNA so that both the retroelement and the integration site now have 5′ overhangs (*Figure 14.39*). These overhangs might not have complementary sequences but they still appear to interact in some way so that the retroelement becomes inserted into the genomic DNA. The interaction results in loss of the retroelement overhangs and filling in of the gaps that are left, which means that the integration site becomes duplicated into a pair of direct repeats, one at either end of the inserted retroelement.

References

Andersson DI, Slechta ES and Roth JR (1998) Evidence that gene amplification underlies adaptive mutability of the bacterial *lac* operon. *Science*, **282**, 1133–1135.

Ashley CT and Warren ST (1995) Trinucleotide repeat expansion and human disease. *Ann. Rev. Genet.*, **29**, 703–728.

Baumann P and West SC (1998) Role of the human RAD51 protein in homologous recombination and double-stranded-break repair. *Trends Biochem. Sci.*, **23**, 247–251.

Blattner FR, Plunkett G, Bloch CA, et al. (1997) The complete genome sequence of *Escherichia coli* K-12. *Science*, **277**, 1453–1462.

Cairns J, Overbaugh J and Miller S (1988) The origin of mutants. *Nature*, **335**, 142–145.

Camerini-Otero RD and Hsieh P (1995) Homologous recombination proteins in prokaryotes and eukaryotes. *Ann. Rev. Genet.*, **29**, 509–552.

Campuzano V, Montermini L, Moltò MD, et al. (1996) Friedreich's ataxia: autosomal recessive disease caused by an intronic GAA triplet repeat expansion. *Science*, **271**, 1423–1427.

Cascalho M, Wong J, Steinberg C and Wabl M (1998) Mismatch repair co-opted by hypermutation. *Science*, **279**, 1207–1210.

Chicurel M (2001) Can organisms speed their own mutation? *Science*, **292**, 1824–1827.

Cooper PK, Nouspikel T, Clarkson SG and Leadon SA (1997) Defective transcription-coupled repair of oxidative base damage in Cockayne syndrome patients from XP group G. *Science*, **275**, 990–993.

Cotton RGH (1997) Slowly but surely towards better scanning for mutations. *Trends Genet.*, **13**, 43–46.

Critchlow SE and Jackson SP (1998) DNA end-joining: from yeast to man. *Trends Biochem. Sci.*, **23**, 394–398.

Cummings CJ and Zoghbi HY (2000) Trinucleotide repeats: mechanisms and pathophysiology. *Ann. Rev. Genomics Hum. Genet.*, **1**, 281–328.

Daniel R, Katz RA and Skalka AM (1999) A role for DNA-PK in retroviral DNA integration. *Science*, **284**, 644–647.

Devine SE and Boeke JD (1996) Integration of the yeast retrotransposon *Ty1* is targeted to regions upstream of genes transcribed by RNA polymerase III. *Genes Devel.*, **10**, 620–633.

Eggleston AK and West SC (1996) Exchanging partners in *E. coli*. *Trends Genet.*, **12**, 20–26.

Foster PL (1999) Mechanism of stationary phase mutation: a decade of adaptive mutation. *Ann. Rev. Genet.*, **33**, 57–88.

Francino MP and Ochman H (1997) Strand asymmetries in DNA evolution. *Trends Genet.*, **13**, 240–245.

Freudenreich CH, Kantrow SM and Zakian VA (1998) Expansion and length-dependent fragility of CTG repeats in yeast. *Science*, **279**, 853–856.

Gacy AM, Goellner GM, Spiro C, et al. (1998) GAA instability in Friedreich's Ataxia shares a common, DNA-directed and intraallelic mechanism with other trinucleotide diseases. *Mol. Cell*, **1**, 583–593.

Goodman MF (2000) Coping with replication 'train wrecks' in *Escherichia coli* using Pol V, Pol II and RecA proteins. *Trends Biochem. Sci.*, **25**, 189–195.

Gowen LC, Avrutskaya AV, Latour AM, Koller BH and Leadon SA (1998) BRCA1 required for transcription-coupled repair of oxidative DNA damage. *Science*, **281**, 1009–1012.

Hanaoka F (2001) SOS polymerases. *Nature*, **409**, 33–34.

Hanawalt PC (2000) The bases for Cockayne syndrome. *Nature*, **405**, 415–416.

Holliday R (1964) A mechanism for gene conversion in fungi. *Genet. Res.*, **5**, 282–304.

Johnson RE, Prakash S and Prakash L (1999) Efficient bypass of a thymine-thymine dimer by yeast DNA polymerase, Polη. *Science*, **283**, 1001–1004.

Johnson RE, Washington MT, Haracska L, Prakash S and Prakash L (2000) Eukaryotic polymerases ι and ζ act sequentially to bypass DNA lesions. *Nature*, **406**, 1015–1019.

Kolodner RD (1995) Mismatch repair: mechanisms and relationship to cancer susceptibility. *Trends Biochem. Sci.*, **20**, 397–401.

Kolodner RDE (2000) Guarding against mutation. *Nature*, **407**, 687–689.

Kunkel TA and Wilson SH (1996) Push and pull of base flipping. *Nature*, **384**, 25–26.

Kwon HJ, Tirumalai R, Landy A and Ellenberger T (1997) Flexibility in DNA recombination: structure of the lambda integrase catalytic core. *Science*, **276**, 126–131.

Lehmann AR (1995) Nucleotide excision repair and the link with transcription. *Trends Biochem. Sci.*, **20**, 402–405.

Lindahl T and Wood RD (1999) Quality control by DNA repair. *Science*, **286**, 1897–1905.

Mandel J-L (1997) Breaking the rule of three. *Nature*, **386**, 767–769.

Mellon I, Rajpal DK, Koi M, Boland CR and Champe GN (1996) Transcription-coupled repair deficiency and mutations in human mismatch repair genes. *Science*, **272**, 557–560.

Meselson M and Radding CM (1975) A general model for genetic recombination. *Proc. Natl Acad. Sci. USA*, **72**, 358–361.

Perutz MF (1999) Glutamine repeats and neurodegenerative diseases: molecular aspects. *Trends Biochem. Sci.*, **24**, 58–63.

Rafferty JB, Sedelnikova SE, Hargreaves D, et al. (1996) Crystal structure of DNA recombination protein RuvA and a model for its binding to the Holliday junction. *Science*, **274**, 415–421.

Richardson C and Jasin M (2000) Frequent chromosomal translocations induced by DNA double-strand breaks. *Nature*, **405**, 697–700.

Roberts RJ and Cheng X (1998) Base flipping. *Ann. Rev. Biochem.*, **67**, 181–198.

Seeberg E, Eide L and Bjørås M (1995) The base excision repair pathway. *Trends Biochem. Sci.*, **20**, 391–397.

Shannon M and Weigert M (1998) Fixing mismatches. *Science*, **279**, 1159–1160.

Shapiro JA (1979) Molecular model for the transposition and replication of bacteriophage Mu and other transposable elements. *Proc. Natl Acad. Sci. USA*, **76**, 1933–1937.

Shen J-C and Loeb LA (2000) The Werner syndrome gene: the molecular basis of RecQ helicase-deficiency diseases. *Trends Genet.*, **16**, 213–220.

Sobol RW, Horton JK, Kühn R, et al. (1996) Requirement of mammalian DNA polymerase-β in base-excision repair. *Nature*, **379**, 183–186.

Sutherland GR, Baker E and Richards RI (1998) Fragile sites still breaking. *Trends Genet.*, **14**, 501–506.

Svejstrup JQ, Vichi P and Egly J-M (1996) The multiple roles of transcription factor/repair factor TFIIH. *Trends Biochem. Sci.*, **21**, 346–350.

Symer DE and Bender J (2001) Hip-hopping out of control. *Nature*, **411**, 146–149.

Walker JR, Corpina RA and Goldberg J (2001) Structure of the Ku heterodimer bound to DNA and its implications for double-strand break repair. *Nature*, **412**, 607–614.

Wood RD, Mitchell M, Sgouros J and Lindahl T (2001) Human DNA repair genes. *Science*, **291**, 1284–1289.

Wu L and Hickson ID (2001) DNA ends RecQ-uire attention. *Science*, **292**, 229–230.

Further Reading

Buermeyer AB, Deschênes SM, Baker SM and Liskay RM (1999) Mammalian DNA mismatch repair. *Ann. Rev. Genet.*, **33**, 533–564.

Harfe BD and Jinks-Robertson S (2000) DNA mismatch repair and genetic instability. *Ann. Rev. Genet.*, **34**, 359–399. — *Comprehensive details of mismatch repair in bacteria and eukaryotes.*

Kowalcsykowski SC (2000) Initiation of genetic recombination and recombination-dependent replication. *Trends Biochem. Sci.*, **25**, 156–165. — *Contains useful summaries of various aspects of recombination.*

Kunkel TA and Bebenek K (2000) DNA replication fidelity. *Ann. Rev.*

Biochem., **69**, 497–529. — *Covers the processes that ensure that the minimum number of errors are made during DNA replication.*

Shinagawa H and Iwasaki H (1996) Processing the Holliday junction in homologous recombination. *Trends Biochem. Sci.*, **21**, 107–111. — *An illuminating description of the central event in recombination.*

Sutton MD, Smith BT, Godoy VG and Walker GC (2000) The SOS response: recent insights into *umuDC*-dependent mutagenesis and DNA damage tolerance. *Ann. Rev. Genet.*, **34**, 479–497.

West SC (1997) Processing of recombination intermediates by the RuvABC proteins. *Ann. Rev. Genet.*, **31**, 213–244. — *Comprehensive information on branch migration and resolution of Holliday junctions.*

STUDY AIDS FOR CHAPTER 14

Key terms

Give short definitions of the following terms:

(6–4) lesion
(6–4) photoproduct photolyase
2-aminopurine
5-bromouracil (5-bU)
Ada enzyme
Alkylating agent
Allele-specific oligonucleotide (ASO) hybridization
AP endonuclease
AP site
Auxotroph
Base analog
Base excision repair
Baseless site
Branch migration
Chi form
Chi site
Co-integrate
Conditional-lethal
Constitutive mutant
Cyclobutyl dimer
Dark repair
Deaminating agent
Delayed-onset
Deletion mutation

Direct repair
D-loop
DNA adenine methylase (Dam)
DNA cytosine methylase (Dcm)
DNA glycosylase
DNA photolyase
DNA repair
Ethidium bromide
Ethylmethane sulfonate (EMS)
Excision repair
Fragile site
Frameshift
Gain-of-function
Gene conversion
General recombination
Haploinsufficiency
Heteroduplex
Holliday structure
Homologous recombination
Hypermutation
Inhibitor resistance
Insertion mutation
Integrase

Intercalating agent
Leaky mutant
Long patch repair
Loss-of-function
MGMT
Minimal medium
Mismatch
Mismatch repair
Missense mutation
Mutagen
Mutasome
Mutation
Mutation scanning
Mutation screening
Nick
Non-homologous end-joining (NHEJ)
Non-penetrance
Nonsense mutation
Non-synonymous mutation
Nucleotide excision repair
Permissive conditions
Phosphodiesterase
Photoproduct
Photoreactivation
Point mutation
Programmed mutation
Proofreading

Prophage
Prototroph
Readthrough
RecA protein
RecBCD enzyme
Reciprocal strand exchange
Recombinase
Recombination
Recombination repair
Regulatory mutant
Replication slippage
Resolution
Restrictive conditions
Short patch repair
Silent mutation
Site-specific recombination
SOS response
Spontaneous mutation
Synonymous mutation
Tautomer
Temperature-sensitive mutation
Transcription-coupled repair
Trinucleotide repeat expansion disease
Triplex
UvrABC endonuclease

Self study questions

1. Distinguish between the terms 'mutation', 'DNA repair' and 'recombination'.
2. Explain how errors in DNA replication can lead to mutations.
3. Giving examples, summarize the key features of trinucleotide repeat expansion diseases.
4. List the various types of chemical and physical agents that have mutagenic properties. Give at least one example of each type of agent and describe the types of mutation that they cause.
5. Describe the various effects that a mutation can have on the coding properties of a genome.
6. Distinguish between the effects of mutations on the somatic and germ cells of multicellular organisms.
7. Name and define the four major types of mutant phenotype recognized in bacteria.
8. Describe, with examples, what is meant by the terms 'hypermutation' and 'programmed mutation'.
9. Distinguish between the various types of DNA repair mechanism that are known.
10. Compare and contrast the direct DNA repair systems of bacteria and eukaryotes.
11. Give detailed descriptions of the base excision and nucleotide excision repair processes of bacteria and eukaryotes.
12. Describe the mismatch repair processes of bacteria and eukaryotes, paying attention to the ways in which the daughter and parent strands are recognized in the two types of organism.
13. Define the term 'non-homologous end-joining' and explain how this process results in the repair of double-strand breaks in DNA molecules.
14. How can DNA damage be bypassed during genome replication in *Escherichia coli* and eukaryotes?
15. Discuss the links between DNA repair and human disease.
16. Draw a fully annotated diagram of the Holliday model for homologous recombination.
17. In what way does the Meselson–Radding modification improve the Holliday model for homologous recombination?
18. Describe the functions of each of the proteins thought to be involved in homologous recombination in *Escherichia coli*.
19. Draw a fully annotated diagram of the double-strand break model for recombination in yeast.
20. Describe how site-specific recombination underlies insertion and excision of the λ genome into and out of the *Escherichia coli* chromosome.
21. Explain how recombination events can result in the replicative or conservative transposition of a DNA sequence.
22. Draw a fully annotated diagram illustrating the transposition mechanism of a retrovirus.

Problem-based learning

1. Explore the current knowledge concerning trinucleotide repeat expansion diseases, including hypotheses that attempt to explain why triplet expansion in these genes leads to a disease.
2. 'Not all mutations have an immediate impact: some are delayed onset and only confer an altered phenotype later in the individual's life. Others display non-penetrance in some individuals, never being expressed even though the individual has a dominant mutation or is a homozygous recessive.' Devise mechanisms to explain how mutations can exhibit delayed onset or non-penetrance.
3. Evaluate the evidence for programmed mutations.
4. The bacterium *Deinococcus radiodurans* is highly resistant to radiation and to other physical and chemical mutagens. Discuss how these special properties of *D. radiodurans* are reflected in its genome sequence. (See White O, Eisen JA, Heidelberg JF, et al. [1999] Genome sequence of the radio-resistant bacterium *Deinococcus radiodurans* R1. *Science*, **286**, 1571–1577.)
5. Assess the general importance of the double-strand break model for gene conversion in yeast. Is there evidence for this type of recombination in organisms other than yeast?

15

How Genomes Evolve

Chapter Contents

15.1 Genomes: the First 10 Billion Years 460

    15.1.1 The origins of genomes 460

15.2 Acquisition of New Genes 465

    15.2.1 Acquisition of new genes by gene duplication 465

    15.2.2 Acquisition of new genes from other species 473

15.3 Non-coding DNA and Genome Evolution 476

    15.3.1 Transposable elements and genome evolution 476

    15.3.2 The origins of introns 477

15.4 The Human Genome: the Last 5 Million Years 479

Learning outcomes

When you have read Chapter 15, you should be able to:

- Speculate on the events that led to evolution of the first genomes

- Distinguish between the various ways in which genomes can obtain new genes

- Using examples, discuss the possible impacts that duplication of whole genomes and of individual genes or groups of genes has had on genome evolution

- Explain how new genes can arise by domain duplication and domain shuffling

- Assess the likely impact of lateral gene transfer on genome evolution in bacteria and in eukaryotes

- Outline how transposable elements may have influenced genome evolution

- Define and evaluate the 'introns early' and 'introns late' hypotheses

- List the differences between the human and chimpanzee genomes and discuss how such similar genomes can give rise to such different biological attributes

MUTATION AND RECOMBINATION provide the genome with the means to evolve, but we learn very little about the evolutionary histories of genomes simply by studying these events in living cells. Instead we must combine our understanding of mutation and recombination with comparisons between the genomes of different organisms in order to infer the patterns of genome evolution that have occurred. Clearly, this approach is imprecise and uncertain but, as we will see, it is based on a surprisingly large amount of hard data and we can be reasonably confident that, at least in outline, the picture that emerges is not too far from the truth.

In this chapter we will explore the evolution of genomes from the very origins of biochemical systems through to the present day. We will look at ideas regarding the **RNA world**, prior to the appearance of the first DNA molecules, and then examine how DNA genomes have gradually become more complex. Finally, in Section 15.4 we will compare the human genome with the genomes of other primates in order to identify the evolutionary changes that have occurred during the last five million years and which must, somehow, make us what we are.

15.1 Genomes: the First 10 Billion Years

Cosmologists believe that the universe began some 14 billion years ago with the gigantic 'primordial fireball' called the Big Bang. Mathematical models suggest that

after about 4 billion years galaxies began to fragment from the clouds of gas emitted by the Big Bang, and that within our own galaxy the solar nebula condensed to form the Sun and its planets about 4.6 billion years ago (*Figure 15.1*). The early Earth was covered with water and it was in this vast planetary ocean that the first biochemical systems appeared, cellular life being well established by the time land masses began to appear, some 3.5 billion years ago. But cellular life was a relatively late stage in biochemical evolution, being preceded by self-replicating polynucleotides that were the progenitors of the first genomes. We must begin our study of genome evolution with these precellular systems.

15.1.1 The origins of genomes

The first oceans are thought to have had a similar salt composition to those of today but the Earth's atmosphere, and hence the dissolved gases in the oceans, was very different. The oxygen content of the atmosphere remained very low until photosynthesis evolved, and to begin with the most abundant gases were probably methane and ammonia. Experiments attempting to recreate the conditions in the ancient atmosphere have shown that electrical discharges in a methane–ammonia mixture result in chemical synthesis of a range of amino acids, including alanine, glycine, valine and several of the others found in proteins (Miller, 1953). Hydrogen cyanide and formaldehyde are also formed, these participating in additional reactions to give other amino acids, as well as purines,

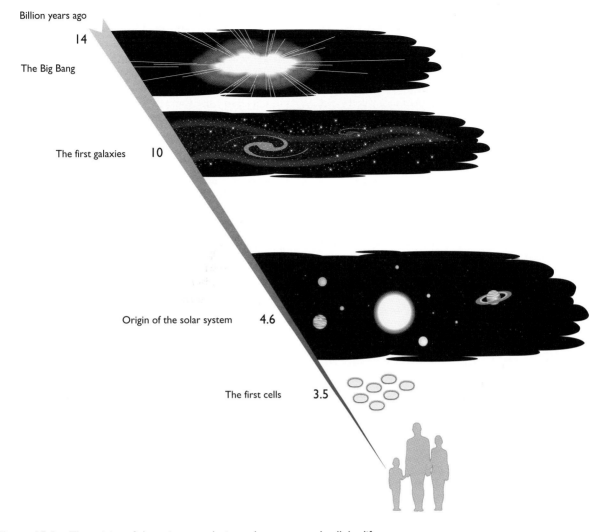

Figure 15.1 The origins of the universe, galaxies, solar system and cellular life.

pyrimidines and, in less abundance, sugars. At least some of the building blocks of biomolecules could therefore have accumulated in the ancient chemosphere.

The first biochemical systems were centered on RNA

Polymerization of the building blocks into biomolecules might have occurred in the oceans or could have been promoted by the repeated condensation and drying of droplets of water in clouds (Woese, 1979). Alternatively, polymerization might have taken place on solid surfaces, perhaps making use of monomers immobilized on clay particles (Wächtershäuser, 1988), or in hydrothermal vents (Wächtershäuser, 1992). The precise mechanism need not concern us; what is important is that it is possible to envisage purely geochemical processes that could lead to synthesis of polymeric biomolecules similar to the ones found in living systems. It is the next steps that we must worry about. We have to go from a random collec-

tion of biomolecules to an ordered assemblage that displays at least some of the biochemical properties that we associate with life. These steps have never been reproduced experimentally and our ideas are therefore based mainly on speculation tempered by a certain amount of computer simulation. One problem is that the speculations are unconstrained because the global ocean could have contained as many as 10^{10} biomolecules per liter and we can allow a billion years for the necessary events to take place. This means that even the most improbable scenarios cannot be dismissed out of hand and a way through the resulting maze has been difficult to find.

Progress was initially stalled by the apparent requirement that polynucleotides and polypeptides must work in harness in order to produce a self-reproducing biochemical system. This is because proteins are required to catalyze biochemical reactions but cannot carry out their own self-replication. Polynucleotides can specify the synthesis of proteins and self-replicate, but it was thought that they could do neither without the aid of proteins. It appeared that the biochemical system would have to spring fully formed from the random collection of biomolecules because any intermediate stage could not be perpetuated. The major breakthrough came in the mid-1980s when it was discovered that RNA can have catalytic activity. Those ribozymes that are known today carry out three types of biochemical reaction:

- Self-cleavage, as displayed by the self-splicing Group I, II and III introns and by some virus genomes (*Table 10.4*, page 298 and Section 10.2.3);
- Cleavage of other RNAs (as carried out by, for example, RNase P; *Table 10.4*, page 298 and Section 10.2.2);
- Synthesis of peptide bonds, by the rRNA component of the ribosome (Section 11.2.3 and Research Briefing 11.1, page 332).

In the test tube, synthetic RNA molecules have been shown to carry out other biologically relevant reactions such as synthesis of ribonucleotides (Unrau and Bartel, 1998), synthesis and copying of RNA molecules (Ekland and Bartel, 1996; Johnston *et al.*, 2001) and transfer of an RNA-bound amino acid to a second amino acid forming a dipeptide, in a manner analogous to the role of tRNA in protein synthesis (Section 11.1; Lohse and Szostak, 1996). The discovery of these catalytic properties solved the polynucleotide–polypeptide dilemma by showing that the first biochemical systems could have been centered entirely on RNA (Bartel and Unrau, 1999).

Ideas about the RNA world have taken shape in recent years (Robertson and Ellington, 1998). We now envisage that RNA molecules initially replicated in a slow and haphazard fashion simply by acting as templates for binding of complementary nucleotides which polymerized spontaneously (*Figure 15.2*). This process would have been very inaccurate so a variety of RNA sequences would have been generated, eventually leading to one or more with nascent ribozyme properties that were able to direct their own, more accurate self-replication. It is possible that a form of natural selection operated so that the most efficient replicating systems began to predominate, as has been shown to occur in experimental systems. A greater accuracy in replication would have enabled RNAs to increase in length without losing their sequence specificity, providing the potential for more sophisticated catalytic properties, possibly culminating in structures as complex as present-day Group I introns (see *Figure 10.26*, page 299) and ribosomal RNAs (see *Figure 11.11*, page 323).

To call these RNAs 'genomes' is a little fanciful, but the term **protogenome** has attractions as a descriptor for molecules that are self-replicating and able to direct simple biochemical reactions. These reactions might have included energy metabolism, based, as today, on the release of free energy by hydrolysis of the phosphate–phosphate bonds in the ribonucleotides ATP and GTP, and the reactions might have become compartmentalized within lipid membranes, forming the first cell-like structures. There are difficulties in envisaging how long-chain unbranched lipids could form by chemical or ribozyme-catalyzed reactions, but once present in sufficient quantities they would have assembled spontaneously into membranes, possibly encapsulating one or more protogenomes and providing the RNAs with an enclosed environment in which more controlled biochemical reactions could be carried out.

The first DNA genomes

How did the RNA world develop into the DNA world? The first major change was probably the development of protein enzymes, which supplemented, and eventually replaced, most of the catalytic activities of ribozymes (Freeland *et al.*, 1999). There are several unanswered questions relating to this stage of biochemical evolution, including the reason why the transition from RNA to protein occurred in the first place. Originally, it was assumed that the 20 amino acids in polypeptides provided proteins with greater chemical variability than the four ribonucleotides in RNA, enabling protein enzymes to catalyze a broader range of biochemical reactions, but this explanation has become less attractive as more and more ribozyme-catalyzed reactions have been demonstrated in the test tube. A more recent suggestion is that protein catalysis is more efficient because of the inherent flexibility of folded polypeptides compared with the greater rigidity of base-paired RNAs (Csermely, 1997). Alternatively, enclosure of RNA protogenomes within membrane vesicles could have prompted the evolution of the first proteins, because RNA molecules are hydrophilic and must be given a hydrophobic coat, for instance by attachment to peptide molecules, before being able to pass through or become integrated into a membrane (Walter *et al.*, 2000).

The transition to protein catalysis demanded a radical shift in the function of the RNA protogenomes. Rather than being directly responsible for the biochemical reactions occurring in the early cell-like structures, the protogenomes became coding molecules whose main function was to specify the construction of the catalytic proteins. Whether the ribozymes themselves became coding molecules, or coding molecules were synthesized by the ribozymes is not known, although the most persuasive theories about the origins of translation and the genetic code suggest that the latter alternative is more likely to be correct (*Figure 15.3*; Szathmáry, 1993). Whatever the mechanism, the result was the paradoxical situation whereby the RNA protogenomes had abandoned their roles as enzymes, which they were good at, and taken on a coding function for which they were less well suited because of the relative instability of the RNA phosphodiester bond, resulting from the indirect effect of the 2'-OH group (Section 1.1.2). A transfer of the coding function to the more stable DNA seems almost inevitable and would not have been difficult to achieve, reduction of ribonucleotides giving deoxyribonucleotides which could then be polymerized into copies of the RNA pro-

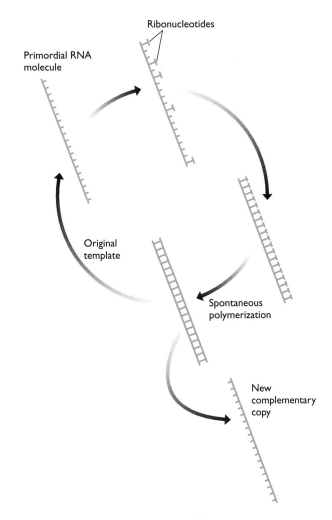

Figure 15.2   Copying of RNA molecules in the early RNA world.

Before the evolution of RNA polymerases, ribonucleotides that became associated with an RNA template would have had to polymerize spontaneously. This process would have been inaccurate and many RNA sequences would have been generated.

togenomes by a reverse-transcriptase-catalyzed reaction (*Figure 15.4*). The replacement of uracil with its methylated derivative thymine probably conferred even more stability on the DNA polynucleotide, and the adoption of double-stranded DNA as the coding molecule was almost certainly prompted by the possibility of repairing DNA damage by copying the partner strand (Sections 14.2.2 and 14.2.3).

According to this scenario, the first DNA genomes comprised many separate molecules, each specifying a single protein and each therefore equivalent to a single gene. The linking together of these genes into the first chromosomes, which could have occurred either before or after the transition to DNA, would have improved the efficiency of gene distribution during cell division, as it is easier to organize the equal distribution of a few large chromosomes than many separate genes. As with most stages in early genome evolution, several different mech-

anisms by which genes might have become linked have been proposed (Szathmáry and Maynard Smith, 1993).

How unique is life?

If the experimental simulations and computer models are correct then it is likely that the initial stages in biochemical evolution occurred many times in parallel in the oceans or atmosphere of the early Earth. It is therefore quite possible that 'life' arose on more than one occasion, even though all present-day organisms appear to derive from a single origin. This single origin is indicated by the remarkable similarity between the basic molecular biological and biochemical mechanisms in bacterial, archaeal and eukaryotic cells. To take just one example, there is no obvious biological or chemical reason why any particular triplet of nucleotides should code for any particular amino acid, but the genetic code, although not universal,

(A) A ribozyme that is also a coding molecule

(B) A ribozyme that synthesizes coding molecules

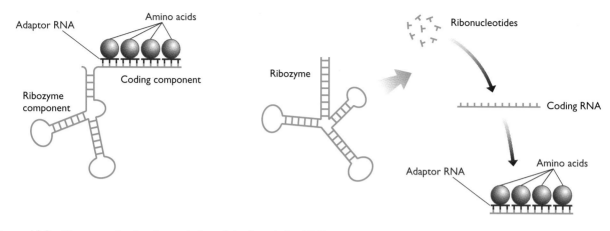

Figure 15.3 Two scenarios for the evolution of the first coding RNA.

A ribozyme could have evolved to have a dual catalytic and coding function (A), or a ribozyme could have synthesized a coding molecule (B). In both examples, the amino acids are shown attaching to the coding molecule via small adaptor RNAs, the presumed progenitors of today's tRNAs.

is virtually the same in all organisms that have been studied. If these organisms derived from more than one origin then we would anticipate two or more very different codes.

If multiple origins are possible, but modern life is derived from just one, then at what stage did this particular biochemical system begin to predominate? The ques-

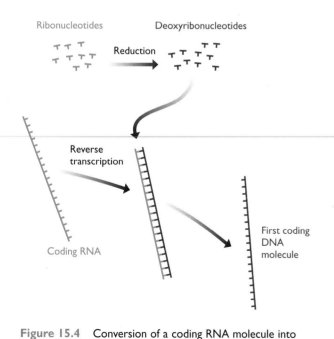

Figure 15.4 Conversion of a coding RNA molecule into the progenitor of the first DNA genome.

tion cannot be answered precisely, but the most likely scenario is that the predominant system was the first to develop the means to synthesize protein enzymes and therefore probably also the first to adopt a DNA genome. The greater catalytic potential and more accurate replication conferred by protein enzymes and DNA genomes would have given these cells a significant advantage compared with those still containing RNA protogenomes. The DNA–RNA–protein cells would have multiplied more rapidly, enabling them to out-compete the RNA cells for nutrients which, before long, would have included the RNA cells themselves.

Are life forms based on informational molecules other than DNA and RNA possible? Orgel (2000) has reviewed the possibility that RNA was preceded by some other informational molecule at the very earliest period of biochemical evolution and concluded that a pyranosyl version of RNA, in which the sugar takes on a slightly different structure, might be a better choice than normal RNA for an early protogenome because the base-paired molecules that it forms are more stable (Beier *et al.*, 1999; Eschenmoser, 1999). The same is true of **peptide nucleic acid (PNA)**, a polynucleotide analog in which the sugar–phosphate backbone is replaced by amide bonds (*Figure 15.5*). PNAs have been synthesized in the test tube and have been shown to form base pairs with normal polynucleotides. However, there are no indications that either pyranosyl RNA or PNA were more likely than RNA to have evolved in the prebiotic soup.

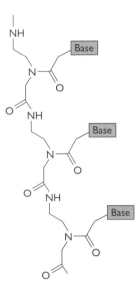

Figure 15.5   A short stretch of peptide nucleic acid.

A peptide nucleic acid has an amide backbone instead of the sugar–phosphate structure found in a standard nucleic acid.

15.2 Acquisition of New Genes

Although the very old fossil record is difficult to interpret, there is reasonably convincing evidence that by 3.5 billion years ago biochemical systems had evolved into cells similar in appearance to modern bacteria. We cannot tell from the fossils what kinds of genomes these first real cells had, but from the preceding section we can infer that they were made of double-stranded DNA and consisted of a small number of chromosomes, possibly just one, each containing many linked genes.

If we follow the fossil record forwards in time we see the first evidence for eukaryotic cells – structures resembling single-celled algae – about 1.4 billion years ago (*Figure 15.6*), and the first multicellular algae by 0.9 billion years ago. Multicellular animals appeared around 640 million years ago, although there are enigmatic burrows suggesting that animals lived earlier than this. The Cambrian Revolution, when invertebrate life proliferated into many novel forms, occurred 530 million years ago and ended with the disappearance of many of the novel forms in a mass extinction 500 million years ago. Since then, evolution has continued apace and with increasing diversification: the first terrestrial insects, animals and plants were established by 350 million years ago, the dinosaurs had been and gone by the end of the Cretaceous, 65 million years ago, and the first hominoids appeared a mere 4.5 million years ago.

Morphological evolution was accompanied by genome evolution. It is dangerous to equate evolution with 'progress' but it is undeniable that as we move up the evolutionary tree we see increasingly complex genomes. One indication of this complexity is gene number, which varies

from less than 1000 in some bacteria to 30 000–40 000 in vertebrates such as humans. However, this increase in gene number has not occurred in a gradual fashion: instead there seem to have been two sudden bursts when gene numbers increased dramatically (Bird, 1995). The first of these expansions occurred when eukaryotes appeared about 1.4 billion years ago, and involved an increase from the 5000 or fewer genes typical of prokaryotes to the 10 000 or more seen in most eukaryotes. The second expansion is associated with the first vertebrates, which became established soon after the end of the Cambrian, with each protovertebrate probably having at least 30 000 genes, this being the minimum number for any modern vertebrate, including the most 'primitive' types.

There are two ways in which new genes could be acquired by a genome:

- By duplicating some or all of the existing genes in the genome (Section 15.2.1);
- By acquiring genes from other species (Section 15.2.2).

Both events have been important in genome evolution, as we will see in the next two sections.

15.2.1 Acquisition of new genes by gene duplication

The duplication of existing genes is almost certainly the most important process for the generation of new genes during genome evolution. There are several ways in which it could occur:

- By duplication of the entire genome;
- By duplication of a single chromosome or part of a chromosome;
- By duplication of a single gene or group of genes.

The second of these possibilities can probably be discounted as a major cause of gene number expansions based on our knowledge of the effects of chromosome duplications in modern organisms. Duplication of individual human chromosomes, resulting in a cell that contains three copies of one chromosome and two copies of all the others (the condition called **trisomy**), is either lethal or results in a genetic disease such as Down syndrome, and similar effects have been observed in artificially generated trisomic mutants of *Drosophila*. Probably, the resulting increase in copy numbers for some genes leads to an imbalance of the gene products and disruption of the cellular biochemistry (Ohno, 1970). The other two ways of generating new genes – whole-genome duplication and duplication of a single or small number of genes – have probably been much more important.

Whole-genome duplications can result in sudden expansions in gene number

The most rapid means of increasing gene number is by duplicating the entire genome. This can occur if an error during meiosis leads to the production of gametes that are diploid rather than haploid (*Figure 15.7*). If two diploid gametes fuse then the result will be a type of

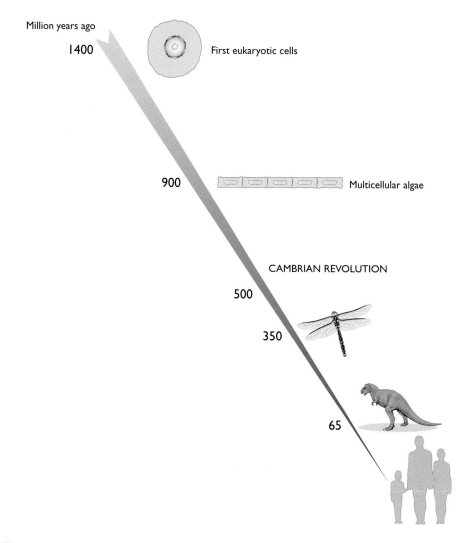

Figure 15.6 The evolution of life.

autopolyploid, in this case a tetraploid cell whose nucleus contains four copies of each chromosome.

Autopolyploidy, as with other types of polyploidy (see page 475), is not uncommon among plants. Autopolyploids are often viable because each chromosome still has a homologous partner and so can form a bivalent during meiosis. This allows an autopolyploid to reproduce successfully, but generally prevents interbreeding with the original organism from which it was derived. This is because a cross between, for example, a tetraploid and diploid would give a triploid offspring which would not itself be able to reproduce because one full set of its chromosomes would lack homologous partners (*Figure 15.8*). Autopolyploidy is therefore a mechanism by which speciation can occur, a pair of species usually being defined as two organisms that are unable to interbreed. The generation of new plant species by autopolyploidy has in fact been observed, notably by Hugo de Vries, one of the rediscoverers of Mendel's experiments. During his work with evening primrose, *Oenothera lamarckiana*, de Vries isolated a tetraploid version of this normally diploid plant, which he named *Oenothera gigas*. Autopolyploidy among animals is less common, especially in those with two distinct sexes, possibly because of problems that arise if a nucleus possesses more than one pair of sex chromosomes.

Autopolyploidy does not lead directly to gene expansion because the initial product is an organism that simply has extra copies of every gene, rather than any new genes. It does, however, provide the *potential* for gene expansion because the extra genes are not essential to the functioning of the cell and so can undergo mutational change without harming the viability of the organism. With many genes, the resulting changes in nucleotide sequence will be deleterious and the end result will be an inactive pseudogene, but occasionally the mutations will lead to a new gene function that is useful to the cell. This aspect of genome evolution is more clearly illustrated by considering duplications of single genes rather than of entire genomes, so we will postpone a full discussion of it until the next section.

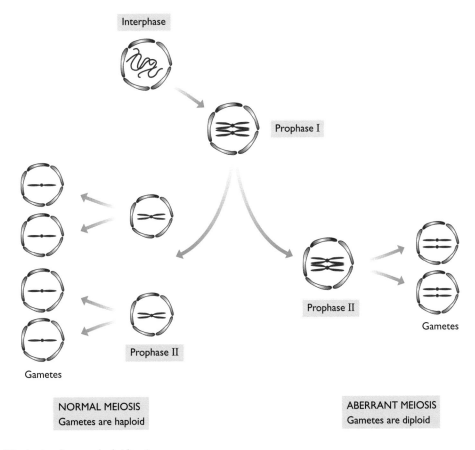

Figure 15.7   The basis of autopolyploidization.

The normal events occurring during meiosis are shown, in abbreviated form, on the left (compare with *Figure 5.15*, page 138). On the right, an aberration has occurred between prophase I and prophase II and the pairs of homologous chromosomes have not separated into different nuclei. The resulting gametes will be diploid rather than haploid.

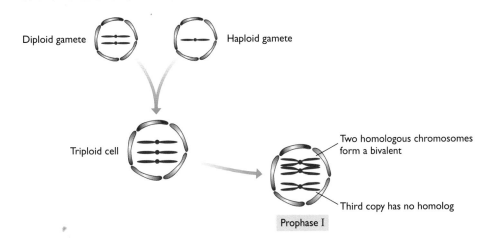

Figure 15.8   Autopolyploids cannot interbreed successfully with their parents.

Fusion of the diploid gamete produced by the aberrant meiosis shown in *Figure 15.7* with a haploid gamete produced by the normal meiosis leads to a triploid nucleus, one that has three copies of each homologous chromosome. During prophase I of the next meiosis, two of these homologous chromosomes will form a bivalent but the third will have no partner. This has a disruptive effect on the segregation of chromosomes during anaphase (see *Figure 5.15*, page 138) and usually prevents meiosis from reaching a successful conclusion. This means that gametes are not produced and the triploid organism is sterile. Note that the bivalent could have formed between any two of the three homologous chromosomes, not just between the pair shown in the diagram.

Segmental duplications in the yeast and human genomes

RESEARCH 15.1 BRIEFING

Examination of the yeast and human genomes reveals evidence of past duplication events.

Duplication of individual genes has been recognized for some time as having played an important role in genome evolution, largely because the results of gene duplication – multigene families – are common in present-day genomes. The evolutionary importance of larger duplications, involving the entire genome or substantial segments of individual chromosomes, has been more controversial because the traces of this type of event are less easy to discern when modern genomes are examined. Recent work with the yeast and human genome sequences suggests that segmental duplications have been relatively common in the evolutionary histories of these organisms.

Indicators of segmental duplications

From what we understand about the way in which genomes change over time, we might anticipate that evidence for a segmental duplication would be quite difficult to obtain. Many of the extra gene copies resulting from such a duplication would be expected to decay into pseudogenes and no longer be visible in the DNA sequence. Those genes that are retained, because their duplicated functions are useful to the organism or because they have evolved new functions, should be identifiable, but it would be impossible to distinguish if they have arisen by a segmental duplication or simply by duplication of individual genes. For a segmental duplication to be signaled it would be necessary to find duplicated *sets* of genes, with the same order of genes in both sets.

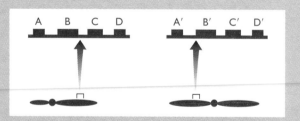

To what extent these duplicated sets are still visible in the genome will depend on how frequently past recombination events have moved genes to new positions.

Evidence for segmental duplications in the yeast genome

To search the *Saccharomyces cerevisiae* genome for evidence of past duplications, homology analysis (Section 7.2.1) was carried out with every yeast gene tested against every other yeast gene. The objective was to identify pairs of genes that might have arisen by duplication events. To be considered as a possible pair, two genes had to display at least 25% identity when the predicted amino acid sequences of their protein products were compared. About 800 gene pairs were identified, but did these arise by many individual gene duplications or by a fewer number of segmental duplications? If the latter, then at least some of the members of different gene pairs should retain their positions close to one another in the genome. The gene distributions were therefore examined to see if any duplicated sets of genes could be identified. Fifty-five duplicate sets were discovered, altogether comprising 376 pairs of genes, with each of these sets containing at least three genes in the same order. An example of a duplicated set on chromosomes VII and XVI is shown here:

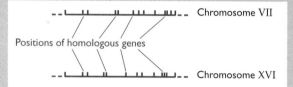

The existence of these 55 duplicate gene sets is strong evidence that segmental duplication has occurred frequently during evolution of the yeast genome. In fact, the duplications are so extensive – covering half the genome – that the possibility that the entire genome has been duplicated cannot be discounted. If a whole-genome duplication is assumed, then application of a molecular clock to the comparative sequence data (Section 16.2.2) indicates that the event took place about 100 million years ago.

Duplications in the human genome

Similar analyses have been carried out with the human genome. No evidence for whole-genome duplication has been found, at least not during the recent evolutionary past, but a substantial amount of large segmental duplication has clearly occurred. This is illustrated by the diagram on the facing page, which depicts the duplication events that are inferred for the last 35 million years of evolution of the long arm of human chromosome 22. As in the yeast study, these duplications have been identified by comparing different parts of the genome, in this case to identify regions of more than 1 kb in length that display 90% or higher sequence similarity. This analysis ignores genome-wide repeat sequences

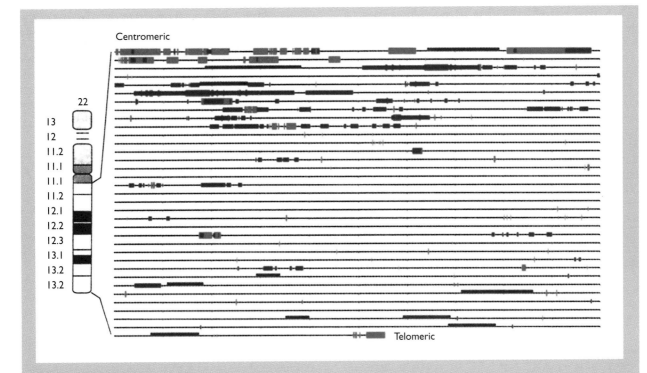

and so locates regions that are probably the products of relatively recent duplication events. The diagram depicts the 34 Mb of the long arm of chromosome 22 as a series of thin horizontal lines, each one representing 1 Mb of DNA sequence, running from the centromere at the top to the telomere at the bottom. This part of the draft human genome sequence contains eleven gaps which are shown as black bars. The red boxes, which make up 3.9% of the 34 Mb, are sequences that are duplicated within this chromosome arm, and the blue boxes (6.4% of the total) are duplications of regions in other chromosomes.

The pattern of duplications in the long arm of chromosome 22 is fairly typical of the genome as a whole. Individual duplications range from 1 to 400 kb and there is a distinct bias towards the areas adjacent to the centromeres, with relatively few duplications in the more distal regions of each chromosome arm. Although the segmental duplications are large, we must bear in mind that the average human gene is 20–25 kb in length and that the genes are scattered sparsely throughout the genome. When the sequences of the duplications are examined it emerges that few of these result in duplication of entire genes, but that several involve parts of genes, and that some of these lead to the upstream exons of one gene being placed alongside downstream exons of a second gene. Some of these new combinations are transcribed, but it is not clear whether the transcripts are functional.

The evolutionary potential of segmental duplications is therefore uncertain. What is clear, however, is that recombination (Section 14.3) between a pair of intrachromosomal duplications can result in deletion of the region between the duplications, and that deletions generated in this way can give rise to genetic disease. An example is Charcot–Marie–Tooth syndrome, which develops 5–15 years after birth and is characterized by degeneration of the peripheral nervous system, leading to weakness and difficulty in walking. The syndrome is caused by recombination between a pair of 24-kb segmental duplications on human chromosome 17, deleting a 1.5 Mb segment of the genome. The loss of the genes contained in this segment gives rise to the disease.

References

Wolfe KH and Shields DC (1997) Molecular evidence for an ancient duplication of the entire yeast genome. *Nature*, **387**, 708–713.

Piskur J (2001) Origin of the duplicated regions in the yeast genome. *Trends Genet.*, **17**, 302–303.

Bailey JA, Yavor AM, Massa HF, Trask BJ and Eichler EE (2001) Segmental duplications: organization and impact within the current human genome project assembly. *Genome Res.*, **11**, 1005–1017.

International Human Genome Sequencing Consortium (2001) Initial sequencing and analysis of the human genome. *Nature*, **409**, 860–921.

Eichler EE (2001) Recent duplication, domain accretion and dynamic mutation of the human genome. *Trends Genet.*, **17**, 661–669.

Are there any indications of genome duplication in the evolutionary histories of present-day genomes? From what we understand about the way in which genomes change over time, we might anticipate that evidence for whole-genome duplication would be quite difficult to obtain. Many of the extra gene copies resulting from genome duplication would be expected to decay into pseudogenes and no longer be visible in the DNA sequence. Those genes that are retained, because their duplicated function is useful to the organism or because they have evolved new functions, should be identifiable, but it would be impossible to distinguish if they have arisen by genome duplication or simply by duplication of individual genes. For a genome duplication to be signaled it would be necessary to find duplicated *sets* of genes, with the same order of genes in both sets. To what extent these duplicated sets are still visible in the genome will depend on how frequently past recombination events have moved genes to new positions. This type of analysis has been applied to the *Saccharomyces cerevisiae* DNA sequence, leading to the suggestion that this genome is the product of a duplication that took place approximately 100 million years ago (Wolfe and Shields, 1997; Research Briefing 15.1), but this hypothesis is still controversial (Piskur, 2001). Comparisons between the *Arabidopsis thaliana* genome sequence and segments

of other plant genomes suggest that the ancestor of the *A. thaliana* genome underwent four rounds of genome duplication between 100 and 200 million years ago (Vision *et al.*, 2000; Bancroft, 2001). The increased number of Hox gene clusters present in some types of fish (see page 472) has been used as an argument for a duplication event in the genomic lineage leading to these organisms (Taylor *et al.*, 2001).

Duplications of individual genes and groups of genes have occurred frequently in the past

If genome duplication has not been a common evolutionary event, then increases in gene number must have occurred primarily by duplications of individual genes and small groups of genes. This hypothesis is supported by DNA sequencing, which has revealed that multigene families are common components of all genomes (Section 2.2.1). By comparing the sequences of individual members of a family (using the techniques described in Chapter 16) it is usually possible to trace the individual gene duplications involved in evolution of the family from a single progenitor gene that existed in an ancestral genome (*Figure 15.9*; Henikoff *et al.*, 1997). There are several mechanisms by which these gene duplications could have occurred:

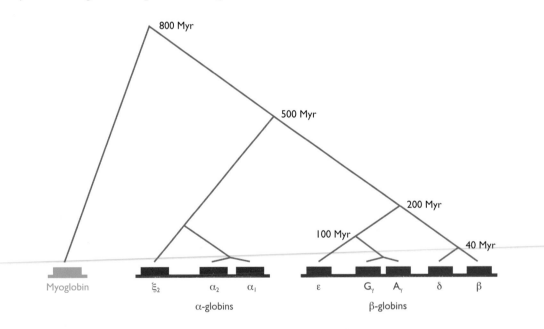

Figure 15.9   Gene duplications during the evolution of the human globin gene families.

Comparisons of their nucleotide sequences enable the evolutionary relationships between the globin genes to be deduced, using the molecular phylogenetics techniques described in Chapter 16. The dates of key duplications are shown. The initial split was between an ancestral gene that gave rise, in one lineage, to the modern gene for the muscle oxygen-binding protein, myoglobin, and, in the other lineage, to the globin genes. This duplication is estimated to have occurred approximately 800 million years ago. The proto-α and proto-β lineages split by a duplication that occurred 500 million years ago and the duplications within the α and β families took place during the last 200 million years. Note that each set of genes is now on a different chromosome: the myoglobin gene is on chromosome 22, the α-globin genes are on chromosome 16, and the β-globin genes are on chromosome 11. See *Figure 2.14*, page 45 for more details about the globin genes. Based on Strachan and Read (1999). Abbreviation: Myr, million years.

- *Unequal crossing-over* is a recombination event initiated by similar nucleotide sequences that are not at identical places in a pair of homologous chromosomes. As shown in *Figure 15.10A*, the result of unequal crossing-over can be duplication of a segment of DNA in one of the recombination products.

- *Unequal sister chromatid exchange* occurs by the same mechanism as unequal crossing-over, but involves a pair of chromatids from a single chromosome (see *Figure 15.10B*).

- *DNA amplification* is sometimes used in this context to describe gene duplication in bacteria and other haploid organisms (Romero and Palacios, 1997), in which duplications can arise by unequal recombination between the two daughter DNA molecules in a replication bubble (*Figure 15.10C*).

- *Replication slippage* (see *Figure 14.5*, page 423) could result in gene duplication if the genes are relatively short, although this process is more commonly associated with the duplication of very short sequences such as the repeat units in microsatellites.

The initial result of gene duplication is two identical genes. As mentioned above with regard to genome duplication, selective constraints will ensure that one of these genes retains its original nucleotide sequence, or something very similar to it, so that it can continue to provide the protein function that was originally supplied by the single gene copy before the duplication took place. The second copy is probably not subject to the same selective pressures and so can accumulate mutations at random. Evidence shows that the majority of new genes that arise by duplication acquire deleterious mutations that inactivate them so that they become pseudogenes (Wagner, 2001). From the sequences of the pseudogenes in the α- and β-globin gene families (*Figure 2.14*, page 45), it appears that the commonest inactivating mutations are frameshifts and nonsense mutations that occur within the coding region of the gene, with

mutations of the initiation codon and TATA box being less frequent.

Occasionally, the mutations that accumulate within a gene copy do not lead to inactivation of the gene, but instead result in a new gene function that is useful to the organism. We have already seen that gene duplication in the globin gene families led to the evolution of new globin proteins that are used by the organism at different stages in its development (see *Figure 2.14*, page 45). We also noted (page 45) that all the globin genes, both the α- and β-types, are related and hence form a gene superfamily that originated with a single ancestral globin gene that split to give the proto-α and proto-β globins about 500 million years ago (see *Figure 15.9*). Further back, about 800 million years ago, this ancestral globin gene itself arose by gene duplication, its sister duplicate evolving to give the modern gene for myoglobin, a muscle protein whose main function, like that of the globins, is the storage of oxygen (Doolittle, 1987). We observe similar patterns of evolution when we compare the sequences of other genes. The trypsin and chymotrypsin genes, for example, are related by a common ancestor approximately 1500 million years ago (Barker and Dayhoff, 1980). Both now code for proteases involved in protein breakdown in the vertebrate digestive tract, trypsin cutting other proteins at arginine and lysine amino acids and chymotrypsin cutting at phenylalanines, tryptophans and tyrosines. Genome evolution has therefore produced two complementary protein functions where originally there was just one.

The most striking example of gene evolution by duplication, whether by duplication of a small group of genes or by whole-genome duplication, is provided by the homeotic selector genes, the key developmental genes responsible for specification of the body plans of animals. As described in Section 12.3.3, *Drosophila* has a single cluster of homeotic selector genes, called HOM-C, which consists of eight genes each containing a homeodomain sequence coding for a DNA-binding motif in the protein product (see *Figure 12.30*, page 376). These eight genes, as

---

Box 15.1: Gene duplication and genetic redundancy

The text adopts the conventional scenario which states that after a duplication, one of the two gene copies can accumulate mutations which either result in inactivation of that gene copy or lead to a new gene function. Implicit in this scenario is the assumption that the cell needs only one gene to perform the original function, so one of the copies can change. This interpretation of gene evolution is complicated by **genetic redundancy**, which occurs when two genes in a single genome perform the same function (Brookfield, 1997). Several examples of genetic redundancy have been discovered in the course of projects using gene inactivation to identify functions of newly sequenced genes (Section 7.2.2).

Occasionally, gene inactivation does not result in any change of phenotype because a second gene exists which is able to take over the function of the one that has been inactivated.

Genetic redundancy raises interesting evolutionary questions. If the functional gene is satisfying the biochemical requirement then there would appear to be no selective pressure on the redundant gene. Without this selective pressure we would anticipate that the sequence of the redundant gene would change because loss of its function would not result in any immediate disadvantage to the cell. In fact, three different situations have been identified which would lead to maintenance of the redundant gene copy (Nowak *et al.*, 1997), the simplest of these being if the redundant gene also has a second function, not satisfied by the first gene, so that its sequence is maintained by a different form of selective pressure.

(A) Unequal crossing over

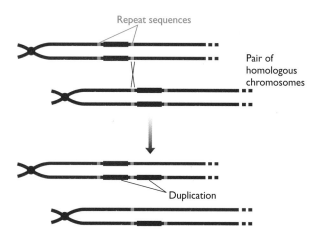

(B) Unequal sister chromatid exchange

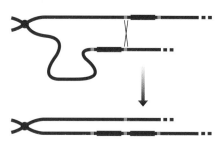

(C) During DNA replication

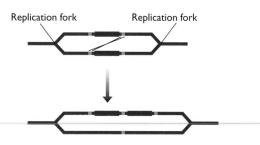

Figure 15.10 Models for gene duplication by (A) unequal crossing-over between homologous chromosomes, (B) unequal sister chromatid exchange, and (C) during replication of a bacterial genome.

In each case, recombination occurs between two different copies of a short repeat sequence, shown in green, leading to duplication of the sequence between the repeats. Unequal crossing-over and unequal sister chromatid exchange are essentially the same except that the first involves chromatids from a pair of homologous chromosomes and the second involves chromatids from a single chromosome. In (C), recombination occurs between two daughter double helices that have just been synthesized by DNA replication.

well as other homeodomain genes in *Drosophila*, are believed to have arisen by a series of gene duplications that began with an ancestral gene that existed about 1000 million years ago. The functions of the modern genes, each specifying the identity of a different segment of the fruit fly, gives us a tantalizing glimpse of how gene duplication and sequence divergence could, in this case, have been the underlying processes responsible for increasing the morphological complexity of the series of organisms in the *Drosophila* evolutionary tree.

Vertebrates have four Hox gene clusters (see *Figure 12.30*, page 376), each a recognizable copy of the *Drosophila* cluster, with sequence similarities between genes in equivalent positions. Not all of the vertebrate Hox genes have been ascribed functions, but we believe that the additional versions possessed by vertebrates relate to the added complexity of the vertebrate body plan. Two observations support this conclusion. The amphioxus, an invertebrate that displays some primitive vertebrate features, has two Hox clusters (Brooke *et al.*, 1998), which is what we might expect for a primitive 'protovertebrate'. Ray-finned fishes, probably the most diverse group of vertebrates with a vast range of different variations of the basic body plan, have seven Hox clusters (Amores *et al.*, 1998).

Gene duplication is not always followed by sequence divergence and the evolution of a family of genes with different functions. Some multigene families are made up of genes with identical or near-identical sequences. The prime examples are the rRNA genes, whose copy numbers range from two in *Mycoplasma genitalium* to 500+ in *Xenopus laevis* (Section 2.2.1), with all of the copies having virtually the same sequence. These multiple copies of identical genes presumably reflect the need for rapid synthesis of the gene product at certain stages of the cell cycle. With these gene families there must be a mechanism that prevents the individual copies from accumulating mutations and hence diverging away from the functional sequence. This is called **concerted evolution**. If one copy of the family acquires an advantageous mutation then it is possible for that mutation to spread throughout the family until all members possess it. The most likely way in which this can be achieved is by gene conversion which, as described in Section 14.3.1, can result in the sequence of one copy of a gene being replaced with all or part of the sequence of a second copy. Multiple gene conversion events could therefore maintain identity among the sequences of the individual members of a multigene family.

Genome evolution also involves rearrangement of existing genes

As well as the generation of new genes by duplication followed by mutation, novel protein functions can also be produced by rearranging existing genes. This is possible because most proteins are made up of structural domains (Section 3.3.3), each comprising a segment of the polypeptide chain and hence encoded by a contiguous series of

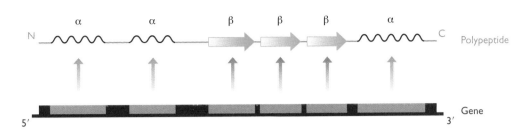

Figure 15.11 Structural domains are individual units in a polypeptide chain coded by a contiguous series of nucleotides.

In this simplified example, each secondary structure in the polypeptide is looked upon as an individual structural domain. In reality, most structural domains comprise two or more secondary structural units.

nucleotides (*Figure 15.11*). There are two ways in which rearrangement of domain-encoding gene segments can result in novel protein functions.

- *Domain duplication* occurs when the gene segment coding for a structural domain is duplicated by unequal crossing-over, replication slippage or one of the other methods that we have considered for duplication of DNA sequences (*Figure 15.12A*). Duplication results in the structural domain being repeated in the protein, which might itself be advantageous, for example by making the protein product more stable. The duplicated domain might also change over time as its coding sequence becomes mutated, leading to a modified structure that might provide the protein with a new activity. Note that domain duplication causes the gene to become longer. Gene elongation appears to be a general consequence of genome evolution, the genes of higher eukaryotes being longer, on average, than those of lower organisms.

- *Domain shuffling* occurs when segments coding for structural domains from completely different genes are joined together to form a new coding sequence that specifies a hybrid or mosaic protein, one that would have a novel combination of structural features and might provide the cell with an entirely new biochemical function (*Figure 15.12B*).

Implicit in these models of domain duplication and shuffling is the need for the relevant gene segments to be separated so that they can themselves be rearranged and shuffled. This requirement has led to the attractive suggestion that exons might code for structural domains. With some proteins, duplication or shuffling of exons does seem to have resulted in the structures seen today. An example is provided by the α2 Type I collagen gene of vertebrates, which codes for one of the three polypeptide chains of collagen. Each of the three collagen polypeptides has a highly repetitive sequence made up of repeats of the tripeptide glycine–X–Y, where X is usually proline and Y is usually hydroxyproline (*Figure 15.13*). The α2 Type I gene, which codes for 338 of these repeats, is split into 52 exons, 42 of which cover the part of the gene coding for the glycine–X–Y repeats. Within this region, each exon encodes a set of complete tripeptide repeats. The number of repeats per exon varies but is 5 (5 exons), 6 (23

exons), 11 (5 exons), 12 (8 exons) or 18 (1 exon). Clearly this gene could have evolved by duplication of exons leading to repetition of the structural domains.

Domain shuffling is illustrated by tissue plasminogen activator (TPA), a protein found in the blood of vertebrates and which is involved in the blood clotting response. The TPA gene has four exons, each coding for a different structural domain (*Figure 15.14*). The upstream exon codes for a 'finger' module that enables the TPA protein to bind to fibrin, a fibrous protein found in blood clots and which activates TPA. This exon appears to be derived from a second fibrin-binding protein, fibronectin, and is absent from the gene for a related protein, urokinase, which is not activated by fibrin. The second TPA exon specifies a growth-factor domain which has apparently been obtained from the gene for epidermal growth factor and which may enable TPA to stimulate cell proliferation. The last two exons code for 'kringle' structures which TPA uses to bind to fibrin clots; these kringle exons come from the plasminogen gene (Li and Graur, 1991).

Type I collagen and TPA provide elegant examples of gene evolution but, unfortunately, the clear links that they display between structural domains and exons are exceptional and are rarely seen with other genes. Many other genes appear to have evolved by duplication and shuffling of segments, but in these the structural domains are coded by segments of genes that do not coincide with individual exons or even groups of exons. Domain duplication and shuffling still occur, but presumably in a less precise manner and with many of the rearranged genes having no useful function. Despite being haphazard, the process clearly works, as indicated by, among other examples, the number of proteins that share the same DNA-binding motifs (Section 9.1.4). Several of these motifs probably evolved *de novo* on more than one occasion, but it is clear that in many cases the nucleotide sequence coding for the motif has been transferred to a variety of different genes.

15.2.2 Acquisition of new genes from other species

The second possible way in which a genome can acquire new genes is to obtain them from another species.

(A) Domain duplication

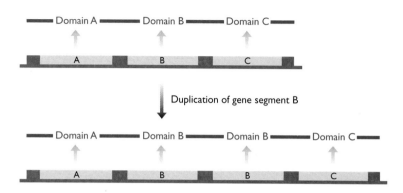

(B) Domain shuffling

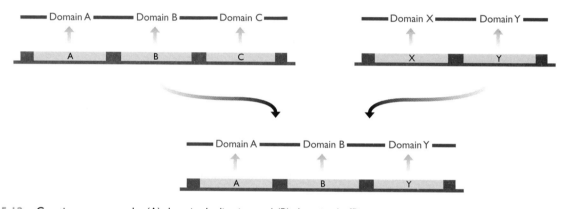

Figure 15.12 Creating new genes by (A) domain duplication and (B) domain shuffling.

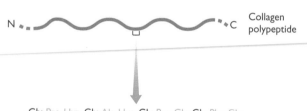

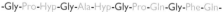

-Gly-Pro-Hyp-Gly-Ala-Hyp-Gly-Pro-Gln-Gly-Phe-Gln-

Figure 15.13 The α2 Type I collagen polypeptide has a repetitive sequence described as Gly–X–Y.

Every third amino acid is glycine, X is often proline and Y is often hydroxyproline (Hyp). See *Table 3.1*, page 84 for other amino acid abbreviations. Hydroxyproline is a post-translationally modified version of proline (Section 11.3.3). The collagen polypeptide has a helical conformation, but one that is more extended than the standard α-helix.

Comparisons of bacterial and archaeal genome sequences suggest that **lateral gene transfer** has been a major event in the evolution of prokaryotic genomes (Section 2.3.2). The genomes of most bacteria and archaea contain at least a few hundred kb of DNA, representing tens of genes, that appears to have been acquired from a second prokaryote.

There are several mechanisms by which genes can be transferred between prokaryotes but it is difficult to be sure how important these various processes have been in shaping the genomes of these organisms. Conjugation (Section 5.2.4), for example, enables plasmids to move between bacteria and frequently results in the acquisition of new gene functions by the recipients. On a day-to-day basis, plasmid transfer is important because it is the means by which genes for resistance to antibiotics such as chloramphenicol, kanamycin and streptomycin spread through bacterial populations and across species barriers, but its evolutionary relevance is questionable. It is true that the genes transferred by conjugation can become

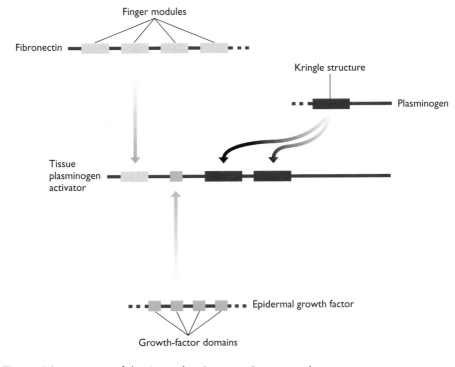

Figure 15.14   The modular structure of the tissue plasminogen activator protein.

See the text for details.

integrated into the recipient bacterium's genome, but usually the genes are carried by composite transposons (see *Figure 2.29B*, page 63), which means that the integration is reversible and so might not result in a permanent change to the genome. A second process for DNA transfer between prokaryotes, transformation (Section 5.2.4), is more likely to have had an influence on genome evolution. Only a few bacteria, notably members of the *Bacillus*, *Pseudomonas* and *Streptococcus* genera, have efficient mechanisms for the uptake of DNA from the surrounding environment, but efficiency of DNA uptake is probably not relevant when we are dealing with an evolutionary time-scale. More important is the fact that gene flow by transformation can occur between any pair of prokaryotes, not just closely related ones (as is the case with conjugation), and so could account for the transfers that appear to have occurred between bacterial and archaeal genomes (Section 2.3.2).

In plants, new genes can be acquired by polyploidization. We have already seen how autopolyploidization can result in genome duplication in plants (see *Figure 15.7*, page 467). **Allopolyploidy**, which results from interbreeding between two different species, is also common and, like autopolyploidy, can result in a viable hybrid. Usually, the two species that form the allopolyploid are closely related and have many genes in common, but each parent will possess a few novel genes or at least distinctive alleles of shared genes. For example, the bread wheat, *Triticum aestivum*, is a hexaploid that arose by

allopolyploidization between cultivated emmer wheat, *T. turgidum*, which is a tetraploid, and a diploid wild grass, *Aegilops squarrosa*. The wild-grass nucleus contained novel alleles for the high-molecular-weight glutenin genes which, when combined with the glutenin alleles already present in emmer wheat, resulted in the superior properties for breadmaking displayed by the hexaploid wheats. Allopolyploidization can therefore be looked upon as a combination of genome duplication and interspecies gene transfer.

Among animals, the species barriers are less easy to cross and it is difficult to find clear evidence for lateral gene transfer of any kind. Several eukaryotic genes have features associated with archaeal or bacterial sequences, but rather than being the result of lateral gene transfer, these similarities are thought to result from conservation during millions of years of parallel evolution. Most proposals for gene transfer between animal species center on retroviruses and transposable elements. Transfer of retroviruses between animal species is well documented, as is their ability to carry animal genes between individuals of the same species, suggesting that they might be possible mediators of lateral gene transfer. The same could be true of transposable elements such as P elements, which are known to spread from one *Drosophila* species to another, and *mariner*, which has also been shown to transfer between *Drosophila* species and which may have crossed from other species into humans (Robertson *et al.*, 1996; Hartl *et al.*, 1997).

15.3 Non-coding DNA and Genome Evolution

So far we have concentrated our attention on the evolution of the coding component of the genome. As coding DNA makes up only 1.5% of the human genome (see Box 1.4, page 23) our view of genome evolution would be very incomplete if we did not devote some time to considering non-coding DNA. The problem is that in many respects there is little that can be said about the evolution of non-coding DNA. We envisage that duplications and other rearrangements have occurred through recombination and replication slippage, and that sequences have diverged through accumulation of mutations unfettered by the restraining selective forces acting on functional regions of the genome. We recognize that some parts of the non-coding DNA, for example the regulatory regions upstream of genes, have important functions, but as far as most of the non-coding DNA is concerned, all we can say is that it evolves in an apparently random fashion.

This randomness does not apply to all components of the non-coding DNA. In particular, transposable elements and introns have interesting evolutionary histories and are of general importance in genome evolution, as described in the following two sections.

15.3.1 Transposable elements and genome evolution

Transposable elements have a number of effects on evolution of the genome as a whole. The most significant of these is the ability of transposons to initiate recombination events that lead to genome rearrangements. This has nothing to do with the transposable activity of these elements, it simply relates to the fact that different copies of the same element have similar sequences and can there-

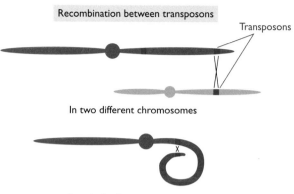

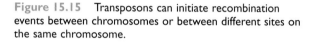

Figure 15.15 Transposons can initiate recombination events between chromosomes or between different sites on the same chromosome.

fore initiate recombination between two parts of the same chromosome or between different chromosomes (*Figure 15.15*). In many cases, the resulting rearrangement will be harmful because important genes will be deleted, but some instances where the result has been beneficial have been documented. Recombination between a pair of LINE-1 elements (Section 2.4.2) approximately 35 million years ago is thought to have caused the β-globin gene duplication that resulted in the Gγ and Aγ members of this gene family (see *Figure 15.9*, page 470; Maeda and Smithies, 1986).

Movement of transposons from one site to another can also have an impact on genome evolution. The transposition of a LINE-1 element can occasionally result in a short piece of the adjacent DNA being transferred along with the transposon, a process called **3′ transduction**, the transferred segment being located at the 3′ end of the element. LINE-1 elements are sometimes found in introns

---

Box 15.2: The origin of a microsatellite

There is no mystery about the origins of microsatellite repeat sequences (Section 2.4.1). A dimeric microsatellite, consisting of two repeat units in tandem array, can easily arise by chance mutational events. Replication slippage (Section 14.1.1) can then increase the copy number to three, four and more.

The birth of a microsatellite within a globin pseudogene of primates has been documented by comparing the nucleotide sequences of various species (Messier et al., 1996). In the orangutan, gibbon and other species some distance from humans, the sequence 5′–ATGTGTGT–3′ occurs at the relevant position. In the gorilla, bonobo, chimpanzee and human a single point mutation has changed this sequence to 5′–ATGTATGT–3′, with subsequent expansion of the 5′–ATGT–3′ repeat motif.

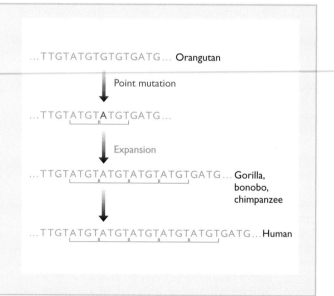

so 3' transduction could conceivably move downstream exons to new sites in the genome (Kazazian, 2000). Transposition has also been associated with altered patterns of gene expression. For example, the efficiency with which DNA-binding proteins that are attached to upstream regulatory sequences can activate transcription of a gene might be affected if a transposon moves into a new site immediately upstream of the gene (*Figure 15.16*). Transcription of the gene might also be influenced by the presence of promoters and/or enhancers within the transposon, so the gene becomes subject to an entirely new regulatory regime (McDonald, 1995). An interesting example of transposon-directed gene expression occurs with the mouse gene *Slp*, which codes for a protein involved in the immune response, the tissue specificity of *Slp* being conferred by an enhancer located within an adjacent retrotransposon (Stavenhagen and Robins, 1988). There are also examples where insertion of a transposon into a gene has resulted in an altered splicing pattern (Purugganan and Wessler, 1992).

15.3.2 The origins of introns

Ever since introns were first discovered in the 1970s their origins have been debated. There are few controversies surrounding the Group I, II and III types (see *Table 10.2*, page 287) as it is generally accepted that all these self-splicing introns evolved in the RNA world and have survived ever since without undergoing a great deal of change. The problems surround the origins of the GU–AG introns, the ones that are found in large numbers in eukaryotic nuclear genomes.

'Introns early' and 'introns late': two competing hypotheses

A number of proposals for the origins of GU–AG introns have been put forward but the debate is generally considered to be between two opposing hypotheses:

- **'Introns early'** states that introns are very ancient and are gradually being lost from eukaryotic genomes.

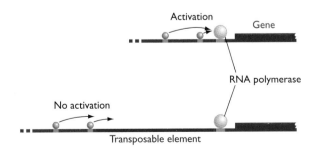

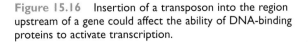

Figure 15.16 Insertion of a transposon into the region upstream of a gene could affect the ability of DNA-binding proteins to activate transcription.

- **'Introns late'** states that introns evolved relatively recently and are gradually accumulating in eukaryotic genomes.

There are several different models for each hypothesis. For 'introns early' the most persuasive model is the one also called the 'exon theory of genes' (Gilbert, 1987) which holds that introns were formed when the first DNA genomes were constructed, soon after the end of the RNA world. These genomes would have contained many short genes, each derived from a single coding RNA molecule and each specifying a very small polypeptide, perhaps just a single structural domain. These polypeptides would probably have had to associate together into larger multidomain proteins in order to produce enzymes with specific and efficient catalytic mechanisms (*Figure 15.17*). To aid the synthesis of a multidomain enzyme it would have been beneficial for the enzyme's individual polypeptides to become linked into a single protein, such as we see today. This was achieved by splicing together the transcripts of the relevant minigenes, a process that was aided by rearranging the genome so that groups of minigenes specifying the different parts of individual multidomain proteins were positioned next to each other. In other words, the minigenes became exons and the DNA sequences between them became introns.

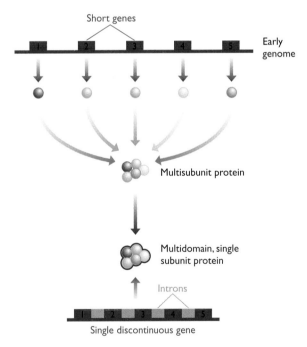

Figure 15.17 The 'exon theory of genes'.

The short genes of the first genomes probably coded for single-domain polypeptides that would have had to associate together to form a multisubunit protein to produce an effective enzyme. Later the synthesis of this enzyme could have been made more efficient by linking the short genes together into one discontinuous gene coding for a multidomain single-subunit protein.

According to the exon theory of genes and other 'introns early' hypotheses, all genomes originally possessed introns. But we know that bacterial genomes do not have GU–AG introns, so if these hypotheses are correct then we must assume that for some reason introns became lost from the ancestral bacterial genome at an early stage in its evolution. This is a stumbling block because it is difficult to envisage how a large number of introns could be lost from a genome without risking the disruption of many gene functions. If an intron is removed from a gene with any imprecision then a part of the coding region will be lost or a frameshift mutation will occur, both of which would be expected to inactivate the gene. The 'introns late' hypothesis avoids this problem by proposing that, to begin with, no genes had introns, these structures invading the early eukaryotic nuclear genome and subsequently proliferating into the numbers seen today. The similarities between the splicing pathways for GU–AG and Group II introns (Section 10.2.3) suggest that the invaders that gave rise to GU–AG introns might well have been Group II sequences that escaped from organelle genomes (Eickbush, 2000). However, the similarity between GU–AG and Group II introns does not prove the 'introns late' view, because it is equally possible to devise an 'introns early' model, different to the exon theory of genes, in which Group II sequences gave rise to GU–AG introns, but at a very early stage in genome evolution.

The current evidence disproves neither hypothesis

One of the reasons why the debate regarding the origin of GU–AG introns has continued for over 20 years is because evidence in support of either hypothesis has been difficult to obtain and is often ambiguous. One prediction of 'introns early' is that there should be a close similarity between the positions of introns in homologous genes from unrelated organisms, because all these genes are descended from an ancestral intron-containing gene (*Figure 15.18*). Early support for 'introns early' came when this was shown to be the case for four introns in animal and plant genes for triosephosphate isomerase (Gilbert *et al.*, 1986). However, when a larger number of species was examined the positions of the introns in this gene became less easy to interpret: it appeared that introns had been lost in some lineages but gained in others. This scenario fits both 'introns early' and 'introns late' as both allow for the loss, gain or repositioning of introns by recombination events occurring in individual lineages. When many genes in many organisms are examined the general picture that emerges is that intron numbers have gradually increased during the evolution of animal genomes, this being put forward as evidence for 'introns late' (Palmer and Logsdon, 1991), despite the fact that animal mitochondrial genomes do not contain Group II introns that could supplement the existing nuclear introns by repeated invasions. Intron numbers must therefore have increased by recombination events, which is possible with both hypotheses.

An alternative approach has been to try to correlate exons with protein structural domains, as the 'introns early' hypothesis predicts that such a link should be evident, even allowing for the fuzzying effects of evolution since the primitive minigenes were assembled into the first real genes. Again, the first evidence to be obtained supported 'introns early'. A study of vertebrate globin proteins concluded that each of these comprises four structural domains, the first corresponding to exon 1 of the globin gene, the second and third to exon 2, and the fourth to exon 3 (*Figure 15.19*; Go, 1981). The prediction that there should be globin genes with another intron that splits the second and third domains was found to be correct when the leghemoglobin gene of soybean was shown to have an intron at exactly the expected position (Jensen

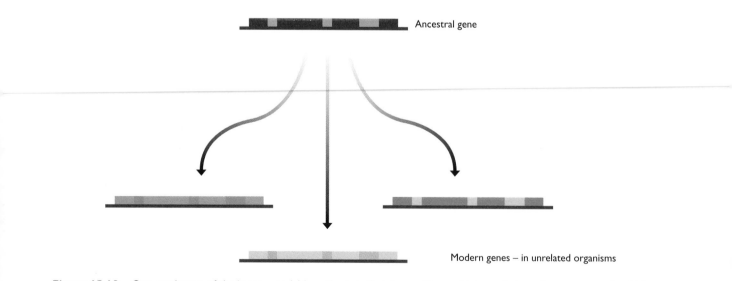

Ancestral gene

Modern genes – in unrelated organisms

Figure 15.18 One prediction of the 'introns early' hypothesis is that the positions of introns in homologous genes should be similar in unrelated organisms, because all these genes are descended from an ancestral intron-containing gene.

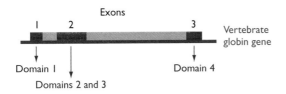

Figure 15.19   A vertebrate globin gene showing the relationship between the three exons and the four domains of the globin protein.

et al., 1981). Unfortunately, as more globin genes were sequenced more introns were discovered – more than ten in all. The positions of the majority of these do not correspond to junctions between domains.

The globin genes therefore conform with the general principle that emerged from our discussion of domain shuffling (Section 15.2.1): that in most cases there are no clear links between gene exons and protein structural domains. But is our definition of 'structural domain' correct? A structural domain within a protein may not simply correspond with a group of secondary structures such as α-helices and β-sheets. A more subtle interpretation might be that a structural domain is a polypeptide segment whose amino acids are less than a certain distance

apart in the protein's tertiary structure. It has been suggested that when this definition is adopted there is a better correlation between structural domain and exon (de Souza *et al.*, 1996).

15.4 The Human Genome: the Last 5 Million Years

Although the evolutionary history of humans is controversial, it is generally accepted that our closest relative among the primates is the chimpanzee and that the most recent ancestor that we share with the chimps lived 4.6–5.0 million years ago (Takahata, 1995). Since the split, the human lineage has embraced two genera – *Australopithecus* and *Homo* – and a number of species, not all of which were on the direct line of descent to *Homo sapiens* (*Figure 15.20*). The result is us, a novel species in possession of what are, at least to our eyes, important biological attributes that make us very different from all other animals. So how different are we from the chimpanzees?

As far as our genomes are concerned the answer is 'about 1.5%', this being the extent of the nucleotide sequence dissimilarity between humans and chimpanzees (Hacia, 2001). Within the coding DNA the difference is less than 1.5%, with many genes having identical sequences in the two genomes, but even in the noncoding regions the dissimilarity is rarely more than 3%. Only a few clear differences have been discovered:

- Humans lack a 92-bp segment of the gene for the *N*-glycolyl-neuraminic acid hydroxylase and so cannot synthesize the hydroxylated form of *N*-glycolyl-

Box 15.3: The role of non-coding DNA

The presence of extensive amounts of non-coding DNA in eukaryotic genomes (see Box 1.4, page 23) is a puzzle for molecular evolutionists. Why is this apparently superfluous DNA tolerated?

One possibility is that the non-coding DNA has a function that has not yet been identified and, as such, must be maintained because without it the cell would be non-viable. Plausible functions are not as difficult to identify as might be imagined. In several places in the preceding chapters the importance of chromatin structure has been stressed, including the attachment of chromatin to sites within the nucleus. Possibly, the non-coding component of genomes is involved in these aspects of genome organization (Cavalier-Smith, 1978). Alternatively, non-coding DNA might have a broadranging control function that so far has eluded discovery by molecular biologists (Zuckerkandl, 1976).

A second possibility is that non-coding DNA has no function but is tolerated by the genome because there is no selective pressure to get rid of it. Possession of non-coding DNA is neither an advantage nor a disadvantage and so the non-coding DNA is simply propagated along with the coding DNA. According to this hypothesis the non-coding DNA could simply be 'junk' or might be parasitic 'selfish DNA' (Orgel and Crick, 1980).

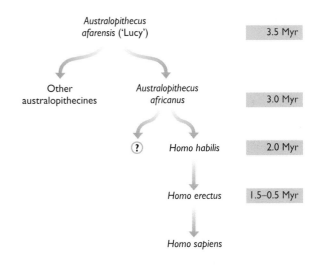

Figure 15.20   One possible scheme for the evolution of modern humans from australopithecine ancestors.

There are many controversies in this area of research and several different hypotheses have been proposed for the evolutionary relationships between different fossils. Abbreviation: Myr, million years.

neuraminic acid, which is present on the surfaces of some chimpanzee cells (Chou *et al.*, 1998; Muchmore *et al.*, 1998). This may have an effect on the ability of certain pathogens to enter human cells, and could possibly influence some types of cell–cell interaction, but the difference is not thought to be particularly significant.

■ Several recent gene duplications have occurred, resulting in gene copies that can be described as human-specific or chimpanzee-specific, as they are present in only one or the other genome. However, as far as gene functions are concerned these new genes are not significant because they have not yet had time to accumulate mutations to any great extent and so, in effect, are simply second copies of the genes from which they were derived.

■ Some components of the non-coding DNA in the two genomes have diverged extensively, illustrating how quickly repetitive DNA can evolve. For example, the alphoid DNA sequences present at human centromeres (Section 2.2.1) are quite different from the

equivalent sequences in chimpanzee and gorilla chromosomes (Archidiacono *et al.*, 1995). The human genome also contains novel versions of the Alu element (Section 2.4.2; Zietkiewicz *et al.*, 1994).

■ Human and chimpanzee genomes have undergone a few rearrangements, as revealed when the chromosome banding patterns are compared. The most dramatic difference is that human chromosome 2 is two separate chromosomes in chimpanzees (*Figure 15.21*), so chimpanzees, as well as other apes, have 24 pairs of chromosomes whereas humans have just 23 pairs. Four other chromosomes – human numbers 5, 6, 9 and 12 – also have visible differences to their chimpanzee counterparts, although the other 18 chromosomes appear to be very similar if not identical (Yunis and Prakash, 1982).

These differences are interesting as far as genome evolution is concerned but none of them reveals anything about the basis of the special biological attributes possessed by humans. This question – what makes us different from chimpanzees and other apes – is perplexing molecular biologists, who have been frustrated by the absence of a sequencing project for any of the ape genomes (Gibbons, 1998). But this is only part of the problem because many of the key differences between humans and apes are likely to lie with subtle changes in the expression patterns of genes involved in developmental processes and in specification of interconnections within the nervous system. Differences in the expression patterns of genes in the brains of humans and chimps have been revealed by microarray analysis (see Technical Note 5.1, page 133; Normile, 2001), but understanding how these differences relate to brain function will not be easy. It is clear, however, that what makes us human is probably not the human genome itself, but the way in which the genome functions.

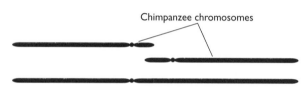
Chimpanzee chromosomes

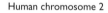
Human chromosome 2

Figure 15.21 Human chromosome 2 is the product of a fusion between two chimpanzee chromosomes.

For more details about the banding patterns of these chromosomes, from which the fusion is deduced, see Strachan and Read (1999).

References

Amores A, Force A, Yan Y-L, et al. (1998) Zebrafish *hox* clusters and vertebrate genome evolution. *Science*, **282**, 1711–1714.

Archidiacono N, Antonacci R, Marzella R, Finelli P, Lonoce A and Rocchi M (1995) Comparative mapping of human alphoid sequences in great apes using fluorescence *in situ* hybridization. *Genomics*, **25**, 477–484.

Bancroft I (2001) Duplicate and diverge: the evolution of plant genome microstructure. *Trends Genet.*, **17**, 89–93.

Barker WC and Dayhoff MO (1980) Evolutionary and functional relationships of homologous physiological mechanisms. *BioScience*, **30**, 593–600.

Bartel DP and Unrau PJ (1999) Constructing an RNA world. *Trends Genet.*, **12**, M9–M13.

Beier M, Reck F, Wagner T, Krishnamurthy R and Eschenmoser A (1999) Chemical etiology of nucleic acid structure: comparing pentopyranosyl-$(2' \rightarrow 4')$ oligonucleotides with RNA. *Science*, **283**, 699–703.

Bird AP (1995) Gene number, noise reduction and biological complexity. *Trends Genet.*, **11**, 94–100.

Brooke NM, Garcia-Fernàndez J and Holland PWH (1998) The ParaHox gene cluster is an evolutionary sister of the Hox gene cluster. *Nature*, **392**, 920–922.

Brookfield JFY (1997) Genetic redundancy. *Adv. Genet.*, **36**, 137–155.

Cavalier-Smith T (1978) Nuclear volume control by nucleoskeletal DNA, selection for cell volume and cell growth rate and the solution to the DNA C-value paradox. *J. Cell Sci.*, **34**, 247–278.

Chou HH, Takematsu H, Diaz S, et al. (1998) A mutation in human CMP-sialic acid hydroxylase occurred after the *Homo–Pan* divergence. *Proc. Natl Acad. Sci. USA*, **95**, 11751–11756.

Csermely P (1997) Proteins, RNAs and chaperones in enzyme evolution: a folding perspective. *Trends Biochem. Sci.*, **22**, 147–149.

de Souza SJ, Long M, Schoenbach L, Roy SW and Gilbert W (1996) Intron positions correlate with module boundaries in ancient proteins. *Proc. Natl Acad. Sci. USA*, **93**, 14 632–14 636.

Doolittle RF (1987) The evolution of the vertebrate plasma proteins. *Biol. Bull.*, **172**, 269–283.

Eickbush TH (2000) Introns gain ground. *Nature*, **404**, 940–943.

Ekland EH and Bartel DP (1996) RNA-catalysed RNA polymerization using nucleoside triphosphates. *Nature*, **382**, 373–376.

Eschenmoser A (1999) Chemical etiology of nucleic acid structure. *Science*, **284**, 2118–2124.

Freeland SJ, Knight RD and Landweber LF (1999) Do proteins predate DNA? *Science*, **286**, 690–692.

Gibbons A (1998) Which of our genes makes us human? *Science*, **281**, 1432–1434.

Gilbert W (1987) The exon theory of genes. *Cold Spring Harbor Symp. Quant. Biol.*, **52**, 901–905.

Gilbert W, Marchionni M and McKnight G (1986) On the antiquity of introns. *Cell*, **46**, 151–153.

Go M (1981) Correlation of DNA exonic regions with protein structural units in hemoglobin. *Nature*, **291**, 90–92.

Hacia JG (2001) Genome of the apes. *Trends Genet.*, **17**, 637–645.

Hartl DL, Lohe AR and Lozovskaya ER (1997) Modern thoughts on an ancyent *marinere*: function, evolution, regulation. *Ann. Rev. Genet.*, **31**, 337–358.

Henikoff S, Greene EA, Pietrokovski S, Bork P, Attwood TK and Hood L (1997) Gene families: the taxonomy of protein paralogs and chimeras. *Science*, **278**, 609–614.

Jensen EQ, Paludan K, Hyldig-Nielsen JJ, Jorgensen P and Markere KA (1981) The structure of a chromosomal leghemoglobin gene from soybean. *Nature*, **291**, 677–679.

Johnston WK, Unrau PJ, Lawrence MS, Glasner ME and Bartel DP (2001) RNA-catalyzed RNA polymerization: accurate and general RNA-templated primer extension. *Science*, **292**, 1319–1325.

Kazazian HH (2000) L1 retrotransposons shape the mammalian genome. *Science*, **289**, 1152–1153.

Li W-H and Graur D (1991) *Fundamentals of Molecular Evolution*. Sinauer, Sunderland, MA.

Lohse PA and Szostak JW (1996) Ribozyme-catalysed amino-acid transfer reactions. *Nature*, **381**, 442–444.

Maeda N and Smithies O (1986) The evolution of multigene families: human haptoglobin genes. *Ann. Rev. Genet.*, **20**, 81–108.

McDonald JF (1995) Transposable elements: possible catalysts of organismic evolution. *Trends Ecol. Evol.*, **10**, 123–126.

Messier W, Li S-H and Stewart C-B (1996) The birth of microsatellites. *Nature*, **381**, 483.

Miller SL (1953) A production of amino acids under possible primitive Earth conditions. *Science*, **117**, 528–529.

Muchmore EA, Diaz S and Varki A (1998) A structural difference between the cell surfaces of humans and the great apes. *Am. J. Phys. Anthropol.*, **107**, 187–198.

Normile D (2001) Gene expression differs in human and chimp brains. *Science*, **292**, 44–45.

Nowak MA, Boerlijst MC, Cooke J and Maynard Smith J (1997) Evolution of genetic redundancy. *Nature*, **388**, 167–170.

Ohno S (1970) *Evolution by Gene Duplication*. George Allen and Unwin, London.

Orgel LE (2000) A simpler nucleic acid. *Science*, **290**, 1306–1307.

Orgel LE and Crick FHC (1980) Selfish DNA: the ultimate parasite. *Nature*, **284**, 604–607.

Palmer JD and Logsdon JM (1991) The recent origin of introns. *Curr. Opin. Genet. Dev.*, **1**, 470–477.

Piskur J (2001) Origin of the duplicated regions in the yeast genome. *Trends Genet.*, **17**, 302–303.

Purugganan MD and Wessler S (1992) The splicing of transposable elements and its role in intron evolution. *Genetica*, **86**, 295–303.

Robertson HM, Zumpano KL, Lohe AR and Hartl DL (1996) Reconstructing the ancient mariners of humans. *Nature Genet.*, **12**, 360–361.

Robertson MP and Ellington AD (1998) How to make a nucleotide. *Nature*, **395**, 223–225.

Romero D and Palacios R (1997) Gene amplification and genomic plasticity in prokaryotes. *Ann. Rev. Genet.*, **31**, 91–111.

Stavenhagen JB and Robins DM (1988) An ancient provirus has imposed androgen regulation on the adjacent mouse sex limited protein gene. *Cell*, **55**, 247–254.

Strachan T and Read AP (1999) *Human Molecular Genetics*, 2nd edition. BIOS Scientific Publishers, Oxford.

Szathmáry E (1993) Coding coenzyme handles: a hypothesis for the origin of the genetic code. *Proc. Natl Acad. Sci. USA*, **90**, 9916–9920.

Szathmáry E and Maynard Smith J (1993) The origin of chromosomes. II. Molecular mechanisms. *J. Theoret. Biol.*, **164**, 447–454.

Takahata N (1995) A genetic perspective on the origin and history of humans. *Ann. Rev. Ecol. System.*, **26**, 343–372.

Taylor JS, Van de Peer Y and Meyer A (2001) Genome duplication, divergent resolution and speciation. *Trends Genet.*, **17**, 299–301.

Unrau PJ and Bartel DP (1998) RNA-catalysed nucleotide synthesis. *Nature*, **395**, 260–263.

Vision TJ, Brown DG and Tanksley SD (2000) The origins of genomic duplications in *Arabidopsis*. *Science*, **290**, 2114–2117.

Wächtershäuser G (1988) Before enzymes and templates: theory of surface metabolism. *Microbiol. Rev.*, **52**, 452–484.

Wächtershäuser G (1992) Groundworks for an evolutionary biochemistry – the iron sulfur world. *Prog. Biophys. Mol. Biol.*, **58**, 85–201.

Wagner A (2001) Birth and death of duplicated genes in completely sequenced eukaryotes. *Trends Genet.*, **17**, 237–239.

Walter P, Keenan R and Schmitz U (2000) SRP – where the RNA and membrane worlds meet. *Science*, **287**, 1212–1213.

Woese CR (1979) A proposal concerning the origin of life on the planet Earth. *J. Mol. Evol.*, **13**, 95–101.

Wolfe KH and Shields DC (1997) Molecular evidence for an ancient duplication of the entire yeast genome. *Nature*, **387**, 708–713.

Yunis JJ and Prakash O (1982) The origin of Man: a chromosomal pictorial legacy. *Science*, **215**, 1525–1530.

Zietkiewicz E, Richer C, Makalowski W, Jurka J and Labuda D (1994) A young *Alu* subfamily amplified independently in human and African great apes lineages. *Nucleic Acids Res.*, **22**, 5608–5612.

Zuckerkandl E (1976) Gene control in eukaryotes and c-value paradox: 'excess' DNA as an impediment to transcription of coding sequences. *J. Mol. Evol.*, **9**, 73–104.

Further Reading

Futuyama DJ (1998) *Evolutionary Biology*, 3rd edition. Sinauer, Sunderland, MA. — *An accessible description of evolutionary biology.*

Gesteland RF, Cech TR and Atkins JF (eds) (1999) *The RNA World*, 2nd edition. Cold Spring Harbor Laboratory Press, Cold Spring Harbor.

Jackson M, Strachan T and Dover GA (1996) *Human Genome Evolution*. BIOS Scientific Publishers, Oxford. — *An advanced treatment of the subject.*

Li W-H (1997) *Molecular Evolution*. Sinauer, Sunderland, MA. — *Detailed descriptions of many of the topics covered in this chapter.*

Maynard Smith J and Szathmáry E (1995) *The Major Transitions in Evolution*. WH Freeman, Oxford. — *A remarkable book that begins with the origin of life and ends with the evolution of human language.*

Otto SP and Whitton J (2000) Polyploid incidence and evolution. *Ann. Rev. Genet.*, **34**, 401–437.

STUDY AIDS FOR CHAPTER 15

Key terms

Give short definitions of the following terms:

3′ transduction
Allopolyploidy
Autopolyploidy
Concerted evolution
Domain duplication
Domain shuffling
Genetic redundancy
Introns early
Introns late

Lateral gene transfer
Peptide nucleic acid (PNA)
Protogenome
RNA world
Trisomy
Unequal crossing-over
Unequal sister chromatid
  exchange

Self study questions

1. Starting at 4.6 billion years ago, outline the key periods in genome evolution.
2. Summarize current thinking regarding the processes that led to evolution of the first genomes. Be careful to distinguish between the RNA world and the DNA world and to indicate how the transition from the former to latter is thought to have occurred.
3. Which periods during the last 1.5 billion years are linked to sudden increases in gene number?
4. Describe how the formation of an autopolyploid could result in an increase in gene number.
5. What indications are there that genome duplication has been important during the evolutionary histories of present-day genomes?
6. Using diagrams, distinguish between the four processes that could lead to gene duplication.
7. Explain how the globin gene superfamily illustrates the importance of gene duplication in genome evolution.
8. Discuss the impact of gene duplication on the evolution of the homeotic selector genes of eukaryotes.
9. Define the term 'concerted evolution' and state why this process is important in the evolution of some multigene families.

10. Describe, with examples, the processes of domain duplication and domain shuffling.
11. Discuss the evidence for lateral gene transfer between prokaryotic organisms. To what extent is lateral gene transfer likely to have contributed to genome evolution in eukaryotes?
12. Outline the impact that transposable elements can have on genome evolution.
13. Distinguish between the 'introns early' and 'introns late' hypotheses. What evidence is there to support these hypotheses?
14. List the ways in which the human genome differs from that of chimpanzees. What are the likely genetic explanations for the important differences between the biological attributes of humans and chimps?

Problem-based learning

1. How unique is life?
2. Are the examples of domain duplication and domain shuffling given in Section 15.2.1 special cases or are they representative of genome evolution in general?
3. 'Among animals, the species barriers are less easy to cross and it is difficult to find clear evidence for lateral gene transfer of any kind.' This statement reflects current thinking but is not secure. Indeed, one of the publications describing the draft human genome sequence lists 31 human genes that might have been acquired from bacteria by lateral gene transfer (IHGSC [International Human Genome Sequencing Consortium] [2001] Initial sequencing and analysis of the human genome. *Nature*, **409**, 860–921). Explore the controversies surrounding the influence of lateral gene transfer on the composition of the human genome.
4. Evaluate the 'introns early' and 'introns late' hypotheses.
5. Examine the differences between the human and chimpanzee genomes and provide a more detailed account of the reasons for the biological differences between humans and chimps.

16

Molecular Phylogenetics

Chapter Contents

16.1 The Origins of Molecular
     Phylogenetics   484

16.2 The Reconstruction of DNA-based
     Phylogenetic Trees   487

   16.2.1 The key features of DNA-based
           phylogenetic trees   487
   16.2.2 Tree reconstruction   489

16.3 The Applications of Molecular
     Phylogenetics   494

   16.3.1 Examples of the use of
           phylogenetic trees   494
   16.3.2 Molecular phylogenetics as a
           tool in the study of human
           prehistory   496

Learning outcomes

When you have read Chapter 16, you should be able to:

■ Recount how taxonomy led to phylogeny and discuss the reasons why molecular markers are important in phylogenetics

■ Describe the key features of a phylogenetic tree and distinguish between inferred trees, true trees, gene trees and species trees

■ Explain how phylogenetic trees are reconstructed, including a description of DNA sequence alignment, the methods used to convert alignment data into a phylogenetic tree, and how the accuracy of a tree is assessed

■ Discuss, with examples, the applications and limitations of molecular clocks

■ Give examples of the use of phylogenetic trees in studies of human evolution and the evolution of the human and simian immunodeficiency viruses

■ Describe how molecular phylogenetics is being used to study the origins of modern humans, and the migrations of modern humans into Europe and the New World

IF GENOMES EVOLVE by the gradual accumulation of mutations, then the amount of difference in nucleotide sequence between a pair of genomes should indicate how recently those two genomes shared a common ancestor. Two genomes that diverged in the recent past would be expected to have fewer differences than a pair of genomes whose common ancestor is more ancient. This means that by comparing three or more genomes it should be possible to work out the evolutionary relationships between them. These are the objectives of **molecular phylogenetics**.

16.1 The Origins of Molecular Phylogenetics

Molecular phylogenetics predates DNA sequencing by several decades. It is derived from the traditional method for classifying organisms according to their similarities and differences, as first practiced in a comprehensive fashion by Linnaeus in the 18th century. Linnaeus was a systematicist not an evolutionist, his objective being to place all known organisms into a logical classification which he believed would reveal the great plan used by the Creator – the *Systema Naturae*. However, he unwittingly laid the framework for later evolutionary schemes by dividing organisms into a hierarchic series of taxonomic categories, starting with kingdom and progressing down through phylum, class, order, family and genus to species. The naturalists of the 18th and early 19th centuries likened this hierarchy to a 'tree of life' (*Figure 16.1*), an analogy that was adopted by Darwin (1859) in *The Origin of Species* as a means of describing the interconnected evolutionary histories of living organisms. The classificatory scheme devised by Linnaeus therefore became reinterpreted as a **phylogeny** indicating not just the similarities between species but also their evolutionary relationships.

Whether the objective is to construct a classification or to infer a phylogeny, the relevant data are obtained by examining variable characters in the organisms being compared. Originally, these characters were morphological features, but molecular data were introduced at a surprisingly early stage. In 1904 Nuttall used immunological tests to deduce relationships between a variety of animals, one of his objectives being to place humans in their correct evolutionary position relative to other primates, an issue that we will return to in Section 16.3.1. Nuttall's work showed that molecular data can be used in phylogenetics, but the approach was not widely adopted until the late 1950s, the delay being due largely to technical limitations, but also partly because classification and phylogenetics had to undergo their own evolutionary changes before the value of molecular data could be fully appreciated. These changes came about with the introduction of **phenetics** and **cladistics** (Box 16.1), two novel

Modern species

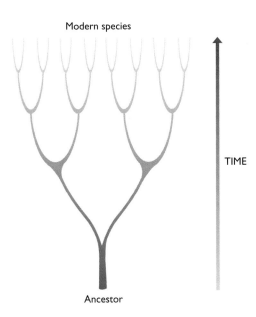

TIME

Ancestor

Figure 16.1   The tree of life.

An ancestral species is at the bottom of the 'trunk' of the tree. As time passes, new species evolve from earlier ones so the tree repeatedly branches until we reach the present time, when there are many species descended from the ancestor.

phylogenetic methods which, although quite different in their approach, both place emphasis on large datasets that can be analyzed by rigorous mathematical procedures. The difficulty in obtaining large mathematical datasets when morphological characters are used was one of the main driving forces behind the gradual shift towards molecular data, which have three advantages compared with other types of phylogenetic information:

- When molecular data are used, a single experiment can provide information on many different characters: in a DNA sequence, for example, every nucleotide position is a character with four **character states**, A, C, G and T. Large molecular datasets can therefore be generated relatively quickly.
- Molecular character states are unambiguous: A, C, G and T are easily recognizable and one cannot be confused with another. Some morphological characters, such as those based on the shape of a structure, can be less easy to distinguish because of overlaps between different character states.
- Molecular data are easily converted to numerical form and hence are amenable to mathematical and statistical analysis.

The sequences of protein and DNA molecules provide the most detailed and unambiguous data for molecular phylogenetics, but techniques for protein sequencing did not become routine until the late 1960s, and rapid DNA sequencing was not developed until 10 years after that.

Early studies therefore depended largely on indirect assessments of DNA or protein variations, using one of three methods:

- **Immunological data**, such as those obtained by Nuttall (1904), involve measurements of the amount of cross-reactivity seen when an antibody specific for a protein from one organism is mixed with the same protein from a different organism. Remember that in Section 12.2.1 we learned that antibodies are immunoglobulin proteins that help to protect the body against invasion by bacteria, viruses and other unwanted substances by binding to these 'antigens'. Proteins also act as antigens, so if human β-globin, for example, is injected into a rabbit then the rabbit makes an antibody that binds specifically to that protein. The antibody will also cross-react with β-globins from other vertebrates, because these β-globins have similar structures to the human version. The degree of cross-reactivity depends on how similar the β-globin being tested is to the human protein, providing the similarity data used in the phylogenetic analysis.
- **Protein electrophoresis** is used to compare the electrophoretic properties, and hence degree of similarity, of proteins from different organisms. This technique has proved useful for comparing closely related species and variations between members of a single species.
- **DNA–DNA hybridization data** are obtained by hybridizing DNA samples from the two organisms being compared. The DNA samples are denatured and mixed together so that hybrid molecules form. The stability of these hybrid molecules depends on the degree of similarity between the nucleotide sequences of the two DNAs, and is measured by determining the melting temperature (see *Figure 5.8*, page 132), a stable hybrid having a higher melting temperature than a less stable one. The melting temperatures obtained with DNAs from different pairs of organisms provide the data used in the phylogenetic analysis.

By the end of the 1960s these indirect methods had been supplemented with an increasing number of protein sequence studies (e.g. Fitch and Margoliash, 1967) and during the 1980s DNA-based phylogenetics began to be carried out on a large scale. Protein sequences are still used today in some contexts, but DNA has now become by far the predominant molecule. This is mainly because DNA yields more phylogenetic information than protein, the nucleotide sequences of a pair of homologous genes having a higher information content than the amino acid sequences of the corresponding proteins, because mutations that result in non-synonymous changes alter the DNA sequence but do not affect the amino acid sequence (*Figure 16.2*). Entirely novel information can also be obtained by DNA sequence analysis because variability in both the coding and non-coding regions of the genome can be

Box 16.1: Phenetics and cladistics

Phenetics, when first introduced (Michener and Sokal, 1957), challenged the prevailing view that classifications should be based on comparisons between a limited number of characters that taxonomists believed to be important for one reason or another. Pheneticists argued that classifications should encompass as many variable characters as possible, these characters being scored numerically and analyzed by rigorous mathematical methods.

Cladistics (Hennig, 1966) also emphasizes the need for large datasets but differs from phenetics in that it does not give equal weight to all characters. The argument is that in order to infer the branching order in a phylogeny it is necessary to distinguish those characters that provide a good indication of evolutionary relationships from other characters that might be misleading. This might appear to take us back to the pre-phenetic approach but cladistics is much less subjective: rather than making assumptions about which characters are 'important', cladistics demands that the evolutionary relevance of individual characters be defined. In particular, errors in the branching pattern within a phylogeny are minimized by recognizing two types of anomalous data.

■ **Convergent evolution** or **homoplasy** occurs when the same character state evolves in two separate lineages. For example, both birds and bats possess wings, but bats are more closely related to wingless mammals than they are to birds (see Figure A). The character state 'possession of wings' is therefore misleading in the context of vertebrate phylogeny.

■ **Ancestral character states** must be distinguished from **derived character states**. An ancestral (or **plesiomorphic**) character state is one possessed by a remote common ancestor of a group of organisms, an example being five toes in vertebrates. A derived (or **apomorphic**) character state is one that evolved from the ancestral state in a more recent common ancestor, and so is seen in only a subset of the species in the group being studied. Among vertebrates, the possession of a single toe, as displayed by modern horses, is a derived character state (see Figure B). If we did not realize this then we might conclude that humans are more closely related to lizards, which like us have five toes, rather than to horses.

Phenetics and cladistics have had an uneasy relationship over the last 40 years. Most of today's evolutionary biologists favor cladistics, even though a strictly cladistic approach throws up some apparently counter-intuitive results, a notable example being the conclusion that the birds should not have their own class (Aves) but be included among the reptiles.

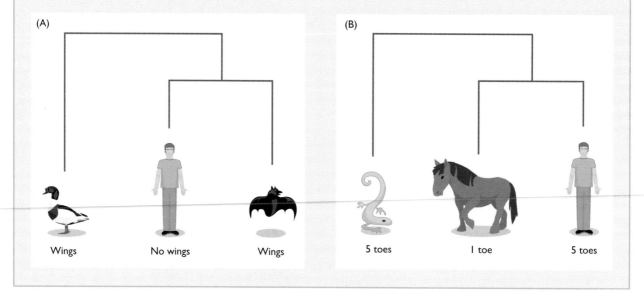

examined. The ease with which DNA samples for sequence analysis can be prepared by PCR (Section 4.3) is another key reason behind the predominance of DNA in modern molecular phylogenetics.

As well as DNA sequences, molecular phylogenetics also makes use of DNA markers such as RFLPs, SSLPs and SNPs (Section 5.2.2), particularly for intraspecific studies such as those aimed at understanding migrations of prehistoric human populations (Section 16.3.2). Later in this chapter we will consider various examples of the use of both DNA sequences and DNA markers in molecular phylogenetics, but first we must make a more detailed study of the methodology used in this area of genome research.

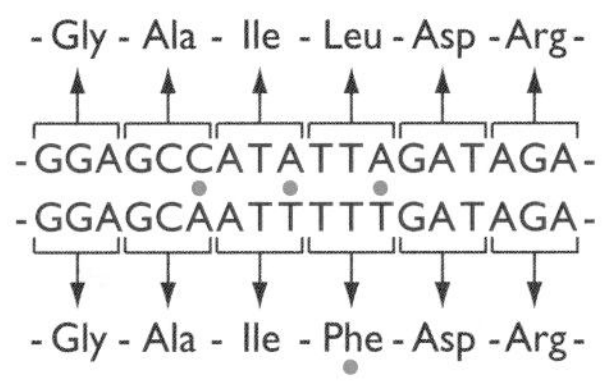

- Gly - Ala - Ile - Leu - Asp - Arg -
- GGAGCCATATTAGATAGA -
- GGAGCAATTTTTGATAGA -
- Gly - Ala - Ile - Phe - Asp - Arg -

Figure 16.2 DNA yields more phylogenetic information than protein.

The two DNA sequences differ at three positions but the amino acid sequences differ at only one position. These positions are indicated by green dots. Two of the nucleotide substitutions are therefore synonymous and one is non-synonymous (see *Figure 14.11*, page 430).

(A) An unrooted tree
A
B
C
D
Branches
Internal nodes
External nodes
(B) Rooted trees
A
B
C
D
1
2
3
4
5

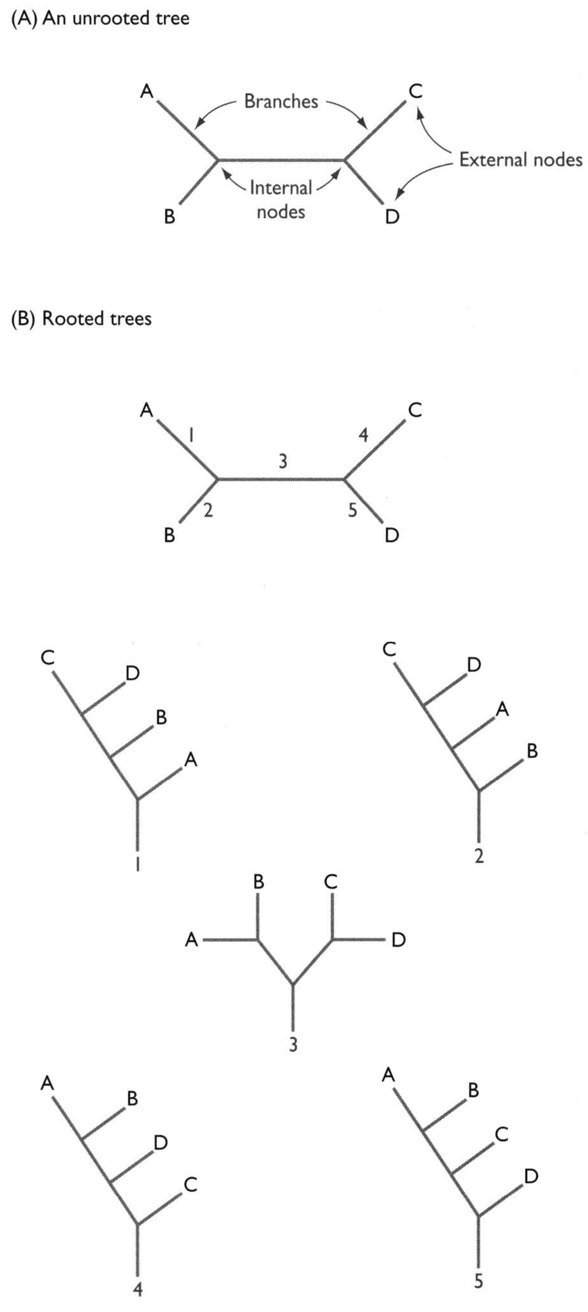

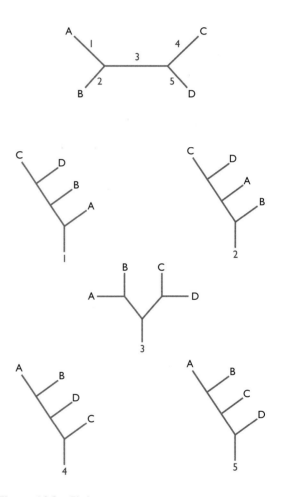

Figure 16.3 Phylogenetic trees.

(A) An unrooted tree with four external nodes. (B) The five rooted trees that can be drawn from the unrooted tree shown in part A. The positions of the roots are indicated by the numbers on the outline of the unrooted tree.

16.2 The Reconstruction of DNA-based Phylogenetic Trees

The objective of most phylogenetic studies is to reconstruct the tree-like pattern that describes the evolutionary relationships between the organisms being studied. Before examining the methodology for doing this we must first take a closer look at a typical tree in order to familiarize ourselves with the basic terminology used in phylogenetic analysis.

16.2.1 The key features of DNA-based phylogenetic trees

A typical phylogenetic tree is shown in *Figure 16.3A*. This tree could have been reconstructed from any type of comparative data, but as we are interested in DNA sequences we will assume that the tree shows the relationships between four homologous genes, called *A*, *B*, *C* and *D*. The **topology** of this tree comprises four **external nodes**, each representing one of the four genes that we have compared, and two **internal nodes** representing ancestral genes. The lengths of the **branches** indicate the degree of difference between the genes represented by the nodes. The degree of difference is calculated when the sequences are compared, as described in Section 16.2.2.

The tree in *Figure 16.3A* is **unrooted**, which means that it is only an illustration of the relationships between *A*, *B*, *C* and *D* and does not tell us anything about the series of evolutionary events that led to these genes. Five different evolutionary pathways are possible, each depicted by a different **rooted** tree, as shown in *Figure 16.3B*. To distinguish between them the phylogenetic analysis must include at least one **outgroup**, this being a homologous gene that we know is less closely related to *A*, *B*, *C* and *D* than these four genes are to each other. The outgroup enables the root of the tree to be located and the correct evolutionary pathway to be identified. The criteria used when choosing an outgroup depend very much on the type of analysis that is being carried out. As an example, let us say that the four homologous genes in our tree come from human, chimpanzee, gorilla and orangutan. We could then use as an outgroup the homologous gene from another primate, such as the baboon, which we know from paleontological evidence branched away from the lineage leading to human, chimpanzee, gorilla and orangutan before the time of the common ancestor of those four species (*Figure 16.4*).

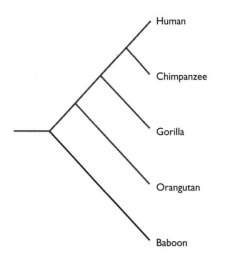

Figure 16.4 The use of an outgroup to root a phylogenetic tree.

The tree of human, chimpanzee, gorilla and orangutan genes is rooted with a baboon gene because we know from the fossil record that baboons split away from the primate lineage before the time of the common ancestor of the other four species. For more information on phylogenetic analysis of humans and other primates see Section 16.3.1.

We refer to the rooted tree that we obtain by phylogenetic analysis as an **inferred tree**. This is to emphasize that it depicts the series of evolutionary events that are inferred from the data that were analyzed, and may not be the same as the **true tree**, the one that depicts the actual series of events that occurred. Sometimes we can be fairly confident that the inferred tree is the true tree, but most phylogenetic data analyses are prone to uncertainties which are likely to result in the inferred tree differing in some respects from the true tree. In Section 16.2.2 we will look at the various methods used to assign degrees of confidence to the branching pattern in an inferred tree, and later in the chapter we will examine some of the controversies that have arisen as a result of the imprecise nature of phylogenetic analysis.

Gene trees are not the same as species trees

The tree shown in *Figure 16.4* illustrates a common type of molecular phylogenetics project, where the objective is to use a **gene tree**, reconstructed from comparisons between the sequences of **orthologous** genes (those derived from the same ancestral sequence; see page 196), to make inferences about the evolutionary history of the species from which the genes are obtained. The assumption is that the gene tree, based on molecular data with all its advantages, will be a more accurate and less ambiguous representation of the **species tree** than that obtainable by morphological comparisons. This assumption is often correct, but it does not mean that the gene tree is the *same* as the species tree. For that to be the case, the internal nodes in the gene and species trees would have to be precisely equivalent. However, they are not equivalent, because:

■ An *internal node in a gene tree* represents the divergence of an ancestral gene into two genes with dif-

---

Box 16.2: Terminology for molecular phylogenetics

The text includes definitions of most of the important terms used in molecular phylogenetics. Here are a few additional definitions that you may find useful when reading research articles on this subject:

■ **Operational taxonomic unit** or **OTU** is the term used to describe the organisms being compared when a tree is constructed from morphological data. 'OTU' is sometimes also used by molecular phylogeneticists as a synonym for 'gene or other nucleotide sequence', but it is perhaps best avoided in this context because a gene is not, strictly speaking, a taxonomic unit. A taxonomic unit is a component of a classification scheme, such as 'species'. To describe a gene as a taxonomic unit runs the risk of confusing a gene tree with a species tree.

■ **Monophyletic** refers to two or more DNA sequences that are derived from a single common ancestral DNA sequence. A group of monophyletic sequences is called a **clade** if it comprises all of the sequences included in the analysis that are descended from the ancestral sequence.

■ **Polyphyletic** refers to a group of DNA sequences that are derived from two or more distinct ancestral sequences.

■ **Parsimony** was originally a philosophical term stating that, when deciding between competing hypotheses, preference should be given to the one that involves the fewest unconnected assumptions. In molecular phylogenetics, parsimony is an approach that decides between different tree topologies by identifying the one that involves the shortest evolutionary pathway, this being the pathway that requires the smallest number of nucleotide changes to go from the ancestral sequence, at the root of the tree, to all the present-day sequences that have been compared.

For definitions of other phylogenetic terms, see Lincoln *et al.* (1998).

ferent DNA sequences: this occurs by mutation (*Figure 16.5A*).

■ An *internal node in a species tree* represents a speciation event (*Figure 16.5B*): this occurs by the population of the ancestral species splitting into two groups that are unable to interbreed, for example, because they are geographically isolated.

The important point is that these two events – mutation and speciation – are not expected to occur at the same time. For example, the mutation event could precede the speciation. This would mean that, to begin with, both alleles of the gene are present in the unsplit population of the ancestral species (*Figure 16.6*). When the population split occurs, it is likely that both alleles will still be present in each of the two resulting groups. After the split, the new populations evolve independently. One possibility is that the results of **random genetic drift** (see Box 16.3, page 497) lead to one allele being lost from one population and the other being lost from the other population. This establishes the two separate genetic lineages that we infer from phylogenetic analysis of the gene sequences

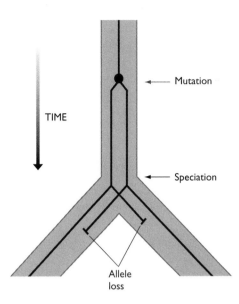

Figure 16.6 Mutation might precede speciation, giving an incorrect time for a speciation event if a molecular clock is used.

See the text for details. Based on Li (1997).

present in the modern species resulting from the continued evolution of the two populations.

How do these considerations affect the equivalence of the gene and species trees? There are various implications, two of which are as follows:

■ If a **molecular clock** (Section 16.2.2) is used to date the time at which the gene divergence took place, then it cannot be assumed that this is also the time of the speciation event. If the node being dated is ancient, say 50 million or more years ago, then the error may not be noticeable. But if the speciation event is recent, as when primates are being compared, then the date for the gene divergence might be significantly different to that for the speciation event.

■ If the first speciation event is quickly followed by a second speciation event in one of the two resulting populations, then the branching order of the gene tree might be different from that of the species tree. This can occur if the genes in the modern species are derived from alleles that had already appeared before the first of the two speciation events, as illustrated in *Figure 16.7*.

16.2.2 Tree reconstruction

In this section we will look at how tree reconstruction is carried out with DNA sequences, concentrating on the four steps in the procedure:

■ Aligning the DNA sequences and obtaining the comparative data that will be used to reconstruct the tree;

■ Converting the comparative data into a reconstructed tree;

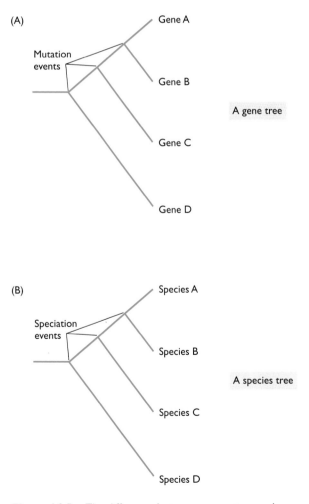

Figure 16.5 The difference between a gene tree and a species tree.

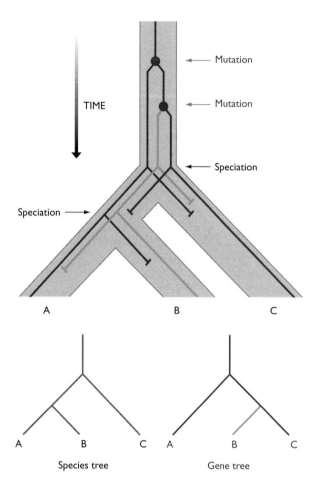

Figure 16.7   A gene tree can have a different branching order from a species tree.

In this example, the gene has undergone two mutations in the ancestral species, the first mutation giving rise to the 'blue' allele and the second to the 'green' allele. Random genetic drift in association with the two subsequent speciations results in the red allele lineage appearing in species A, the green allele lineage in species B and the blue allele lineage in species C. Molecular phylogenetics based on the gene sequences will reveal that the red–blue split occurred before the blue–green split, giving the gene tree shown on the right. However, the actual species tree is different, as shown on the left. Based on Li (1997).

■   Assessing the accuracy of the reconstructed tree;
■   Using a molecular clock to assign dates to branch points within the tree.

Sequence alignment is the essential preliminary to tree reconstruction

The data used in reconstruction of a DNA-based phylogenetic tree are obtained by comparing nucleotide sequences. These comparisons are made by aligning the sequences so that nucleotide differences can be scored. This is the critical part of the entire enterprise because if the alignment is incorrect then the resulting tree will definitely not be the true tree.

The first issue to consider is whether the sequences being aligned are homologous. If they are homologous then they must, by definition, be derived from a common ancestral sequence (Section 7.2.1) and so there is a sound basis for the phylogenetic study. If they are not homologous then they do not share a common ancestor. The phylogenetic analysis will find a common ancestor because the methods used for tree reconstruction always produce a tree of some description, even if the data are completely erroneous, but the resulting tree will have no biological relevance. With some DNA sequences – for example, the β-globin genes of different vertebrates – there is no difficulty in being sure that the sequences being compared are homologous, but this is not always the case, and one of the commonest errors that arises during phylogenetic analysis is the inadvertent inclusion of a non-homologous sequence.

Once it has been established that two DNA sequences are indeed homologous, the next step is to align the sequences so that homologous nucleotides can be compared. With some pairs of sequences this is a trivial exercise (*Figure 16.8A*), but it is not so easy if the sequences are relatively dissimilar and/or have diverged by the accumulation of insertions and deletions as well as point mutations. Insertions and deletions cannot be distinguished when pairs of sequences are compared so we refer to them as **indels**. Placing indels at their correct positions is often the most difficult part of sequence alignment (*Figure 16.8B*).

Some pairs of sequences can be aligned reliably by eye. For more complex pairs, alignment might be possible by the **dot matrix** method (*Figure 16.9*). The two sequences

(A) A simple sequence alignment

```
AGCAATGGCCAGACAATAATG
AGCTATGGACAGACATTAATG
*** **** ****** *****
```

(B) A more difficult sequence alignment

```
GACGACCATAGACCAGCATAG
GACTACCATAGA-CTGCAAAG
*** ********* ** *** **
```
Two possible positions for the indel

```
GACGACCATAGACCAGCATAG
GACTACCATAGACT-GCAAAG
*** ********* *** **
```

Figure 16.8   Sequence alignment.

(A) Two sequences that have not diverged to any great extent can be aligned easily by eye. (B) A more complicated alignment in which it is not possible to determine the correct position for an indel. If errors in indel placement are made in a multiple alignment then the tree reconstructed by phylogenetic analysis is unlikely to be correct. In this diagram, the red asterisks indicate nucleotides that are the same in both sequences.

are written out on the x- and y-axes of a graph, and dots placed in the squares of the graph paper at positions corresponding to identical nucleotides in the two sequences. The alignment is indicated by a diagonal series of dots, broken by empty squares where the sequences have nucleotide differences, and shifting from one column to another at places where indels occur.

More rigorous mathematical approaches to sequence alignment have also been devised. The first of these is the **similarity approach** (Needleman and Wunsch, 1970), which aims to maximize the number of matched nucleotides – those that are identical in the two sequences. The complementary approach is the **distance method** (Waterman *et al.*, 1976), in which the objective is to minimize the number of mismatches. Often the two procedures will identify the same alignment as being the best one.

Usually the comparison involves more than just two sequences, meaning that a **multiple alignment** is required. This can rarely be done effectively with pen and paper so, as in all steps in a phylogenetic analysis, a computer program is used. For multiple alignments, Clustal is often the most popular choice (Jeanmougin *et al.*, 1998). Clustal and other software packages for phylogenetic analysis are described in Technical Note 16.1.

Converting alignment data into a phylogenetic tree

Once the sequences have been aligned accurately, an attempt can be made to reconstruct the phylogenetic tree. To date nobody has devised a perfect method for tree reconstruction, and several different procedures are used routinely. Comparative tests have been run with artificial

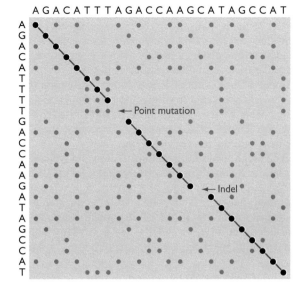

Figure 16.9   The dot matrix technique for sequence alignment.

The correct alignment stands out because it forms a diagonal of continuous dots, broken at point mutations and shifting to a different diagonal at indels.

data, for which the true tree is known, but these have failed to identify any particular method as being better than any of the others (Felsenstein, 1988).

The main distinction between the different tree-building methods is the way in which the multiple sequence

---

Phylogenetic analysis

Software packages for construction of phylogenetic trees.

Few sets of DNA sequences are simple enough to be converted into phylogenetic trees entirely by hand. Virtually all research in this area is carried out by computer with the aid of any one of a variety of software packages designed specifically for one or other of the steps in tree reconstruction.

One of the easiest to use and most popular packages is Clustal, which was originally written in 1988 and has undergone several upgrades in the intervening years (Jeanmougin et al., 1998). Clustal is primarily a program for carrying out multiple alignments of protein or DNA sequences, which it is able to do very effectively provided that the sequences being compared do not contain extensive internal repeat motifs. Clustal is usually used in conjunction with NJplot, a simple program for tree reconstruction by the neighbor-joining method. One

important advantage of Clustal and NJplot is that they do not require huge amounts of computer memory and so can be run on small PCs or Macintosh computers.

More comprehensive software packages enable the researcher to choose between a variety of different methods for tree reconstruction and to carry out more sophisticated types of phylogenetic analysis. The most widely used of these packages are PAUP (Swofford, 1993) and PHYLIP (Felsenstein, 1989). The tree-building programs in PAUP are often looked upon as the most accurate ones currently available and are able to handle relatively large datasets (Eernisse, 1998). PHYLIP has the advantage of including a number of software tools not readily available from other sources. Other popular packages include PAML (Yang, 1997), MacClade, and HENNIG86.

alignment is converted into numerical data that can be analyzed mathematically in order to reconstruct the tree. The simplest approach is to convert the sequence information into a **distance matrix**, which is simply a table showing the evolutionary distances between all pairs of sequences in the dataset (*Figure 16.10*). The evolutionary distance is calculated from the number of nucleotide differences between a pair of sequences and is used to establish the lengths of the branches connecting these two sequences in the reconstructed tree.

The **neighbor-joining method** (Saitou and Nei, 1987) is a popular tree-building procedure that uses the distance matrix approach. To begin the reconstruction, it is initially assumed that there is just one internal node from which branches leading to all the DNA sequences radiate in a star-like pattern (*Figure 16.11A*). This is virtually impossible in evolutionary terms but the pattern is just a starting point. Next, a pair of sequences is chosen at random, removed from the star, and attached to a second internal node, connected by a branch to the center of the star, as shown in *Figure 16.11B*. The distance matrix is then used to calculate the total branch length in this new 'tree'. The sequences are then returned to their original positions and another pair attached to the second internal node, and again the total branch length is calculated. This operation is repeated until all the possible pairs have been examined, enabling the combination that gives the tree with the shortest total branch length to be identified. This pair of sequences will be neighbors in the final tree; in the interim, they are combined into a single unit, creating a new star with one branch fewer than the original one. The whole process of pair selection and tree-length calculation is now repeated so that a second pair of neighboring sequences is identified, and then repeated again so that a third pair is located, and so on. The result is a complete reconstructed tree.

The advantage of the neighbor-joining method is that the data handling is relatively easy to carry out, largely because the information content of the multiple alignment has been reduced to its simplest form. The disadvantage is that some of the information is lost, in particular that pertaining to the identities of the ancestral

Multiple alignment

1	AGGCCAAGCCATAGCTGTCC
2	AGGCAAAGACATACCTGACC
3	AGGCCAAGACATAGCTGTCC
4	AGGCAAAGACATACCTGTCC

Distance matrix

	1	2	3	4
1	–	0.20	0.05	0.15
2		–	0.15	0.05
3			–	0.10
4				–

Figure 16.10 A simple distance matrix.

The matrix shows the evolutionary distance between each pair of sequences in the alignment. In this example the evolutionary distance is expressed as the number of nucleotide differences per nucleotide site for each sequence pair. For example, sequences 1 and 2 are 20 nucleotides in length and have four differences, corresponding to an evolutionary difference of 4/20 = 0.2. Note that this analysis assumes that there are no **multiple substitutions** (also called **multiple hits**). Multiple substitution occurs when a single site undergoes two or more changes (e.g. the ancestral sequence . . .A**T**GT . . . gives rise to two modern sequences: . . .A**G**GT . . . and . . . A**C**GT . . .). There is only one nucleotide difference between the two modern sequences, but there have been two nucleotide substitutions. If this multiple hit is not recognized then the evolutionary distance between the two modern sequences will be significantly underestimated. To avoid this problem, distance matrices for phylogenetic analysis are usually constructed using mathematical methods that include statistical devices for estimating the amount of multiple substitution that has occurred.

(A) The starting point for the neighbor-joining method

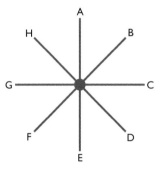

(B) Removal of two sequences from the star

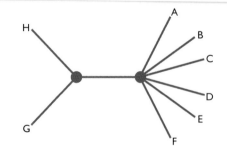

Figure 16.11 Manipulations carried out when using the neighbor-joining method for tree reconstruction.

See the text for details.

and derived nucleotides (equivalent to ancestral and derived character states, defined in Box 16.1, page 486) at each position in the multiple alignment. The **maximum parsimony** method (Fitch, 1977) takes account of this information, utilizing it to recreate the series of nucleotide changes that resulted in the pattern of variation revealed by the multiple alignment. The assumption, possibly erroneous, is that evolution follows the shortest possible route and that the correct phylogenetic tree is therefore the one that requires the minimum number of nucleotide changes to produce the observed differences between the sequences. Trees are therefore constructed at random and the number of nucleotide changes that they involve calculated until all possible topologies have been examined and the one requiring the smallest number of steps identified. This is presented as the most likely inferred tree.

The maximum parsimony method is more rigorous in its approach compared with the neighbor-joining method, but this increase in rigor inevitably extends the amount of data handling that is involved. This is a significant problem because the number of possible trees that must be scrutinized increases rapidly as more sequences are added to the dataset. With just five sequences there are only 15 possible unrooted trees, but for ten sequences there are 2 027 025 unrooted trees and for 50 sequences the number exceeds the number of atoms in the universe (Eernisse, 1998). Even with a high-speed computer it is not possible to check every one of these trees in a reasonable time, if at all, so often the maximum parsimony method is unable to carry out a comprehensive analysis. The same is true with many of the other more sophisticated methods for tree reconstruction.

Assessing the accuracy of a reconstructed tree

The limitations to the methods used in phylogenetic reconstruction lead inevitably to questions about the veracity of the resulting trees. Statistical tests of the accuracy of a reconstructed tree have been devised (Hillis, 1997; Whelan *et al.*, 2001) but these are necessarily complex because a tree is geometric rather than numeric, and the accuracy of one part of the topology may be greater or lesser than the accuracy of the other parts.

The routine method for assigning confidence limits to different branch points within a tree is to carry out a **bootstrap analysis**. To do this we need a second multiple alignment that is different from, but equivalent to, the real alignment. This new alignment is built up by taking columns, at random, from the real alignment, as illustrated in *Figure 16.12*. The new alignment therefore comprises sequences that are different from the original, but it has a similar pattern of variability. This means that when we use the new alignment in tree reconstruction we do not simply reproduce the original analysis, but we should obtain the same tree.

In practice, 1000 new alignments are created so 1000 replicate trees are reconstructed. A **bootstrap value** can then be assigned to each internal node in the original tree, this value being the number of times that the branch pat-

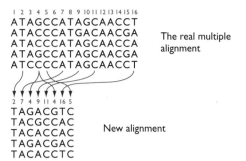

Figure 16.12 Constructing a new multiple alignment in order to bootstrap a phylogenetic tree.

The new alignment is built up by taking columns at random from the real alignment. Note that the same column can be sampled more than once.

tern seen at that node was reproduced in the replicate trees. If the bootstrap value is greater than 700/1000 then we can assign a reasonable degree of confidence to the topology at that particular internal node.

Molecular clocks enable the time of divergence of ancestral sequences to be estimated

When we carry out a phylogenetic analysis our primary objective is to infer the pattern of the evolutionary relationships between the DNA sequences that are being compared. These relationships are revealed by the topology of the tree that is reconstructed. Often we also have a secondary objective: to discover when the ancestral sequences diverged to give the modern sequences. This information is interesting in the context of genome evolution, as we discovered when we looked at the evolutionary history of the human globin genes (see *Figure 15.9*, page 470). The information is even more interesting on occasions when we are able to equate a gene tree with a species tree, because now the times at which the ancestral sequences diverged approximate to the dates of speciation events.

To assign dates to branch points in a phylogenetic tree we must make use of a molecular clock. The molecular clock hypothesis, first proposed in the early 1960s, states that nucleotide substitutions (or amino acid substitutions if protein sequences are being compared) occur at a constant rate. This means that the degree of difference between two sequences can be used to assign a date to the time at which their ancestral sequence diverged. However, to be able to do this the molecular clock must be calibrated so that we know how many nucleotide substitutions to expect per million years. Calibration is usually achieved by reference to the fossil record. For example, fossils suggest that the most recent common ancestor of humans and orangutans lived 13 million years ago. To calibrate the human molecular clock we therefore compare human and orangutan DNA sequences to determine the amount of nucleotide substitution that has occurred, and then divide this figure by 13, followed by 2, to obtain a rate of substitution per million years (*Figure 16.13*).

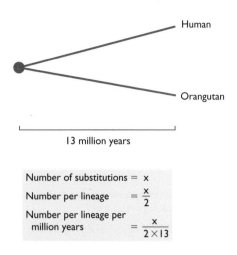

Figure 16.13   Calculating a human molecular clock.

The number of substitutions is determined for a pair of homologous genes from human and orangutan: call this number 'x'. The number of substitutions per lineage is therefore $\frac{x}{2}$, and the number per million years is $\frac{x}{2 \times 13}$.

At one time it was thought that there might be a universal molecular clock that applied to all genes in all organisms (Ochman and Wilson, 1987). Now we realize that molecular clocks are different in different organisms and are variable even within a single organism (Strauss, 1999). The differences between organisms might be the result of generation times, because a species with a short generation time is likely to accumulate DNA replication errors at a faster rate than a species with a longer generation time. This probably explains the observation that rodents have a faster molecular clock than primates (Gu and Li, 1992). Within an organism the variations are as follows:

■ Non-synonymous substitutions occur at a slower rate than synonymous ones. This is because a mutation that results in a change in the amino acid sequence of a protein might be deleterious to the organism, so the accumulation of non-synonymous mutations in the population is reduced by the processes of natural selection (see Box 16.3, page 497). This means that when gene sequences in two species are compared, there are usually fewer non-synonymous than synonymous substitutions.

■ The molecular clock for mitochondrial genes is faster than that for genes in the nuclear genome. This is probably because mitochondria lack many of the DNA repair systems that operate on nuclear genes (Section 14.2; Gibbons, 1998).

Despite these complications, molecular clocks have become an immensely valuable adjunct to tree reconstruction, as we will see in the next section when we look at some typical molecular phylogenetics projects.

16.3 The Applications of Molecular Phylogenetics

Molecular phylogenetics has grown in stature since the start of the 1990s, largely because of the development of more rigorous methods for tree building, combined with the explosion of DNA sequence information obtained initially by PCR analysis and more recently by genome projects. The importance of molecular phylogenetics has also been enhanced by the successful application of tree reconstruction and other phylogenetic techniques to some of the more perplexing issues in biology. In this final section we will survey some of these successes.

16.3.1 Examples of the use of phylogenetic trees

First, we will consider two projects that illustrate the various ways in which conventional tree reconstruction is being used in modern molecular biology.

DNA phylogenetics has clarified the evolutionary relationships between humans and other primates

Darwin (1871) was the first biologist to speculate on the evolutionary relationships between humans and other primates. His view – that humans are closely related to the chimpanzee, gorilla and orangutan – was controversial when it was first proposed and fell out of favor, even among evolutionists, in the following decades. Indeed, biologists were among the most ardent advocates of an anthropocentric view of our place in the animal world (Goodman, 1962).

From studies of fossils, paleontologists had concluded prior to 1960 that chimpanzees and gorillas are our closest relatives but that the relationship was distant, the split, leading to humans on the one hand and chimpanzees and gorillas on the other, having occurred some 15 million years ago. The first detailed molecular data, obtained by immunological studies in the 1960s (Goodman, 1962; Sarich and Wilson, 1967) confirmed that humans, chimpanzees and gorillas do indeed form a single clade (see Box 16.2, page 488) but suggested that the relationship is much closer, a molecular clock indicating that this split occurred only 5 million years ago. This was one of the first attempts to apply a molecular clock to phylogenetic data and the result was, quite naturally, treated with some suspicion. In fact, an acrimonious debate opened up between paleontologists, who believed in the ancient split indicated by the fossil evidence, and biologists, who had more confidence in the recent date suggested by the molecular data. This debate was eventually 'won' by the molecular biologists, whose view that the split occurred about 5 million years ago became generally accepted.

As more and more molecular data were obtained, the difficulties in establishing the exact pattern of the evolutionary events that led to humans, chimpanzees and gorillas became apparent. Comparisons of the mitochondrial genomes of the three species by restriction mapping (Section 5.3.1) and DNA sequencing suggested that the chimpanzee and gorilla are more closely related to each other than either is to humans (*Figure 16.14A*), whereas DNA–DNA hybridization data supported a closer relationship between humans and chimpanzees (*Figure 16.14B*). The reason for these conflicting results is the close similarity between DNA sequences in the three species, the differences being less than 3% for even the most divergent regions of the genomes (Section 15.4). This makes it difficult to establish relationships unambiguously.

The solution to the problem has been to make comparisons between as many different genes as possible and to target those loci that are expected to show the greatest amount of dissimilarity. By 1997, 14 different molecular datasets had been obtained, including sequences of variable loci such as pseudogenes and non-coding sequences (Ruvolo, 1997). Analysis of these datasets confirmed that the chimpanzee is the closest relative to humans, with our lineages diverging 4.6–5.0 million years ago. The gorilla is a slightly more distant cousin, its lineage having diverged from the human–chimp one between 0.3 and 2.8 million years earlier (*Figure 16.14C*).

(A) Mitochondrial DNA data

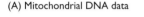

(B) DNA–DNA hybridization data

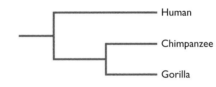

(C) Combined molecular datasets

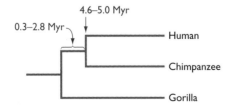

Figure 16.14 Different interpretations of the evolutionary relationships between humans, chimpanzees and gorillas.

See the text for details. Abbreviation: Myr, million years.

The origins of AIDS

The global epidemic of acquired immune deficiency syndrome (AIDS) has touched everyone's lives. AIDS is caused by human immunodeficiency virus 1 (HIV-1), a retrovirus (Section 2.4.2) that infects cells involved in the immune response. The demonstration in the early 1980s that HIV-1 is responsible for AIDS was quickly followed by speculation about the origin of the disease. Speculation centered around the discovery that similar immunodeficiency viruses are present in primates such as the chimpanzee, sooty mangabey, mandrill and various monkeys. These simian immunodeficiency viruses (SIVs) are not pathogenic in their normal hosts but it was thought that if one had become transferred to humans then within this new species the virus might have acquired new properties, such as the ability to cause disease and to spread rapidly through the population.

Retrovirus genomes accumulate mutations relatively quickly because reverse transcriptase, the enzyme that copies the RNA genome contained in the virus particle into the DNA version that integrates into the host genome (see Section 2.4.2), lacks an efficient proofreading activity (Section 13.2.2) and so tends to make errors when it carries out RNA-dependent DNA synthesis. This means that the molecular clock runs rapidly in retroviruses, and genomes that diverged quite recently display sufficient nucleotide dissimilarity for a phylogenetic analysis to be carried out. Even though the evolutionary period we are interested in is less than 100 years, HIV and SIV genomes contain sufficient data for their relationships to be inferred by phylogenetic analysis.

The starting point for this phylogenetic analysis is RNA extracted from virus particles. RT-PCR (see Technical Note 4.4, page 120) is therefore used to convert the RNA into a DNA copy and then to amplify the DNA so that sufficient amounts for nucleotide sequencing are obtained. Comparison between virus DNA sequences has resulted in the reconstructed tree shown in *Figure 16.15* (Leitner *et al.*, 1996; Wain-Hobson, 1998). This tree has a number of interesting features. First it shows that different samples of HIV-1 have slightly different sequences, the samples as a whole forming a tight cluster, almost a star-like pattern, that radiates from one end of the unrooted tree. This star-like topology implies that the global AIDS epidemic began with a very small number of viruses, perhaps just one, which have spread and diversified since entering the human population. The closest relative to HIV-1 among primates is the SIV of chimpanzees, the implication being that this virus jumped across the species barrier between chimps and humans and initiated the AIDS epidemic. However, this epidemic did not begin immediately: a relatively long uninterrupted branch links the center of the HIV-1 radiation with the internal node leading to the relevant SIV sequence, suggesting that after transmission to humans, HIV-1 underwent a latent period when it remained restricted to a small part of the global human population, presumably in Africa, before beginning its rapid spread to other parts of the world. Other primate

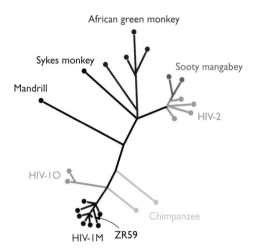

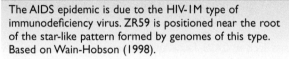

Figure 16.15 The phylogenetic tree reconstructed from HIV and SIV genome sequences.

The AIDS epidemic is due to the HIV-1M type of immunodeficiency virus. ZR59 is positioned near the root of the star-like pattern formed by genomes of this type. Based on Wain-Hobson (1998).

SIVs are less closely related to HIV-1, but one, the SIV from sooty mangabey, clusters in the tree with the second human immunodeficiency virus, HIV-2. It appears that HIV-2 was transferred to the human population independently of HIV-1, and from a different simian host. HIV-2 is also able to cause AIDS, but has not, as yet, become globally epidemic.

An intriguing addition to the HIV/SIV tree was made in 1998 when the sequence of an HIV-1 isolate from a blood sample taken in 1959 from an African male was sequenced (Zhu *et al.*, 1998). The RNA was highly fragmented and only a short DNA sequence could be obtained, but this was sufficient for the sequence to be placed on the phylogenetic tree (see *Figure 16.15*). This sequence, called ZR59, attaches to the tree by a short branch that emerges from near the center of the HIV-1 radiation. The positioning indicates that the ZR59 sequence represents one of the earliest versions of HIV-1 and shows that the global spread of HIV-1 was already underway by 1959. A later and more comprehensive analysis of HIV-1 sequences has suggested that the spread began in the period between 1915 and 1941, with a best estimate of 1931 (Korber *et al.*, 2000). Pinning down the date in this way has enabled epidemiologists to begin an investigation of the historic and social conditions that might have been responsible for the start of the AIDS epidemic.

16.3.2 Molecular phylogenetics as a tool in the study of human prehistory

Now we will turn our attention to the use of molecular phylogenetics in intraspecific studies: the study of the evolutionary history of members of the same species. We could choose any one of several different organisms to illustrate the approaches and applications of intraspecific studies, but many people look on *Homo sapiens* as the most interesting organism so we will investigate how molecular phylogenetics is being used to deduce the origins of modern humans and the geographic patterns of their recent migrations in the Old and New Worlds.

Intraspecific studies require highly variable genetic loci

In any application of molecular phylogenetics, the genes chosen for analysis must display variability in the organisms being studied. If there is no variability then there is no phylogenetic information. This presents a problem in intraspecific studies because the organisms being compared are all members of the same species and so share a great deal of genetic similarity, even if the species has split into populations that interbreed only intermittently. This means that the DNA sequences that are used in the phylogenetic analysis must be the most variable ones that are available. In humans there are three main possibilities.

- *Multiallelic genes*, such as members of the HLA family (Section 5.2.1), which exist in many different sequence forms;
- *Microsatellites*, which evolve not through mutation but by replication slippage (Section 14.1.1). Cells do not appear to have any repair mechanism for reversing the effects of replication slippage, so new microsatellite alleles are generated relatively frequently.
- *Mitochondrial DNA* which, as mentioned in Section 16.2.2, accumulates nucleotide substitutions relatively rapidly because mitochondria lack many of the repair systems that slow down the molecular clock in the human nucleus. The mitochondrial DNA variants present in a single species are called **haplotypes**.

It is important to note that it is not the potential for change that is critical to the application of these loci in phylogenetic analysis, it is the fact that different alleles or haplotypes of the locus coexist in the population as a whole. The loci are therefore **polymorphic** (see Box 16.3) and information pertaining to the relationships between different individuals can be obtained by comparing the combinations of alleles and/or haplotypes that those individuals possess.

The origins of modern humans – out of Africa or not?

It seems reasonably certain that the origin of humans lies in Africa because it is here that all of the oldest pre-human fossils have been found. The paleontological evidence reveals that hominids first moved outside of Africa over 1 million years ago, but these were not modern humans, they were an earlier species called *Homo erectus*. These were the first hominids to become geographically dispersed, eventually spreading to all parts of the Old World.

Box 16.3: Genes in populations

New alleles and haplotypes appear in a population because of mutations that occur in the reproductive cells of individual organisms. This means that many genes are **polymorphic**, two or more alleles being present in the population as a whole, each with its own **allele frequency**. Allele frequencies change over time as a result of **natural selection** and **random genetic drift**. Natural selection occurs because of differences in **fitness** (the ability of an organism to survive and reproduce) that result in the 'preservation of favourable variations and the rejection of injurious variations' (Darwin, 1859). Natural selection therefore decreases the frequencies of alleles that reduce the fitness of an organism, and increases the frequency of alleles that improve fitness. In reality, few new alleles that arise in a population have a significant impact on fitness so most are not affected by natural selection, but their frequencies still change because of random genetic drift caused by the arbitrary nature of birth, death and reproduction.

Either because of natural selection or random drift, one allele can begin to predominate in a population and eventually achieve a frequency of 100%. This allele has now become **fixed**. Mathematical models predict that over time different alleles become fixed in a population, resulting in a series of **gene substitutions**.

If a species splits into two populations that do not interbreed extensively, then the allele frequencies in the two populations will change differently so that after a few tens of generations the two populations will have distinctive genetic features. Eventually, different gene substitutions occur in the two populations, but even before this happens we can distinguish between them because of the differences in allele frequencies. These differences can be used to date the time when the population split occurred and even to determine if one or both populations experienced a **bottleneck**, a period when the population size became substantially reduced.

The events that followed the dispersal of *Homo erectus* are controversial. From comparisons using fossil skulls and bones, paleontologists have concluded that the *Homo erectus* populations that became located in different parts of the Old World gave rise to the modern human populations of those areas by a process called **multiregional evolution** (*Figure 16.16A*). There may have been a certain amount of interbreeding between humans from different geographic regions, but, to a large extent, these various populations remained separate throughout their evolutionary history.

Doubts about the multiregional hypothesis were first raised by re-interpretations of the fossil evidence and were subsequently brought to a head by publication in 1987 of a phylogenetic tree reconstructed from mitochondrial RFLP data obtained from 147 humans representing populations from all parts of the World (Cann *et al.*, 1987). The tree (*Figure 16.17*) confirmed that the ancestors of modern humans lived in Africa but suggested that they were still there about 200 000 years ago. This inference was made by applying the mitochondrial molecular clock to the tree, which showed that the ancestral mitochondrial DNA, the one from which all modern mitochondrial DNAs are descended, existed between 140 000 and 290 000 years ago. The tree showed that this mitochondrial genome was located in Africa, so the person who possessed it, the so-called mitochondrial Eve (she had to be female because mitochondrial DNA is only inherited through the female line), must have been African.

The discovery of mitochondrial Eve prompted a new scenario for the origins of modern humans. Rather than evolving in parallel throughout the world, as suggested by the multiregional hypothesis, **Out of Africa** states that *Homo sapiens* originated in Africa, members of this species then moving into the rest of the Old World between 100 000 and 50 000 years ago, displacing the descendents of *Homo erectus* that they encountered (see *Figure 16.16B*).

Such a radical change in thinking inevitably did not go unchallenged. When the RFLP data obtained by Cann *et al.* (1987) were examined by other molecular phylogeneticists it became clear that the original computer analysis had been flawed, and that several quite different trees could be reconstructed from the data, some of which did not have a root in Africa. These criticisms were countered by more detailed mitochondrial DNA sequence datasets, most of which are compatible with a relatively recent African origin and so support the Out of Africa hypothesis rather than multiregional evolution (e.g. Ingman *et al.*, 2000). An interesting complement to 'mitochondrial Eve' has been provided by studies of the Y chromosome, which suggest that 'Y chromosome Adam' also lived in Africa some 200 000 years ago (Pääbo, 1999). Of course, this Eve and Adam were not equivalent to the biblical characters and were by no means the only people alive at that time: they were simply the individuals who carried the ancestral mitochondrial DNA and Y chromosomes that gave rise to all the mitochondrial DNAs and Y chromosomes in existence today. The important point is that these ancestral DNAs were still in Africa well after the spread of *Homo erectus* into Eurasia.

The mitochondrial DNA and Y chromosome studies appear to provide strong evidence in support of the Out of Africa theory. But complications have arisen from studies of nuclear genes other than those on the Y chromosome. For example, β-globin sequences give a much earlier date, 800 000 years ago, for the common ancestor (Harding *et al.*, 1997), and studies of an X chromosome gene, *PDHA1*, place the ancestral sequence at 1 900 000 years ago (Harris and Hey, 1999). Molecular anthropologists are currently debating the significance of

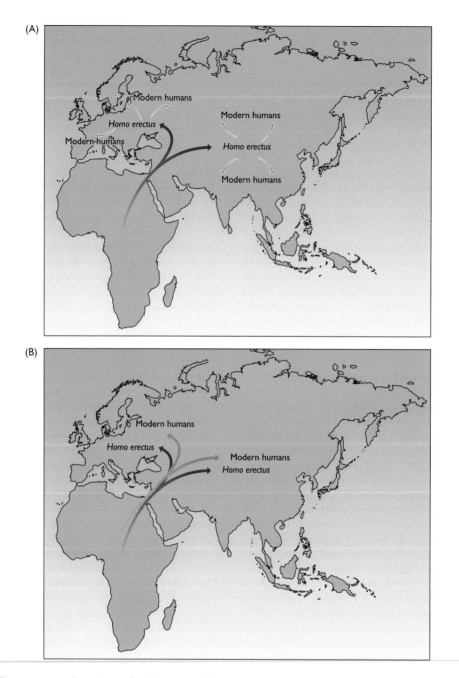

Figure 16.16 Two competing hypotheses for the origins of modern humans.

(A) The multiregional hypothesis states that *Homo erectus* left Africa over 1 million years ago and then evolved into modern humans in different parts of the Old World. (B) The Out of Africa hypothesis states that the populations of *Homo erectus* in the Old World were displaced by new populations of modern humans that followed them out of Africa.

these results (Pääbo, 1999). More datasets, and hopefully some sort of Grand Synthesis, are eagerly awaited.

The patterns of more recent migrations into Europe are also controversial

By whatever evolutionary pathway, modern humans were present throughout most of Europe by 40 000 years ago. This is clear from the fossil and archaeological records. The next controversial issue in human prehistory concerns whether these populations were displaced about 30 000 years later by other humans migrating into Europe from the Middle East.

The question centers on the process by which agriculture spread into Europe. The transition from hunting and gathering to farming occurred in the Middle East some

RESEARCH

16.1

BRIEFING

Neandertal DNA

Sequence analysis of 'ancient DNA' extracted from a fossil bone between 30 000 and 100 000 years old provides support for the Out of Africa hypothesis.

Neandertals are extinct hominids who lived in Europe between 300 000 and 30 000 years ago. They were descended from the *Homo erectus* populations who left Africa about 1 million years ago and, according to the Out of Africa hypothesis, were displaced when modern humans reached Europe about 50 000 years ago. Therefore, one prediction of the Out of Africa hypothesis is that there is no genetic continuity between Neandertals and the modern humans that live in Europe today. Bearing in mind that the last Neandertal died out 30 000 years ago, is there any way that we can test this hypothesis?

Ancient DNA could provide an answer. It has been known for some years that DNA molecules can survive the death of the organism in which they are contained, being recoverable centuries and possibly millennia later as short degraded molecules preserved in bones and other biological remains. There is never very much ancient DNA in a specimen, possibly no more than a few hundred genomes in a gram of bone, but that need not concern us because we can always use PCR to amplify these tiny amounts into larger quantities from which we can obtain DNA sequences.

The field of ancient DNA has been plagued with controversies over the last 10 years. In the early 1990s there were many reports of ancient human DNA being detected in bones and other archaeological specimens, but often it turned out that what had been amplified by PCR was not ancient DNA at all, but contaminating modern DNA left on the specimen by the archaeologist who dug it up or by the molecular biologist who carried out the DNA extraction. The worldwide success of the film *Jurassic Park* led to reports of DNA in insects preserved in amber and even in dinosaur bones, but all of these claims are now doubted. Many biologists started to wonder if ancient DNA existed at all, but gradually it became clear that if the work is carried out with extreme care it is sometimes possible to extract authentic ancient DNA from specimens up to about 50 000 years old. This is just old enough to include a few Neandertal bones.

DNA from a Neandertal bone

The Neandertal specimen selected for study is believed to be between 30 000 and 100 000 years old. DNA extraction was carried out with a fragment of bone weighing about 400 mg and a technique called **quantitative PCR** was used to determine if human DNA molecules were present and, if so, how many. The results indicated that the bone fragment contained about 1300 copies of the Neandertal mitochondrial genome, enough to suggest that sequence analysis was worth attempting.

PCRs were directed at what was expected to be the most variable part of the Neandertal mitochondrial genome. Because it was anticipated that the DNA would be broken into very short pieces, a sequence was built up in sections by carrying out nine overlapping PCRs, none amplifying more than 170 bp of DNA but together giving a total length of 377 bp. This sequence provided the data used to test the Out of Africa hypothesis.

A phylogenetic tree was constructed to compare the sequence obtained from the Neandertal bone with the sequences of six mitochondrial DNA haplotypes from modern humans. The Neandertal sequence was positioned on a branch of its own, connected to the root of the tree but not linked directly to any of the modern human sequences.

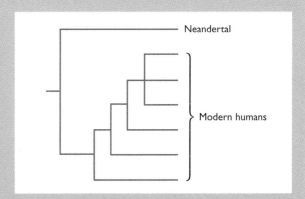

Next, a huge multiple alignment was made in order to compare the Neandertal sequence with 994 sequences from modern humans. The differences were striking. The Neandertal sequence differed from the modern sequences at an average of 27.2 ± 2.2 nucleotide positions whereas the modern sequences, which came from all over the world, not just Europe, differed from each other at only 8.0 ± 3.1 positions. Similar results were obtained when mitochondrial DNA from a second Neandertal skeleton was examined (Ovchinnikov *et al.*, 2000). The degree of difference between Neandertal and modern European DNA is incompatible with the notion that modern Europeans are descended from Neandertals and strongly supports the Out of Africa hypothesis. Supporters of multiregional evolution are not convinced, however, and the debate over modern human origins continues.

References

Krings M, Stone A, Schmitz RW, Krainitzki H, Stoneking M and Pääbo S (1997) Neandertal DNA sequences and the origin of modern humans. *Cell*, **90**, 19–30.
Ovchinnikov IV, Götherström A. Romanova GP, Kharitonov VM, Lidén K and Goodwin W (2000) Molecular analysis of Neanderthal DNA from the northern Caucasus. *Nature*, **404**, 490–493.

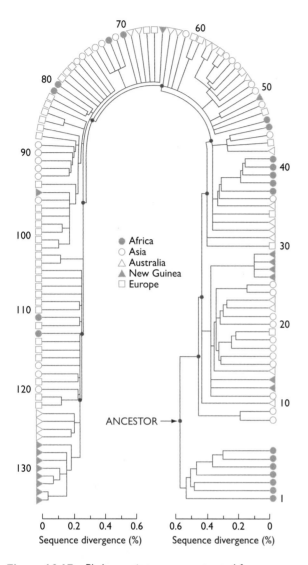

Africa
Asia
Australia
New Guinea
Europe

ANCESTOR →

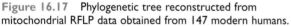

0   0.2   0.4   0.6   0.6   0.4   0.2   0
Sequence divergence (%)   Sequence divergence (%)

Figure 16.17 Phylogenetic tree reconstructed from mitochondrial RFLP data obtained from 147 modern humans.

The ancestral mitochondrial DNA is inferred to have existed in Africa because of the split in the tree between the seven modern African mitochondrial genomes placed below the ancestral sequence and all the other genomes above it. Because this lower branch is purely African it is deduced that the ancestor was also African. The scale bars at the bottom indicate sequence divergence from which, using the mitochondrial molecular clock, it is possible to assign dates to the branch points in the tree. The clock suggests that the ancestral sequence existed between 140 000 and 290 000 years ago. Reprinted with permission from Cann et al. (1987) Nature, **325**, 31–36. Copyright 1987 Macmillan Magazines Limited.

9000–10 000 years ago, when early Neolithic villagers began to cultivate crops such as wheat and barley. After becoming established in the Middle East, farming spread into Asia, Europe and North Africa. By searching for evidence of agriculture at archaeological sites, for example

by looking for the remains of cultivated plants or for implements used in farming, it has been possible to trace the expansion of farming along two routes through Europe, one around the coast to Italy and Spain and the second through the Danube and Rhine valleys to northern Europe (*Figure 16.18*).

How did farming spread? The simplest explanation is that farmers migrated from one part of Europe to another, taking with them their implements, animals and crops, and displacing the indigenous pre-agricultural communities that were present in Europe at that time. This **wave of advance** model was initially favored by geneticists because of the results of a large-scale phylogenetic analysis of the allele frequencies for 95 nuclear genes in populations from across Europe (Cavalli-Sforza, 1998). Such a large and complex dataset cannot be analyzed in any meaningful way by conventional tree building but instead has to be examined by more advanced statistical methods, ones based more in population biology than phylogenetics. One such procedure is **principal component analysis**, which attempts to identify patterns in a dataset corresponding to the uneven geographic distribution of alleles, these uneven distributions possibly being indicative of past population migrations. The most striking pattern within the European dataset, accounting for about 28% of the total genetic variation, is a gradation of allele frequencies across Europe (*Figure 16.19*). This pattern implies that a migration of people occurred either from the Middle East to northeast Europe, or in the opposite direction. Because the former coincides with the expansion of farming, as revealed by the archaeological record, this first principal component was looked upon as providing strong support for the wave of advance model.

The analysis looked convincing but two criticisms were raised. The first was that the data provided no indication of when the inferred migration took place, so the link between the first principal component and the spread of agriculture was based solely on the pattern of the allele gradation, not on any complementary evidence relating to the period when this gradation was set up. The second criticism arose because of the results of a second study of European human populations, one that did include a time dimension (Richards *et al.*, 1996). This study looked at mitochondrial DNA haplotypes in 821 individuals from various populations across Europe. It failed to confirm the gradation of allele frequencies detected in the nuclear DNA dataset, and instead suggested that European populations have remained relatively static over the last 20 000 years. A refinement of this work led to the discovery that eleven mitochondrial DNA haplotypes predominate in the modern European population, each with a different time of origin, thought to indicate the date at which the haplotype entered Europe (*Figure 16.20*; Richards *et al.*, 2000). The most ancient haplotype, called U, first appeared in Europe approximately 50 000 years ago, coinciding with the period when, according to the archaeological record, the first modern humans moved into the continent as the ice sheets withdrew to the north at the end of the last major glaciation. The youngest haplotypes,

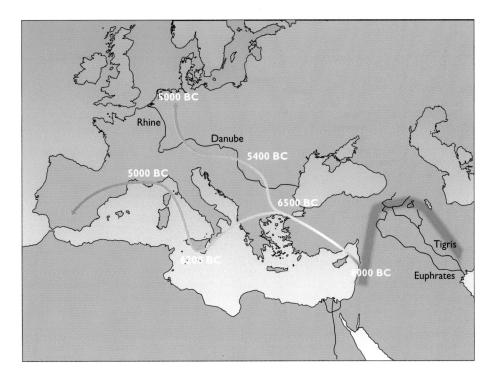

Figure 16.18   The spread of agriculture from the Middle East to Europe.

The dark-green area is the 'Fertile Crescent', the area of the Middle East where many of today's crops – wheat, barley, etc. – grow wild and where these plants are thought to have first been taken into cultivation.

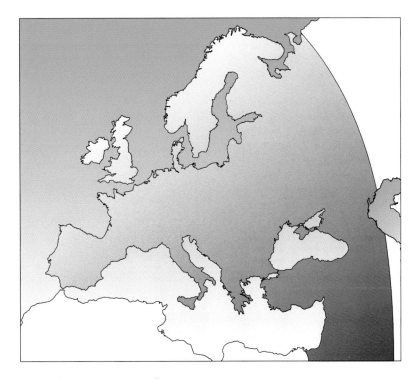

Figure 16.19   A genetic gradation across modern Europe.

See the text for details.

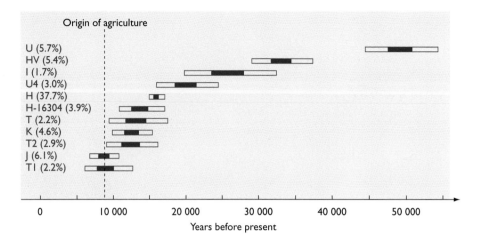

Figure 16.20 The eleven major European mitochondrial haplotypes.

The calculated time of origin for each haplotype is shown, the closed and open parts of each bar indicating different degrees of confidence. The percentages refer to the proportions of the modern European population with each haplotype. All the haplotypes except J and T1 had entered Europe before the origin of agriculture 9000–10 000 years ago. Redrawn from Richards *et al.* (2000).

J and T1, which at 9000 years in age could correspond to the origins of agriculture, are possessed by just 8.3% of the modern European population, suggesting that the spread of farming into Europe was not the huge wave of advance indicated by the principal component study. Instead, it is now thought that farming was brought into Europe by a smaller group of 'pioneers' who interbred with the existing pre-farming communities rather than displacing them.

Prehistoric human migrations into the New World

Finally we will examine the completely different set of controversies surrounding the hypotheses regarding the patterns of human migration that led to the first entry of people into the New World. There is no evidence for the spread of *Homo erectus* into the Americas, so it is presumed that humans did not enter the New World until after modern *Homo sapiens* had evolved in, or migrated into, Asia. The Bering Strait between Asia and North America is quite shallow and if the sea level dropped by 50 meters it would be possible to walk across from one continent to the other. It is believed that this was the route taken by the first humans to venture into the New World (*Figure 16.21*).

The sea was 50 meters or more below its current level for most of the last Ice Age, between about 60 000 and 11 000 years ago, but for most of this time the route would have been impassable because of the build-up of ice. Also, the northern parts of America would have been arctic during much of this period, providing few game animals for the migrants to hunt and very little wood with which they could make fires. These considerations, together with the absence of archaeological evidence of humans in North America before 11 500 years ago, led to the adoption of 'about 12 000 years ago' as the date for the first entry of humans into the New World. Recent discoveries

of evidence of human occupation at sites dating to 20 000 years ago, both in North and South America, has prompted some rethinking, but it is still generally assumed that a substantial population migration into North America, possibly the one from which all modern Native Americans are descended, occurred about 12 000 years ago.

What information does molecular phylogenetics provide? The first relevant studies were carried out in the late 1980s using RFLP data. These indicated that Native

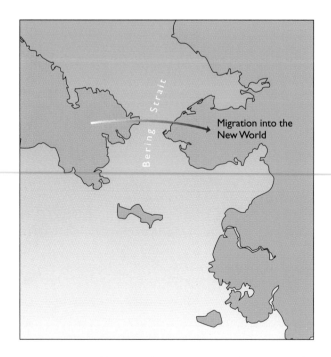

Figure 16.21 The route by which humans first entered the New World.

Americans are descended from Asian ancestors and identified four distinct mitochondrial haplotypes among the population as a whole (Wallace *et al.*, 1985; Schurr *et al.*, 1990). Linguistic studies had already shown that American languages can be divided into three different groupings, suggesting that modern Native Americans are descended from three sets of people, each speaking a different language. The inference from the molecular data that there may in fact have been four ancestral populations was not too disquieting. The first significant dataset of mitochondrial DNA sequences was obtained in 1991, enabling the rigorous application of a molecular clock. This indicated that the migrations into North America occurred between 15 000 and 8000 years ago (Ward *et al.*, 1991), which is consistent with the archaeological evidence that humans were absent from the continent before 11 500 years ago.

These early phylogenetic analyses confirmed, or at least were not too discordant with, the complementary evidence provided by archaeological and linguistic studies. However, the additional molecular data that have been acquired since 1992 have tended to confuse rather than clarify the issue. For example, different datasets have provided a variety of estimates for the number of migrations into North America. The most comprehensive analysis, based on mitochondrial DNA (Forster *et al.*, 1996), puts this figure at just one migration, and suggests that it occurred between 25 000 and 20 000 years ago, much earlier than the traditional date. Studies of Y chromosomes have assigned a date of approximately 22 500 years ago to the 'Native American Adam', the carrier of the Y chromosome that is ancestral to most, if not all, of the Y chromosomes in modern Native Americans (De Mendoza and Braginski, 1999). The implication from these studies is that humans became established in North America about 20 000 years ago, much earlier than indicated by the archaeological and early genetic evidence. This hypothesis is still being evaluated by other molecular biologists and archaeologists.

References

Cann RL, Stoneking M and Wilson AC (1987) Mitochondrial DNA and human evolution. *Nature*, **325**, 31–36.

Cavalli-Sforza LL (1998) The DNA revolution in population genetics. *Trends Genet.*, **14**, 60–65.

Darwin C (1859) *The Origin of Species by Means of Natural Selection, or the Preservation of Favoured Races in the Struggle for Life.* Penguin Books, London.

Darwin C (1871) *The Descent of Man, and Selection in Relation to Sex.* Princeton University Press, Princeton, NJ.

De Mendoza DH and Braginski R (1999) Y chromosomes point to Native American Adam. *Science*, **283**, 1439–1440.

Eernisse DJ (1998) A brief guide to phylogenetic software. *Trends Genet.*, **14**, 473–475.

Felsenstein J (1988) Phylogenies from molecular sequences: inference and reliability. *Ann. Rev. Genet.*, **22**, 521–565.

Felsenstein J (1989) PHYLIP – Phylogeny Inference Package (Version 3.20). *Cladistics*, **5**, 164–166.

Fitch WM (1977) On the problem of discovering the most parsimonious tree. *Am. Nat.*, **111**, 223–257.

Fitch WM and Margoliash E (1967) Construction of phylogenetic trees. A method based on mutation distances as estimated from cytochrome *c* sequences is of general applicability. *Science*, **155**, 279–284.

Forster P, Harding R, Torroni A and Bandelt HJ (1996) Origin and evolution of native American mtDNA variation: a reappraisal. *Am. J. Hum. Genet.*, **59**, 935–945.

Gibbons A (1998) Calibrating the molecular clock. *Science*, **279**, 28–29.

Goodman M (1962) Immunochemistry of the primates and primate evolution. *Ann. NY Acad. Sci.*, **102**, 219–234.

Gu X and Li W-H (1992) Higher rates of amino acid substitution in rodents than in humans. *Mol. Phylogenet. Evol.*, **1**, 211–214.

Harding RM, Fullerton SM, Griffiths RC, et al. (1997) Archaic African and Asian lineages in the genetic ancestry of modern humans. *Am. J. Hum. Genet.*, **60**, 772–789.

Harris EE and Hey J (1999) X chromosome evidence for ancient human histories. *Proc. Natl Acad. Sci. USA*, **96**, 3320–3324.

Hennig W (1966) *Phylogenetic Systematics.* University of Illinois Press, Urbana, IL.

Hillis DM (1997) Biology recapitulates phylogeny. *Science*, **276**, 218–219.

Ingman M, Kaessmann H, Pääbo S and Gyllensten U (2000) Mitochondrial genome variation and the origin of modern humans. *Nature*, **408**, 708–713.

Jeanmougin F, Thompson JD, Gouy M, Higgins DG and Gibson TJ (1998) Multiple sequence alignment with Clustal X. *Trends Biochem. Sci.*, **23**, 403–405.

Korber B, Muldoon M, Theiler J, et al. (2000) Timing the ancestor of the HIV-1 pandemic strains. *Science*, **288**, 1789–1796.

Leitner T, Escanilla D, Franzen C, Uhlen M and Albert J (1996) Accurate reconstruction of a known HIV-1 transmission history by phylogenetic tree analysis. *Proc. Natl Acad. Sci. USA*, **93**, 10864–10869.

Li W-H (1997) *Molecular Evolution.* Sinauer, Sunderland, MA.

Lincoln R, Boxshall G and Clark P (1998) *A Dictionary of Ecology, Evolution and Systematics*, 2nd edition. Cambridge University Press, Cambridge.

Michener CD and Sokal RR (1957) A quantitative approach to a problem in classification. *Evolution*, **11**, 130–162.

Needleman SB and Wunsch CD (1970) A general method applicable to the search of similarities in the amino acid sequences of two proteins. *J. Mol. Biol.*, **48**, 443–453.

Nuttall GHF (1904) *Blood Immunity and Blood Relationship.* Cambridge University Press, Cambridge.

Ochman H and Wilson AC (1987) Evolution in bacteria: evidence for a universal substitution rate in cellular genomes. *J. Mol. Evol.*, **26**, 74–86.

Pääbo S (1999) Human evolution. *Trends Genet.*, **15**, M13–M16.

Richards M, Côrte-Real H, Forster P, et al. (1996) Paleolithic and Neolithic lineages in the European mitochondrial gene pool. *Am. J. Hum. Genet.*, **59**, 185–203.

Richards M, Macauley V, Hickey E, et al. (2000) Tracing European founder lineages in the Near Eastern mtDNA pool. *Am. J. Hum. Genet.*, **67**, 1251–1276.

Ruvolo M (1997) Molecular phylogeny of the hominoids: inferences from multiple independent DNA sequence data sets. *Mol. Biol. Evol.*, **14**, 248–265.

Saitou N and Nei M (1987) The neighbor-joining method: a new method for reconstructing phylogenetic trees. *Mol. Biol. Evol.*, **4**, 406–425.

Sarich VM and Wilson AC (1967) Immunological time scale for hominid evolution. *Science*, **158**, 1200–1203.

Schurr TG, Ballinger SW, Gan YY, et al. (1990) Amerindian mitochondrial DNAs have rare Asian mutations at high frequencies suggesting they are derived from four primary maternal lineages. *Am. J. Hum. Genet.*, **46**, 613–623.

Strauss E (1999) Can mitochondrial clocks keep time? *Science*, **283**, 1435–1438.

Swofford DL (1993) *PAUP: Phylogenetic Analysis Using Parsimony.* Illinois Natural History Survey, Champaign, IL.

Wain-Hobson S (1998) 1959 and all that. *Nature,* **391,** 531–532.

Wallace DC, Garrison K and Knowler WC (1985) Dramatic founder effects in Amerindian mitochondrial DNAs. *Am. J. Phys. Anthropol.,* **68,** 149–155.

Ward RH, Frazier BL, Dew-Jager K and Pääbo S (1991) Extensive mitochondrial diversity within a single Amerindian tribe. *Proc. Natl Acad. Sci. USA,* **88,** 8720–8724.

Waterman MS, Smith TF and Beyer WA (1976) Some biological sequence metrics. *Adv. Math.,* **20,** 367–387.

Whelan S, Liò P and Goldman N (2001) Molecular phylogenetics: state-of-the-art methods for looking into the past. *Trends Genet.,* **17,** 262–272.

Yang Z (1997) PAML: a program package for phylogenetic analysis by maximum likelihood. *CABIOS,* **13,** 555–556.

Zhu T, Korber BT, Nahmias AJ, Hooper E, Sharp PM and Ho DD (1998) An African HIV-1 sequence from 1959 and implications for the origin of the epidemic. *Nature,* **391,** 594–597.

Further Reading

Avise JC (1994) *Molecular Markers, Natural History and Evolution.* Chapman & Hall, New York. — *A detailed description of the use of molecular data in studies of evolution.*

Doolittle WF (1999) Phylogenetic classification and the universal tree. *Science,* **284,** 2124–2128. — *Discusses the strengths and weaknesses of molecular phylogenetics as a means of inferring species trees.*

Futuyama DJ (1998) *Evolutionary Biology,* 3rd edition. Sinauer, Sunderland, MA.

Hall BG (2001) *Phylogenetic Trees Made Easy: A How-To Manual for Molecular Biologists.* Sinauer, Sunderland, MA.

Hartl DL and Clark AG (1997) *Principles of Population Genetics,* 3rd edition. Sinauer, Sunderland, MA. — *A good introduction to population genetics, emphasizing the relevance of this subject to evolutionary biology.*

Hillis DM, Moritz C and Mable BK (eds) (1996) *Molecular Systematics,* 2nd edition. Sinauer, Sunderland, MA. — *Comprehensive coverage of techniques for phylogenetic tree reconstruction.*

Li WH and Graur D (1991) *Fundamentals of Molecular Evolution.* Sinauer, Sunderland, MA. — *Another excellent introduction.*

Nei M (1996) Phylogenetic analysis in molecular evolutionary genetics. *Ann. Rev. Genet.,* **30,** 371–403. — *Brief review of tree-building techniques.*

Thornton JW and Desalle R (2000) Gene family evolution and homology: genomics meets phylogenetics. *Ann. Rev. Genomics Hum. Genet.,* **1,** 41–73. — *Stresses the changes that need to occur in molecular phylogenetics in order to deal with genomic sequences.*

STUDY AIDS FOR CHAPTER 16

Key terms

Give short definitions of the following terms:

Allele frequency
Ancestral character state
Ancient DNA
Apomorphy
Bootstrap analysis
Bootstrap value
Bottleneck
Branch
Character states
Cladistics
Convergent evolution
Derived character state
Distance matrix
Distance method
Dot matrix
External node
Fitness
Gene substitution
Gene tree
Haplotype
Homoplasy
Indel

Inferred tree
Internal node
Maximum parsimony
Molecular clock
Molecular phylogenetics
Monophyletic
Multiple alignment
Multiple hits
Multiple substitutions
Multiregional evolution
Natural selection
Neighbor-joining method
Operational taxonomic unit (OTU)
Orthologous genes
Outgroup
Out of Africa
Parsimony
Phenetics
Phylogeny
Plesiomorphy
Polymorphic loci

Polyphyletic
Principal component analysis
Quantitative PCR
Random genetic drift
Rooted tree
Similarity approach

Species tree
Topology
True tree
Unrooted tree
Wave of advance

Self study questions

1. Describe how taxonomy gradually led to phylogeny.

2. Define the terms 'phenetics' and 'cladistics' and outline the important features of each of these approaches to phylogenetics.

3. List the various types of molecular data that have been used in phylogenetics, indicating how each type of data is obtained.

4. Explain why DNA sequences are the principal type of molecular data used in modern molecular phylogenetics.

5. Draw and annotate a typical unrooted tree.

6. Explain how an outgroup can be used to convert an unrooted tree into a rooted one.

7. Distinguish between the terms 'inferred tree', 'true tree', 'gene tree' and 'species tree'. Explain why gene trees and species trees are not equivalent.

8. Describe how alignment of DNA sequences is used as a preliminary to tree reconstruction.
9. Outline the key features of the neighbor-joining and maximum parsimony methods of tree reconstruction.
10. How is the accuracy of a reconstructed tree assessed?
11. Describe how a molecular clock is calibrated and explain why there is no universal molecular clock.
12. Outline how molecular phylogenetics has contributed to our understanding of the evolutionary relationships between humans and other primates.
13. Describe how molecular phylogenetics has been used to investigate the origins of AIDS.
14. What types of variable loci are used when molecular phylogenetics is applied to intraspecific studies?
15. Distinguish between the multiregional and Out of Africa hypotheses for the origins of modern humans. What evidence is there for either hypothesis?
16. Describe how molecular phylogenetics has been used to trace the migrations of modern humans into Europe.
17. Describe the current models for the migration of modern humans into the New World.

Problem-based learning

1. Can a gene tree ever be equivalent to a species tree?
2. How reliable are molecular clocks?

3. Write a report on the science described in Ruvolo M (1997) Molecular phylogeny of the hominoids: inferences from multiple independent DNA sequence data sets. *Mol. Biol. Evol.*, **14**, 248–265.
4. Evaluate the genetic evidence in support of the Out of Africa hypothesis.
5. Explore how molecular phylogenetics has been used to study the mitochondrial DNA haplotypes present in modern European populations. What equivalent work has been done on mitochondrial DNA haplotypes among Native Americans?
6. Phylogenetic studies of mitochondrial DNA assume that this genome is inherited through the maternal line and that there is no recombination between maternal and paternal genomes. Assess the validity of this assumption and describe how the hypotheses regarding the origins and migrations of modern humans would be affected if recombination between maternal and paternal genomes was shown to occur. Possible starting points for your research into this problem are: Ladoukakis ED and Zouros E (2001) Recombination in animal mitochondrial DNA: evidence from published sequences. *Mol. Biol. Evol.*, **18**, 2127–2131; Meunier J and Eyre-Walker A (2001) The correlation between linkage disequilibrium and distance: implications for recombination in hominid mitochondria. *Mol. Biol. Evol.*, **18**, 2132–2135.

Appendix

Keeping up to Date

GENOME RESEARCH MOVES at a fast pace. If you check through the reading lists at the end of each chapter you will see that many of the papers cited in this second edition of *Genomes* were published in 2001, the year in which the book was written. Important discoveries are likely to be made over the next 2–3 years and keeping up to date is a big challenge for students taking advanced courses in molecular biology.

In the past, the only way to learn about new advances was to read about them in research journals and other publications. This is still a good way to keep up to date, but today it is also possible to find out about the latest research, almost as it happens, through the Internet. This Appendix gives you some guidance about how to make the best use of these written and electronic resources.

Keeping up to Date by Reading the Literature

If you have made use of the reading lists at the end of each chapter, then by now you will be familiar with the best sources of up-to-date information on genome research. You will know that it is not necessary to tackle the research papers that describe the latest results – the 'primary' publications – because most of the important papers are summarized in more accessible reviews and news articles. These articles can be found in two types of journal (*Table A1*):

- **Review journals** are devoted to review articles and do not themselves contain any primary research papers. Most are published monthly and are therefore able to report new discoveries soon after they are announced. Others, such as the *Annual Reviews* series, come out once a year and aim to provide comprehensive surveys of the current status of various research areas.
- A few **research journals** publish not only primary papers, but also review articles that place these papers in context and explain their significance in a way that is easily understood by the general reader. They describe papers published in other journals as well as in their own pages, and so give a comprehensive coverage of the latest results. They are so useful and important that there are few scientists, in any

Table A1 Journals that publish review articles relevant to genome research

Journal	Frequency	Level*
Review journals		
Annual Review of Biochemistry	Yearly	3
Annual Review of Genetics	Yearly	3
Annual Review of Genomics and Human Genetics	Yearly	3
Bioessays	Monthly	2
Current Biology	Twice monthly	2
Current Opinion in Cell Biology	Bimonthly	2
Current Opinion in Genetics and Development	Bimonthly	2
Current Opinion in Structural Biology	Bimonthly	2
New Scientist	Weekly	1
Scientific American	Monthly	1
Trends in Biochemical Sciences	Monthly	2
Trends in Biotechnology	Monthly	2
Trends in Genetics	Monthly	2
Research journals containing some review articles		
Cell	Biweekly	3
Nature	Weekly	2
Nature Biotechnology	Monthly	2
Nature Genetics	Monthly	2
Nature Structural Biology	Monthly	2
Science	Weekly	2

*Level 1 journals are aimed at a general readership and their articles are less detailed than the text of *Genomes*; the review articles in level 2 journals are roughly equivalent to *Genomes*; level 3 journals are more advanced.

area of research, who do not read *Science* and/or *Nature* every week.

Keeping up to Date Using the Internet

Many research labs have their own Internet sites where they describe their ongoing research projects and latest discoveries. There are too many of these to list here, but they are easily found if you know the name of the university or research institute in which the group is located, which is always noted in their written publications. Simply find the home page for the university or institute (using a search engine) and usually you will see links to the research pages you are looking for.

Even more exciting is the possibility of examining new results by logging on to the various Internet sites that researchers use to keep themselves up to date. There is at least one site devoted to each organism whose genome is being or has been sequenced, as well as many sites specializing in more general areas of genome research (*Table A2*). These sites are designed for research use and the information they contain is sometimes difficult to understand if you are not an expert in that particular area, but many include at least some general information and all of them are interesting. You rarely need a password to gain access because most genome research operates on the basis that everyone should be able to see the latest results and data, so feel free to log on and see what there is.

Table A2 Internet addresses relevant to genome research

Description	Address
Sites relevant to the Human Genome Project	
CEPH (various maps)	http://www.cephb.fr/bio/ceph-genethon-map.html
dbEST (EST sequences)	http://www.ncbi.nlm.nih.gov/dbEST/
dbSNP (SNPs)	http://www.ncbi.nlm.nih.gov/SNP/
dbSTS (STS sequences)	http://www.ncbi.nlm.nih.gov/dbSTS/
Généthon (various maps)	http://www.genethon.fr/
Radiation hybrid database	http://www.ebi.ac.uk/RHdb
Sites relevant to other eukaryotic genome projects	
Various eukaryotes	http://www.sanger.ac.uk/Projects/
	http://www.mips.biochem.mpg.de/
Microbial eukaryotes	
Candida albicans	http://alces.med.umn.edu/Candida.html
Dictyostelium discoideum	http://glamdring.ucsd.edu/others/dsmith/dictydb.html
Neurospora crassa	http://www.mips.biochem.mpg.de/
Plasmodium falciparum	http://ben.vub.ac.be/malaria/mad.html
	http://www.wehi.edu.au/biology/malaria/who.html
Saccharomyces cerevisiae	http://genome-www.stanford.edu/Saccharomyces/
Invertebrates	
Bombyx mori	http://www.ab.a.u-tokyo.ac.jp/sericulture/shimada.html
Caenorhabditis elegans	http://www.ddbj.nig.ac.jp/htmls/c-elegans/html/CE_INDEX.html
Drosophila melanogaster	http://flybase.bio.indiana.edu/
Mosquito	http://klab.agsci.colostate.edu/
Vertebrates	
Chicken	http://www.ri.bbsrc.ac.uk/chickmap/chickgbase/manager.html
Cow	http://locus.jouy.inra.fr/cgi-bin/bovmap/intro2.pl
Dog	http://mendel.berkeley.edu/dog.html
Mouse	http://www.informatics.jax.org/
Pufferfish	http://fugu.hgmp.mrc.ac.uk
Rat	http://ratmap.gen.gu.se/
Zebrafish	http://zfish.uoregon.edu/

Table A2 (cont)   Internet addresses relevant to genome research

Description	Address
Plants	
Various plants	http://ars-genome.cornell.edu/
Arabidopsis thaliana	http://www.arabidopsis.org/
	http://www.tigr.org/tdb/at/at.html
Beans	http://www.ba.cnr.it/Beanref/
Cotton	http://algodon.tamu.edu/
Forest trees	http://dendrome.ucdavis.edu/
Maize	http://aestivum.moulon.inra.fr/imgd/
Sites relevant to prokaryotic genome projects	
Various species	http://www.tigr.org/tdb/mdb/mdb.html
Other interesting sites	
Sequence databases	
European Bioinformatics Institute	http://www.ebi.ac.uk
Genbank	http://www.ncbi.nlm.nih.gov
DNA Database of Japan	http://www.ddbj.nig.ac.jp/
Bioinformatics resources	http://www.ncgr.org/
Codon usage data	http://www.kazusa.or.jp/codon/
Eukaryotic promoters	http://www.epd.isb-sib.ch/
Restriction endonucleases	http://www.neb.com/rebase/rebase.html
RNA editing	http://www.rna.ucla.edu/
Telomeres	http://www.genlink.wustl.edu/teldb/index.html
Vertebrate gene evolution	http://pbil.univ-lyon1.fr/databases/hovergen.html

Glossary

2 μm circle   A plasmid found in the yeast *Saccharomyces cerevisiae* and used as the basis for a series of cloning vectors.

2-aminopurine   A base analog that can cause mutations by replacing adenine in a DNA molecule.

−25 box   A component of the bacterial promoter.

3′-OH terminus   The end of a polynucleotide that terminates with a hydroxyl group attached to the 3′-carbon of the sugar.

3′ transduction   Transfer of a segment of genomic DNA from one place to another caused by movement of a LINE element.

3′-untranslated region   The untranslated region of an mRNA downstream of the termination codon.

30 nm chromatin fiber   A relatively unpacked form of chromatin consisting of a possibly helical array of nucleosomes in a fiber approximately 30 nm in diameter.

5-bromouracil   A base analog that can cause mutations by replacing thymine in a DNA molecule.

5′-P terminus   The end of a polynucleotide that terminates with a mono-, di- or triphosphate attached to the 5′-carbon of the sugar.

5′-untranslated region   The untranslated region of an mRNA upstream of the initiation codon.

(6–4) lesion   A dimer between two adjacent pyrimidine bases in a polynucleotide, formed by ultraviolet irradiation.

(6–4) photoproduct photolyase   An enzyme involved in photoreactivation repair.

α-helix   One of the commonest secondary structural conformations taken up by segments of polypeptides.

β-*N*-glycosidic bond   The linkage between the base and sugar of a nucleotide.

β-sheet   One of the commonest secondary structural conformations taken up by segments of polypeptides.

β-turn   A sequence of four amino acids, the second usually glycine, which causes a polypeptide to change direction.

γ-complex   A component of DNA polymerase III comprising subunit γ in association with δ, δ′, χ and ψ.

κ-homology domain   A type of RNA-binding domain.

π–π interactions   The hydrophobic interactions that occur between adjacent base pairs in a double-stranded DNA molecule.

Acceptor arm   Part of the structure of a tRNA molecule.

Acceptor site   The splice site at the 3′ end of an intron.

Acidic domain   A type of activation domain.

Acridine dye   A chemical compound that causes a frameshift mutation by intercalating between adjacent base pairs of the double helix.

Activation domain   The part of an activator that makes contact with the initiation complex.

Activator   A DNA-binding protein that stabilizes construction of the RNA polymerase II transcription initiation complex.

Acylation   The attachment of a lipid sidechain to a polypeptide.

Ada enzyme   An *Escherichia coli* enzyme that is involved in the direct repair of alkylation mutations.

Adaptor   A synthetic, double-stranded oligonucleotide used to attach sticky ends to a blunt-ended molecule.

Adenine   A purine base found in DNA and RNA.

Adenosine deaminase acting on RNA (ADAR)   An enzyme that edits various eukaryotic mRNAs by deaminating adenosine to inosine.

Adenylate cyclase   The enzyme that converts ATP to cyclic AMP.

A-DNA   A structural configuration of the double helix, present but not common in cellular DNA.

Affinity chromatography   A column chromatography method that makes use of a ligand that binds to the molecule being purified.

Agarose gel electrophoresis   Electrophoresis carried out in an agarose gel and used to separate DNA molecules between 100 bp and 50 kb in length.

Alkaline phosphatase   An enzyme that removes phosphate groups from the 5′ ends of DNA molecules.

Alkylating agent   A mutagen that acts by adding alkyl groups to nucleotide bases.

Allele   One of two or more alternative forms of a gene.

Allele frequency   The frequency of an allele in a population.

Allele-specific oligonucleotide (ASO) hybridization   The use of an oligonucleotide probe to determine which of two alternative nucleotide sequences is contained in a DNA molecule.

Allopolyploid   A polyploid nucleus derived from fusion between gametes from different species.

Alphoid DNA   The tandemly repeated nucleotide sequences located in the centromeric regions of human chromosomes.

Alternative splicing   The production of two or more mRNAs from a single pre-mRNA by joining together different combinations of exons.

Alu   A type of SINE found in the genomes of humans and related mammals.

Alu-PCR   A clone fingerprinting technique that uses PCR to detect the relative positions of Alu sequences in cloned DNA fragments.

Amino acid   One of the monomeric units of a protein molecule.

Amino terminus   The end of a polypeptide that has a free amino group.

Aminoacyl or A site   The site in the ribosome occupied by the aminoacyl-tRNA during translation.

Aminoacylation   Attachment of an amino acid to the acceptor arm of a tRNA.

Aminoacyl-tRNA synthetase   An enzyme that catalyzes the aminoacylation of one or more tRNAs.

Ancestral character state   A character state possessed by a remote common ancestor of a group of organisms.

Ancient DNA   DNA preserved in ancient biological material.

Annealing   Attachment of an oligonucleotide primer to a DNA or RNA template.

Anticodon   The triplet of nucleotides, at positions 34–36 in a tRNA molecule, that base-pairs with a codon in an mRNA molecule.

Anticodon arm   Part of the structure of a tRNA molecule.

Antigen   A substance that elicits an immune response.

Antitermination   A bacterial mechanism for regulating the termination of transcription.

Antiterminator protein   A protein that attaches to bacterial DNA and mediates anti-termination.

AP (apurinic/apyrimidinic) site   A position in a DNA molecule where the base component of the nucleotide is missing.

AP endonuclease   An enzyme involved in base excision repair.

Apomorphic character state   A character state that evolved in a recent ancestor of a subset of organisms in a group being studied.

Apoptosis   Programmed cell death.

Archaea   One of the two main groups of prokaryotes, mostly found in extreme environments.

Ascospore   One of the haploid products of meiosis in an ascomycete such as the yeast *Saccharomyces cerevisiae*.

Ascus   The structure which contains the four ascospores produced by a single meiosis in the yeast *Saccharomyces cerevisiae*.

Attenuation   A process used by some bacteria to regulate expression of an amino acid biosynthetic operon in accordance with the levels of the amino acid in the cell.

AU–AC intron   A type of intron found in eukaryotic nuclear genes: the first two nucleotides in the intron are 5′–AU–3′ and the last two are 5′–AC–3′.

Autonomously replicating sequence (ARS)   A DNA sequence, particularly from yeast, that confers replicative ability on a non-replicative plasmid.

Autopolyploid   A polyploid nucleus derived from fusion of two gametes from the same species, neither of which is haploid.

Autoradiography   The detection of radioactively labeled molecules by exposure of an X-ray-sensitive photographic film.

Autosome   A chromosome that is not a sex chromosome.

Auxotroph   A mutant microorganism that can grow only when supplied with a nutrient that is not needed by the wild type.

B chromosome   A chromosome possessed by some individuals in a population, but not all.

Bacteria   One of the two main groups of prokaryotes.

Bacterial artificial chromosome (BAC)   A high-capacity cloning vector based on the F plasmid of *Escherichia coli*.

Bacteriophage   A virus that infects a bacterium.

Bacteriophage P1 vector   A high-capacity cloning vector based on bacteriophage P1.

Barr body   The highly condensed chromatin structure taken up by an inactivated X chromosome.

Basal promoter   The position within a eukaryotic promoter where the initiation complex is assembled.

Basal promoter element   Sequence motifs that are present in many eukaryotic promoters and set the basal level of transcription initiation.

Basal rate of transcription   The number of productive initiations of transcription occurring per unit time at a particular promoter.

Base analog   A compound whose structural similarity to one of the bases in DNA enables it to act as a mutagen.

Base excision repair   A DNA repair process that involves excision and replacement of an abnormal base.

Base pair   The hydrogen-bonded structure formed by two complementary nucleotides. When abbreviated to 'bp', the shortest unit of length for a double-stranded DNA molecule.

Base ratio   The ratio of A to T, or G to C, in a double-stranded DNA molecule. Chargaff showed that the base ratios are always close to 1.0.

Baseless site   A position in a DNA molecule where the base component of the nucleotide is missing.

Base-pairing   The attachment of one polynucleotide to another, or one part of a polynucleotide to another part of the same polynucleotide, by base pairs.

Base-stacking   The hydrophobic interactions that occur between adjacent base pairs in a double-stranded DNA molecule.

Basic domain   A type of DNA-binding domain.

B-DNA   The commonest structural conformation of the DNA double helix in living cells.

Beads-on-a-string   An unpacked form of chromatin consisting of nucleosome beads on a string of DNA.

Bioinformatics   The use of computer methods in studies of genomes.

Biolistics   A means of introducing DNA into cells that involves bombardment with high-velocity microprojectiles coated with DNA.

Biological information   The information contained in the genome of an organism and which directs the development and maintenance of that organism.

Biotechnology   The use of living organisms, often, but not always microbes, in industrial processes.

Biotinylation   Attachment of a biotin label to a DNA or RNA molecule.

Bivalent   The structure formed when a pair of homologous chromosomes lines up during meiosis.

Blunt end   An end of a double-stranded DNA molecule where both strands terminate at the same nucleotide position with no single-stranded extension.

Bootstrap analysis   A method for inferring the degree of confidence that can be assigned to a branch point in a phylogenetic tree.

Bootstrap value   The statistical value obtained by bootstrap analysis.

Bottleneck   A temporary reduction in the size of a population.

Branch   A component of a phylogenetic tree.

Branch migration   A step in the Holliday model for homologous recombination, involving exchange of polynucleotides between a pair of recombining double-stranded DNA molecules.

Buoyant density   The density possessed by a molecule or particle when suspended in an aqueous salt or sugar solution.

C terminus   The end of a polypeptide that has a free carboxyl group.

CAAT box   A basal promoter element.

Cap   The chemical modification at the 5′ end of most eukaryotic mRNA molecules.

Cap binding complex   The complex that makes the initial attachment to the cap structure at the beginning of the scanning phase of eukaryotic translation.

CAP site   A DNA-binding site for the catabolite activator protein.

Capping   Attachment of a cap to the 5′ end of a eukaryotic mRNA.

Capsid   The protein coat that surrounds the DNA or RNA genome of a virus.

Carboxyl terminus   The end of a polypeptide that has a free carboxyl group.

CASPs (CTD-associated SR-like proteins)   Proteins thought to play regulatory roles during splicing of GU–AG introns.

Catabolite activator protein   A regulatory protein that binds to various sites in a bacterial genome and activates transcription initiation at downstream promoters.

Catabolite repression   The means by which extracellular glucose levels dictate whether genes for sugar utilization are switched on or off in bacteria.

cDNA   A double-stranded DNA copy of an mRNA molecule.

cDNA capture or cDNA selection   Repeated hybridization probing of a pool of cDNAs with the objective of obtaining a subpool enriched in certain sequences.

Cell cycle   The series of events occurring in a cell between one division and the next.

Cell cycle checkpoint   A period before entry into S or M phase of the cell cycle, a key point at which regulation is exerted.

Cell senescence   The period in a cell lineage when the cells are alive but no longer able to divide.

Cell-free protein synthesizing system   A cell extract containing all the components needed for protein synthesis and able to translate added mRNA molecules.

Cell-specific module   Sequence motifs present in the promoters of eukaryotic genes that are expressed in just one type of tissue.

Centromere   The constricted region of a chromosome that is the position at which the pair of chromatids are held together.

Chain termination method   A DNA sequencing method that involves enzymatic synthesis of polynucleotide chains that terminate at specific nucleotide positions.

Chaperonin   A multi-subunit protein that forms a structure that aids the folding of other proteins.

Character state   One of at least two alternative forms of a character used in phylogenetic analysis.

Chemical degradation sequencing   A DNA sequencing method that involves the use of chemicals that cut DNA molecules at specific nucleotide positions.

Chemical shift   The change in the rotation of a chemical nucleus, used as the basis of NMR.

Chi form   An intermediate structure seen during recombination between DNA molecules.

Chi site   A repeated nucleotide sequence in the *Escherichia coli* genome that is involved in the initiation of homologous recombination.

Chimera   An organism composed of two or more genetically different cell types.

Chloroplast   One of the photosynthetic organelles of a eukaryotic cell.

Chloroplast genome   The genome present in the chloroplasts of a photosynthetic eukaryotic cell.

Chromatid   The arm of a chromosome.

Chromatin   The complex of DNA and histone proteins found in chromosomes.

Chromatosome   A subcomponent of chromatin made up of a nucleosome core octamer with associated DNA and a linker histone.

Chromosome   One of the DNA–protein structures that contains part of the nuclear genome of a eukaryote. Less accurately, the DNA molecule(s) that contain(s) a prokaryotic genome.

Chromosome scaffold   A component of the nuclear matrix which changes its structure during cell division, resulting in condensation of the chromosomes into their metaphase forms.

Chromosome theory   The theory, first propounded by Sutton in 1903, that genes lie on chromosomes.

Chromosome walking   A technique that can be used to construct a clone contig by identifying overlapping fragments of cloned DNA.

***Cis*-displacement**   Movement of a nucleosome to a new position on a DNA molecule.

Clade   A group of monophyletic organisms or DNA sequences that include all of those in the analysis that are descended from a particular common ancestor.

Cladistics   A phylogenetic approach that stresses the importance of understanding the evolutionary relevance of the characters that are studied.

Cleavage and polyadenylation specificity factor (CPSF)   A protein that plays an ancillary role during polyadenylation of eukaryotic mRNAs.

Cleavage stimulation factor (CstF)   A protein that plays an ancillary role during polyadenylation of eukaryotic mRNAs.

Clone   A group of cells that contain the same recombinant DNA molecule.

Clone contig   A collection of clones whose DNA fragments overlap.

Clone contig approach   A genome sequencing strategy in which the molecules to be sequenced are broken into manageable segments, each a few hundred kb or few Mb in length, which are sequenced individually.

Clone fingerprinting   Any one of several techniques that compare cloned DNA fragments in order to identify ones that overlap.

Clone library   A collection of clones, possibly representing an entire genome, from which individual clones of interest are obtained.

Cloning vector   A DNA molecule that is able to replicate inside a host cell and therefore can be used to clone other fragments of DNA.

Closed promoter complex   The structure formed during the initial step in assembly of the transcription initiation complex. The closed promoter complex consists of the RNA polymerase and/or accessory proteins attached to the promoter, before the DNA has been opened up by breakage of base pairs.

Cloverleaf   A two-dimensional representation of the structure of a tRNA molecule.

Coactivator   A protein that works in conjunction with an activator to stabilize construction of the RNA polymerase II transcription initiation complex.

Coding RNA   An RNA molecule that codes for a protein; an mRNA.

Codominance   The relationship between a pair of alleles which both contribute to the phenotype of a heterozygote.

Codon   A triplet of nucleotides coding for a single amino acid.

Codon–anticodon recognition   The interaction between a codon on an mRNA molecule and the corresponding anticodon on a tRNA.

Codon bias Refers to the fact that not all codons are used equally frequently in the genes of a particular organism.

Cohesin The protein that holds sister chromatids together during the period between genome replication and nuclear division.

Cohesive end An end of a double-stranded DNA molecule where there is a single-stranded extension.

Co-integrate An intermediate in the pathway resulting in replicative transposition.

Commitment complex The initial structure formed during splicing of a GU–AG intron.

Comparative genomics A research strategy that uses information obtained from the study of one genome to make inferences about the map positions and functions of genes in a second genome.

Competent Refers to a culture of bacteria that have been treated, for example, by soaking in calcium chloride, so that their ability to take up DNA molecules is enhanced.

Complementary Refers to two nucleotides or nucleotide sequences that are able to base-pair with one another.

Complementary DNA (cDNA) A double-stranded DNA copy of an mRNA molecule.

Composite transposon A DNA transposon comprising a pair of insertion sequences flanking a segment of DNA usually containing one or more genes.

Concatamer A DNA molecule made up of linear genomes or other DNA units linked head to tail.

Concerted evolution The evolutionary process that results in the members of a multigene family retaining the same or similar sequences.

Conditional-lethal mutation A mutation that results in a cell or organism able to survive only under permissive conditions.

Conjugation Transfer of DNA between two bacteria that come into physical contact with one another.

Conjugation mapping A technique for mapping bacterial genes by determining the time it takes for each gene to be transferred during conjugation.

Consensus sequence A nucleotide sequence that represents an 'average' of a number of related but non-identical sequences.

Conservative replication A hypothetical mode of DNA replication in which one daughter double helix is made up of the two parental polynucleotides and the other is made up of two newly synthesized polynucleotides.

Conservative transposition Transposition that does not result in copying of the transposable element.

Constitutive control Control over bacterial gene expression that depends on the sequence of the promoter.

Constitutive heterochromatin Chromatin that is permanently in a compact organization.

Constitutive mutation A mutation that results in continuous expression of a gene or set of genes that is normally subject to regulatory control.

Context-dependent codon reassignment Refers to the situation whereby the DNA sequence surrounding a codon changes the meaning of that codon.

Contig A contiguous set of overlapping DNA sequences.

Contour clamped homogeneous electric fields (CHEF) An electrophoresis method used to separate large DNA molecules.

Conventional pseudogene A gene that has become inactive because of the accumulation of mutations.

Convergent evolution   The situation that occurs when the same character state evolves independently in two lineages.

Core enzyme   The version of *Escherichia coli* RNA polymerase, subunit composition $\alpha_2\beta\beta'$, that carries out RNA synthesis but is unable to locate promoters efficiently.

Core octamer   The central component of a nucleosome, made up of two subunits each of histones H2A, H2B, H3 and H4, around which DNA is wound.

Core promoter   The position within a eukaryotic promoter where the initiation complex is assembled.

Co-repressor   A small molecule that must be bound to a repressor protein before the latter is able to attach to its operator site.

cos **site**   One of the cohesive, single-stranded extensions present at the ends of the DNA molecules of certain strains of λ phage.

Cosmid   A high-capacity cloning vector consisting of the λ cos site inserted into a plasmid.

Cotransduction   Transfer of two or more genes from one bacterium to another via a transducing phage.

Cotransformation   Uptake of two or more genes on a single DNA molecule during transformation of a bacterium.

CpG island   A GC-rich DNA region located upstream of approximately 56% of the genes in the human genome.

CREB   An important transcription factor.

Crossing-over   The exchange of DNA between chromosomes during meiosis.

Cryptic splice site   A site whose sequence resembles an authentic splice site and which might be selected instead of the authentic site during aberrant splicing.

Cryptogene   One of several genes in the trypanosome mitochondrial genome which specify abbreviated RNAs that must undergo pan-editing in order to become functional.

CTD-associated SR-like protein (CASP)   A type of protein thought to play a regulatory role during splicing of GU–AG introns.

C-terminal domain (CTD)   A component of the largest subunit of RNA polymerase II, important in activation of the polymerase.

C-value paradox   The non-equivalence between genome size and gene number that is seen when comparisons are made between some eukaryotes.

Cyanelle   A photosynthetic organelle that resembles an ingested cyanobacterium.

Cyclic AMP (cAMP)   A modified version of AMP in which an intramolecular phosphodiester bond links the 5' and 3' carbons.

Cyclin   A regulatory protein whose abundance varies during the cell cycle and which regulates biochemical events in a cell-cycle-specific manner.

Cyclobutyl dimer   A dimer between two adjacent pyrimidine bases in a polynucleotide, formed by ultraviolet irradiation.

Cys$_2$His$_2$ finger   A type of zinc-finger DNA-binding domain.

Cytochemistry   The use of compound-specific stains, combined with microscopy, to determine the biochemical content of cellular structures.

Cytosine   One of the pyrimidine bases found in DNA and RNA.

D arm   Part of the structure of a tRNA molecule.

Dark repair   A type of nucleotide excision repair process that corrects cyclobutyl dimers.

De novo **methylation**   Addition of methyl groups to new positions on a DNA molecule.

Deadenylation-dependent decapping   A process for degradation of eukaryotic mRNAs that is initiated by removal of the poly(A) tail.

Deaminating agent   A mutagen that acts by removing amino groups from nucleotide bases.

Degeneracy   Refers to the fact that the genetic code has more than one codon for most amino acids.

Degradosome   A multienzyme complex responsible for degradation of bacterial mRNAs.

Delayed-onset mutation   A mutation whose effect is not apparent until a relatively late stage in the life of the mutant organism.

Deletion mutation   A mutation resulting from deletion of one or more nucleotides from a DNA sequence.

Denaturation   Breakdown by chemical or physical means of the non-covalent interactions, such as hydrogen bonding, that maintain the secondary and higher levels of structure of proteins and nucleic acids.

Density gradient centrifugation   A technique in which a cell fraction is centrifuged through a dense solution, in the form of a gradient, so that individual components are separated.

Deoxyribonuclease   An enzyme that cleaves phosphodiester bonds in a DNA molecule.

Derived character state   A character state that evolved in a recent ancestor of a subset of organisms in a group being studied.

Development   A coordinated series of transient and permanent changes that occurs during the life history of a cell or organism.

Diauxie   The phenomenon whereby a bacterium, when provided with a mixture of sugars, uses up one sugar before beginning to metabolize the second sugar.

Dicer   The ribonuclease that plays a central role in RNA interference.

Dideoxynucleotide   A modified nucleotide that lacks the 3′ hydroxyl group and so terminates strand synthesis when incorporated into a polynucleotide.

Differential centrifugation   A technique that separates cell components by centrifuging an extract at different speeds.

Differential splicing   The production of two or more mRNAs from a single pre-mRNA by joining together different combinations of exons.

Differentiation   The adoption by a cell of a specialized biochemical and/or physiological role.

Dihybrid cross   A sexual cross in which the inheritance of two pairs of alleles is followed.

Dimer   A protein or other structure that comprises two subunits.

Diploid   A nucleus that has two copies of each chromosome.

Direct readout   The recognition of a DNA sequence by a binding protein that makes contacts with the outside of a double helix.

Direct repair   A DNA repair system that acts directly on a damaged nucleotide.

Direct repeat   A nucleotide sequence that is repeated twice or more frequently in a DNA molecule.

Directed evolution   A set of experimental techniques that is used to obtain novel genes with improved products.

Discontinuous gene   A gene that is split into exons and introns.

Dispersive replication   A hypothetical mode of DNA replication in which both polynucleotides of each daughter double helix are made up partly of parental DNA and partly of newly synthesized DNA.

Displacement replication A mode of replication which involves continuous copying of one strand of the helix, the second strand being displaced and subsequently copied after synthesis of the first daughter strand has been completed.

Distance matrix A table showing the evolutionary distances between all pairs of nucleotide sequences in a dataset.

Distance method A rigorous mathematical approach to alignment of nucleotide sequences.

Disulfide bridge A covalent bond linking cysteine amino acids on different polypeptides or at different positions on the same polypeptide.

D-loop An intermediate structure formed during the Meselson–Radding model for homologous recombination. Also an intermediate formed during displacement replication.

DNA Deoxyribonucleic acid, one of the two forms of nucleic acid in living cells; the genetic material for all cellular life forms and many viruses.

DNA adenine methylase (Dam) An enzyme involved in methylation of *Escherichia coli* DNA.

DNA bending A type of conformational change introduced into a DNA molecule by a binding protein.

DNA-binding motif The part of a DNA-binding protein that makes contact with the double helix.

DNA-binding protein A protein that attaches to a DNA molecule.

DNA chip A high-density array of DNA molecules used for parallel hybridization analyses.

DNA cloning Insertion of a fragment of DNA into a cloning vector, and subsequent propagation of the recombinant DNA molecule in a host organism.

DNA cytosine methylase (Dcm) An enzyme involved in methylation of *Escherichia coli* DNA.

DNA-dependent DNA polymerase An enzyme that makes a DNA copy of a DNA template.

DNA-dependent RNA polymerase An enzyme that makes an RNA copy of a DNA template.

DNA glycosylase An enzyme that cleaves the β-*N*-glycosidic bond between a base and the sugar component of a nucleotide as part of the base excision and mismatch repair processes. The name is a misnomer and should be *DNA glycolyase*, but the incorrect usage is now embedded in the literature.

DNA gyrase A Type II topoisomerase of *Escherichia coli*.

DNA ligase An enzyme that synthesizes phosphodiester bonds as part of DNA replication, repair and recombination processes.

DNA marker A DNA sequence that exists as two or more readily distinguished versions and which can therefore be used to mark a map position on a genetic, physical or integrated genome map.

DNA methylation Refers to the chemical modification of DNA by attachment of methyl groups.

DNA methyltransferase An enzyme that attaches methyl groups to a DNA molecule.

DNA photolyase A bacterial enzyme involved in photoreactivation repair.

DNA polymerase   An enzyme that synthesizes DNA.

DNA polymerase α   The enzyme that primes DNA replication in eukaryotes.

DNA polymerase δ   The main eukaryotic DNA replicating enzyme.

DNA polymerase γ   The enzyme responsible for replication of the mitochondrial genome.

DNA polymerase I   The bacterial enzyme that completes synthesis of Okazaki fragments during genome replication.

DNA polymerase II   A bacterial DNA polymerase involved in DNA repair.

DNA polymerase III   The main DNA replicating enzyme of bacteria.

DNA repair   The biochemical processes that correct mutations arising from replication errors and the effects of mutagenic agents.

DNA replication   Synthesis of a new copy of the genome.

DNase I hypersensitive site   A short region of eukaryotic DNA that is relatively easily cleaved with deoxyribonuclease I, possibly coinciding with positions where nucleosomes are absent.

DNA sequencing   The technique for determining the order of nucleotides in a DNA molecule.

DNA shuffling   A PCR-based procedure that results in directed evolution of a DNA sequence.

DNA topoisomerase   An enzyme that introduces or removes turns from the double helix by breakage and reunion of one or both polynucleotides.

DNA transposon   A transposon whose transposition mechanism does not involve an RNA intermediate.

DNA tumor virus   A virus with a DNA genome, able to cause cancer after infection of an animal cell.

Domain   A segment of a polypeptide that folds independently of other segments; also the segment of a gene coding for such a domain.

Domain duplication   Duplication of a gene segment coding for a structural domain in the protein product.

Domain shuffling   Rearrangement of segments of one or more genes, each segment coding for a structural domain in the gene product, to create a new gene.

Dominant   The allele that is expressed in a heterozygote.

Donor site   The splice site at the 5′ end of an intron.

Dot matrix   A method for aligning nucleotide sequences.

Double helix   The base-paired double-stranded structure that is the natural form of DNA in the cell.

Double heterozygote   A nucleus that is heterozygous for two genes.

Double homozygote   A nucleus that is homozygous for two genes.

Double restriction   Digestion of DNA with two restriction endonucleases at the same time.

Double-strand break repair   A DNA repair process that mends double-stranded breaks.

Double stranded   Comprising two polynucleotides attached to one another by base-pairing.

Double-stranded RNA-binding domain (dsRBD)   A common type of RNA-binding domain.

Downstream   Towards the 3′ end of a polynucleotide.

Dynamic allele-specific hybridization (DASH)   A solution hybridization technique used to type SNPs.

E site   A position within a bacterial ribosome to which a tRNA moves immediately after deacylation.

Electrophoresis   Separation of molecules on the basis of their net electrical charge.

Electrostatic interactions   Ionic bonds that form between charged chemical groups.

Elongation factor   A protein that plays an ancillary role in the elongation step of transcription or translation.

Elongator   A yeast protein, possibly with histone acetyltransferase activity, involved in the elongation phase of transcription.

Embryonic stem (ES) cell   A totipotent cell from the embryo of a mouse or other organism.

End-labeling   The attachment of a radioactive or other label to one end of a DNA or RNA molecule.

End-modification   The chemical alteration of the end of a DNA or RNA molecule.

End-modification enzyme   An enzyme used in recombinant DNA technology that alters the chemical structure at the end of a DNA molecule.

Endogenous retrovirus (ERV)   An active or inactive retroviral genome integrated into a host chromosome.

Endonuclease   An enzyme that breaks phosphodiester bonds within a nucleic acid molecule.

Endosymbiont theory   A theory that states that the mitochondria and chloroplasts of eukaryotic cells are derived from symbiotic prokaryotes.

Enhanceosome   A structure formed by DNA bending that comprises a collection of proteins involved in activation of the RNA polymerase II transcription initiation complex.

Enhancer   A regulatory sequence that increases the rate of transcription of a gene or genes located some distance away in either direction.

Episome   A plasmid that is able to integrate into the host cell's chromosome.

Episome transfer   Transfer between cells of some or all of a bacterial chromosome by integration into a plasmid.

Ethidium bromide   A type of intercalating agent that causes mutations by inserting between adjacent base pairs in a double-stranded DNA molecule.

Ethylmethane sulfonate (EMS)   A mutagen that acts by adding alkyl groups to nucleotide bases.

Euchromatin   Regions of a eukaryotic chromosome that are relatively uncondensed, thought to contain active genes.

Eukaryote   An organism whose cells contain membrane-bound nuclei.

Excision repair   A DNA repair process that corrects various types of DNA damage by excising and resynthesizing a region of polynucleotide.

Exit site   A position within a bacterial ribosome to which a tRNA moves immediately after deacylation.

Exon   A coding region within a discontinuous gene.

Exonic splicing enhancer (ESE)   A nucleotide sequence that plays a positive regulatory role during splicing of GU–AG introns.

Exonic splicing silencer (ESS)   A nucleotide sequence that plays a negative regulatory role during splicing of GU–AG introns.

Exon–intron boundary   The nucleotide sequence at the junction between an exon and an intron.

Exon skipping   Aberrant splicing in which one or more or exons are omitted from the spliced RNA.

Exon theory of genes   An 'introns early' hypothesis that holds that introns were formed when the first DNA genomes were constructed.

Exon trapping   A method, based on cloning, for identifying the positions of exons in a DNA sequence.

Exonuclease   An enzyme that removes nucleotides from the ends of a nucleic acid molecule.

Exosome   A multiprotein complex involved in degradation of mRNA in eukaryotes.

Exportin   A protein involved in transport of molecules out of the nucleus.

Expressed sequence tag (EST)   A cDNA that is sequenced in order to gain rapid access to the genes in a genome.

Extein   The functional component of a discontinuous protein.

External node   The end of a branch in a phylogenetic tree, representing one of the organisms or DNA sequences being studied.

Extrachromosomal gene   A gene in a mitochondrial or chloroplast genome.

F plasmid   A fertility plasmid that directs conjugal transfer of DNA between bacteria.

Facultative heterochromatin   Chromatin that has a compact organization in some, but not all cells, thought to contain genes that are inactive in some cells or at some periods of the cell cycle.

FEN1   The 'flap endonuclease' involved in replication of the lagging strand in eukaryotes.

Fiber-FISH   A specialized form of FISH that enables high marker resolution.

Field inversion gel electrophoresis (FIGE)   An electrophoresis method used to separate large DNA molecules.

Fitness   The ability of an organism or allele to survive and reproduce.

Fixation   Refers to the situation that occurs when a single allele reaches a frequency of 100% in a population.

Flow cytometry   A method for the separation of chromosomes.

FLpter value   The unit used in FISH to describe the position of a hybridization signal relative to the end of the short arm of the chromosome.

Fluorescence recovery after photobleaching (FRAP)   A technique used to study the mobility of nuclear proteins.

Fluorescent *in situ* hybridization (FISH)   A technique for locating markers on chromosomes by observing the hybridization positions of fluorescent labels.

Flush end   An end of a double-stranded DNA molecule where both strands terminate at the same nucleotide position with no single-stranded extension.

fMet   *N*-formylmethionine, the modified amino acid carried by the tRNA that is used during the initiation of translation in bacteria.

Folding domain   A segment of a polypeptide that folds independently of other segments.

Folding pathway   The series of events, involving partially folded intermediates, that results in an unfolded protein attaining its correct three-dimensional structure.

Footprinting   A range of techniques used for locating bound proteins on DNA molecules.

Fosmid   A high-capacity vector carrying the F plasmid origin of replication and a λ *cos* site.

Fragile site   A position in a chromosome that is prone to breakage because it contains an expanded trinucleotide repeat sequence.

Frameshift mutation   A mutation resulting from insertion or deletion of a group of nucleotides that is not a multiple of three and which therefore changes the frame in which translation occurs.

Frameshifting   The movement of a ribosome from one reading frame to another at an internal position within a gene.

Functional analysis   The area of genome research devoted to identifying the functions of unknown genes.

Functional domain   A region of eukaryotic DNA around a gene or group of genes that can be delineated by treatment with deoxyribonuclease I.

Fusion protein   A protein that consists of a fusion of two polypeptides, or parts of polypeptides, normally coded by separate genes.

G1 phase   The first gap period of the cell cycle.

G2 phase   The second gap period of the cell cycle.

Gain-of-function mutation   A mutation that results in an organism acquiring a new function.

Gamete   A reproductive cell, usually haploid, that fuses with a second gamete to produce a new cell during sexual reproduction.

GAPs (GTPase activating proteins)   A set of proteins that are intermediates in the Ras signal transduction pathway.

Gap genes   Developmental genes that play a role in establishing positional information within the *Drosophila* embryo.

Gap period   One of two intermediate periods within the cell cycle.

GC box   A type of basal promoter element.

GC content   The percentage of nucleotides in a genome that are G or C.

Gel electrophoresis   Electrophoresis performed in a gel so that molecules of similar electrical charge can be separated on the basis of size.

Gel retardation analysis   A technique that identifies protein-binding sites on DNA molecules by virtue of the effect that a bound protein has on the mobility of the DNA fragments during gel electrophoresis.

Gel stretching   A technique for preparing restricted DNA molecules for optical mapping.

Gene   A DNA segment containing biological information and hence coding for an RNA and/or polypeptide molecule.

Gene cloning   Insertion of a fragment of DNA, containing a gene, into a cloning vector, and subsequent propagation of the recombinant DNA molecule in a host organism.

Gene conversion   A process that results in the four haploid products of meiosis displaying an unusual segregation pattern.

Gene expression   The series of events by which the biological information carried by a gene is released and made available to the cell.

Gene flow   The transfer of a gene from one organism to another.

Gene fragment   A gene relic consisting of a short isolated region from within a gene.

General recombination   Recombination between two homologous double-stranded DNA molecules.

General transcription factor (GTF)   A protein or protein complex that is a transient or permanent component of the initiation complex formed during eukaryotic transcription.

Gene substitution   The replacement of an allele that at one time was fixed in the population by a second allele, this second allele arising by mutation and increasing in frequency until itself reaching fixation.

Gene superfamily   A group of two or more evolutionarily related multigene families.

Genes-within-genes   Refers to a gene whose intron contains a second gene.

Gene therapy   A clinical procedure in which a gene or other DNA sequence is used to treat a disease.

Genetic code   The rules that determine which triplet of nucleotides codes for which amino acid during protein synthesis.

Genetic footprinting   A technique for the rapid functional analysis of many genes at once.

Genetic linkage   The physical association between two genes that are on the same chromosome.

Genetic mapping   The use of genetic techniques to construct a genome map.

Genetic marker   A gene that exists as two or more readily distinguished alleles and whose inheritance can therefore be followed during a genetic cross, enabling the map position of the gene to be determined.

Genetic profile   The banding pattern revealed after electrophoresis of the products of PCRs directed at a range of microsatellite loci.

Genetic redundancy   The situation that occurs when two genes in the same genome perform the same function.

Genetics   The branch of biology devoted to the study of genes.

Gene tree   A phylogenetic tree that shows the evolutionary relationships between a group of genes or other DNA sequences.

Genome   The entire genetic complement of a living organism.

Genome expression   The series of events by which the biological information carried by a genome is released and made available to the cell.

Genome-wide repeat   A sequence that recurs at many dispersed positions within a genome.

Genomic imprinting   Inactivation by methylation of a gene on one of a pair of homologous chromosomes.

Genotype   A description of the genetic composition of an organism.

Gigabase pair   1 000 000 kb; 1 000 000 000 bp.

Global regulation   A general down-regulation in protein synthesis that occurs in response to various signals.

Glutamine-rich domain   A type of activation domain.

Glycosylation   The attachment of sugar units to a polypeptide.

GNRPs (guanine nucleotide-releasing proteins)   A set of proteins that are intermediates in the Ras signal transduction pathway.

Green fluorescent protein   A protein that is used to label other proteins and whose gene is used as a reporter gene.

Group I intron   A type of intron found mainly in organelle genes.

Group II intron   A type of intron found in organelle genes.

Group III intron   A type of intron found in organelle genes.

GTPase activating proteins (GAPs)   A set of proteins that are intermediates in the Ras signal transduction pathway.

GU–AG intron   The commonest type of intron in eukaryotic nuclear genes. The first two nucleotides of the intron are 5′–GU–3′ and the last two are 5′–AG–3′.

Guanine   One of the purine nucleotides found in DNA and RNA.

Guanine methyltransferase   The enzyme that attaches a methyl group to the 5′ end of a eukaryotic mRNA during the capping reaction.

Guanine nucleotide releasing proteins (GNRPs)   A set of proteins that are intermediates in the Ras signal transduction pathway.

Guanylyl transferase   The enzyme that attaches a GTP to the 5′ end of a eukaryotic mRNA at the start of the capping reaction.

Guide RNA   A short RNA that specifies the positions at which one or more nucleotides are inserted into an abbreviated RNA by pan-editing.

Hammerhead   An RNA structure with ribozyme activity that is found in some viruses.

Haploid   A nucleus that has a single copy of each chromosome.

Haploinsufficiency   The situation where inactivation of a gene on one of a pair of homologous chromosomes results in a change in the phenotype of the mutant organism.

Haplotype   A collection of alleles that are usually inherited together.

Helicase   An enzyme that breaks base pairs in a double-stranded DNA molecule.

Helix–loop–helix motif   A dimerization domain commonly found in DNA-binding proteins.

Helix–turn–helix motif   A common structural motif for attachment of a protein to a DNA molecule.

Helper phage   A phage that is introduced into a host cell in conjunction with a related cloning vector, in order to provide enzymes and other proteins required for replication of the cloning vector.

Heterochromatin   Chromatin that is relatively condensed and is thought to contain DNA that is not being transcribed.

Heteroduplex   A DNA–DNA or DNA–RNA hybrid.

Heteroduplex analysis   Transcript mapping by analysis of DNA–RNA hybrids with a single-strand-specific nuclease such as S1.

Heterogenous nuclear RNA (hnRNA)   The nuclear RNA fraction that comprises unprocessed transcripts synthesized by RNA polymerase II.

Heteropolymer   An artificial RNA comprising a mixture of different nucleotides.

Heterozygosity   The probability that a person chosen at random from the population will be heterozygous for a particular marker.

Heterozygous   A diploid nucleus that contains two different alleles for a particular gene.

High mobility group N (HMGN) protein   A group of nuclear proteins that influence chromatin structure.

High-performance liquid chromatography (HPLC)   A column chromatography method with many applications in biochemistry.

Histone   One of the basic proteins found in nucleosomes.

Histone acetylation   Modification of chromatin structure by attachment of acetyl groups to core histones.

Histone acetyltransferase (HAT)   An enzyme that attaches acetyl groups to core histones.

Histone code   The hypothesis that the pattern of chemical modification on histone proteins influences various cellular activities.

Histone deacetylase (HDAC)   An enzyme that removes acetyl groups from core histones.

Holliday structure   An intermediate structure formed during recombination between two DNA molecules.

Holocentric chromosome   A chromosome that does not have a single centromere but instead has multiple kinetochores spread along its length.

Holoenzyme   The version of the *Escherichia coli* RNA polymerase, subunit composition $\alpha_2\beta\beta'\sigma$, that is able to recognize promoter sequences.

Homeodomain   A DNA-binding motif found in many proteins involved in developmental regulation of gene expression.

Homeotic mutation   A mutation that results in the transformation of one body part into another.

Homeotic selector gene   A gene that establishes the identity of a body part such as a segment of the *Drosophila* embryo.

Homologous chromosomes   Two or more identical chromosomes present in a single nucleus.

Homologous genes   Genes that share a common evolutionary ancestor.

Homologous recombination   Recombination between two homologous double-stranded DNA molecules, i.e. ones which share extensive nucleotide sequence similarity.

Homology searching   A technique in which genes with sequences similar to that of an unknown gene are sought, the objective being to gain an insight into the function of the unknown gene.

Homoplasy   The situation that occurs when the same character state evolves independently in two lineages.

Homopolymer   An artificial RNA comprising just one nucleotide.

Homopolymer tailing   The attachment of a sequence of identical nucleotides (e.g. AAAAA) to the end of a nucleic acid molecule, usually referring to the synthesis of single-stranded homopolymer extensions on the ends of a double-stranded DNA molecule.

Homozygous   A diploid nucleus that contains two identical alleles for a particular gene.

Horizontal gene transfer   Transfer of a gene from one species to another.

Hormone response element   A nucleotide sequence upstream of a gene that mediates the regulatory effect of a steroid hormone.

Housekeeping protein   A protein that is continually expressed in all or at least most cells of a multicellular organism.

Hsp70 chaperone   A family of proteins that bind to hydrophobic regions in other proteins in order to aid their folding.

Human Genome Project   The publicly funded project responsible for one of the draft human genome sequences and which continues to study the functions of human genes.

Hybridization   The attachment to one another, by base-pairing, of two complementary polynucleotides.

Hybridization probing   A technique that uses a labeled nucleic acid molecule as a probe to identify complementary or homologous molecules to which it base-pairs.

Hydrogen bond   A weak electrostatic attraction between an electronegative atom such as oxygen or nitrogen and a hydrogen atom attached to a second electronegative atom.

Hydrophobic effects   Chemical interactions that result in hydrophobic groups becoming buried inside a protein.

Hypermutation   An increase in the mutation rate of a genome.

Immunocytochemistry   A technique that uses antibody probing to locate the position of a protein in a tissue.

Immunoelectron microscopy   An electron microscopy technique that uses antibody labeling to identify the positions of specific proteins on the surface of a structure such as a ribosome.

Immunoscreening   The use of an antibody probe to detect a polypeptide synthesized by a cloned gene.

Importin   A protein involved in transport of molecules into the nucleus.

In *vitro* mutagenesis   Techniques used to produce a specified mutation at a predetermined position in a DNA molecule.

In *vitro* packaging   Synthesis of infective λ phages from a preparation of λ proteins and a concatamer of λ DNA molecules.

Incomplete dominance   Refers to a pair of alleles, neither of which displays dominance, the phenotype of a heterozygote being intermediate between the phenotypes of the two homozygotes.

Indel   A position in an alignment between two DNA sequences where an insertion or deletion has occurred.

Inducer   A molecule that induces expression of a gene or operon by binding to a repressor protein and preventing the repressor from attaching to the operator.

Inferred tree   A tree obtained by phylogenetic analysis.

Informational problem   The problem tackled by the early molecular biologists concerning the nature of the genetic code.

Inhibition domain   The part of a eukaryotic repressor that makes contact with the initiation complex

Inhibitor-resistant mutant   A mutant that is able to resist the toxic effects of an antibiotic or other type of inhibitor.

Initiation codon   The codon, usually but not exclusively 5′–AUG–3′, found at the start of the coding region of a gene.

Initiation complex   The complex of proteins that initiates transcription. Also the complex that initiates translation.

Initiation factor   A protein that plays an ancillary role during initiation of translation.

Initiation of transcription   The assembly upstream of a gene of the complex of proteins that will subsequently copy the gene into RNA.

Initiation region   A region of eukaryotic chromosomal DNA within which replication initiates at positions that are not clearly defined.

Initiator (Inr) sequence   A component of the RNA polymerase II core promoter.

Initiator tRNA   The tRNA, aminoacylated with methionine in eukaryotes or *N*-formylmethionine in bacteria, that recognizes the initiation codon during protein synthesis.

Inosine   A modified version of adenosine, sometimes found at the wobble position of an anticodon.

Insertion mutation   A mutation that arises by insertion of one or more nucleotides into a DNA sequence.

Insertion sequence   A short transposable element found in bacteria.

Insertion vector   A λ vector constructed by deleting a segment of non-essential DNA.

Insertional editing   A less extensive form of pan-editing that occurs during processing of some viral RNAs.

Insertional inactivation   A cloning strategy whereby insertion of a new piece of DNA into a vector inactivates a gene carried by the vector.

Instability element   A sequence present in yeast mRNAs that affects degradation.

Insulator   A segment of DNA that acts as the boundary point between two functional domains.

Integrase   A Type I topoisomerase that catalyzes insertion of the λ genome into *Escherichia coli* DNA.

Integron   A set of genes and other DNA sequences that enable plasmids to capture genes from bacteriophages and other plasmids.

Intein   An internal segment of a polypeptide that is removed by a splicing process after translation.

Intein homing   The conversion of a gene coding for a protein that lacks an intein into one coding for an intein-plus protein, catalyzed by the spliced component of the intein.

Intercalating agent   A compound that can enter the space between adjacent base pairs of a double-stranded DNA molecule, often causing mutations.

Intergenic region   The regions of a genome that do not contain genes.

Internal node   A branch point within a phylogenetic tree, representing an organism or DNA sequence that is ancestral to those being studied.

Internal ribosome entry site (IRES)   A nucleotide sequence that enables the ribosome to assemble at an internal position in some eukaryotic mRNAs.

Interphase   The period between cell divisions.

Interphase chromosome   A chromosome, present in a cell during the period between cell divisions, which adopts a relatively uncondensed chromatin structure.

Interspersed repeat   A sequence that recurs at many dispersed positions within a genome.

Interspersed repeat element PCR (IRE-PCR)   A clone fingerprinting technique that uses PCR to detect the relative positions of genome-wide repeats in cloned DNA fragments.

Intrinsic terminator   A position in bacterial DNA where termination of transcription occurs without the involvement of Rho.

Intron   A non-coding region within a discontinuous gene.

Intron homing   The conversion of a gene lacking an intron into one that contains an intron, catalyzed by a protein coded by that intron.

Introns early   The hypothesis that introns evolved relatively early and are gradually being lost from eukaryotic genomes.

Introns late   The hypothesis that introns evolved relatively late and are gradually accumulating in eukaryotic genomes.

Inverted repeat   Two identical nucleotide sequences repeated in opposite orientations in a DNA molecule.

Iron-response element   A type of response module.

Isoaccepting tRNAs   Two or more tRNAs that are charged with the same amino acid.

Isochore   A segment of genomic DNA that has a uniform base composition which differs from that of the adjacent segments.

Isotope   One of two or more atoms that have the same atomic number but different atomic weights.

Janus kinase (JAK)   A type of kinase that plays an intermediary role in some types of signal transduction involving STATs.

Junk DNA   One interpretation of the intergenic DNA content of a genome.

Karyogram   The entire chromosome complement of a cell, with each chromosome described in terms of its appearance at metaphase.

Karyopherin   A protein involved in transport of RNA out of or into the nucleus.

Kilobase pair (kb)   1000 base pairs.

Kinetochore   The part of the centromere to which spindle microtubules attach.

Klenow polymerase   A DNA polymerase enzyme, obtained by chemical modification of *Escherichia coli* DNA polymerase I, used primarily in chain termination DNA sequencing.

Knockout mouse   A mouse that has been engineered so that it carries an inactivated gene.

Kornberg polymerase   The DNA polymerase I enzyme of *Escherichia coli*.

Kozak consensus   The nucleotide sequence surrounding the initiation codon of a eukaryotic mRNA.

Lactose operon   The cluster of three genes that code for enzymes involved in utilization of lactose by *Escherichia coli*.

Lactose repressor   The regulatory protein that controls transcription of the lactose operon in response to the presence or absence of lactose in the environment.

Lagging strand   The strand of the double helix which is copied in a discontinuous fashion during genome replication.

Lariat   Refers to the lariat-shaped intron RNA that results from splicing a GU–AG intron.

Lateral gene transfer   Transfer of a gene from one species to another.

Leader segment   The untranslated region of an mRNA upstream of the initiation codon.

Leading strand   The strand of the double helix which is copied in a continuous fashion during genome replication.

Leaky mutation   A mutation that results in partial loss of a characteristic.

Lethal mutation   A mutation that results in death of the cell or organism.

Leucine zipper   A dimerization domain commonly found in DNA-binding proteins.

Ligase   An enzyme that synthesizes phosphodiester bonds as part of DNA replication, repair and recombination processes.

LINE (long interspersed nuclear element)   A type of genome-wide repeat, often with transposable activity.

LINE-1   One type of human LINE.

Linkage   The physical association between two genes that are on the same chromosome.

Linkage analysis   The procedure used to assign map positions to genes by genetic crosses.

Linker DNA   The DNA that links nucleosomes: the 'string' in the 'beads-on-a-string' model for chromatin structure.

Linker histone   A histone, such as H1, that is located outside of the nucleosome core octamer.

Locus   The chromosomal location of a genetic or DNA marker.

Locus control region (LCR)   A DNA sequence that maintains a functional domain in an open active configuration.

Lod score   A statistical measure of linkage as revealed by pedigree analysis.

Long patch repair   A nucleotide excision repair process of *Escherichia coli* that results in excision and resynthesis of up to 2 kb of DNA.

Loss-of-function mutation   A mutation that reduces or abolishes a protein's activity.

LTR element   A type of genome-wide repeat typified by the presence of long terminal repeats (LTRs).

Lysogenic infection cycle   The type of bacteriophage infection that involves integration of the phage genome into the host DNA molecule.

Lysozyme   A protein used to destabilize the bacterial cell wall prior to DNA purification.

Lytic infection cycle   The type of bacteriophage infection that involves lysis of the host cell immediately after the initial infection, with no integration of the phage DNA molecule into the host genome.

M phase   The stage of the cell cycle when mitosis or meiosis occurs.

Macrochromosome   One of the larger gene-deficient chromosomes seen in the nuclei of chickens and various other species.

MADS box   A DNA-binding domain found in several transcription factors involved in plant development.

Maintenance methylation   Addition of methyl groups to positions on newly synthesized DNA strands that correspond with the positions of methylation on the parent strand.

Major groove   The larger of the two grooves that spiral around the surface of the B-form of DNA.

Major histocompatibility complex (MHC)   A mammalian multigene family coding for cell surface proteins and including several multiallelic genes.

Map   A chart showing the positions of genetic and/or physical markers in a genome.

MAP kinase   A signal transduction pathway.

Mapping reagent   A collection of DNA fragments spanning a chromosome or the entire genome and used in STS mapping.

Marker   A distinctive feature on a genome map. Also a gene, carried by a cloning vector, that codes for a distinctive protein product and/or phenotype and so can be used to determine if a cell contains a copy of the cloning vector.

Mass spectrometry   An analytical technique in which ions are separated according to their charge-to-mass ratios.

Maternal-effect gene   A *Drosophila* gene that is expressed in the parent and whose mRNA is subsequently injected into the egg, after which it influences development of the embryo.

Mating type   The equivalent of male and female for a eukaryotic microorganism.

Mating-type switching   The ability of yeast cells to change from a to α mating type, or vice versa, by gene conversion.

Matrix-assisted laser desorption ionization time-of-flight (MALDI-TOF)   A type of mass spectrometry used in proteomics.

Matrix-associated region (MAR)   An AT-rich segment of a eukaryotic genome that acts as an attachment point to the nuclear matrix.

Maturase   A protein, coded by a gene in an intron, thought to be involved in splicing.

Maximum parsimony method   A method for construction of phylogenetic trees.

Mediator   A protein complex that forms a contact between various activators and the C-terminal domain of the largest subunit of RNA polymerase II.

Megabase pair (Mb)   1000 kb; 1 000 000 bp.

Meiosis   The series of events, involving two nuclear divisions, by which diploid nuclei are converted to haploid gametes.

Melting   Denaturation of a double-stranded DNA molecule.

Melting temperature (T_m)   The temperature at which the two strands of a double-stranded nucleic acid molecule or base-paired hybrid detach as a result of complete breakage of hydrogen bonding.

Meselson–Stahl experiment   The experiment which showed that cellular DNA replication occurs by the semiconservative process.

Messenger RNA (mRNA)   The transcript of a protein-coding gene.

Metaphase chromosome   A chromosome at the metaphase stage of cell division, when the chromatin takes on its most condensed structure and features such as the banding pattern can be visualized.

Methyl-CpG-binding protein (MeCP)   A protein that binds to methylated CpG islands and may influence acetylation of nearby histones.

MGMT (O^6-methylguanine-DNA methyltransferase)   An enzyme involved in the direct repair of alkylation mutations.

Microarray   A low-density array of DNA molecules used for parallel hybridization analysis.

Microsatellite   A type of simple sequence length polymorphism comprising tandem copies of, usually, di-, tri- or tetranucleotide repeat units. Also called a simple tandem repeat (STR).

Minichromosome   One of the smaller, gene-rich chromosomes seen in the nuclei of chickens and various other species.

Minigene   The name given to the pair of exons carried by a cloning vector used in the exon-trapping procedure.

Minimal medium   A medium that provides only the minimum nutritional requirements for growth of a microorganism.

Minisatellite   A type of simple sequence length polymorphism comprising tandem copies of repeats that are a few tens of nucleotides in length. Also called a variable number of tandem repeats (VNTR).

Minor groove   The smaller of the two grooves that spiral around the surface of the B-form of DNA.

Mismatch   A position in a double-stranded DNA molecule where base-pairing does not occur because the nucleotides are not complementary; in particular, a non-base-paired position resulting from an error in replication.

Mismatch repair   A DNA repair process that corrects mismatched nucleotide pairs by replacing the incorrect nucleotide in the daughter polynucleotide.

Missense mutation   An alteration in a nucleotide sequence that converts a codon for one amino acid into a codon for a second amino acid.

Mitochondrial genome   The genome present in the mitochondria of a eukaryotic cell.

Mitochondrion   One of the energy-generating organelles of eukaryotic cells.

Mitosis   The series of events that results in nuclear division.

Model organism   An organism which is relatively easy to study and hence can be used to obtain information that is relevant to the biology of a second organism that is more difficult to study.

Modification assay   A range of techniques used for locating bound proteins on DNA molecules.

Modification interference   A technique used to identify nucleotides involved in interactions with a DNA-binding protein.

Modification protection   A technique used to identify nucleotides involved in interactions with a DNA-binding protein.

Molecular biologist   A person who studies the molecular life sciences.

Molecular chaperone   A protein that helps other proteins to fold.

Molecular clock   A device based on the inferred mutation rate that enables times to be assigned to the branch points in a gene tree.

Molecular combing   A technique for preparing restricted DNA molecules for optical mapping.

Molecular evolution   The gradual changes that occur in genomes over time as a result of the accumulation of mutations and structural rearrangements resulting from recombination and transposition.

Molecular life sciences   The area of research comprising molecular biology, biochemistry and cell biology, as well as some aspects of genetics and physiology.

Molecular phylogenetics   A set of techniques that enable the evolutionary relationships between DNA sequences to be inferred by making comparisons between those sequences.

Monohybrid cross   A sexual cross in which the inheritance of one pair of alleles is followed.

Monophyletic   Refers to two or more organisms or DNA sequences that are derived from a single ancestral organism or DNA sequence.

mRNA surveillance   A RNA degradation process in eukaryotes.

Multicopy   A gene, cloning vector or other genetic element that is present in multiple copies in a single cell.

Multicysteine zinc finger   A type of zinc-finger DNA-binding domain.

Multigene family   A group of genes, clustered or dispersed, with related nucleotide sequences.

Multiple alignment   An alignment of three or more nucleotide sequences.

Multiple alleles   The different alternative forms of a gene that has more than two alleles.

Multiple hit or multiple substitution   The situation that occurs when a single nucleotide in a DNA sequence undergoes two mutational changes, giving rise to two new alleles, both of which differ from each other and from the parent at that nucleotide position.

Multipoint cross   A genetic cross in which the inheritance of three or more markers is followed.

Multiregional evolution   A hypothesis that holds that modern humans in the Old World are descended from *Homo erectus* populations that left Africa over 1 million years ago.

Mutagen   A chemical or physical agent that can cause a mutation in a DNA molecule.

Mutagenesis   Treatment of a group of cells or organisms with a mutagen as a means of inducing mutations.

Mutant   A cell or organism that possesses a mutation.

Mutasome   A protein complex that is constructed during the SOS response of *Escherichia coli.*

Mutation An alteration in the nucleotide sequence of a DNA molecule.

Mutation scanning A set of techniques for detection of mutations in DNA molecules.

Mutation screening A set of techniques for determining if a DNA molecule contains a specific mutation.

N terminus The end of a polypeptide that has a free amino group.

Natural selection The preservation of favorable alleles and the rejection of injurious ones.

N-degron An N-terminal amino acid sequence that influences the degradation of a protein in which it is found.

Neighbor-joining method A method for construction of phylogenetic trees.

Nick A position in a double-stranded DNA molecule where one of the polynucleotides is broken as a result of the absence of a phosphodiester bond.

Nitrogenous base One of the purines or pyrimidines that form part of the molecular structure of a nucleotide.

N-linked glycosylation The attachment of sugar units to an asparagine in a polypeptide.

Non-coding RNA An RNA molecule that does not code for a protein.

Non-homologous end-joining (NHEJ) Another name for the double-strand break repair process.

Non-penetrance The situation whereby the effect of a mutation is never observed during the lifetime of a mutant organism.

Non-polar A hydrophobic (water-hating) chemical group.

Nonsense-mediated RNA decay (NMD) A process for degradation of eukaryotic mRNAs that is initiated by the presence of an internal termination codon.

Nonsense mutation An alteration in a nucleotide sequence that changes a triplet coding for an amino acid into a termination codon.

Non-synonymous mutation A mutation that converts a codon for one amino acid into a codon for a second amino acid.

Northern blotting The transfer of RNA from an electrophoresis gel to a membrane prior to northern hybridization.

Northern hybridization A technique used for detection of a specific RNA molecule against a background of many other RNA molecules.

Nuclear genome The DNA molecules present in the nucleus of a eukaryotic cell.

Nuclear magnetic resonance (NMR) spectroscopy A technique for determining the three-dimensional structure of large molecules.

Nuclear matrix A proteinaceous scaffold-like network that permeates the cell.

Nuclear pore complex The complex of proteins present at a nuclear pore.

Nuclear receptor superfamily A family of receptor proteins that bind hormones as an intermediate step in modulation of genome activity by these hormones.

Nuclease An enzyme that degrades a nucleic acid molecule.

Nuclease protection experiment A technique that uses nuclease digestion to determine the positions of proteins on DNA or RNA molecules.

Nucleic acid The term first used to describe the acidic chemical compound isolated from the nuclei of eukaryotic cells. Now used specifically to describe a polymeric molecule comprising nucleotide monomers, such as DNA and RNA.

Nucleic acid hybridization Formation of a double-stranded hybrid by base-pairing between complementary polynucleotides.

Nucleoid The DNA-containing region of a prokaryotic cell.

Nucleolus   The region of the eukaryotic nucleus in which rRNA transcription occurs.

Nucleoside   A purine or pyrimidine base attached to a five-carbon sugar.

Nucleosome   The complex of histones and DNA that is the basic structural unit in chromatin.

Nucleosome remodeling   A change in the conformation of a nucleosome, associated with a change in access to the DNA to which the nucleosome is attached.

Nucleotide   A purine or pyrimidine base attached to a five-carbon sugar, to which a mono-, di-, or triphosphate is also attached. The monomeric unit of DNA and RNA.

Nucleotide excision repair   A repair process that corrects various types of DNA damage by excising and resynthesizing a region of a polynucleotide.

Nucleus   The membrane-bound structure of a eukaryotic cell in which the chromosomes are contained.

Octamer module   A basal promoter element.

Okazaki fragment   One of the short segments of RNA-primed DNA synthesized during replication of the lagging strand of the double helix.

Oligonucleotide   A short synthetic single-stranded DNA molecule.

Oligonucleotide-directed mutagenesis   An *in vitro* mutagenesis technique in which a synthetic oligonucleotide is used to introduce a predetermined nucleotide alteration into the gene to be mutated.

Oligonucleotide hybridization analysis   The use of an oligonucleotide as a hybridization probe.

O-linked glycosylation   The attachment of sugar units to a serine or threonine in a polypeptide.

Open promoter complex   A structure formed during assembly of the transcription initiation complex consisting of the RNA polymerase and/or accessory proteins attached to the promoter, after the DNA has been opened up by breakage of base pairs.

Open reading frame (ORF)   A series of codons starting with an initiation codon and ending with a termination codon. The part of a protein-coding gene that is translated into protein.

Operational taxonomic unit (OTU)   One of the organisms being compared in a phylogenetic analysis.

Operator   The nucleotide sequence to which a repressor protein binds to prevent transcription of a gene or operon.

Operon   A set of adjacent genes in a bacterial genome, transcribed from a single promoter and subject to the same regulatory regime.

Optical mapping   A technique for the direct visual examination of restricted DNA molecules.

ORF scanning   Examination of a DNA sequence for open reading frames in order to locate the genes.

Origin of replication   A site on a DNA molecule where replication initiates.

Origin recognition complex (ORC)   A set of proteins that binds to the origin recognition sequence.

Origin recognition sequence   A component of a eukaryotic origin of replication.

Orphan family   A group of homologous genes whose functions are unknown.

Orthogonal field alternation gel electrophoresis (OFAGE)   An electrophoresis system in which the field alternates between pairs of electrodes set at an angle of 45°, used to separate large DNA molecules.

Orthologous   Refers to homologous genes located in the genomes of different organisms.

Outgroup   An organism or DNA sequence that is used to root a phylogenetic tree.

Out of Africa   A hypothesis which holds that modern humans evolved in Africa, moving to the rest of the Old World between 100 000 and 50 000 years ago, displacing the descendants of *Homo erectus* that they encountered.

Overlapping genes   Two genes whose coding regions overlap.

P1-derived artificial chromosome (PAC)   A high-capacity vector that combines features of bacteriophage P1 vectors and bacterial artificial chromosomes.

Pair-rule genes   Developmental genes that establish the basic segmentation pattern of the *Drosophila* embryo.

Pan-editing   The extensive insertion of nucleotides into an abbreviated RNA, resulting in a functional molecule.

Paralogous   Refers to two or more homologous genes located in the same genome.

Paranemic   Refers to a helix whose strands can be separated without unwinding.

Parental genotype   The genotype possessed by one or both of the parents in a genetic cross.

Parsimony   An approach that decides between different phylogenetic tree topologies by identifying the one that involves the shortest evolutionary pathway.

Partial linkage   The type of linkage usually displayed by a pair of genetic and/or physical markers on the same chromosome, the markers not always being inherited together because of the possibility of recombination between them.

Partial restriction   Digestion of DNA with a restriction endonuclease under limiting conditions so that not all restriction sites are cut.

Pedigree   A chart showing the genetic relationships between the members of a human family.

Pedigree analysis   The use of pedigree charts to analyze the inheritance of a genetic or DNA marker in a human family.

P element   A DNA transposon of *Drosophila*.

Pentose   A sugar comprising five carbon atoms.

Peptide bond   The chemical link between adjacent amino acids in a polypeptide.

Peptide nucleic acid (PNA)   A polynucleotide analog in which the sugar–phosphate backbone is replaced by amide bonds.

Peptidyl or P site   The site in the ribosome occupied by the tRNA attached to the growing polypeptide during translation.

Peptidyl transferase   The enzyme activity that synthesizes peptide bonds during translation.

Permissive conditions   Conditions under which a conditional-lethal mutant is able to survive.

PEST sequences   Amino acid sequences that influence the degradation of proteins in which they are found.

Phage   A virus that infects a bacterium.

Phage display   A technique for identifying proteins that interact with one another.

Phage display library   A collection of clones carrying different DNA fragments, used in phage display.

Phagemid   A cloning vector comprising a mixture of plasmid and phage DNA.

Phenetics   A classificatory approach based on the numerical typing of as many characters as possible.

Phenotype   The observable characteristics displayed by a cell or organism.

Phosphodiesterase   A type of enzyme that can break phosphodiester bonds.

Phosphodiester bond   The chemical link between adjacent nucleotides in a polynucleotide.

Phosphorimaging   An electronic method for determining the positions of radioactive markers in a microarray or on a hybridization membrane.

Photobleaching   A component of the FRAP technique for studying protein mobility in the nucleus.

Photolithography   A technique that uses pulses of light to construct an oligonucleotide from light-activated nucleotide substrates.

Photolyase   An *Escherichia coli* enzyme involved in photoreactivation repair.

Photoproduct   A modified nucleotide resulting from treatment of DNA with ultraviolet radiation.

Photoreactivation   A DNA repair process in which cyclobutyl dimers and (6–4) photoproducts are corrected by a light-activated enzyme.

Phylogeny   A classification scheme that indicates the evolutionary relationships between organisms.

Physical mapping   The use of molecular biology techniques to construct a genome map.

Pilus   A structure involved in bringing a pair of bacteria together during conjugation; possibly the tube through which DNA is transferred.

Plaque   A zone of clearing on a lawn of bacteria caused by lysis of the cells by infecting bacteriophages.

Plasmid   A usually circular piece of DNA often found in bacteria and some other types of cell.

Plectonemic   Refers to a helix whose strands can only be separated by unwinding.

Plesiomorphic character state   A character state possessed by a remote common ancestor of a group of organisms.

Point mutation   A mutation that results from a single nucleotide change in a DNA molecule.

Polar   A hydrophilic (water-loving) chemical group.

Polyacrylamide gel electrophoresis   Electrophoresis carried out in a polyacrylamide gel and used to separate DNA molecules between 10 and 1500 bp in length.

Polyadenylate-binding protein   A protein that aids poly(A) polymerase during polyadenylation of eukaryotic mRNAs, and which plays a role in maintenance of the tail after synthesis.

Polyadenylation   The addition of a series of As to the 3′ end of a eukaryotic mRNA.

Polyadenylation editing   A form of editing that occurs with many animal mitochondrial RNAs, resulting in a termination codon being formed by adding a poly(A) tail to an mRNA that ends with the nucleotides U or UA.

Poly(A) polymerase   The enzyme that attaches a poly(A) tail to the 3′ end of a eukaryotic mRNA.

Poly(A) tail   A series of A nucleotides attached to the 3′ end of a eukaryotic mRNA.

Polymer   A compound made up of a long chain of identical or similar units.

Polymerase chain reaction (PCR)   A technique that results in exponential amplification of a selected region of a DNA molecule.

Polymorphic   Refers to a locus that is represented by a number of different alleles or haplotypes in the population as a whole.

Polynucleotide   A single-stranded DNA or RNA molecule.

Polynucleotide kinase   An enzyme that adds phosphate groups to the 5′ ends of DNA molecules.

Polypeptide   A polymer of amino acids.

Polyphyletic   A group of DNA sequences that are derived from two or more distinct ancestral sequences.

Polyprotein   A translation product consisting of a series of linked proteins which are processed by proteolytic cleavage to release the mature proteins.

Polypyrimidine tract   A pyrimidine-rich region near the 3′ end of a GU–AG intron.

Polysome   An mRNA molecule that is being translated by more than one ribosome at the same time.

Positional cloning   A procedure that uses information on the map position of a gene to obtain a clone of that gene.

Positional effect   Refers to the different levels of expression that result after insertion of a gene at different positions in a eukaryotic genome.

Post-replication complex (post-RC)   A complex of proteins, derived from a pre-RC, that forms at a eukaryotic origin of replication during the replication process and ensures that the origin is used just once per cell cycle.

POU domain   A DNA-binding motif found in a variety of proteins.

Pre-initiation complex   The structure comprising the small subunit of the ribosome, the initiator tRNA plus ancillary factors that forms the initial association with the mRNA during protein synthesis. Also the structure that forms at the core promoter of a gene transcribed by RNA polymerase II.

Pre-mRNA   The primary transcript of a protein-coding gene.

Prepriming complex   A complex of proteins formed during initiation of replication in bacteria.

Pre-replication complex (pre-RC)   A protein complex that is constructed at a eukaryotic origin of replication and enables initiation of replication to occur.

Pre-RNA   The initial product of transcription of a gene or group of genes, subsequently processed to give the mature transcript(s).

Pre-rRNA   The primary transcript of a gene or group of genes specifying rRNA molecules.

Pre-spliceosome complex   An intermediate in the splicing pathway for a GU–AG intron.

Pre-tRNA   The primary transcript of a gene or group of genes specifying tRNA molecules.

Pribnow box   A component of the bacterial promoter.

Primary structure   The sequence of amino acids in a polypeptide.

Primary transcript   The initial product of transcription of a gene or group of genes, subsequently processed to give the mature transcript(s).

Primase   The RNA polymerase enzyme that synthesizes RNA primers during bacterial DNA replication.

Primer   A short oligonucleotide that is attached to a single-stranded DNA molecule in order to provide a start point for strand synthesis.

Primosome   A protein complex involved in genome replication.

Principal component analysis   A procedure that attempts to identify patterns in a large dataset of variable character states.

Prion   An unusual infectious agent that consists purely of protein.

Processed pseudogene   A pseudogene that results from integration into the genome of a reverse-transcribed copy of an mRNA.

Processivity   Refers to the amount of DNA synthesis that is carried out by a DNA polymerase before dissociation from the template.

Programmed frameshifting   The controlled movement of a ribosome from one reading frame to another at an internal position within a gene.

Programmed mutation   The possibility that under some circumstances an organism can increase the rate at which mutations occur in a specific gene.

Prokaryote   An organism whose cells lack a distinct nucleus.

Proliferating cell nuclear antigen (PCNA)   An accessory protein involved in genome replication in eukaryotes.

Proline-rich domain   A type of activation domain.

Promiscuous DNA   DNA that has been transferred from one organelle genome to another.

Promoter   The nucleotide sequence, upstream of a gene, to which RNA polymerase binds in order to initiate transcription.

Promoter clearance   The completion of successful initiation of transcription that occurs when the RNA polymerase moves away from the promoter sequence.

Promoter escape   The stage in transcription during which the polymerase moves away from the promoter region and becomes committed to making a transcript.

Proofreading   The 3'→5' exonuclease activity possessed by some DNA polymerases which enables the enzyme to replace a misincorporated nucleotide.

Prophage   The integrated form of the genome of a lysogenic bacteriophage.

Protease   An enzyme that degrades protein.

Proteasome   A multi-subunit protein structure that is involved in the degradation of other proteins.

Protein   The polymeric compound made of amino acid monomers.

Protein electrophoresis   Separation of proteins in an electrophoresis gel.

Protein engineering   Various techniques for making directed alterations in protein molecules, often to improve the properties of enzymes used in industrial processes.

Protein folding   The adoption of a folded structure by a polypeptide.

Protein interaction map   A map showing the interactions between all or some of the proteins in a proteome.

Protein–protein crosslinking   A technique that links together adjacent proteins in order to identify proteins that are positioned close to one another in a structure such as a ribosome.

Proteome   The collection of functioning proteins synthesized by a living cell.

Proteomics   A variety of techniques used to study proteomes.

Protogenome   An RNA genome that existed during the RNA world.

Protoplast   A cell from which the cell wall has been completely removed.

Prototroph   An organism that has no nutritional requirements beyond those of the wild type and which can grow on minimal medium.

Pseudogene   An inactivated and hence non-functional copy of a gene.

Punctuation codon   A codon that specifies either the start or the end of a gene.

Punnett square   A tabular analysis for predicting the genotypes of the progeny resulting from a genetic cross.

Purine   One of the two types of nitrogenous base found in nucleotides.

Pyrimidine   One of the two types of nitrogenous base found in nucleotides.

Pyrosequencing   A novel DNA sequencing method in which addition of a nucleotide to the end of a growing polynucleotide is detected directly by conversion of the released pyrophosphate into a flash of chemiluminescence.

Quantitative PCR   A PCR method that enables the number of DNA molecules in a sample to be estimated.

Quaternary structure   The structure resulting from the association of two or more polypeptides.

RACE (rapid amplification of cDNA ends)   A PCR-based technique for mapping the end of an RNA molecule.

Radiation hybrid   A collection of rodent cell lines that contain different fragments of a second genome, constructed by a technique involving irradiation and used as a mapping reagent, for example in studies of the human genome.

Radioactive marker   A radioactive atom incorporated into a molecule and whose radioactive emissions are subsequently used to detect and follow that molecule during a biochemical reaction.

Radiolabeling   The technique for attaching a radioactive atom to a molecule.

Random genetic drift   The process that leads to alleles gradually changing their frequency in a population.

Ras   A protein involved in signal transduction.

Reading frame   A series of triplet codons in a DNA sequence.

Readthrough mutation   A mutation that changes a termination codon into a codon specifying an amino acid, and hence results in readthrough of the termination codon.

RecA   An *Escherichia coli* protein involved in homologous recombination.

RecBCD enzyme   An enzyme complex involved in homologous recombination in *Escherichia coli*.

Recessive   The allele that is not expressed in a heterozygote.

Reciprocal strand exchange   The exchange of DNA between two double-stranded molecules, occurring as a result of recombination, such that the end of one molecule is exchanged for the end of the other molecule.

Recognition helix   An α-helix in a DNA-binding protein, one that is responsible for recognition of the target nucleotide sequence.

Recombinant   A progeny member that possesses neither of the combinations of alleles displayed by the parents.

Recombinant DNA molecule   A DNA molecule created in the test tube by ligating pieces of DNA that are not normally joined together.

Recombinant DNA technology   The techniques involved in the construction, study and use of recombinant DNA molecules.

Recombinant protein   A protein synthesized in a recombinant cell as the result of expression of a cloned gene.

Recombinase   A diverse family of enzymes that catalyze site-specific recombination events.

Recombination   A large-scale rearrangement of a DNA molecule.

Recombination frequency   The proportion of recombinant progeny arising from a genetic cross.

Recombination hotspot   A region of a chromosome where crossovers occur at a higher frequency than the average for the chromosome as a whole.

Recombination repair   A DNA repair process that mends double-stranded breaks.

Regulatory control   Control over bacterial gene expression that depends on the influence of regulatory proteins.

Regulatory mutant   A mutant that has a defect in a promoter or other regulatory sequence.

Release factor   A protein that plays an ancillary role during termination of translation.

Renaturation   The return of a denatured molecule to its natural state.

Repetitive DNA   A DNA sequence that is repeated two or more times in a DNA molecule or genome.

Repetitive DNA fingerprinting   A clone fingerprinting technique that involves determining the positions of genome-wide repeats in cloned DNA fragments.

Repetitive DNA PCR   A clone fingerprinting technique that uses PCR to detect the relative positions of genome-wide repeats in cloned DNA fragments.

Replacement vector   A λ vector designed so that insertion of new DNA is by replacement of part of the non-essential region of the λ DNA molecule.

Replica plating   A technique for transfer of colonies from one Petri dish to another, such that their relative positions on the surface of the agar medium are retained.

Replication factor C (RFC)   A multi-subunit accessory protein involved in eukaryotic genome replication.

Replication factory   A large structure attached to the nuclear matrix; the site of genome replication.

Replication fork   The region of a double-stranded DNA molecule that is being opened up to enable DNA replication to occur.

Replication licensing factors (RLFs)   A set of proteins that regulate genome replication, in particular by ensuring that only one round of genome replication occurs per cell cycle.

Replication mediator protein (RMP)   A protein responsible for detachment of single-strand binding proteins during genome replication.

Replication origin   A site on a DNA molecule where replication initiates.

Replication protein A (RPA)   The main single-strand binding protein involved in replication of eukaryotic DNA.

Replication slippage   An error in replication that leads to an increase or decrease in the number of repeat units in a tandem repeat such as a microsatellite.

Replicative form   The double-stranded form of the M13 DNA molecule found within infected *Escherichia coli* cells.

Replicative transposition   Transposition that results in copying of the transposable element.

Replisome   A complex of proteins involved in genome replication.

Reporter gene   A gene whose phenotype can be assayed and which can therefore be used to determine the function of a regulatory DNA sequence.

Resolution   Separation of a pair of recombining double-stranded DNA molecules.

Response module   A sequence motif found upstream of various genes that enables transcription initiation by RNA polymerase II to respond to general signals from outside of the cell.

Restriction endonuclease   An enzyme that cuts DNA molecules at a limited number of specific nucleotide sequences.

Restriction fragment length polymorphism (RFLP)   A restriction fragment whose length is variable because of the presence of a polymorphic restriction site at one or both ends.

Restriction mapping   Determination of the positions of restriction sites in a DNA molecule by analyzing the sizes of restriction fragments.

Restrictive conditions   Conditions under which a conditional-lethal mutant is unable to survive.

Retroelement   A genetic element that transposes via an RNA intermediate.

Retroposon   A retroelement that does not have LTRs.

Retrotransposition   Transposition via an RNA intermediate.

Retrotransposon   A genome-wide repeat with a sequence similar to an integrated retroviral genome and possibly with retrotransposition activity.

Retroviral-like element (RTVL)   A truncated retroviral genome integrated into a host chromosome.

Retrovirus   A virus with an RNA genome that integrates into the genome of its host cell.

Reverse transcriptase   A polymerase that synthesizes DNA on an RNA template.

Reverse transcriptase PCR (RT-PCR)   PCR in which the first step is carried out by reverse transcriptase, so RNA can be used as the starting material.

Rho   A protein involved in termination of transcription of some bacterial genes.

Rho-dependent terminator   A position in bacterial DNA where termination of transcription occurs with the involvement of Rho.

Ribbon–helix–helix motif   A type of DNA-binding domain.

Ribonuclease   An enzyme that degrades RNA.

Ribonuclease D   An enzyme involved in processing pre-tRNA in bacteria.

Ribonuclease MRP   An enzyme involved in processing eukaryotic pre-rRNA.

Ribonuclease P   An enzyme involved in processing pre-tRNA in bacteria.

Ribonucleoprotein (RNP) domain   A common type of RNA-binding domain.

Ribose   The sugar component of a ribonucleotide.

Ribosomal protein   One of the protein components of a ribosome.

Ribosomal RNA (rRNA)   The RNA molecules that are components of ribosomes.

Ribosome   One of the protein–RNA assemblies on which translation occurs.

Ribosome binding site   The nucleotide sequence that acts as the attachment site for the small subunit of the ribosome during initiation of translation in bacteria.

Ribosome recycling factor (RRF)   A protein responsible for disassembly of the ribosome at the end of protein synthesis in bacteria.

Ribozyme   An RNA molecule that has catalytic activity.

RNA   Ribonucleic acid, one of the two forms of nucleic acid in living cells; the genetic material for some viruses.

RNA-dependent DNA polymerase   An enzyme that makes a DNA copy of an RNA template; a reverse transcriptase.

RNA-dependent RNA polymerase   An enzyme that makes an RNA copy of an RNA template.

RNA editing   A process by which nucleotides not coded by a gene are introduced at specific positions in an RNA molecule after transcription.

RNA interference   An RNA degradation process in eukaryotes.

RNA polymerase   An enzyme that synthesizes RNA on a DNA or RNA template.

RNA polymerase I   The eukaryotic RNA polymerase that transcribes ribosomal RNA genes.

RNA polymerase II   The eukaryotic RNA polymerase that transcribes protein-coding and snRNA genes.

RNA polymerase III   The eukaryotic RNA polymerase that transcribes tRNA and other short genes.

RNA transcript   An RNA copy of a gene.

RNA world   The early period of evolution when all biological reactions were centered on RNA.

Rolling circle replication   A replication process that involves continual synthesis of a polynucleotide which is 'rolled off' of a circular template molecule.

Rooted   Refers to a phylogenetic tree that provides information on the past evolutionary events that have led to the organisms or DNA sequences being studied.

S phase   The stage of the cell cycle when DNA synthesis occurs.

S value   The unit of measurement for a sedimentation coefficient.

S1 nuclease   An enzyme that degrades single-stranded DNA or RNA molecules, including single-stranded regions in predominantly double-stranded molecules.

SAP (stress activated protein) kinase   A stress-activated signal transduction pathway.

Satellite DNA   Repetitive DNA that forms a satellite band in a density gradient.

Scaffold   A series of sequence contigs separated by sequence gaps.

Scaffold attachment region (SAR)   An AT-rich segment of a eukaryotic genome that acts as an attachment point to the nuclear matrix.

Scanning   A system used during initiation of eukaryotic translation, in which the pre-initiation complex attaches to the 5′-terminal cap structure of the mRNA and then scans along the molecule until it reaches an initiation codon.

Second messenger   An intermediate in a certain type of signal transduction pathway.

Secondary structure   The conformations, such as α-helix and β-sheet, taken up by a polypeptide.

Sedimentation analysis   The centrifugal technique used to measure the sedimentation coefficient of a molecule or structure.

Sedimentation coefficient   The value used to express the velocity at which a molecule or structure sediments when centrifuged in a dense solution.

Segment polarity genes   Developmental genes that provide greater definition to the segmentation pattern of the *Drosophila* embryo established by the action of the pair-rule genes.

Segregation   The separation of homologous chromosomes, or members of allele pairs, into different gametes during meiosis.

Selectable marker   A gene carried by a vector and conferring a recognizable characteristic on a cell containing the vector or a recombinant DNA molecule derived from the vector.

Selective medium   A medium that supports the growth of only those cells that carry a particular genetic marker.

Selfish DNA   DNA that appears to have no function and apparently contributes nothing to the cell in which it is found.

Semiconservative replication   The mode of DNA replication in which each daughter double helix is made up of one polynucleotide from the parent and one newly synthesized polynucleotide.

Sequenase   An enzyme used in chain termination DNA sequencing.

Sequence contig   A contiguous DNA sequence obtained as an intermediate in a genome sequencing project.

Sequence skimming   A method for rapid sequence acquisition in which a few random sequences are obtained from a cloned fragment, the rationale being that if the fragment contains any genes then there is a good chance that at least some of them will be revealed by these random sequences.

Sequence tagged site (STS)   A DNA sequence that is unique in the genome.

Serial analysis of gene expression (SAGE)   A method for studying the composition of a transcriptome.

Sex cell   A reproductive cell; a cell that divides by meiosis.

Sex chromosome   A chromosome that is involved in sex determination.

Shine–Dalgarno sequence   The ribosome binding site upstream of an *Escherichia coli* gene.

Short interfering RNA (siRNA)   An intermediate in the RNA interference pathway.

Short patch repair   A nucleotide excision repair process of *Escherichia coli* that results in excision and resynthesis of about 12 nucleotides of DNA.

Short tandem repeat (STR)   A type of simple sequence length polymorphism comprising tandem copies of, usually, di-, tri- or tetranucleotide repeat units. Also called a microsatellite.

Shotgun approach   A genome sequencing strategy in which the molecules to be sequenced are randomly broken into fragments which are then sequenced individually.

Shuttle vector   A vector that can replicate in the cells of more than one organism (e.g. in *Escherichia coli* and in yeast).

Signal peptide   A short sequence at the N terminus of some proteins that directs the protein across a membrane.

Signal transduction   Control of cellular activity, including genome expression, via a cell-surface receptor that responds to an external signal.

Silencer   A regulatory sequence that reduces the rate of transcription of a gene or genes located some distance away in either direction.

Silent mutation   A change in a DNA sequence that has no effect on the expression or functioning of any gene or gene product.

Similarity approach   A rigorous mathematical approach to alignment of nucleotide sequences.

Simple sequence length polymorphism (SSLP)   An array of repeat sequences that display length variations.

SINE (short interspersed nuclear element)   A type of genome-wide repeat, typified by the Alu sequences found in the human genome.

Single-copy DNA   A DNA sequence that is not repeated elsewhere in the genome.

Single nucleotide polymorphism (SNP)   A point mutation that is carried by some individuals of a population.

Single orphan   A single gene, with no homolog, whose function is unknown.

Single-strand binding protein (SSB)   One of the proteins that attach to single-stranded DNA in the region of the replication fork, preventing base pairs forming between the two parent strands before they have been copied.

Single stranded   A DNA or RNA molecule that comprises just a single polynucleotide.

Site-directed hydroxyl radical probing   A technique for locating the position of a protein in a protein–RNA complex, such as a ribosome, by making use of the ability of Fe(II) ions to generate hydroxyl radicals which cleave nearby RNA phosphodiester bonds.

Site-directed mutagenesis   Techniques used to produce a specified mutation at a predetermined position in a DNA molecule.

Site-specific recombination   Recombination between two double-stranded DNA molecules that have only short regions of nucleotide sequence similarity.

Slippage   The translocation of a ribosome along a short non-coding nucleotide sequence between the termination codon of one gene and the initiation codon of a second gene.

SMAD family   A group of proteins involved in signal transduction.

Small cytoplasmic RNA (scRNA)   A type of short eukaryotic RNA molecule with various roles in the cell.

Small nuclear ribonucleoprotein (snRNP)   Structures involved in splicing GU–AG and AU–AC introns and in other RNA processing events, comprising one or two snRNA molecules complexed with proteins.

Small nuclear RNA (snRNA)   A type of short eukaryotic RNA molecule involved in splicing GU–AG and AU–AC introns and in other RNA processing events.

Small nucleolar RNA (snoRNA)   A type of short eukaryotic RNA molecule involved in chemical modification of rRNA.

Somatic cell   A non-reproductive cell; a cell that divides by mitosis.

Sonication   A procedure that uses ultrasound to cause random breaks in DNA molecules.

SOS response   A series of biochemical changes that occur in *Escherichia coli* in response to damage to the genome and other stimuli.

Southern hybridization   A technique used for detection of a specific restriction fragment against a background of many other restriction fragments.

Species tree   A phylogenetic tree that shows the evolutionary relationships between a group of species.

Spliceosome   The protein–RNA complex involved in splicing GU–AG or AU–AC introns.

Splicing   The removal of introns from the primary transcript of a discontinuous gene.

Splicing pathway   The series of events that converts a discontinuous pre-mRNA into a functional mRNA.

Spontaneous mutation   A mutation that arises from an error in replication.

SR-like CTD-associated factor (SCAF)   Proteins thought to play regulatory roles during splicing of GU–AG introns.

SR protein   A protein that plays a role in splice-site selection during splicing of GU–AG introns.

STAT (signal transducer and activator of transcription)   A type of protein that responds to binding of an extracellular signaling compound to a cell surface receptor by activating a transcription factor.

Stem cell   A progenitor cell that divides continually throughout the lifetime of an organism.

Stem-loop structure   A structure made up of a base-paired stem and non-base-paired loop, which can form in a single-stranded polynucleotide that contains an inverted repeat.

Steroid hormone   A type of extracellular signaling compound.

Steroid receptor   A protein that binds a steroid hormone after the latter has entered the cell, as an intermediate step in modulation of genome activity.

Sticky end   An end of a double-stranded DNA molecule where there is a single-stranded extension.

Strong promoter   A promoter that directs a relatively large number of productive initiations per unit time.

Structural domain   A segment of a polypeptide that folds independently of other segments. Also, a loop of eukaryotic DNA, predominantly in the form of the 30 nm chromatin fiber, attached to the nuclear matrix.

STS mapping   A physical mapping procedure that locates the positions of sequence tagged sites (STSs) in a genome.

Stuffer fragment   A DNA fragment contained within a λ vector that is replaced by the DNA to be cloned.

Substitution mutation   Commonly used as a synonym for a point mutation.

Supercoiling   A conformational state in which a double helix is overwound or underwound so that superhelical coiling occurs.

Superwobble   The extreme form of wobble that occurs in vertebrate mitochondria.

Suppressor mutation   A mutation in one gene that reverses the effect of a mutation in a second gene.

Syncytium   A cell-like structure comprising a mass of cytoplasm and many nuclei.

Synonymous mutation   A mutation that changes a codon into a second codon that specifies the same amino acid.

Synteny   Refers to a pair of genomes in which at least some of the genes are located at similar map positions.

TψC arm   Part of the structure of a tRNA molecule.

T4 polynucleotide kinase   An enzyme that adds phosphate groups to the 5′ ends of DNA molecules.

TAF and initiator-dependent cofactor (TIC)   A type of protein involved in initiation of transcription by RNA polymerase II.

Tandemly repeated DNA   DNA sequence motifs that are repeated head to tail.

Tandem repeat   Direct repeats that are adjacent to each other.

TATA-binding protein (TBP)   A component of the general transcription factor TFIID, the part that recognizes the TATA box of the RNA polymerase II promoter.

TATA box   A component of the RNA polymerase II core promoter.

Tautomeric shift   The spontaneous change of a molecule from one structural isomer to another.

Tautomers   Structural isomers that are in dynamic equilibrium.

TBP-associated factor (TAF)   One of several components of the general transcription factor TFIID, playing ancillary roles in recognition of the TATA box.

TBP domain   A type of DNA-binding domain.

T-DNA   The portion of the Ti plasmid that is transferred to the plant DNA.

Telomerase   The enzyme that maintains the ends of eukaryotic chromosomes by synthesizing telomeric repeat sequences.

Telomere   The end of a eukaryotic chromosome.

Telomere binding protein (TBP)   A protein that binds to and regulates the length of a telomere.

Temperature-sensitive mutation   A type of conditional-lethal mutation, one that is expressed only above a threshold temperature.

Template   The polynucleotide that is copied during a strand synthesis reaction catalyzed by a DNA or RNA polymerase.

Template-dependent DNA polymerase   An enzyme that synthesizes DNA in accordance with the sequence of a template.

Template-dependent DNA synthesis   Synthesis of a DNA molecule on a DNA or RNA template.

Template-dependent RNA polymerase   An enzyme that synthesizes RNA in accordance with the sequence of a template.

Template-dependent RNA synthesis   Synthesis of an RNA molecule on a DNA or RNA template.

Template-independent DNA polymerase   An enzyme that synthesizes DNA without the use of a template.

Template-independent RNA polymerase   An enzyme that synthesizes RNA without the use of a template.

Template strand   The polynucleotide that acts as the template for RNA synthesis during transcription of a gene.

Terminal deoxynucleotidyl transferase   An enzyme that adds one or more nucleotides to the 3' end of a DNA molecule.

Termination codon   One of the three codons that mark the position where translation of an mRNA should stop.

Termination factor   A protein that plays an ancillary role in termination of transcription.

Terminator sequence   One of several sequences on a bacterial genome involved in termination of genome replication.

Tertiary structure   The structure resulting from folding the secondary structural units of a polypeptide.

Test cross   A genetic cross between a double heterozygote and a double homozygote.

Thermal cycle sequencing   A DNA sequencing method that uses PCR to generate chain-terminated polynucleotides.

Thermostable   Able to withstand high temperatures.

Thymine   One of the pyrimidine bases found in DNA.

Ti plasmid   The large plasmid found in those *Agrobacterium tumefaciens* cells able to direct crown gall formation on certain species of plants.

T_m   Melting temperature.

Tn3-type transposon   A type of DNA transposon that does not have flanking insertion sequences.

Topology   The branching pattern of a phylogenetic tree.

Totipotent   Refers to a cell that is not committed to a single developmental pathway and can hence give rise to all types of differentiated cell.

Trailer segment   The untranslated region of an mRNA downstream of the termination codon.

Transcript   An RNA copy of a gene.

Transcription   The synthesis of an RNA copy of a gene.

Transcription bubble   The non-base-paired region of the double helix, maintained by RNA polymerase, within which transcription occurs.

Transcription-coupled repair   A nucleotide excision repair process that results in repair of the template strands of genes.

Transcription factory   A large structure attached to the nuclear matrix; the site of RNA synthesis.

Transcription initiation   The assembly, upstream of a gene, of the complex of proteins that will subsequently copy the gene into RNA.

Transcriptome   The entire mRNA content of a cell.

Transcript-specific regulation   Regulatory mechanisms that control protein synthesis by acting on a single transcript or a small group of transcripts coding for related proteins.

Trans-**displacement**   Transfer of a nucleosome from one DNA molecule to another.

Transduction   Transfer of bacterial genes from one cell to another by packaging in a phage particle.

Transduction mapping   The use of transduction to map the relative positions of genes in a bacterial genome.

Transfection   The introduction of purified phage DNA molecules into a bacterial cell.

Transfer-messenger RNA (tmRNA)   A bacterial RNA involved in protein degradation.

Transfer RNA (tRNA)   A small RNA molecule that acts as an adaptor during translation and is responsible for decoding the genetic code.

Transformant   A cell that has become transformed by the uptake of naked DNA.

Transformation   The acquisition by a cell of new genes by the uptake of naked DNA.

Transformation mapping   The use of transformation to map the relative positions of genes in a bacterial genome.

Transforming principle   The compound, now known to be DNA, responsible for transformation of an avirulent *Streptococcus pneumoniae* bacterium into a virulent form.

Transgenic mouse   A mouse that carries a cloned gene.

Transition   A point mutation that replaces a purine with another purine, or a pyrimidine with another pyrimidine.

Translation   The synthesis of a polypeptide, the amino acid sequence of which is determined by the nucleotide sequence of an mRNA in accordance with the rules of the genetic code.

Translational bypassing   A form of slippage in which a large part of an mRNA is skipped during translation, elongation of the original protein continuing after the bypassing event.

Translocation   The movement of a ribosome along an mRNA molecule during translation.

Transposable element   A genetic element that can move from one position to another in a DNA molecule.

Transposable phage   A bacteriophage that transposes as part of its infection cycle.

Transposase   An enzyme that catalyzes transposition of a transposable genetic element.

Transposition   The movement of a genetic element from one site to another in a DNA molecule.

Transposon A genetic element that can move from one position to another in a DNA molecule.

Transposon tagging A gene isolation technique that involves inactivation of a gene by movement of a transposon into its coding sequence, followed by the use of a transposon-specific hybridization probe to isolate a copy of the tagged gene from a clone library.

Transversion A point mutation that involves a purine being replaced by a pyrimidine, or vice versa.

Trinucleotide repeat expansion disease A disease that results from the expansion of an array of trinucleotide repeats in or near to a gene.

Triplet binding assay A technique for determining the coding specificity of a triplet of nucleotides.

Triplex A DNA structure comprising three polynucleotides.

Trisomy The presence of three copies of a homologous chromosome in a nucleus that is otherwise diploid.

tRNA nucleotidyltransferase The enzyme responsible for the post-transcriptional attachment of the triplet 5'–CCA–3' to the 3' end of a tRNA molecule.

***trp* RNA-binding attenuation protein (TRAP)** A protein involved in attenuation regulation of some operons in bacteria such as *Bacillus subtilis*.

True tree A phylogenetic tree that depicts the actual series of evolutionary events that led to the group of organisms or DNA sequences being studied.

Truncated gene A gene relic that lacks a segment from one end of the original, complete gene.

Tus The protein that binds to a bacterial terminator sequence and mediates termination of genome replication.

Twintron A composite structure made up of two or more Group II and/or Group III introns embedded in each other.

Two-dimensional gel electrophoresis A method for separation of proteins used especially in studies of the proteome.

Type 0 cap The basic cap structure, consisting of 7-methylguanosine attached to the 5' end of an mRNA.

Type 1 cap A cap structure comprising the basic 5'-terminal cap plus an additional methylation of the ribose of the second nucleotide.

Type 2 cap A cap structure comprising the basic 5'-terminal cap plus methylation of the riboses of the second and third nucleotides.

Ubiquitin A 76-amino-acid protein which, when attached to a second protein, acts as a tag directing that protein for degradation.

Unequal crossing-over A recombination event that results in duplication of a segment of DNA.

Unequal sister chromatid exchange A recombination event that results in duplication of a segment of DNA.

Unit factor Mendel's term for a gene.

Unrooted Refers to a phylogenetic tree that merely illustrates relationships between the organisms or DNA sequences being studied, without providing information about the past evolutionary events that have occurred.

Upstream Towards the 5' end of a polynucleotide.

Upstream control element A component of an RNA polymerase I promoter.

Upstream promoter element   Components of a eukaryotic promoter that lie upstream of the position where the initiation complex is assembled.

Uracil   One of the pyrimidine bases found in RNA.

U-RNA   A uracil-rich nuclear RNA molecule including the snRNAs and snoRNAs.

UvrABC endonuclease   A multienzyme complex involved in the short patch repair process of *Escherichia coli*.

V loop   Part of the structure of a tRNA molecule.

van der Waals forces   A particular type of attractive or repulsive non-covalent bond.

Variable number of tandem repeats (VNTR)   A type of simple sequence length polymorphism comprising tandem copies of repeats that are a few tens of nucleotides in length. Also called a minisatellite.

Vegetative cell   A non-reproductive cell; a cell that divides by mitosis.

Wave of advance   A hypothesis which holds that the spread of agriculture into Europe was accompanied by a large-scale movement of human populations.

Weak promoter   A promoter that directs relatively few productive initiations per unit time.

Whole-genome shotgun approach   A genome sequencing strategy which combines random shotgun sequencing with a genome map, the latter used to aid assembly of the master sequence.

Wild type   A gene, cell or organism that displays the typical phenotype and/or genotype for the species and is therefore adopted as a standard.

Winged helix–turn–helix   A type of DNA-binding domain.

Wobble hypothesis   The process by which a single tRNA can decode more than one codon.

X inactivation   Inactivation by methylation of most of the genes on one copy of the X chromosome in a female nucleus.

X-ray crystallography   A technique for determining the three-dimensional structure of a large molecule.

X-ray diffraction   The diffraction of X-rays that occurs during passage through a crystal.

X-ray diffraction pattern   The pattern obtained after diffraction of X-rays through a crystal.

Yeast artificial chromosome (YAC)   A high-capacity cloning vector constructed from the components of a yeast chromosome.

Yeast two-hybrid system   A technique for identifying proteins that interact with one other.

Z-DNA   A conformation of DNA in which the two polynucleotides are wound into a left-handed helix.

Zinc finger   A common structural motif for attachment of a protein to a DNA molecule.

Zoo blotting   A technique that attempts to determine if a DNA fragment contains a gene by hybridizing that fragment to DNA preparations from related species, on the basis that genes have similar sequences in related species and so give positive hybridization signals, whereas the regions between genes have less similar sequences and so do not hybridize.

Zygote   The cell resulting from fusion of gametes during meiosis.

Index

Note: Entries which are simply page numbers refer to the main text. Other entries have the following abbreviations immediately after the page number: B, Box; F, Figure; G, Glossary; RB, Research Briefing; T, Table; TN, Technical Note.

−10 box, 255, 256F, 256T, 258, 258F, 262
16S rRNA, 295, 296F, 321, 322F, 323F, 324
18S rRNA, 44, 255T, 295, 321, 322F
2 μm circle, 119, 511G
2-aminopurine, 426, 511G
2′-deoxyadenosine 5′-triphosphate, 10, 11F
2′-deoxycytidine 5′-triphosphate, 10, 11F
2′-deoxyguanosine 5′-triphosphate, 10, 11F
2′-deoxyribose, 9, 11F
2′-deoxythymidine 5′-triphosphate, 10, 11F
2-methylcytosine, 437
23S rRNA, 295, 296F, 321, 322F, 332RB
−25 box, 256, 257F, 259, 259T, 261F, 267B, 511G
28S rRNA, 44, 255T, 295, 298T, 321, 322F
3′→5′ exonuclease, 100, 101F, 166B, 395
³H, 98TN
3-methyladenine, 437, 439T
3′-RACE, 194F
3′ splice site, 288, 288F
3′ terminus, 10, 11F, 511G
3′ transduction, 476–7, 511G
3′-untranslated region (3′-UTR), 20, 511G
30 nm chromatin fiber, 37, 38F, 225F, 229, 231B, 511G
³²P, 8, 10F, 98TN
−35 box, 255, 256F, 256T, 257–8, 258F, 262
³⁵S, 8, 10F, 98TN
4-thiouridine, 301T
4, 6-diamino-2-phenylindole dihydrochloride (DAPI), 150–1
40S ribosome subunit, 321–2, 322F
5′→3′ exonuclease, 100, 101F, 166B, 395
5-azacytidine, 233RB
5-bromo-4-chloro-3-indolyl-β-D-galactopyranoside, 111

5-bromouracil (5-bU), 425, 425F, 511G
5-hydroxycytosine, 437
5-hydroxyuracil, 439T
5-methylcytosine, 149F
5′-RACE, 194F
5S rRNA, 44, 255T, 295, 296F, 321, 322F
5′ splice site, 288, 288F
5′ terminus, 10, 11F, 511G
5′-untranslated region (5′-UTR), 20, 511G
5.8S rRNA, 44, 255T, 295, 321, 322F
50S ribosome subunit, 321–2, 322F
(6–4) lesion, 427, 427F, 436, 511G
(6–4) photoproduct photolyase, 436, 511G
7-methylguanosine, 283, 285F, 301T, 437
7SK RNA, 74B
7SL RNA, 74B
70S ribosome subunit, 321–2, 322F
8-oxoguanine, 439T
80S ribosome subunit, 321–2, 322F

α2 Type I collagen gene, 473, 474F
α-galactosidase, 304T
α-globin genes, 44–5, 45F, 470F, 471
α-helix, 81, 81F, 511G
α-spin, 248, 248F
α subunit of bacterial RNA polymerase, 275, 279RB
α T-cell receptor locus, 20
β-barrel dimer, 249T
β-galactosidase, 111, 112F, 206T
β-globin genes
    evolution, 470F, 471, 476
    gene organization, 44–5, 45F
    introns, 287F, 288T
    locus control region, 228, 228F
    used to study the origins of modern humans, 497
β-glucuronidase, 206T
β-N-glycosidic bond, 10, 11F, 427, 428F, 511G
β-sheet, 81, 81F, 511G
β-spin, 248, 248F
β subunit of bacterial RNA polymerase, 275–6, 278–9RB

β′ subunit of bacterial RNA polymerase, 275, 278–9RB
β T-cell receptor locus, 18, 19F, 20–1
β-turn, 249, 511G
γ-complex, 399, 511G
δ globin, 45, 45F
ε globin 45, 45F
κ chain, 362, 362T
κB site, 267B
κ-homology domain, 250B, 511G
λ chain, 362, 362T
λ insertion vector, 112, 114F, 115F
λ replacement vector, 112, 114F, 114T
π–π interactions, 14, 511G
σ subunit of bacterial RNA polymerase, 241T, 255, 257–8, 258F, 262, 263F, 275, 366–9, 368F, 369F
σ³² subunit, 262, 263F, 264
σ⁵⁴ subunit, 262
σ⁷⁰ subunit, 262, 263F
σ^A subunit, 366
σ^E subunit, 366–7, 368F
σ^F subunit, 366–7, 368F
σ^G subunit, 367, 369F
σ^H subunit, 366
σ^K subunit, 367, 369F

A

ABC model, 375B
abdominal A gene, 376F
Abdominal B gene, 375, 376F
Aberrant splicing, 289, 290F
ABF1 protein, 393, 393F
ABF2 protein, 255B
ABO series, 129
Acceptor arm, 315, 315F, 316F, 511G
Acceptor site, 288, 288F, 511G
Ac/Ds elements, 64
Ace1p protein, 352F, 353
Acetylation of histones, 228–9
Acetylation of proteins, 341, 341F, 341T
Acidic domain, 266, 511G
Acquired immune deficiency syndrome (AIDS), 495–6, 496F
Acrylamide, 165TN
ACTH, 341F

Actin, 89

Activation domain, 211, 213F, 266, 512G

Activator, 211, 213F, 241T, 264, 265F, 265–6, 266F, 268–9, 512G

Acute lymphoblastic leukemia cells, 80

Acute myeloid leukemia, 80, 284

Acylation of proteins, 341T, 342, 512G

ADA complex, 229

Ada enzyme, 436, 512G

Adaptor, 107, 107F, 512G

Adaptor role of tRNA, 314, 315F

ADAR, 304

Adenine, 10, 11F, 512G

Adenosine 5′-triphosphate, 10, 13F

Adenosine deaminases, 241T

Adenosine deaminase acting on RNA (ADAR), 304, 512G

Adenoviruses, 119

Adenylate cyclase, 354, 355F, 512G

Adrenocorticotropic hormone (ACTH), 341F

Aegilops squarrosa, 475

Aepyrum pernix, 57F

Affinity chromatography, 245, 246F, 512G

A-form of DNA, 15–16, 16T, 17F, 253, 512G

African green monkey, 496F

Agarose gel electrophoresis, 37TN, 103, 105F, 120TN, 149, 150F, 242, 242F, 512G

Agriculture, 500, 501F

Agrobacterium tumefaciens, 51T, 119

AIDS, 495–6, 496F

AKAP149 gene, 197F

Alanine, 82, 83F, 84T

Alcohol dehydrogenase, 33

Alkaline phosphatase, 98TN, 107, 512G

Alkylating agent, 426, 512G

Alkylation repair, 436

Allele, 129, 512G

Allele frequency, 497B, 512G

Allele-specific oligonucleotide (ASO) hybridization, 429TN, 512G

Allolactose, 263, 264F, 355F

Allopolyploidy, 475, 512G

Alphoid DNA, 38, 59–60, 480, 512G

Alternative splicing, 20, 20F, 22B, 77, 77F, 291–4, 292F, 293F, 512G

Alu, 24, 24T, 63, 158, 512G

*Alu*I, 103T

Alu-PCR, 179, 179F, 512G

Aly protein, 309

Amino acid
  definition, 80, 512G
  diversity of amino acid types, 82–3
  relevance to protein function, 88–9
  spontaneous chemical synthesis, 460
  structures, 81F, 83F

Aminoacylation, 315–17, 317F, 318F, 512G

Aminoacyl site, 324F, 325T, 329–30, 330F, 331F, 512G

Aminoacyl-tRNA synthetase, 241T, 316–17, 317T, 512G

Amino terminus, 81, 81F, 512G

Ammonium persulfate, 165TN

Amoeba dubia, 33

Amphioxus, 472

Ampicillin resistance, 109–10

Amylotrophic lateral sclerosis, 215T

Anabaena, 31, 297

Anaphase, 41F, 137F, 138F

Ancestral character state, 486B, 513G

Anchor protein, 371, 371F

Ancient DNA, 499RB, 513G

Animals, origins, 465, 466F

Antennapedia, 249T, 328, 376F

Antennapedia complex (ANT-C), 375–6, 376F

Antennapedia module, 267B

Anthrax, 366

Antibody, 89, 485

Anticodon, 315F, 316, 316F, 513G

Anticodon arm, 315F, 316, 316F, 513G

Antigen, 361, 485, 513G

Antirrhinum, 375B

Antitermination, 277, 280B, 280F, 513G

Antiterminator protein, 277, 513G

AP endonuclease, 437, 438F, 513G

APETELA proteins, 375B

Apocytochrome *b* gene, 47F

Apolipoprotein B, 304, 304T, 306F

Apomorphic character state, 486B, 513G

Apoptosis, 42, 413, 513G

AP site, 427, 428F, 437, 438F, 442, 444, 513G

Apurinic/apyrimidinic site, 427, 428F, 437, 438F, 442, 444, 513G

Aquifex aeolicus, 56, 56F, 57F

Arabidopsis thaliana
  centromere DNA, 38–9
  flower development, 375B
  gene catalog, 43F
  gene density, 41, 42F
  genome duplication, 470
  genome sequence, 31T
  genome size, 33T
  mitochondrial genome, 48T
  promiscuous DNA, 47–9
  protein domains, 44T
  transposable elements, 450B
  web site, 509T

Arc repressor protein, 249T

Archaea
  accessing the archaeal genome, 223F
  aminoacylation, 317
  definition, 31, 513G
  DNA topoisomerases, 388T, 389
  genome replication, 403B

  genomes, 33–5, 49–59
  introns, 287T, 302B
  lateral gene transfer, 474–5
  proteasome, 344
  RNA polymerase, 255
  transcription initiation, 258B

Archaeoglobus fulgidus, 31T, 57F

AR gene, 424T

Arginine, 82, 83F, 84T

ARS, 392, 393F, 513G

ARS1, 393F

ARS binding factor 1 (ABF1), 393, 393F

Ascospore, 79F, 513G

Ascus, 79F, 447, 513G

A site, 324F, 325T, 329–30, 330F, 331F, 512G

ASO hybridization, 429TN, 512G

Asparagine, 83F, 84T

Aspartic acid, 82, 83F, 84T

Aspergillus nidulans, 33T, 48T

Aspergillus oryzae, 102T

Ataxia telangiectasia, 215T, 444

ATM protein, 413

ATP, 10, 13F

ATPase subunit genes, 47F

ATR protein, 413

Attenuation, 277–81, 280F, 281F, 282F, 513G

att sites, 449, 451F

ATX gene, 444

AU–AC intron, 287T, 294–5, 513G

Australopithecus, 479, 479F

Automated DNA sequencing, 170–2, 171F

Autonomously replicating sequence (ARS), 392, 393F, 513G

Autopolyploid, 466, 467F, 513G

Autoradiography, 98TN, 104, 513G

Autosome, 4, 513G

Auxotroph, 145, 432, 432F, 513G

Avery, O, 6

B

Baboon, 487, 488F

BAC, 114T, 117, 176, 182, 513G

Bacillus
  σ subunits, 262
  attenuation, 281, 282F, 283F
  cloning vectors, 118–19
  diauxie, 354
  gene catalog, 59
  lateral gene transfer, 57F
  operons, 56
  sporulation, 366–9, 370RB
  transformation, 475

Bacillus anthracis, 366

Bacillus megaterium, 317

Bacillus subtilis, 31, 56, 281, 282F, 283F, 354, 366–9, 370RB

Bacteria
  accessing the bacterial genome, 223F
  capsules, 6, 7F
  definition, 31, 513G

Bacteria *continued*
  genes are made of DNA, 6–8
  genetic mapping, 144–5, 144F, 146F
  genomes, 33–5, 49–59
  ribosomes, 321–3, 322F, 323F, 324F, 331F
  wobble, 319
Bacterial artificial chromosome (BAC), 114T, 117, 176, 182, 513G
Bacterial RNA synthesis
  antitermination, 277, 280B, 280F
  attenuation, 277–81, 280F, 281F, 282F
  choice between elongation and termination, 277–81
  elongation, 274–5, 275F, 278–9RB
  termination, 275–7, 276F, 277F, 278–9RB
Bacteriophage
  definition, 8, 513G
  genes are made of DNA, 8
  lysogenic infection cycle, 9F, 112, 113F
  lytic infection cycle, 9F, 112
  structure, 8, 9F
  transduction, 144F, 145, 146F
Bacteriophage 434, 250F
Bacteriophage λ
  antitermination, 280B
  DNA replication, 390F, 390–1
  integration, 448–50, 451F
Bacteriophage φX174, 44B
Bacteriophage cloning vector
  λ insertion vectors, 112, 114F, 115F
  λ replacement vectors, 112, 114F, 115F
  M13 vectors, 168, 168F, 193, 205TN
  P1 vectors, 114T, 117, 513G
Bacteriophage M13, 168F, 168–9
Bacteriophage P1, 114T, 117
Bacteriophage SP6, 120TN
Bacteriophage T1, 435RB
Bacteriophage T2, 8, 9F
Bacteriophage T3, 120TN
Bacteriophage T4, 105, 297, 333, 333F
Bacteriophage T7, 101T, 120TN, 166B, 202RB
*Bam*HI, 103, 103T, 104F
Barley, 501F
Barr body, 235, 513G
Basal promoter, 256, 514G
Basal promoter element, 267B, 514G
Basal rate of transcription initiation, 262, 514G
Base, 10, 11F
Base analog, 424, 425F, 425–6, 514G
Base excision repair, 437, 438F, 514G
Base isomerization, 301T
Base pair, 16B, 514G
Base-pairing, 14, 15F, 14B, 320F, 422F, 514G

Base ratio, 13, 514G
Base-stacking, 14, 514G
Basic domain, 249T, 251, 514G
Basic Local Alignment Search Tool (BLAST), 196
B-cell lymphoma, 80
B chromosome, 39B, 513G
B'-DNA, 15
Beads-on-a-string structure, 36, 38F, 226, 514G
Bean genome, 509T
Bering Strait, 502, 502F
B-form of DNA, 14, 16, 16T, 17F, 252, 514G
*Bgl*I, 103T
bicoid gene, 375, 376F
Bicoid module, 267B
Bicoid protein, 250B, 372–4, 373F
Bidirectional DNA replication, 391, 391F
Big Bang, 460, 461F
Biochemical markers, 129, 129T
Bioinformatics, 188–91, 196–8, 509T, 514G
Biolistics, 119, 514G
Biological information, 4, 514G
Biotinylation of proteins, 341T, 342, 514G
Bis, 165TN
Bithorax complex (BX-C), 375–6, 376F
Bivalent, 138, 138F, 514G
BLAST, 196
Blistering of cuticle phenotype, 203RB
Blood clotting, 473
Blood sugar levels, 339
Bloom's syndrome, 215, 215T, 295, 444
Blunt end, 103, 104F, 514G
B lymphocyte, 361
Body morphology phenotype, 203RB
Bohr, N, 6
Bombyx mori, 33T, 508T
Bond energies, 82B
Bonobo, 476B
Bootstrap analysis, 493, 493F, 514G
Bootstrap value, 493
Borrelia burgdorferi, 34, 51, 53, 53T, 57T
Bottleneck, 497B, 514G
Bragg, L, 246
Bragg, W, 246
Brain medulloblastoma, 79F
Branch, in a phylogenetic tree, 487, 487F, 514G
Branch migration, 445, 445F, 514G
Brassica oleracea mitochondrial genome, 48T
BRCA1 gene, 444
Bread wheat, 475
Breadmaking properties, 475
Breakage and reunion model for DNA replication, 385
Breast cancer, 79F, 444

Brenner, S, 85RB, 369
*Bsm*FI, 207, 208F
*Bsr*BI, 103, 103T
Buoyant density, 42TN, 60F, 514G
Buttonhead protein, 374, 374F
BX-C, 375–6, 376F
bZIP leucine zipper, 254F

C
C1b5p protein, 412
C1b6p protein, 412
CAAT box, 267B, 515G
Caenorhabditis elegans
  appearance, 369F
  complexity, 21
  gene catalog, 43, 43F
  genome sequence, 30, 31T
  genome size, 33T
  holocentric chromosomes, 39B
  model organism, 369
  protein domains, 44T
  RNA interference, 200, 201F, 202–3RB
  vulva development, 369–72
  web site, 508T
  zinc fingers, 250
Cajal body, 223, 223F
Calcium channel, 359, 360F
Calcium chloride, 109
Calmodulin, 360, 360F
Cambrian Revolution, 465, 466F
cAMP, 354, 355F, 518G
Campylobacter jejuni, 35
Canavanine, 129T
Candida albicans, 508T
Candida cylindracea, 87T
CAP, 264, 354, 355F, 515G
Cap binding complex, 326F, 515G
Capping, 282–3, 285F, 515G
Capsid, 8, 515G
Cap structure, 76, 325–8, 326F
Capsules, 6, 7F
Carboxyl terminus, 81, 81F, 515G
Carcinogen, 424, 424T
Cascade pathway, 358, 358F
CASP, 291, 515G
Catabolite activator protein (CAP), 264, 354, 355F, 515G
Catabolite repression, 354, 355F, 515G
Caudal protein, 373–4, 373F
C-banding, 39T
CCdA-phosphate-puromycin, 332RB
Cdc6p protein, 409–12, 412F
Cdc7pDbf4p protein, 412
C-DNA, 15
C'-DNA, 15
C''-DNA, 15
cDNA, 22F, 101T, 102, 154, 155F, 515G
cDNA capture, 193, 515G
cDNA library, 207, 245–6
cDNA selection, 193, 515G
cDNA sequencing, 192–3
Celera Genomics, 16

Cell cycle, 409, 409F, 515G
Cell cycle arrest, 413
Cell cycle checkpoint, 409, 413, 515G
Cell-free protein synthesizing system, 85RB, 515G
Cell fusion, 155, 158F
Cell senescence, 407–9, 408F, 515G
Cells, origins, 460, 461F, 462
Cell-specific module, 267B, 515G
Cell surface receptors, 354, 356F, 356T
CENP-A, 39
Centre d'Études du Polymorphisme Humaine (CEPH), 144, 181, 508T
Centrifugation techniques, 42TN
Centromere
  component of yeast artificial chromosome, 114, 116F
  definition, 38, 38F, 515G
  DNA content, 38
  gene density, 39, 41, 42F
  human and chimpanzee differences, 480
  in genome sequences, 18
  position in a chromosome, 38
  proteins, 39
  satellite DNA, 59–60
CEPH, 144, 181, 508T
Cesium chloride density gradient centrifugation, 42TN
Cetyltrimethylammonium bromide (CTAB), 110TN
cGMP, 359
Chain termination DNA sequencing
  automated sequencing, 170–2, 171F
  DNA polymerases for sequencing, 166B
  outline, 165–8, 167F, 515G
  primer, 169F, 169–70
  template, 168–9
  thermal cycle sequencing, 170, 171F
Chaperonin, 337–8, 338F, 515G
Character state, 485, 515G
Charcot–Marie–Tooth syndrome, 469RB
Chargaff, E, 13
Chase, M, 8
CHEF, 150, 517G
Chemical degradation DNA sequencing, 164, 170B, 515G
Chemical modification of proteins, 335, 340–2, 341F, 341T, 342F
Chemical modification of RNA, 76
Chemical shift, 248, 515G
Chemiluminescence, 98TN, 172, 172F
Chicken, 39B, 508T
Chi form, 445, 445F, 515G
Chimera, 199
Chimpanzee, 21, 183, 476B, 479–80, 480F, 487, 488F, 494–5, 495F, 496F

Chi site, 446, 447F, 515G
Chk1 protein, 413
Chk2 protein, 413
Chlamydomonas reinhardtii, 46, 48T
Chloroplast genome, 46–9, 48T, 49F, 515G
Chloroplast RNA polymerase, 255B
Chondrus crispus, 48T
Chromatid, 138, 516G
Chromatin, 35–6, 516G
Chromatin domains, 224–8
Chromatin modifications
  high mobility group N (HMGN) proteins, 231B
  histone acetylation, 228–9
  histone code, 229
  histone deacetylation, 231–4
  methylation, 232–3RB, 234–6
  nucleosome remodeling, 229–31, 230F
Chromatosome, 36, 516G
Chromomycin A$_3$, 159
Chromosome
  contains genes, 6
  banding, 38, 39T, 40F
  definition, 516G
  differences between human and chimpanzee, 480, 480F
  DNA packaging, 35–7
  duplications, 465
  metaphase chromosomes, 37–41
  multiradiate, 233RB
  sizes, 35
  staining techniques, 38, 39T
  unusual types, 39B
Chromosome scaffold, 223, 516G
Chromosome theory, 6, 516G
Chromosome walking, 176–8, 177F, 516G
Chymotrypsin, 471
Circular DNA, 42TN
Cis, 263B
Cis-displacement, 230, 230F, 516G
Cladistics, 484, 486B, 516G
Class I aminoacyl-tRNA synthetase, 316, 317T
Class II aminoacyl-tRNA synthetase, 316, 317T
Class 1 tRNA, 316
Class 2 tRNA, 316
Clastogen, 424T
Clay particles, 461
Cleavage and polyadenylation specificity factor (CPSF), 241T, 286, 286F, 287F, 516G
Cleavage stimulation factor, 241T, 286, 286F, 516G
Clone contig, 127, 128F, 159–60, 176–9, 516G
Clone fingerprinting, 178–79, 179F, 516G
Clone library, 113, 114T, 117TN, 516G
Cloning vector
  definition, 516G

Escherichia coli bacteriophages, 112–13, 168
E. coli plasmids, 109–11, 168–9
for long pieces of DNA, 113–17
for organisms other than *E. coli*, 117–19
multicopy, 201
Closed promoter complex, 257, 257F, 516G
Clostridium, 88T
Clouds, 461
Cloverleaf structure, 315F, 315–16, 516G
Clustal, 491, 491B
Coactivator, 266, 516G
Cochlea, 294
Cockayne syndrome, 284, 444
Codominance, 135, 516G
Codon, 84, 516G
Codon–anticodon interaction, 317–19, 319F, 516G
Codon bias, 189–91, 517G
Codon usage data, 509T
Cohesin, 137, 404, 405F, 517G
Cohesive end, 103, 104F, 517G
Co-integrate, 451–2, 452F, 517G
Cold Spring Harbor, 8
Colinearity between mRNA and polypeptide, 84, 84F
Collagen, 89, 341T, 473, 474F
Collins, F, 182
Colon cancer, 79F, 80, 215T
Col plasmid, 51T
Columbia University, 6, 13
Combinatorial screening, 177–8, 178F
Commitment complex, 289, 291F, 292F, 517G
Comparative genomics, 182, 213–15, 517G
Complementary DNA (cDNA), 22F, 101T, 102, 154, 155F, 517G
Composite transposon, 63F, 64, 517G
Concatamer, 107F, 112, 115F, 517G
Concerted evolution, 472, 517G
Conditional-lethal mutation, 432–3, 517G
Confocal microscopy, 133TN, 209F
Conjugation, 57B, 144F, 145, 146F, 474–5, 517G
Connective tissue cells, 407–8
Connexin domain, 44T
Consensus sequence, 191, 517G
Conservative replication, 386–98, 386F, 387F, 517G
Conservative transposition, 63F, 63–4, 451F, 451–4, 452F, 517G
Constitutive control of bacterial transcription initiation, 261–2, 263F, 517G
Constitutive heterochromatin, 39T, 40F, 182, 225, 517G
Constitutive mutation, 433, 433F, 517G

Context-dependent codon reassignment, 86–8, 87T, 517G
Contour clamped homogeneous electric fields (CHEF), 150, 517G
Conventional pseudogene, 22, 517G
Convergent evolution, 486B, 518G
Copper, 129T, 352F, 352–3
Core octamer, 36, 229, 230, 231B, 260RB, 518G
Co-repressor, 263, 265F, 518G
Core promoter, 256, 257F, 267B, 518G
Cortex protein, 372
Corticotropin-like intermediate lobe protein (CLIP), 341F
cos site, 112, 518G
Cosmid, 113–14, 114T, 115F, 518G
COSY, 248
Cotton, 509T
Covalent bonds, 14
Cow, 88T, 508T
CpG island, 191, 234, 235F, 518G
CPSF, 241T, 286, 286F, 287F, 516G
CRE, 267B, 518G
CREB protein, 267B, 359, 518G
Crick, F, 8, 83–4, 314, 318, 384–5
Cricket, 33
Cro protein, 280B
Crossing-over, 138–40, 138F, 139F, 444, 518G
Crosslinking studies, 278–9RB
Cross-talk, 227, 227F
Crown gall disease, 51T
Cryptic splice site, 289, 290F, 293F, 294, 518G
Cryptogene, 305B, 518G
CSB protein, 284, 285T
C segment, 362, 362T, 363F
CstF, 241T, 286, 286F, 516G
CTAB, 110TN
CTD-associated SR-like protein (CASP), 291, 518G
C-terminal domain (CTD) of RNA polymerase II, 259, 259B, 261F, 266, 266F, 283, 286, 287F, 518G
C terminus, 81, 81F, 515G
CTP, 10, 13F
Cucumis melo, 48T
curly leaf gene, 375B
C-value paradox, 31, 33, 518G
Cyanelle, 47, 518G
Cyanobacteria, 31, 47
Cyanophora paradoxa, 47
Cyclic AMP (cAMP), 354, 355F, 518G
Cyclic AMP response module (CRE), 267B
Cyclic GMP (cGMP), 359
Cyclin, 412, 412F, 518G
Cyclin-dependent kinases, 412, 413
Cyclin-independent protein kinase, 412
Cyclobutyl dimer, 426, 427F, 436, 437, 442, 444, 518G
Cycloheximide, 129T

Cys$_2$His$_2$ finger, 249T, 250, 251F, 518G
Cysteine, 81, 83F, 84F, 84T
Cystic fibrosis, 121, 215, 215T
Cytidine 5′-triphosphate, 10, 13F
Cytochemistry, 6, 518G
Cytochrome *c* oxidase genes, 47F
Cytokine, 89, 356
Cytosine, 10, 11F, 518G
Cytosine glycol, 439T

D

DAG, 359, 360F
Dam, 440, 520G
DAPI, 150–1
Dark repair, 437, 518G
D arm, 315F, 316, 316F, 518G
Darwin, C, 434, 435RB, 484
dATP, 10, 11F
dbEST, 508T
D box, 302, 303F
dbSNP, 508T
dbSTS, 508T
Dcm, 440, 520G
dCTP, 10, 11F
D-DNA, 15
Deacetylation of histones, 231–4, 236
Deadenylation-dependent decapping, 306–7, 307F, 519G
Deaminating agent, 426, 426F, 519G
Deamination of nucleotides, 301T, 304
Death domain, 42, 44T
Deformed gene, 376F
Deformed (Dfd) protein, 363–5
Degeneracy, 86, 519G
Degradative plasmids, 51T
Degradosome, 306, 519G
Deinococcus radiodurans, 53, 53T, 57F, 150F, 151
Deiodinases, 88T
Delayed development phenotype, 203RB
Delayed-early transcripts of bacteriophage λ, 280B
Delayed-onset mutation, 432, 519G
Delbrück, M, 6, 385, 435RB
Deletion cassette, 198, 199F
Deletion mutation, 418, 430, 430F, 519G
Delta sequence, 32
Denaturation of DNA, 152, 519G
Denaturation of proteins, 101, 336, 336F, 519G
De novo methylation, 232–3RB, 234, 234F, 518G
Density gradient centrifugation, 42TN, 387F, 388, 410RB, 519G
Dentatoribral-pallidoluysian atrophy, 424T
Deoxyribodipyrimidine photolyase, 436
Deoxyribonuclease, 7, 8F, 519G
Deoxyribonuclease I (DNase I), 102T, 226–8, 243, 243F

Deoxyribonuclease I hypersensitive region, 228, 228F, 521G
Deoxyribonuclease I sensitive region, 226, 226F
Derived character state, 486B, 519G
Desulfomicrobium baculatum, 88T
Development
 Bacillus sporulation, 366–9, 370RB
 Caenorhabditis elegans vulva development, 369–72
 definition, 348, 519G
 Drosophila melanogaster, 372–7
 plants, 375B
Developmental regulators, 267B
de Vries, H, 466
Dfd protein, 363–5
dGTP, 10, 11F
Diacylglycerol (DAG), 359, 360F
Diauxie, 354, 355F, 519G
Dicer, 308, 519G
Dictyostelium discoideum, 508T
Dideoxynucleotide, 156RB, 166–8, 167F, 519G
Differential centrifugation, 42TN, 519G
Differential splicing, 20, 20F, 22B, 77, 77F, 291–4, 292F, 293F, 519G
Differentiation, 348, 519G
Dihydrouracil, 439T
Dihydrouridine, 301T, 316
Dimethyl sulfate (DMS), 170B
Dimethyl sulfate modification interference assay, 245, 245F
Dimethyl sulfate modification protection assay, 244F
DinB, 444
Dinoflagellate, 46
Dinosaurs, 465, 466F
Dioxetane, 98TN
Diploid, 4, 519G
Directed mutagenesis, 204
Direct readout, 249, 252–3, 253F, 519G
Direct repair, 434–7, 436F, 519G
Direct repeat, 450, 451F, 519G
Discontinuous gene
 definition, 519G
 examples, 18, 19, 19F
 in *Drosophila melanogaster* genome, 33
 in *Saccharomyces cerevisiae* genome, 32
 number of exons, 19
Dispersive replication, 386–8, 386F, 387F, 520G
Displacement replication, 390, 390F, 519G
Display phage, 212F
Distance matrix, 492, 492F, 520G
Distance method, 491, 520G
Distinctiveness genes, 58–9
Disulfide bridge, 81, 520G
D-loop, 446, 520G
DMPK gene, 424T
DMS, 170B

DMS modification interference assay, 245, 245F

DMS modification protection assay, 244F

DNA
base-pairing, 14
base ratios, 13–14
base-stacking, 13
definition, 520G
density, 13
discovery, 5
double helix, 11–16, 15F
fibers, 13
genetic material, 6–8
origins of life, 462–3, 464F
structure, 8–10, 11–16
units of length, 16B
X-ray diffraction pattern, 13

DNA adenine methylase (Dam), 440, 520G

DNA amplification, 471, 472F

DnaA protein, 370RB, 391, 392F

DNA bending, 253, 520G

DnaB protein, 392, 392F, 397, 398F, 399F

DnaBC protein, 392, 392F

DNA-binding motif, 249–52, 249T, 520G

DNA-binding protein
definition, 520G
individual motifs, 248–52
interaction with DNA, 16
interaction with proteins, 252–4
locating binding positions, 242–5
purifying, 245–6
role, 240
special features, 248–52
studying DNA–protein complexes, 246–8

DNA chip
description, 133TN, 520G
in DNA sequencing, 172
in transcriptome studies, 78, 207–8, 209F
typing SNPs, 132

DNA cloning
definition, 520G
development, 97
in *Escherichia coli*, 109–17, 168–9
in organisms other than *E. coli*, 117–19
outline, 108F, 108–9
vectors, 109–19, 168–9

DNA cytosine methylase (Dcm), 440, 520G

DNA Database of Japan, 509T

DNA-dependent RNA polymerase, 74, 520G

DNA–DNA hybridization in molecular phylogenetics, 485, 495, 495F

DNA fiber, 13

DNA glycosylase, 241T, 437, 438F, 439T, 520G

DNA gyrase, 50, 388T, 389, 520G

DNA helicase II, 438, 439F, 440, 441F

DNA labeling, 98TN, 101T

DNA ligase, 97, 99F, 241T, 435, 436F, 437, 438F, 439, 439F, 440, 440F, 441F, 520G

DNA ligase IV, 442, 443F

DNA ligation, 105, 106F

DNA manipulation
cloning, 109–19
enzymes, 97–108
polymerase chain reaction, 119–22

DNA marker
definition, 129, 520G
restriction fragment length polymorphism, 130, 130F, 131F
simple sequence length polymorphism, 130, 132F
single nucleotide polymorphism, 130–2, 131B, 132F, 134F
use in molecular phylogenetics, 486

DNA methylation
5′–CG–3′ sequences, 149, 149F
definition, 520G
genome modification, 232–3RB, 234–6
mismatch repair, 440, 441F
transposition, 450B

DNA methyltransferase, 232–3RB, 234–5, 520G

DNA methyltransferase 1, 232–3RB

DNA methyltransferase 2, 232RB

DNA methyltransferase 3a, 232–3RB

DNA methyltransferase 3b, 232–3RB

DNA packaging, 35–7

DNA photolyase, 436, 520G

DNA-PK$_{CS}$ protein kinase, 442, 443F

DNA polymerase
activity, 97, 99F, 100F
definition, 521G
DNA-binding motif, 249T
DNA-binding proteins, 241T
DNA labeling, 98TN, 101T
mode of action, 10, 12F, 98–100, 100F, 101F
types, 395–6, 396T
types used in research, 100–2, 101T

DNA polymerase I, 100, 101T, 166B, 395, 396T, 400, 400F, 437, 438F, 439, 439F, 440, 441F, 521G

DNA polymerase II, 395, 396T, 444, 521G

DNA polymerase III, 395, 396F, 396T, 397–9, 399F, 400F, 332, 333F, 521G

DNA polymerase IV, 396T, 444

DNA polymerase V, 396T, 442, 444F

DNA polymerase α, 395, 396F, 396T, 397, 400, 521G

DNA polymerase β, 396T

DNA polymerase γ, 395, 396T, 521G

DNA polymerase δ, 395, 396F, 396T, 397, 400–1, 402F, 437, 521G

DNA polymerase ε, 396T, 413

DNA polymerase ζ, 396T, 444

DNA polymerase η, 396T, 444

DNA polymerase ι, 396T, 444

DNA polymerase κ, 396T, 404

DNA purification, 110TN

DNA repair
defects in repair, 444
definition, 418, 419F, 521G
direct repair systems, 434–7, 436F
double-strand break repair, 441–2, 442F, 443F
excision repair, 437–9, 438F, 439F, 440F
link to DNA replication, 413
mismatch repair, 440–1, 441F
mitochondria, 494
recombination repair, 441–2, 442F
single-strand break repair, 442, 443F
types of repair system, 435, 436F

DNA replication
accuracy, 420, 422F
archaea, 403B
definition, 521G
elongation, 393–401
initiation, 391–3
outline, 384–5, 385F
regulation, 409–13
telomere replication, 404–9
termination, 401–4
topological issues, 384–91

DNase I, 102T, 226–8, 243, 243F

DNase I hypersensitive region, 228, 228F, 521G

DNase I sensitive region, 226, 226F

DNA sequencing
assembling a contiguous sequence, 172–81
definition, 521G
methodology, 164–72
molecular phylogenetics, 485
of the human genome, 182
outline of strategies, 126–28, 128F

DNA synthesis, 394F

DNA topoisomerase, 50, 241T, 388T, 388–9, 389F, 398F, 521G

DNA transposon, 18, 19F, 24, 24T, 63–4, 451F, 451–4, 521G

dnaX gene, 332, 333F

Dnmt1, 232–3RB, 234

Dnmt2, 232RB

Dnmt3a, 232–3RB, 234

Dnmt3b, 232–3RB, 23–5

Dog, 88T, 508T

Domain duplication, 473, 474F, 521G

Domain shuffling, 472, 474F, 475F, 521G

Dominant, 134, 134F, 521G

Donor site, 288, 288F, 521G

Dot matrix method, 490–1, 491F, 521G

Double-bond saturation of nucleotides, 301T
Double Helix, 13
Double helix
    definition, 521G
    evidence for, 13–14
    importance, 8
    key features, 14
    structural flexibility, 14–16
Double heterozygote, 141, 521G
Double homozygote, 141, 521G
Double restriction, 147, 148F, 521G
Double-strand break recombination model, 447–8, 450F
Double-strand break repair, 436F, 441–2, 442F, 443F, 521G
Double-stranded RNA binding domain (dsRBD), 250B, 521G
Downstream, 21F, 522G
Drosophila melanogaster
    alternative splicing, 293F, 294
    complexity, 21
    development, 372–7
    developmental regulators, 267B
    DNA-binding proteins, 249T
    DNA methylation, 233RB
    early genetical studies, 129, 138–40
    errors in genome sequence, 181
    establishment of segmentation pattern, 374F, 374–5
    gene catalog, 43F
    gene density, 33T
    gene inactivation, 199
    gene regulation by feedback loop, 363–5
    gene silencing, 363, 365F
    genetic code, 87T
    genome project, 180–1
    genome sequence, 30, 31T
    genome size, 33T
    genome-wide repeat content, 33T
    homeotic selector genes, 375–6, 376F, 471–2
    homology analysis, 196–7
    insulators, 226, 226F
    introns per gene, 33T
    lateral gene transfer, 475
    maternal effect genes, 372–4, 373F
    mitochondrial genome, 48T
    organization, 32–3, 34F
    protein domains, 44T
    RNA-binding proteins, 250B
    segmentation pattern, 374, 374F
    sex determination, 293F, 294
    telomeres, 408B
DRPLA gene, 424T
D segment, 362, 362T, 363F
dsRBD, 250B, 521G
DSX protein, 293F, 294
dTTP, 10, 11F
Dystrophin gene, 288T

E

Early replication origin, 411RB, 412–13

*Eco*RI, 102, 103T
E-DNA, 15
eEF-1, 327T, 329–30
eEF-2, 327T, 330
EF-G, 325T, 330, 330F
EF-Ts, 325T, 330, 330F
EF-Tu, 325T, 329–30, 330F
Egg-laying defects phenotype, 203RB
eIF-1, 326, 327T
eIF-1A, 326, 327T
eIF-2, 326, 326F, 327T, 328
eIF-3, 326–7, 326F, 327T
eIF-4A, 326F, 327, 327T
eIF-4B, 326F, 327, 327T
eIF-4E, 326F, 327, 327T
eIF-4F, 327, 327T
eIF-4G, 327, 327T
eIF-5, 327T, 328
eIF-6, 327T, 328
Electroendosmosis value, 37TN
Electron density map, 246, 247F
Electron microscopy, 36, 46, 222–3, 223F, 322
Electrophoresis, 37TN, 522G
Electrostatic interactions, 82B, 253, 522G
ELL protein, 284, 285T
Elongation factors
    bacterial translation, 325T
    definition, 522G
    eukaryotic translation, 327T
    RNA polymerase I, 295
    RNA polymerase II, 284, 285T
    RNA polymerase III, 295
Elongation of translation, 328–33, 330F
Elongator protein, 284, 285T, 522G
Elongin protein, 285T
Embryonic lethal phenotype, 203RB
Embryonic stem (ES) cell, 199, 232RB, 522G
Emmer wheat, 475
Empty spiracles protein, 374, 374F
EMS, 426, 522G
End-modification enzymes, 97, 99F, 522G
End-modification of RNA, 76, 522G
Endogenous retrovirus (ERV), 62, 522G
Endonuclease, 102, 522G
Endoplasmic reticulum, 319
Endorphin, 341F
Endosymbiont theory, 46–9, 522G
engrailed gene, 376
Enhanceosome, 266, 522G
Enhancer, 264, 265F, 267B, 450B, 477, 522G
Enterobacter aerogenes, 88T
Enzymes for DNA manipulation
    DNA ligases, 105–7
    DNA polymerases, 98–102
    end-modification enzymes, 107–8
    nucleases, 102–4
Ephrin domain, 44T

Epidermal growth factor, 473, 475F
Epinephrine, 356T
Episome transfer, 57B, 145, 522G
EPM1 locus, 424T
eRF-1, 327T, 334, 335F
eRF-3, 327T, 334–5
ERVs, 62, 522G
ERV class I, 24T
ERV(K) class II, 24T
ERV(L) class III, 24T
Erythroid module, 267B
ES cell, 199, 232RB, 522G
Escherichia coli
    16S rRNA, 323F
    accuracy of genetic map, 145
    bacteriophage λ integration, 448–50, 451F
    bacteriophage cloning vectors, 111–3
    base ratios, 13
    catabolite repression, 354, 355F
    chromosome, 50
    cosmids, 113–14
    DNA-binding proteins, 249T, 250F
    DNA ligases, 105–7
    DNA polymerases, 100–2, 101T, 395, 396T
    DNA purification, 110TN
    DNA repair, 436, 437–9, 440
    DNA replication, 396–400, 402–3
    elongation of translation, 328–33, 330F
    end-modification enzymes, 107–8
    gene catalog, 58T, 58–9
    gene number, 33
    gene organization, 54F
    genetic mapping, 144–5, 144F, 146F
    genome organization, 34–5, 34F
    genome sequence, 30, 31T
    genome size, 33
    genomes of different strains, 56–7
    homologous recombination, 446–7, 447F, 447B
    host for bacteriophages, 8
    host for single-stranded DNA synthesis, 168–9
    identified genes, 195
    initiation of translation, 324–5, 326F, 328
    lac gene sequences, 190F
    lateral gene transfer, 57F
    molecular chaperones, 337–8, 338F
    non-coding DNA, 54
    nucleases, 102, 102T, 103T
    nucleoid, 50, 51F, 52RB
    ORF scanning, 189, 190F
    origin of replication, 391–2, 391F
    plasmid cloning vectors, 109–11
    plasmids, 51T
    pre-rRNA processing, 296F
    pre-tRNA processing, 297F

Escherichia coli continued
  programmed mutations, 433–4, 435RB
  promoter, 255–6, 256F, 256T
  replication error rate, 420–2
  ribosome binding site, 325T
  RNA polymerase, 255
  RNA synthesis, 274–82
  selenoproteins, 88T
  semiconservative DNA replication, 386–8, 387F
  termination of replication, 402–3, 403F, 404F
  termination of translation, 333–5, 334F
  transcription initiation, 257–8, 258F, 261–4, 263F, 264F, 265F
  unusual translation elongation events, 331, 333F
  vectors for long pieces of DNA, 117
ESE, 290–1, 292F, 293F, 294, 522G
E site, 330, 331F, 522G
ESS, 290–1, 523G
EST, 154, 232RB, 523G
Estradiol, 353F
Estradiol receptor, 353F
Estrogen receptor, 354F
Estrogen response element, 353F
Ethenoadenine, 439T
Ethenocytosine, 439T
Ethidium bromide, 37TN, 426, 426F, 522G
Ethylmethane sulfonate (EMS), 426, 522G
Euchromatin, 225, 522G
Eukaryotes
  definition, 30, 522G
  genomes, 31–3, 35–49
  origins, 465, 466F
Euplotes, 87T
European Bioinformatics Institute, 509T
even-skipped gene, 376
Excinuclease, 437
Excision repair, 436F, 437–9, 438F, 439F, 440F, 522G
Excisionase, 450
Exit site, 330, 331F, 522G
Exon, 18, 19, 21F, 473, 475F, 522G
Exon duplication, 473
Exonic splicing enhancer (ESE), 290–1, 292F, 293F, 294, 522G
Exonic splicing silencer (ESS), 290–1, 523G
Exon–intron boundary, 191, 194–5, 195F, 288, 288F, 523G
Exon shuffling, 473, 475F
Exon skipping, 289, 290F, 293F, 294, 523G
Exon theory of genes, 477, 477F, 523G
Exon trapping, 194, 195F, 523G
Exonuclease, 102, 523G
Exosome, 306, 523G

Exportin, 308, 523G
Exportin-t, 308
Expressed sequence tag (EST), 154, 232RB, 523G
External node, 487, 487F, 523G
Extrachromosomal gene, 46, 48T, 523G

F
Factor VIII gene, 288T
FACT protein, 284, 285T
Facultative heterochromatin, 225, 523G
Fanconi's anemia, 444
Farming, 500, 501F
Feedback loop, 363–5, 365F
FEN1 endonuclease, 400–1, 401F, 402F, 424, 523G
Ferritin, 89, 328, 329F
Fertile Crescent, 501F
Fertility plasmids, 51T
Fiber-FISH, 153, 523G
Fibrin, 473
Fibroblast culture, 407–8
Fibronectin, 473, 475F
Field inversion gel electrophoresis (FIGE), 150, 523G
Finger module, 473, 475F
Fish, 470, 472
FISH, *see Fluorescent* in situ *hybridization*
Fixation, 497B, 523G
Flap domain of bacterial RNA polymerase, 276, 279RB
Flap endonuclease, 400–1, 401F, 402F, 424
floricaula gene, 375B
Flow cytometry, 158F, 159, 523G
Flower development, 375B
FLpter value, 153, 523G
Fluorescence recovery after photobleaching (FRAP), 223, 224TN, 523G
Fluorescent *in situ* hybridization (FISH)
  definition, 145, 523G
  improved resolution, 153
  methodology, 152, 152F
  uses, 153
Fluorescent labeling, 133TN, 152–3, 156RB, 158F, 159, 170–1, 171F, 206, 209F, 224TN
Fluorescent marker, 98TN
Folding pathway, 337, 337F, 523G
Forest tree genomes, 509T
Formaldehyde, 460
Formamide, 152, 152F
Formamidopyrimidine, 439T
Formate dehydrogenase, 88T
Fosmid, 114T, 117, 524G
F plasmid, 51T, 117, 523G
Fragile site, 423, 524G
Fragile X syndrome, 424T
Fragile XE mental retardation, 424T
Frameshifting, 330–1, 524G

Frameshift mutation, 423, 524G
Franklin, R, 13
FRAP, 223, 224TN, 523G
Friedrich's ataxia, 424, 424T
Fritillaria assyriaca, 33T
FRM genes, 424T
Frog, 88T
Fruit fly, *see Drosophila melanogaster*
Fucose, 342F
Fugu rubripes, 214, 508T
Functional domain, 226, 524G
fushi tarazu gene, 376, 376F
Fusion protein, 211, 212F, 213F, 524G

G
G1 phase, 409, 409F, 524G
G2 phase, 409, 409F, 524G
GABP protein, 249T
Gain of function, 201, 432, 524G
GAL4 protein, 249T
Galactose, 55, 55F, 342F
galE gene, 325T
Gamete, 4, 137, 138F, 449F, 524G
GAP, 359, 359F, 524G
Gap 1 phase, 409, 409F, 524G
Gap 2 phase, 409, 409F, 524G
Gap gene, 374, 374F, 524G
GATA-1 protein, 267B
G-banding, 39T, 40F, 41
GC box, 267B, 524G
GCN4 protein, 249T
GCN5 protein, 229
Gel electrophoresis, 37TN, 524G
Gel retardation, 242, 242F, 524G
Gel stretching, 150–1, 151F, 524G
Genbank, 509T
Genes
  as markers, 129, 129T
  definition, 524G
  determining functions of, 195–206
  in populations, 497B
  locating in a genome sequence, 188–94
  made of DNA, 6–8
  origins, 463, 477, 477F
  unit factors, 5
Gene *60,* 333, 333F
Gene catalogs, 21F, 24–5, 43, 43F, 58T, 58–9
Gene cloning, *see DNA cloning*
Gene conversion, 360–1, 361F, 447–8, 449F, 524G
Gene densities, 33T, 41, 42F
Gene duplication, 470F, 470–2, 471B, 472F, 480
Gene families, 43–5
Gene flow, 56, 47B, 475, 524G
Gene fragment, 22, 23F, 524G
Gene inactivation, 198–200, 471B
Gene numbers, 31T
Gene overexpression, 200–1, 204F
Gene rearrangement, 20–1, 472–3, 473F, 474F, 475F
Gene segments, 18, 19F, 20–1
Gene substitutions, 497B, 525G

Gene superfamily, 45, 525G
Gene therapy, 119, 525G
Gene tree, 488–9, 489F, 490F, 525G
General recombination, 444–7, 445F, 446F, 447F, 448F, 525G
General transcription factor (GTF), 258–9, 259T, 261F, 525G
Genes-within-genes, 44B, 302, 303F, 525G
Généthon, 508T
Genetic code
    basic features, 84–6
    codon meanings, 86F
    decoding by wobble, 319, 321F
    definition, 525G
    elucidation, 85RB
    link between transcriptome and proteome, 84
    not universal, 86
    origin and evolution, 87B
    role, 84–6
    unusual codes, 86, 87B, 87T
Genetic drift, 489, 497B
Genetic footprinting, 199, 525G
Genetic linkage, 134, 525G
Genetic mapping
    definition, 128, 525G
    DNA markers, 129–32
    genes as markers, 129
    linkage analysis, 134–45
    of the human genome, 182
    resolution and accuracy, 145, 147F
Genetic marker, 129, 129T, 525G
Genetic profile, 60, 61F, 525G
Genetic redundancy, 471B, 525G
Genome activation, 228–31
Genome anatomies
    eukaryotic nuclear genomes, 35–45
    eukaryotic organelle genomes, 46–9
    overview, 30–5
    prokaryotic genomes, 49–59
    repetitive DNA, 59–64
Genome duplication, 465–70, 467F, 468–9RB
Genome evolution
    acquisition of new genes, 465–75
    gene duplication, 470–2
    gene rearrangement, 472–3
    genome duplication, 465–70
    human genome, 479–80
    introns, 477–9
    lateral gene transfer, 474–5
    non-coding DNA, 476–9
    origins of genomes, 450–63
    segmental duplications, 468–9RB
    transposable elements, 476–7
    web site, 509T
Genome regulation
    Bacillus sporulation, 366–9, 370RB
    Caenorhabditis elegans vulva development, 369–72
    chromatin structure changes, 362–3

Drosophila melanogaster, 372–7
    during development, 365–77
    feedback loops, 363–5
    genome rearrangements, 360–2
    permanent changes, 360–5
    plants, 375B
    transient changes, 350–60
Genome replication
    archaea, 403B
    elongation, 393–401
    initiation, 391–3
    regulation, 409–13
    telomere replication, 404–9
    termination, 401–4
    topological issues, 384–91
Genome sequences, 30, 31T
Genome silencing, 234–6
Genome sizes, 31, 33, 33T
Genome-wide repeats
    amount in different genomes, 33T
    blocking in a hybridization probe, 152, 153F
    centromere DNA, 38–9
    definition, 525G
    in *Saccharomyces cerevisiae* genome, 32
    in the human genome, 22–4, 24T
    problems in sequencing, 126, 127F
    types, 60–4
Genomic imprinting, 235, 236F, 525G
Genomic library, 114T
GFP, 206T, 224TN, 525G
Giant protein, 374F
Giemsa stain, 39T, 40F
Gigabase pair, 16B, 525G
Gliadin, 78, 89
Global regulation of protein synthesis, 328, 525G
Globin genes
    evolution, 470F, 471, 476, 478–9, 479F
    gene organization, 45F
    gene superfamily, 45
    multigene family, 44–5
Glucagon, 356T
Glucocorticoid receptor, 354F
Glucose, 55, 55F
Glutamate receptor, 304T
Glutamic acid, 82, 83F, 84T
Glutamine, 83F, 84T
Glutamine-rich domain, 266, 525G
Glutathione peroxidase, 88T
Glutenin genes, 475
Glycan, 342F
Glycine, 82, 83F, 84T
Glycine reductase, 88T
Glycolytic pathway, 55
Glycoprotein, 342, 342F
Glycosylation, 341T, 342, 342F, 525G
GNRP, 359, 359F, 525G
Gorilla, 476B, 480 487, 488F, 494–5, 495F
G-protein coupled receptor, 356T
Gramineae, 215

Grasses, 475
Grauzone protein, 372
Green fluorescent protein (GFP), 206T, 224TN, 525G
Griffith, F, 6
GroEL/GroES complex, 337–8, 338F
Group I intron, 287T, 296–8, 297F, 298F, 298T, 343, 462, 525G
Group II intron, 287T, 298T, 302B, 462, 478, 526G
Group III intron, 287T, 298T, 302B, 462, 526G
Growth-factor domain, 473, 475F
Growth hormone gene, 268, 268F, 269F
GrpE protein, 337
GTF, 258–9, 259T, 261F, 525G
GTP
    component of RNA, 10, 13F
    energy for protein synthesis, 325, 325T, 326, 326F, 328, 329–30, 330F, 334
GTPase activating protein (GAP), 359, 359F, 526G
GU–AG intron
    aberrant splicing, 289, 290F
    alternative splicing, 291–4, 292F, 293F
    definition, 526G
    conserved sequence motifs, 288, 288F
    origins, 478
GU–AG intron *continued*
    outline of splicing pathway, 288–9, 289F
    role of snRNAs and other proteins, 289–91, 290F, 291F, 292F
Guanidylate cyclase, 359
Guanine, 10, 11F, 526G
Guanine methyltransferase, 283, 285F, 526G
Guanine nucleotide releasing protein (GNRP), 359, 359F, 526G
Guanosine 5′-triphosphate, 10, 13F
Guanylyl transferase, 283, 285F, 526G
G–U base pair, 319, 320F, 321F
Guide RNA, 305B, 526G
Gyrase, 388T, 389

H
H1 isochore, 41
H1 histone, 36
H1a–e histones, 36
H1t histone, 36
H-16304 haplotype, 501F
H19 gene, 235, 236F
H19 RNA, 74B
H2 isochore, 41
H2A histone, 36, 38F, 230F, 236 284
H2B histone, 36, 38F, 229, 230F, 284
H3 histone, 36, 38F, 39, 229, 230F, 260RB, 341, 341F

H3 isochore, 41
H4 histone, 36, 38F, 230F, 236, 260RB
H5 histone, 36
Haemophilus influenzae
    gene catalog, 58T, 58–9
    genome project, 173–6, 174F
    lateral gene transfer, 57F
    restriction endonuclease, 102,
        103T
Hair cell, 294
Hammerhead structure, 298T, 526G
Hamster cells, 155, 158F
Haploid, 4, 526G
Haploinsufficiency, 432, 526G
Haplotype, 496, 500–1, 502F, 526G
hAT, 24T
HAT medium, 155
HDAC proteins, 231–4, 235, 235F,
    527G
HD gene, 423, 424T
Heat, mutagenic effects, 427, 428F
Heat shock, 262, 262F, 264
Heat-shock module, 267B
Heavy chain, 361–2, 362F, 362T
Heavy isotope labeling, 386–8, 387F
Heavy nitrogen labeling, 42TN
Hedgehog, 342
HeLa cell, 223F
Helicase, 259T, 277, 392, 397, 399F,
    526G
Helicobacter pylori, 57F, 211–12
Helix–loop–helix motif, 254, 526G
Helix–turn–helix, 249T, 249–50, 250F,
    526G
Hemoglobin, 44, 89, 247
HENNIG86, 491B
Hepatitis delta virus, 298T
Hereditary non-polyposis colorectal
    cancer (HNPCC), 444
Hermaphrodite, 371
Hershey, A, 8
HeT-A, 408B
Heterochromatin, 224–5, 363, 365F,
    526G
Heteroduplex, 445F, 445–6, 446F,
    526G
Heteroduplex analysis, 193–4, 195F,
    526G
Heterogenous nuclear RNA, 76,
    526G
Heteropolymer, 85RB, 526G
Heterozygous, 134, 134F, 526G
Hexaploid wheat, 475
Hexose-1-phosphate
    uridyltransferase, 325T
H haplotype, 501F
High incidence of males phenotype,
    203RB
High mobility group (HMG)
    domain, 249T
High mobility group N (HMGN)
    protein, 231B, 526G
High-performance liquid
    chromatography (HPLC),
    429TN, 526G

*Hinf*I, 102, 103, 103T
HIS2 gene, 212, 213F
his2 gene, 212, 213F
his10 gene, 212, 213F
Histidine, 82, 83F, 84T
Histones, 35–7, 228–9, 231–4, 241T,
    249T, 260RB, 341T, 526G
Histone acetylation, 228–9, 527G
Histone acetyltransferase, 284, 527G
Histone code, 229, 527G
Histone deacetylase (HDAC), 231–4,
    235, 235F, 527G
Histone deacetylation, 231–4, 236
Histone fold, 249T, 259
HIV-1, 304, 304T, 495–6, 496F
HIV-2, 496, 496F
HLA-B, 129
HLA-DRB1, 129
HLA genes, 129, 496
HMG domain, 249T
HMG14 protein, 285T
HMGN protein, 231B, 526G
HML loci, 361, 361F, 362
HNPCC, 444
hnRNA, 76, 526G
Hoechst 33258, 159
Holley, R, 315
Holliday model, 445F, 445–6
Holliday structure, 445F, 445–6,
    527G
Holocentric chromosomes, 39B,
    527G
HOM-C, 376F, 377, 471–2
homeless, 197F
Homeobox, 44T
Homeodomain, 249, 249T, 250B,
    251F, 376, 471, 527G
Homeotic gene, 363, 365F
Homeotic gene complex (HOM-C),
    376F, 377, 471–2
Homeotic mutant, 375, 527G
Homeotic selector genes
    definition, 527G
    Drosophila, 375–6, 376F
    evolution, 470, 471–2
    vertebrates, 376F, 377, 377F
Homo erectus, 479F, 496–7, 498F,
    499RB, 502
Homo habilis, 479F
Homologous chromosomes, 138,
    138F, 527G
Homologous genes, 191, 196, 527G
Homologous recombination
    definition, 527G
    description, 444–7, 445F, 446F,
        447F, 448F
    in DNA cloning, 118F, 119
    in gene inactivation, 198F, 198–9,
        199F
Homology, 191, 196
Homology analysis, 196–8, 197F,
    468RB
Homology search, 191, 527G
Homoplasy, 486B, 527G
Homopolymer, 85RB, 527G

Homopolymer tailing, 107, 107F,
    527G
Homozygous, 134, 134F, 527G
Hormone, 89
Hormone response element, 353,
    353F, 527G
Housekeeping gene, 234
Housekeeping protein, 80, 527G
Hox gene, 376F, 377, 377F, 470, 472
HP1 protein, 229
HPLC, 429TN, 526G
HPRT, 155
H-Ras protein, 359
Hsp40 protein, 337
Hsp70 chaperone, 267B, 337–8, 338F,
    527G
hsp70 gene, 226, 226F
Hucklebein protein, 374, 374F
HU/IHF motif, 249T
Humans
    evolutionary position, 484, 487,
        488F, 494–5, 495F
    migrations into Europe, 498–502,
        501F, 502F
    migrations into the New World,
        502F, 502–3
    origins of modern humans,
        496–8, 498F, 499RB, 500F
    pedigree analysis, 142–4, 143F
    prehistory, 496–503, 498F, 499RB,
        500F, 501F, 502F
    selenoproteins, 88T
    transcriptome, 79F, 80
Human genome
    base ratios, 13F
    content, 18–24
    disease genes, 215
    gene catalog, 41–3, 43F, 45F
    gene density, 33T, 41
    gene functions, 21, 21F
    gene location, 41
    gene number, 31
    genome-wide repeats, 18, 19F,
        22–4, 24T, 33T
    how many genes? 22B
    Human Genome Project, 16–18,
        24–5
    imprinting, 235, 236F
    introns per gene, 33T
    isochores, 41
    karyogram, 40F, 41
    lateral gene transfer, 475
    locus control region, 228, 228F
    microsatellites, 18–19, 19F, 24, 25T
    non-coding RNA, 74B
    organization, 23B
    protein domains, 44T
    pseudogenes and other
        evolutionary relics, 18, 19F, 22
    recent evolution, 479–80, 480F
    segmental duplication, 468–9RB
    size, 18, 18F, 33T
    wobble, 319, 321F
Human Genome Project
    accuracy, 182

Human Genome Project *continued*
definition, 527G
ethical considerations, 183
future, 182–3
human diversity projects, 183
mapping phase, 181–2
sequencing phase, 182
Human immunodeficiency virus 1
(HIV-1), 304, 304T, 495–6, 496F
Human immunodeficiency virus 2
(HIV-2), 496, 496F
Human mitochondrial genome
copy number, 46
genetic code, 86, 87T
human migrations into Europe,
500–2, 502F
human migrations into the New
World, 503
in intraspecific studies, 496
molecular clock, 494
origins of modern humans, 497,
500F
outline, 24, 25F
overlapping genes, 44B
primate evolution, 495, 495F
size, 46, 48T
Hunchback protein, 373–4, 373F
Huntington's disease, 423, 424T
HU protein, 50, 249T
HV haplotype, 501F
Hybridization analysis, 117TN,
131–2, 132F, 134F, 192, 192F,
193F
Hybridization probe, 104, 152, 153F
Hybridization probing, 176, 177F,
527G
Hydrogen bond, 14, 82B, 528G
Hydrogen cyanide, 460
Hydrogenosome, 47
Hydrophobic group, 337
Hydrophobic interaction, 14, 82B,
254F, 528G
Hydroxylation of proteins, 341T
Hydroxyproline, 473, 474F
Hyperactive phenotype, 203RB
Hyperediting, 304
Hypermethylation, 235, 236F, 450B
Hypermutation, 433, 434F, 528G
Hypoxanthine, 426, 426F, 437, 439T
Hypoxanthine phosphoribosyl
transferase (HPRT), 155
Hypoxanthine, aminopterin and
thymidine (HAT) medium, 155
hyRNA, 74B

I

Ice Age, 502
ICF, 233RB, 235
IF-1, 325, 325T, 326F
IF-2, 325, 325T, 326F
IF-3, 324, 325T, 326F
Igf2 gene, 235, 236F
IGH locus, 362T, 363F
IGK locus, 362T
IGL locus, 362T

I haplotype, 501F
IHF protein, 249T
Immediate-early transcripts of
bacteriophage λ, 280B
Immune response, 477
Immunocytochemistry, 206, 207,
528G
Immunodeficiency, centromere
instability and facial
anomalies (ICF), 233RB, 235
Immunoelectron microscopy, 322,
528G
Immunoglobulin, 20, 304T, 328,
361–2, 362F, 362T, 433, 434F, 442
Immunoglobulin diversity, 361–2,
362T, 363F
Immunological data in molecular
phylogenetics, 484, 485, 494
Immunological screening, 117TN,
528G
Importin, 308, 528G
Importin β, 309
Imprinted gene, 235, 236F
Incomplete dominance, 135, 528G
Indel, 490F, 490, 491F, 528G
Inducer, 263, 264F, 528G
Inferred tree, 488, 528G
Informational link between mRNA
and protein, 314
Informational problem, 84, 528G
Inhibition domain, 268
Inhibitor-resistant mutation, 433,
528G
Initiation codon, 86, 86F, 528G
Initiation complex for eukaryotic
translation, 327, 528G
Initiation factors
bacteria, 324, 325T, 326F
definition, 528G
eukaryotes, 327T
Initiation of translation
bacteria, 324–5, 326F, 328
eukaryotes, 325–8, 326F
regulation, 328, 329F
Initiation region, 393, 528G
Initiator (Inr) sequence, 156, 257F,
267B, 528G
Initiator tRNA, 324, 325F, 326, 326F,
528G
Inosine, 301T, 304, 319, 320F, 321F,
528G
Inositol-1,4,5-trisphosphate
$(\mathrm{Ins}(1,4,5)\mathrm{P}_3)$, 359, 360F
Insertional editing, 305B, 529G
Insertional inactivation, 109, 529G
Insertion mutation, 418, 430, 430F,
529G
Insertion sequence, 35, 63F, 64, 529G
Insertion vector, 112, 114F, 115F,
529G
In situ hybridization, 152
Instability element, 306–7, 529G
Insulator, 226–7, 226F, 227F, 529G
Insulin, 89, 267B, 288T, 339, 340F,
356T

Integrase, 449–50, 454F, 455, 529G
Integration host factor (IHF) protein,
249T
Integron, 53, 529G
Intein homing, 343, 343F, 529G
Intein splicing, 335, 342–3, 343F,
529G
Intercalating agent, 426, 426F, 529G
Interferons, 356
Intergenic region, 22, 529G
Interleukin, 89, 356
Internal node, 487, 487F, 529G
Internal ribosome entry site (IRES),
328, 529G
Internet resources, 507–8, 508T, 509T
Interphase, 137F, 138F, 153, 529G
Interspersed repeat, 23, 23F, 529G
Interspersed repeat element PCR
(IRE-PCR), 179, 179F, 529G
Intrinsic terminator, 276, 276F, 529G
Introns
definition, 18, 21F, 529G
genes-within-genes, 44B
human β-globin gene, 287F
lengths, 288T
number in *Saccharomyces cerevisiae*
genome, 32
number per gene, 19, 33T, 288T
origins, 287, 477F, 477–9, 478F,
479F
pre-RNA, 296–8, 298F
pre-tRNA, 298, 300F
problems in ORF scanning,
189–90F
splicing, 287–95
types, 287T
Intron homing, 343, 529G
Intron splicing
AU–AC introns, 294–5
differential splicing, 20, 20F, 22B,
77, 77F, 291–4, 292F, 293F
GU–AG introns, 287–94
location in nucleus, 223, 223F
pre-rRNA introns, 296–8, 297F,
298F
pre-tRNA introns, 298, 300F
synthesis of snoRNA, 302, 303F
Intron–exon boundary, 191, 194–5,
195F, 288, 288F
Introns early, 287, 477F, 477–9, 478F,
529G
Introns late, 287, 477–9, 529G
In vitro mutagenesis, 204, 205TN,
528G
In vitro packaging, 112, 528G
Ion channels, 356T
Ionizing radiation, 427
IRE-PCR, 179, 179F, 529G
IRES, 328, 529G
Iron-response element, 328, 329F,
529G
IS1, 34F, 35, 64
IS186, 34F, 35, 64
Islets of Langerhans, 339
Isoaccepting tRNAs, 315, 530G

Isochore, 41, 530G
Isoleucine, 83F, 84T

J

Jacob, F, 96, 262
Janus kinase (JAK), 357, 357F, 530G
J haplotype, 501, 501F
J segment, 362, 362T, 363F
Junk DNA, 22, 478B, 530G

K

Karyogram, 38, 40F, 530G
Karyopherin, 308, 530G
Kendrew, J, 247
K haplotype, 501F
Killer plasmid, 51T
Kilobase pair, 16B, 530G
KinA protein, 370RB
KinB protein, 370RB
KinC protein, 370RB
Kinetochore, 39, 39B, 41F, 530G
Kings College, London, 13
Klebsiella pneumoniae, 262
Klenow polymerase, 100–1, 101T, 166B, 530G
Knirps protein, 374, 374F
Knockout mice, 119, 199, 232–3RB, 530G
Kornberg, A, 395
Kornberg polymerase, 100, 101T, 530G
Kozak consensus sequence, 326F, 327, 530G
K-Ras protein, 359
Kringle structure, 473, 475F
Krüppel protein, 374F
Ku protein, 442, 442F, 443F

L

L1 isochore, 41
L1 ribosomal protein, 329F
L10 ribosomal protein, 325T
L11 ribosomal protein, 329F
L2 isochore, 41
Labeling RNA, 120TN
Labeling with heavy nitrogen, 42TN
Labeling with metallic beads, 169, 169F
labial palps gene, 375, 376F
Lactoferrin, 352
Lactose, 55–6, 55F
Lactose operator, 262–3, 264F
Lactose operon, 190F, 256F, 256T, 262–3, 264F, 325T, 333F, 354, 355F, 433F, 433–4, 530G
Lactose repressor, 249, 249T, 262–3, 264F, 433F, 530G
lacZ gene, 55, 55F, 206T
lacZ' gene, 111, 112F
Lagging strand, 395, 395F, 399F, 399–401, 400F, 402F, 406F, 530G
Lamarckian evolution, 435RB
Lariat structure, 289, 289F, 530G
Larval lethal phenotype, 203RB

Late replication origin, 411RB, 412–13
Lateral gene transfer, 57–8, 57F, 59F, 474–5, 530G
LCR, 227–8, 228F, 531G
Leading strand, 395, 395F, 399F, 399–401, 406F, 530G
leafy gene, 375B
Leaky mutation, 433, 530G
Leder, P, 85RB
Leghemoglobin, 478
Lethal mutation, 433, 530G
Leucine, 82, 83F, 84T
Leucine zipper, 254, 254F, 530G
Leukemia, 80
Ligase, *see DNA ligase*
Light chain, 361–2, 362F, 362T
Light microscopy, 222
LIN-3 protein, 371
LINE, 18, 19F, 24, 24T, 62, 62F, 530G
LINE-1, 24T, 62, 476, 530G
LINE-2, 24T
LINE-3, 24T
Linear DNA, 42TN
Linguistic studies, 503
Linkage analysis
    by human pedigree analysis, 142–4
    by planned breeding experiments, 140–2
    definition, 530G
    in bacteria, 144–5
    principles, 134–40
Linker, 107, 107F
Linker histone, 36–7, 38F, 530G
Linnaeus, C, 484
Liposome, 200, 201F
Lipotropin, 341F
Liverwort, 48T
Locus control region (LCR), 227–8, 228F, 531G
Locust, 33T
Lod score, 144, 531G
Long interspersed nuclear element (LINE), 18, 19F, 24, 24T, 62, 62F, 530G
Long patch excision repair, 439, 531G
Long patch mismatch repair, 440
Long terminal repeat, 453F, 454
Loss of function, 200, 431–2, 432F, 531G
LTR element
    description, 61–2, 531G
    in *Saccharomyces cerevisiae* genome, 32
    in maize genome, 33
    in the human genome, 18, 19F, 24, 24T
Luciferase, 206T
Lucy, 479F
Luria, S, 435RB
lux gene, 206T
Lymphoid cell module, 267B
Lysine, 82, 83F, 84T

Lysogenic infection cycle, 9F, 112, 113F, 531G
Lysozyme, 110TN, 531G
Lytic infection cycle, 9F, 112, 531G

M

Maaløe, O, 8
Mac1p protein, 352F, 353
MacClade, 491B
Machado–Joseph disease, 424T
MacLeod, C, 6
Macrochromosome, 39B, 531G
MacroH2A1 histone, 236
MADS box, 375B, 531G
Maintenance methylation, 232–3RB, 234, 234F, 531G
Maize, 33, 33T, 34F, 48T, 87T, 509T
Major groove, 16, 16T 252–3, 253F, 253–4, 531G
Malaria parasite, *see Plasmodium falciparum*
MALDI-TOF, 210F, 211, 531G
MaLR, 24T
Mandrill, 496F
Mannose, 342F
MAP kinase system, 358F, 358–9, 531G
Mapping genomes
    FISH, 152–3
    genetic mapping, 128–45
    physical mapping, 145–60
    reasons, 126–8
    restriction mapping, 145–52
    STS mapping, 153–60
Mapping reagent, 155–60, 531G
MAR, 225, 225F, 531G
Marchantia polymorpha, 48T
mariner, 475
Mass spectrometry, 208, 210F, 531G
Mass-to-charge ratio, 210F, 211
MAT gene, 361, 361F
Maternal-effect gene, 372–4, 373F, 531G
Mating type, 360, 531G
Mating-type switching, 360–1, 361F, 362, 531G
Matrix-assisted laser desorption ionization time-of-flight (MALDI-TOF), 210F, 211, 531G
Matrix-associated region (MAR), 225, 225F, 531G
Matthaei, H, 85RB
Maturase protein, 47F, 531G
Maximum parsimony, 493, 532G
MBD4 protein, 439T
McCarty, M, 6
McClintock, B, 64
MCM family of proteins, 412
MCM1 protein, 361
Measles, 304
Mechanically stretched chromosomes, 153
MeCP, 235, 235F, 532G
Mediator, 259, 266, 266F, 532G
Megabase pair, 16B, 532G

Mega-YAC, 114T
Meiosis
    aberrant, 465–6, 467F
    definition, 532G
    description, 137–8, 138F
    role in sporulation, 79F
Meischer, JF, 5
Mek protein, 358, 358F
Melanotropin, 341F
Mellitin, 339, 340F, 341T
Melting temperature, 132F, 532G
Mendel, G, 5, 129, 134–6
Mendel's experiments, 134–5, 134F
Mendel's Laws, 135, 135F
Meselson, M, 388
Meselson–Radding modification,
    446, 446F
Meselson–Stahl experiment, 386–8,
    387F, 532G
Messenger RNA, *see mRNA*
Metallic bead, 169, 169F
Metaphase, 137F, 138F
Met-encephalin, 341F
Methanobacterium thermoautrophicum,
    57F
Methanococcus, 88T
Methanococcus jannaschii, 31T, 56, 57F
Methionine, 83F, 84T
Methylation
    of histones, 229
    of DNA, 232–3RB, 234–6
    of nucleotides, 301T
    of proteins, 341T
    of rRNA, 302, 303F
Methyl-CpG-binding protein
    (MeCP), 235, 235F, 532G
Methylguanine, 244F, 245, 245F
MetJ repressor protein, 249T, 252F
Metridium senile, 48T
MF1 endonuclease, *see FEN1*
    endonuclease
MGMT, 436, 532G
Microarray, 78, 133TN, 208, 209F,
    410–11RB, 532G
Micrococcus, 87T
Microsatellite
    description, 60, 532G
    in genetic mapping, 130, 132F,
        142–4, 143F
    in intraspecific studies, 496
    in the human genome, 18–19, 24,
        25T
    origins, 476B
Microsome, 321
Microtubule, 41F, 137F
Migrations of humans, 498–502,
    501F, 502F
Minichromosome, 39B, 532G
Minigene, 194, 195F, 532G
Minimal genome content, 58–9
Minimal medium, 432F, 532G
Minisatellite, 60, 130, 132F, 532G
Minor groove, 16, 16T, 252–3, 253F,
    532G
MIR, 24, 24T

MIR3, 24, 24T
Mismatch repair, 436F, 440–1, 441F,
    532G
Missense mutation, 430, 532G
Mitochondrial Eve, 497
Mitochondrial genome
    definition, 532G
    genetic code, 86, 87B, 87T
    genetic content, 46, 47F, 48T
    human migrations into Europe,
        500–2, 502F
    human migrations into the New
        World, 503
    in intraspecific studies, 496
    molecular clock, 494
    origins of mitochondrial
        genomes, 46–9
    origins of modern humans, 497,
        500F
    overlapping genes, 44B
    physical features, 46
    primate evolution, 495, 495F
    RNA editing, 305B
    sizes, 46, 48T
    summary, 4
    superwobble, 319
    tRNAs, 316
    why have them? 46
Mitochondrial RNA polymerase,
    255B
Mitogen activated protein (MAP)
    kinase system, 358F, 358–9,
    531G
Mitosis, 137, 137F, 409, 409F, 532G
Mitotic cyclins, 412, 412F
Mnt repressor protein, 249T
Model organism, 199, 532G
Modification interference assay,
    244–5, 245F, 533G
Modification protection assay, 242–4,
    243F, 244F, 533G
Modified nucleotides, 301T
Molecular biology, 6
Molecular chaperone, 89, 337–8,
    338F, 533G
Molecular clock, 489, 493–4, 494F,
    494–6, 497, 533G
Molecular combing, 150, 151, 151F,
    533G
Molecular life sciences, 25, 533G
Molecular phylogenetics
    applications, 494–503
    definition, 484, 533G
    origins, 484–6
    phenetics and cladistics, 486B
    problems with lateral gene
        transfer, 58, 59F
    software, 491B
    tree reconstruction, 487–94
Molting phenotype, 203RB
Monod, J, 96, 262, 354
Monophyly, 488B, 533G
Morgan, TH, 137–40, 372
Mosquito, 508T
Mot1 protein, 268

Mother cell, 366, 367F
Motility proteins, 89
Mouse
    genome size, 33T
    mitochondrial genome, 48T
    radiation hybrid mapping,
        156–7RB
    selenoproteins, 88T
    web sites, 508T
MPG protein, 439T
M phase, 409, 409F, 531G
mRNA degradation
    bacteria, 305–6, 306F
    eukaryotes, 306–8, 307F
    role in genome expression, 71,
        72F
mRNA stability, 286–7
mRNA surveillance, 307, 307F, 533G
mtTF1 protein, 255B
Multicopy cloning vector, 201
Multicysteine zinc finger, 249T,
    250–1, 251F, 533G
Multigene family, 43–5, 196, 533G
Multipartite genome, 34
Multiple alignment, 491–2, 492F,
    533G
Multiple alleles, 129, 533G
Multiple hits, 492F, 533G
Multiple substitutions, 492F, 533G
Multiple vulvas phenotype, 203RB
Multipoint crosses, 141B, 533G
Multiradiate chromosome, 233RB
Multiregional evolution, 497, 498F,
    533G
Muscle development, 363
Muscular dystrophy, 290
Mutagen
    definition, 418, 424, 424T, 533G
    general effect on replication, 421F
    types, 424–7, 425F, 426F, 427F,
        428F
Mutasome, 442, 533G
Mutation
    causes, 420–8
    creation of pseudogene, 22
    definition, 418, 419F, 534G
    detection, 429TN
    effects on genomes, 428–31
    effects on microorganisms, 432–3
    effects on multicellular
        organisms, 431–2
    hypermutation, 433
    programmed mutations, 433–4,
        435RB
Mutation scanning, 429TN, 534G
Mutation screening, 429TN, 534G
MutH protein, 440, 441F
MutHY protein, 440
MutL protein, 440
MutS protein, 440, 441F
Mycobacterium tuberculosis, 30, 31T,
    57F
Mycoplasma, 86, 87B, 87T
Mycoplasma genitalium
    gene catalog, 58T, 58–9

Mycoplasma genitalium continued
  genome project, 176
  genome sequence, 31T
  lateral gene transfer, 57F
  rRNA genes, 472
Mycoplasma pneumoniae, 57F
Myoblast module, 267B
MyoD protein, 267B, 363
Myoglobin, 247, 470F, 471
Myosin, 89
Myotonic dystrophy, 215T, 424T

N

N-2-acetylaminofluorene, 444
N-acetylgalactosamine, 342F
N-acetylglucosamine, 342F
Nanos protein, 373–4, 373F
National Institutes of Health, 85RB
Native Americans, 502–3
Natural selection, 497B, 534G
NC2 protein, 268
N-CoR protein, 268, 268F, 269F
N-degron, 344, 534G
Neandertals, 499RB
Neighbor-joining method, 492F,
  492–3, 534G
Neisseria meningitidis, 35
Neolithic, 500
Neuroblastoma, 79F
Neurodegenerative diseases, 423
Neurofibromatosis, 44B, 304T, 215T
Neurospora crassa, 46, 508T
NF-1 protein, 267B
NF-κB protein, 249T, 267B, 357–8
N-formlylation of proteins, 341T
N–formylmethionine, 317, 318F, 324,
  326F
NF-Y protein, 267B
N-glycolyl-neuraminic acid, 479–80
N-glycolyl-neuraminic acid
  hydroxylase, 479
NHEJ, 442, 443F, 444, 534G
Nicks, 436, 436F, 534G
Nicotiana tabacum, 48T
NiFeSe hydrogenase, 88T
Nirenberg, M, 85RB
Nitrocellulose membrane, 103–4
Nitrogen fixation, 197, 262
Nitrous acid, 426
NJplot, 491B
N-linked glycosylation, 341T, 342,
  342F, 534G
NMD, 307, 307F, 534G
NMR spectroscopy, 248, 248F, 534G
N-myristoylation of proteins, 341T
N, N–methylenebisacrylamide (bis),
  165TN
N, N, N', N'–tetramethylethylene-
  diamine (TEMED), 165TN
NOESY, 248
Non-coding RNA
  chemical modifications, 298–304
  definition, 19, 534G
  differences between human and
    chimpanzee, 480

  evolution, 476B, 476F, 476–9,
    477F, 478F, 479F
  introns in non-coding RNA,
    296–8
  role, 479B
  synthesis and processing, 295–6
  types, 72–3, 74B
Non-Hodgkin lymphoma, 80
Non-homologous end-joining
  (NHEJ), 442, 443F, 444, 534G
Non-penetrance, 432, 534G
Non-radioactive marker, 98TN
Nonsense mutation, 430, 430F,
  433–4, 534G
Nonsense-mediated RNA decay
  (NMD), 307, 307F, 534G
Non-synonymous mutation, 429,
  430F, 494, 534G
Northern hybridization, 120TN, 192,
  192F, 534G
*Not*I, 103T, 149
N protein, 280B
N-Ras protein, 359
N terminus, 81, 81F, 534G
NTH1 protein, 439T
Nuclear magnetic resonance (NMR)
  spectroscopy, 248, 248F, 534G
Nuclear matrix, 223, 223F, 225F,
  534G
Nuclear pore complex, 308, 308F,
  534G
Nuclear receptor superfamily, 353,
  534G
Nuclease, 97, 99F, 102T, 102–4, 241T,
  534G
Nuclease footprinting, 243, 243F
Nuclease protection experiment, 35,
  36F, 322, 534G
Nucleic acid, 5, 534G
Nucleoid, 50, 51F, 52RB, 241T, 534G
Nucleolus, 223, 223F, 535G
Nucleoprotein, 6
Nucleoside, 10, 11F, 535G
Nucleosome, 36–7, 38F, 229–31, 230F,
  231B, 259, 260RB, 535G
Nucleosome remodeling, 229–31,
  230F, 535G
Nucleosome remodeling complex,
  230–1
Nucleotidase, 172, 172F
Nucleotide, 9–10, 11F, 535G
Nucleotide excision repair, 437–9,
  439F, 440F, 535G
Nucleotide modifications, 301T
Nucleotide replacement, 301T
Nucleus, 35, 36F, 222–4, 223F, 224–8,
  308F, 308, 309, 535G
NuRD complex, 231–4, 235
Nylon membrane, 103–4

O

O⁶-methylguanine-DNA
  methyltransferase (MGMT),
  436, 532G
Oct-1 protein, 249T, 267B

Oct-2 protein, 249T
Octamer module, 267B, 535G
Odorants, 356T
Oenothera gigas, 466
Oenothera lamarckiana, 466
OFAGE, 149–50, 150F, 535G
OGG1 protein, 439T
Okazaki fragment, 397, 399–401,
  400F, 402F, 404–5, 406F, 408F,
  535G
Oligonucleotide, 99, 535G
Oligonucleotide-directed
  mutagenesis, 205TN, 535G
Oligonucleotide hybridization
  analysis, 131–2, 132F, 134F,
  172, 245–6, 535G
Oligonucleotide synthesis, 133TN
O-linked glycosylation, 341T, 342,
  342F, 535G
Oncogen, 424T
Open promoter complex, 257, 257F,
  535G
Open reading frame (ORF), 189,
  189F, 535G
Operational taxonomic unit (OTU),
  488B, 535G
Operator, 262–3, 264F, 265F, 535G
Operon
  definition, 35, 535G
  discovery, 96
  lactose operon, 55–6, 55F
  tryptophan operon, 55F, 56
  unusual ones, 56, 56F
Optical mapping, 150–2, 150F, 535G
Orangutan, 476B, 487, 488F, 494F
ORC, 392, 393F, 409, 535G
ORF, 189, 189F, 535G
ORF scanning, 189–91, 190F, 535G
Organelle genomes
  genetic content, 46, 47F, 48T, 49F
  origins, 46–9
  physical features, 46
Organelle RNA polymerase, 255B
oriC, 391, 392F
Origin firing, 411RB, 412–13
Origin of replication
  archaea, 403B
  component of yeast artificial
    chromosome, 114, 116F
  definition, 535G
  Escherichia coli, 391–2, 391F
  higher eukaryotes, 393
  in cloning vectors, 109
  Saccharomyces cerevisiae, 392–3,
    393F
Origin of Species, 484
Origin recognition protein, 241T
Origin replication complex (ORC),
  392, 393F, 409, 535G
Origins of life, 450–63
Orphan families, 197F, 198, 535G
Orthodenticle protein, 374, 374F
Orthogonal field alternation gel
  electrophoresis (OFAGE),
  149–50, 150F, 535G

Orthologous genes, 196, 488, 536G
Oryza sativa, 33T, 46, 48T, 49F, 215
Osteoporosis, 201
OTU, 488B, 535G
Outgroup, 487, 488F, 536G
Out of Africa hypothesis, 497, 499RB, 536G
Ovarian cancer, 444
Overlapping genes, 44B, 536G
Oxytricha, 405T

P
P1-derived artificial chromosome, 114T, 117, 536G
p300/CBP protein, 229, 359
p53 protein, 413
p55 protein, 229
PADP, 286, 286F, 327, 537G
Paired homeodomain, 249T
Pair-rule gene, 374, 536G
PAML, 491B
Pancreas, 339
Pancreatic cancer, 80
Pan-editing, 305B, 536G
Papillomavirus E2 protein, 249T
pap operon, 306
Paralogous genes, 196, 536G
Paralyzed phenotype, 203RB
Paramecium, 46
Paramyxovirus, 305B
Paranemic, 385, 536G
Parental genotype, 139, 536G
PARP1 proteins, 442F
Parsimony, 488B, 536G
Partial linkage
    basis to genetic mapping, 139–40
    definition, 536G
    discovery, 136F, 136–7
    explanation for, 137–9
Partial restriction, 147, 148F, 536G
Pauling, L, 13
PAUP, 491B
Pax proteins, 249T
pBIN19, 119, 119F
pBR322, 109, 109F
PCNA, 396, 400, 413, 539G
PCR, *see Polymerase Chain Reaction*
PDHA1 gene, 497
Pea, 33T, 48T, 134–7, 134F, 136F
Pedigree analysis, 142–4, 536G
P element, 199, 475, 536G
Pentose, 9, 11F, 536G
Peptide bond, 81, 81F, 536G
Peptide hormone, 340, 341F
Peptide nucleic acid (PNA), 464, 465F, 536G
Peptidyl site, 329–30, 330F, 331F, 536G
Peptidyl transferase, 298T, 330, 332RB, 536G
Permissive conditions, 432, 536G
Perutz, M, 247
PEST sequence, 344, 536G
Phage, *see Bacteriophage*
Phage display, 211, 212F, 536G

Phage display library, 211, 212F, 536G
Phagemid, 169, 536G
Phenetics, 484, 486B, 537G
Phenotype, 129, 200B, 537G
Phenylalanine, 82, 83F, 84T
Phosphate group, 10, 11F
Phosphatidylinositol-4,5-bisphosphate (PtdIns(4,5)P$_2$), 359, 360F
Phosphodiester bond, 10, 11F, 105, 537G
Phosphodiesterase, 427, 438F, 537G
Phospholipase, 359, 360F
Phosphorimaging, 98TN, 133TN, 537G
Phosphorylation of proteins, 269, 341T
Photobleaching, 224TN, 537G
Photolithography, 133TN, 537G
Photolyase, *see DNA photolyase*
Photoproduct, 426–7, 427F, 436, 442, 444, 537G
Photoreactivation, 436, 537G
PHYLIP, 491B
Phylogenetic Analysis by Maximum Likelihood (PAML), 491B
Phylogenetic Analysis Using Parsimony (PAUP), 491B
Phylogenetic tree
    bootstrap analysis, 493, 493F
    gene trees and species trees, 488–9, 489F, 490F
    key features, 487F, 487–8
    molecular clock, 489, 493–4, 494F, 494–6, 497
    outgroup, 487, 488F
    tree reconstruction, 489–93, 490F, 491F, 492F
Phylogeny, 484, 537G
Phylogeny Inference Package (PHYLIP), 491B
Physical gap, 173–6, 175F, 180, 181F
Physical mapping
    comparison of mouse and human maps, 157RB
    definition, 128, 537G
    FISH, 152–3
    of the human genome, 182–3
    reasons for, 145
    restriction mapping, 145–52
    STS mapping, 153–60
Picornaviris, 328
PiggyBac, 24T
Pilus, 144F, 537G
Piperidine, 170B, 244F, 245F
pistillata gene, 375B
Pisum sativum, 33T, 48T
Pit-1 protein, 249T, 267B, 268, 268F, 269F
Pituitary cell module, 267B
Pituitary gland, 340
Plant cloning vectors, 119, 119F
Plant mitochondrial genomes, 46, 48T

Plaque, 113, 115F, 537G
Plasmid
    component of bacterial genomes, 51–3, 53T
    definition, 33, 35F, 537G
    features, 51T
    functions, 33–4
    role in DNA cloning, 96, 97F, 108F, 109–11
Plasmid cloning vector
    cosmids, 113–14, 114T, 115F
    for single-stranded DNA, 168
    pBR322, 109, 109F
    phagemids, 169
    pUC8, 111, 112F
    recombinant selection, 109–11, 111F, 112F
Plasmid transfer, 474–5
Plasminogen, 473, 475F
Plasmodium falciparum
    genome sequence, 30, 31T
    mitochondrial genome, 46, 47T
    optical mapping, 151
    web site, 508T
Plectonemic, 385, 537G
Plesiomorphic character state, 486B, 537G
pMB1 plasmid, 109
PNA, 464, 465F, 536G
Pneumonia, 6
PNPase, 305–6, 306F
Point mutation, 131B, 418, 419B, 420–2, 421F, 422F, 537G
pol gene, 454–5
polA gene, 395
Poliovirus, 328
Polyacrylamide gel electrophoresis, 37TN, 165, 165TN, 166F, 537G
Polyadenylate-binding protein (PADP), 286, 286F, 327, 537G
Polyadenylation, 286–7, 286F, 287F, 537G
Polyadenylation editing, 305B, 537G
Poly(A) polymerase, 74, 286, 537G
Poly(A) tail, 76, 193, 194F, 286F, 286–7, 306–7, 325–8, 537G
Polycomb protein, 363, 365F, 376
Polycomb response element, 363, 365F
Polyethylene glycol, 155
Polyglutamine codons, 423
Polymerase chain reaction (PCR)
    analysing results, 121, 122F
    applications, 121
    carrying out a PCR, 120–1
    definition, 538G
    development, 97
    in chromosome walking, 176–8, 177F
    in DNA sequencing, 169, 169F, 170, 171F
    in radiation hybrid analysis, 156RB
    in RFLP analysis, 130, 131F
    in RNA interference, 202RB

Polymerase chain reaction (PCR)
continued
in sequence assembly, 175F, 176
in SSLP analysis, 130, 132F
in transcript mapping, 193, 194F
of RNA, 120TN
outline, 121F, 122F
studying a clone library, 117TN
Polymerase cleft, 249T
Polymerase I and transcript release factor (PTRF), 295
Polymorphic loci, 496, 497B, 538G
Polynucleotide, 9–10, 11F, 461–3, 463F, 464F, 538G
Polynucleotide kinase, *see T4 polynucleotide kinase*
Polynucleotide phosphorylase (PNPase), 305–6, 306F
Polypeptide, 80, 538G
Polyphyly, 488B, 538G
Polyploidy, 466, 467F, 475
Polyprotein, 339–40, 340F, 538G
Polypyrimidine tract, 288, 288F, 538G
Population genetics, 497B
Positional cloning, 178, 538G
Positional effect, 226, 226F, 538G
Post-embryonic phenotype, 203RB
Post-replication complex (post-RC), 409, 538G
POU domain, 249T, 250, 268, 268F, 269F, 538G
Pre-initiation complex
eukaryotic translation, 326F, 326–7, 538G
transcription, 259, 261F, 538G
Pre-mRNA, 19–20, 76, 538G
Pre-mRNA introns, 288–95
pre-mRNA processing
capping, 283, 285F
intron splicing, 287–95
polyadenylation, 286–7, 286F, 287F
RNA editing, 303–4
Prepriming complex, 392, 538G
Preproinsulin, 339, 340F
Pre-replication complex (pre-RC), 409–12, 538G
Pre-RNA, 74–5, 76, 538G
Pre-rRNA, 76, 538G
Pre-rRNA introns, 296–8, 297F, 298F
Pre-rRNA processing
chemical modifications, 299–303, 303F
cutting events, 295–6, 296F
introns, 296–8, 297F, 298F
Pre-spliceosome complex, 290, 291F, 538G
Prespore, 366, 367F
Pre-tRNA, 76, 538G
Pre-tRNA introns, 287T, 298, 300F
Pre-tRNA processing
chemical modifications, 298, 301T
cutting events, 295–6, 297F
introns, 298, 300F

PriA protein, 397
Primary level of protein structure, 81, 81F
Primary vulva cell fate, 371, 371F
Primase, 396F, 397, 538G
Primer, 99, 100F, 166, 167F, 169–70, 538G
Priming of DNA synthesis, 396F, 396–7
Primosome, 397, 538G
Principal component analysis, 500, 539G
proboscipedia gene, 376F
Processed pseudogene, 22, 22F, 539G
Processivity, 166B, 539G
Progesterone receptor, 354F
Programmed frameshifting, 331, 333F, 539G
Programmed mutations, 433–4, 435RB, 539G
Progressive myoclonus epilepsy, 424T
Proinsulin, 339, 340F
Prokaryote
definition, 30–1, 539G
genomes, 33–5, 49–59
Prolactin gene, 268, 268F, 269F
Proliferating cell nuclear antigen (PCNA), 396, 400, 413, 539G
Proline, 83F, 84T
Proline-rich domains, 266, 539G
Promiscuous DNA, 47–9, 539G
Promoter
bacterial, 255–6, 256F, 256T
definition, 74, 539G
eukaryotic, 256–7
RNA polymerase I, 256, 257F
RNA polymerase II, 256, 257F, 267B
RNA polymerase III, 256–7, 257F
web site, 509T
Promoter clearance, 257, 257F, 283, 284F, 539G
Promoter escape, 283, 284F, 539G
Proofreading, 100, 101F, 395, 422, 422F, 495, 539G
Pro-opiomelanocortin, 340, 341F
Prophage, 449, 539G
Prophase, 137F, 138F
Protease, 8, 8F, 539G
Proteasome, 344, 539G
Protection assay, 242–4, 243F, 244F
Protective protein, 89
Protein
content of cell, 80–9
copy numbers, 80
definition, 539G
domains, 88
housekeeping proteins, 80
importance of amino acid sequence, 88–9
link between transcriptome and proteome, 83–8
link with biochemistry of cell, 88–9

multiplicity of protein function, 89
structure, 80–3
Protein degradation, 71, 72F, 343–4
Protein domains, 42, 44T, 472–3, 473F, 474F, 475F
Protein electrophoresis, 208, 210F, 485, 539G
Protein engineering, 204, 539G
Protein folding
definition, 335, 539G
molecular chaperones, 337–8, 338F
outline, 88–9, 89F
role in genome expression, 71, 72F
test-tube folding, 335–7, 336F
Protein interaction map, 211–14, 214F, 539G
Protein kinase A, 359
Protein mobility, 223–4, 224TN
Protein processing
chemical modification, 340–2
folding, 335–8
intein splicing, 342–3
proteolytic cleavage, 339–40
role in genome expression, 71, 72F
Protein sequencing in molecular phylogenetics, 485
Protein synthesis
role of ribosome, 319–35
role of tRNA, 314–19
Protein–protein crosslinking, 322, 539G
Proteolytic cleavage
cleavage of ends, 339, 340F
definition, 335
polyprotein processing, 339–40, 340F
Proteome
definition, 25, 70, 71F, 539G
folding, 335–8
processing, 339–43
protein degradation, 343–4
studying, 208–11, 210F
synthesis, 314–35
Proteomics, 208–11, 210F, 539G
Protogenome, 462, 539G
Protoplast, 119, 539G
Prototroph, 432, 539G
Protovertebrate, 465, 472
Protruding vulva phenotype, 203RB
Pseudogene, 22, 22F, 45F, 539G
Pseudomonas, 31T, 51T, 118–19, 475
Pseudouridine, 299, 301T, 316
Pseudouridinylation of rRNA, 302
P site 329–30, 330F, 331F, 536G
*Pst*I, 103, 103T, 104F
PtdIns(4,5)P$_2$, 359, 360F
PTRF, 295
pUC8, 111, 112F
Puffer fish, 33T, 214, 508T
Punctuation codon, 86, 86F, 540G
Purine, 10, 11F, 460, 540G

Pyranosyl RNA, 464
Pyrimidine, 10, 11F, 460, 540G
Pyrococcus abyssi, 403B
Pyrococcus horikoshii, 57F
Pyrophosphate, 12F, 172, 172F
Pyrosequencing, 172, 172F, 540G

Q

Q-banding, 39T
Q protein, 280RB
Quantitative PCR, 499RB, 540G
Quaternary level of protein
structure, 81–2, 540G
Queosine, 298, 301T
Quinacrine, 39T

R

R1 plasmid, 109
R6.5 plasmid, 109
RACE, 193, 194F, 540G
RAD27 gene, 242
RAD51 protein, 448
Radiation hybrid, 155–9, 156–7RB,
158F, 508T, 540G
Radioactive labeling, 8, 10F, 152,
170–1, 540G
Radioactive marker, 98TN, 540G
Raf protein, 358, 358F, 359F
Random genetic drift, 489, 497B,
540G
Rapid amplification of cDNA ends
(RACE), 193, 194F, 540G
Ras system, 359, 359F, 540G
Rat, 88T, 508T
Ray-finned fish, 472
R-banding, 39T
Rbk plasmid, 51T
RDE-1 nuclease, 201F, 308
Reading frame, 189, 540G
Readthrough, 430, 430F, 540G
Reb1p protein, 266, 295, 296F
RecA protein, 241T, 442, 444F, 446,
447F, 448, 540G
RecBCD pathway, 446–7, 447F, 540G
RecE pathway, 446, 447B
Recessive, 134, 134F, 540G
RecF pathway, 446, 447F
RecG protein, 447
Reciprocal strand exchange, 445F,
446, 540G
Reclinomonas americana, 46, 47T
Recognition helix, 249, 252B, 540G
Recombinant cloning vector, 109
Recombinant DNA technology, 96,
540G
Recombinant genotype, 139
Recombinant selection, 109–10, 111,
111F, 112F
Recombinase, 449, 540G
Recombination
between transposable elements,
476, 476F
definition, 418, 419F, 541G
double-strand break model,
447–8, 450F

during meiosis, 138
gene conversion, 447–8, 449F
homologous recombination,
444–7, 445F, 446F, 447F, 448F
in gene duplication, 471, 472F
retrotransposition, 453F, 454F,
454–5
site-specific recombination,
448–50, 451F
transposition, 450–5, 451F, 452F,
453F, 454F
Recombination frequency, 139, 140F,
541G
Recombination hotspot, 139, 541G
Recombination repair, 436F, 441–2,
442F, 541G
Regulatory control of bacterial
transcription initiation, 262–4,
264F, 265F, 541G
Regulatory mutation, 433, 541G
Release factors
bacteria, 325T, 334, 334F
definition, 541G
eukaryotes, 327T, 334, 335F
Rel homology domain, 249T
Rel protein, 357–8
Repetitive DNA, 59–64, 541G
Repetitive DNA fingerprint, 178,
179F, 541G
Repetitive DNA PCR, 179, 179F,
541G
Repetitive DNA, blocking, 152, 153F
Replacement vector, 112, 114F, 114T,
541G
Replica plating, 110, 111F, 541G
Replication error, 420–4, 421F, 422F,
423F
Replication factor C (RFC), 400, 413,
541G
Replication factory, 401, 402F, 541G
Replication fork, 385F, 397–401, 541G
Replication licensing factor (RLF),
412, 541G
Replication mediator protein (RMP),
399, 541G
Replication origin
archaea, 403B
definition, 541G
Escherichia coli, 391F, 391–2
higher eukaryotes, 393
Saccharomyces cerevisiae, 392–3,
393F
Replication protein A (RPA), 398,
398F, 541G
Replication slippage, 423, 423F, 471,
541G
Replicative form, 168, 541G
Replicative transposition, 63F, 63–4,
451F, 451–4, 452F, 541G
Replisome, 400, 541G
Reporter gene, 206, 206F, 206T, 541G
Rep protein, 397
Repressor
bacterial, 262–3, 264F, 265F
DNA-binding proteins, 241T

eukaryotic, 266–9
Research journals, 507, 507T
Resistance plasmid, 51T
Resolution of the Holliday structure,
445, 445F, 541G
Resonance frequency, 248
Respiratory complex, 24
Response module, 267B, 542G
Restriction endonuclease
activity, 102, 102F, 104F
definition, 102T, 542G
DNA-binding proteins, 241T
examples, 103T
in RFLP analysis, 130
recognition sequences, 103T
restriction digest, 103–4, 105F
types of end, 103, 104F
web site, 509T
Restriction fragment length
polymorphism (RFLP), 130,
130F, 131F, 146–7, 147F, 181,
486, 497, 500F, 502–3, 542G
Restriction mapping
definition, 145, 542G
direct examination of DNA
molecules, 150–2
increasing the scale, 147–50
methodology, 147, 148F
Restrictive conditions, 433, 542G
Retinoblastoma protein, 231
Retroelement, 61–3, 408B, 453F, 454F,
454–5, 542G
Retroposon, 62–3, 408B, 542G
Retrotransposition, 453F, 454F,
454–5, 542G
Retrotransposon, 62, 62F, 450B, 542G
Retrovirus, 61, 62F, 119, 475, 495–6,
496F, 542G
Reverse transcriptase, 101T, 101–2,
155F, 193, 454, 542G
Reverse transcriptase-PCR (RT-
PCR), 120TN, 193, 495, 542G
Reverse transcription, 22F, 61, 61F
Review journals, 507, 507T
RF-1, 325T, 334, 334F
RF-2, 325T, 334, 334F
RF-3, 325T, 334, 334F
RFC, 400, 413, 541G
RFLP, 130, 130F, 131F, 146–7, 147F,
181, 486, 497, 500F, 502–3,
542G
Rhinovirus, 328
Rho-dependent termination, 276–7,
277F, 542G
Rho protein, 276–7, 277F, 542G
Rhombomere, 377F
Ribbon–helix–helix motif, 249T,
251–2, 252F, 542G
Ribonuclease, 8F, 241T, 247F, 335–7,
336F, 542G
Ribonuclease II, 305, 306F
Ribonuclease III, 296F, 305, 306F
Ribonuclease D, 297F, 542G
Ribonuclease E, 297F, 305–6, 306F
Ribonuclease F, 296F, 297F

Ribonuclease H, 401, 402F, 453F, 454
Ribonuclease M16, 296F
Ribonuclease M23, 296F
Ribonuclease M5, 296F
Ribonuclease MRP, 74B, 295, 542G
Ribonuclease P, 74B, 296F, 297F, 298T, 462, 542G
Ribonucleoprotein (RNP) domain, 250B, 542G
Ribose, 9, 10, 13F, 542G
Ribosomal protein, 241T, 322, 322F, 328, 329F, 542G
Ribosomal RNA, *see rRNA*
Ribosome
    definition, 542G
    ribozyme, 262
    role in attenuation, 277–81, 280F, 281F, 282F
    role in translation, 323–35
    rRNA content, 44
    sedimentation coefficients, 42TN
    structure, 321–3, 322F, 323F, 324F, 331F
    subunits, 321–2, 322F
    summary of role, 319
Ribosome binding site, 324, 325F, 325T, 542G
Ribosome recycling factor (RRF), 325T, 334, 334F, 542G
Ribozyme
    definition, 542G
    origins of genomes, 462, 453F, 464F
    rRNA, 319, 330, 332RB
    types, 298, 298T
Rice, 33T, 46, 48T, 49F, 215
Rickettsia prowazekii, 57F
RLF, 412, 541G
RMP, 399, 541G
RNA
    content of cell, 72–80, 73F
    definition, 542G
    half-life, 304–5
    origins, 461
    processing, 74–7
    structure, 10–11, 13F
    studying RNA, 120TN
    synthesis, 74, 75F
    transcriptome, 78–80
    units of length, 16B
RNA-binding motifs, 250B
RNA degradation
    bacteria, 305–6, 306F
    eukaryotes, 306–8, 307F
    role in genome expression, 71, 72F
RNA-dependent DNA polymerase, 101–2, 543G
RNA-dependent RNA polymerase, 74, 543G
RNA editing
    definition, 76, 77F, 543G
    effect on gene number estimates, 22B
    effect on genetic code, 86

    process, 303–4, 304T, 305B, 306F
    web site, 509T
RNA interference, 200, 201F, 202–3RB, 308, 543G
RNA labeling, 120TN
RNA polymerase
    archaeal, 255
    bacterial, 241T, 255, 257–8, 258F, 275–81, 278–9RB
    chloroplast, 255B
    definition, 74, 543G
    DNA-binding motif, 249T
    mitochondrial, 255B
    mode of action, 10
RNA polymerase I, 254–5, 255T, 259–61, 266, 295, 296F, 543G
RNA polymerase II
    definition, 543G
    elongation, 284, 285T
    initiation of transcription, 254–5, 255T, 258–9, 261F, 266, 266F, 267B
    mRNA synthesis, 283–7
    termination, 286–7, 287F
RNA polymerase III, 254–5, 255T, 261, 266, 295, 543G
RNA processing
    capping, 283, 285F
    chemical modifications, 298–304
    cutting events, 295–6, 303F
    intron splicing, 287–95, 296–8
    polyadenylation, 286–7, 286F, 287F
    pre-rRNA, 299–303
    pre-tRNA, 298, 301T
    role in genome expression, 71, 72F
RNA sequencing, 120TN
RNA synthesis
    bacteria, 274–82
    chemical modification, 298–304
    eukaryotic mRNA, 283–95
    eukaryotic rRNA, 295–8
    eukaryotic tRNA, 295–8
    intron splicing, 287–95, 296–8
    non-coding RNA, 295–8
    RNA editing, 303–4
    RNA polymerase I, 295
    RNA polymerase II, 283–7
    RNA polymerase III, 295
RNA transport, 308F, 308–9
RNA transposon, 61–3
RNA world, 296, 460, 462, 453F, 464F, 543G
RNP domain, 250B, 542G
Ro RNA, 74B
Robots, 171
Rolling circle replication, 390F, 390–1, 543G
Rooted tree, 487, 487F, 543G
RPA, 398, 398F, 541G
RpAp46 protein, 231
RpAp48 protein, 231
Rpd3 protein, 231
rplJ gene, 325T

RRF, 325T, 334, 334F, 542G
rRNA
    B chromosomes, 39B
    base-pairing, 14B
    chromosomal location, 40F
    content of cell, 73
    definition, 73, 74B
    genes, 44, 472
    transcription, 255T
Rsk protein, 358, 358F
RT-PCR, 120TN, 193, 495, 542G
Ruv proteins, 447, 448F

S
S1 nuclease, 102T 194, 195F, 543G
S12 ribosomal protein, 433
SII protein, 285T
S5 ribosomal protein, 323
Saccharomyces cerevisiae
    accuracy of genetic map, 145, 147F
    alternative splicing, 291
    artificial chromosomes, 114–7
    cloning vectors, 114T, 114–7, 116F, 118F, 119
    copper-regulated gene expression, 352F, 352–3
    DNA-binding proteins, 249T, 251F
    double-strand break recombination model, 447–8, 450F
    extrachromosomal genes, 46
    gene catalog, 43F
    gene conversion, 447–8, 449F
    gene density, 33T
    gene inactivation, 198–9, 199F, 200B, 201F
    gene number, 31
    gene organization, 47F
    genetic code, 87T
    genome duplication, 468RB, 470
    genome organization, 32, 34F
    genome replication, 410–11RB
    genome sequence, 30, 31T
    genome size, 33T
    genome-wide repeat content, 33T
    histone acetylation, 229
    histone deacetylation, 231
    homologs of human disease genes, 215, 215T
    homology analysis, 197F, 197–8
    identified genes, 195
    inteins, 342
    introns, 288
    introns per gene, 33T
    linkage analysis, 140
    mating type switching, 360–1, 361F, 362
    mitochondrial genome number, 46
    mitochondrial genome size, 46, 48T
    mitochondrial transcription, 255T
    mRNA degradation, 306–7

Saccharomyces cerevisiae continued
    origin of replication, 392–3, 393F
    protein domains, 44T
    protein interaction map, 212–13, 214F
    regulation of DNA replication, 409–12
    sporulation, 78–9, 79F
    transcriptome, 78–80
    two-hybrid system, 211, 213F
    web site, 508T
SAGA complex, 229, 265
SAGE, 207, 208F, 544G
Salmon sperm, 5
Sanger, F, 86
*Sap*I, 149
SAP kinase system, 359, 543G
SAR, 225, 225F, 543G
Satellite DNA, 59–60, 60F, 543G
*Sau*3AI, 103, 103T, 104F
SCA genes, 424T
SCAF, 291, 545G
Scaffold, 180, 181, 543G
Scaffold-attachment region (SAR), 225, 225F, 543G
Scanning, 325–6, 326F, 543G
Schizosaccharomyces pombe, 268
Schrödinger, E, 5
scRNA, 73, 74B, 255T, 545G
scs, 226, 226F
scs′, 226, 226F
Sda protein, 370RB
Sea urchin, 33T
Second messenger, 359–60, 360F, 543G
Secondary level of protein structure, 80, 81, 81F
Secondary vulva cell fate, 371, 371F
Sedimentation analysis, 295, 543G
Sedimentation coefficient, 42TN, 543G
Segment polarity gene, 374–5, 543G
Segmental duplication, 468–9RB
Selectable marker, 109, 544G
Selenocysteine, 82, 84F, 86–8, 87T, 88T, 317, 318F, 340
Selenoprotein, 88T
Selenoprotein P, 88T
Selenoprotein W, 88T
Selfish DNA, 478B, 544G
Self-splicing, 462
Self-spicing intron, 297–8, 298T
Semiconservative replication, 386F, 386–8, 387F, 389–91, 544G
Sendai virus, 155
Senescence, 407–9, 408F
SEPALLATA protein, 375B
Septum, 366, 367F
Sequenase, 101, 101T, 166B, 544G
Sequence alignment, 490F, 490–91, 491F
Sequence contig, 173, 180, 181F, 544G
Sequence database, 509T
Sequence gap, 173, 175F, 180, 181F

Sequence tagged site (STS), 154, 544G
Sequence tagged site (STS) content mapping, 179, 179F
Sequence tagged site (STS) mapping
    definition, 145, 546G
    in Human Genome Project, 182
    mapping reagents, 155–60
    rationale, 153–4, 154F
    types of STS, 154–5
Serial analysis of gene expression (SAGE), 207, 208F, 544G
Serine, 82, 83F, 84T
Serine-threonine kinase receptor, 356T, 364RB
Serum albumin, 89, 288T
Serum response factor, 267B
Serum response module, 267B
Sex cell, 4, 544G
Sex chromosome, 4, 544G
Sex combs reduced gene, 376F
Sex determination in *Drosophila melanogaster*, 293F, 294
SF1 protein, 289, 291F, 292F
*Sgf*I, 149
SGS1 protein, 215, 215T, 295
Shine–Dalgarno sequence, 324, 325F, 544G
Short interfering RNA (siRNA), 201F, 308, 544G
Short interspersed nuclear element (SINE), 18, 19F, 24, 24T, 62F, 62–3, 544G
Short patch excision repair, 437, 439F, 544G
Short patch mismatch repair, 440
Short tandem repeat (STR), *see Microsatellite*
Shotgun sequencing, 126, 127F, 173–6, 174F, 544G
Sialic acid, 342F
Signal peptide, 339, 340F, 544G
Signal transducer and activator of transcription (STAT), 89, 355–8, 357F, 545G
Signal transduction
    definition, 354, 544G
    human genes for, 21
    MAP kinase system, 358F, 358–9
    Ras system, 359, 359F
    SAP kinase system, 359
    second messengers, 359–60, 360F
    SMADs, 364RB
    STATs, 355–8, 357F
Signal transmission, 351–60
Silencers, 264, 544G
Silent mating-type cassettes, 361
Silent mutation, 428, 544G
Silica, 110TN
Silkworm, 33T, 508T
Simian immunodeficiency virus (SIV), 495–6, 496F
Similarity approach, 491, 544G
Simple sequence length polymorphism (SSLP), 130, 132F, 181, 155, 486, 544G

Sin3 complex, 231–4, 235
SINE, 18, 19F, 24, 24T, 62F, 62–3, 544G
Single nucleotide polymorphism (SNP), 18, 130–2, 131B, 132F, 134F, 486, 545G
Single orphan, 197F, 198, 545G
Single-strand binding protein (SSB), 241T, 398, 398F, 399F, 545G
Single-strand break repair, 441, 442F
siRNA, 201F, 308, 544G
Sir proteins, 223F, 231–4
Site-directed hydroxyl radical probing, 323, 323F, 545G
Site-directed mutagenesis, 204, 205TN, 545G
Site-specific recombination, 448–50, 451F, 545G
SIV, 495–6, 496F
SL1 protein, 259
Slippage, 331, 333F, 545G
Slo proteins, 294, 294F
Slp gene, 477
SMA/MAD related (SMAD) proteins, 364RB, 545G
*Sma*I, 149
Small cytoplasmic RNA (scRNA), 73, 74B, 255T, 545G
Small nuclear ribonucleoprotein (snRNP), 241T, 289–91, 290F, 291F, 292F
Small nuclear RNA (snRNA)
    definition, 73, 74B, 545G
    RNA transport, 309
    role in splicing, 289–91, 290F, 291F, 292F
    transcription, 255T
Small nucleolar RNA (snoRNA), 44B, 73, 74B, 255T, 302, 303F, 545G
SMUG1 protein, 439T
SNP, 18, 130–2, 131B, 132F, 134F, 486, 545G
Sodium bisulfite, 426
Sodium dodecyl sulfate, 110TN
Solar system, 460, 461F
Solenoid model, 38F
Solution hybridization, 132, 134F
Somatic cell, 4, 545G
Sonication, 172, 545G
Sooty mangabey, 496, 496F
SOS response, 427, 442–4, 444F, 545G
Southern hybridization, 103–4, 105F, 130, 131F, 545G
Soybean, 478
Specialized chromatin structure (scs), 226, 226F
Species concept, 56–8
Species tree, 488–9, 489F, 490F, 545G
Sperm typing, 140
S phase, 409, 409F, 412–13, 543G
S-phase cyclin, 412, 412F
Spinal and bulbar muscular atrophy, 424T
Spinocerebellar ataxia, 424T

Spliceosome, 289–91, 290F, 291F, 292F, 545G
Splicing pathway, 288–9, 289F, 291, 545G
SpoOA protein, 366, 368F, 370RB
SpoOB protein, 370RB
SpoOF protein, 370RB
SpoIIAA protein, 366, 368F
SpoIIAB protein, 366–8, 368F
SpoIIE protein, 366–8
SpoIIGA protein, 368, 368F
SpoIIR protein, 368, 368F
SpoIVB protein, 368, 369F
SpoIVF protein, 368, 369F
Spontaneous mutation, 420, 545G
Sporulation, 366, 367F
SR protein, 290–1, 292F, 293F, 294, 545G
SR-like CTD-associated protein (SCAF), 291, 545G
SRS2 protein, 295
SRY protein, 249T, 266
SSB, 241T, 398, 398F, 399F, 545G
SSLP, 130, 132F, 181, 155, 486, 544G
Ssn6-Tup1 repressor, 268
Stahl, F, 388
STAT, 89, 355–8, 357F, 545G
Staufen protein, 372
Stem cell, 405–7, 546G
Sterile phenotype, 203RB
Sterile progeny phenotype, 203RB
Steroid hormone, 353, 352F, 354F, 546G
Steroid receptor, 249T, 353, 352F, 354F, 546G
Steroid receptor zinc finger, 251, 251F
Sticky end, 103, 104F, 546G
Storage protein, 89
STR, *see Microsatellite*
Streptococcus, 6–8, 31T, 475
Streptomyces, 51, 118–19
Streptomycin, 433
Stress activated protein (SAP) kinase system, 359, 543G
Strong promoter, 262, 546G
Strongylocentrotus purpuratus, 33T
Structural domain, 225–6, 472, 473F, 477, 546G
Structural protein, 89
STS, 154, 544G
STS content mapping, 179, 179F
STS mapping
    definition, 145, 546G
    in Human Genome Project, 182
    mapping reagents, 155–60
    rationale, 153–4, 154F
    types of STS, 154–5
Stuffer fragment, 112, 546G
Sturtevant, A, 139
Substitution mutation, 419B, 546G
Sucrose density gradient centrifugation, 42TN
Suillus grisellus, 48T
Sulfur substitution of nucleotides, 301T

Sulfurylase, 172, 172F
Supercoiled DNA, 42TN, 50, 50F, 52RB, 110TN, 546G
Superwobble, 319, 546G
Suppressor mutation, 85RB, 370RB, 546G
Suppressor of dnaA1 (Sda) protein, 370RB
Sutton, WS, 6
Svedberg, T, 42TN
Svedberg unit, 42TN
SWI5 protein, 251F
Swi/Snf complex, 230–1, 234, 265
SXL protein, 293F, 294
Sykes monkey, 496F
Syncytium, 372, 373F, 546G
Synechocystis, 57F
Synonymous mutation, 429, 430F, 494, 546G
Synteny, 214, 546G
Synthesis phase, 409, 409F, 412–13
Systema Naturae, 484

T
T1 haplotype, 501, 501F
T2 haplotype, 501F
T4 polynucleotide kinase, 98TN, 107–8, 538G, 546G
TψC arm, 315F, 316, 316F, 546G
TAF, 258–9, 259T, 260RB, 261F, 546G
TAF$_{II}$42, 260RB
TAF$_{II}$62, 260RB
TAF- and initiator-dependent cofactor (TIC), 259, 546G
Tailless protein, 374, 374F
Takifugu rubripes, 33T
Tandemly repeated DNA, 23, 23F, 59–60, 126, 127F, 546G
Taq DNA polymerase, 101, 101T, 120–1, 121F
TART, 408B
TATA-binding protein (TBP), 241T, 249T, 252, 258–9, 259T, 261F, 268, 546G
TATA box, 256, 257F, 259, 259T, 261F, 267B, 546G
Tautomer, 420, 546G
Tautomeric shift, 420, 422F, 546G
TBP, 241T, 249T, 252, 258–9, 259T, 261F, 268, 546G
TBP-associated factor (TAF), 258–9, 259T, 260RB, 261F, 546G
TBP domain, 249T, 252, 546G
Tc-1, 24T
T-cell receptor diversity, 361–2
TDG protein, 439T
T-DNA, 15, 119, 119F, 547G
Telomerase, 405, 407F, 408F, 547G
Telomerase RNA, 74B
Telomere
    component of yeast artificial chromosome, 114, 116F
    definition, 38F, 39, 547G
    DNA content, 39, 41F
    Drosophila, 408B

in genome sequences, 18
    maintaining their lengths, 404–5, 406F, 407F
    minisatellite DNA, 60
    proteins, 39–41
    repeat sequences, 405T, 407F
    senescence and cancer, 405–9
    web site, 509T
Telophase, 137F, 138F
TEMED, 165TN
Temperature-sensitive mutation, 432–3, 547G
Template-dependent DNA polymerase, 98–100, 547G
Template-dependent DNA synthesis, 14, 547G
Template-dependent RNA synthesis, 14, 74F, 547G
Template-independent DNA polymerase, 107, 547G
Template-independent RNA polymerase, 74, 286, 286F, 547G
Template switching, 453F, 454
Teratogen, 424T
Terminal deoxynucleotidyl transferase, 98TN, 107, 107F, 194F, 547G
Termination codon, 85RB, 86, 86F, 88, 307, 307F, 547G
Termination factor, 266, 547G
Termination of replication, 401–4
Termination of transcription
    bacteria, 275–81, 276F, 277F, 278–9RB, 280B, 280F, 282F, 282F
    RNA polymerase II, 286–7, 287F, 295, 296F
Termination of translation, 333–5, 334F
Terminator sequence, 403, 404F, 547G
Ternary complex, 326
Tertiary level of protein structure, 81, 82F, 547G
Test cross, 141–2, 142F, 547G
Tetracycline resistance, 109–10
Tetrahymena, 33, 33T, 229, 296–8, 297F, 298F, 405T
TFIIA, 259, 259T, 261F
TFIIB, 259, 259T, 261F
TFIID, 258–9, 259T, 260RB, 261F
TFIIE, 229, 259, 259T, 261F
TFIIF, 229, 259T, 261F
TFIIH, 259, 259T, 261F, 439
TFIIIA, 249T
TFIIIB, 261
Thalassemia, 121, 228
T haplotype, 501F
Thermal cycle DNA sequencing, 170, 171F, 547G
Thermatoga maritima, 57F, 58
Thermostable protein, 101
Thermus aquaticus, 101, 101T, 120
Thermus thermophilus, 120TN, 324F, 331F

Threonine, 82, 83F, 84T
Thymidine kinase (TK), 155
Thymidylate synthase gene, 297
Thymine, 10, 11F, 547G
Thymine glycol, 437, 439T
TIC, 259, 546G
Ti plasmid, 51T, 119, 119F, 547G
TIF-1B, 259
TIF-IA, 259
TIF-IC, 259
Tissue plasminogen activator (TPA), 473, 475F
Titin, 19
TK, 155
T lymphocyte, 361
T$_{m'}$ 132F, 547G
tmRNA, 73
Tn3-type transposon, 63F, 64, 451, 547G
Tobacco chloroplast genome, 48T
TOCSY, 248
TOL plasmid, 51T
TPA, 473, 475F
Trans, 263B
Transcript mapping, 193–4, 194F, 195F
Transcription bubble, 275, 275F, 548G
Transcription-coupled repair, 439, 444, 548G
Transcription factory, 402F, 548G
Transcription initiation
    archaea, 258B
    definition, 548G
    Escherichia coli, 257–8, 258F
    regulation in bacteria, 261–4
    regulation in eukaryotes, 265–9
    RNA polymerase I, 259–61
    RNA polymerase II, 258–9, 261F
    RNA polymerase III, 261
Transcription initiation complex
    assembly, 257–69
    assembly position, 74
    role in genome expression, 70, 72F
Transcriptome
    definition, 25, 70, 71F, 548G
    link with proteome, 83–8
    studying, 78–80, 207–8, 208F, 209F
Transcript-specific regulation of protein synthesis, 328, 548G
Trans-displacement, 230, 230F, 548G
Transduction, 57B, 144F, 145, 146F, 548G
Transesterification, 289, 290, 289F, 297, 297F
Transfection, 112, 548G
Transfer-messenger RNA (tmRNA), 73, 548G
Transfer RNA, *see* tRNA
Transferrin, 328
Transformation, 6, 57B, 109, 144F, 145, 146F, 475, 548G
Transforming growth factor-β, 364RB

Transforming principle, 6, 548G
Transgenic mouse, 201, 204F, 548G
Transition, 419B, 548G
Translation factor, 241T
Translation
    archaea, 334B
    bacteria, 324–5, 328–35
    definition, 70, 548G
    elongation, 328–33
    eukaryotes, 325–35
    initiation, 323–8
    regulation, 328
    termination, 333–5
    unusual events, 330–3
Translational bypassing, 331–3, 333F, 548G
Translational slippage, 331, 333F
Translocation, 330, 330F, 548G
Transport protein, 89
Transposable element, 23–4, 475, 476F, 476–7, 477F, 548G
Transposable phage, 63F, 64, 451, 548G
Transposase, 64, 548G
Transposition, 61, 63F, 63–4, 450–5, 451, 452F, 453F, 454F, 548G
Transposon, 61, 549G
Transposon tagging, 199, 201F, 549G
Transversion, 419B, 549G
TRAP, 281, 282F, 283F, 549G
Tra protein, 293F, 294
Tree of life, 484, 485F
Tree reconstruction, 489–93, 490F, 491F, 492F
Tree topology, 487, 487F
Treponema pallidum, 53, 57F
TRF1, 39
TRF2, 39
Trichomonad, 47
Trichothiodystrophy, 444
Trimethylpsoralen, 52RB
Trinucleotide repeat expansion disease, 423–4, 424T, 549G
Triosephosphate isomerase, 478
Triplet binding assay, 85RB, 549G
Triplet expansion, 423–4
Triplex DNA helix, 446–7, 447F, 549G
Triploid nucleus, 467F
Trisomy, 465, 549G
Triticum aestivum, 33T, 475
Triticum turgidum, 475
tRNA
    adaptor molecule, 83–4
    aminoacylation, 315–17, 317F, 318F
    base-pairing, 14B
    codon–anticodon interactions, 317–19, 319F
    definition, 73, 74B, 548G
    in *Saccharomyces cerevisiae* genome, 32
    structure, 315F, 315–16, 316F, 335F
    summary of role, 314, 315F
    transcription, 255T
    wobble, 318–19, 320F, 321F

tRNA nucleotidyltransferase, 296, 297F, 549G
tRNA primer for retrotransposition, 453F, 454
tRNA^{Phe} ribozyme, 298T
trp RNA-binding attenuation protein (TRAP), 281, 282F, 283F, 549G
True tree, 488, 549G
Truncated gene, 22, 23F, 549G
TRY4, 18, 19F
TRY5, 18, 19F
Trypanosome, 305B
Trypsin, 211, 471
Tryptophan, 82, 83F, 84T
Tryptophan operon, 256F, 256T, 263–4, 265F, 277–81, 280F, 281F, 282F, 333F
Tryptophan repressor, 249T, 263, 264F
TTF-1 protein, 266, 295, 296F
Tth DNA polymerase, 120TN
tudor domain, 196–7, 197F
Tus protein, 403, 404F, 549G
Twintron, 287T, 302B, 549G
Two-dimensional gel electrophoresis, 209, 210F, 549G
Two-step gene replacement, 204, 206F
Ty1, 201F
Ty1/copia family, 62, 62F, 454
Ty2, 32
Ty3/gypsy family, 62, 454
Type 0 cap, 284, 285F, 549G
Type 1 cap, 284, 285F, 549G
Type I collagen gene, 473, 474F
Type 1 deiodinase, 88T
Type I DNA topoisomerases, 388T, 388–9, 389F
Type 1 neurofibromatosis, 215T
Type I restriction endonucleases, 102
Type 2 cap, 284, 285F, 549G
Type 2 deiodinase, 88T
Type II DNA topoisomerases, 388T, 388–9, 389F, 448
Type II restriction endonucleases, 102–4
Type 3 deiodinase, 88T
Type III restriction endonucleases, 102
Type VII collagen gene, 288T
Tyrosine, 82, 83F, 84T, 247F
Tyrosine-kinase-associated receptor, 356T
Tyrosine kinase receptor, 356T, 357

U
U1–snRNP, 289–90, 290F, 291F, 294
U11/U12–snRNP, 294
U16 snoRNA, 303F
U2AF^{35}, 289, 291F, 292F
U2AF^{65}, 289, 291F, 292F, 293F, 294
U2–snRNP, 289–90, 291F, 294–5
U24 snoRNA, 303F
U4 haplotype, 501F

U4/6–snRNP, 289–90, 291F, 295
U4atac/U6atac-snRNP, 294
U5–snRNP, 289–90, 291F, 294
U6–snRNA, 255T
UBF protein, 259
Ubiquitin, 229, 344, 549G
Ubiquitination of histones, 229
UCE, 256, 257F, 259
U haplotype, 500, 502F
uidA gene, 206T
Ultrabithorax gene, 376F
Ultracentrifugation, 42TN, 321–2
Ultraviolet radiation visualization of
  DNA, 37TN
UmuD'$_2$C complex, 442, 444F
Uncoordinated phenotype, 203RB
Unequal crossing-over, 471, 472F,
  549G
Unequal sister chromatid exchange,
  471, 472F, 549G
UNG protein, 439T
Unipartite genome, 34
Unit factor, 5, 549G
Universal primer, 169, 169F
University of Colorado, 52RB
Unrooted tree, 487, 487F, 549G
Upstream, 21F, 549G
Upstream control element, 256, 257F,
  259, 549G
Upstream promoter element, 256,
  257F, 550G
Uracil, 10, 13F, 437, 439T, 550G
Urea, 336F, 336–7
Uridine 5'-triphosphate, 10, 13F
Urokinase, 473
UTP, 10, 13F
UV radiation, 426–7, 427F
UvrABC endonuclease, 437, 439F,
  550G

V
V28 gene segment, 18, 19F, 20, 362
V29-1 gene segment, 18, 19F, 20, 362
Valine, 82, 83F, 84T
van der Waals forces, 82B, 253F, 550G
Variable number of tandem repeat
  (VNTR), *see Minisatellite*
Vault RNA, 74B
Venter, C, 179, 182
Very short patch mismatch repair,
  440
Vibrio cholerae, 31T, 51–3, 53T
Virulence plasmid, 51T
V loop, 315F, 316, 316F, 317T, 550G
VNTR, *see Minisatellite*
V segment, 362, 362T, 363F, 433, 434F
Vsr endonuclease, 440
Vulva development, 369–72, 371F

W
Watson, J, 8, 384–5
Wave of advance model, 500, 550G
Werner's syndrome, 215, 215T, 295,
  444
What is Life? 5
Wheat, 33T, 78, 214–15, 501F
Whole-genome shotgun sequencing,
  126–7, 128T, 179–81, 550G
Wild type, 145, 550G
Wilms tumor-1, 304T
Wilson's disease, 215T
Winged helix–turn–helix, 249T, 250,
  550G
Wobble, 318–19, 320F, 321F, 550G
Wyosine, 298

X
X25 locus, 424T
Xanthine, 426

Xenopus laevis, 472
Xeroderma pigmentosum, 444
X-gal, 111, 112F
X inactivation, 235–6, 362, 550G
X-irradiation, 155, 158F
xis gene, 450
Xist, 74B, 235
X-ray crystallography, 246–7, 247F,
  260RB, 278–9RB, 298, 316, 323,
  332RB, 447, 550G
X-ray diffraction, 13, 246, 247F, 550G
X-ray diffraction pattern, 246, 247F,
  550G
X-ray sensitive film, 98TN
XRCC4 protein, 442, 443F

Y
Y chromosome, 497, 503
Yeast artificial chromosome (YAC),
  114T, 114–17, 116F, 176, 182,
  182F, 550G
Yeast mating type, 360–1, 361F
Yeast two-hybrid system, 211, 213F,
  364RB, 550G
Yersinia pestis, 31T
YIp5, 118F, 119
Yra1p protein, 309

Z
Z-DNA, 16, 16T, 17F, 253, 550G
Zea mays, 33T, 48T, 87T, 509T
Zebrafish, 508T
zerknüllt gene, 376F
Zinc, 352
Zinc binuclear cluster, 249T
Zinc finger, 42, 44F, 249T, 250–51,
  251F, 252B, 550G
Zoo-blotting, 192, 193F, 550G
Zygote, 449F, 550G